Einstein

Einstein
A Life in Science and Music

Brian Foster
Donald H Perkins Professor Emeritus of Experimental Physics, University of Oxford
Alexander von Humbolt Professor Emeritus, University of Hamburg/DESY

Great Clarendon Street, Oxford, OX2 6DP,
United Kingdom

Oxford University Press is a department of the University of Oxford.
It furthers the University's objective of excellence in research, scholarship,
and education by publishing worldwide. Oxford is a registered trademark of
Oxford University Press in the UK and in certain other countries.

Published in the United States of America by Oxford University Press
198 Madison Avenue, New York, NY 10016, United States of America

British Library Cataloguing in Publication Data
Data available

Library of Congress Control Number: 2025941214

ISBN 9780198794875

DOI: 10.1093/oso/9780198794875.001.0001

Printed and bound by
CPI Group (UK) Ltd, Croydon, CR0 4YY

The manufacturer's authorized representative in the EU for product safety is
Oxford University Press España S.A. of Parque Empresarial San Fernando de Henares,
Avenida de Castilla, 2 – 28830 Madrid (www.oup.es/en or product.safety@oup.com).
OUP España S.A. also acts as importer into Spain of products made by the manufacturer.

Pray can you tell me – that is, without anger, before I write my chapter upon straight lines – by what mistake – who told them so – or how has it come to pass, that your men of wit and genius have all along confounded this line, with the line of Gravitation?

The Life and Opinions of *Tristram Shandy* by Laurence Sterne – Chapter 3.LXXXIII

To my family: Sabine, Paul, & Mark, with love.

Foreword

A distinguished theoretical physicist recently expressed the view that there should be a moratorium on biographies of Einstein. Obviously, I do not agree but nevertheless I recognise his rationale. Indeed, one of the daunting tasks facing any new biographer is to assimilate the immense extant literature on almost all aspects of Einstein's life. This is a task in which I admit to having failed. While I have certainly read, or at least dipped into, hundreds of Einstein biographies, there are certainly hundreds more with which I am unacquainted. Therefore, I admit from the outset that the current volume is incomplete. However, all volumes, even the fattest and most comprehensive, are incomplete. Mostly strikingly for me, and the reason why I embarked on the present work, is that almost none pays anything more than the most cursory attention to arguably the second most important thing in Einstein's life other than physics: music.

The genesis of this volume lies in a group of lectures that I developed with my friend and violin teacher, Jack Liebeck. The centenary of Einstein's *annus mirabilis* of 1905 led to 2005 being designated 'World Year of Physics'. As an enthusiast for public engagement with science, I looked around for a theme on which to hang a presentation. When Jack reminded me that Einstein had been a violinist, the idea of the *Superstrings* and *Einstein's Universe* lectures was born. The presentations married Einstein's theories with modern developments in physics, particularly at the Large Hadron Collider at CERN. Each lecture was punctuated by musical interludes played by Jack of solo violin music with a particular connection to Einstein. In the intervening twenty years, we have given these lectures hundreds of times to a live audience totalling by now more than 30,000.

As I learned more and more about Einstein, I was startled to observe the very little attention paid to Einstein's musical activities in the literature. Slowly the idea of rectifying this omission began to gather strength and I launched into what I intended to be a rather specialised and brief biographical essay. I soon realised however that Einstein's relationship with music was intimately bound up with his personality and emotions, so that it could not be understood in isolation but only in the context of a highly complex life. The specialised essay became a 'full' biography in which special attention was given to all things musical and the emotional landscape from which this welled, in particular his family, lovers, and friends. The middle of these categories is the reason why I used the word 'arguably' when I characterised music as the second most important factor in his life. As readers will see, his romantic adventures, with which music was often bound up, might indeed have edged music into third place. Thus, all three of these preoccupations are given priority in this book.

If a biography of Einstein is to remain within the practical bounds of book production, the addition of some topics must necessarily result in the omission of others. This book pays almost no attention to Einstein's considerable activity in the philosophy of science, his views on religion, and relatively little to his political activities except in far as they touched on his major preoccupations such as pacificism, anti-fascism, and Zionism and the Jewish people. These were to Einstein more reflections of his deeply held beliefs than 'political' activity as such. I have done my best to explain the basics of his work in his priority, physics, within the constraints of my public presentations, that is, using no equations other than $E = mc^2$. Clearly therefore the explanations are simplified up to but hopefully not beyond the point of becoming misleading. To paraphrase what Einstein may have said, 'Explanations of science should be as simple as possible, but not simpler'. Those interested enough to seek a much more rigorous exposition of Einstein's physics are recommended to that classic of Einstein works, *Subtle is the Lord* by Abraham Pais.

Einstein's last interview before his death in March 1955 was to Bernard Cohen:

> [Einstein] told me most emphatically that he thought the worst person to document any ideas about how discoveries are made is the discoverer. He said many people had asked him how he had come to think of this or that, but he had always found himself a very poor source of information about the genesis of his own ideas. Einstein thought that the historian is likely to have better insight into the thought processes of the scientist than the scientist himself.[1]

I have been guided by these wise words. Einstein was indeed a most unreliable witness of his own story. This is particularly evident in the origins of special relativity, where he contradicted himself on several occasions. Einstein wasn't particularly interested in the history of his own ideas but was obsessed with their further development. He also tended to wishful thinking; what he thought he ought to have done morphed, over the years, into what he thought he had done. An example here is his involvement in the development of atomic weapons. I have therefore relied most heavily on the record of what others observed and wrote about his life and times. Almost always, this has been achieved by going back to the original records, as catalogued in the Einstein Archive at the Hebrew University in Jerusalem – an institution to which Einstein was greatly attached and which he played a vital part in founding.

The Einstein Archive is voluminous. The first few decades are available in the output of the Einstein Papers Project, based in California Institute of Technology. The so-far extant sixteen volumes, covering up to May 1929, are essential sources for the first part of the book. When one goes beyond May 1929, it is as if the overhead lights in the archive have been switched off and one is left to wander around with a guttering candle. The information is still there but the effort required to extract it is orders of magnitude greater. This is a tribute to the astonishing level of scholarship of the editors of the Einstein Papers Project, now led by Diana Kormos Buchwald but originally by John Stachel, Martin J. Klein, A. J. Kox, Michel Janssen, Jürgen Renn, Robert Schulmann, and many others. To them and their fellow editors I and all Einstein biographers owe a profound debt of gratitude.

A word on translation: each volume in the Einstein Papers project has a companion volume in which (most) items are translated into English. These translations are excellent; however, as 'translations of record' they are of necessity somewhat literal. I have preferred to use my own translations from German, French, and Italian throughout, to give a more colloquial flavour. This implies less accuracy, but I believe makes them more readable and I do not think I have introduced any material inaccuracy. If I have, inadvertently, then the responsibility is of course mine alone.

In addition to the contents of the Einstein Archive, I have also relied on newspaper reports. After the eclipse of 1919, Einstein was rarely if ever out of the headlines. In particular, I have found the columns of *The New York Times*, the *Los Angeles Times*, *The Times* of London, and the *Jewish Telegraph Agency* and its publications vital sources. While Einstein lived in Germany, the German 'quality' dailies and magazines, such as *Jüdische Rundschau* and *Berliner Tageblatt*, were another invaluable resource.

The biographies written by contemporaries who knew Einstein have also been mined. Particularly valuable for his early life is the biography written by his sister, Maja Winteler-Einstein, which is available in full in manuscript from the Einstein Archive. Also important are the books

[1] Cohen, I. Bernard. April 3rd, 1955, Scientific American, 193(1): 71.

by Einstein's sons-in-law, Rudolf Kayser (writing under the pen-name Anton Reiser) and Dimitri Marianoff. The latter is highly unreliable in many respects, but the author lived in the same apartment as Einstein for a considerable time and his recollections of Einstein's character and behaviour have a ring of truth. Another younger contemporary was Carl Seelig. His books were the product of intensive research and correspondence not only with Einstein but also many of his early friends. Seelig's earliest book has the rare distinction, along with that of Kaiser's book and that of Einstein's colleague and successor at Prague, Philipp Frank, of gaining Einstein's approval.

The first 'modern' biography by an author not an acquaintance of Einstein is that of Ronald Clark. It too was the product of considerable scholarship and still contains much useful information. It is now however more than fifty years old and Einstein scholarship has moved on greatly in the interim. I have used modern biographies where the authors' original research and interviews have turned up information not available elsewhere. Of these, the most useful were the books of Highfield and Carter, Overbye, and Brian. Finally, for a comprehensive study of the subject, look no further than the immensely readable book by Isaacson.

Some specialist books about Einstein's family and friends have also thrown light on aspects of his story and character. *Einsteins Schwester* by Rogger is an invaluable source on the person probably closest to him, his sister Maja. Heinrich Zangger's correspondence with Einstein stretched over almost forty years; Robert Schulmann did a heroic job not only by reading Zangger's almost illegible handwriting, which Einstein himself often gave up on, but also by annotating and editing their correspondence with impressive scholarship in *Seelenverwandte*. The life and tragic times of Einstein's first wife, Mileva Marić, are recorded, although not always reliably, in *Im Schatten Albert Einsteins* by Desanka Trbuhović-Gjurić. Martin Klein's sadly incomplete biography *Paul Ehrenfest: The Making of a Theoretical Physicist* details the life of the physicist probably closest personally to Einstein. Although Einstein's correspondence with his two oldest and closest friends, Michele Besso and Maurice Solovine, are also available in edited volumes, I have generally preferred to use the original letters in the Einstein Archive.

A brief note on nomenclature: in principle I could have referred to Einstein as 'Albert' but as a physicist I felt uncomfortable with pretending familiarity with such a towering figure in my subject. I settled therefore on 'Einstein'. Einstein's sons' given names were Albert and Eduard. Albert was always called this by both parents. For obvious reasons of avoiding confusion, in this book I use his own subsequent adoption of Hans Albert for the elder. The younger was never called Eduard by either parent; they used a variety of diminutives, of which the most common was Tetel, closely followed by Tete. I chose to use the latter. I have generally omitted accents in favour of anglicisations in Slavic names as well as in German names where there are simpler anglicised alternatives, for example, Zurich rather than Zürich.

I am greatly indebted to many institutions and people. Most important are the various branches of the Albert Einstein Archive. The central repository at the Hebrew University of Jerusalem hosted me for several long visits and answered many questions both in person and by email. I am grateful to the Curator, Roni Grosz and to Barbara Wolff, Chaya Becker, and Anna Rabin for the hospitality and unfailing helpfulness in responding to often obscure queries. The Einstein Papers Project at the California Institute of Technology is the centre of Einstein scholarship. I am very grateful in particular to Diana Kormos-Buchwald, the director, for agreeing to host me for several visits, including a particularly valuable extended one in which my interaction with both staff and visiting scholars helped me to understand the basis of Einstein scholarship. I also greatly appreciate my interactions with Ze'ev Rosenkranz and Emily de Araújo. I am grateful to the Princeton University Library and Special Collections for hospitality during an extended visit.

Another institution with a wealth of Einstein material is the Zurich Stadtarchiv. I appreciate their hospitality during a visit. Also the ETH-Bibliothek were helpful and responsive to all my questions. I thank Andrew Neilson at the Menuhin Archives of the Royal Academy of Music for his help and hospitality. I am grateful to the Nuffield College Library Archives for access to the Cherwell Papers and to the staff of Rijksmuseum Boerhaave in Leiden for access to the Ehrenfest papers. The library of Deutsches Elektronen Synchrotron (DESY) in Hamburg has loaned me several Einstein books on (very) long-term loan. For more general research I have made extensive use of the world-class facilities at both the British Library and the Bodleian Library, the latter of which also contains the Deneke Papers. The online facilities afforded by *archive.org* are outstanding; the pettiness of some of the major publishers in withdrawing their books from the archive is deeply reprehensible.

At a personal level, I am grateful for several pieces of advice and information from Prof Diana Kormos Buchwald. Jack Liebeck's friendship and artistry on the violin inspired the enterprise. Several colleagues and friends have read and commented on parts of this book. These include Dr Jason McFall, Evan Quinter and Prof Ling Lin. Those who have read and given highly constructive comments on the entire book have my deep gratitude as well as amazement at their stamina: Prof Norman McCubbin, Prof Nick Trefethen, and Prof Dr Albrecht Wagner. Their enthusiasm kept me going. Remaining errors are of course my responsibility.

Finally, I express my appreciation of the support of my various employers, the University of Oxford's Department of Physics, and DESY and the University of Hamburg. They gave me the space within which I could, eventually, complete this book, although it drifted back into my (nominal) retirement from both. My sons Paul and Mark, well launched on their own lives, were mostly unaware but their love was always a support. My wife Sabine was both well aware of the project and enormously patient; I thank her for all her love and support and in particular the many hours in which we puzzled over Elsa Einstein's handwriting, Einstein's own particular brand of obscure colloquialisms, and for her ability to read the old German *Kurrentschrift* that often defeated me.

Contents

List of Figures

Chapter 1

Origins and beginnings (1879–1895)

When Albert Einstein's biographer, Carl Seelig, asked about the origin of his scientific abilities, Einstein responded that 'the question of my heritage is irrelevant'. While his abilities indeed seem unique, Einstein modestly maintained that his achievements arose not from any particular talent, but simply from being passionately curious. Perhaps his ancestry was irrelevant to his scientific prowess. However, with regard to his more general outlook on life, particularly his taste in the arts, especially music, he was very much a product of his ancestry and environment. Einstein was a German Jew, a decisive fact in the turbulent twentieth century that he did more than most to form.

Of course, German Jew is much too crude a characterisation of his ancestry. His father's family can be traced back to the small town of Bad Buchau in Baden-Württemberg in southwest Germany, where Einsteins have been recorded as early as the seventeenth century. There had been a Jewish community there since the fifteenth century, residing in a ghetto that was dissolved only at the beginning of the nineteenth century. The community thrived and many of its members became entrepreneurs in the textile industry; by the end of the nineteenth century, a third of the population of Bad Buchau was Jewish. Naftali Einstein, Einstein's great-great-grandfather, was born there in 1733, and his son Rupert lived his whole life in Buchau, dying in 1834. Rupert had many children, one of whom was Einstein's grandfather Abraham. Another, Rafael, was the grandfather of Elsa, Einstein's second wife. The location of the house that Abraham and his wife Helene née Moss (known as Hindel in the family)[1] built on the site of a former palace garden in Buchau is today marked by a commemorative plaque. All of their six children were born there, including Einstein's father, Hermann, who was born in 1847, and Jakob, born three years later. Jakob was to play an important role in the lives of both Hermann and Albert.

Einstein's mother Pauline Koch came from a similar background to that of Hermann. Her parents, Julius Derzbacher and Jette Bernheim, were both born in Jebenhausen, a small village about 40 km southeast of Stuttgart; they married in 1847. Julius began life as a humble baker but sometime around 1853 he and his wife moved to Cannstatt, directly across the river from Stuttgart, the capital of the Kingdom of Württemberg. This move from small villages to cities was one followed by many of their co-religionists. For example, there were 550 Jews recorded in Jebenhausen in 1843; this declined to 85 by 1880. Together with Derzbacher's earlier change of name to Koch, which carries no Jewish connotations, this move implies a plan to make his way in a wider, typically anti-Semitic, world. Julius and his brother became successful grain merchants, eventually receiving a royal warrant. The two brothers lived together with their wives in a single home for decades, perhaps a result of Julius's parsimonious attitudes, as recorded by Einstein's sister Maja. She also implies that their home was cultured, since once he became successful, Julius liked to consider himself a patron of the arts. Certainly Pauline, who was born in 1858, took piano lessons

[1] Their portraits in oils hung in Einstein's drawing room in his house in Princeton.

Einstein. Brian Foster, Oxford University Press. © Brian Foster (2026). DOI: 10.1093/oso/9780198794875.003.0001

and became a good pianist. Her love of music was to be one of the critical factors in Einstein's formative years.

Hermann Einstein was an intelligent boy who had a clear bent towards mathematics. The educational possibilities of Buchau being strictly limited, at the age of fourteen he was sent the roughly 100 km to Stuttgart to attend secondary school. No doubt he relied on traditional Jewish hospitality by lodging with a Jewish family, just as years later he extended hospitality to poor Jewish students in Munich. Hermann did well and obtained a graduation certificate that would normally have been a stepping stone to university. However, he did not become a student and instead became a merchant. The main reason for this decision was undoubtedly, as Maja records, financial; his family were not well off and he had many siblings. However, poverty was not an insuperable impediment to attending university. Almost a third of all Jewish students attending Prussian universities in the 1880s were officially declared to be paupers and excused fees for attending lectures. Less than a decade after Hermann's decision, that more Jews from poor families attended university than did Catholics from a similarly deprived background led to serious anti-Semitic disturbances in German universities. Whether early manifestations of these disturbances helped to convince the placid and equable Hermann to forego university matriculation is unknown. Whatever the reasons, Hermann in later life seemed to regret his choice and found vicarious fulfilment by encouraging his son's enthusiasm for mathematics and science.

To understand Einstein's family life, it is necessary to appreciate the social milieux in which his parents and he grew up. The German-speaking states of central Europe in the mid-nineteenth century were remnants of the Holy Roman Empire broken up by Napoleon as he conquered Europe, riding the tsunami of the French Revolution. The new revolutionary ethos not only swept away the mediaeval borders but also many archaic customs. For centuries, Jews were universally discriminated against and, if allowed to reside in cities, mostly confined within ghettos. The great scholar Moses Mendelssohn, known as the 'German Socrates' and grandfather of the composers Felix and Fanny, entered Berlin as a penniless boy from Dessau in 1743 through the only gate open to Jews and cattle. At this time, despite the caricature of Jews as rich money-lenders, roughly 80% of German Jews lived in abject poverty, often as wandering vagabonds, excluded from guilds and crafts, at the very bottom of society. Moses, endeavouring to escape this fate, changed his name, as did Einstein's grandfather Julius. He became Moses Mendelssohn rather than Moishe Ben Mendel Dessau in order to improve his chances of receiving a first-class education. There was no university in Berlin at that time, but had there been, Moses would not have been allowed to enter it. The contemporary attitude was reflected when a young Jew applied to the Berlin Singakademie but was rebuffed by being informed that it was mathematically impossible that a Jew could compose music. Jews were admitted to universities in Prussia only after the Napoleonic Wars. They could study law but not practice or teach it; only the practice of medicine was allowed. Nevertheless, the legal discrimination against Jews was gradually relaxed. The Kingdom of Württemberg, within which both Pauline and Hermann Einstein were born, recognised Jews as full citizens in 1828. However, several German states only finally emancipated Jews after the foundation of the German Empire in 1871.

Thus the Kochs and the Einsteins raised their families in an environment in which the legal discrimination against Jews was gradually relaxed, but where anti-Semitism among the populace was never far below the surface. Einstein himself remarked on this in 1920:

> Among the children anti-Semitism was alive particularly at elementary school. It was based on the children's remarkable awareness of racial characteristics and on impressions left from religious instruction. Active attacks and verbal abuse on the way to and from school were frequent but usually not all that serious. They sufficed, however, to establish an acute feeling of alienation already in childhood.

Most aspiring Jewish families reacted to this undercurrent by attempting to be as similar to other Germans as possible. Since strict Jewish dietary laws made the sharing of food, and thereby most social occasions, with gentiles almost impossible, many Jews ceased to be observant or even religious at all. This was the path taken by both the Einsteins and the Kochs. The possibilities of the Industrial Revolution, which spread from Great Britain in the nineteenth century, opened up avenues for social advancement to many of the former poorer classes. Jews seemed quicker than most to realise that education was the key to exploiting the new-found social mobility. This was amplified by the traditional Jewish respect for rabbinical learning and the written word. Consequently, Jewish students flocked to the newly accessible universities in disproportionate numbers.

Perhaps the most common social bond of this period was music. In a world in which travel was at the pace of a horse, mass communication consisted of a local newspaper and such things as organised sport were unknown; almost the only respectable entertainments were music and reading. Since reading is predominantly enjoyed alone, music was the mainstay both of family and community enjoyment. Any aspiring family who could afford one owned a piano, around which families would gather to sing or play chamber music. Making pianos was a major industrial activity. For example, Martha Novak Clinkscale's monumental book contains 415 *pages* listing known piano manufacturers, mainly in the US and Europe, between 1820 and 1860. Any families who could not afford a piano would be likely to have other musical instruments, often (in the case of Jewish families) a violin. Being easily portable and cheap to manufacture, the violin was ideal for an itinerant race often fleeing violent persecution. Julius Koch's success as grain supplier to the Court of Württemberg meant that he could afford a piano and the tuition by which his daughter Pauline became a proficient and enthusiastic musician. In her turn, she ensured that both her children had the same opportunities.

How Pauline and Hermann met is not documented, but it may be assumed that Julius's prosperity was reflected in a hospitable open house for his co-religionists of which the impecunious Hermann, during his time as a student and apprentice merchant in Stuttgart, took advantage. If this was indeed the case, he not only satisfied his hunger but also charmed the daughter of the house. Certainly the young Pauline was deeply in love with the good-natured Hermann, and they were married in Cannstatt on 8 August 1876. Hermann by this time had completed his apprenticeship and joined a bedding business run by two of his cousins, Israel and Levy. It was based in the city of Ulm on the Danube River, which here formed the border between Württemberg and Bavaria. Hermann and Pauline registered as citizens of Ulm on 6 September 1877. Initially, they lived in a small apartment just south of Cathedral Square. When Pauline became pregnant, they moved to a larger apartment at what became 20 Bahnhofstrasse. Both locations were within a few minutes' walking distance of Hermann's shop in the Weinhof. It was in Bahnhofstrasse, on 14 March 1879, that their first child, Albert, was born.

Einstein's birth was a difficult one. Not only was his head large and misshapen, but also he was a very large baby who caused his grandmother, on her first visit to the newly born child, to exclaim that he was much too fat. Einstein told the story that he didn't talk until the age of three. In fact, a letter from his maternal grandmother to Pauline Einstein makes it clear that he was already quite voluble at the age of just over two. Einstein's sister Maria was born in November 1881. Despite his initial disappointment that she had no wheels, Maja, as she was always known, became a treasured companion and friend to her elder brother.

The growing family was now living in a comfortable house in Munich, having moved there in June 1880 at the instigation of Hermann's younger brother Jakob. Jakob became one of the most significant people in Einstein's early life. As energetic as Hermann was placid, Jakob had followed Hermann to Stuttgart in 1867 and trained as an engineer at the Polytechnikum. In the summer of

1870, he entered the army as an engineer and served in the Franco–Prussian War. The war came to a rapid conclusion in favour of Prussia and its allies, including Württemberg. This led to the foundation of the German Empire under the Kings of Prussia in January 1871. Although active operations ceased after the suppression of the Paris Commune by the French government at the end of May 1871, it is likely that Jakob served a full five-year term in the military, since he is next heard of in 1876 in Munich, where he founded a gas-fitting and plumbing firm. It was this firm that his brother joined when he moved to Munich. Hermann, Pauline, and Albert lived in the same building that housed the company, in Müllerstrasse 3, just outside the historic city centre. Jakob was living alone at the time. On his marriage in 1882, he also moved into a separate apartment in the same building. Jakob's wife Ida was an Einstein cousin whose family also originated in Bad Buchau, although she herself was born in Munich. The intimacy of the Einstein clan is illustrated by Albert also eventually following the same path of marriage inside the extended family.

Einstein's mother was keen to encourage his traits for concentration and self-reliance. The Müllerstrasse apartment, which was on the third floor, was near to the centre of a city buzzing with traffic. Trams with their distinctive blue wagons, the 'city omnibus' as well as countless hackney coaches, all modes of transport drawn by horses, clattered up and down streets thronged with people. At three or four years old, Albert was shown a route once, perhaps to a local shop, and how to cross the busy intersections safely. Subsequently he was asked to retrace the journey. Surreptitiously observed by his mother, he carried out his instructions and returned without incident.

No doubt Jakob's curiosity and receptivity had been sharpened by his experiences in the Franco–Prussian war. The illumination of the trenches in the Siege of Paris by a brilliant arc light, driven by a generator, on the Moulin de la Galette at the summit of Montmartre, was a sensation for the besieged civilians and both sets of combatants. Jakob was always on the lookout for new ideas and became interested in the development of electricity as a transformative social and industrial force, part of what later became known as the Second Industrial Revolution. He followed the development of the first commercial generator by Siemens in 1866 and the invention of the light bulb by Edison in 1877. This invention opened the floodgates for schemes to electrify the private homes of the rich and subsequently public spaces. As it happens, Stuttgart was the scene of the first electrical lighting installation in Germany. In 1882 Paul Reißer, like Jakob a former student of the Polytechnikum, built a generator and used it to power thirty light bulbs. Jakob turned his engineering knowledge towards this fledgling industry. Many of the installations had substantial mechanical engineering components, where he could use the skills he had learned at Stuttgart and in the army. He quickly immersed himself in the techniques of the new electrical technology. He had many ideas on how to produce industrial quantities of electric power. Already in 1882, the Einstein company exhibited at a Munich Electrical Exhibition. Their products included dynamos and arc lights, which were to be mainstays of their future activity. In 1885 Jakob persuaded Hermann that the time was ripe to switch their activities completely into this new area and on 6 May they registered J. Einstein & Co. as an electro-technical factory, at the same time dissolving their two-thirds interest in Keissling and Co., a firm that produced boilers.

Jakob's motives in teaming up with Hermann included not only fraternal affection and his mercantile knowledge, but also his access to capital. Hermann's father-in-law, Julius, was a wealthy man. Hermann's brother-in-law, Rudolf Einstein, who was also his first cousin, was another source of financial support for his activities. Rudolf owned a large textile factory in Hechingen, a small town in Swabia to which reference will be made several times in this story. Hermann was able to convince both Julius and Rudolf to invest heavily in J. Einstein & Co. For a considerable time, this looked like a good investment. Munich was ripe for electrification; there was a successful experiment in lighting a part of the Central Station in the autumn of 1879, while the Residenztheater became the first public building to be lit by electricity in 1883.

While Hermann and Jakob's business prospered, Albert generally ignored the rough games of the growing band of children (Jakob's son Robert was born in 1884, his daughter Edith in 1888) in the household and played quietly with his set of building bricks.[2] At first the growing household filled the apartment in Müllerstrasse; after the foundation of the electro-technical business, both families moved about 1 km southwest, to a large villa in Rengerweg on a substantial plot, part of which was annexed for the company factory. By the time of the move, Einstein was at a crucial point in his development.

Ambitious for her son to succeed, Pauline had tried to give him a head start when he was five years old by employing a private tutor in advance of his starting school when he reached his sixth birthday. Most crucially for his subsequent life, she was determined that her children should have the opportunity to share her love of music. Music was always ringing through the house. Pauline loved to play with others and is said to have delighted in playing piano duets with engineers drafted in from the Einstein factory next door. When Einstein was around five or six years old, a typical age to begin, she took the fateful decision to start him on the violin, no doubt in anticipation of being able to play duets with him on the piano. The violin is a challenging instrument for a young child to learn since, in contrast to the piano, it is very difficult to make any sort of pleasant sound until after a considerable investment of time and effort. Einstein, who throughout his life was exceedingly reluctant to devote time to repetitive practice methods such as scales, clearly found the learning experience both frustrating and irritating. Maja reports that he hurled a stool at his teacher in a fit of rage, upon which she fled and was never seen again, although whether this incident was related to violin playing or his general tuition is unclear. At any rate, his violin lessons continued, although they did not give him any pleasure. As he recorded in 1940 in a draft letter to his friend and biographer Philipp Frank: 'I took violin lessons between the age of six and fourteen, but had no luck with my teachers, to whom music was nothing more than a mechanical exercise.'

Maja also reported on the beginning of another musical activity that he pursued throughout his life, indeed even after the infirmities of old age forced him to discontinue playing the violin. He loved to sit at the piano and improvise, picking out chords and arpeggios in harmonics of his own invention. Curiously, he appears never to have had formal piano lessons; his mother must have introduced him to at least the basics of keyboard technique.

Another seminal moment in Einstein's life occurred at around the same time as he began to play the violin. He recorded in his 'Autobiographical Notes' in 1947 that, at four or five years of age, he was transfixed when his father showed him a compass. The fact that the needle turned to always align itself with some unseen force made a strong impression on the boy. For the first time he had experienced something outside the world of clear cause and effect, of 'touch' and reaction. He realised that there were hidden forces in the world that, although sources of wonder, might possibly be understood.

The move to the house in Rengerweg in 1885 gave the growing Einstein much more space and a secure childhood base on which both he and his sister looked back with nostalgia. The house was improved to become a substantial three-storey building with a balcony, set back to avoid the noise

---

[2] These Ankersteinbaukasten, mentioned by Maja in her memoir of Einstein, were sold at auction by Christies in 2016 for $83,000.

from the busy Munich streets and looking out over a spacious yard containing several majestic old trees. The Einsteins were often visited by their extended family, including Einstein's cousins Alfred and Robert Koch, the latter subsequently becoming a schoolfellow of Einstein's and a rival in love. On Robert's death in 1952, Einstein wrote to Alfred recalling their playing together in the yard of Rengerweg in a large crate. In this vessel, Einstein and Robert were the crew, Alfred the captain.

Einstein particularly looked forward to visits from his uncle, Caesar Koch, whom he considered to be the family member closest to him. Much more regular visits were paid by his young cousins, including Elsa, daughter of Rudolf Einstein and Pauline's sister Fanny, making her Einstein's cousin from both sides of the family and later to be his second wife. Part of Elsa's attraction for Einstein in later years was this shared ancestry and the 'gemütlichkeit' (the nearest translation is 'coziness') that went with it. They spoke a broad Schwäbisch dialect together; Einstein never entirely lost his Schwäbisch accent, although it was overlaid with several other influences, including Swiss German. Elsa's visits to Munich were often combined with a trip to the opera. Their son-in-law Rudolf Kayser, writing under the pseudonym Anton Reiser, asserted that:

> Since earliest childhood, Albert and Elsa Einstein had been attached to each other. They had played together in the home of Albert's parents in Munich, and had, at the same time, had their first artistic experiences at the Munich Opera. Albert's attractive little cousin sat in the orchestra, while he himself sat in the highest gallery.

Elsa recalled joining the family on Sunday picnics to one of the Bavarian Lakes, or jaunts into the mountains:

> where the small inns served the famous *Weisswurst* [Einstein] is still so fond of. That was the only diversion he ever had. He was not a solemn little boy, not at all . . . he would not run with us or play with us, but he loved to laugh – only he loved more to think.

Elsa was not quite right about his only diversion being trips to the mountains; he loved word games. In a letter to a friend in 1934, Maja recalled:

> It was our custom to commemorate any important family events with the composition of doggerel verses at family festivities. My brother was a bit of a poet, which he still is today. He is prouder of his verses than any of his other accomplishments.

Indeed, this pleasure in the composition of short poems, humorous in a rather heavy Germanic way, lasted throughout Einstein's life.

In 1885, the year of the move to Rengerweg, Einstein entered primary school. He attended the Petersschule on Blumenstrasse, about a twenty-minute walk from his home. The fact that he began in the second grade indicates that the private tuition that his parents had arranged had indeed given him a head start. However, his experiences at school began his long career of resistance to authority wherever he found it, both in society and in science. He characterised his teachers, whose methods of drilling by rote were reinforced by raps across the knuckles with a ruler, as corporals. He refused to memorise the large doses of 'facts' so beloved of the unimaginative teacher. Furthermore, he was unable and unwilling to immediately snap back his answer to the teachers' questions as his fellow pupils did. Even in arithmetic he did not excel, although he could always solve even difficult problems in his own time despite being prone to calculational errors.

Like most primary schools in Munich, the Petersschule was a Catholic school in which the pupils were given religious instruction. Einstein, the only Jew in his class, listened to this instruction with equanimity. His parents were uninterested in religion, and he had never been exposed to any particular system of belief. Despite their agnosticism and perhaps in reaction to the compulsory

Catholic instruction, his parents did arrange for a distant relative to outline to him the tenets of the Jewish faith. Einstein became interested and indeed for a considerable time was sufficiently religious to follow Jewish ritual practices, including eating kosher, which must have caused his mother considerable difficulties in a non-kosher household. His growing interest and feeling for music led him to compose little songs glorifying God that he sometimes quietly sang to himself. It was only at about twelve years of age that his exposure to scientific books for young people gradually destroyed his faith and indeed imbued him with a conviction that the state was deliberately deceiving its citizens.

Einstein's not particularly happy time at the Petersschule came to an end in October 1888, when he transferred to the Luitpold Gymnasium. Gymnasia, then as now, were the 'academic' branch of German school education, rather similar to the English grammar school. The Luitpold Gymnasium was expanding rapidly, from 684 pupils when Einstein enrolled to 1,330 six years later. It was also rather socially exclusive, and it was unusual for the son of a merchant to attend. The curriculum was dominated by Classics. Einstein appreciated the clear and logical structure of Latin grammar and obtained good marks in both that and Greek, despite Maja's report that an exasperated Greek teacher had once exclaimed that he would never achieve anything with his life!

As Einstein made his way home from his new school in the twilight of the winter afternoons of 1888, his way would have been lighted, if at all, by the dimly glowing mantels of gas lamps. His family's firm was trying its best to replace them with electricity. The autumn of 1888 was a particularly exciting time in the compound in Rengerweg, which had been redesignated Adlzreiterstraße in 1886. Having already cut their teeth with lighting for the Oktoberfest and various breweries in 1885 and small Italian towns, including Varese in 1887, the Einstein Company in October 1888 was awarded the contract to electrify the Munich suburb of Schwabing (still independent at the time). Work began in November and was completed in time for a grand opening on 26 February 1889. The work of the Einstein firm was widely praised and the Bavarian dignitaries who attended the opening were impressed. Afterwards they were transported in 150 carriages to the Salvator Brewery, which had also been electrified by J. Einstein & Co. in 1888. In this establishment, the happy illumination of the streets of Schwabing was toasted until the early hours. On 2 March, the Corporation of Schwabing paid almost 44,000 marks[3] to J. Einstein & Co. This financial success must have gratified not only Hermann and Jacob but also Grandfather Julius, who had invested significant amounts of his own money in the company. He had moved into the Adlzreiterstraße house after the death of his wife Jette in 1886. The home to which Einstein returned after his first months at the Luitpold Gymnasium was both crowded and happy.

Einstein's attitude to the Luitpold was scarcely more favourable than it had been to the Petersschule. By Alexander Moszkowski's account, he promoted his teachers there from corporals to lieutenants, by which he meant that their discipline was enforced more by moral authority than by physical violence, although it was nonetheless objectionable for that. It was therefore his stimulating home environment that must have been responsible for the majority of his intellectual development. Given the fact that he lived in close proximity to Jakob and his company, Einstein could hardly fail to be influenced by both; for the influence of the former there is a great deal of evidence. Einstein's first steps in mathematics were surely encouraged by Hermann but catalysed by Jakob, who introduced him to algebra at an early age by describing it as a sort of hunting expedition: 'It's an amusing science – if you can't catch the creature, you give him the name 'x' and carry on hunting until you track him down.' Moszkowski gave a slightly different version of the same story and also reported that Jakob told Einstein the form of the Pythagorean theorem,

[3] This corresponds to a value of around $6M in 2024 prices.

which immediately intrigued him. The boy worked intensively on the problem for three weeks and then came up with his own proof. The joy and satisfaction that he obtained from this was immense. Together with the episode of the compass referred to earlier, this episode marks the very first steps of Einstein the theoretical physicist.

Jakob was also instrumental in awakening Einstein's interest in geometry. Einstein recalled that 'At the age of 12 I experienced a second wonder of a totally different nature: in a little book dealing with Euclidean plane geometry, which came into my hands at the beginning of a schoolyear.' Excited by his first steps in mathematical investigations, Einstein applied to his parents for some of the textbooks that he was to use when he started to study mathematics at the Luitpold the following year. This request was apparently transferred to Jakob, who obliged by presenting Einstein with a copy of Adolf Sickenberger's *Leitfaden der Elementaren Mathematik, Zweiter Teil, Planimetrie*. This thin volume had an enormous effect on the young Einstein, who afterwards referred to it as the 'holy geometry book'.

The influence of the family firm was less explicit but just as real. In the years between 1886 and 1893, J. Einstein & Co. took out at least seven patents on various aspects of electrical engineering. Such legalities would inevitably have involved both Jakob and Hermann and as Lewis Pyenson remarks 'we can be confident that all facets of electrical generation, distribution and utility were topics of daily conversation in the Einstein household just when Albert was undergoing an intellectual awakening.' Living from an early age with patent drawings on the kitchen table surely laid the foundations for Einstein's future satisfaction in his work at the Bern Patent Office and indeed for his fascination with inventions and gadgets. The location of the factory next door to the Einsteins' house means that he had many opportunities to help out in Jakob's office on the upper story. An internal walkway in front of the office looked out directly over the busy and noisy shop floor where the firm's dynamos were constructed. The industrial application of electricity and magnetism was part of Einstein's daily life.

Somewhat before the arrival of the 'holy geometry book', another and probably even more decisive influence than Jakob arrived in Einstein's life. In the late autumn of 1889, a young Polish medical student, Max Talmud (he changed his surname to Talmey when he moved to the US) began to visit Adlzreiterstraße. His brother had previously been hosted by the Einsteins, who thereby fulfilled a traditional duty towards their fellow Jews by offering the hospitality of their table for lunch once a week. The student discovered a lively intelligence in the young boy and the two became fast friends. Talmud awoke Einstein's interest in physics and subsequently philosophy, while Jakob continued to provide mathematics books and problems. Although Talmud gave him a much thicker newly published book on geometry by Spieker after Einstein had devoured the 'holy geometry book', the three books on mathematics by Lübsen, which Einstein worked through to understand calculus, were all Jakob's property. Meanwhile, Talmud introduced him to Bernstein's popular and colourfully illustrated series of books on science, which Einstein afterwards remembered with pleasure and gratitude.

Bernstein's first volume discussed the aims of science and, among other things, the speed of light and the aether through which it was thought to propagate. At this stage in the development of physics, it was thought that all observations could be reduced at a fundamental level to mechanics as formulated by Newton. Although Newton had assumed that light consisted of particles ('corpuscles'), Huygens had shown convincingly that it was in fact a wave.[4] Waves,

---

[4] In perhaps his most revolutionary work, Einstein in 1905 showed that both Newton and Huygens were right—light behaves sometimes as a particle and sometimes as a wave.

it was thought, were disturbances that required a mechanical medium to support their passage, as is the case for waves in water. Hence there must be a medium throughout the universe that supported light in its passage. The problem was that the void between the stars seemed empty so that the mechanical properties of the aether must be incredible weak and attenuated. Unsuccessful theories to reconcile the vital support of the aether for the propagation of light with its practical undetectability littered the last decades of the nineteenth century. With great foresight, Bernstein opined that it had been given a name, the aether, even though no one had ever, or would ever, see it. However, in a subsequent volume Bernstein declared that the degeneration of the orbit of a comet had made the existence of the aether completely indubitable. He also suggested that, since each kind of light had exactly the same speed, the law of the speed of light could be considered the most general of all Nature's laws. Such statements must have sparked the curiosity of the young Einstein on questions central to the special relativity paper of 1905.

As Einstein worked his way through Spieker and Lübsen, rapidly completing almost all of the exercises, he left Talmud floundering in his wake. Indeed, at fourteen years old, Einstein already understood integral and differential calculus and analytical geometry, far beyond the mathematics syllabus of the Luitpold. Clearly unwilling to give up his stimulating conversations with the precocious young man, Talmud turned to discussing philosophy:

> I recommended to him the reading of Kant. At that time, he was still a child, only thirteen years old, yet Kant's works, incomprehensible to ordinary mortals, seemed to be clear to him. Kant became Albert's favourite philosopher after he had read through his *Critique of Pure Reason* and the works of other philosophers.

At the same time as these first steps in mathematics, science, and philosophy was the discovery of something of almost equal importance to Einstein's life. The 1940 draft letter to Frank quoted earlier continues: 'I really only started to learn [the violin] at the age of 13, after I fell in love with Mozart's sonatas. The effort to reproduce to some degree their artistic content and unique grace forced me to acquire the necessary technique, without ever doing systematic practice.' Confusingly, Talmud recounts that Einstein only had violin lessons for two years, but presumably he means for two years after the time he first met him. In any case, it is clear that Einstein derived little if any benefit from his lessons and decided to dispense with them and teach himself. Such a course is fraught with danger, as violin playing is a highly technical art and without correction from the expert eye of a competent teacher it is very easy to develop idiosyncrasies that significantly hinder subsequent playing ability. For the moment, however, Einstein pursued his joyous discovery of Mozart: 'The feeling of profound emotion he experienced in reading books of geometry is perhaps to be compared only with his experience as a 14-year-old boy when for the first time he was able to take an active part in chamber-music performance.'

Whatever the technical limitations that Einstein's decision to stop violin lessons may eventually have caused, the awakening of his passion for music in general and Mozart in particular was one of the two stars, the other being physics, by which he would steer the course of the rest of his life. His father remarked in a letter to his brother-in-law Jakob in Genoa that 'Albert does such bad work in school. He only wants to play the violin. I'm afraid he will come to no good end.' From now on, his enthusiasm grew with the years until it became almost an obsession.

While Einstein and his mother happily played Mozart and Beethoven sonatas in the drawing room of the comfortable house in Adlzreiterstraße, the outlook for J. Einstein & Co. was darkening. Despite being one of the leading German companies exhibiting in the Frankfurt Electrical Exhibition of 1891, it was also one of the smallest. By 1893, competition in the electrical lighting business had become intense. Although the Einsteins were successful in bids for various minor contracts in Bavaria subsequent to the Schwabing project, they were increasingly unable to compete with the growing might of companies such as AEG, Schuckert & Co., and Siemens & Halske. A turning point was the loss of a major contract for the electrification of a central region of Munich. All four of the aforementioned companies bid for the contract; J. Einstein & Co, whose bid was 374,000 marks, was the most expensive. After political pressure against the 'non-Bavarian' Siemens & Halske, who were the cheapest bidder, the contract was eventually awarded to Schuckert in July 1893.

This was a major blow to the Einstein company, but it was more of a last straw than a singular catastrophe. The root cause of the company's failure was that it had insufficient capital and no access to unsecured capital from banks. The scale of electrification projects was increasing enormously and only access to major capital could permit the growth that promoted the economies of scale essential to produce winning bids. Whereas Siemens and AEG could access many millions of marks, J. Einstein & Co. was never able to raise a loan from a bank except via mortgages on property which it had acquired, mostly through Einstein's grandfather Julius Koch, or by loans from his Uncle Rudolf in Hechingen. Indeed, throughout the life of the firm, Julius is often recorded as injecting small to medium infusions of capital. Pyenson believed that the company needed at least 1 million marks (roughly $120 million in 2024 prices) to compete effectively. The extent of their problem is shown by their efforts to raise capital in this period, which produced a grand total of 74,000 marks, via two mortgages on Adlzreiterstraße 14 – at 10% interest.

There were several other contributory factors to the firm's failure, most of which were Jakob's responsibility. J. Einstein & Co. only ever produced direct-current (DC) devices. In Bavaria there was great potential for hydroelectric power but mostly in the southern mountains, a great distance from the major population centres. Alternating current (AC) is a much more efficient way of transporting power over long distances. In Jakob's defence, it must be said that many others made the same mistake, including Thomas Edison! Another factor was that Jakob's personality was not suited to leading a technology-based factory. He spread the company's resources too widely over many good ideas and patents. He was also extremely irascible, waging several long-running vendettas, particularly with successive Bavarian officials responsible for certifying electrical equipment. Finally, the loss of the Munich contract coincided with a general economic depression in Germany that forced many undercapitalised companies to the wall.

Maja thought that the responsibility for the company's failure was shared between Jakob and Hermann; however, she gives the impression that Hermann's easy-going and indecisive disposition was an important factor. His errors however seem to be rather of omission than commission. What he could in fairness be criticised for was not realising that the company's capital was insufficient and that they needed to specialise to markets in which they could compete. He also did not effectively resist Jakob's wilder ideas and tendency to pick damaging fights. He always deferred to Jakob in matters of the firm's strategic direction. Nevertheless, at this period, few foresaw that many small firms similar to that of the Einsteins would be insolvent in a few years.

Wherever the blame lies, it was clear by 1893 that the company was in severe difficulties. The Einstein brothers explored selling the firm as a going concern to Siemens & Halske for 600,000 marks, with them continuing to manage it. These negotiations failed and the firm closed its doors at the end of June 1894.

With the failure of J. Einstein & Co., the fairly idyllic period of Einstein's childhood and early youth ended abruptly. Maja makes it very clear that the dissolution of the Adlzreiterstraße home was a traumatic experience for her:

> The beautiful property with the villa in which Albert Einstein had spent a happy childhood was sold to a builder, who immediately turned the beautiful plot into a building site, cut down the majestic trees and built a row of horrible dwellings for rent. Until the family moved, the children were forced to witness the destruction of their fondest memories from the windows.

The effect on Einstein must have been even more profound, since the break-up of his close-knit family changed his circumstances even more than Maja's. Jakob once again persuaded Hermann to throw in his lot with him and move to Italy. The company had had a good name in the country after their lighting schemes in Varese and Susa had been a success; furthermore, at that time Italy was not suffering from the economic depression in Germany. As optimistic as ever, Jakob saw a bright future and greener grass on the other side of the Alps. As preparations were made for the move, Einstein's parents decided that he would have to stay behind to complete his schooling. He had little facility for languages, and he spoke no Italian, so transplanting to an Italian school seemed unfeasible. Maja however, to whom foreign languages came easily, could in any case attend the International School of the Protestant Families in Milan. Unfortunately for Einstein, this took pupils only up to age thirteen. So, lodgings were found for him and distant relatives agreed to keep an eye on him. Another member of the family also didn't make the journey to Italy. Grandfather Julius, who must have lost a very considerable amount of money with the liquidation of J. Einstein and Co., returned to Cannstatt and died there a year later. On 1 June, Hermann, Pauline, and Maja left Munich. Although they probably spent the summer with Albert, at the start of October they were in Milan, leaving the disconsolate youth to return to Munich for the start of term at the Luitpold.

The break-up of a happy home and extended family, the growing realisation that his father was unlikely to be a successful man of business (certainly as long as he followed in the wake of Uncle Jakob), and his general unhappiness with the atmosphere at the Luitpold combined to depress Einstein's spirits. The Einstein family had enjoyed walks as a family; however, never gregarious, it is likely that Albert began to take long solitary walks in the environs of Munich, confirming a taste for both walking and the rolling pre-Alpine landscape that was an important factor in his life while he remained in Europe. His violin was also a consolation. After his discovery of Mozart, he began to stretch himself with Beethoven sonatas. He also studied the music of Johann Sebastian Bach. Probably around the time of his solitary sojourn in Munich he began to explore Schubert's violin and piano duets.[5]

There is evidence for another musical activity around this time from an unimpeachable musical source. Distinguished author, musicologist, and a distant cousin of Albert, Alfred Einstein was in the class below him at the Luitpold. Many years later, when both lived in Berlin and Alfred was music critic of the *Berliner Tageblatt*, they were often in contact when their correspondence was misdirected. They became close friends. In his autobiographical fragment, Alfred recalled that he had stood in front of Albert during school choir practice. When Albert learned that his name was also Einstein, he had pulled the hair of his younger namesake in a friendly way but then, in typical schoolboy fashion, ignored the younger boy completely!

However, no musical diversion could significantly mitigate Einstein's growing unhappiness and loneliness, particularly when he contemplated the fact that he would soon be forced to enter the

---

[5] As evinced by a disintegrating and well-thumbed copy of an edition containing Deutsch catalogue numbers D934, D574, D895, D802 in Einstein's sheet music collection in the Albert Einstein Archive in Jerusalem. The piano part has an ink stamp from the Max Hieber Music Shop in Munich, which opened in 1885.

German system for compulsory military training. This was certainly one of the critical factors in the decision that he took late in 1894 to leave his school and indeed Germany as soon as possible. Any German male who had not enrolled for military training after sixteen years of age would be considered a deserter by the authorities and subject to military punishment. Einstein's sixteenth birthday fell on 14 March 1895. His determination to depart was confirmed sometime before the end of the Christmas term when his class teacher, Joseph Degenhart, sent for Einstein to tell him that he wanted him to leave the school. When asked what his transgression had been, Degenhart replied that Einstein's mere presence in the classroom undermined his authority, an indication of the unwillingness of Einstein to conform to the norms of the school and his determination to assert his own intellectual independence. His abilities in mathematics were also seen as subversive: 'The talent manifested was that of genius. To the teachers, it was not always pleasing. Many questions the boy asked embarrassed them to the point of silence.'

At around the time of Degenhart's request, Einstein asked the family's physician for a certificate testifying that he had had a mental breakdown and needed six months without school to recover. This may not have been so far from the truth of his mental state at that time; at any rate the doctor, who was a friend of the family, obliged. Presented with this, and presumably having had the report of Degenhart, the headmaster granted Einstein's request to be released from attendance at the Luitpold. Armed with a letter from his mathematics teacher that his knowledge of the subject was up to university standard, but without any other school certificate, Einstein left the school officially on 29 December 1894. His parents' somewhat irregular letters had painted an attractive picture of Italy, so he hurried to the station to catch the first train to Milan.

The surprise and concern with which his parents greeted Einstein when he arrived unannounced at their home in Milan can be imagined. The letter carefully obtained from his maths teacher implies that he had not run away spontaneously. He did have a plan for his future that involved entering a university, not all of which required the completion of the school-leaving certificate. It is possible that Einstein knew even before his departure from Munich that the Polytechnikum in Zurich (now ETH Zurich) was such an institution. Equally, he may also just have thought that once he had shaken the dust of Germany from his feet, everything would work out somehow. In the family conferences that followed his arrival in Milan, the Polytechnikum – German-speaking but not German and only 300 km or so north of Milan – would have been an obvious possibility. There was another factor. Gustav Maier, an old friend of the family who had been Director of the Reichsbank in Ulm and a fellow non-observant Jew, was in Zurich. He had become so disillusioned with the militarism of Wilhelmine Germany that he moved to Zurich in 1893 to found societies and publish pamphlets on socialism and pacifism. He was friendly with Albin Herzog, the Director of the Polytechnikum. Once the Einsteins had decided that the Polytechnikum was the best option for Albert, they hoped to prevail upon Maier to write to Herzog on behalf of the teenager. Some string-pulling was necessary since Einstein was two years below the minimum age to be eligible to take the entrance examination.

Another part of the 'deal' by which Einstein's parents were reconciled to his arrival was that he should give up what Hermann termed his 'philosophical nonsense' and get a sensible trade as an engineer. He feared that the new Italian venture was going to be an uphill struggle and that he might therefore be unable to support his son financially. Einstein promised to knuckle down to prepare himself for the examination in October and bought several thick physics textbooks, including the newly published set by Violle, which he worked his way through. He had also agreed to prepare himself for a 'sensible trade' by helping out in the offices of Officine Elettrotecniche Nazionale Einstein-Garrone, which had been incorporated in Pavia in March 1894 by the Einstein brothers in partnership with Garrone, an Italian engineer with whom they had worked in the Italian electrification projects of the Munich firm. In March 1895, the family moved to Pavia, where a factory had been constructed the previous autumn. This imposing building, with its lofty chimney, was situated on the banks of the canal (it can still be seen today, converted into apartments and without the chimney, on the corner of Viale Venezia and Viale Partigiani). The required capital came principally from Rudolf Einstein, as well as the proceeds of the sale of the Munich properties and Pauline's share of Julius Koch's estate. About eighty people worked there – a considerably smaller number than the 200 of J. Einstein & Co. in its prime. The major reason for the location in Pavia is likely to have been Garrone's intimation that the contract for the electrification of the city was pending and the company would be handily placed to bid for it.

The possibility of a successful bid for the electrification contract must have been increased by the considerable influence that the Einsteins and Garrone seemed to have had in both Milan and Pavia. This is demonstrated by the ease with which the Einsteins were received into the cities socially. They were able to rent one of the prime locations in Pavia, now Via Ugo Foscolo 11, 100 metres from the Town Hall and Botanical Gardens and a ten-minute walk from the new factory. This mediaeval house has an imposing frontage that now carries plaques not only commemorating Foscolo, the poet, but also Albert Einstein. The impressive double doors open into a courtyard typical of mediaeval Italian palaces of the nobility, surrounded by a colonnade and stairs leading to the upper story, the *piano nobile*, where the reception rooms containing fifteenth-century frescoes are situated. Jakob found accommodation for his family elsewhere in town.

From this comfortable base in Via Foscolo, Einstein studied Violle's *Lehrbuch der Physik*. He also regularly visited the factory and helped as much as he could in Jakob's office. He had access there to various journals in electrical engineering to which the company had a subscription, some of which also published articles that were essentially pure physics[6]. Einstein surely read through these with interest as he lent a hand in the factory. One of Einstein's contributions has been preserved through a statement of Jakob to an apprentice in Einstein, Garrone & Co.: 'You know, my nephew is amazing. My engineer and I had been racking our brains for days about a particular problem and then along came the lad and in fifteen minutes he had solved the whole thing. He'll make a name for himself one day!'

As Einstein leafed through the *Electrotechnische Zeitschrift* and *Lehrbuch der Physik*, the ground fertilised by Bernstein and Sickenberger's books began for the first time to bear fruit. He began to try to emulate the papers in Jakob's journals and during his Pavia stay he produced two pieces of work. One of these was written in the late spring of 1895. It concerned the areas of physics to which Einstein had had most exposure: electricity and magnetism. It was entitled 'On the Investigation of the State of the Aether in Magnetic Fields' and begins with an apology for his presumption in attacking such a difficult theme. He postulated that when a current passes, the aether is excited and generates a magnetic field. As might be expected, particularly from Einstein's reading of Bernstein's confident statement on the aether mentioned above, all is couched in terms of the aether as an elastic medium carrying electrical disturbances. It contains no mathematics and has no real conclusions, but it is a remarkable document from a sixteen-year-old boy. It is interesting that he enclosed the work in a letter that he sent to his Uncle Caesar. It might have been expected that he would show it to his Uncle Jakob, who was a scientist of sorts, whereas Caesar was a grain merchant. If he didn't show it to Jakob, it may have been that the young scholar felt nervous about exposing himself to his uncle's acerbic wit. In the accompanying letter, Einstein told Caesar of his

[6] Bracco points out that a paper in '*La lumière électrique*' concerning Helmholtz's Least Action Principle in electrodynamics follows a paper presenting Jakob Einstein's arc lamps.

intention to enter the Polytechnikum and promised to write again when the result was known. Einstein's second piece of work from this period involves a new development in the adolescent Einstein's life – girls.

The Einstein who had disembarked from the Munich train and made his way to his parents' house at the Palazzo Ricordi, in the shadow of Milan's great cathedral, was a troubled, lonely boy whose doctor's certificate attesting that he had had a nervous breakdown was not too far from the truth. One of his first acts after he unpacked was to ask for his father's help to renounce his citizenship of the Kingdom of Württemberg, symbolically cutting his ties with Germany and also ensuring his escape from compulsory military service before the legal age limit of seventeen. Since his parents retained German citizenship and since he had just arrived in Italy, he had no other option, when his citizenship was revoked in January 1896, but to be stateless for several years.[7] At the same time he decided to renounce his religion, so that the certificate releasing him from his citizenship records his religious denomination as 'none'.

Having thus decided in the winter of 1894–1895 to cut himself adrift from all his ties with his homeland and religious background, the young man was open to the influence of his newly adopted country. When the damp and chilly fogs of the northern Italian winter gave way to spring and the family move to Pavia was complete, Einstein began to regain his spirits. His sister observed this with some wonderment: 'Although limited at first by his only knowing Milan and Pavia, Italy made a big impression on him. The way of life, the landscape, the art, all attracted him and from distant lands in later life he looked back at this time with longing.' She also concluded: 'The free life and independent study changed a quiet, dreamy boy into a happy, popular and outgoing young man.'

As Einstein's state of mind improved, he began to make friends, although perhaps typically not among boys his own age. His closest friend had a strong similarity to Max Talmud. Otto Neustätter was Jewish, born in Munich in 1870 where he graduated as a physician in 1894. Shortly thereafter he moved to Pavia, where he was an ophthalmologist at a hospital. Quite how he became friendly with Einstein is not clear. He may well have known the Einsteins while they were in Munich since he and Talmud were almost exact contemporaries as medical students. Alternatively, the German-speaking Jewish community in Pavia would have been relatively small and they may well have met socially. Whatever the circumstances, they hit it off and became inseparable in that glorious Pavese summer. Neustätter had a friend, Ernesta Marangoni, usually called Ernestina. She was nineteen, a chemistry student at the Technical Institute, and came from a distinguished Italian Jewish family. She met the Einsteins when Pauline arranged a Saturday musical afternoon for Maja and Ernestina, beginning a friendship that was to endure until Maja's death. The first meeting with Einstein took place when Ernestina and her mother walked down to the broad Ticino River, near to one of the glories of Pavia, the Ponte Coperto. This bridge, dating from 1354, had a colonnaded walkway. It was near to the Lido, a favourite spot for the youth of Pavia to bathe in the hot summer weather. Here Ernestina came across her friend Otto, who was with a young man lazily floating on his back in the river,[8] whom he introduced as Alberto. Ernestina was clearly struck by the appearance of the teenager, whom she described as delicate but healthy, rather pale, with brown eyes and brown

---

[7] In stark contrast to the current situation, the lack of a passport did not stop Einstein from crossing the Swiss–Italian border many times in the next few years. As he wrote to his biographer Carl Seelig, more than half a century later, 'No normal person in those days had any idea what a passport was, as such a thing was not necessary to travel'.

[8] Floating on his back may have been the limit of his aquatic abilities since Einstein always maintained that he had never learned to swim.

hair. A strong attachment grew up between them, which seems likely to have become romantic. Part of the nostalgia with which Einstein looked back at those Pavia summers may well have been tinged with the bittersweet feelings of first love. Having passed through puberty and filled out physically, the shy and lonely Munich schoolboy became a happy young man on the banks of the Ticino.

Otto, Alberto, and Ernestina were inseparable that summer. Einstein had already begun to learn Italian while still in Munich, presumably in preparation for his trip. Ernestina considered his Italian good, but noted that he told jokes in German, which his sister translated. Ernestina's family home was in Casteggio, a small town about twenty miles south of Pavia. The Villa Marangoni was a monumental structure that towered over the old town from the Pistornile hill. Then it was rural on one side; today it lies in the centre of the town. In a letter to Ernestina written fifty years later, Einstein recalled his frequent visits with nostalgic pleasure. He had clearly also become intimate with her family; in the letter he refers with great fondness to her mother and also to her father as a second Leonardo da Vinci. A vitally important part of Einstein and Ernestina's friendship was their shared love of music. Ernestina, who played the piano, described their weekly musical gatherings in Casteggio, at which Einstein played with passion on an excellent violin that she borrowed for him. They mostly played the German classical composers and were joined by Maja and other musician friends.

That summer, Einstein and Otto decided to set off on an expedition from Casteggio and over the next four days they walked the roughly 90 km between nearby Voghera and Genoa. There Einstein stayed with his uncle Jakob Koch, whose children's stories of their home had first awakened Einstein's interest in Italy many years earlier in the garden on the Adlzreiterstraße in Munich. Writing to congratulate Einstein on his fiftieth birthday, Otto, by then a distinguished ophthalmologist in Berlin, reminisced about their trip: 'Do you remember our hike in the Italian mountains when at night we watched the stars and you spoke of the enormous impression that such a vast vista provoked? Perhaps even then you were starting to have those grand visions which you have now presented to the world.'

During the hottest part of the summer, the Einstein family escaped for a few weeks to the mountains, to the small village of Airolo, nestling at the foot of the St Gotthard Pass. Here, their influential Italian connections gave the Einsteins a link to the Jewish Italian ex-minister Luigi Luzzatti, who in the following year rejoined the government as Minister of the Treasury and eventually became Prime Minister. Luzzatti developed a liking for the young Einstein. The holidays over, the family returned to Pavia. Ernestina remembered beautiful days spent in the hills at harvest time with Einstein, who arrived in September to help with the harvest in Fontanone, a village to the east of Casteggio. A family friend, Ernestina's future father-in-law, owned a vineyard there. At some point this year or possibly early in the next, Einstein wrote his second piece of physics. This manuscript, a copy of which in his mother's hand was presented to Ernestina by Einstein, was entitled 'On Electricity and the Electric Current.' The following year, he asked his mother to try to retrieve it because it was incorrect. Unfortunately, the manuscript is now lost.

As summer drew to an end, Einstein's parents contacted their friend Maier, who on 24 September wrote to the Director of the Polytechnikum. Maier must surely have met Einstein by then since, without personal experience, he would hardly have been willing to write in such glowing terms to Herzog. He called Einstein a '*Wunderkind*' and requested special permission for the age requirement to sit the Polytechnikum entrance exam to be waived. Herzog replied cautiously, recommending that Einstein should obtain a school leaving certificate first, but agreed that in the event that his advice was rejected, Einstein could indeed take the examination in October, subject to written confirmation of his abilities from the Headmaster of the Luitpold Gymnasium. For the latter, presumably Einstein's letter from his maths teacher would have had to suffice. When

Maier reported Herzog's permission to Pavia, the fog surrounding Einstein's future seemed to be about to clear.

Einstein travelled to Zurich for the entrance examination, which took place over several days starting on 8 October. He stayed with Maier and is quite likely to have been accompanied by his mother. He took papers in two groups: general and scientific. The majority of the subjects were examined orally, although a written German essay and technical and free-hand drawings were also required. He did well in physical science and mathematics, but in particular his knowledge of history and modern languages was considered inadequate, and he was failed. When the results were announced on 14 October, they were no surprise to Einstein, who had emerged crestfallen from the oral examinations. As he remembered almost sixty years later, 'The examination gave me painful evidence of the patchy nature of my education, despite the examiners being patient and understanding. I thought my failure was thoroughly deserved.'

All was not lost, however. Einstein's poring over his books when he could have been playing the violin with Ernestina had not been in vain. His examiner in physics had been Professor Heinrich Weber, who had been impressed enough to suggest that if Einstein remained in Zurich, he could attend his physics lectures. Presumably, in typical Einstein fashion, his study hours had been devoted to those subjects that he enjoyed; others, such as geography and French, had been put aside. The examiners recognised his precocious ability in maths and physics and Herzog wrote to Hermann and Pauline suggesting that Einstein should attend the final year of a Swiss school. If he graduated successfully, he would be admitted to the Polytechnikum the following autumn. Even the question of which school could be settled quickly. Maier was an old friend of Jost Winteler, a teacher at the school in Aarau. Winteler shared Maier's radical politics. He had a large family and a comfortable home in Aarau and often took in lodgers who attended the school. Maier wrote to Winteler, and he agreed to take Einstein into the family home. When word reached Pavia, Einstein hurriedly packed his books and clothes and, violin in hand, caught the first train to Zurich. Time was of the essence since the academic year in Aarau, which was divided into four quarters, began in April and Einstein would have to fit in as best he could half-way through. He arrived in Aarau just before 26 October and took an entrance assessment. He was judged worthy of entrance, although his status as a Wunderkind was somewhat tarnished by his receiving only the second-highest possible mark for physics and maths. His chemistry was found seriously wanting and his French execrable; remedial teaching would be arranged. On 26 October 1895, Einstein was admitted to the Aargau Kantonsschule and began a new phase in his life.

Chapter 2

School and first loves (1895–1896)

It was an autumnal day in 1895 when Einstein walked the short distance from the Aarau railway station to Rössligut, the Winteler's large and comfortable house. As he passed the Cantonal School, he must have dreaded returning to the iron discipline and rote learning that had characterised his experience in Munich. However, the Cantonal School was a very different sort of establishment to the Luitpold Gymnasium. Founded in 1802 by public subscription, it occupied a three-storey building built some years earlier as a hospital. It offered pupils either a traditional Gymnasium-like course, emphasising the Classics, or a 'trade school', which included substantial elements of science and mathematics. The Cantonal School had gained sufficient reputation that, after the Polytechnikum in Zurich was founded in 1855, its exit diploma qualified students for direct admission without taking the Polytechnikum's entrance examination. By the time Einstein arrived in Aarau, after scraping through his entrance test, it was relatively easy to 'mix and match' courses from any of the strands of study a pupil could follow. The classrooms were still the original dark and gloomy eighteenth-century structures. However, he would have seen a new building taking shape, four storeys high surmounted by a clock tower, which was completed the following spring.

Not only was the Cantonal School much more congenial than the Luitpold, but Aarau was also much less tainted with anti-Semitism than Munich. In the first part of the nineteenth century, the town contained the largest Jewish population in Switzerland, attracted by the unusually early granting of full citizenship rights to Jews in 1824. The town was small, having 7,000 inhabitants by 1900, and dominated by agricultural activity. Only thirty minutes from Zurich via the frequent train service, it is set in the valley of the River Aare as it cuts its way through the rolling hill scenery of the Jura, carrying the icy waters of the Aare Glaciers to the Rhine at the Swiss border town of Koblenz.

The Winteler household could not have been better suited to Einstein's needs. Here he found a surrogate family and surroundings much calmer and more relaxed than his recent domestic circumstances, perturbed by the ever-present business worries and changes of address. The patriarch Jost Winteler was a remarkable man, with achievement in subjects as diverse as politics, ornithology, and poetry. However, it was in linguistics that he was most distinguished, his work in phonetics being considered a milestone in the subject. Born in central Switzerland in 1846, he studied philology in Jena, where he conceived a deep detestation for Prussian militarism. He was happy to return to Switzerland as a teacher, but his career was characterised by a variety of indiscretions and conflicts with the authorities, which meant that he passed through three schools in nine years. He arrived in Aarau in 1884, where he found the tolerant and relaxed atmosphere much more to his taste. He remained for twenty-five years, teaching a variety of subjects but concentrating on history. His wide interests and hobbies, as well as his liberal and enlightened views on politics, made Rösslígut a stimulating environment for his seven children, as well as his young lodger. His wife Pauline was the stable centre around which family life revolved, beloved of all who came into contact with her, not least Einstein. Pauline and Jost had married in November

Einstein. Brian Foster, Oxford University Press. © Brian Foster (2026). DOI: 10.1093/oso/9780198794875.003.0002

1871 while Jost was studying in Jena, Pauline's native city; that their eldest child Anna was born in March 1872 indicates that there must have been considerable haste in arranging the marriage ceremony. Children followed in regular succession, a further six between Anna and the youngest, Paul, who was born in 1882.

Einstein slotted in between Mathias and Jost Junior in age and found 'a welcome and understanding for his idiosyncrasies and immediately felt completely at home there'. Indeed, within a day or two of arriving, he had written to tell his parents how well he had settled down. The free and wide-ranging discussions at the Winteler dinner table had made a strong impression upon him. In addition, there was a link with his real family; arriving at the same time and also put under the supervision of Winteler, although he lodged with a neighbour, was his cousin Robert Koch,[1] his almost exact contemporary and son of his uncle Jakob Koch (not to be confused with Einstein's other uncle Jakob, his father's brother and partner). Robert must also have been a regular at the Wintelers' family gatherings.

Happy and stimulated in his domestic environment, Einstein soon settled into his new routine. For the first time in his life, school, if not an unalloyed pleasure, was at least not a torment. In particular, he was lucky in his physics teacher and the Rector of the School, August Tuchschmid:

> In his inspired teaching Tuchschmid aimed for depth rather than breadth. He insisted before his classes that the concepts of the infinitesimal and the differential were keys to all advanced physics and mathematics. A former pupil of his recalled that Tuchschmid's introduction to electric current was a shining piece of experimental physics which equalled university lectures. In all his courses Tuchschmid strove for clarity, intelligibility, and precision. He omitted those parts of physics not based on rigorous scientific foundations or those areas giving play to theoretical speculations. He never engaged in chatter on physical subjects and spared no effort to convey fundamental laws and phenomena.

Two other teachers at the school impressed Einstein and were impressed by him. Friedrich Mühlberg was a distinguished researcher in geology whose research output was comparable to that of a university professor. He was the leading expert on the geology of the Jura mountains. Although Einstein did not take classes in geography, since these were taken in the first two years of the curriculum, which he had missed, he interacted with Mühlberg through lessons on natural history. Mühlberg had written the authoritative work on the flora of the Aargau canton. He also organised school outings to areas of geological interest. Mühlberg must have succeeded in awakening Einstein's interest in geology, since he was subsequently to take an elective course on this at the Polytechnikum. Heinrich Ganter, Einstein's mathematics teacher, had written an influential textbook on geometry. He treated his pupils as adults and was concerned as much with the development of their characters as with imparting the elements of mathematics.

The most important activity for Einstein in Aarau, as indeed it had been both in Munich and Pavia, was music. He played at school, at Rössligut, and anywhere else he could. The Wintelers were no exception to the cultured middle-class tradition of domestic music. Several of the Winteler children played the piano and also sang. Einstein lost no time in opening his violin case and joining in. His favourite partner was Marie, two years older, who was an accomplished, as well as an exceedingly pretty, pianist. However, he also found music partners throughout the town and in the school itself. Music was an important school activity and also formed part of the curriculum. Einstein received instrumental instruction, restarting the lessons that he had abandoned for self-tuition in Munich. He did very well in his music examinations; indeed, at the end of his first

[1] Why Robert also arrived half-way through the term together with Einstein, given Einstein's eccentric route to Aarau, is unknown but illustrates yet again the strong links within the extended Einstein clan.

quarter his marks for music, specifically his violin playing, were higher than for any of his other subjects, including physics, and only equalled by mathematics. There was a school orchestra, of which Einstein became an enthusiastic member.

A school friend from that time, Hans Byland, who although a year older was in the same class as Einstein, played the double bass in the orchestra and was also a pianist. He gave a vivid picture of the intellectual and musical Einstein of these early months in Aarau:

> He walked about with his grey felt hat pushed back on his head of luxuriant black hair at the rapid, indeed I could almost say manic, pace of a restless spirit who carried around a world of thoughts inside his head. Nothing escaped the piercing glance of those large bright brown eyes. Whoever approached him was in the presence of an intellectually curious personality. The quizzical curl of his top lip over his protruding lower lip warned off the Philistine. Unrestricted by any convention, the smiling philosopher took his stand against the world, his caustic wit flaying unmercifully anyone vain or artificial. In conversation he was assured, his taste informed by travel – his parents lived at that time in Milan – which gave him confidence in his judgements. He was not shy in giving his opinion, whomever it might offend. This courage in defending the truth as he saw it was stamped on his whole personality and eventually impressed any opponent. The at that time exuberant club and beer culture left him cold: 'Beer makes you stupid and lazy' he quoted Bismarck. As a substitute, the sixteen-year-old became otherwise inebriated, with Kant's *Critique of Pure Reason* and decided on theoretical physics as being his future field of research [. . .] One day we met in the impressive dining-room of the school to play Mozart Sonatas. As he began to play, the surroundings seemed to recede. For the first time I was hearing the real Mozart in all the Grecian beauty of its clear lines, now full of elfin grace, now powerful. 'That's heavenly', he said, 'we must do it again.' His playing was full of energy! I hardly knew him any longer as this brilliant cynic who offended everyone. He couldn't help being one of those people who have a dual character; who know how to project a rough exterior to protect the delicate workings of an intensely emotional sensibility. Fate had decreed that this precise thinker should pitch his tent with the romantic Winteler family, where he felt very happy. Then, as now, Schumann lieder were a necessity, for example *Der Nussbaum*, *Die Lotosbluete*, etc. Heine, his favourite poet, would have appreciated this music.[2] However, often before his last notes had died away, he would destroy the mood, pitching us out of our heavenly trance by making a joke. He hated to be thought sentimental and quickly damped down his highly combustible emotions within his controlled and cool personality.

A picture of the domestic Einstein in this period is given by the eldest Winteler child, Anna, who remembered him as not going out a great deal and often working on his own, but nevertheless certainly not unsociable. He regularly sat around the dining table in the evening and took a full part in family discussions. He was a keen participant in the regular Sunday family walks in the beautiful Jura scenery that surrounded Aarau and would get involved in long philosophical discussions with Jost Winteler, to whom he would also explain his latest ideas in physics. As Einstein's thoughts on physics began to deepen, one such question was on the nature of light. He began to ask himself questions—the 'thought experiments' that would characterise his future physics research. At this time he was particularly interested in the behaviour of light emitted from sources moving with respect to the aether. He tried to imagine what would happen if one could travel at the same speed as a ray of light and was puzzled by the seeming contradictions on the propagation of a wave – he wrote: 'One would see a time-independent wave field. But such a thing does not appear to exist!'[3]. The seeds that had been sown by Bernstein's books in Munich and the electricity journals

[2] *Die Lotosbluete* is a setting of a Heine poem.

[3] It should be noted that several authors have doubted the accuracy of Einstein's recollection here. Given what he knew in Aarau, there is no reason why he would have believed that a stationary light wave could

in Uncle Jakob's office were beginning to germinate in the fecund atmosphere of Aarau; they would eventually grow into the theory of relativity.

The biggest musical excitement of Einstein's first months in Aarau was the appearance of the great violinist Joseph Joachim, the friend and collaborator of Johannes Brahms. Byland's memoir is quoted by Seelig: 'When it was known that the great violinist Josef Joachim would appear in Aarau and play the G-major sonata of Brahms in 1896, Einstein diligently prepared himself by practising it for months beforehand.' While Byland's memory of Einstein's excitement at Joachim's visit rings true, after thirty years the details had become hazy. In fact, Joachim's concert with the pianist Robert Freund took place in the afternoon of Sunday 17 November 1895. Joachim and Freund played the Brahms Sonata and some violin miniatures by Schumann. Joachim played the Chaconne from J. S. Bach's D Minor Partita for solo violin, BWV1004. The rest of the programme was taken up by pieces for mixed choir by Schumann, for women's voices by Brahms, and for men's choir by Max Bruch, sung by the Caecilien Society choir. This was one of Aarau's most venerable institutions and was responsible for organising the subscription concert season. Since the concert took place less than a month after Einstein's arrival, clearly Byland's memory of Einstein 'playing [the Brahms Sonata] for months' is somewhat of an exaggeration. Given Einstein's well-documented and admitted dislike of practice of any kind, what probably happened is that he played it through a few times with Byland in the weeks before the concert.

Also in November, a staff meeting of the Cantonal School noted that Einstein was still receiving remedial lessons in French, chemistry, and natural history, and that his request to be excused singing and gymnastics had been granted. Clearly his musical activities did not extend as far as using his voice. As a foreigner, he was also excused from the military training that all prospective Swiss citizens were expected to undergo. The remedial lessons seemed to have some effect as his marks at the end of the first quarter in December in at least natural history were good; in chemistry, he received high marks for effort, but it was remarked that he hadn't yet been assessed in class as he was still catching up. His French remained fairly poor, certainly not as good as his Italian, strengthened as it had been by his interactions with Ernestina, with whom he was still in written communication. It was probably at some time during this six months' absence from Pavia that he sent Ernestina his second work of physics, copied by his mother (referred to in Chapter 1) and which she identified as having received in 1896.

As autumn turned into winter in Aarau, Einstein felt sufficiently at home at Rössligut that he decided that he would remain there over the Christmas vacation. In fact, the magnet holding him in Aarau was Marie. Sometime just before Christmas, their musical partnership became a romantic one. In fact, Einstein had a rival from his own family; both he and his cousin Robert were contending for Marie's favour. On the Saturday before Christmas, Einstein and Robert travelled into Zurich, to make several visits and presumably to shop for Christmas and New Year's gifts for the Wintelers. Jost expressed some surprise that they had not made time to visit their mentor, Gustav Meier. Ernestina sent Einstein seasonal greetings on a postcard on 25 December.

After the Christmas festivities and as a greeting for New Year, both of Einstein's parents wrote to the Wintelers, thanking Jost for his pastoral care and the news he sent of their son's progress. The letter from Pauline Einstein has a section addressed to Marie Winteler that makes it clear that, by this time, Einstein had beaten his rival Robert for Marie's affections. Not only had Einstein

---

not exist – indeed, it is predicted in theories of the aether. It may well be that he conflated a thought-train relating to a different type of theory of light, known as 'emission theories', which puzzled him later in his search to understand light propagation; or, of course, he could have misunderstood something he had read and been led to the right conclusion for the wrong reasons!

and Marie become romantically involved, but also they proudly informed both sets of parents, although it could hardly have been hidden from the Wintelers. Pauline's words make it clear that she is overjoyed with this news and apparently Marie's parents, with typical tolerance, also viewed it benignly. Once again, music had brought Einstein into intimacy with an attractive girl. She was charmed by the not yet seventeen-year-old violinist, two years younger than she, and fell deeply in love with him. He too felt passionately for Marie and this holiday period was a particularly happy time. They exchanged many letters over the winter, since Marie was away for most of the week at the school in nearby Niederlenz, where she was standing in for an absent teacher. The tone of these letters contradicts Byland's description, written many years after the event, of Einstein's 'controlled and cool personality'. They are passionate, affectionate, and indeed at times mawkish; Einstein was clearly very much in love.

Einstein's musical activities were central to his relationship with Marie. His letters to her are full of references to their playing together. In February he told her that 'Music has bound our souls together so wonderfully.' At the end of the winter semester and just before his return to Pavia, the playing of Einstein and sixteen of his schoolfellows had been examined by a local musical luminary. After commenting that there was a general tendency to stiffness in the bowing arm, the examiner was nevertheless pleased with the technical accomplishments and intonation of the violinists. Einstein was picked out for special praise; he 'gave a brilliant and deeply insightful performance of an Adagio from a sonata by Beethoven.' He also played in public: the Bach Chaconne,[4] as well as sonatas by Händel and Mozart and Tartini's 'Devil's Trill' Sonata. The school's musical director, Franz Rödelberger, organised an evening concert at the Aarau church, for which he had made an arrangement of the Bach 'Air'.[5] Einstein played first violin. His rich tone and accurate keeping of time impressed the second violinist, Hans Wohlwend, who was in a less senior class. He asked him, admiringly, whether he counted the beats. 'Oh no', laughed Einstein, 'it's in my blood.' During the weekends, Einstein often accompanied Wohlwend to his parents' house in Lenzburg, a small town about a ten-minute train journey from Aarau. Here they played Schubert and Schumann lieder with Hans's mother, who had an excellent voice, as well as playing chamber music with others from the town in the Wohlwends' musical salon.

Einstein's school life continued on a satisfactory, if not outstanding, course. It is rather difficult to be sure of his progress because of the very confusing school marking system. When Einstein arrived at the Cantonal School, the best mark in assessments was 1 and the worst 6. After the end of the school year in April 1896, this was inverted, to match that used for the school leaving certificate. In addition, some subjects, such as Italian, were examined at the end of the third year of study (for Einstein, April 1896, so that he completed the third year's work in six months) and carried forward to the final certificate. Certainly, Pyenson was misled by this confusion into believing that Einstein had done very poorly initially and made a dramatic improvement in performance after April 1896; in fact, he continued to excel in physics and mathematics and to perform consistently well in all other subjects except languages and chemistry, with some improvement in the latter subject towards the end. This ambiguity extends to the *Collected Papers of Albert Einstein* in which his mark of five in Italian is interpreted as a 'low' five rather than a 'high' five. Given that it had been six months since he had been at home in Pavia, where he exercised his Italian in conversations with Ernestina, it had almost certainly deteriorated, so either interpretation is possible. On the other hand, given the fact that the '5' appears in his final certificate without any qualification alongside

[4] Presumably the Chaconne from the D Minor Partita referred to above.

[5] Presumably the Air from the Third Orchestral Suite, BWV 1068.

all other marks which are clearly 'high' 5s and 6s, it would be impossibly misleading if it were a 'low' five.

The school year ended on 8 April and Einstein travelled back to Pavia for the first time in six months. His first absence from Marie was marked by a letter of extravagant affection, beginning with 'Beloved Darling' and ascending from there. As Highfield and Carter remark:

> Einstein's excursions into the language of love always tended towards the treacly, and this letter [. . .] sets a tone that is echoed in his future correspondence with [his future wife] Mileva. Einstein tended to be most affectionate when the objects of his love were at a safe distance, as if he could then make of them what he wanted.

While Marie was temporarily at a safe distance, Ernestina was close at hand and it is clear that they remained, at the very least, friends. Shortly after his arrival in Aarau the previous autumn, Einstein had visited the photographer Wolfsgruber to have his portrait taken. A print was sent to the Marangoni family. Einstein inscribed it in the heavily playful manner that became his custom with female intimates: 'To Mrs but not Miss Marangoni from your friend Albert Einstein' (Figure 2.1). A certain uneasiness can be discerned in his letter to Marie:

> My mother . . . makes fun of me because the girls of whom I was previously enamoured no longer please me [. . .] How different it is to play a simple, sweet song with my darling than to master an, admittedly difficult, sonata with stiff, elegantly dressed ladies from Pavia whose ideal is 'as fast as possible while making as few mistakes as possible' or 'to extract oneself from a situation as genteelly as possible'.

He goes on to make disparaging remarks about various aspects of life and the people in Pavia, entirely at odds with his later recollections or his contemporary feelings as recorded by Maja. Einstein's discomfort at making such an early start on his long career of 'infidelity', if such a word can be used of a youth of seventeen, is plain to see. The letter implies that he continued to play chamber music in the Villa Marangoni, managing to hold in check the homesickness and pining of which he complained to Marie. At any rate, he had not long to suffer since term began again on 29 April, only eight days after his letter was posted. The visit to Pavia had however been coloured by a real sadness and concern: it was clear that his father's business was once more sliding into trouble.

The major reason for the foundation of the factory in Pavia had been the prospect of the contract for the electrification of the city. In January of 1896, the Einstein & Garrone Company had placed an advertisement in the local newspaper announcing their intention to use hydroelectric power by diverting water from the canal that ran alongside their factory; the connection fees were to be announced in a subsequent advertisement. This never appeared, since the company got into difficulties over the water rights they were expecting to exploit. In February, the Volta Society, with whom the Einsteins had been collaborating, also published a notice in the newspaper in which they announced that they would be competing with Einstein & Garrone and that they had applied for the water rights to extract water from the canal. Clearly something had gone badly awry, probably due to lapses of communication with the Volta Society and some carelessness and unwarranted assumptions by Einstein & Garrone on what they were allowed to do with the canal water.[6] The firm's situation deteriorated rapidly thereafter.

---

[6] See also Note 21 to Maja Einstein's 'Beitrag für sein Lebensbild', which, while generally agreeing with the conclusions stated here, has a different timeline for the events.

Figure 2.1 Einstein at age 17.

Since the business situation already appeared grave when Einstein left Pavia, he would have been relieved both to escape from the prevailing gloom and to be reunited with Marie. He returned a few days before teaching began in order to witness a joyous occasion, as 26 April saw the dedication of the new school block that had been taking shape since his arrival. It was a gorgeous sunny spring Sunday as Tuchschmid presided over the ceremony that inaugurated, among other fine facilities, his own physics laboratory with some excellent equipment, including a dynamo and precise galvanometer. Einstein's lessons began again in these new facilities on 29 April 1896.

As the term wound down and the final examinations loomed, Einstein joined one of Prof Mühlberg's famous school outings to the mountains. On 24 June in glorious sunshine, the twenty students took the 140-km train journey eastwards to Appenzell. From there they gradually climbed 700 metres, following a beautiful, glaciated valley to the mountain hut at Meglisalp. On the next morning they set out for their goal, the summit of the Säntis at 2500 metres above sea level. This involved a further 1000-m climb, mostly along a steep ridge with precipitous drops to either side. Throughout the trip, Mühlberg pointed out interesting geology to his pupils. At one point, he asked Einstein whether he thought some rock strata were rising or falling. 'It is all the same to me, Professor', replied Einstein. Further evidence of his rebellious attitude was that, despite the very clear instructions that Mühlberg would have issued, Einstein was not wearing appropriate climbing boots. When a violent thunderstorm that afternoon rendered the rock surfaces slippery, Einstein lost his footing and was about to disappear over a snowy precipice when one of the party extended his alpenstock for Einstein to grasp, arresting his slide just in time before gravity exerted a terminal influence on its future master.

A couple of weeks later, back in Aarau, Einstein was prominent in another adventure. A colourful account of Einstein at the end of his time in the Kantonschule is given by his classmate Emil Ott. It shows that, despite his thoughts mostly dwelling on matters such as whether light waves could be stationary, Einstein was still a young man capable of high jinks. Ott's account is of the Maienzug, a festivity dating back to the sixteenth century, in which the youth of Aarau marked the beginning of the summer holidays with fun, parades, a festive dinner, and flowers. The nine pupils in Einstein's final-year class decked themselves out in bizarre costumes consisting of black gym vests on which white skulls were painted, completed by black trousers and top hats. Thus attired, the class formed into a company of three ranks of three, and together with the other pupils charged into a mock battle with an ad hoc group of townspeople. The merry discharge of blanks from an assortment of rifles was only halted by a summer downpour. Ott depicts Einstein as also firing a rifle, despite rarely (if ever) having touched one in his life, and having great fun. Given his long-standing pacifist beliefs, which had already begun to crystallise in Munich, this was probably the only time in his life that he handled a gun.

Einstein's movements after the Maienzug are unclear; he is next heard of on 7 September, when he wrote to the Aarau authorities from Rössligut with a declaration of his background and the statutory request to be entered for the Matura examination. His whereabouts from mid-July until early September are unknown. His family usually visited a mountain resort of some type for several weeks during this period, but the critical situation of Einstein & Garrone may well have precluded this in 1896. It seems likely that Einstein returned to Pavia to do whatever he could to help in the winding up of the firm. This latest setback was too much for Uncle Jakob. With Hermann's blessing, he dissolved their partnership and left for a job with a large electrical concern elsewhere in Italy.[7] There seems to have been little or no subsequent contact between Einstein and his uncle; indeed, there is some evidence that he may have blamed Jakob at least in part for his father's failures. Maja, on the other hand, blamed Garrone and reported that, with the collapse of Einstein & Garrone, all of her mother's inheritance was lost, as well as substantial sums from various relatives.

Hermann, however, was not ready to lose his independence and work for another firm. He determined to try again, motivated by his awareness of how hard Pauline would be hit by this loss in status. The Einstein clan, in particular 'rich' Rudolf Einstein from Hechingen, was still ready to offer Hermann financial assistance, despite the vociferous opposition of his son to further independent business ventures. This opposition, remarkably for someone who had just written in his Matura examination that he had no practical sense, extended to Einstein taking the train to Hechingen on the spur of the moment – a 400-km round trip from Aarau – to try to dissuade his

[7] Working inside a large organization seemed to suit Jakob. By 1900 he had moved to Aachen as the manager of a power plant. He then moved to Vienna in 1906 where he had several responsible positions, ending as manager for the Vienna branch of the well-known Berlin manufacturer of electrical measuring equipment, Dr Aron GmbH. He was divorced from his wife Ida in 1909 and died in 1912.

uncle Rudolf from advancing any new capital[8]. He convinced neither Hermann nor Rudolf and the decision was made to begin a new dynamo factory in Milan. By the time Einstein returned to Aarau in September, Einstein & Garrone had gone out of business and been absorbed by Carlo Monti & Co.

A snapshot of Einstein's intellectual status in the summer of 1896 is provided by his examination for the Matura, the culmination of his time in Aarau and his passport to higher education at the Polytechnikum. His year's work had already been evaluated in his final report, dated 5 September. This presents an impressive record of achievement, with only his French letting him down badly. Some of the marks, for example for geography and Italian, were carried directly through to the Matura. Fortunately for us, this was not true of French, for which the examination contained an exercise in prose composition entitled '*Mes projects d'avenir*' in which Einstein sets out his future path in stumbling French. It is worth quoting this in full:

> 'A happy man is too satisfied with the present to dwell too much upon the future. But on the other hand, young people especially like to contemplate bold projects. Also, it is natural for a serious young man to envision his desired goals with the greatest possible precision.
>
> If I am lucky and successfully pass my examinations, I shall enrol in the polytechnical school in Zurich. I shall stay there four years to study mathematics and physics. I suppose I will become a teacher of these branches of natural science, opting for the theoretical part of these sciences.
>
> Here are the reasons that have induced me to this plan. They are, most of all, my individual inclination for abstract and mathematical thinking, lack of imagination and of practical sense. My desires have also led me to the same decision. This is quite natural; everybody likes to do that for which he has a talent. Besides, I am also much attracted by a certain independence offered by the scientific profession.

There is much of interest too in his answers to the other parts of the examination, the written parts of which took place between 18 and 21 September, while the final oral examination was held on 30 September. In his physics answer he gave a careful discussion of electricity and magnetism, the design of a compass to measure magnetic field, and a galvanometer to measure electric current. As Pyenson notes, at no point does he mention the concept of an electric field, central to Maxwell's work but much too novel to appear in Tuchschmid's lessons. Pyneson summarises his Matura answers in the following way:

> One is struck by the clear and unpretentious quality of Einstein's examination essays, strong points present as well in his later published writings. The essays begin with a statement of a generalisation of physical phenomena in terms of theoretical concepts, such as electromagnetic lines of force. Then the essays consider the problem at hand by omitting distracting albeit interesting detail in favour of the fundamentals.

The results of the Matura were decided directly after the oral examinations on 30 September; Einstein's school leaving certificate is dated Saturday 3 October and its beautiful calligraphy indicates an excellent performance, with Einstein heading the class. He received top marks for all his mathematics and physics subjects as well as history. Now he could look forward to fulfilling his earlier aim, deferred by his year in Aarau, of pursuing his love of physics at the Zurich Polytechnikum.

It is likely that Einstein returned to Pavia after completing his Matura in early October in order to help the family move back to Milan. Their address there certainly does not indicate the reduction

---

[8] It may also be possible that Einstein attended the wedding of his cousin Elsa to Max Löwenthal, which took place sometime in 1896, and took this opportunity to have a quiet word with his uncle on the subject.

in the family's resources reported by Maja. A monumental archway led to the inner courtyard of 21 Via Bigli, around which stretched eleven rooms, one of which was Hermann's office. The house, which was owned by a prince, had once belonged to Countess Maffei, a famous writer and ardent supporter of the independence and unity of Italy. Her Via Bigli salon was a famous meeting place for prominent figures in that struggle, known as the Risorgimento. They would have circulated around two impressive reception rooms, which, according to Ernestina Marangoni, who was a regular visitor to the new Einstein home, communicated 'by means of a great arc'. It was probably from this impressive building, his family's base for the next six years, that Einstein left Milan for Zurich. On 29 October, he registered himself as a resident and began his new life as a student of physics and mathematics at the Polytechnikum.

Chapter 3

Student at the Polytechnikum in Zurich (1896–1900)

In October 1896, Einstein walked into Zurich from the main train station. It was a city that he already knew quite well, having made several visits there during the previous year. While a metropolis in comparison to Aarau, with around 120,000 inhabitants, it was still provincial compared to the other cities in which he had lived, Munich and Milan. Nevertheless, it was growing rapidly and had already established itself as the business capital of Switzerland. The growth was mostly concentrated in the suburbs; the old city itself wound down along the River Limmat to the Zurichsee, merchants' houses surrounding a few impressive mediaeval churches. The new building of the Polytechnikum loomed at the top of the old town, which was flanked by the heights of the Uetilberg. Across the lake, the snowy peaks of the Alps were visible in the distance. There was a great deal of beauty and little of the squalor of an industrial city.

For the first two years of his stay, Einstein took lodgings in a large and unprepossessing apartment house run by Frau Kägi in the Hottingen district, Unionstrasse 4. The area was popular with students and a pleasant fifteen-minute walk to the Poly, ending through the Universitätsspital park. Given the parlous nature of the family finances, Einstein was constrained to live extremely simply. He supported himself on a monthly allowance of 100 Swiss francs from his Genoa relatives, of which he saved 20 francs towards the fee required to obtain Swiss citizenship when he eventually satisfied the residence requirement. The remaining money was barely sufficient to pay his rent and to eat. He was fortunate to have several acquaintances in the city who could be relied upon to provide a wholesome meal as well as support and company. These included Gustav Maier, an old friend of his parents. Many years later, Einstein wrote to the Maiers on the occasion of their golden wedding anniversary, recalling 'Your welcoming home was always open to me when I was a student, even when I arrived with boots dirty from a descent of the Uetilberg.' Another acquaintance was Albert Karr, an old schoolfriend and a distant relative of his mother. His wife was a fine singer, whom Einstein often accompanied at the frequent musical events held at the Karr home.

As Einstein registered for his courses and picked up his timetable, he would have noted that it was filled with mathematics. Many physics students today complain in their first year that although they have signed up to do physics, they seem to have wandered into a mathematics degree instead. The necessity to take on board a large slice of pure and applied mathematics in order to approach many of the most important problems in physics was at least as marked in Einstein's day, particularly as his course was designed to produce teachers in physics and mathematics. Although he was soon to develop a distaste for mathematics courses, he began his work well and studiously. He attended a number of elective courses that matched the interests he had developed in Munich, including the philosophy of Kant and the works of Goethe. He met the other four students who had enrolled for his course, of which three were men: Marcel Grossmann, Louis Kollros, and

Einstein. Brian Foster, Oxford University Press. © Brian Foster (2026). DOI: 10.1093/oso/9780198794875.003.0003

Jakob Ehrat. Unusually for that period and fatefully for Einstein, there was one woman: Mileva Marić.

Einstein tried to concentrate on his exercises on differential equations but found himself distracted by a number of issues both from his real and his adopted family. The founding of the new factory and the winding up and merger of the Einstein & Garrone company rumbled on until the end of the year, with the Einstein family moving back to Milan in October.

More immediately disturbing to Einstein, because she was nearby, was Marie Winteler. She had left Aarau just after Einstein, taking up a temporary teaching position in Olsberg, forty kilometres away, near the Rhine and the German border. She was still clearly in love with Einstein and bombarded him with playful and affectionate letters. She also reinforced the bond by doing his laundry, which he dispatched to her by train at regular intervals. There is a heart-rending quality to a letter written from Olsberg in November 1896 in which she tells Einstein that she received his laundry basket and looked in vain for a little note from him but was very grateful to see his handwriting on the address label. 'I thank you so much, Albert, that you want to come to Aarau . . . with your beloved child, your violin, on one side and your other child[1] (also beloved?) on the other.' The lack of any personal note with the laundry basket indicates that, by then, Marie was much more engaged with the relationship than was Einstein. Indeed, later in her letter, after recounting her ninety-minute walk through the woods in pouring rain to the rail station to return the washing, she asks what he meant by telling her, in a letter now lost, that they should no longer correspond. Apparently, Einstein was happy to continue to have his clothes washed by Marie but wished in other matters to distance himself from her. However, he relented sufficiently to break his silence at the end of November, complaining in his letter about a teapot that she had sent him. A wish to avoid awkward intimacy with Marie, after she returned home from Olsberg in mid-December, probably influenced Einstein's decision to spend the short Christmas holidays in the new apartment in Milan rather than in Aarau. He caught the train back to Milan around 20 December.

At some time in the winter of 1896–1897, Einstein made a new friend, as usual through his passion for music. Michele Angelo Besso would become as close to him as anyone except Maja; he would play an important role in the formulation of both special and general relativity and remain a friend for life. Besso, who came from an Italian Sephardic Jewish family, was born near Zurich six years before Einstein and like him, gravitated from the family home in Italy to the Polytechnikum, where he too took the physics and mathematics course, graduating in 1895 with higher marks than Einstein was eventually to achieve. He took a position in an electrical factory in a suburb of Zurich. Einstein had lost no time after arriving in Zurich in establishing musical connections. As well as playing sonatas with some of the Polytechnikum academic staff, he also frequented an informal salon hosted by Selina Caprotti, where sympathetic students were given sustenance – in Einstein's case much needed – and entertained themselves and their hostess with music. Einstein was still in contact with both his musical friends from Aarau, Wohlwend and Byland, and played sonatas at Caprotti's with Wohlwend.

Having met at Caprotti's, Einstein and Besso sat down and began to talk; immediately they became absorbed in a discussion of the nature of light and the contrasting views of Newton, who contended that light consisted of small particles ('corpuscles'), and Huygens, who believed that it was a wave. The perfect foil for Einstein, Besso was highly intelligent and unorthodox, a dedicated follower of the philosophy of Ernest Mach (to which he introduced Einstein), but never ambitious,

[1] This is meant to refer to Marie herself.

particularly for academic success. Indeed, he often exasperated Einstein, hardly himself the most worldly of people, by his scatter-brained attitude to everyday activities. As they sat in a corner in the comfortable middle-class house in Zurich, trying to make themselves heard over the amateur music-making, they immersed themselves a conversation about physics and its philosophy that would last for the rest of their lives.

Since Besso was also a Jewish physicist and a violinist, he and Einstein could hardly have had more in common. Indeed, it transpired that they also shared another enthusiasm: for Jost Winteler's daughters. Einstein subsequently introduced Besso to Anna Winteler, who was probably acting as Einstein's pianist, at one of Caprotti's evenings; Besso and Anna Winteler married in 1898.

It was typical of Einstein's questioning and sceptical attitude to physics that he and Besso should have spent time in Frau Caprotti's salon discussing what was then considered a solved problem. As Einstein began his study, the foundations of physics seemed reassuringly solid. Physicists at that time considered that the controversy over the nature of light had been decided long ago in favour of Huygens.[2] However, this was a rare setback for Newton's ideas, which elsewhere reigned supreme. He had established his laws of motion, which are still used today to describe the overwhelming majority of the situations experienced in everyday life (e.g. the behaviour of moving vehicles, the collisions of pool balls, the necessary strength and composition of building materials). Newton also promulgated his law of gravitation, which explained the motion of the planets around the sun. Thus he set what he passionately believed to be God's creation in a framework of absolute time and space, within which cause produced effect in a perfectly predictable way. As far as Newton was concerned, 'absolute' space was not empty, but filled with God, who was omnipresent.[3] However, quite how distant planets were affected by the sun's gravity was puzzling to Newton, who wrote:

> That Gravity should be innate, inherent and essential to Matter, so that one body may act upon another at a distance thro' a Vacuum, without the Mediation of any thing else, by and through which their Action and Force may be conveyed from one to another, is to me so great an Absurdity that I believe no Man who has in philosophical Matters a competent Faculty of thinking can ever fall into it. Gravity must be caused by an Agent acting constantly according to certain laws; but whether this Agent be material or immaterial, I have left to the Consideration of my readers.

In fact, he made a variety of suggestions as to what this medium might be, essentially a mixture of various sorts of 'aether', each type with mechanical properties that separately supported disturbances (such as light, electricity, gravity).

Over the next several centuries, natural philosophers assimilated Newton's entire world view. The developments in physics from the seventeenth to the end of the nineteenth century can be characterised by increasing application of Newton's laws and a reduction in the centrality of the aether. The problem of an aether essential to the propagation of light but without detectable mechanical properties would play a key part in the story of special relativity. Einstein himself, writing many years later, recalled that the greatest impression that the new student of

---

[2] More or less as Besso and Einstein spoke, Wilhelm Wien was writing a paper in which he proposed an approximation that described the spectrum of radiation from a heated 'black' (perfectly absorbing) body at short wavelengths. However, it failed at long wavelengths. It was the attempt to modify Wien's Law that a few years later led Max Planck to propose the idea that energy was discretised in units called 'quanta'.

[3] It is amusing to note that things have now come full circle. Modern theories of particle physics consider the vacuum to be filled with an omnipresent field, the Higgs field, that generates mass and the 'God particle', a name coined by Leon Ledermann, otherwise known as the Higgs boson.

nineteenth-century physics received was surprise at the success of the application of mechanical principles in areas without any apparent connection to mechanics; he cited in particular the theory of gas properties, such as the internal resistance to their flow (known as viscosity), how different gases mix when placed in contact (known as diffusion), and how they conduct heat. These properties of gases are obtained from statistical averages of countless collisions of atoms treated as real physical objects, rather than simply as parameterisations of chemical processes. The success of this approach was due mainly to the work of Ludwig Boltzmann and James Clerk Maxwell, before Einstein was born; it was bitterly resisted by many influential scientists, in particular Mach, and Wilhelm Ostwald. Nevertheless, the atomistic theory and the concomitant sway of Newtonian mechanics won out. Flushed with success, proponents were convinced, as Maxwell had been, that his equations describing electromagnetic disturbances could also be reduced to mechanical interactions. Einstein mentions explicitly how Maxwell's equations and their description by an 'electromagnetic field' was the crucial concept that allowed him to shake off the dead hand of Newton and to look towards a new formulation of physics. He was encouraged in this by Mach's writings, in which he continued to fulminate against Newton and all his works, or at least his philosophical assumptions, such as the reality of absolute space and time. Mach's book *Die Mechanik in Ihrer Entwicklung* poured scorn on the very idea that anything that could not be measured, such as the existence of an absolute space, could have any importance in science. This book, recommended to Einstein by Besso in one of their earliest conversations, was among the most important influences on Einstein's early development as a physicist.

As Einstein laboured over his mathematics textbooks and enjoyed Geiser's lectures on infinitesimal geometry, which he characterised as masterworks of the pedagogical art, cracks were beginning to appear in the imposing façade of nineteenth-century physics. In 1895, Wilhelm Röntgen had discovered X-rays, the first indication that atoms not only exist but have an internal structure. Albert Michelson and Edward Morley had in 1887 carried out their famous experiment in the United States, in which light travelling along paths perpendicular to each other can be combined and differences in velocity along the two arms accurately determined. If the earth was moving through a stationary aether that supported the transmission of light, then the velocity of light would be changed parallel to the earth's motion but unchanged perpendicular to it. Michelson and Morley failed to detect any change whatsoever. Following Wien's publication in 1897, the shape of the radiation spectrum from heated bodies was proving particularly difficult to explain in classical theory. These fissures were not particularly apparent to the leaders of the field. Lord Kelvin is often wrongly quoted as having announced to the British Association that 'There is nothing new to be discovered in physics now. All that remains is more and more precise measurement.' However, Arthur Michelson did state that:

> While it is never safe to affirm that the future of Physical Science has no marvels in store even more astonishing than those of the past, it seems probable that most of the grand underlying principles have been firmly established . . . An eminent physicist remarked that the future truths of physical science are to be looked for in the sixth place of decimals.[4]

[4] A prediction perhaps only surpassed in its spectacular wrong-headedness by that of William Pitt the Younger, Prime Minister of Great Britain, who told the House of Commons in February 1792 that 'unquestionably there never was a time in the history of this country when, from the situation of Europe, we might more reasonably expect fifteen years of peace, than we may at the present moment.' Two months later, France declared war on Austria and Prussia, thus precipitating more than twenty years of sanguinary conflict that redrew the map of Europe.

Never safe, indeed. In fact, physics was on the verge of a revolution more profound than anything in its history. Albert Einstein, walking quietly to the Polytechnikum for his classes every morning, was destined to be one of the chief revolutionaries.

The spring break between semesters in March 1897 allowed Einstein to return once more to Milan. His mother wrote a gossipy, intimate letter to Marie Winteler on 24 March that makes it clear that Einstein had not at that point told her anything about the cooling in their relationship. His mother reported on how well her son was looking, his high spirits, the size of the appetite that he had developed, and his continuing obsession with music. She almost seemed relieved to record that she had now handed over her role as accompanist to Maja, who had become a fine pianist. Marie's feelings on receiving this letter must have been mixed: pleasure at hearing about her adored boyfriend's activities together with pain at not hearing from the man himself. On Einstein's return for the new semester, matters came to a head. The pattern of events is not entirely clear, but it seems most likely that Einstein ended their relationship in the second half of April or early in May. This can be deduced from two sources: letters that Einstein wrote to Pauline Winteler in May and June, and the testimony of Julia Niggli, yet another female musician attracted into Einstein's orbit.

Julia Niggli was the daughter of Aarau Town Clerk Arnold Niggli, who was also a distinguished man of letters. Among several other works, mostly about music, he had written a short biography of Schubert that became very popular, particularly in Vienna. Six years older than Einstein, Julia was a highly intelligent and creative woman, whose education included stays abroad and a certificate from the University of Zurich qualifying her to teach English and French. She had been working in Italy while Einstein had been at the Cantonal school and thus only met him for the first time on her return in 1897. Her memoir, written almost sixty years later, gives nevertheless a very sharp picture of that first meeting and of the Winteler household in 1897.

Julia and Einstein met one morning at the Saturday market, which was first held in Aarau in 1283. It took place in the Graben, a loop of a road following the line of the old city walls and enclosing a space planted with plane trees under which the stalls were set out. Julia's description of the encounter between Frau Winteler and herself makes it likely that it took place after the winter, with circumstantial evidence pointing to a weekend in April, perhaps directly after Einstein returned from Milan. Frau Winteler, who knew Julia well, introduced Einstein, who was carrying her basket as she did her shopping. As she bought some fruit, Einstein began the conversation. As ever, his opening gambit concerned music: 'I hear that you play the piano. We could play something together – I have brought my violin with me.' Returning to her basket with her purchased goods, Frau Winteler confirmed an invitation to coffee and cake. A few hours later, Julia arrived at Rössligut with some trepidation, as she did not consider herself a particularly good pianist. She was relieved when Einstein remarked that his favourite pieces were Schubert songs, with which she was well acquainted through her father's enthusiasm for the composer. They played pieces from *Die schöner Müllerin.* Julia was entranced:

> Albert Einstein played the song melodies on his violin with an intimacy and fiery passion that tore at my heartstrings. By some mysterious process he matched his violin tone to the so-familiar text in such a way that he enhanced it with a power beyond that of the words themselves. It was as if something only implied by the words had been called into life. I was particularly impressed by his rendition of the song 'Trockene Blumen' [. . .] As the last note died away, Albert Einstein wrapped up his violin in its silk cloth as a loving mother might wrap her child and carefully placed it in its case. 'Now come and have some coffee,' called the Frau Professor, the considerate hostess, through the open door of the neighbouring room, 'You've played well.'

Julia's report of the talk around the coffee table makes no mention of the fact that Marie and Einstein were romantically involved, and there seems to have been no awkwardness. Presumably the young lovers were still formally together but the fact that Einstein had been so keen to play with Julia rather than with Marie speaks volumes.

The parting of the ways between Einstein and Marie at around this time is circumstantially confirmed by two letters that Einstein wrote to Pauline Winteler just before, and during, Pentecost, which, in 1897, fell in early June. They give an interesting reflection on his character at the time. They are intensely emotional and are eloquent testimony to his deep affection for Frau Winteler, who has clearly filled a place left partially vacant by his sometimes-difficult relationship with his mother. In the first letter, he announced that he would not after all be coming to Aarau for the Pentecost holiday. The letter is undated but the most likely time for its composition is sometime in May, and after the meeting with Julia Niggli described above. At that time, apparently, Einstein had no difficulty in interacting socially, if distantly, with Marie. Now, however, Einstein wrote that he really didn't believe that it was fair on Marie if he were to return, as he would have liked, to Rössligut. He admitted the pain that he had already caused Marie. He felt less guilty since his sacrifice in giving up his anticipated visit to Aarau had caused him some small fraction of the pain he knew Marie felt. His regret is evident, as is the homesickness for the daily routine in Rösslígut. At the same time, he told Pauline Winteler that his solace was to be found in his scientific studies of God's works in the universe, although he admitted that this may have been just his way of hiding from the harsh realities of life. He told Pauline of his other unfailing consolation, describing his Saturday evening music-making with Byland at Selina Caprotti's salon. He concluded by remarking that Byland had read aloud to him various works by Gerhard Hauptmann; Einstein recommended a new play to Pauline, *Hannele's Ascension to Heaven*, which had won Hauptmann the Grillparzer Prize in the previous year and by which Einstein had been deeply moved. It seems likely that he formally ended his relationship with Marie at this time, perhaps at the end of the weekend described in the previous paragraph.

The second letter was written on the Whit Monday holiday that he had hoped to spend in Aarau but during which he was now enjoying some peace and quiet in his Zurich lodgings. Most of his letter is filled with reminiscences about Rössligut. He also informed Pauline Winteler about his activities over the holiday weekend. Unsurprisingly, they involved days and nights of making music, with an elderly lady, probably Selina Caprotti's mother. Einstein remarked that she at least didn't involve him in other more dangerous interactions. His usual accompanist Byland had returned home to Lenzburg for the holidays. Einstein had asked him to make the short trip to Aarau to call in on the Wintelers. This planned visit seemed to give Einstein a vicarious pleasure.

Although the relationship between Einstein and Marie was over for now, it continued to strongly affect Marie's life (as will be documented in Chapter 8). Einstein seemed initially to find no difficulty in moving on, while Marie, in contrast, didn't find it easy to part with Einstein; indeed, if Marie's account in old age is to be believed, Einstein's mother arrived in Aarau around this time to see if she could do anything to fix up the relationship, of which she had always approved. This bizarre visit, and indeed Einstein's attitude, may have been related to the first stirrings of a more enduring relationship of which Pauline Einstein certainly did not approve: that with Einstein's fellow physics student, Mileva Marić.

Mileva Marić was born a citizen of the Austro-Hungarian Empire but considered herself a Serb. Austria–Hungary was a complex state, forged from remnants of the Holy Roman Empire after the French Revolutionary Wars, bolted together with the hereditary possessions of the Habsburgs: Austria, Hungary, and Bohemia. The addition of Hungary to the name of the Austrian Empire in 1867 was the result of decades of agitation among the Magyar nobility for equal status with the German ruling classes. Austria had historically been the bulwark protecting Europe from the Turk, who had pushed up to the walls of Vienna on several occasions, the last of which being as recently as 1683. The Serbs had been at the forefront of that long struggle and nursed centuries of resentment at the forces of Islam that had conquered their homeland and subjected them to fierce persecution. As the Turkish Empire fell into the decline and the decadence that led to it being known as the 'Sick Man of Europe', both Austria–Hungary and Russia moved into the vacuum, dismembering the European part of the Turkish Empire via a mixture of territorial acquisition and the creation of an array of client states. The area surrounding Titel, the small town in which Mileva was born in 1875, had been allocated to the region of influence of the Hungarian part of the Dual Monarchy in 1867. The Magyar influence was steadily growing, displacing Serbian speakers with those whose native language was Hungarian.

Born into this melting pot of races and cultures in 1875, Mileva had a comfortable upbringing. Her father had made his career originally in the army but had left it early to become a prosperous civil servant. He owned substantial property in the region inherited from his wife's parents. Mileva was born with a congenital defect, a displaced hip that was not noticed until she began to walk with a limp, by which time it was too late to cure. Eight years elapsed between her birth and that of her younger sister and brother, so that she had for that period the undivided parental attention received by an only child, and she remained her father's favourite. She loved music and began piano lessons at eight years of age, initially making excellent progress. She was a talented artist. At school, it quickly became clear that she had a facility for mathematics and as she progressed, she became fascinated by science in general and physics in particular. When her father was appointed to an important administrative position in the High Court at Zagreb in 1891, he pulled strings to get her admitted to the exclusively male Royal Classical Gymnasium in 1892. She herself requested permission to be admitted to the physics class in 1894. The mental, and probably physical, torments that she must have had to endure in such an environment consisting exclusively of boys would have crushed a lesser spirit. That she flourished is a tribute to her strength of character and determination, and it emboldened her to pursue an academic career.

As well as her ability in music, science, and maths, Mileva was something of a linguist. Her father had learned German, the *lingua franca* of the polyglot Austro–Hungarian army, in his military days and it was spoken at home. She also studied French and would have had to learn Latin; indeed, her father tried to pull more strings to excuse her from the obligatory Greek at the Zagreb Royal Classical, but failed. Mileva took Greek in her stride. She graduated with the highest marks in physics and mathematics. However, around the same time she fell quite seriously ill with a lung infection. Her fluency in German opened up an opportunity to move to Zurich, whose mountain air was thought to be beneficial for lung diseases such as tuberculosis. Again, her courage and determination are impressive. She left her comfortable home at the age of eighteen and relocated more than 1000 kilometres to an unknown country. The Canton of Zurich was a pioneer in female higher education, which offered opportunities to slake Mileva's thirst for further knowledge. In November 1894, she enrolled in the Zurich Girls High School and began studying for the Matura exam that would be, as it had been for Einstein, a passport for entry into the university or the Polytechnikum. In the spring of 1896, she passed. She elected to study medicine and enrolled on a medical degree course at the university. It didn't go well, and she quickly decided to cut her losses and return to her first loves, mathematics and physics. In October of that year she, together with Einstein, Grossman, Kohlross and Ehrat, registered for the Mathematics and Physics Teaching Diploma course at the Polytechnikum.

Mileva and Einstein would surely have met at one or other of their classes. Trbuhović-Gjurić recounts an occasion in their first semester when Mileva helped Einstein with a mathematics problem that had defeated him, but there is no evidence of any other interaction between them until the summer of that academic year. As the weather warmed up following his Pentecostal break-up with Marie, Einstein often drank coffee with his colleagues, presumably sometimes including Mileva, on the outside tables of the student coffee houses on the Bahnhofstrasse. Once, when asked why he was late, he replied 'I can tell you precisely. My landlady, who irons clothes for a living, told me that she can iron much better while I am playing my violin – so I carried on playing for a while.' He also began sailing, which became a lifelong pastime. The Zurichsee is an excellent place to sail, and Einstein took up the sport with enthusiasm. Walking in the hills and mountains of Switzerland remained a delight for Einstein and it was on a shared walk that the first indication of intimacy with Mileva was revealed. Again, the sheer pluck of Mileva and her determination not to miss out on any aspect of life can only awake admiration for the young girl with a pronounced limp, accompanying Einstein, a fit and experienced walker, on such a climb. The students returned to their homes at the end of the semester, Mileva to Kac and Einstein to Milan – probably going on to the family's usual vacation somewhere in the Alps.

While Einstein was spending August with his parents, Switzerland was hosting two events that would have some significance for his future career. Zurich became briefly the centre of the world of mathematics and theoretical physics when the first International Congress of Mathematicians was held there from 9–11 August. There were 242 delegates, 38 of whom were women. It was organised in great part by Einstein's Polytechnikum teachers, including Geiser and Hermann Minkowski. Among the delegates was Besso, then living in nearby Winterthur. The keynote speech was due to be given by Poincaré, but he was unable to attend due to illness, and it was read out to the assembled delegates. His subject was the relationship between pure mathematics and theoretical physics.[5] He explained that mathematics is the only language in which physicists can communicate. He used Maxwell as one example, pointing out that he could formulate his equations twenty years in advance of any confirmatory experiments because he was deeply aware of the symmetries of mathematics and was used to 'thinking in vectors'. No doubt Besso and Einstein discussed the congress on Einstein's return to Zurich in the autumn. It is a pity that Einstein had not heard Poincaré's extremely elegant and inspiring contribution, since his first year of mathematical grind at the Polytechnikum had left him with both a distaste for pure mathematics and a deep thirst to get into the laboratory and study physics. Many years later he wrote that '. . . it was not clear to me as a young student that access to a more profound knowledge of the basic principles of physics depends on the most intricate mathematical methods. This dawned upon me only gradually after years of independent scientific work.'

[5] Contrary to the statement of many authors, Poincaré's contribution published in the proceedings makes no mention of the non-existence of absolute space or time. The confusion probably arose from Einstein's fellow student, Louis Kollros, whose article 'Albert Einstein en Suisse – Souvenirs' at the Congress to celebrate the fiftieth anniversary of relativity in 1956 mentions Poincaré's contribution at the Zurich meeting but quotes from Poincaré's 1902 *Science and Hypothesis*, which contains an elegant summary of his thoughts on these matters.

The second event of relevance to Einstein's future life took place at the end of August in nearby Basel. This was the First Zionist Congress, convened by Theodore Herzl. It adopted the Basel Program, which began with a declaration of intent: 'Zionism aims at establishing for the Jewish people a publicly and legally assured home in Palestine.' Although Chaim Weizmann, who subsequently drew Einstein into Zionist work, was unable to be present, he was already an active Zionist and attended all subsequent congresses.

On Einstein's return to the Polytechnikum for the start of the new semester in October, he discovered to his surprise that Mileva was nowhere to be found. On asking her friends, he discovered that she had decided not to return to Zurich but to travel to Heidelberg, where she spent the winter semester of 1897–1898 attending undergraduate lectures at the physics department.

The reasons for Mileva's decision to leave the Polytechnikum are unclear. Although Einstein's friendship was clearly important enough for her to have mentioned him to her father, who gave her some tobacco as a gift for him, it is difficult to believe that her decision had much if anything to do with Einstein. Their relationship was still only that of friends. Once he had established that Mileva had left Zurich, Einstein wrote a rather long letter to her. Although this letter has been lost, somewhat surprisingly since it is likely to have been the first she received from the man who was to become the love of her life, her reply has survived. Mileva waited a considerable time before sending this and addresses Einstein in her letter with the formal 'Sie'. Although there is affectionate banter in the letter, it is far from intimate. The health considerations that were her original reason for coming to Zurich were presumably no longer a consideration, since the Heidelberg climate is certainly less healthy than Zurich; indeed, Mileva's letter comments on the thick fog that rises from the Neckar River. The most likely reason for her decision is also the most straightforward; she was not particularly enjoying her course in Zurich, which had contained little physics in the first year. Perhaps she was already beginning to struggle with her studies, problems that were to become only too evident in the coming years. Heidelberg at the turn of the nineteenth century had a strong reputation for physics, building on the work of Kirchhoff, who was succeeded by Georg Quincke, a greatly admired lecturer. It had some able faculty members, such as future Nobel Laureate (and inveterate opponent of Einstein) Philipp Lenard. Mileva probably hoped that a new start and new scenery would renew her enthusiasm. This, as well as her naivety, is evident in her description of one of Lenard's lectures on the kinetic theory of gases. As she tried to come to terms with the ideas of physics that winter, she and Einstein kept in touch via infrequent, at least on his part, letters.

As one friend left Einstein's life, temporarily as it turned out, another arrived. Friedrich Adler, son of the founder of the Social Democratic Party of Austria, arrived at the Polytechnikum in October 1897 to study chemistry. After one year, however, he switched to physics and became a close friend of Einstein. The start of the semester saw Einstein finally able to satisfy his thirst for physics; Heinrich Weber began his introductory lecture course, which spanned the entire academic year. Einstein took copious and relatively accurate notes and wrote to Mileva in February 1898 that he was greatly enjoying the course and indeed looked forward to every new lecture. She for her part had decided that the grass was not particularly greener in Germany compared to Switzerland and had written, probably in late January, to tell Einstein that she had decided to return to Zurich for the spring semester. Einstein, despite his delayed reply, was enthusiastic, promising to lend her his notes and those of Grossmann and reassuring her that she would quickly be able to catch up with what she had missed.

Einstein's pleasure in Weber's lectures did not endure. They were clear and elegant, as far as they went, but the problem was that they did not go far enough. Indeed, they stopped somewhere around 1870. A contemporary of Einstein at the Aargau Cantonal School, Adolf Fisch, who had saved him from falling off the Santis during the school expedition, wrote

> Weber was a typical spokesman of classical physics. Anything that came after Helmholtz was simply ignored. At the end of our studies we knew everything about the history of physics but nothing about current work or its future. We therefore had no other option than to study new publications on our own.

Decades later, Einstein recalled that, as 1898 wore on, he skipped many of his lectures, preferring to work at home. However, the ETH regulations were very strict, requiring written explanations if more than one day's attendance was missed, so that Einstein's recollection must be somewhat exaggerated. Indeed, several of Grossmann's and Einstein's classmates were reprimanded or worse for absences. Einstein's recollection of truancy probably refers mostly to subjects he studied that were not formally marked at the end of the year, particularly in mathematics. This is borne out by one of his mathematics professor's remarks after Einstein had become famous. Minkowski taught courses in a variety of mathematical areas, including function theory and partial differential equations, during the last three years of Einstein's studies. He moved to Göttingen in 1902 and subsequently took up the theory of special relativity, to which he made a significant contribution. He remarked to Max Born that, when he had read Einstein's paper, 'It came as an enormous surprise to me, because Einstein [at ETH] had been a complete wastrel. He had never bothered himself at all about mathematics.'

In contrast, Maja certainly believed that Einstein was a conscientious student:

> His studies were carried out diligently. Among students it is customary to miss the odd lecture, particularly if the evening before had seen too much alcohol consumed or when they want to use the time to enjoy themselves. Albert Einstein however, who neither knew nor wanted to know such distractions, attended his lectures conscientiously and indeed also went to others that were not in his study plan.

No doubt he was often present in body, but not in spirit, perhaps reading about Maxwell's equations from a book hidden from the lecturer's view. For his revision for his intermediate and final examinations, he came to rely more and more on Grossmann's comprehensive notes, which survive and where the occasion marginal annotation can be ascribed to Einstein. They met weekly for coffee at the Café Metropol to discuss not only physics but also whatever was on their minds; they also made up half of a walking party with fellow students Ehrat and Walter Leich. Grossmann didn't seem to mind being Einstein's amanuensis; he was already convinced that his friend had a future in physics. He remarked to his parents shortly after their first meeting that 'This Einstein is going to be a very great man one day!' Einstein became a regular guest at the comfortable Grossman home in Thalwil, a few kilometres along the shores of the Zurichsee, where he was an enthusiastic participant in their frequent music-making.

In addition to working at home, Einstein increasingly began to frequent the room of Mileva Marić. She had returned to Zurich shortly after Einstein's February letter, before the start of the semester in April. There is some uncertainty about the date of her return, since she didn't register again with the Zurich authorities until 16 April, when the semester began. However, letters from her Slav friends imply that she was already back at the Pension Bächtold by 24 February, but that she was spending most of her time with them at the Pension Engelbrecht nearby. On 21 May, Mileva introduced Einstein to her friend Milana Bota, who wrote to her parents 'Marić has introduced me to her good friend, who is a German called Einstein. He plays the violin wonderfully, he is a real artist, so I will be able to play music with someone again.' A subsequent letter implies that her hopes were fulfilled and that she and Einstein spent a whole afternoon making music together. When Einstein's visits became frequent enough to cause problems with Frau Bächtold,

Mileva used Ružica Dražić's rooms in the Pension Engelbrecht to work and meet Einstein in while she waited until the autumn for a room in the floor above to become vacant.

Matters at home in Milan continued to be a distraction for Einstein. After a promising start, his father's dynamo business had failed. He wrote to Maja almost despairingly:

> If my advice had been followed, Papa would have looked for a job two years ago and spared us all this aggravation [. . .] of course, the unhappiness of my poor parents hits me hardest; after so many years they hardly have a happy moment anymore. Even harder for me to bear is that fact that I, as a grown man, can't do the least thing to help. I am just a burden to my family. It would indeed be better if I were dead. The only thing that keeps me going and stops me despairing is that I tried everything within my limited influence and that, year in, year out, I allowed myself no amusements or distractions except those offered to me by my studies.

After the crisis had abated somewhat, he wrote in a more optimistic, indeed too optimistic, frame of mind: 'I have quite a lot to do, but not so much that I can't find time for a few hours walk through Zurich's wonderful surroundings. During my walks, I am comforted by the thought that my parents' worst worries are over.' By early 1899, Hermann had decided to return to his earlier activity, building electrical generating stations and he registered yet another company with the Milanese authorities, with himself as director, to take advantage of a concession he had obtained to build the electrical plant for the small town of Canneto sull'Oglio, halfway between Cremona and Mantua.

There is almost no documentary evidence for the period between the summer of 1898 and the early spring of 1899, during which Einstein and Mileva's relationship changed from friendship to romantic involvement. That such a change took place is clear from the tone of the surviving letters. Two that cannot be dated exactly but are likely to be from the first half of 1898 begin with the rather formal greeting 'Geehrtes Fräulein' ('Honoured Miss') while one sent in the spring of 1899 begins 'Dear' followed by something indecipherable. The whole tone of this latter letter, which was sent just after Einstein arrived back in Milan for the spring break, is utterly different to its predecessors. It is full of slang and is chatty and indiscrete as well as intimate. In one section Einstein reminisces flirtatiously about a dinner they had had together as guests of Professor Stern of Zurich University. Crucially, he also recounts that he had shown Mileva's photograph, which he is clearly now carrying around with him, to his mother. He records that she spent a very long time looking at the photograph. Given her subsequent attitude to the relationship, it can be assumed that this long contemplation was not an approving one and that she was puzzled by what her wonderful son could see in this rather dowdy, dark, and clearly older, woman. Perhaps sensing this, Einstein cheerily remarked that Mileva was a very clever little thing. This probably cut no ice with his 'old lady', as he referred to his mother in his letter. The evidence therefore points to an intimacy between Einstein and Mileva that developed after Einstein returned to Zurich from Milan for the start of the 1898–1899 semester; by the spring of 1899 it had developed into a fully fledged romance.

Einstein returned to Zurich on 17 September 1898 to prepare for his mid-term oral examination, which began on 3 October. However, Mileva postponed taking the examination until the following October, presumably because of her stay in Heidelberg, indicating that Einstein's assurance in his letter earlier that year that she could easily catch up had not been the case. Einstein did very well in the examination, receiving the highest mark of the five students in his course.[6]

---

[6] The fifth student, replacing Mileva, was Louis-Gustave Du Pasquier. He scored higher than Einstein in physics but was weak in the mathematics courses and scored the second lowest of the five marks

When he returned to Zurich in September, Einstein moved lodgings from Unionstrasse to Klosbachstrasse 87, where Frau Stephanie Markwalder-Bürgin became his landlady. The new lodgings were only about 100 metres away from his previous ones. Suzanne Markwalder was a young primary school teacher living in her mother's house when Einstein moved in with them. Seelig interviewed Susanne many years later and reconstructed a charming picture of Einstein's life in the Klosbachstrasse during the 1898–1899 academic year. As usual, much of it centred around music:

> In the evenings, there was often music, with Einstein playing the violin. He preferred Mozart in which I did my best to accompany him on the piano. I still have the copy of Mozart's piano sonatas that Herr Einstein presented me with, on which he had written 'in friendship and admiration'. He was very patient with my limited technique. The worst he would say was: 'There you are again, stuck like a donkey facing a slope.' He would point with his bow to my entry and we would carry happily on. One day a widow who lived above us asked my mother if she and her two daughters could come down to our lounge to listen in on our practice session. With some hesitation, she agreed. When the three music lovers had installed themselves on the sofa, they began to knit, disturbing our practice with the clattering of their long needles as well as the occasional whisper when one dropped a stitch. Einstein ostentatiously closed my music and returned his violin to its case. When the widow asked 'What, have you finished playing already?' he replied sarcastically, 'We really don't want to disturb your work.'
>
> [...]
>
> He sometimes went sailing with his friends on the Zurichsee. He took me with him once. I still remember that, when becalmed with the sail hanging like a crinkled leaf, he would take out a little notebook and scribble in it until the breeze returned. As soon as it did, he was immediately ready to carry on sailing. To the pleasure of the inhabitants, he once improvised a serenade with a schmaltzy Italian tenor in the garden outside our house. He often played at the house of Professor Stern, a historian, to whom he had been introduced by friends from Milan and where he was always a welcome dinner guest. On one such occasion he got into a heated argument with another physicist during which his opponent attacked his theoretical premises. Einstein wasn't perturbed but at the end of the meal pointed to his violin which he had brought with him and suggested 'Let's go to the music room! We can play your favourite – Händel!'[7]
>
> [...]
>
> At home [...] we would chat or read aloud from Heine's *Harzreise*. One summer's day, as he was about to get his violin and close the balcony doors, he heard someone playing a Mozart piano sonata in a house nearby. 'Do you know who that is playing the piano?' he asked. I explained that it was probably the piano teacher who lived in an attic. He immediately put his violin under his arm and hurried out, without collar or tie. 'Herr Einstein, you can't go out like that!' I called after him but either he didn't hear or didn't want to hear. Shortly afterwards the garden gate closed and before long we heard a violin playing an accompanying tune to the Mozart sonata. When Einstein returned, he said enthusiastically 'What a delightful lady. From now on I will play with her often.' We subsequently met old Fraulein Wegelin, who appeared a few hours later in a black silk dress and asked uncertainly who that extraordinary young man was? We reassured her that he was a harmless student. She told us that it had given her such a shock when this unknown musician had rushed into her room shouting 'Carry on playing!'

The spring and early summer of 1899 saw a blossoming of the relationship between Einstein and Mileva. After their work in the Polytechnikum was over for the day, he usually accompanied her to her room, where they discussed physics and their plans for the future. Increasingly Einstein

---

overall. Undeterred, he went on to become a professor of mathematics and made several distinguished contributions to number theory.

[7] This illustrates Einstein's habitual fondness for puns – händel in German also means 'argument'.

used Mileva as a sounding board. At around this time, he told Julia Niggli, with whom he kept in touch and whom he often met and played music with on his trips to Aarau, that 'I often work with [Mileva]. She almost knows more maths than I do. Getting to know her is a real plus for me. She isn't particularly pretty but she is very clever.'

Milana Bota many years later recalled her impressions of Einstein and Mileva in this period:

> I lived on the third floor in a very expensive pension. All nations were represented there: the Germans, the Italians, the English, the Americans, and we, the Serbs too. Miza (this is how friends called Mrs. Einstein), Helene and I met early on and became friends. Miza lived above us on the fourth floor. In her room, in total quiet, we always found Miza and Albert Einstein, and we had not the slightest inkling that one day he would become such a famous man.
>
> [...]
>
> We went together for walks, to concerts, to social events. Albert did not pay much attention to his appearance. Whenever you met him, he was without a necktie; his pants were not ironed; his vest was not buttoned. He was very slovenly and had long hair like a poet. When we were returning to the pension, Albert, I don't know why, always walked three steps ahead of Miza and whistled. He would jump over two or three steps, as if he were chasing somebody. His face had an ironic look, and he liked to laugh at other people's expense. Whenever he was not busy, he played the violin. I accompanied him on the piano, while Miza played the tamburitza. Albert loved to listen to songs from Vojvodina, for which Miza was very talented. He was a real virtuoso with the violin. He liked most to play Bach, Schubert, Schumann and Beethoven.

Einstein appears to have stayed on in Zurich until the end of July, since he had arranged to join his mother and Maja for a holiday in the aptly named Hotel Paradies in Mettmenstetten, just a few kilometres west of Zurich. Newly built, it had already acquired an excellent reputation as a fashionable health resort and looked down on the little town and a landscape of rolling hills. Einstein had already invited Julia Niggli to visit him there and although she accepted on the assurance that his mother and sister would be chaperones, she had little doubt that such a trip would be forbidden by her puritan mother. In fact, she had already written to Einstein asking him to visit her in Aarau before she left for a holiday, as she was greatly troubled by the attentions of a much older man and wished to ask Einstein's advice. It can be deduced that she found these attentions not unwelcome, despite the fact that the older man had admitted that marriage played no part in his future plans. Einstein had been unable to visit Aarau or indeed even to write to her since an accident in the physics laboratories resulted in an injury sufficiently serious that it had required stitches at the hospital; Einstein could neither write nor, more importantly, play the violin. He wrote to Julia that he was sure that his violin would be unable to understand why she remained in her case, admitting that he was suffering greatly from no longer being able to express his innermost feelings through the medium of music. However, he did promise that, since he was intending to do some scientific work with a physics teacher from the other secondary school in Aarau in September, there would be ample opportunity for them to make music together.

In his next letter, written about halfway through his holiday, Einstein answered Julia's reiterated request for advice about her older man with a long description of the faithlessness of men that seems to be inspired by his own behaviour to Marie Winteler. He also approved of Julia's plan to escape amorous advances and her parents' censoriousness by leaving Switzerland. He offered her a position as a tutor to Alice, the seven-year-old daughter of Einstein's Aunt Julie, who was staying with them at the Hotel Paradies. It was she who provided the financial support for his study. Julia left in the autumn for Frankfurt, from whence, just after Christmas of that year, she wrote to Rosa Winteler asking for news of her adopted brother Albert. In the letter, Julia chastised Rosa for not giving her any news, despite her knowing that she was very interested in him. She asked Rosa to send her Einstein's address, as she wanted to send him a postcard. It seems that Einstein's charm and violin had made another conquest and that Julia's confidences may have been intended as much to deepen her intimacy with Einstein as to solicit the advice of a man who was, after all, a student several years her junior.[8]

Einstein typically had no compunction about encouraging Julia Niggli's long-distance flirtation while carrying on correspondence with Mileva. Indeed, he was also flirting with another girl at the hotel, the seventeen-year-old sister-in-law of the manager who accompanied Einstein on walking tours in the region. Anna Schmid (known usually as Anneli) kept a commonplace book in which Einstein wrote one of the little poems of which he was so fond, the gist of which was that the price of his inscription was a kiss and that if she were offended by that, the best punishment was to give him another in return.

While Einstein was engaged in such pleasantries at the Hotel Paradies, Mileva had returned to her parents in Kac. She was worried by her impending intermediate examination and Einstein's letters over the summer are full of reassurances that she would sail through it. She was, rightly as it turned out, not so confident. The letters they exchanged are interesting both from the point of view of their deepening intimacy and from his hints as to how his thinking on physics was developing. He discussed 'their' household in Zurich and the pleasure he got from their reading and discussing physics together. He reported on his attempts to understand Helmholtz's theory of the electrodynamics of moving bodies, and that he believed that there must be a simpler way to describe it, rather than postulating an aether whose state of motion and physical properties were indeterminate. His conception of electricity and magnetism being generated by actual particles of some description carrying electric charge was far from common at the time, even though J. J. Thomson had announced the reality of the electron experimentally at a Royal Institution lecture in London two years previously.[9] Einstein was convinced by Helmholtz's argument that Faraday's laws of electrolysis clearly indicated the existence of particles carrying various amounts of electric charge. He returned to this theme in another letter written just before his return to Milan, in which he described an idea for an experiment, similar to that of Fizeau, on how light propagates through transparent bodies and its relation to the aether. In the period between his letters, he had spent some time in Aarau, staying with the Wintelers (Marie was not at home) and presumably working with Wüest, the physics teacher with whom he was exchanging ideas.

Einstein remained in Milan until 15 October. He may have been reunited there with his old friend Besso, who had got a job as an engineer working on the long-distance transmission of electricity – and thereby becoming in some sense a competitor of Hermann Einstein! Besso moved to Milan with his wife Anna and their baby son sometime in 1899. At the end of September, Einstein sent Mileva a letter that mentions a recent paper he had read by Wien that discussed various theories of the aether and how its properties might be measured. Among the experimental results referred to is that of Michelson and Morley. Einstein had written to Wien about an idea to build apparatus, presumably for his Diplomarbeit, a short thesis required of every undergraduate in the final year of study. His proposal sounds like a variant of the Michelson–Morley experiment. Weber, however, was not encouraging and Einstein gave up the idea, particularly as he received

---

[8] Whatever Julia's intentions towards Einstein, they lost contact for many years after a final letter from Einstein on 11 September as he was leaving Mettmenstetten to return to Milan with his mother and sister.

[9] It was a combination of Thomson's subsequent experiments in 1899, which demonstrated that the electron charge was equal to the unit of electrolytic charge and the fact that Lorentz supported Thomson's interpretation, that led to the general acceptance of the electron as a particle.

Figure 3.1 The extended Winteler family. From left to right: Marie Winteler, Maja Einstein, Paul Winteler, Anna Winteler, Jost Winteler, Pauline Winteler, Rosa Winteler.

no recorded response from Wien, possibly because Einstein wrote to him at his former address in Aachen, but he had a few months previously moved to Gießen.

In addition to physics, Einstein's letter also informed Mileva about Maja's move to Aarau, where she was to attend the Ladies Teacher Training Seminar, a dependent establishment of the Cantonal school intended for girls, who were not allowed to attend gymnasia at that time in either Germany or Switzerland. Although she lodged at the Töchterheim, the hostel attached to the school, she found a welcome and a substitute home, just as Einstein had, with the Wintelers (see Figure 3.1). Einstein continued his letter by reassuring Mileva that he wouldn't be going backwards and forwards all the time to Aarau to visit his sister because Marie was returning. With an amazing tactlessness, he reminded his current girlfriend – relatively plain, unsure of herself, and just about to take a critical examination – that he had been madly in love with Marie. He wrote that he was worried: 'But if I were to see that girl a couple more times, I would certainly go mad, I know it and I fear it like fire.' What sort of reception he got when he returned to Zurich and saw Mileva again on 15 October is not recorded. She was at least relieved at having passed her intermediate examination, albeit with an average grade of 5.05, putting her fifth of the six candidates.

On 19 October, Einstein registered with the authorities at his new address back at Frau Hägi's house at Unionstrasse 4. Having by now satisfied the residence requirements, on the same day he set in train the long process of gaining Swiss citizenship by writing to the Swiss President and Federal Council. Given that he had managed perfectly adequately to commute back and forth between Milan and Switzerland for the previous four years as a stateless person, Maja's statement that he wished to become a Swiss citizen out of conviction is credible:

> Already in his school in Aarau he had become familiar with a freer type of citizen spirit than he had been used to in Germany, and had been impressed by a participation in democracy similar to that in North America, with its combination of independence and community spirit. He saw in Switzerland an ideal that very much chimed in with his own democratically minded outlook.

However, even the unworldly Einstein probably realised that possession of Swiss citizenship would open up a variety of positions as a Beamter, a tenured civil servant, both in the education and other branches of Swiss officialdom, that otherwise would be closed to him. Since he eventually took up such a position in the Patent Office in Bern, the cost of 600 Swiss francs, from a total of 800 he had saved laboriously at the rate of 20 francs per month from his 100 Franc Genovese allowance, was well worth it. He finally became a Swiss citizen on 21 February 1901.

There is little documentary evidence for the remainder of Einstein's final year at the Polytechnikum. Reiser notes that:

> At Zurich, he frequented the Opera. He became ever fonder of playing [the violin] himself, especially fantasies, a kind of free monologue addressed entirely to his own inner self. From that time on, he elicited from his violin not only classical and modern music, but spontaneous improvisations which had no other meaning than that of the moment.

No doubt these improvisations helped him return with renewed vigour to the highly uncongenial task of learning by heart various parts of Grossmann's voluminous lecture notes for regurgitation at his final examination. It is clear that most of his and Mileva's time in the new year of 1900 was taken up with preparing for their final examination. One requirement was that they had each to write a short thesis on a topic of their choice to be agreed with their professor. It has already been noted that Einstein's first suggestions, including something similar to the Michelson–Morley experiment, had not met with Weber's approval. In the end, both Einstein and Mileva pursued relatively straightforward work relating to heat conduction, one of Weber's favourite subjects. The selection of her thesis topic is one of the things she mentions to her friend Helene Kaufler (who had also lived in Plattenstrasse when she was a history student at the university) in one of the few letters that survive from this period. Although Mileva implies that Einstein was also excited about his topic in the same area of heat conduction, he later recalled that it held no interest for him whatever.

Although the absence of dates in most of the extant letters makes establishing time sequences difficult, it is likely that Einstein went back to Milan for the spring break; it could be during this absence that Mileva wrote a charming little letter in which she sends him a kiss and asks if he likes her as much as she likes him. The importance of this is that it is the first time that she uses the familiar *Du* form of address. Among adults at that time, *Du* was only used between very close friends or family, so that this usage indicates another ratchet upwards in their level of intimacy.

It may be during this stay in Milan that Einstein tidied up a loose end from an earlier relationship. Ernestina Marangoni had continued to visit Einstein's Milan home since the family's departure from Pavia. She recalled being taken to hear Toscanini at La Scala and taking the Einstein family to hear the famous poet Gabriele D'Annunzio give an address at the inauguration of the Università Popolare di Milano.[10] She also recalled that '... before his graduation, Albert told me that he had

[10] Ernestina was ambiguous but it seems likely that it was Einstein who took her to La Scala, which was just around the corner from their house, whereas she took his parents to hear D'Annunzio. Toscanini first

sold his violin, as it was not possible to study both music and physics. I grieved a lot, but luckily it was not a definitive separation.' It seems unlikely that Einstein sold his violin or that he ever believed that it was not possible to play music and study physics at the same time. The fact that Einstein told Ernestina this gives the impression that he was trying to find an excuse to put some distance between them, since their relationship had always pivoted around their making music together. This is another indication that Einstein was by now in love with Mileva.

A letter written around the time of their final examination in the summer sheds an interesting light on relationships both between Mileva and Einstein and between them and others. Mileva was answering her friend Helene, who had written to Mileva describing a visit that Helene had made to Einstein's parents in Milan. Helene confirmed Mileva's fears that Pauline was very opposed to her son's interest in her. Mileva also documented the break-up of the friends who had lodged at the Pension Engelbricht, both in terms of domicile and intimacy. She notes that both of her friends were about to leave and that they seemed to be annoyed with her for some reason; the cause seems to have been her relationship with Einstein, which clearly crowded out everything else in her life. Milana Bota, who only two years earlier had swooned to her parents about Einstein's wonderful violin playing and how she looked forward to making music with him, now wrote to them that she saw Mileva rarely, which she ascribed to Einstein, whom she hated. The couple's relationship with Helene Kaufler was, however, still very good. In a letter to her mother on 7 June, Helene described a charming break they took from their work towards their examinations: 'After dinner we first took a walk by the lakeside, and then I accompanied Mr. Einstein on the piano. I wish you could hear him play – he plays with such a deep passion and full sound and, at the same time, in a measured, controlled way.'

Both Einstein and Mileva must have worked hard in preparation for the final examination, which took place towards the end of July. Neither particularly distinguished themselves; Einstein's average of 4.91 put him last of the four who graduated, the others being Kollros, Grossman, and Ehrat, in that order. At least he got his diploma, however, whereas Mileva, whose average was 4.0, did not. Both she and Einstein did rather poorly on their theses; and a warning sign of the breakdown of the relations between Einstein and Weber appeared when the latter insisted that Einstein produce another copy of his thesis written on the regulation paper before he would assess it. However, it was Mileva's marks in astronomy and in particular the mathematics course on the theory of functions taught by Minkowski that let her down. Her mark in the latter, 2.5, was little short of disastrous. Their years of shared intimacy seem to have had a detrimental effect on both their exam performances, in particular on Mileva's mathematics, in which she had also struggled in her intermediate examination. Since Einstein had specifically remarked on her ability in maths early in their relationship, it must be suspected that they had spent too much time bouncing wild ideas about physics off each other and insufficient time on the basic groundwork; whereas Einstein could still scrape through the examinations because of his extraordinary abilities, Mileva could not.

Einstein and Mileva must have learned their final marks and discussed the consequences in somewhat of a rush, probably the same day they left Zurich on 27 July – Einstein for his traditional Alpine holiday with his mother and sister, Mileva once again for Kac. Mileva decided to return the following year and make another attempt to gain her teaching diploma. As later developments indicate, they must also have discussed their future together and marriage.

conducted at La Scala in 1896 and he became Principal Conductor in 1898. The opening speech referred to took place on 1 March 1901, whereas Einstein did not arrive in Milan until around 20 March.

Although Einstein now had his diploma, he had found the examinations neither pleasant nor easy. Many years later he recalled that 'This necessity (revision) had such a discouraging effect on me that after I passed my final examinations I couldn't bear to face scientific problems for a whole year.' However, as often seems to be the case, Einstein's recollections in old age are misleading and, in fact, a matter of weeks after his examinations were over, he was once again deep in his studies of physics; and, of course, playing his violin.

Chapter 4

Mileva and first steps in research (1900–1902)

The newly qualified specialist teacher in mathematics and physics left Zurich without any clear idea of his future – except that it was to be with Mileva. He had every expectation of obtaining an assistantship with one of the Polytechnikum professors, which almost all successful graduates were offered. However, when he left on 27 July 1900, in the company of his Aunt Julie, to rendezvous with his mother and sister, nothing had been arranged. They met at the station of Sarnen at the head of the eponymous lake, around 10 km from their final destination, the Hotel Kurhaus-Melchtal. The hotel had been built in the 1880s in the tiny hamlet of Melchtal, hemmed in by mountains and leading up to the beautiful and isolated Melchsee, at a height of around 2000 m at the head of the valley. From here, many scenic walking routes fanned out. On the drive up to the hotel, Einstein and Maja decided to leave Pauline and Aunt Julie and walk the final stretch up to the village. They talked, Maja warning Einstein to be cautious in discussing the sensitive subject of Mileva with his mother.

As soon as they had all unpacked, Einstein knocked on the door of his mother's room. In response to Einstein reporting Mileva's examination results, Pauline asked what Mileva would do now. Einstein, immediately on the defensive, replied that she would be his wife. The ensuing scene of weeping – Einstein reported that she cried like a child – led to a series of entreaties and concerns, including that Einstein would be in a pretty pickle if Mileva got pregnant. Bridling at this, Einstein repudiated in the strongest terms[1] the idea that they were having sex. Her tirade was only cut short by the arrival of a family friend in the room, which turned the conversation to more trivial matters. This was only a temporary respite however, as, after an evening of music-making, Pauline returned to the attack. She complained that Mileva was a bookworm just like Einstein but what he needed was a wife. She went on to opine that by the time Einstein was thirty, Mileva would be an old witch. Once Pauline had exhausted her querulous concerns, some calm returned. Einstein complained about the poor weather and the interminable meals at the hotel, which bored both him and Maja to such an extent that he had taken up physics again and was reading a work by Kirchhoff. He continued to be solicitous to his mother, soothing her with his assiduous violin playing in partnership with other hotel guests.

The weather improved as the holiday continued. Einstein and Maja walked in the beautiful environs, climbing a mountain, perhaps the 1900-m Heitlistock or the 2700-m Huetstock, from which they had an excellent view of the snowfields of the 3200-m Titlis, the highest mountain to the immediate south of the valley. Maja returned shortly afterwards to Aarau, where she had fallen in love with Paul Winteler, the youngest son of the family; the Einsteins and their associates seemed

[1] Since this letter was being sent to Mileva and gives no inkling that the statement was anything other than the truth, we can conclude that their relationship had not at this stage progressed beyond mutual caresses. That it was to do so within the next few months is however documented in the clearest possible way by the eventual realisation of Pauline's worst fear—becoming a grandmother.

Einstein. Brian Foster, Oxford University Press. © Brian Foster (2026). DOI: 10.1093/oso/9780198794875.003.0004

unable to resist the charms of the Winteler clan. Einstein continued in Melchtal until the second week in August, when he began the journey towards Milan. He had decided to turn down the prospect of a temporary clerical job in favour of letting his father initiate him into his electrical business so that he could stand in should Hermann Einstein become ill. This was probably also a ploy to please his father, hoping thereby to soften his concerns, which were as strong as his mother's, about a marriage to Mileva. On the way to Milan, he made a stop in Zurich to investigate his chances of obtaining an assistantship at the Polytechnikum. Largely because of the possibility that some of his fellow new graduates would not take up their offers, the situation was unclear. It must however have begun to register with Einstein that his prospects were not as good as he had originally hoped. He stayed in Zurich for just over a week, desultorily calling in at the Polytechnikum, where the faculty would have been almost all on vacation, and writing to Mileva to tell her how much he missed her. His letters from this period are touching and passionate; he clearly did miss her greatly, especially when in Zurich.

Einstein spent the remainder of August and September with his family. He listened patiently to his father's concerns about a prospective marriage for someone with no visible means of support until, at the start of September, Einstein accompanied him on a tour of his company's electricity-generating stations. One was at Canneto sull'Oglio, halfway between Cremona and Mantua, the other at Isola della Scala, just south of Verona. The latter being only 130 km west of Venice, the two spent some time exploring the beauties and canals of the city together. In a postcard to his mother, Einstein reported enjoying sea-bathing and music in beautiful sunshine.

It was during his stay in Milan that Einstein renewed his intimacy with Michele Besso, who was working with a Milanese electrical firm. Despite his recollections in old age, reported in the previous chapter, Einstein just a few weeks after his final examinations was positively bubbling with enthusiasm for physics and with new ideas in several directions. Indeed, it is in this first year after graduation that Einstein the physicist can be discerned for the first time as a brilliant and creative researcher, brimming with ideas. He and Besso spent long evenings deep in discussion of physics; Einstein looked forward to getting out of his parents' house to the somewhat bourgeois but comfortable rooms where Michele and his wife Anna, formerly Winteler, were bringing up their toddler son, with whom Einstein enjoyed playing. He not only discussed his own ideas on physics with Besso but also responded to his friend's interests. One such was to investigate the energy radiating from a cable carrying an alternating current, presumably a work-related issue that Besso was grappling with. This investigation is rather poignant, since it was his father's persistent failure to take up alternating-current rather than direct-current electricity generation that had led to many of his business problems.

In September 1900, Einstein went hill walking in the vicinity of Lake Maggiore, did some physics, and pursued his usual musical diversions. He eagerly anticipated making music with Luigi Ansbacher, a law student whose Milanese Jewish family were close friends of the Einsteins. Ansbacher was an enthusiastic musician who subsequently became the lawyer and friend of his idol Arturo Toscanini. He also did some legal work for Rudolf Einstein relating to the Canneto sull'Oglio generating station. The physics studies related to two books by Boltzmann, which convinced Einstein that the properties of gases were indeed derivable from the myriad collisions of atoms and molecules treated in a statistical way. This inspired him to push forward with speculations on the behaviour of liquids in narrow tubes, which he had begun before his final examinations. He discussed his ideas with Besso and looked forward to filling in Mileva on his work, which postulated that the attraction between molecules was simply given by two numbers specific to each type of molecule and a force depending in an unknown way on their separation. These speculations were ultimately to lead to Einstein's first paper in December. The lack of any allowance for the size of the molecule, except what could be absorbed into the unknown molecule-specific constants, was the flaw in Einstein's theory; subsequently he discounted it as worthless. Nevertheless, at the time he was enchanted and energised by the fact that a simple hypothesis such as the nature of the intermolecular force allowed him to explain what appeared at first sight to be very disparate liquid properties. Still excited by a revelation that would become in many ways the guiding principle of his scientific career, a few months later he wrote to Marcel Grossmann that 'It is a wonderful feeling to discern the essential unity of a complex collection of phenomena that appear from the direct appreciation of the senses to be entirely unconnected.'

Einstein's work with Besso was interrupted by a visitor, Besso's brother-in-law, Fritz Winteler. Einstein was irritated by Fritz, who had become an assistant at the Polytechnikum after graduation and now held an assistantship in physical chemistry at Darmstadt. No doubt he had his own views on molecular forces, but it is likely that the animosity arose from Fritz having successfully obtained assistantships at two institutions. Einstein, in contrast, had his worst suspicions confirmed on his return to Zurich at the beginning of October, when he learned that he was not going to be appointed to an assistantship at the Polytechnikum. He had written in September to his old mathematics professor, Hurwitz, applying for a position vacated by a Dr Matter, who had taken up a teaching job. Since Einstein had latterly absented himself from all mathematics classes except those that were compulsory, this was always a long shot and in October it became clear that it had missed the mark. Furthermore, Heinrich Weber patently had no intention of employing the iconoclastic Einstein; indeed, rather than do so and in the absence of any other suitable physicists, he appointed two engineers to his vacant assistantships.

Einstein's disappointment was assuaged by Mileva's return to Zurich around 9 October, accompanied by her sister Zorka, who stayed for a short time to see a little of the wider world in which her sister moved. Now without any means of support after his Aunt Julie's termination of his allowance, his parents' lack of funds, and their disapproval of his being in Zurich with Mileva, Einstein determined to put his new diploma to use by taking private pupils. He also was paid to carry out some calculations on sunspots and other astronomical phenomena for a former teacher of his, Professor Alfred Wolfer, now Director of the Swiss Federal Observatory. He began to work on a doctoral thesis based in Weber's laboratory in the general area of electricity generated by differences in temperature. It is difficult to believe that the renewal of this acquaintanceship gave pleasure to either party, but Einstein knew that Weber's laboratory was the best equipped in the institution; nonetheless, either because of conflict with Weber or more likely because his interests drifted away from areas that could be researched in this laboratory and towards more theoretical investigations, Einstein soon lost interest in his initial topic in favour of his work of the properties of liquids.

Over the autumn, Einstein wrote up his paper on capillary effects in consultation with Mileva, who was inordinately proud of 'their' paper, even though it appeared in *Annalen der Physik* under Einstein's name only. Given the subsequent controversy over, in particular, the paper on special relativity that was to appear in 1905, it is important to try to disentangle their contributions to this first paper. Although their correspondence shows that Mileva and Einstein had many discussions about the paper, the tenor of the letters implies that the key ideas were his, developed while he was in Milan over the summer. Einstein apparently did not feel it necessary either to include her as an author or to acknowledge her contribution to the paper when it was published in March 1901.

Mileva was concerned in this period much more by what she considered to be the likely imminent departure of Einstein from Zurich. She wrote in some distress to Helene Savić, saying that if Einstein left her, half her world would depart with him. Mileva knew that he intended to spend Christmas in Milan and was very concerned that his parents would succeed in pressuring him to stay there and to cut his ties with her. An additional level of uncertainty arose from his active attempts to obtain a position as an assistant elsewhere in the German-speaking world; Einstein's habitual optimism would have given Mileva the idea that at any moment he would be successful.

Einstein did spend the Christmas period in Milan. A glimpse of Einstein's family life there at this time is given by a letter sent to him in 1931 by Edward Lebegott, a piano prodigy born in Italy in 1889 who came to Milan to study at the Conservatory. The Einsteins' traditional hospitality to fellow Jews that had benefitted Max Talmud in Munich presumably explains how Lebegott became intimate with them, spending every Wednesday evening at Via Bigli making music. He recalled a Christmas when Einstein returned home:

> I remember you very clearly . . . playing the violin on one of these evenings. Among other things we had at dinner a delicious cake baked by your mother herself. We spent a happy and exhilarating evening sitting around the table with its festive decorations in the beautiful dining room. I have never forgotten those times, which were among the best and happiest of my life.[2]

Einstein returned to Zurich on 3 January 1901. He continued to work out his physics ideas, sharing coffee with Mileva and trying without success to find a position in the academic world. He also busied himself with completing the formalities for the grant of Swiss citizenship, which was finally made official on 21 February. He was puzzled by his difficulty in gaining academic employment, although a realistic view of the prospects of someone who was on bad terms with the professor who would write his letters of recommendation and who had graduated lowest of all those in his class to get a diploma, might have given him pause for thought. It is also surely the case that, even in relatively tolerant and liberal Switzerland, anti-Semitism may well have played a part in his lack of success.

Once he received the offprints of his first paper in March 1901, Einstein forwarded copies to various scientific luminaries, including Wilhelm Ostwald, whose data he had cited therein. Despite two letters, Ostwald did not reply. Since he was deeply unsympathetic to the molecular treatment of liquid properties that Einstein championed in his paper, this is unsurprising, particularly considering the likely reaction of eminent professors to unsolicited letters begging employment. Eventually, unbeknownst to Einstein, his father also wrote to Ostwald pressing his son's case. The letter sheds light on Einstein's mood and situation in the spring of 1901:

> My son is terribly unhappy about his current lack of a position and as every day goes by he becomes more and more convinced that his career has become derailed and he has no chance of making an onwards connection. He is also depressed by the thought that he is only a burden to us, people of limited means. Distinguished Professor, I know that my son has a higher opinion of you than of any other of the active leaders of physics and I therefore am emboldened to request you most earnestly to read his paper that was published in *Annalen für* (sic) *Physik*; if you could send him a few lines of encouragement it would

---

[2] The Christmas to which Lebegott is referring is difficult to reconstruct. He mentions in the letter that Einstein was then living in Winterthur, but since he only lived there for a few months in 1901, not including a Christmas, the most likely explanation from other circumstantial evidence in the letter is that the Christmas referred to is that of 1900 and that thirty years later Lebegott remembered Winterthur from subsequent conversations in the apartment at Via Bigli. Lebegott subsequently became a professional musician and prolific composer in the US, where he was musical director of the Los Angeles Oratorio Society and the Mark Strand Orchestra, which played at the Strand Theatre in Manhattan in conjunction with silent movie presentations. By 1940, the lack of success of his musical career, which he noted in his letter to Einstein, resulted in him becoming a real-estate broker.

> restore his confidence in his abilities and his enjoyment of life. If in addition it might be possible for you to find him a position as an assistant, either now or in the autumn, my gratitude would be unbounded.

Even this level of flattery could not elicit a reply from the distinguished professor.

The above picture of Einstein's mental state can be complemented by his physical condition as testified by his Military Service Book, which he received and was obliged to maintain following the grant of Swiss citizenship. All Swiss male citizens are required to undergo compulsory military training; when he reported for his medical on 13 March, he was recorded as being 1.71 m high, with a chest measurement of 87 cm and upper arm 28 cm. It was not, however, his almost emaciated condition that ruled him out of further active military service but a combination of varicose veins, flat feet, and excessive foot perspiration. The latter gives some physiological reason, rather than the usual one of absent-mindedness, for his preference in later life not to wear socks. Einstein was by all accounts mortified by this official verdict on his physical condition; while he was still repelled by all things military, he was willing to do his duty in defending his democratic and egalitarian new country.

It was presumably lack of funds, as much as his traditional visit for the spring break between semesters, that drove Einstein to return to Milan towards the end of March 1901, from whence he would continue his campaign for a position. The situation Einstein encountered in Milan when he returned was depressing. His parents were weighed down by their usual financial problems and pressure from their main creditor, 'rich Uncle Rudolf'; indeed, several weeks later they were forced to ask Einstein, hardly in a sound financial situation himself, to send some funds to Aarau to support Maja. A few days after Einstein's arrival in Milan, Maja returned, causing a further worsening of the atmosphere. Her position with the Wintelers was extremely complicated. She had fallen in love with Paul Winteler, but her courting took place within a poisonous atmosphere caused by Einstein's ex-girlfriend Marie, who was still deeply upset by his desertion. This caused Maja considerable problems, since Marie was naturally reminded of her pain whenever they encountered each other. When Maja turned to Einstein for advice and help, he absolutely refused to discuss such 'women's matters' with her and continued to avoid any further encounter or communication with Marie. He complained to Mileva about Maja's poisonous mood. This was probably the moment when Maja's life-long devotion to her brother was most sorely tested.

The arrival in Milan of Jost Winteler, Paul's father, for the Easter vacation did not help the situation for either sibling. Einstein was dragged from his contemplation of physics problems to act as a guide; a few weeks later he was livid to discover that Jost, hardly the most conventional of men, nevertheless considered that Einstein was carrying on a scandalous way of life with Mileva in Zurich. This was particularly galling since it is clear that, at this point, Einstein would have been unable truthfully to deny any further accusations from his mother as to his and Mileva's sexual activities. As for Maja, although Jost fell in with Einstein's refusal to discuss romantic involvements, his presence continually reminded Maja of her problems with Marie. She became sufficiently concerned that she wrote to Pauline Winteler, who, like Einstein, she also considered an adoptive mother. Maja asked whether she should avoid Aarau in future and continue her education in Zurich rather than return and bring further strife to Rössligut. Fortunately, Frau Winteler would not hear of this and Maja could resume her courting with Paul; a year later the problem was solved when Marie left for a teaching position elsewhere in the Aargau.

When Einstein arrived back in Milan, he spent time working on physics at the excellent library of the venerable Istituto Lombardo, which was around the corner from the Einstein family's apartment. However, with Besso absent visiting his parents in Trieste and with Mileva still in Zurich, Einstein was starved of physics conversation. He was therefore pleased to welcome Besso back on 2 April, immediately going to Besso's flat and spending four hours discussing a wide range of subjects. These included the definition of absolute space, which was constantly at the back of Einstein's mind, despite his current enthusiasm for the properties of liquids and gases. He also exploited Besso's contacts, including an uncle who was a professor of mathematics, to present his case for an assistantship at Italian universities.

Einstein tried to entertain Mileva by writing an amusing anecdote about Besso, no doubt also hoping to disarm the jealousy he suspected she felt over his intimate physics conversations with his friend, whom she considered to be usurping her rightful role. The anecdote originated from Besso's employer, who was a friend of the Einsteins, with whom he had recently spent an evening of music-making. He recounted how Besso had been asked to visit a nearby town to inspect some electrical cables that had recently been laid. He decided to set off the night before to save some time but missed the train. The following day he forgot all about his mission until it was too late. On the day after, although Besso arrived in time for the train, he realised that he had completely forgotten what he was supposed to do when he arrived, so he sent a hurried postcard to the office before he boarded the train asking them to send his instructions to his destination by telegram. Einstein concluded the anecdote by wondering whether Besso was completely normal, and signed off by encouraging Mileva in her determination to go to lots of concerts in his absence – in particular, asking her to tell him whether she enjoyed Bach's Mass in B minor. Whatever his opinion about Besso's powers of organisation, he never doubted the quality of his physics intuition; this became a vital element in the great flowering of Einstein's creativity four years later. However, they now began a period of separation as Besso and his family returned to Trieste, relocating there after his father died in October 1901. In any case, Einstein's days in Milan were also drawing to a close.

On 13 April, some good news on Einstein's future finally arrived in the shape of a letter from Grossman, who reported that his father had recommended Einstein to his friend Friedrich Haller, the Director of the Federal Patent Office in Bern. Einstein's gratitude at being remembered by his old friend was heartfelt; clearly his father's characterisation to Ostwald of his low spirits was accurate. Even though the position in question, which was yet far from secured, was not academic, by now Einstein was desperate enough to grasp at any straw. After expressing his gratitude, he went on to tell Grossman about life in Milan in the glorious spring weather and how he was retaining his sanity by playing music with friends. In addition to the usual duets at home with Maja and his mother, Einstein spent considerable time playing at the Ansbacher home; their niece, who was a music student, was staying there during this period. He was also thinking deeply about physics.

More good news arrived on the following day. Professor Jakob Rebstein, a mathematics teacher at the Technikum, a technical school in Winterthur about 25 km northeast of Zurich, had been called up for two months of military service starting on 15 May. He was looking for a stand-in teacher and wondered whether Einstein would be interested. Einstein jumped at the chance, despite the fact that he would be teaching for thirty hours a week. His letter to Mileva, in which he announced both pieces of news, was simply electric with excitement and happiness, teasing Mileva by saying that the prospect of teaching geometry, which Mileva had failed disastrously in her examination, held no fears for her 'valiant Swabian'. His next letter was even more ecstatic in tone, pledging love and brimming with plans for their future together. After some prevarication from Mileva, owing to her spirits being depressed by a chiding letter she had received from her parents, they agreed to meet in Como in the early morning of 5 May 1901.

Later that month, Mileva wrote an account of their trip to Savić – one of the most joyous letters that Mileva ever sent. She detailed their trip, beginning from their loving reunion on the station platform. They spent half a day in Como, then went northwards on the ferry, first to the northernmost part of the lake at Colico and then went to Cadennabia on the west bank where they visited the gardens of the Villa Carlotta, resplendent at that time of year with blooming azaleas. On the following day, they scaled the Splügen Pass in a horse-drawn sledge, necessary since the snow was still deep on the ground in the highest reaches; indeed, Mileva reported six-metre drifts. Leaving the sledge, they continued their journey by walking over the pass into Switzerland via the Hinterrhine valley with the spectacular gorge at Via Mala. This walk of more than 40 km was some achievement for a girl with a limp. Indeed, after his arrival in Winterthur, Einstein solicitously enquired as to whether her feet had yet recovered. She was also, it turned out, pregnant.[3]

Mileva's suspicions about her condition were confirmed within a couple of weeks; it is likely that she informed Einstein of the news on one of what became their regular Sunday meetings, this one in Zurich on 26 May. He wrote to her on 28 May enquiring after her health and that of 'the boy'. He seems unconcerned by this news that fulfilled his mother's worst fears. His long periods in Milan at Besso's house, where he reportedly enjoyed playing with Besso's young son, probably made him look forward to the prospect of family life. Indeed, Maja confirmed that 'He really loves children. Already when, as a student, he was on vacation in Italy, the street children all knew him and ran after him until he talked to them.' However, with typical self-absorption, Einstein gave little indication of considering the implications for Mileva – physically, socially, and emotionally – as she prepared to retake her examinations at the Polytechnikum a few weeks later.

After their brief holiday but before he learned of Mileva's pregnancy, the two had parted in Zurich, where Einstein picked up his belongings from his former landlady and travelled to Winterthur. Rebstein met him at the station on 9 May and took him to the Technikum, where he monitored the lessons to get a feel for the environment in which he would be teaching. He visited his old friend and music partner from Aarau, Hans Wohlwend, who by chance worked in a local import-export firm. The two friends were overjoyed to find themselves neighbours again and Wohlwend lost no time in arranging accommodation for Einstein owned by his landlady and near to his own. Einstein was delighted with the room, which he described in detail to Mileva, together with the fine view and the wonderful garden, full of blooms. It is clear that he and Wohlwend spent much time in each other's company, no doubt playing violin duets after dining together each evening at Wohlwend's lodgings. Einstein also accompanied Wohlwend on visits to his family home in Lenzburg, refilling his place with violin in their music salon. Einstein's old accompanist Hans Byland was still in the neighbourhood; however, he was out of favour since Einstein had heard that Byland had agreed with some of Jost Winteler's criticism of Einstein's 'debauched' lifestyle in Zurich. Completing this set of reunions, Einstein visited Grossmann at his parents' home near Zurich.

Einstein's stay in Winterthur was generally happy. He learned from his parents that Jost Winteler, despite his reservations about Einstein's lifestyle, had nevertheless thought of him when he had heard of a vacant teaching position in Burgdorf, near Bern. Einstein wrote to Winteler in thanks, waxing lyrical about his pleasure in teaching the Winterthur boys and how he now considers school teaching to be his vocation. The delight of reuniting with old friends, re-establishing musical partnerships, and making new ones, as well as enjoying teaching and progressing with his scientific endeavours was only marred by separation from Mileva and a growing concern about their prospects with a baby on the way. Einstein's letters to Mileva from this period are peppered with various schemes for gaining employment, including applying for the Burgdorf position,

[3] The various dates that can be reconstructed indicate that the baby was conceived just before Einstein left Zurich for Milan, around 20 March.

which was unlikely as he clearly did not fulfil the requirements of the job description; he even contemplated working in insurance in Besso's father's firm if it would enable him to support Mileva and their child. None of these possibilities came to anything; he continued to hope for an academic position, as illustrated by an exchange of letters with distinguished physicist Paul Drude, at that time Professor of Physics in Gießen.

Einstein had been developing a theory of the properties of metals based on the motion of electrons, similar to Drude's published work, which had initially impressed him. However, he believed that he had discovered some errors in these papers, so he wrote to Drude setting out what he considered to be obvious problems. It is likely that these were related to the fundamental basis of Boltzmann's statistical explanation of gas properties, which underlay Drude's theories. In his letter, Einstein also took care to inform the eminent professor that he was looking for a job as an assistant. Of course, no offer was made, although Drude did at least reply, essentially ignoring the criticisms; not unreasonably, Drude considered that Boltzmann, one of the great icons of nineteenth-century physics, was much more likely to be correct about such matters than an unknown Polytechnikum graduate. It is again a measure of Einstein's naivety that he imagined that a letter from an unknown neophyte criticising Drude's and Boltzmann's work would result in a job offer. Einstein was incensed by Drude's cavalier dismissal of his concerns, which added to his and Mileva's conviction that they were fighting against the entire physics establishment.

As soon as he arrived in Winterthur, Einstein's insatiable appetite for music-making led to a quest for musical partners, in addition to those he had renewed with old friends. One such new partner was Marie Barthelts, a thirty-six-year-old music teacher who was probably recommended to him by Wohlwend. She wrote to Einstein many years later, chiming in with the world-wide congratulations on his fiftieth birthday, reminding him of their former happy times together:

> . . . [in] looking for an accompanist to play with, you came unexpectedly to me! Since I was out [when you called] you arranged to come in the evening; your youth encouraged me, although I was somewhat concerned when you first of all blew out the candles on the music stand, saying that you knew the music by heart. We played Mozart sonatas – inexplicably, at that time unknown to me – and it went so well that you arranged regular evenings with me, which for my part I was pleased to reserve. I don't precisely remember the other things we played, but I do recall that you liked to play the vocal part of Mendelssohn songs – I can still hear "The sun is already climbing over the mountain, through the wood . . . ." We had so much fun playing together; my mother would often ask you to have your supper with us, just our normal meal, but you would never take anything except fruit and a drink after we had played. You ate a cup of soup and a sausage beforehand [. . .], which was always enough for you. Sometimes you played on Hans Wohlwend's instrument, which was better than yours; sometimes he came with you and we played Trios – do you remember?

There are some very evocative images of a bygone age here from Barthelts, who in her sixties still lived with her ninety-year-old mother in the same house in Winterthur: the brash young man blowing out the candles on the music stand – despite the efforts of Hermann, Jakob, and their competitors, domestic lighting by electricity was then still only available to the rich; Einstein's habitual simple evening meal of soup and a sausage; Einstein swapping his cheap violin for the better one of his friend; all three playing trios by candlelight.

After his morning of concentrated teaching, Einstein worked hard in the afternoons in the local library, absorbed in puzzling through elements of the statistical theories of Boltzmann and the connection with his own work on liquid properties in terms of molecular dynamics. He decided to write a doctoral thesis and associate himself with Alfred Kleiner, the professor of physics at the University of Zurich, which was necessary as the Polytechnikum had no power to award doctorates.

Einstein seems to have joined Wohlwend as a violinist in the Orchestergesellschaft Winterthur. Founded in 1885, this orchestra was composed of amateur players, as opposed to the professionals of the Stadtorchester, the oldest orchestra in Switzerland, which traced its history back to 1629. The Orchestergesellschaft rehearsed in the Technikum. It was presumably at rehearsals that Einstein met Alfred Schnauder, who later became conductor from 1914–1920. Schnauder was also a composer who kept in touch after Einstein left Winterthur and subsequently sent him several of his compositions.

Einstein's appointment and his generally happy stay in Winterthur came to an end on 15 July. He was still applying for various positions; one in the mathematics department at the Cantonal school in Frauenfeld had been advertised but once again Einstein lost out – this time to his friend Grossmann. Einstein did not take up Mileva's suggestion that they should go together to visit her parents in Kac, where a somewhat awkward interview would have awaited him. Indeed, after leaving Winterthur he decided to join his mother and sister in their annual holiday, returning to the Hotel Paradies in Mettmenstetten. This is a difficult decision to understand, since Mileva's second attempt to pass her final examination was imminent and she would surely have derived strength and support in her pregnant condition from her lover's presence by her side. His only contribution was a brief couple of sentences in a letter in which he wished her luck and said they should go on a celebratory outing after her ordeal was over. Sadly, no celebrations were called for; in any case, when Mileva's marks were released on 27 July, Einstein had left the Hotel Paradies for a few days and was mountain walking in the Urnerboden region south of Zurich. Mileva had made essentially no progress in her average score since the previous year. Given her delicate condition and various preoccupations, as well as the marked deterioration of her relationship with Weber, it is a great credit to her that her performance had not also deteriorated. Nevertheless, this constituted the end of her attempts to make her own way in physics; her future contributions would be in assisting Einstein's career.

Packing her bags to return to her parents' home, Mileva was naturally concerned about the reception she would get when her condition became undisguisable. She asked Einstein to write to her father, presumably repeating the promises he had already made to her that they would be married as soon as Einstein had a secure position. She also asked him to either send her a copy or to meet briefly before her departure. She had discovered in the railway timetable that the train from Zurich called at Mettmenstetten and then returned after a halt of fifteen minutes at Zug; if Einstein could catch it early in the morning, they could steal an hour together on the return journey between Mettmenstetten and Zug without his mother being aware. It is not known if Einstein took her up on this ingenious subterfuge; a few days later she returned to Vojvodina.

Einstein's movements over the rest of the summer are unclear; he intended to, and most probably did, return to Milan for a while. However, by September he was back in Winterthur, as evinced by a letter to Grossmann congratulating him on his appointment to the teaching position in Frauenfeld. This letter also illuminates his physics interests in this period. He wrote that he was working hard on trying to reformulate some of Boltzmann's work to make it more self-consistent and also that his long-standing interest in the motion of bodies through the aether was continuing. He had thought of a particularly simple way to investigate this using interference, probably a variation on the Michelson–Morley experiment. He also informed Grossmann of the good news that he had succeeded in obtaining a position as a tutor to an English pupil in a private school in Schaffhausen, about 30 km north of Winterthur, starting on 15 September. He was to lodge with the head of the Lehr- und Erziehungsanstalt, Dr Jakob Nüesch, and his family.

It is likely that a friend from Zurich, either Jakob Ehrat or Conrad Habicht, played a part in Einstein's obtaining the position with Nüesch. Both were natives of Schaffhausen.[4] Ehrat had graduated one place higher than Einstein and had successfully obtained an assistantship at the Polytechnikum. Habicht had gone on to study in Munich and Berlin and had only recently returned home when Einstein appeared on the scene. Habicht was not only a student of physics and mathematics – having switched from initial studies in philosophy – but also he was a highly proficient violinist. It is therefore unsurprising that he and Einstein became close friends.

If Winterthur had been an almost idyllic introduction to the joys of pedagogy, the position at Schaffhausen was much less satisfactory. The fault lay not so much in the pupil,[5] a nineteen-year-old Englishman named Louis Cahen who wished to gain entry to the Polytechnikum to study engineering, as in Einstein's personality clash with Nüesch. The terms under which Einstein had accepted the position stipulated that he should live and dine with the Nüeschs and their four children, as well as with other lodgers. The noisy and untidy household made it difficult for Einstein to concentrate on his physics and the meals seemed like an extension of his workday. When he requested that he should be given the cost of his upkeep to be used at a tavern, Nüesch exploded. Although he eventually gave way, according to Einstein, intimidated by his threats to leave, the tension in the relationship remained. Eventually, Einstein moved out of the Nüesch household entirely, first to lodge with another family but by December into the Cardinal restaurant and guest house, an imposing four-storey structure on the Bahnhofstrasse.

While Einstein was negotiating his way through social minefields in Schaffhausen, Mileva was having an uncomfortable stay at home. Her parents had reacted as might be expected to both news of her pregnancy and that there was no immediate prospect of the father regularising their relationship. She no longer had a goal in physics to work towards and her interest in the subject, except for Einstein's endeavours, palpably waned. Her academic failure also greatly grieved her father, who had always believed in and supported her studies. In addition, Pauline Einstein had sent a highly abusive letter to Mileva's parents in which she had fulminated against Mileva for ruining her son's future prospects. It is unclear what precipitated this missive, but it would be explained if at some point that autumn Einstein had finally told his parents of Mileva's condition. Very few letters are extant from the period between Mileva's leaving Zurich and November, by when she seems to have had enough of the poisonous atmosphere at home. She needed to seek solace and encouragement from her beloved Albert, so she caught the train back to Switzerland; in order not to cause any scandal resulting from her by now obviously delicate condition, she installed herself in the Hotel Steinerhof, 20 km down the river at Stein am Rhein.

Any solace for Mileva seems however to have been in short supply. Einstein learned of his beloved's arrival in the neighbourhood from a bunch of flowers that she sent him; far from rushing into her arms, he made excuse after excuse to avoid visiting her. Instead he sent her a bundle of books to keep her entertained. Mileva grew increasingly exasperated, especially over his excuse that he had no money for the train fare, offering to send him money if his monthly salary and full

[4] Although Seelig states that Habicht was the person who recommended Einstein, and implies that they both studied at the Polytechnikum, other authorities, including Einstein himself, say that they only met after Einstein arrived in Schaffhausen. Perhaps Seelig became confused between Habicht and Ehrat, although it is equally possible that Einstein forgot that he and Habicht first met in Zurich; certainly they were both in Zurich in Einstein's student years, although Habicht initially studied philosophy at the university.

[5] Reiser mentions two pupils, but there seems to be no other evidence for this.

board was insufficient to leave him a few francs in his pocket. Clearly Einstein did not relish being reminded by the sight of Mileva's swelling belly that he was not able to devote himself single-mindedly to physics; he had to find a secure position to allow him to support a wife and child. Eventually they met a couple of times and Einstein was able to bring her up to date with his physics activities. He had completed two papers that he intended to submit to the university as a doctoral thesis. One was his reformulation of some of Boltzmann's work; the other concerned molecular forces in gases. He submitted them to Professor Kleiner for his consideration on 23 November; the following month, he made an excursion to Zurich to discuss his thesis but discovered that Kleiner had not yet read it. Unperturbed, he explained his ideas on the electrodynamics of moving bodies to Kleiner, who seemed both interested and appreciative and who offered to act as Einstein's referee for future applications for positions, thereby relieving Einstein from his dependence on his nemesis, Weber. Einstein told Mileva that Kleiner was not as stupid as he had thought and was a decent man!

Although the tensions between Einstein and his employer meant that his sojourn in Schaffhausen was not particularly happy, neither was it the social desert that his letter to Mileva in mid-December implied. He wrote that he led a life of complete solitude and talked to no one except his pupil Louis. In fact, only the week before he had played in a concert given by the Musikkollegium Schaffhausen. Habicht was a member of the orchestra and introduced Einstein also into the violin section. Habicht was by all accounts a first-class violinist; indeed, he was musical director of the Musikkollegium for twenty-three years and its leader for even longer, having returned to Schaffhausen in 1915 as a physics teacher in the Cantonal school, where he remained until his retirement in 1948. The concert in which Einstein played took place in the Inthurneum, the music school of the town, on 11 December. The programme was typical for the time, not dissimilar to that of Joachim that Einstein had listened to, enthralled, at the start of his life in Switzerland. It began with an overture by Weber, followed by a selection of songs from the baritone Otto Watter from Zurich and Spohr's Violin Concerto No. 9 with the leader of the orchestra, Klein, as soloist. The second half continued with a similar mixture of orchestral works and lieder, ending with Dvořák's Symphony No. 9 in E Minor (the 'New World' Symphony). In addition to the rehearsals leading up to the performance, Einstein also attended other concerts as an audience member and spent many evenings playing violin duets with Habicht. Einstein's family also paid visits: his cousin Robert stayed with him for several days in November during Mileva's stay in Stein am Rhein, and Maja was only around an hour's train journey away in Aarau. Einstein visited her there. No doubt realising that Mileva was having a miserable time in Novi Sad, he did not dwell in his letters to her on the compensations of his 'monastic' lifestyle in Schaffhausen.

On the same day that Einstein appeared in the Musikkollegium orchestra, he received a letter from Grossmann informing him that the position in the Patent Office that he had been trying to secure for Einstein through his father's influence was about to be advertised and that he was confident that his friend would indeed be appointed to it. Einstein enthusiastically told Mileva that this would allow them to marry but asked her to consult her father about finessing a baby's impending appearance in their new household. In all the excitement of this news and his concert, he entirely forgot, not for the first time, to congratulate Mileva on her birthday on 19 December. Einstein decided not to go to Milan for the Christmas holidays, but spent Christmas Eve and Christmas Day with Maja in his favourite haunt, the Hotel Paradies, before returning to Schaffhausen.

Shortly after Grossmann's letter, Haller wrote to Einstein asking him to apply for the vacant position. The cordiality of Haller's letter makes it likely that, rather than taking the reference from Grossmann at face value, he had already met and made his own assessment of Einstein. Although there is no record of a meeting in Einstein's correspondence, both Maja and Reiser mention an interview taking place in 1901. Einstein was examined for two hours by being given various patents to read, on which he had to form an immediate opinion. Although it was clear that he lacked knowledge of many technical areas, Haller must have been sufficiently impressed to convince himself that Einstein would be a suitable candidate for the next vacancy.

January 1902 turned out to be an eventful month. Towards its end, Einstein's febrile relationship with his employer finally ended 'with a bang' and, although he had signed on for a full year, he packed his bags and left town with immediate effect. No doubt the prospect of the position in the Patent Office emboldened him to be even more free and frank in discussions with Nüesch, who had already given evidence of scant patience with his impudent assistant. At some time earlier in the month, Einstein had probably paid a visit to Zurich to hear Kleiner's verdict on his thesis. It was not positive. Nevertheless, Kleiner must have been impressed since he did not reject the thesis outright and remained on good terms with Einstein. He convinced him to withdraw the thesis. Einstein picked up the entire 230-franc fee refund from the university on 1 February. This cash must have come in very handy as he boarded the train to begin a new phase of his life in Bern, the capital of the Swiss Federation and site of the Patent Office and his anticipated position. When he arrived in the sleepy mediaeval city, however, he had no job, lodgings, income, nor any immediate prospects of gaining his doctorate nor any other academic advancement. Meanwhile, at some point in the month of January, Mileva had, with great difficulty, given birth to their daughter.

Chapter 5

Marriage and the Patent Office in Bern (1902–1904)

Mileva had a difficult labour, to the extent that for many days she was exhausted and unable to write to Einstein with the news of the arrival of a healthy daughter, Lieserl. The news eventually reached him via a letter written by Mileva's father and forwarded from Schaffhausen to Bern. In his reply to Mileva, Einstein recounted how worried he was to see an envelope in her father's handwriting; fearing the worst, he was delighted to discover that both mother and baby were well, even though the mother was keeping to her bed. He plied Mileva with many questions about the baby's doings and whether a photograph could be sent to him. If it was, it was the nearest he ever got to seeing Lieserl. It was mentioned in Chapter 4 that Einstein had suggested that Mileva ask her father how they might keep the baby with them. Either the question was never asked, or Herr Marić was unable to come up with an idea, for the baby never joined the couple when Mileva eventually came to Bern. The only other mention of her in Einstein's correspondence is in September 1903, when Einstein wrote to Mileva, who had returned to Vojvodina for a visit to her parents. Liserl's fate is unknown.

The second half of Einstein's letter in response to his becoming a father sang the praises of the city in which he had just taken up residence. He loved the quaint arcades in the historical centre that protected pedestrians from rain and was very proud of the room that he had rented. He sent Mileva a sketch of it, including where the various pieces of furniture were placed. He was also happy to find he was not without acquaintances in Bern. A school friend from Aarau, Hans Frösch, was studying in the city. He took Einstein to a medical class that interested him so much that he planned to attend every Saturday. His friend Conrad Habicht was working towards his doctorate in mathematics at the university; although he spent much time away from Bern at his parents' house in Schaffhausen, where he and Einstein had become intimate, he was often in Bern during term time. Paul Winteler, now his sister's boyfriend, studied jurisprudence at the university until March 1903, when he left for a year abroad. For a while, he and Einstein often shared communal meals of the cheapest and simplest kind. Indeed, Einstein had so little money in these first months in Bern that he often went hungry. Although he had been hard pressed for funds in Zurich, he had always been able, when in desperate need of sustenance, to dine with family friends or feed himself from the buffet after playing his violin in music salons, such as that of Selina Caprotti. In Bern, he had as yet no such possibilities and no allowance from his Genoa relatives. In later life, he blamed his chronic digestive problems on this period of deprivation.

Despite hardships and shortages of funds, Einstein began to enjoy himself enormously, indeed too much for Mileva's taste, who was still recuperating and looking after Lieserl in Novi Sad. Einstein had to reassure her within two weeks of his arrival that she should not be jealous of Habicht and Frösch, who meant nothing to him in comparison to her. She was no doubt uncomfortably aware, however, that her place as his physics confidant was being slowly usurped by others.

Immediately after arriving in Bern, Einstein sought to earn some money to support himself until the expected Patent Office position materialised. He placed an advertisement in the

Einstein. Brian Foster, Oxford University Press. © Brian Foster (2026). DOI: 10.1093/oso/9780198794875.003.0005

local paper offering 'Private lessons in Mathematics and Physics for undergraduates and school pupils taught thoroughly by Albert Einstein, holder of the specialist teaching diploma of the Federal Polytechnikum. Gerechtigkeitsgasse 32, first floor. Trial lessons without charge.' This had an immediate response and by around 10 February 1902, he was teaching two pupils, one an architect and the other an engineer, from whom he expected to earn a total of 4 francs per lesson. Since this was utterly inadequate to live on, even assuming that these pupils continued their lessons – and they are never mentioned again – Einstein re-advertised during the Easter vacation.

It was this advertisement that caught the eye of Romanian student Maurice Solovine, who was about four years older than Einstein and of Jewish extraction. He had been perfunctorily studying a variety of subjects at the university philosophy faculty since 1900 but had recently decided that he was really interested only in physics, so he knocked on Einstein's door. Encouraged by a deep 'Come in', he opened the door to be met by the impression of Einstein's eyes shining at him from the gloomy interior. The trial lesson gradually extended into a two-hour discussion on various aspects of physics, only interrupted by Einstein needing to leave the house; even the farewell was extended by a further thirty minutes of discussion in the street before they arranged to meet the following day. At their third meeting, Einstein indicated that they should forget about lessons and just spend their time discussing the big open questions in physics that clearly fascinated them both. Solovine left a description of Einstein's pedagogical method at that time:

> I was amazed at his unique ability to tackle and solve problems in physics. He wasn't an inspiring speaker and neither did he use startling analogies. He laid out his arguments slowly and monotonously, but what he said was wonderfully clear. To make an abstract concept more understandable, he sometimes used an example drawn from everyday experience. Einstein, who used his mathematical armoury with exceptional skill, often argued against the misuse of mathematics in physical problems. 'Physics', he said, 'is in its essence a precise and clear science. The role of mathematics is merely to put the tools into our hands whereby we can express the laws that govern the phenomena.'

Solovine had the idea that they should extend their activities in a form of a 'journal club', reading some of the major works of physics and philosophy and eventually also literature. He suggested that they start with the *Grammar of Science* by the English mathematician Karl Pearson, who was among the first to apply statistical methods to a wide range of scientific problems, particularly in biology. Assuming that the friends read the German translation of the second edition, published in English in 1900 and considerably extended from the first edition, then the book ranged from philosophical discussion of time and space to Darwin's theory of evolution. There is a delicious irony in one extract:

> We have in the law of gravitation an excellent example of a scientific law. [. . .] at last Newton reaches [a formula] so simple and so wide-embracing that many have thought nothing further can be achieved in this direction. 'Here', says Paul du Bois-Raymond, 'is the limit to our possible knowledge.'

Unsurprisingly, it took them several weeks to discuss their way through this. Shortly afterwards, they were joined by Habicht, back for the start of the summer semester, who joined their discussions with enthusiasm. However, Paul Winteler felt somewhat out of his depth in the meetings of the 'Olympia Academy', as the three friends, with deliberate portentousness, named their regular gatherings. He gradually ceased to partake in the communal meals with Einstein; in any case, he had got into some trouble with the authorities by joining a demonstration against a Bern professor who offended Swiss students by opining that Switzerland was really intellectually a part of Germany. Paul thereafter decamped to France for the summer.

Solovine described the development of the Olympia Academy over the next few months:

> It was important to Einstein that we ate together. These meals were of exemplary simplicity. The menu was typically a salami sausage, some Swiss cheese, some fruit, a little honey and one or two cups of tea – but we were so happy! One could indeed apply the words of Epicurus: 'Contented poverty is an honourable estate.' [. . .] In order to live, Einstein at that time had to give private lessons but it was difficult to find pupils and impossible to charge very much. One day when we were discussing what other sources of income he could find to survive on, he told me that the easiest thing for him to do would be to busk around the streets with his violin. I told him that if he really was going to do that, I would learn the guitar in order to accompany him.

Having worked their way through Pearson, the young men read Mach's *The Analysis of Sensations* and *Mechanics*, the latter having already been read by Einstein, J. S. Mills's *A System of Logic*, Hume's *A Treatise of Human Nature*, Spinoza's *Ethics*, lectures by Helmholtz, a few chapters from Ampère's *Essay on the Philosophy of the Sciences*, Riemann's *On the Hypotheses that Lie at the Bases of Geometry*, sections of *Critique of Pure Experience* by Avenarius, Clifford's *On the Nature of Things-in-Themselves*, and Dedekind on number theory. The main thrust of their initial reading list was more philosophy than physics. They also read Poincaré's *Science and Hypothesis*, which Solovine noted was a work that made a great impression on them, keeping them busy for several weeks. As they worked through it, several of the issues that would dominate Einstein's scientific work for the rest of his life cropped up: non-Euclidean geometries, Brownian motion, Lorentz's electron theory, Fizeau's experiment, the principle of relativity, absolute space and time. Einstein developed a great sympathy with Spinoza's philosophy. In later life, he often characterised his religious beliefs as akin to those of the Dutchman; Spinoza's staunch belief in determinism and that free will was an illusion were things that Einstein hung onto tenaciously. They coloured his views on future developments in physics, in particular the indeterminism inherent in the behaviour of the atom as described by quantum mechanics.

At the end of his life Einstein still recalled the impression that these months of concentrated discussion and study had made upon him: 'We occupied ourselves mostly with D. Hume (in an excellent German translation). Reading this had considerable influence on the development of my thinking – along with Poincaré and Mach.' Poincaré's relativity principle simply recasts the conventional view of so-called Galilean relativity, that it is not possible to ascertain the velocity of a moving reference frame by any measurement made solely within it. However, he does make clear that he refutes the Newtonian concept of absolute space, which indeed Mach criticises in much more intemperate terms. Poincaré states 'Thus, absolute space, absolute time, and even geometry are not conditions that are imposed on mechanics', and concludes his book with a description of Lorentz's recent and successful theory of electrodynamics, with considerable foresight and perspicacity:

> . . . the edifice of electro-dynamics seemed, at any rate in its broad lines, definitively constructed [. . .] fresh experiments have been undertaken. What result will they give? I shall take care not to risk a prophecy which might be falsified between the day this book is ready for the press and the day on which it is placed before the public.

The falsification took somewhat longer than Poincaré feared, but its seeds were already germinating in Einstein's mind.

Not everything in the Olympia Academy was dry academic discussion. Solovine reported many light-hearted and indeed frivolous activities. They expanded their reading to include works of literature, including plays such as Sophocles's *Antigone*, Racine's *Andromache*, and novels such as Dickens's *Christmas Stories*, and a large part of Cervantes's *Don Quixote*. Solovine also reported that Einstein would sometimes brighten up their meetings with his violin. Given their previous musical collaboration, Habicht would certainly have joined in some duets.

Music could also be a distraction in other ways. Solovine noted that:

> Bern was fascinating because the greatest violinists, cellists and pianists who were travelling from one end of Europe to the other would always give a couple of concerts in the Swiss capital, at Europe's centre. We took advantage of many of these opportunities. One day I noticed a poster on a wall advertising a concert by a famous Czech string quartet. The programme was very attractive: quartets from Beethoven, Smetana and Dvorak! When I arrived at Einstein's that night for our Academy meeting, I told him the good news and that I would book three tickets for us. 'I think it would be better', said Einstein, 'to skip the concert and read some Hume, which is incredibly enlightening.' 'OK, fine', I replied – but when I passed the concert hall at 5 p.m. on the day of the concert and saw the programme again, I was tempted and went automatically to the ticket office to ask if there were tickets available.
>
> The lady looked at the plan and offered me several tickets in different parts of the hall; finally she found a couple of available tickets at reduced price for those on small incomes. I lost my head and bought a ticket.
>
> Now on this evening our meeting was to be at my place. I rushed home to prepare dinner. Since I knew my friends were fond of hard-boiled eggs, I added four eggs on a side plate to the usual meal and covered them with a note on which I wrote: *Amicis carissimis ova dura et salutem* – 'To my dearest friends, hard-boiled eggs and greetings.'
>
> I asked my landlady to present my apologies to my guests and tell them that I was called away on urgent business. When they turned up for dinner and heard this fairy tale they knew immediately what was going on. They ate everything on the table. Knowing that I hated tobacco in all its forms, they began smoking as if possessed by the devil. Einstein puffed on his pipe, while Habicht smoked strong cigars. They filled the plates with pipe ash and cigar butts and piled up the crockery and all the furniture on the bed, leaving a note on the wall saying: *Amico carissimo fumum spissum et salutem* - 'To our dearest friend, thick smoke and greetings.'
>
> It was my habit after a concert to take a short walk. The music ran around inside my head and this allowed me digest what I had just heard. I did so that night. I strolled around the streets until 1 a. m. When I got home and opened my door I was met by a miasma of tobacco smoke. Thinking I was going to suffocate, I tore the window open and busied myself in moving the tower of furniture from my bed, which reached almost to the roof. When I lay down to try to sleep, I couldn't close an eye because of the awful stink of tobacco smoke that had impregnated the sheets and pillows. Only as dawn broke did I get any rest.
>
> When that evening I sought out Einstein to have our evening meal and Academy meeting, he greeted me with beetle brows and the words 'You reprobate! You miss our meeting to hear a couple of violinists! You barbarian, you bumpkin! If you ever perpetrate another such offence, you will be drummed out of the academy in disgrace!'
>
> Since the previous evening's meeting had been cancelled because of my absence, we carried on that evening until 1 a. m.'

This is one of very few recorded occasions when Einstein turned down the chance to attend a concert!

Sometimes tiredness overtook them before they could finish their discussion, which was often precipitated by not much more than the first sentence of the passage they were studying. In this case, the friends would sometimes meet in the street and restart a discussion of the previous night or, after Einstein began work at the Patent Office, during his lunch hour. In the balmy summer evenings after their Academy meeting, they would often climb the Gurten, a hill rising about 300 metres above Bern, to watch the sunrise. They were moved by the glorious views of the stars to discuss astronomy, until the sight of the sun illuminating the white alpine peaks with a roseate glow moved them to silence. Once the mountain restaurant opened, they would partake of a hearty breakfast, returning to the city by 9 a. m., tired but happy. Other places they visited included Thun, about thirty kilometres away, at the head of the Thunersee; they typically set off to walk at 6 a.m. and arrived at Thun at midday. The glorious views of the Alps inspired them to talk about geology; after a good meal and much more talk, they caught the train back to Bern.

This bohemian lifestyle and constant exchange of views and ideas were exactly what Einstein needed to spur him on to new endeavours in physics. He always worked best when he had a good sounding board, and in the Olympia Academy, he had two. Indeed, he sometimes had more, as they were frequently joined by Habicht's younger brother Paul, who was a pupil at a school in Bern, and by Lucien Chavan, who was significantly older than the others. He had taken up an appointment in 1900 at the Federal Postal and Telegraph Administration, which had its offices in the same building as the Patent Office, and began to take physics lessons from Einstein sometime in 1903. Like Solovine before him, he was soon attracted into the Academy's orbit. Chavan's wife also became one of Mileva's few close friends in Bern. The families remained close after both had left the city.

Einstein's research activities in Bern continued even before the Academy had taken shape. His work on molecular forces that had resulted in his first paper on capillarity was developed to examine the properties of very dilute solutions. After establishing the utility of the second law of thermodynamics in discussing the dynamics of mixtures of molecules in a liquid, he used it to derive relations on the potential developed in an electrolytic cell consisting of electrodes of the same metal as the solute. He found that the potential depended only on the concentration of the metal in the solution and the nature of the solvent and suggested experiments that could be carried out to verify his underlying ideas on molecular forces. Since he admitted that the experimentation would be laborious, it is no surprise that his suggestions were not taken up by other physicists. The paper was completed in late April, received by *Annalen der Physik* on 30 April, and published on 10 July 1902.

That thermodynamics played a large part in Einstein's second paper was symptomatic of a shift in his interests away from molecular forces towards the basic formulation of thermodynamics and statistical physics that had formed a substantial part of his now-withdrawn thesis. He was concerned to derive the basic laws of thermodynamics from statistical methods with as few underlying assumptions about the mechanical properties of the systems as possible. This was by no means a retreat from his by-now firm conviction of the utility and indeed reality of atoms and molecules in the description of physical phenomena; rather it was an attempt to free the theory from ad hoc assumptions about the details of their properties and interactions that some authors had found necessary. Indeed, he subsequently characterised all his early publications in this area with the comment: 'My main purpose for doing this was to find facts which, as far as possible, would attest to the existence of atoms of a particular, finite, size.'

The programme Einstein was setting out on was ambitious. During it he made several important advances which significantly influenced his subsequent work and his basic outlook on physics. It transpires that much of the work he undertook in this area had already been published independently, some by Boltzmann, but in particular in a book by the great American physicist Josiah Willard Gibbs. It is unsurprising that Einstein was unfamiliar with this work as it was published in English only in 1902 and no German translation appeared until 1905.

A brief review of the science of thermodynamics may be helpful before embarking on the details of Einstein's contributions. The modern study of thermodynamics originated in the construction of steam engines and was formulated by, among others, Boyle, Hooke, Gay-Lussac, and in particular Sadi Carnot, whose 1824 book *Réflexions sur la puissance motrice du feu et sur les machines propres à développer cette puissance* (*Reflections on the Motive Power of Fire and on the Appropriate Machines to Improve this Power*) can be considered as the founding text of modern thermodynamics. This 'classical' thermodynamics characterised systems, usually gases, in terms of macroscopic, average, quantities such as volume, temperature, pressure, etc. The systems were considered continuous and their internal structure, if any, irrelevant. The idea that heat is actually a form of motion of internal constituents, or atoms, goes back at least to the seventeenth century but was only gradually considered to be important, until it was brought to the fore in the mid-nineteenth century by the work of Clausius, Maxwell, and Boltzmann. They explained many of the 'classical' laws in terms of the statistical behaviour of atoms and molecules. Nevertheless, when Einstein began to publish in this area, the 'statistical mechanics' view of matter as consisting of countless constituents in perpetual random motion – except at the absolute zero of temperature, −273 °C, where all motion ceases – was far from generally accepted.

The first of a series of three of Einstein's papers in this area uses the simplest thermodynamical system, called by Gibbs a 'microcanonical ensemble', which consists of a macroscopic system with many degrees of freedom (equivalent to many constituent particles) which is isolated and has a constant energy. Using this, Einstein constructed what Gibbs called the 'canonical ensemble', a collection of microcanonical ensembles that is in thermal equilibrium but can exchange energy both internally and with its environment. He then derived the laws of thermodynamics and calculated the entropy, a measure of the level of non-random structure within a thermodynamic system. Einstein had probably drafted this paper the previous year but the final version was sent to *Annalen* in June 1902 and published in September.

In April, just before the submission of this paper, an unexpected visitor whom Einstein had neither seen nor heard of for many years arrived in Bern. Max Talmud, the friend from his childhood in Munich who had fed his interest in mathematics and science, had been in Milan and called in to see Einstein's parents. He wrote that:

> Only Mr and Mrs Einstein were present, and he was evidently in bad health. There was an atmosphere of seriousness and gloom all around. The cheerfulness of the Einstein home in Munich was no more. My hosts were rather reticent, when I tried to learn something about their circumstances and the fortunes of their son.

This mood is confirmed by Maja, who wrote to Paul Winteler during a visit to Milan at the end of August:

> I am back home for a couple of weeks but under very sad circumstances. Father is in bed, is very weak and can't take pleasure in anything. This is all Albert's fault. Can you imagine, he has completely broken with my parents, he no longer writes to them [. . .] You can imagine how much this is getting me down.

She was glad to escape briefly from the gloom of Via Bigli by paying a visit to her old friend Ernestina Marangoni in Casteggio, where she found life unchanged from their former happy times there. In terms of Maja's unfair accusation against Einstein – their father in fact had a serious heart complaint – it seems he had for the moment given up the fight to gain acceptance for Mileva and was hoping that a period of icy silence from him would eventually break down his parents' opposition.

Having been informed by Einstein's parents of his whereabouts, Talmud hurried away to Bern to visit his erstwhile friend, noting that 'His environment betrayed a good deal of poverty. He lived in a small, poorly furnished room.' Talmud was by then a rather prosperous doctor, so it is unsurprising that he looked on Einstein's accommodation with a more jaundiced eye than had its occupant in his first report to Mileva. After they spent the day together, Einstein presented Talmud with a copy of his first paper; he must have ordered a large number of offprints given the quantity that had already been sent to the tenured physicists of Europe to solicit employment. Talmud passed out of Einstein's life again for many years until they were to meet once more on a different continent, when Einstein had acquired world fame and Talmud a new name.

By the end of May, there was finally movement within the Swiss bureaucracy to fill two vacant places in the Patent Office. A total of six candidates applied, although two were judged insufficiently qualified at an initial stage. The remainder were examined both orally and in writing. It is unclear whether Einstein was examined again or whether his earlier interview sufficed; Director Friedrich Haller had a great deal of latitude in making the appointments. Einstein and Heinrich Schenk were appointed on 2 June 1902 at an annual salary of 3,500 francs, with their employment intended to start not later than 16 June. However, the bureaucratic wheels turned insufficiently quickly, so that Einstein only received his letter of appointment on 19 June with the instruction that he could begin work any time before 1 July.

Einstein's appointment to the Patent Office marked the removal of the most important obstacle that had prevented his marriage to Mileva, but there were other issues. There was firstly the desirability of bringing both families around to the idea. Next was the question of what to do about Lieserl. Also, protest as he might in his letters to Mileva, Einstein was greatly enjoying his time with his friends in Bern. No doubt sensing this and expecting Einstein's imminent appointment to the Patent Office, although no letter from him intimating this survives, Mileva made the decision to leave Lieserl behind and return to Switzerland. It seems likely that she obtained a visa that allowed her to stay for two months and that she arrived in Bern on 6 June; she was certainly nearby when Einstein wrote to her on 28 June arranging to meet two days later. During her stays in Bern she lodged near to but separately from Einstein. The remainder of the year 1902 is very difficult to reconstruct. Only one letter, from Einstein to Hans Wohlwend, has survived. Einstein seems to have moved home twice, once at the beginning of June to Thunstrasse, across the River Aare from Gerechtigkeitsgasse and away from the old city with Einstein's beloved arcades. Since Thunstrasse is further from the Patent Office, the move was probably precipitated by Mileva's imminent arrival; perhaps the accommodation was not quite so cramped and shabby as that which had made so unfavourable an impression on Talmud. Whatever the reason, this was also considered unsuitable, presumably by Mileva after her arrival in Bern. Einstein therefore moved again, to Archivstrasse 8, once more slightly further from the Patent Office, on 15 August. Mileva went back to Vojvodina on 11 July, having agreed with Einstein that they would marry imminently. When she returned on 9 September, having arranged for the marriage bans to be published in Novi Sad in October, on her visa she gave 'to be married' as the reason for her visit.

Sometime in that summer, Einstein wrote to Wohlwend, hoping to confirm tentative arrangements for him to visit Bern. To tempt him, he mentioned that his new landlady had a wonderful violin, no doubt available for Wohlwend's use in their beloved violin duets. He also waxed lyrical about the pleasures of Bern, although he mentioned that he was very busy, with eight hours per day in the Patent Office followed by giving at least one private lesson. Nevertheless, he was enjoying his work, which he considered varied and thought provoking. Even more, he liked having a regular salary! The only negative issue was that his father was unwell.

As described, Maja had visited Milan at the end of August. She would certainly have told Einstein, even had his mother not done so earlier, that his father was ill, particularly since she blamed his silence for Hermann's decline. In October, news arrived that Hermann's health had deteriorated further. Einstein hastened to Milan, to find his father visibly sinking. The long years of disappointment with his business affairs and anxiety over his substantial debts, principally to family members but also to others, had worn down the formerly easy-going and affable man to a shadow of his former self. His heart was failing, and he had no desire to continue struggling. Einstein sat by his bedside and obtained his father's blessing for his marriage to Mileva. With Maja's charge that he was responsible for his father's condition preying on his mind, he felt both anguished and guilty as Hermann drifted away. On 10 October, when Hermann felt the end to be near, he asked everyone to leave the room so that he could die alone. Einstein was distraught. Indeed, he later told his friend Janos Plesch that the death of his father was one of only two occasions in his life when he had been truly moved; the inevitable 'if only' thoughts preyed on his mind for years to come. Hermann was buried in the Civico Mausoleo Palanti in Milan.

With her plans for marriage disrupted by the sad events in Milan, Mileva returned to Vojvodina in early November, having agreed with Einstein that they would finally be married in January. She returned just before Christmas and they published the banns for their marriage both in Bern and Zurich (considered by the Swiss bureaucracy to be Einstein's hometown) on 17 and 18 December, respectively. The marriage date was fixed for 6 January 1903.

While it is no surprise that Pauline Einstein did not attend the festivities, it is significant that Maja also seems to have been absent. She had arrived in Bern in July 1902 and had successfully passed the entrance examination to the University of Bern, hoping to join her beloved Paul as a student, as well as to be near to Einstein. However, she did not matriculate, presumably because of a combination of worries about her father's health and a lack of any funds to support her studies. Instead she decided that, at least for a time, she must get a job and then hope to return to study once she had built up some savings. Michele Besso had returned to the family home in Trieste when his father had died and it was to this home that Maja went as a tutor to Besso's young sister, Beatrice, known as Bice. A photograph exists, said to be taken in Trieste in the summer of 1902, showing Einstein punting a group of people, one of whom seems to be Maja and another who may well be Bice. Einstein may have taken Maja to Trieste to introduce her to Bice, incidentally taking the opportunity before he started work at the Patent Office of spending time with his old friend Besso, probably early in the summer of 1902. Maja returned to Trieste to tutor Bice after her father's burial and the break-up of the Via Bigli home. Hermann had left considerable debts, which both Einstein and Maja felt obliged to try to clear, although those to 'rich Uncle Rudolf' were presumably mostly written off. Pauline Einstein returned to Germany, first to stay with friends in Heilbronn. In 1903, she moved to Hechingen, more or less equidistant between Stuttgart and Ulm, to stay with Rudolf and her sister Fanny. Maja's presumed absence from the ceremony on 6 January indicates that she was not yet reconciled with her brother, since she had always been friendly with Mileva and had often tried to act as a mediator in the long-running family dispute.

Einstein and Mileva's wedding was just about as low-key an event as possible. Only the principals and two witnesses, Habicht and Solovine, are known to have been present. There was no honeymoon and the festivities were limited:

> The evening was spent with the groom's friends, who celebrated enthusiastically until late in the night. When the young couple wanted to enter their home for the first time, it was discovered that the groom had forgotten the key! There was nothing else to do but press the doorbell of the owner, who lived in the house! Only the chair itself was lacking for the absent-minded Professor!'

Only four days after the wedding, Herr and Frau Einstein registered in a new home in Tillierstrasse. This attic flat had beautiful views towards the Swiss Alps and was very close to the River Aare. Mileva set up house and Einstein carried on his life more or less as before. His walk to the Patent Office was not much different to that from Archivstrasse and his lunches and evening discussions with his friends also continued. Mileva's life, however, changed completely. She now had a home and a husband to take care of; as a married woman, she could hold her head up with her neighbours as she did the shopping for the evening meal. Both Einstein and Mileva were content with their new arrangements. Einstein told Besso in late January that 'I am now a married man and lead a very nice, cosy life with my wife. She looks after everything excellently, is a good cook and always in good spirits.' Mileva wrote to Helene Savić that 'I am, if possible, even more attached to my dear treasure than I already was in the Zurich days, he is my only companion and society and I am happiest when he is beside me.' From these letters, it can be deduced that they spent much time alone together. Given their previous collaboration in Zurich, even considering Mileva's disillusionment with physics following her failed examinations and her distraction over the future of Lieserl, most of their conversation would have been about physics. Einstein was beginning to bubble with ideas. His paper on the foundations of thermodynamics had been sent to *Annalen* just after the wedding and others were to appear in the next few months. At this early stage in his career at the Patent Office, he would necessarily have had to give his full attention during working hours to keeping on top of his work, so that much of his paper writing and thinking must perforce have been carried out at home. That Mileva was still an important collaborator and sounding board for Einstein is borne out by his sister, who speaks of the couple in this period as being comrades in study.

Various glimpses of the first few months of the Einstein marriage can be obtained. Solovine reported that nothing changed with respect to the meetings of the Olympia Academy. When they met at the Einstein's flat, 'Mileva was clever and reserved; she listened attentively, but never contributed to our discussions.' This last is interesting, since Mileva's reluctance to contribute was not caused by unease with the company; Mileva was fond of both the Habichts and Solovine. She clearly preferred to expose her thoughts only to Einstein in the privacy of their own apartment.

The Einstein home was the scene of much gaiety in the early years of their stay in Bern. Not only did the often-boisterous Academy continue to meet there, but there were also many musical activities, as Maja described:

> All through the week music lovers were to be found there, making up a very fine string quartet who played the most beautiful classical quartets with real ardour. Among them the cellist was the most noteworthy, who was a passionate musician. He was a real original. He was a Bavarian by birth but had been living in Switzerland for decades and spoke the most extraordinary mixture of both dialects.

She went on to observe that Einstein and Mileva spoke a similar dialect mixture, which can indeed be observed from their correspondence. Maja blamed this for their son, Hans Albert, who would be conceived that summer, being unable to speak 'proper' Bernese Swiss German until the age of five.

Maja also gave a portrait of Einstein's work in the Patent Office once he had settled in:

> His job was therefore reduced in the main to producing a description, from what was often provided in an inadequate form, that was adequate technically, legally, logically and in understandable language. Anyone who knows what manner of possible and impossible things are invented in this day and age and in what sort of obscure way they are often described by the old codgers who invented them, can imagine the amount of head scratching or downright guffaws of laughter this task could produce. This was the reason that candidates for this job at that time were particularly examined on their ability to think logically. The young patent clerk didn't use his own subject, theoretical physics, a great deal, but his capacity for logical thought in rewriting the descriptions was in great demand. The Director, usually difficult to please, realized early on that his new patent clerk was a prize indeed, and closed one or indeed both of his eyes to the realization that Albert Einstein was using a lot of his time to work on problems of theoretical physics that obsessed him day and night. The chief had no reason to complain

> since, despite his private preoccupations, he easily carried out the average workload of a patent clerk. He also was happy in the office, his colleagues liked his light and friendly manner and in later years he looked back on this time with affection and said that he owed his ability to write clearly and succinctly to his work in the Patent Office under the strong, down-to-earth criticism of Director Haller. His time in the Patent Office was therefore a peaceful, fruitful and blessed period of his life.

As Maja noted, Einstein got on well with his colleagues at the office. In particular, he became friendly with Dr Josef Sauter, who had been Professor's Weber's principal assistant at the Polytechnikum in Zurich. He and Einstein had overlapped there briefly but only became friends in Bern. Sauter retained his interest in physics and in particular discussed Einstein's ideas on thermodynamics with him intensively during his first months in the Patent Office. At one point, he discovered an error in one of Einstein's papers, which the author admitted without great concern. Sauter lived on the outskirts of the city, in a region almost surrounded by a great loop of the Aare. Einstein and he often walked to the end of his road, where the countryside began and where there was a private astronomical observatory. Sauter was well connected with the other scientists, both amateur and professional, of Bern. He was a member of the Naturforschende Gesellshaft Bern, the Bern Scientific Researchers' Society, which served as a sort of unofficial Swiss national scientific academy. Sauter brought Einstein to Society meetings in the autumn of 1902 and on 2 May 1903, he was elected a member. The regular rendezvous was in the Spitalgasse, just 100 metres down the road from the Patent Office. Einstein attended the Club's discussion meetings regularly and gave talks there on several occasions during his time in Bern. The secretary of the Society was Dr Paul Gruner, who at that time taught at a Bern Gymnasium and was a Privatdozent at the university. This lowest independent academic rank, although above the assistantship that Einstein had never acquired, was unsalaried. It gave the holder little more than an association with the university and the ability to collect fees from any student who could be convinced to attend lectures that the Dozent chose to give. On making Gruner's acquaintance, Einstein investigated the possibility of also becoming a Privatdozent. However, without a doctorate he would be forced to rely on a regulation that specified that this could be waived if the candidate had other exceptional qualifications. While Einstein considered his *Annalen* papers were ample qualification, this view was not shared by the university authorities. Einstein gave up yet another scheme for academic preferment in disgust, raging against the 'pigsty' that was Bern University.

In July, the happy camaraderie of the Academy began to change when Conrad Habicht obtained his doctorate from the University of Bern. He returned to his beloved Schaffhausen until February 1904, when he was appointed to a teaching position in Schiers, a long way from Bern. Thus, his contacts with Einstein were reduced, but they exchanged irregular letters and occasional visits. His brother Paul, who finished school in Bern that summer, also left to attend Einstein's former place of employment, the Technikum in Winterthur, but kept in touch with the group; he did not officially register his departure with the authorities until January 1905. Even while still at school, Paul was full of ideas for new gadgets. He subsequently became a gifted electrical engineer and some of his ideas awoke Einstein's interest and collaboration. Solovine remained in Bern, enjoying Einstein's friendship and tuition and continuing his studies desultorily at the university until 1904, when he wandered off, first to Strasbourg and then in August 1904 to the University of Lyon and more academic dilettantism.

In late August of 1903, Mileva returned to Kac to visit her parents and Lieserl. She felt very unwell when she changed trains in Budapest, probably the first indications of her pregnancy. Three weeks later, Einstein replied to a lost letter in which Mileva had indicated that she believed that she was pregnant. In this reply he reassured her that he wasn't at all annoyed, but rather pleased because he had been thinking for some time 'that I should arrange that you should get a new

Lieserl, in order that you shouldn't be deprived of what really is every woman's right.' He went on to react to news that Lieserl had contracted scarlet fever. He enquired as to how she had been registered, presumably under which surname, and indeed given name, since Lieserl is normally a diminutive of Elisabeth. No record corresponding to the baby has ever been found. Clearly, at this point, Lieserl was still alive and the two had already decided that she should be adopted, otherwise there would be no sense in producing a 'new Lieserl'. The only plausible reason for the decision to adopt was the fear of the scandal that the appearance of Lieserl in Bern would bring, which could jeopardise Einstein's still unconfirmed position at the Patent Office. The probability is that the arrangements were made by Mileva during this trip. Since it is almost impossible to believe that Mileva would have lost contact with the baby and never mentioned her again had she still been alive, the most likely scenario[1] is that she died, perhaps of the complications from the scarlet fever to which Einstein alluded. Mileva's anguish at the loss of her daughter was only to deepen as the years passed.

At least Mileva had the compensation of feeling new life growing within her after she returned from her Balkan trip. Shortly after her return, on 29 October, the Einsteins moved again, back into the old centre of the city, with its arcades and clocks. Kramgasse is one of the loveliest of Bern's many beautiful streets. Cobbled, its quiet was disturbed by the regular passage of trams. It is dominated at its western end by the clock tower, the fabulous Zytglogge, built in 1530. At the striking of each hour, knights, bears, and other figures chase each other around and around beneath the bells. The apartment was on the second floor, small but just large enough for them and the coming baby. A grandfather clock stood between the two windows, which opened wide onto the street. Einstein's music stand was normally placed in front of one window. A dresser flanked the right-hand wall, while a houseplant stood against the other. A small table stood in the centre where the Einsteins both worked and ate. It was here that the papers that revolutionised twentieth-century physics were born. Before that, however, Hans Albert Einstein came into the world on 14 May 1904. That it was another difficult birth is attested by Mileva's letter to Helene Savić a month later in which she admitted to still feeling very weak.

Einstein loved children. From Hans Albert's birth, he doted on the baby and was always willing to look after him. Maja described the household in 1904:

> Albert's salary was only just about sufficient for the necessities of life. They couldn't afford a maid so it was self-evident to the young husband that he had to give his wife and study comrade a hand with the housework: he brought the coal up from the cellar, chopped and stacked wood, did the shopping on his way home from work (although often he would forget to do it) and he often had to get up during the night and relieve his delicate wife from the task of nursing the child [. . .] What an ideal father he was for his own kids! He carried and took his first child on walks, played ball and trains with him, and never lost his patience even when the boy made tremendous noise while he was thinking hard.

Marcel Grossmann had also recently had a son. Einstein congratulated him in a brief note and compared their scientific work, drawing an analogy with his most recent paper, sent a short time before to *Annalen*. In this, he had derived a simple thermodynamic relationship between the temperature and the characteristic wavelength of the thermal radiation from a 'black body' (an idealised object that absorbs all radiation impinging on it and re-emits it with characteristics that depend only on its temperature). The relationship did not assume any particular characteristics for

[1] Another possibility is that she was adopted by Mileva's friend Helene Savić and took the identity of her own child who might have died. While there are some intriguing hints in Mileva and Helene's correspondence, there is only the flimsiest evidence for such a scenario.

the motion of the atoms involved. Grossmann, meanwhile, was working on problems in geometry. It is also interesting to note that Einstein sent greetings to Grossmann and his family from 'his student', meaning Mileva, another indication that they continued to work together on Einstein's research, although with Mileva in the role of admiring acolyte. She was nevertheless essential for one activity of Einstein's that began in 1904, when he was asked to review and précis various papers in the general area of thermodynamics for the journal *Beiblätter zu den Annalen der Physik*. Since he reviewed several papers in English but there is no indication that he knew any English at all, Mileva, who had studied English to some extent, must have helped him to puzzle through them.

It is interesting to note that Einstein's new paper only contained two references, one of which is to Einstein's own earlier publication and the other is to his by-now admired Boltzmann's *Lectures on Gas Theory*. This is remarkable for at least two reasons. Firstly, in the introduction he used the term 'Elementarquanta', first used by Planck in 1900 and still considered highly novel and speculative. Secondly, he subsequently derived properties of the spectrum of black-body radiation that use several numerical values that can be traced back to Planck's paper. It seems from this early example that Einstein, very unusually, only considered it necessary to refer to papers that had directly influenced his line of thought, rather than those that merely provide incidental information such as numerical values. Most scientists consider it not only a courtesy to cite the papers of colleagues but also feel more comfortable building upon their work, rather than striking out into the great unknown on their own. Einstein, clearly, had no need for such supports.

Even with all of the commitments entailed by a new baby, there was always time for music. When Habicht was available, he and Einstein played violin duos by Spohr, the Bach Double Concerto, and pieces by Schumann and Schubert. When on his own, he would play through the violin parts of his first love, Mozart's sonatas and concertos. He also played the famous 'Air on a G String', arranged from J. S. Bach's 'Orchestral Suite No. 3'. The Bach Solo Sonatas and Partitas were a favourite study, particularly the famous Chaconne, one of the most challenging pieces ever written for solo violin.

Einstein had two weeks of vacation at the end of July and beginning of August. He seems to have spent them partly with Solovine, who came back for a flying visit before decamping to Lyon. The two friends left Mileva to take care of Hans Albert and set off for a few days walking. They went first to their favourite haunt of Thun and thence up into the Bernese Oberland to visit the famous Blue Lake and then to the valley of the Kandersteg. They presumably stayed there for a while and enjoyed the beautiful views up to the peak of the Blüemlisalphorn. The extent to which Einstein was missing the companionship and critical faculties of both Solovine and particularly Habicht, whom he had not seen since the autumn, is palpable from the tone of his postcards in this period. He urges Habicht to come and stay with them and utters dire threats, in ostensible jest, when he fails to keep the appointments. Unfortunately, it seems that Habicht did not receive Einstein's postcards in time to respond to his urgings since he sent them to Habicht's address in Schiers. It is typical that it did not occur to Einstein that since the school vacation had begun, Habicht would have returned home to Schaffhausen.

The hole left by the departure of the Habichts and Solovine had however been filled by the arrival of an even older and closer friend. Besso had been freelancing as an engineer since he had returned to Trieste to wind up his father's affairs. He had not made much of a success of engineering, hardly surprising given Einstein's estimation of his powers of organisation (see Chapter 4). Early in 1904, encouraged by Einstein, Besso applied for a vacant position at the Patent Office as Technical Expert Second Class, to which he was appointed in March. His salary was considerably in excess of Einstein's but then, as Haller pointed out, he was a trained engineer with much more experience whereas Einstein was only a physicist. Any disappointment Einstein might have felt at being passed over for promotion was assuaged when in September, upon recommendation by his strong supporter Haller, his position at the Patent Office was made permanent and he received an increase in salary that took his annual income to 3,900 francs.

Besso arrived to begin work more or less when Solovine left for Strasburg and before long he and Einstein had fallen back into their usual in-depth discussions of physics. Solovine himself, restless as ever, returned to Bern in October. Maja left Trieste, presumably when Besso moved to Bern. She had saved up enough money to support her further education and in November she returned to the Lehrerinseminar in Aarau to complete her studies for her full teacher's certificate. With much of his system of friends again in place, with his position at the Patent Office secure, and with Mileva at home to support his family life and his physics work, the stage was set for the *annus mirabilis* of 1905.

Chapter 6

Annus mirabilis (1905)

The only thing that seemed to distinguish the Christmas of 1904 from the previous festive season that the Einsteins had celebrated in the Kramgasse was the presence of Hans Albert. This was certainly a revolution in their family life and one that gave them both enormous pleasure. Other than that, however, Einstein's routine carried on as usual. After the festivities were over, he resumed his walk along the arcade under the Zytglogge past the railway station to the Patent Office. Sometimes at lunch he would meet Solovine, Chavan, and others. Almost every day he could be found deep in discussion over physical problems with his fellow inhabitant of that secular monastery, as he called it, Michele Besso. He and Einstein were again inseparable; Einstein took him to a meeting of the Scientific Research Society in February and on 4 March 1905, Besso joined Einstein as a member.

Now that he was thoroughly on top of his ordinary work, Einstein's extraordinary speculations on physics could be carried out additionally, more or less surreptitiously, at the office. In the evenings, however, while rocking Hans Albert's cradle, making him toys, or bringing up the coal from the bunker, he continued to think about physics and to discuss this work with Mileva. His violin case remained close at hand and was opened as often as ever for playing through Mozart and Bach when alone, or a variety of composers in musical evenings with many of their friends. He would bring his violin to the office if he had a musical rendezvous after work. In order to get back to his own work, of whatever description, he had his own unique way of terminating tedious discussions with his colleagues over obscure issues of patent practice. He unpacked his violin and played a few notes; never one to miss the chance to reuse an old joke, he would insist that they should rather play Händel than 'händel' (argue).

This continuity and routine, however, was deceptive. Intellectually, Einstein was preparing a revolution, to be promulgated in his four great papers of 1905, that would eventually shake the foundations of physics. Even that revolution, however, was not a sudden, inspirational creation; rather it had gradually been building up, brick by brick, in Einstein's intellectual architecture since at least his days at the Aarau Cantonal School. While something similar happens for almost every scientist, what makes Einstein unique is that the fruits of his intellectual development were so profound and broad and ripened over such a relatively brief period.

The origin of several of the 1905 papers can be discerned in his previous four, dealing with the foundations of molecular thermodynamics and their application. Einstein was now, after at least five years of reading and thought, a true master of the application of thermodynamics and the use of its principles to tackle disparate questions in physics. The first paper of 1905 was despatched to *Annalen* on 17 March. It is entitled 'On a Heuristic Point of View Concerning the Production and Transformation of Light'.

Max Planck had postulated in 1900 that the oscillating electric charges thought to give rise to the electromagnetic radiation given off by black (i.e. perfectly absorbing) bodies could not absorb or emit an arbitrary amount of energy. Rather, the allowed energy exchanges were 'quantised'; integer multiples of a fundamental constant, subsequently given Planck's name. This postulate

Einstein. Brian Foster, Oxford University Press. © Brian Foster (2026). DOI: 10.1093/oso/9780198794875.003.0006

allowed Planck to explain how radiation is not emitted with equal brightness across all wavelengths, the distance between successive peaks of the wave, but follows a characteristic pattern called a spectrum. Planck was able to obtain excellent agreement with the data, in contrast to the complete failure of previous attempts by others. That this spectrum shifts away from the red towards the blue as the temperature increases is well known, for example, from the common characterisation of heated metals as 'red hot' and 'white hot'. Einstein had been familiar with the quantum postulate since shortly after Planck's publication and had puzzled over it in letters and conversations with Mileva. In 1902, Philipp Lenard, who had impressed Mileva in her brief stay in Heidelberg in 1897, had produced some seminal experimental results on the behaviour of electrons (at the time still called cathode rays) ejected from the surface of metals by light. Lenard observed that the maximum speed of the emitted cathode rays was independent of the intensity of the light falling on the metal. All that changed as the light intensity was increased was that greater numbers of electrons were emitted. This was an extremely difficult result for physicists to understand, since the picture of light as a wave implied that an electron would absorb energy continuously until it had acquired sufficient to overcome the barrier at the metal surface and escape. Increasing the intensity should mean that the electric field carried by the light wave would be larger and that therefore the electron would extract more energy from the field, assuming, as was widely believed, that there was a characteristic time before it escaped. Lenard found some evidence that the maximum electron velocity depended on the colour, and hence the wavelength, of the light used to illuminate the metal. However, the evidence for this latter effect was qualitative and Lenard made no attempt, then or subsequently, to investigate this dependence.

Einstein's train of thought in writing his first 1905 paper is evident from his previous paper, in which he applied, for the first time, the thermodynamic methods that he had been developing from the statistical treatment of atomic and molecular collisions to the properties of radiation. He had introduced a method of calculation based on considerations of the fluctuations in energy inherent in any dynamical system, an approach that he utilised to great effect in subsequent work. Using this technique, he proved, on very general grounds, an approximate relationship between the characteristic wavelength of radiation, confined in a cavity of dimensions comparable to the wavelength, and its temperature. It is ironic that the proportionality to the inverse temperature that Einstein derived had been found experimentally almost twenty years earlier at the Polytechnikum by his old adversary, Professor Weber. Einstein's closing sentence sets out a future agenda: the fact that properties of electromagnetic radiation normally derived by other means could also be obtained from thermodynamics 'should not be considered a coincidence'.

It is not certain whether, when Einstein wrote the above words, he already had in mind the application of his thermodynamic methods to the problem of black-body radiation and Planck's hypothesis, although it seems likely. However, by the beginning of 1905 he had the problem at the forefront of his mind. The opening section of the new paper sets out transparently what the problem was and how he intended to approach it. He contrasted the atomic theory of matter, with a large but finite number of discrete entities, with the continuous nature of the fields in Maxwell's equations. A light wave, emanating from a tiny source such as an atom, is nevertheless considered to have energy that is spread continuously through space as it travels. While admitting the success of the wave theory of light in optics, he pointed out that optical phenomena are macroscopic and involve averaging over long time periods. The actual production and absorption of light, which occur instantaneously in comparison with the macroscopic propagation of light, may be governed by different laws. He put forward the proposition that:

> . . . observations such as black-body radiation, photoluminescence, the production of cathode rays via ultraviolet light and other phenomena involving the production or transformation of light may be better understood under the assumption that the energy of light is distributed discontinuously through space. On this assumption, the energy from a light wave emanating from a point source is not continuously distributed over a larger and larger volume, but is isolated into a finite number of energy quanta localised at points of space. These move without further subdivision and can only be absorbed and emitted in these quanta.

This is truly an astonishing proposal, going far beyond Planck's ad hoc and somewhat tentative assumption. Einstein was flying against 200 years of tradition and returning to Newton's discredited corpuscular theory of light. Not that Einstein considered that he was returning to Newton. Far from it, he was setting out on uncharted seas, leaving Sir Isaac on his beach examining pebbles. He first turned to Planck's earlier work on black-body radiation, in which Maxwell's equations and a simple model for the radiation process had been used. Einstein used a well-established statistical principle – that every possible excited state of a system would receive an equal share of its total energy – to obtain a formula for the black-body spectrum. Unfortunately, it predicted that the energy summed over all wavelengths becomes infinite. Since Einstein's formula agreed with both Planck's, and the data, at long wavelengths, he concluded that the use of the Maxwellian picture of radiation from black bodies was only applicable for long wavelengths. In contrast, Wien had developed a different treatment of the problem by looking at the entropy, the measure of the extent to which the radiation field was orderly. This was a statistical approach close to Einstein's heart. Wien's Law was known to fail at long wavelengths but to be approximately valid for short wavelengths.

Einstein developed Wien's statistical treatment to show that the entropy of the electromagnetic radiation at low density in a cavity is identical to that of an ideal gas. Since an ideal gas in the atomic picture consists of a collection of tiny immutable particles, the implication is obvious. Einstein crystallised this by applying Boltzmann's definition that the entropy of a particular state is proportional to the probability of its formation. From this, he deduced that the probability of finding the electromagnetic energy in a particular configuration obeys a distribution identical to what would be expected if it consisted of particles or 'quanta'. With hindsight, these may be given the modern name 'photons'. The energy of each photon is inversely proportional to its wavelength. In the final sections of the paper, Einstein explored the consequences for photoluminescence, for the ionisation of gases by ultraviolet light and, most importantly, for what is now called the photoelectric effect. He referred to Lenard's experimental work in the latter area as 'pioneering' but ignored Lenard's own speculations on its meaning; Einstein showed that his own picture implies that the velocity distribution of the emitted electrons is independent of the light intensity, but the number of emitted electrons is proportional to it. He went on to show that, on the assumption that all of the photon's energy is at least sometimes transferred to the electron,[1] the value of the attractive voltage required to be applied to the metal to stop any electron leaving it must vary linearly with the frequency, which is proportional to the inverse of the wavelength, of the light.

Einstein's paper made little impact. For decades, Planck, the originator of the quantum hypothesis, refused to accept the idea that electromagnetic radiation in transit was quantised. His view was echoed by almost all German physicists, who were convinced that the explanation for Lenard's results was the 'triggering mechanism'. This was postulated to be a resonant effect whereby, when

[1] Einstein implicitly assumed that this transfer dominates the electron's final energy.

the wavelength or equivalently energy of the light matched that of an electron within the material, it was ejected without absorbing any of the energy from the light; its energy came from a distribution of electron energy intrinsic to the interior of the metal. Lenard was further convinced that cathode rays were not particles at all but vortices in the aether and that X-rays were simply extremely energetic cathode rays. Although these and his other preconceptions now seem with the benefit of hindsight to be bizarre, at the time they seemed entirely plausible and indeed were held by the great majority of German (although not British) physicists. They certainly seemed much more plausible than a crazy idea of a patent clerk from Bern that 200 years of wave theory and optics were based on an incorrect premise. One of the reasons that Lenard showed no interest in pursuing Einstein's suggestion that the maximum velocity of his cathode rays depended linearly on the frequency of the incident light was that he 'knew' that it depended on the unknown dynamics going on inside the metal. The verification of Einstein's prediction had to await experiments carried out by Robert Millikan in 1911.

For several years after the publication of Einstein's paper on 9 June 1905, Lenard and Einstein had extremely cordial relations. However, the cumulative effect of Einstein's 1905 papers was eventually to demolish entirely the conceptual basis for Lenard's understanding of physics – which he was not beyond defending by a variety of unscientific and unsavoury tactics, including the suppression of inconvenient data. It was this systematic demolition of Lenard's scientific shibboleths, at least as much as his rabid antisemitism, that formed the background to the conflict between the two by-then Nobel Laureates that was to explode in the 1920s and 1930s.

It is unclear why Einstein decided, just after the completion of the photoelectric-effect paper, to try once more to obtain a doctorate. Still presumably disgusted with the University of Bern, he determined to try once again at Zurich. By now he had gained enough knowledge of the world to realise that his thesis should be as uncontroversial as possible, He had been working for many years on the properties of liquids and gases and in the previous year or so had devised a method to determine the size of molecules that used only familiar techniques from fluid dynamics. Of course, the very existence of molecules was still controversial in some quarters, but Einstein had long become wedded to this idea, which permeates all of his papers. He probably contacted Professor Kleiner in Zurich and sent him the thesis, which was very short, for an informal opinion before he formally submitted it. Having once before had to go through the process of submitting and then withdrawing a thesis, this would have been a wise precaution. The inevitable delay for Kleiner to get around to casting his eye over it would explain the fact that, although it was completed on 30 April, the thesis was not formally submitted to the Dean of the Faculty until 20 July. The likelihood of the thesis having been sent informally in advance is increased by the fact that Einstein many years later recalled that it had at first been returned by Kleiner with the comment that it was too short. Einstein added one line and then returned it, after which it was accepted without further comment. Reverting back to type in comparison to the photoelectric-effect paper, which had a veritable plethora of seven references – two to Planck, four to Lenard, and one to Stark – Einstein's thesis contained only one – to Kirchoff's lectures on mechanics.[2] It was dedicated to his friend Grossmann, as his previous thesis attempt in 1901 had been, in gratitude for Grossmann's help in obtaining the position at the Patent Office.

The thesis uses once again the basic molecular thermodynamics that Einstein was exploiting. The problem of determining molecular size was addressed by examining the behaviour of molecules,

---

[2] The version subsequently published in *Annalen* contained an additional reference – to Einstein's own paper on Brownian motion.

assumed spherical, in a dilute solution whose solvent's molecular size was so much smaller that it could be neglected in comparison to that of the solute (the dissolved substance). The problem was approached from two directions. In one, the macroscopic properties of the solution were calculated assuming that, over distance scales large compared to the average separation of the solute molecules, it could be treated as a homogeneous 'average' of solute and solvent. In the other method, estimates of macroscopic quantities such as the viscosity were obtained by treating the problem at the molecular level, where viscosity can be thought of as originating from the intermolecular attraction of the solvent molecules. The calculation of macroscopic quantities from the two viewpoints must agree if both are valid descriptions. This is true if the viscosity of the solution is a factor $(1+ \Phi)$ bigger than that of the solvent, where Φ is the fraction of the volume taken by the solute molecules compared to that of the solution. The value of Φ is proportional both to the molecular volume (i.e. its radius cubed) and Avogadro's number. Avogadro's number, usually represented by the mathematical symbol N, is essentially the scaling factor to go between molecules and macroscopic quantities. It was originally defined as the number of atoms in one gram of hydrogen and has the value around 6.022×10^{23}.

In order to calculate both the molecular radius and N, Einstein required a second relationship, which he constructed from the 'osmotic pressure' (i.e. the tendency for concentrations of a solute in two solutions in contact with each other to equalise) and diffusion (i.e. the tendency for all molecules to mix). He explored how solute molecules react to a translational force when moving from a region of high concentration to one of a lower concentration. If the system is in equilibrium, the flow due to the diffusion of the solute molecules generated by the thermal fluctuations of the system must balance the flow of the molecules under the force applied. Stokes's Law relates the velocity of a particle (a solute molecule in this case) in a fluid to the applied force; it depends on the radius of the solute molecule and the fluid's viscosity. By considering an equilibrium, Einstein could obtain a relationship for the diffusion coefficient, a measurable quantity, in terms of the product of the molecular radius and Avogadro's number. Having obtained two equations, he could now determine both quantities of interest.

Unfortunately, Einstein made a numerical mistake in his calculations, which was eventually tracked down several years later, at his request, by a student of his. The multiplicative factor to obtain the solution's viscosity from that of the solvent should be $1 + 2.5\ \Phi$, Thus, when he used his thesis result to calculate Avogadro's number and the size of the molecules, his results were somewhat in disagreement with other methods. Once the numerical error had been corrected, his result turned out to be highly useful for a variety of applications. In a sense this is fortuitous, since Einstein made a large number of simplifying assumptions, such as the solvent molecules being much smaller than the solute, the solute molecules being spherical, the applicability of Stokes's Law, etc. It transpires that, in many situations of practical interest, these assumptions are valid. Therefore, although relatively unknown outside specialised areas of hydrodynamics, Einstein's thesis is his most highly cited publication; the second-most highly cited is the erratum, published in 1911, in which the correct formula was obtained.

Einstein's thesis was evaluated by Kleiner and his colleague from the mathematics department, Professor Heinrich Burkhardt. Since Kleiner was an experimentalist with little knowledge of the sort of theory that Einstein was doing, Burkhardt was drafted in to check the mathematics. Kleiner's view was that the calculations in the thesis were 'among the most difficult in hydrodynamics', while Burkhardt considered that Einstein 'demonstrated fundamental mastery of the relevant mathematical methods'. Clearly, he did not read it very carefully since he not only failed to pick up the mistake alluded to above, but also he missed a variety of minor errors in formulae typical of Einstein's papers. The extremely short time between Martin passing on the thesis to

Kleiner and Burkhardt on 22 July and the delivery of two favourable reviews the following day, is another indication that Kleiner, not known for his rapid attention to Einstein's work, is likely to have had sight of the thesis before formal submission. After these favourable reports were received by Martin, he passed the thesis on to the Mathematics and Physics Faculty, who accepted it unanimously on 24 July 1905. Having formally received his doctorate from the University of Zurich on 15 January 1906, Einstein could finally refer to himself as 'Herr Dr Albert Einstein'.

Another contributory factor to the delay in Einstein submitting his thesis was his distraction by a great deal of other work. At the same time as working on his thesis, he was applying similar techniques to address a problem whose name and characteristics he was not even sure of but which he tentatively identified with Brownian motion. This phenomenon had been described by the botanist Robert Brown in 1828, when he examined suspensions of a variety of small granules in liquids. By careful experimentation, he was able to exclude a large number of possible reasons for the random jiggling motion of the particles when viewed through a microscope, without being able to suggest a plausible explanation. Neither in the subsequent eighty years had anyone else, although a significant amount of work and speculation had taken place. The most likely explanation, the result of the impact of molecules carrying thermal energy, had been seemingly excluded by order-of-magnitude calculations of the effect being by far too small to produce the observed motion. This sort of problem was grist to Einstein's mill. As pointed out by Jürgen Renn, the pioneering papers of 1905 are characterised by their interdisciplinary approach, in the sense of being at the boundaries of the three major fields of physics at that time: thermodynamics, mechanics, and electrodynamics. Einstein's unusual background and current situation outside academic life gave him both a wider perspective and a reduced imperative to prove himself an expert in a narrow field. Thus, having spent the previous couple of years honing techniques in statistical mechanics, he was ready to apply them in areas that did not occur to other physicists.

The Brownian motion paper – Einstein's lack of knowledge of the phenomenon is evinced by his not using this name in the title – had its genesis directly in his thesis work. There, he considered the implications of molecular thermodynamics for dilute solutes approximated as spheres and how this could be used to infer the size of molecules. The question of what happened as these spheres were made larger and larger must have occurred to him, particularly as, by then, the previous chemical distinction between solutions and suspensions had become blurred by the invention of new instruments that made solute molecules detectable. He would surely have registered a glaring contradiction between the 'classical' thermodynamics of Boyle and Carnot and the statistical mechanics of Boltzmann and Gibbs that he, Einstein, was championing. In the former, a system of macroscopic spheres suspended in a classical liquid would reach thermal equilibrium after an appropriate time and there could be no further relative motion of the liquid and suspension. In the latter, however, the potential for continuous movement due to random collisions with molecules was there. Chapter 5 mentioned that the Olympia Academy had read Poincaré's *Science and Hypothesis* and spent many weeks working through it. Poincaré mentions Brownian motion and the work of Guoy, who considered that this phenomenon violated Carnot's principle – one of the basic tenets of classical thermodynamics. It may be that Einstein's interest was piqued by this and that this memory came back to him as he was writing his thesis.

Whatever its genesis, the paper is a tour de force. Einstein immediately made it clear that he was playing for high stakes. In the second paragraph, he states:

> If the motion discussed here and the associated laws are indeed observable, then classical Thermodynamics can no longer be considered exactly valid even for microscopically distinguishable volumes and an exact determination of molecular size becomes possible. If, on the other hand, the predicted motion

were to be proven wrong, this would be a strong argument against the molecular-kinetic interpretation of heat.

The first section asserted that there was a significant difference in the interpretation of the dynamics of suspended particles between classical and molecule-based thermodynamics. In the following section, Einstein used the machinery developed in his previous thermodynamic papers to demonstrate this and to show that there is no difference between the behaviour of solute molecules and the much larger particles typical of suspensions, since the characteristics are generated by the motion of the solvent molecules, considered to be much smaller than either. In the third section, Einstein again employed the method first used in his thesis (and described briefly above) of balancing two flows: one due to an applied force and the other due to diffusion. This led to the same result as obtained in his thesis that, at fixed temperature and pressure, diffusion is sensitive only to the viscosity of the liquid and the size of the suspended particles.

In the fourth section, Einstein began to go beyond the considerations already discussed in his thesis by carefully setting up the initial conditions by which he could proceed with calculations of Brownian motion itself. These are that each particle in the suspension acts independently of all the others and that the states of motion of a given individual particle separated by a sufficiently long time interval are independent of its previous motion. He then considered in detail the motion of a particle, for simplicity's sake only in one dimension, over time intervals much smaller than the 'macroscopic' time of observation of the particle through, for example, a microscope, but long enough that its motion between two such intervals is independent of any previous collision. This results in a calculation of the diffusion coefficient that depends only on the probability that a particle is displaced by a tiny amount. Since the diffusion coefficient governs a process that results in macroscopic movement, as for example the dispersion of a drop of red dye in a vessel containing water, the connection between Brownian motion and molecular collisions is already implicitly made. In the remainder of the paper, Einstein made it explicit and calculated the average displacement that can be observed over a minute in a water solution using a suspension of particles of diameter 1 μm. The result, about 6 μm, should have been observable with a microscope of the quality available at that time. Since he had calculated that the diffusion coefficient depended on both the mean distance travelled by the particles and N, measuring the mean distance travelled would in turn allow a determination of Avogadro's number.

The reason why Einstein succeeded in showing that Brownian motion had a molecular origin was that his studies in his previous papers on thermodynamics gave him the correct mathematical tools and philosophy to approach the problem in a novel way. His insight was that, in the molecular interpretation of thermodynamics, macroscopic particles were no different to solutes in a dilute solution, except that they were bigger. Therefore, the same molecular dynamics that governed diffusion also governed Brownian motion: the particles must move, randomly, albeit more slowly than solute molecules in a solution. Previous researchers had calculated the velocities of the suspended particles subject to molecular collisions; they found that they were tiny and randomly oriented and thus had decided that a molecular approach was doomed to failure. No less an authority than Boltzmann himself had come to this conclusion. What they failed to realise was that, if a large object is struck by billions of tiny random blows, its instantaneous velocity will be negligible but its chances of remaining in the same position over a sufficiently long period are also vanishingly small, a phenomenon known as a 'random walk'.

In contrast to the photoelectric-effect paper, the Brownian-motion paper awoke substantial interest from some important figures in European science. Wilhelm Röntgen, the discoverer of X-rays, corresponded with Einstein on the subject, as did Richard Lorenz, a chemist from the

Polytechnikum. Paul Langevin took up the subject in 1908, stimulated by both of Einstein's works, which he extended in several subsequent papers on the subject. Marian Smoluchowski independently developed ideas rather similar to Einstein's. The last sentence in Einstein's paper expresses a wish that 'Soon, a researcher will be able to decide on the correctness of the questions contained herein, which are of great importance for the theory of heat.' Although his appeals for experimental verification in previous papers had been ignored, this time several experimentalists enthusiastically took up the task of verifying his Brownian motion predictions. Eventually, it was Jean Baptiste Perrin and his students who obtained beautiful results between 1908 and 1909 that verified Einstein's predictions. Over the next few years, it was the experimental evidence in favour of Einstein's theoretical explanation of Brownian motion, combined with the agreement of estimates of Avogadro's number from many disparate techniques and physical systems, that essentially decided the controversy over the composition of matter in favour of the atomists.

Einstein's work on Brownian motion brought him into contact with Heinrich Zangger, a distinguished physician who in 1905 became Professor of Forensic Medicine at the University of Zurich. He was a gifted polymath who became one of the foremost European experts on the medical aspects of disasters such as explosions in coal mines, which linked him with Poincaré, who for many years had investigated the physical causes of such catastrophes. Zangger's interests included physics, chemistry, toxicology, anatomy, welfare, and many others. In the course of the research that he was carrying out on milk, a colloid suspension, he devised a simple experimental method to observe the path of falling droplets. His colleague Laurel Stodola, Professor of Mechanical Engineering in Zurich, knew of Einstein through Michele Besso, who had been a student of his. Stodola, who also shared both Einstein's and Besso's love of music, advised Zangger to contact Einstein for advice on how to determine Avogadro's number with his setup. At some point, either in the late summer of 1905 or early in 1906, Zangger travelled to Bern to meet Einstein. Although Zangger did eventually publish a determination of Avogadro's number, the importance of their meeting was not scientific, but personal. They became friends, a friendship which deepened with the years to the extent that Zangger became a trusted confidant and subsequently a vital go-between in Einstein's soon-to-be dysfunctional family life.

Einstein's final two papers in 1905 both concern the electrodynamics of moving bodies. As discussed, Einstein had thought deeply about this issue for many years, since at least his days at the Cantonal School in Aarau. His letters to Mileva while they were at the Polytechnikum are peppered with references to 'our work on relative motion'. There is little documentary evidence from the period after he left Zurich that he worked in this area, although the fact that he was still pondering questions on electromagnetism is evident from a talk he gave to the Bern Scientific Research Society in December 1903 on 'The Theory of Electromagnetic Waves'. Having produced his first three 1905 papers, for the moment he had exhausted ideas on the applications of his new methods of thermodynamics. Driven by the momentum of an unprecedented period of creativity, he switched his attention back to the perennial problem of light, motion, electrodynamics, and the aether.

An enormous amount has been written on the genesis of the theory of special relativity, including quite a lot of material by Einstein himself. Unfortunately, the latter was mostly produced many years after the event and Einstein's recollections were not always particularly reliable. While the large literature on the subject from other authors is also often contradictory, there are points of agreement. Einstein's earliest speculations on catching up with a light beam were never far from his mind. They induced in him a scepticism about the existence of the aether; subsequent years hardened this scepticism into a conviction. Indeed, in his lecture on the origins of relativity at

Kyoto University in 1922, he began his account by noting his scepticism about the existence of the aether and that 'This insight actually provided the first route that led me to what we now call the principle of special relativity.'[3]

The immediate question that occurs once the aether has been abandoned is how then does light propagate? If nothing supports it, the light wave must 'pull' itself along by its own bootstraps, carrying within it the mechanism for its own propagation. Maxwell's equations describe how this occurs. A light wave consists of coupled electric and magnetic fields oscillating perpendicular to each other. The change in one can be considered to generate the other, as codified by Faraday's law, one of the principal inputs that Maxwell used to build his equations. If light waves are self-perpetuating in a vacuum, then it is inescapable that their velocity therein should be a universal constant. However, Einstein knew that this led to further difficulties. If all inertial reference frames (i.e. three-dimensional volumes moving at constant velocity) were equivalent (which Einstein by now firmly believed), then Maxwell's equations imply that the velocity of light must be the same, irrespective of the frame. Otherwise, experiments could be carried out to distinguish one frame from the other. Thus, an observer travelling with constant velocity parallel to a beam of light at half its speed would still observe the beam travelling at light speed. However, the traditional Newtonian/Galilean law of addition of velocities, known for centuries, implies that the speed of light relative to the observer should be reduced to one half. Something was amiss deep within Newton's conception of space and time, but for years Einstein could not pinpoint what it was.

In an attempt to circumvent this roadblock, Einstein was perhaps inspired by his work on quanta of light, which threw doubt on the applicability of Maxwell's equations in all situations. He toyed with the idea of an 'emission' theory of light. Such emission theories, reminiscent of Newton's corpuscular theory, would retain the constancy of the velocity of light only with respect to its source; for other observers, the Newtonian/Galilean addition of velocities would continue to apply. However, the complications that ensued in formulating an alternative to Maxwell's equations proved this attempted cure to be worse than the disease.

Having abandoned emission theories, Einstein was forced back to revising the Newtonian/Galilean addition of velocities. He knew that this had been verified extensively for centuries, so departures could only exist for extreme conditions not attainable in the laboratory, such as velocities close to that of light. He was familiar with a theory that contained such modifications. A decade earlier, H. A. Lorentz had constructed a model that attempted to reconcile the null result of the Michelson–Morley experiment with the existence of the aether, to which he was deeply intellectually committed. He, and independently Irish physicist G. F. Fitzgerald, postulated that bodies moving with respect to the aether suffered a contraction in their length along the direction of motion. The contraction cancelled out the effect expected from motion through the aether in the Michelson–Morley experiment. The physical source of this effect was at that stage unclear. Lorentz made qualitative arguments based on the compaction along the direction of motion of the forces that kept apart the charged constituents of material bodies, which he thought were electrons. Such an effect had recently been calculated for the electromagnetic field by Oliver Heaviside. Subsequently, Lorentz set up a 'transformation' that showed how to change both time and space

[3] In the Japanese record of his lecture, which is the only existing source, Einstein went on to talk about the result of the Michelson–Morley experiment. There is some ambiguity in the translation, but most interpret Einstein as saying that he knew about Michelson–Morley as a student; however, it is possible that he said 'had he known about Michelson–Morley as a student . . . '. Elsewhere, Einstein denied knowing about this experiment in any detail in 1905. However, he was certainly familiar with it from papers he is known to have read.

coordinates so that Maxwell's equations had the same form in any system moving at constant velocity. He was aware that his ansatz involved a modification of time: he called it 'local time' but considered it to be a purely mathematical construction. Poincaré had polished and modified Lorentz's work and, in a paper published in 1905 while Einstein was completing his, went beyond Lorentz by showing that his transformations formed what is known as a 'group' in mathematics. In such a group, an arbitrary number of separate transformations can be replaced by a single transformation that produces the same effect. Poincaré named the group, and the associated transformations, after Lorentz.

Einstein was unaware both of Lorentz's most recent work and Poincaré's tidying up of it, but it would have made no difference had he known about either. He found Lorentz's formulation deeply unsatisfactory. Although it could account for the null result in the Michelson–Morley experiment and explain other effects such as the Fizeau experiment,[4] he considered it an ad hoc construction in which poorly motivated concepts were introduced to 'fix up' the agreement between theory and experiment. Einstein hankered after a much deeper solution, in which the various experimental results would emerge naturally from one or two guiding principles. He was motivated by his earlier work on the foundations of thermodynamics, in which a small number of postulates gave rise to far-reaching explanations of disparate physical phenomena. In order to produce such a theory, he was ready to sacrifice any Newtonian assumption on the altar of simplicity and beauty.

As he bent his mind unceasingly to the problem in that Bernese spring of 1905, Einstein turned to his old friend, sounding board, and fellow patent clerk, Besso. Einstein knew that the pieces of the jigsaw were spread around him: the aether must go, as no inertial reference frames could be preferred; for the same reason, the speed of light must be the same in all inertial frames, so Newtonian/Galilean addition of velocities must be incorrect when speeds approached that of light; finally, Lorentz had shown that experimental results such as that of Michelson and Morley could be understood provided that lengths and times departed from the Newtonian prescription. These were the jigsaw pieces, but how to complete the puzzle? Einstein must have laid this before Besso on numerous occasions during their long talks together. Besso made many suggestions and remarks. Back in the Kramgasse, Einstein mused over Besso's suggestions, perhaps in conversation with Mileva. Suddenly, sometime in mid-May 1905, the penny dropped. The problem was indeed with the concept of time.

Einstein realised that all measurements of time involved a decision about simultaneity. To state that 'the train left the station at 2:00 p.m.' required the observation of the train leaving to be simultaneous with the position of the hands of, for example, a watch denoting this time. However, was simultaneity independent of the relative motion between events? Imagine two observers who meet and synchronise their stop-watches and then separate by walking 100 metres from each other. They exchange images of their watches using light beams to confirm that they remain synchronised. Equidistant from each, a flashlight clearly visible to both flashes. Each stops the watch when they see the flash. When they walk together to compare their watches, they are unsurprised to find that they show the same time. Consider now the same situation, but in which the observers are both travelling on a train. They synchronise watches at the midpoint of the train, separate by the agreed distance, confirm that their watches remain synchronised via a light beam and wait. When the midpoint of the train passes the source of light, it flashes. On this occasion, when the observers

[4] Fizeau's experiment was rather analogous to the Michelson–Morley experiment except that it attempted to measure the difference in the velocity of light produced by passing it through moving water, rather than the aether; this experiment too showed no effect on the velocity of the light.

meet to compare their watches, they are slightly different. The observer in the rear part of the train was moving towards the source when the flash went off, the observer at the front was moving away. Although, according to Einstein's postulate, the speed of light was the same for both observers, one had moved a small distance towards the flash before it arrived whereas the other had moved a small distance away. The flash therefore appeared to have happened earlier to the rear observer than to the front observer. The flash no longer appeared simultaneous to the two observers because they were in a frame of reference moving with respect to the event.

As is the case for so many things in this story, this relativity of simultaneity was not a completely new idea. Again, as so often, Poincaré had come to the same conclusion, that time and simultaneity were relative, as long ago as 1898. He explicitly discussed the effect in *Science and Hypothesis*. Having made a decisive step away from the Newtonian world, however, on the next page he returned to it by declaring that he would 'admit absolute time and Euclidean geometry'. Later, in *The Foundations of Science*, published in the same year as Einstein's paper, he discussed precisely the situation addressed by Einstein, of the relativity of simultaneity in a stationary compared to a moving frame of reference. Both Lorentz and Poincaré, like Einstein, had the pieces of the intellectual jigsaw in front of them. Unlike Einstein, they were not driven by the need for a simple over-arching picture that could produce a new synthesis. They were temperamentally unsuited to overthrowing the Newtonian universe in which they had grown up. Einstein had no such compunction.

As he worked through the implications of his insight about simultaneity, Einstein realised that all the problems that had hitherto plagued him now fell away. All he had to do was replace Newton's framework with his own! Having glimpsed the road ahead, he plunged down it. He excitedly told Besso that, thanks to their conversation of the previous day, he had now solved the problem. He worked with support from Mileva over the next six weeks. However, nothing was ever simple in the Einstein household; in the middle of drafting a paper that would revolutionise our concept of space and time, his domestic arrangements were also shaken up. For unknown reasons, the Einsteins left Kramgasse and thereby the historic centre of Bern and its arcades and migrated westwards to the Mattenhof district. This part of the city had only been developed in the mid-nineteenth century. Around 16 May, the family moved into Besenscheuerweg 28, which was subsequently renamed Tscharnerstrasse. Despite the upheaval this move must have caused, Einstein's concentration was unbroken. He continued to draft the paper and on 29 June, it was posted to *Annalen der Physik*.

This period of revelation is captured in a letter that Einstein wrote to Conrad Habicht. Einstein had already revealed how much he missed his old friend Habicht by sending two postcards in rapid succession early in March imploring him to visit Bern and attend the Academy sessions, thereby increasing the membership by 50%. This time, Einstein took the precaution of writing both to Schiers and to Schaffhausen. However, neither communication had the desired effect, so he wrote again in the second half of May, upbraiding Habicht for his silence and dilatoriness as a member of the Academy, his failure to turn up in Bern when promised, etc. Einstein went on to promise him four pieces of work in exchange for Habicht's often-promised but as-yet-undelivered thesis:

> ... of these the first I can send you soon, because I am expecting my free offprints in the near future. It's about radiation and the energy relationships of light and is very revolutionary, as you will see, provided you send me your thesis *first*! The second is a determination of the true sizes of atoms via diffusion and viscosity of dilute neutral solutions. The third shows that, assuming the molecular theory of heat, particles of size around 1/1000 of a mm suspended in a liquid must show a random motion whose

> origin is thermal motion; such an unexplained motion of small suspended inanimate particles has been observed by physiologists, which they called 'Brownian Motion.' The fourth piece of work on the electrodynamics of moving bodies is only in outline and involves a modification of the theories of space and time; just the kinematics of this paper will be sure to interest you.
>
> [...]
>
> Solo[vine] is still giving private lessons and can't bring himself to take his examinations; I am very sorry for him because he leads a miserable life. He also looks as if he is in a bad way. I don't think though that one can influence him to improve his way of life – you know how he is!

The paper 'On the Electrodynamics of Moving Bodies' begins with a brief introduction and continues in two main sections, the first dealing with 'kinematics' and the second with 'electrodynamics.' It is a rather long paper by Einstein's usual standards: thirty pages in its final published form in *Annalen der Physik*, compared to sixteen for the photoelectric-effect paper and eleven for the Brownian-motion paper. However, it is extremely concise, with precisely zero references and much mathematical detail left for readers to puzzle out for themselves. The lack of references is hardly surprising; it conformed to Einstein's by now established practice and indeed, despite the numerous earlier publications that had dealt with aspects of the problem, this of all his papers could be considered the least influenced by the work of others.

The introduction gives the clearest picture of what were the most important issues in Einstein's mind in 1905 that led to his new synthesis of time and space. He began with an example drawn from the electrodynamics of moving bodies: the observed effects when a charged wire moves through a magnetic field depend only on their relative motion, but the universally accepted explanation given by Maxwell for what happens when the wire moves and the magnet is stationary is qualitatively different from the situation in which the magnet moves and the wire is stationary. Only after highlighting this did Einstein move on to mention other examples 'of a similar type' and the problem of the aether as motivations for his paper. All of these problems, in his opinion, pointed to the fact that there is no state of absolute rest and that all inertial frames are equivalent. He elevated this hypothesis to one of two postulates on which his paper is based. The other is that the velocity of light is constant, irrespective of the state of motion of the source. He gave warning that from these two postulates alone, he would develop a new system of mechanics in which there is neither room nor necessity for an aether. He concluded the introduction by pointing to the source of the current state of confusion in this area, that insufficient attention had been paid in the past to the relationship between clocks and coordinate systems. He then launched into his analysis of the logical consequences of his two postulates.

In the first main section of the paper, Einstein defined his terms by clarifying what he understood by time and simultaneity of events taking place at different spatial coordinates. He set out the rules for synchronisation of clocks, for example, if clock A and clock B are synchronised, as are A and C, then B and C are also synchronous. In these definitions, Einstein implicitly assumed that space is homogeneous. He defined the speed of light in terms of a frame in which all clocks and sources of light are at rest with respect to each other. In fact, his terminology is rather inconsistent in that he persistently talks about 'systems at rest' as if there were an absolute frame of rest; this is, however, nevertheless a useful shorthand for a more cumbersome explanation. He then proceeded to restate his two postulates and to use them to measure the length of a moving rod. Two operations were discussed, one in which the length was measured by an observer co-moving with the rod, the other when there was relative motion between the rod and the observer. Einstein stated that the two lengths were different and demonstrated, by an argument similar to that using the train (discussed earlier in this chapter), that events that are simultaneous in one frame are not simultaneous in another frame moving with constant velocity with respect to the first.

In the next section, Einstein derived the equations necessary to transform from one frame to another moving with respect to it. He now explicitly assumed the homogeneity of space and time, meaning that the necessary equations are linear. He used the same physical set-up as he had in the first section, of a light source and a mirror, which reflects the pulse back to its origin. From this he set up equations that relate three events: the emission of the pulse, its reflection and its receipt back at the origin and discusses them in a frame in which the system moves with a given velocity along one of the coordinate axes. By assuming that the distances involved are infinitesimally small, he arrives at a relation that is the Lorentz transformation between frames. The algebraic manipulation involved in going from the equation given at the bottom of page 898 of the *Annalen* paper to that at the top of page 899 is not trivial but is omitted in the paper. One part involves using a 'Taylor expansion', a mathematical operation useful to obtain successively better approximations to a quantity that can be expressed as a power of a large and a much smaller term summed together. There are several simpler ways, not requiring the use of such techniques, to obtain this result. Indeed, only two years later, Einstein himself used a much simpler derivation in his 1907 review article 'On the Relativity Principle' for the *Yearbook of Radioactivity and Electronics*. It seems quite likely that this section of the paper is the source of the following anecdote. In 1944, while Einstein was transcribing the paper in his own hand to the dictation of his secretary Helen Dukas for sale at auction to raise money[5] for War Bonds, he asked whether he had really written what she had just read to him. When this was confirmed, Einstein remarked that he could have said that more simply. Having derived the general form of the transformations up to multiplicative factors that he later showed were equal to unity, he demonstrated that light indeed travels with a constant velocity in any inertial reference frame.

In the next section, Einstein demonstrated that moving bodies contract in the direction of motion and that the velocity of light is an upper bound to velocity. He further showed that moving clocks run slow compared to clocks at rest; hence, clocks at the equator of the earth run slightly slower than clocks at the poles. Finally, he derived the law of addition of relativistic velocities and showed, as, unbeknownst to him, Poincaré had before him, that the Lorentz transformations form a mathematical group.

The second part of the paper is significantly more technical and applies the relativity principle to the interpretation of Maxwell's equations and electromagnetic theory. Einstein first derived the form of the electric and magnetic fields under the Lorentz transformation. He then demonstrated that there is no longer an asymmetry in the explanation of moving electric and magnetic fields in Faraday's Law, as indeed is inevitable given Einstein's basic postulate that all inertial frames are indistinguishable. The forces on charges moving through electric or magnetic fields are simply those from the static forces of fields whose values are Lorentz transformed to the frame in which the charges are at rest. Einstein went on to derive the Doppler effect (e.g. when the sound of a speeding ambulance siren has a lower tone after it passes) for the change in wavelength with velocity and the law of stellar aberration (the apparent motion of the stars when viewed from the earth, a moving reference frame). By calculating the distortion of a sphere from a frame moving with respect to it, he showed that the energy of light contained in an arbitrary sphere appears to change in exactly the same way as does its frequency. In a masterpiece of understatement, he wrote that it was remarkable that the energy and frequency of light varied according to the motion of the observer with the same law. Thus, he not only supported his conjecture in

[5] It raised the enormous sum of $6.5M, being purchased by the Kansas City Insurance Company, who donated it to the Library of Congress.

the photoelectric-effect paper some months previously but also showed that it was true independent of the motion of the observer. Einstein proceeded to use the calculation of the radiation pressure of light on a mirror to demonstrate that all problems in optics in which an optical element is in motion can be solved in the same way (i.e. by making a Lorentz transformation to the rest frame of the element in question and then solving using normal optical techniques). Finally, he showed that a more general form of Maxwell's equations including a displacement electrical current (caused by a varying electric field) are invariant under Lorentz transformations and derived expressions for the dynamics of a moving electron. He demonstrated that the quantity of electric charge is unchanged irrespective of the motion of an observer. In a frame in which the electron is instantaneously at rest (a necessary assumption since an electron in a magnetic field is being constantly accelerated, which cannot be explicitly treated in special relativity), Einstein showed that the Lorentz force on the electron arises naturally from the relativistic treatment.

The last paragraph in the paper states: 'In concluding, I want to point out that my friend and colleague M. Besso stood faithfully by me throughout my work on the problems discussed herein and I wish to thank him for several valuable suggestions.'

As a physicist, it is impossible even today to read the relativity paper without a feeling of astonishment. It is true, as Einstein implied in his letter to Habicht, that the photoelectric-effect paper is in a sense more revolutionary. Nevertheless, the scope of the subjects treated in the relativity paper and the number of paradoxes and open questions that are resolved by it are breathtaking. From two apparently simple postulates, which were forged in the furnace of many years of intellectual toil and frustration, a new world view appeared. In this, Newton's laws are a wonderful, economical, and almost exact approximation in the everyday world, but as velocities approach that of light, a new system, based on the application of the symmetry of nature and the undetectability of constant motion, is required. There are no stationary electromagnetic waves, such as had haunted the imagination of the Aarau schoolboy. There is no aether with its wraith-like properties and bizarre attachment to the hurtling earth. Maxwell's equations look the same no matter how fast a source or observer moves.

Michelmore reported[6] that Einstein was so exhausted after the paper was completed that he took to his bed for two weeks. Afterwards, the routine of life carried on as before: Einstein walked to and from the Patent Office with Besso,[7] chopped the wood, did the shopping (when he remembered), and looked after young Hans Albert. Above all, he continued to work on his theories. He had also for a while been dabbling in some experimental work. The induction of Besso into the Scientific Researchers Society led to them becoming very friendly with two teachers who were also members, Drs Rudolf Huber and Hans Rothenbühler. In the evenings, the four friends often repaired to a school physics classroom to try out experimental ideas. These evenings of physics experimentation were inevitably concluded by music-making – Rothenbühler was a great lover of music and Huber, like Einstein and Besso, was a dedicated violinist. Huber also conducted his school's orchestra,

[6] Michelmore's source was Hans Albert's memory of what Mileva had told him. Trbuhović-Gjurić repeats the story with additional details, but her source is likely to have been Michelmore's book; there is no other corroboration.

[7] The move to Besenscheuerweg may have been at least partly influenced by being nearer to Besso, whose wife Anna was also friendly with Mileva. Besso lived in an adjoining street, Schwarzenburgstrasse. With this move began the long walks and discussions on the way home from the Patent Office that Einstein recalled with such affection.

which Einstein was often pleased to reinforce with his own contribution, sitting down without any fuss in the middle of the young violinists.

In the second half of June, Einstein continued his reviewing work for *Annalen* by summarising six papers. This somewhat tedious work did not stop him thinking about relativity. The applications that he had described in the final section of the paper were only the beginning of the ramifications of his new world view. He realised that it was a clear implication of relativity that mass and energy were interchangeable. The likelihood is that this came to him during an extended holiday that the family took in the summer. For the first time, Einstein visited Mileva's homeland.

The trip began with a stop with friends in Zurich, from where they took the train to Belgrade to visit several of Mileva's Zurich friends who now lived there, particularly Helen Savić (neé Kaufler), probably her closest friend. Einstein particularly appreciated the city's situation at the confluence of the Danube and the wooded hills on its outskirts, where the family spent a week. They then moved on to Novi Sad. Einstein was greeted with warmth by Mileva's family; now that they were a happily married couple with a young son, all the earlier troubles, including apparently their daughter Lieserl, seem to have been forgotten. Mileva was proud to show off her husband; she remarked to her father that 'We have just finished a very important piece of work, which will make my husband world famous.' Einstein's relaxed, informal and friendly manner, carrying Hans Albert on his shoulders through the city, made a favourable impression on the locals. They visited Mileva's maternal relations and attended a family wedding, where the quantity of food and drink available amazed Einstein.

The family returned to Bern and Einstein hurried to compose a paper on the implications of relativity that had occurred to him over the summer. It was received by *Annalen* on 27 September. This three-page paper, 'Does the Inertia of a Body Depend on Its Energy Content?' is very straightforward. Einstein began by referring to his own earlier paper as the basis for a very interesting consequence. He pointed out that his work was based on the relativity principle, which stated that the change in properties of systems should not depend on whether they were measured in one frame or one moving with respect to it with a constant velocity. Note that three months' contemplation had completely removed his former anxiety about the applicability of Maxwell's equations. In a footnote, he referred to his second postulate in the June paper, that of the constancy of the velocity of light, as being naturally contained within Maxwell's equations.

Einstein used the results of his earlier paper to derive the energy of a system of light waves when measured in a frame moving with respect to the one in which it was emitted. He considered a system initially at rest that emitted equal quantities of light in opposite directions, thereby remaining at rest. He then considered this process from the point of view of a frame moving with respect to the first frame, using the expression from his earlier paper to transform the light energy between the frames. Since the principle of relativity requires that the conservation of energy holds in each of the two frames, he could obtain an expression for the change in kinetic energy of the system, as measured in the second frame, in terms of the emitted light energy. By assuming that the velocity of the moving frame was small compared to that of light, he could relate this change in kinetic energy to the Newtonian[8] formula for the kinetic energy of an object. Einstein showed that the energy lost by the emitting body was equivalent to it having lost mass. Since nothing in his argument would be changed if the energy were lost in ways other than by light emission, he concluded that this was a general property of matter and that 'The mass of a body is a measure of its energy content; if the energy is changed by an amount L, the mass changes in the same way by

[8] A complete relativistic treatment leads to the same result.

an amount L/9.10^{20}'. In modern notation, this can be re-expressed so that the mass m changes by energy E divided by c^2, the velocity of light squared. If Einstein's thesis is his most cited paper, the one sent off in September is his most quoted: the rearrangement of the formula discussed above, $E = mc^2$, is the only mathematical equation that almost everyone on the planet is likely to recognise. In conclusion, Einstein suggested that the mass loss due to the energy released by radioactivity in radium might be measurable and would be a test of his conclusion.[9]

Sometime in the period before the completion of the second relativity paper, Einstein wrote again to Habicht. He accused him of having become very dull and suggested that this was a result of his current teaching position in Schiers. He tried to interest Habicht in working with him in the Patent Office, remarking that although eight hours per day had to be devoted to this activity, there remained another eight hours in every day as well as Sundays for more enjoyable activities. The affection Einstein had for Habicht shines clearly through the letter: 'I would be really pleased if you were here.' Habicht, who became a dedicated schoolteacher, never took up this invitation. Einstein went on to inform him that he had worked out that mass and energy were equivalent and that light carried mass: 'The idea is amusing and plausible – but who knows – perhaps God is laughing at me and has led me a merry dance.'

Despite his joking with Habicht, Einstein had not a shadow of doubt that his analysis was correct. Having completed this momentous addendum to special relativity, Einstein and Mileva awaited the publication of the original paper with great anticipation. This appeared in *Annalen* on 26 September. Maja described the feelings in the Einstein household once the paper had been accepted:

> The celebrations in the little apartment were great. The author thought that publication in the visible and much read journal [*Annalen*] would gain immediate recognition. Indeed, he expected strong contradictions and criticism. He was however deeply disappointed. An icy silence followed publication. Subsequent issues of the journal devoted no words to his work. The experts were reserving judgement.

As the autumn turned into winter and the new year of 1906, Einstein laboured on at the Patent Office and continued to submit reviews and summaries of other people's papers to *Annalen der Physik*. He and Besso walked to and from work through the falling snow, carrying on their usual discussions of physics as if the papers of 1905 had not been published. Physics, however, would never be the same again.

[9] In fact, Planck showed in 1907 that the mass loss from this process was too small to be measured.

Chapter 7

Recognition and first steps in academia (1906–1909)

The winter and early spring of 1906 saw further changes to Einstein's domestic arrangements. In late April the family relocated yet again, crossing the river to Aegertenstrasse in the Kirchenfeld district. The move may have been related to some news that Einstein received in March. Thanks to the good offices of Friedrich Haller, he was promoted to Technical Expert, Second Class. Haller wrote to the Swiss Federal Council recommending this promotion in glowing terms, praising Einstein's ability to deal with the most difficult and technical of patent applications and remarking that he was one of the most valued members of the office. The case for promotion was strengthened by the award of his doctorate; Haller exhibited some premonition that Einstein's ambitions stretched beyond life as a Patent Clerk by telling the Federal Council that his loss would be grievously felt. The promotion came with an increase of his annual salary to 4,500 francs. On receiving his pay slip at the end of April, Einstein is said to have wondered what on earth he would do with all that money. Perhaps Mileva saw her chance to engineer a move to better lodgings.

Meanwhile, Einstein's scientific work did not slacken after the monumental achievements of his *annus mirabilis*. Neither did it show any signs of a narrower scope or concentration; rather, the preoccupations of the 1905 papers were further explored and expanded upon. At the end of 1905 he had written a paper on some further aspects of Brownian motion, in particular an estimate of the rotational motion to be expected. In March 1906 he returned to the photoelectric effect. In May it was again the turn of relativity, this time applied to the motion of centres of gravity, which led to a further proof of his mass–energy relationship; in August he outlined a method to distinguish relativity from other competing theories using measurements of the properties of cathode rays. This pattern continued for the next few years.

Even before Einstein's latest house move, the world of physics was starting to take notice of the relativity paper. Max Planck had almost certainly seen it when it was submitted to *Annalen der Physik*, of which he was an editor. It appealed to his own beliefs about how physics should be done, via a few simple premises from which disparate phenomena could be deduced. He liked the centrality of the speed of light in Einstein's formulation and quickly began to see the implications. Indeed, he was the first to publish a paper using Einstein's ideas. Planck read it to a meeting of the German Physical Society on 23 March and it was subsequently published in the *Proceedings*. He derived, among other things, the relativistic relation between momentum and velocity. Maja remarked on the reaction in the Einstein household to a letter from Planck sometime in late 1905 or early in 1906 with some questions about the relativity paper:

> After the long period of waiting, this was the first sign that his work had been read at all. The joy of the young researcher was all the greater in that the response came from one of the most eminent physicists of his time. Subsequently a substantial correspondence between the two physicists built up, from which, despite the differences in their age and background, a solid friendship developed.

Einstein. Brian Foster, Oxford University Press. © Brian Foster (2026). DOI: 10.1093/oso/9780198794875.003.0007

Planck also interested his students and researchers in relativity by discussing it in a seminar in Berlin in November 1905; one in particular was Max Laue, a young Privatdozent who had just completed his Habilitation (postdoctoral thesis). Although busy with his military service and the administration of Planck's research group, he took up Einstein's ideas with enthusiasm. Soon he was writing to Einstein with questions about applications of relativity. He arranged to spend a vacation mountain climbing in Switzerland with the express purpose of subsequently visiting Einstein in Bern. Many years later he described their meeting to Carl Seelig:

> Having arranged things by letter, I went to meet him at the Patent Office. In the waiting room I was told to walk down the corridor and Einstein would meet me coming in the opposite direction. I did so, but I couldn't believe that the young man who came towards me could be the father of relativity, so I walked past him. Only when he returned from the waiting room did we make each other's acquaintance.

As they walked around the city, they enjoyed the view of the Bernese Oberland while smoking cigars provided by Einstein. The more sophisticated Laue judged his to be so awful that he surreptitiously threw it into the river. Einstein, perhaps discussing Laue's recent climbing vacation, remarked that he didn't understand how people could walk around up there, indicating that his still-frequent mountain walking trips were limited to below the snow line. Although Laue did not recall this, Maja recorded that their walk across the river was directed to Einstein's home: 'Thus in the cramped apartment the young elegantly dressed Berliner with the fashionable Wilhelmine moustache discussed the greatest problems in physics with this shirt-sleeved figure with unkempt hair and stubble!' This was the beginning of a long association and friendship that led, for example, to Laue writing the first monograph on relativity in 1910–1911. As well as Planck, other leaders in German physics took an interest in Einstein's work. Wilhelm Röntgen, needing to prepare a talk on electron motion, wrote in September to request copies of the two 1905 relativity papers, as well as that on Brownian motion. Arnold Sommerfeld attended Röntgen's talk and was particularly impressed by what he heard of relativity, so much so that he prepared a seminar on it and subsequently taught a course on relativity at the Technical University of Munich in 1908–1909, which he later claimed was the first ever given.

The development of new contacts through his research publications coincided with a weakening of those bonds that until now had anchored Einstein to life outside physics. Now they were no longer neighbours, his long walks home with Michele Besso ceased, although he still saw him at the Patent Office. Old friends such as Maurice Solovine, who was studying in his usual desultory way in France, gradually faded out of his life except for the occasional letter or postcard. Habicht, never a reliable correspondent, was absorbed in his teaching career and made only fleeting visits to Bern. His school friend Hans Wohlwend was thousands of miles away in Karachi where he was working for the Winterthur merchants Gebrüder Volkart. Although Jakob Ehrat spent the Christmas of 1907 at the Einstein house, he was an infrequent visitor.

Most importantly, it was around this time that his relationship with Mileva began discernibly to deteriorate. The causes can only be guessed at. Their bond as allies fighting a lone struggle against the physics establishment could hardly survive the beginnings of Einstein's recognition by it. Their son, contrary to the experience of most married couples, did not bring them closer together; their interests lay in different areas of Hans Albert's development. Einstein loved to play with the boy, no doubt fascinated by the slow but perceptible growth in his understanding. Mileva loved him deeply but concentrated on the more mundane areas of ensuring a good meal appeared on the table and worrying about and planning his education. A letter that she wrote to her old friend Helene Savić as year's end approached illustrates her preoccupations. She asked in some detail about Helene's methods of bringing up her own three children and whether she used any particular methodology.

Mileva also asked whether Helen thought that she and Albert had changed in the years since their student days in Zurich, saying that she looked back on those times as the happiest in her life.

A bond to Einstein's youth that was irreparably broken at this time was that to Pauline Winteler, the wife of Jost, who had been an adopted mother to Einstein during his schooldays in Aarau. Her son, Julius (known as Jost Jnr.), loved travel and had criss-crossed the world, ending up in the United States, where he had worked as a cook. His letters from there had already given warning that all was not well. He became convinced that secret societies in the US were plotting to kill him. His father arranged with the Swiss consul in Chicago that Julius should return home. Packed in his trunk, unbeknownst to his family, was a revolver. On 1 November 1906, a chance remark by his sister Rosa's husband, Ernst Bandi, convinced Julius that he was a member of the dreaded secret society; he shot Bandi, also killing his mother before turning the gun on himself. Marie Winteler's mother died in her arms. Einstein wrote a touching letter to his old mentor in which he recalled Pauline's kindness to him and regretted the upset he had caused her in his youth. Einstein would surely have attended the funeral, and it was perhaps at this point that he re-established contact with his old love, Marie Winteler.

Links with the Winteler family continued to deepen. Not only was Besso married to Anna Winteler, but also her brother Paul was in love with Einstein's sister Maja. After achieving her teacher's certificate, she had attended seminars on Romance languages in Berlin and arrived in Bern in April 1907 to continue her studies. Maja lodged with her brother for almost a year; her room in the attic had a wonderful view of the Bernese Alps. As might be expected with two women in a household, both of whom adored the same man, there were tensions that soured their relationship. During much of Maja's stay in Bern, Paul Winteler was also studying there. In December 1908, Maja's studies culminated in a thesis and the award of a doctorate, making her the second Dr Einstein in the family.

Music continued to be of the highest importance to Einstein. Early in 1907, Alfred Schnauder, an old friend from Einstein's days with the Winterthur orchestra, wrote enclosing some of his compositions for violin. One that remained in Einstein's possession until his death was *Gedichte*, a collection of pieces for violin and piano that was published in 1905 by the well-known Hug imprint. Einstein's reply thanking Schnauder for the music has become well known because of his colourful description of his activities in Bern: 'I am a dignified Federal ink-shitter with a decent salary. In addition, I continue to ride my mathematical-physical hobby-horse and to scrape on my violin, within the narrow confines for such inessentials allowed to me by my two-year-old lad.' In fact, Einstein was often able to combine these activities, scribbling away in his notebook about physics while playing with the child with the other hand, as well as playing the violin to keep him quiet.

The first part of the letter to Schnauder contains an apology that Einstein has not been able to play the pieces he sent with an accompanist because his 'rascally young pianist' had been absent since the New Year. Clearly, despite Mileva's love of music and the fact that she had learnt the piano when young and was said to have made excellent progress, she and her husband were not musical partners. Indeed, there is no recorded instance of Mileva accompanying Einstein on the piano. Nor does it seem that, at this stage, they had a piano in their home, despite the fact that one could be rented at that time for around 2–3 francs a month. The only reference to Mileva playing the piano as a young adult is from Helene Savić's grandson, who in his introduction to Mileva's letters to his grandmother stated that in their student days in Zurich: 'Mileva played the tamburitza (a mandolin-like instrument), and later the piano; Helene the piano; and Einstein the violin.' Quite what he meant by 'later the piano' is unclear. It may be that he meant 'many years later' since, in a letter to Helene Savić sent in 1909, Mileva commented with respect to her son Hans Albert that:

> He also loves music a lot and seems to have a gift for it; in recent years I have learned to play the piano a little, and he can quite accurately sing or whistle the tune of a Schubert song that I am playing; it sounds very funny in a small boy like this.

From this letter it can be concluded that by 1909 the Einsteins either owned or rented a piano, mostly for the benefit of Hans Albert. It also seems that despite Mileva's apparent aptitude for the piano as a child, she, like so many others, ceased playing and took it up again only as a mother concerned for her child's musical development. It seems unlikely therefore that she and her husband ever made music together. This was unfortunate since, given Einstein's obsession with music, playing together might have bridged some of the differences that by now had so clearly begun to grow between them. The arrival of Maja in April 1907 provided Einstein with a congenial musical partner whose abilities matched his own.

Einstein's growing correspondence with his colleagues in various German universities, and in particular an extended exchange of letters with Planck, encouraged him to take a first tentative step towards entering their academic world. His lecture on Brownian motion to the Bern Scientific Society in March 1907 led to a discussion in which Besso and Hermann Sahli, Professor of Medicine at the University of Bern, participated. In the early summer he wrote to the Cantonal authorities for permission to petition for his Habilitation at the University of Bern, with a view to obtaining an appointment as a Privatdozent. Einstein expected to be able to combine his work at the Patent Office with such a position. He despatched reprints of all his published papers to the Philosophical Faculty. It was more usual to present a written unpublished thesis on a particular topic rather than a collection of disparate work, but Einstein no doubt thought that his seventeen publications would convince the faculty of his qualifications. For whatever reason, and the strong desire to follow normal procedures should not be discounted, the faculty saw no reason to bend their requirements to accommodate a relatively unknown patent clerk, and in October Einstein was told that he should submit a thesis. Einstein subsequently blamed this rejection on the unfriendly attitude of Aimé Forster, the Professor of Physics at Bern. This is borne out by Seelig, who wrote that Forster returned Einstein's submission, and specifically the 1905 relativity paper, with the comment that he could not understand a word of it!

The summer of 1907 was a busy one for Einstein. His correspondence, predominantly on aspects of relativity, continued to grow. In particular, he entered into an extended correspondence with eminent Professor of Physics at the University of Würzburg, Wilhelm Wien, who was concerned about the possibility of velocities faster than light, particularly in transparent media. This is indeed a complex question and Einstein struggled in a series of letters to answer it satisfactorily, without success.[1] In addition to fielding questions from senior colleagues, he was busy writing papers across the broad range of his interests, including statistical mechanics, the quantum theory, Brownian motion and relativity. It is interesting to note how, as time progressed, Einstein began to refer to the 'relativity principle' without further explanation, indicating his confidence that at least the majority of his readers had by now at least some acquaintance with his 1905 paper. His musings on Brownian motion and the existence of an analogous effect in random fluctuations in the flow of small amounts of electricity led him into an unusually practical train of thought in his work on physics. He began to consider how best to measure such tiny fluctuations. Drawing on his by now

[1] Editors of physics journals appreciate the amount of work required to address the wide variety of questions and obscure situations posed by the many submitted papers 'proving Einstein wrong'. Answering such questions must have been particularly taxing for Einstein since almost all were being posed for the first time.

extensive experience of inventions from the Patent Office, he began to sketch possible devices. He recalled that Paul Habicht, Conrad's brother, was an engineer. His desultory attempts to convince Conrad to visit him in Bern now took on a new urgency and on 15 July he wrote a postcard to them both at their parents' home in Schaffhausen, where they were presumably staying during the school vacation, suggesting an urgent meeting. Since Maja was currently away, they could stay with him, but he preferred to meet them while he was on vacation with Hans Albert and Mileva in the Simmental, about eighty kilometres southwest of Bern and sitting under the massif of Les Diablerets, Wildhorn, and Wildstrubel, which forms the boundary with the Valais to the south. Since Mileva subsequently sent a card telling the Habichts their exact itinerary, it seems likely that they did indeed meet in the Hotel Hirsch in Lenk in Simmental. Certainly, on 16 August, now back in Bern, Einstein sent another card marvelling at the speed with which Paul had turned his ideas into a real device and, his excitement clearly discernible, promising to be with them the following Sunday. In September he wrote to the brothers expressing his 'murderous curiosity' to know how they were getting on, and pleaded for news. His excitement was premature. The story of the Maschinchen, as the collaborators christened it, had a long way to run to a not particularly successful conclusion.

As this hectic summer drew to a close, Einstein must have been very gratified to receive his first request to write a review article. This came from Johannes Stark, then at the Technical University of Hanover but about to move to Greifswald. Stark was Editor of the *Yearbook of Radioactivity and Electronics* and the request was to write a review of relativity. Einstein, pointing out that he could only access a library with difficulty given his working hours, requested Stark to provide him with a list of relevant papers. Stark supplied him with the very brief list of papers concerned with relativity, including one by Planck with which Einstein was unfamiliar. The writing of this review, which also contained new and original work, was one of Einstein's major preoccupations in the autumn of 1907. It was completed and despatched to Stark early in December.

The *Yearbook* review is interesting in charting the development of Einstein's thought since the 1905 paper. The introduction in particular reads very like most modern textbooks of relativity in its concentration on the null result of the Michelson–Morley experiment as a central element in the development of the theory. Einstein records the efforts of Lorentz to account for this, his introduction of a 'local time' and the Lorentz–Fitzgerald contraction. With one grand gesture Einstein was able to sweep away this entire ad hoc construction, as he called it, by equating Lorentz's 'local time' with real time and postulating that physics and specifically the velocity of light were the same in all reference systems. He then went on to derive the Lorentz transformations using a much simpler method than in his original paper, using two reference systems aligned along one axis and comparing the position of a spherical light front in the two coordinate systems, given that the velocity of light was identical in both. The Lorentz transformations now pop out with almost no calculational effort. He went on to consider various applications, showing that the addition of two velocities always gives a velocity less than that of light and applying relativity to various problems in optics. He demonstrated the Lorentz invariance of Maxwell's equations and then considered the electrodynamics of a slowly accelerated electron. So far, the content, if not the treatment, is similar to the 1905 paper. New things included in the review are, for example, the results from experiments by Walter Kaufmann, professor at the University of Bonn, carried out in 1905, measuring the deflection of electrons with respect to electric and magnetic fields. Although the results qualitatively agreed with the predictions of relativity, there were systematic deviations. Einstein diplomatically praised Kaufmann's experimental techniques but left open whether or not the apparent disagreement was related to unknown experimental effects or a real failure of relativity. Einstein acknowledged that theories competing with relativity gave predictions closer to

Kaufmann's data. However, in an aside remarkable for evincing the self-confidence of the young man as well as his approach to theoretical physics, he remarked that he did not believe these theories likely to be correct because they were too narrow in scope, and did not form part of a larger theoretical scheme that could give predictions for other phenomena. Indeed, Einstein's self-belief was justified since subsequent more precise experiments, in particular those by Alfred Bucherer, contradicted Kaufmann's results and agreed well with the predictions of relativity.

It is the fifth and final section of the *Yearbook* article, 'The Principle of Relativity and Gravitation,' that is of most interest with regard to Einstein's future direction in physics. In this section he considered the question of how acceleration might be treated in the context of relativity. Einstein himself vividly recalled the first time, in November 1907, that the equivalence of gravitation and acceleration occurred to him. In his talk on the origins of relativity in Kyoto in 1922, Einstein said:

> I was sitting in a chair in the Patent Office in Bern when all of a sudden I was struck by a thought: 'If a person falls freely, he will certainly not feel his own weight.' I was startled. This simple thought made a really deep impression on me. My excitement motivated me to develop a new theory of gravitation.

In the weeks following this revelation, Einstein worked feverishly on trying to develop its consequences. In the final section of the *Yearbook* article, he immediately posed the question to the reader of whether the principle of relativity could also apply to systems accelerated with respect to one another. Unsurprisingly, he answered the question in the affirmative; he reached the conclusion by considering two systems, one accelerated along an axis and the other in a uniform gravitational field aligned along the same axis. Einstein now made the crucial assertion for the first time of what he later called the Equivalence Principle:

> As far as we know, the laws of physics are identical [in the two systems] since all objects are accelerated identically in a gravitational field. Our experience therefore gives us no reason to suppose that [the two systems] differ from each other in any respect and we will therefore assume the complete physical equivalence of the gravitational field and a corresponding acceleration in what follows.

This assumption was one of the most startling and rich in consequences of all of Einstein's many scientific leaps. It owed something to his study of Mach's works in the Olympia Academy and seemed both simple and logical; in a way typical of Einstein, once he had made the assumption, he built upon this foundation one of the greatest constructs of the human intellect, certainly in physics and arguably in all of human thought. He expanded upon its genesis both in the Kyoto lecture and in his article 'Fundamental Ideas and Methods of the Theory of Relativity, Presented in their Development,' written originally for publication in 1920 in a special edition of *Nature* on relativity, but for which it became much too lengthy. In this article, he emphasised, as he had in the 1905 paper, the importance in his thinking of the appearance of two apparently distinct phenomena that were in fact aspects of the same phenomenon. Special relativity had shown that the properties of the electric and magnetic fields in induction were manifestations of a unified electromagnetic field. In general relativity, as his theory of gravity eventually came to be called, the two phenomena were gravity and acceleration. As Einstein said in his article:

> There then came to me the happiest thought of my life, in the following form. The gravitational field has only a relative existence, analogous to the electric field produced in magneto-electrical induction. *So, if an observer falls freely from the roof of a house, no gravitational field exists during his fall* (Einstein's italics) – at least in his near neighbourhood. Indeed, if this observer drops various objects, they remain either at rest with respect to him or in uniform motion, independent of their particular chemical or physical nature. The observer is therefore justified in considering himself to be at rest.

From the fact that all bodies are acted upon identically by gravity, independent of their composition, Einstein concluded that, rather than an inexplicable foible of nature, this was a manifestation of relativity. Were a particular substance to be acted upon by gravity in a way different to everything else, then an observer could drop something made of this substance and determine that he was in a gravitational field, rather than an accelerated one. The relativity of uniformly accelerated frames therefore implies Galileo's centuries-old observation. In 1907, Einstein was still many years of toil and false starts away from the final theory of general relativity, but already, in the month between his 'happiest thought' occurring to him and the completion of his review, he had made great strides towards defining at least the ansatz of what eventually became his primary programme of work for the next decade.

As Einstein remarked in the final paragraph of Section 17 of the *Yearbook* article, the importance of the Equivalence Principle at this stage was that it allowed him to consider the properties of the gravitational field as if it were an accelerated reference frame, in which he could hope to apply some of the techniques of special relativity. The remainder of the final chapter sets out how such techniques could be applied in some restricted physical situations. He considered the behaviour of clocks in three frames: one stationary, one moving with a constant velocity with respect to the first, and a third with constant acceleration. By considering a time in which the moving frame was instantaneously coincident with the accelerated frame, Einstein could link measurements in the latter with those in the moving frame. Although he was neglecting effects which grow rapidly as the two frames separate, this first-order approximation to a full theory was sufficient already to give him some important insights. For example, he deduced that clocks run slower in a gravitational field. By identifying the periods of oscillation of atoms, whose quantised energy levels give rise to sharp lines in their emission spectra, as manifestations of the clocks in his formulation, he was able to show that, to first approximation, light travelling through a gravitational field was subject to a 'red shift', that is, towards longer wavelengths.

Einstein went on to consider Maxwell's equations in the accelerated frame. By applying the same formulation and approximations, he was able to show that light must be bent in a gravitational field, although by an amount far too small to be detected by effects on earth. Although he concluded from this that there was no prospect of confronting his results with experiment, he subsequently realised that the deflection of starlight by the sun could be detectable during a total solar eclipse. Indeed, it was this detection that captured the imagination of the world in 1919 and gave the crucial evidence for the general theory of relativity. It is possible from the wording of the final paragraph of this review to feel something of Einstein's joy in discovery in physics at this time. He points out that his previous calculations imply a very remarkable result: that the energy in the gravitational field gives rise to a mass E/c^2, mirroring exactly the relationship in special relativity between energy and inertial mass, clearly supporting his assumption that acceleration and gravity are equivalent.[2]

Abraham Pais concludes his account of the 1907 *Yearbook* article in his magisterial book *Subtle is the Lord* by remarking on the fact that this article is a high point at least as important as the 1905 relativity paper. Einstein's courage in putting to one side the beautiful and newly minted framework of special relativity in favour of a new endeavour in which he could already tell that special relativity would be inadequate is astounding.

---

[2] The mass that appears in special relativity is by definition that which governs movement and for which Newton's laws of motion were formulated, known as the 'inertial mass'.

Einstein did not let his new ideas in physics interfere with his other favourite activity, music. Throughout the winter months of the three years between 1907 and his return to Zurich in 1909, he played every week with a group of fellow string players: a lawyer, a maths teacher, a master bookbinder, and a pensioner. Einstein played second violin in this quintet. They played works by Haydn, Mozart, and Beethoven. Since Joseph Haydn wrote no string quintets (although his brother Michael did write five) and Beethoven wrote only two, the group must have played these composers' quartets.

Just before the end of 1907, Einstein wrote to his friend Conrad Habicht, who was laid up after an accident that had ruined his enjoyment of Christmas. As well as giving him news of progress with their Maschinchen, he reported on his completion of the *Yearbook* review and that he was using his new insights on gravitation to try to explain a curious fact about the orbit of the planet Mercury. This orbit is elliptical; the axis joining the furthest point from the sun (aphelion) with the closest point (perihelion) rotates gradually with time with respect to the sun. This rotation, or precession as it is called in astronomy, had been a puzzle for many years. Its existence was predicted in Newtonian gravity from the gravitational pull of the other planets; however, the effect was considerably smaller than observed. This was even more puzzling since Newtonian gravity gave the correct answer for the effect in other planets.[3] However, Einstein's hopes of getting the correct answer were dashed, as he explained in a postscript. A few weeks later, in a letter discussing special relativity, he also informed Sommerfeld, Professor of Physics at Munich University, of his new ideas on gravity.

Early in 1908, Einstein wrote to his old friend Marcel Grossman. The Christmas and New Year holiday gave Einstein time to consider his situation. Working an eight-hour day, six days a week in the Patent Office gave him very little time for his physics work. Having more time for research was a reason he cited to apply for a teaching position at Winterthur, where he had previously had a temporary position. He asked Grossman's advice as to how best to go about this, in particular whether he should seek a personal interview with someone, worrying that his non-Swiss-German, Semitic appearance would count against him. He hoped that Grossmann was enjoying his new position as Professor of Descriptive and Projective Geometry at the Polytechnikum. He must have written this somewhat ruefully, since his own recent attempt to obtain a much lowlier university position in Bern had failed. He had no more success with the position in Winterthur; his friend Adolf Gasser, who had originally alerted him to the possible opening, wrote later in January that someone else had been earmarked to fill the vacancy. Einstein had also written to Gasser about finding a skilled technician to construct the Maschinchen, seeming to be somewhat disillusioned with Paul Habicht's efforts but also indicating the enthusiasm with which he viewed the development of this instrument. Indeed, in another letter to Sommerfeld, he tried to interest one of his assistants in working on it. Irrespective of the difficulties of developing a working device, Einstein thought the general principle used was worthy of publication and he submitted it to the *Physikalische Zeitschrift* in February.

Despite the discouraging news from Gasser, on 20 January Einstein applied to the Canton of Zurich for one of four positions as a maths teacher, pointing out that he would also be willing to teach physics. His application was unsuccessful. However, in another twist, only a couple of weeks later a chance encounter in the library with Paul Gruner, Professor of Physics at Bern, as well as

[3] The data for Venus and the earth were difficult to measure with the instruments available at that time because their orbits are almost circular. In addition, the deviation from the Newtonian prediction for these two planets is much smaller than for Mercury and is negligible for Mars and the outer planets.

pressure from various friends, changed his mind yet again. He determined to write the Habilitationsschrift required to obtain his Habilitation and submitted it to the authorities sometime in mid-February. He was invited to defend his work on 27 February before a committee consisting of the Dean, Gruner, and five other professors, not including Forster, who had a perhaps-diplomatic illness, although he wrote to agree to Einstein's candidature. Einstein's colloquium was entitled 'On the Limits of Validity of Classical Thermodynamics.' His *venia docendi*, as the Habilitation was known in Switzerland, was granted on 28 February. Einstein lost no time in preparing his first lecture course. Even before his *venia docendi* arrived, he wrote to Conrad Habicht to try to trace the whereabouts of his copy of Boltzmann's book on the theory of gases, and took the opportunity yet again to urge him to visit Bern. Einstein began his lecture course on the molecular theory of heat on 21 April. Because of his regular job at the Patent Office, he was forced to lecture at the extraordinary hours of 7:00–8:00 a.m. on Tuesdays and Saturdays. Unsurprisingly, he attracted a small audience: his friends Besso and Lucien Chavan and his colleague from the Patent Office, Heinrich Schenk.

At the end of January, Einstein received two letters from Jakob Laub, a Jew from an area of Poland which, at that time, was a part of the Austro–Hungarian Empire. Laub had converted to Catholicism and just completed his doctorate at the University of Würzburg under the supervision of Wien. He was an enthusiast for relativity and asked permission to spend some time with Einstein in Bern. Einstein agreed and Laub, marvelling that Einstein could produce the theory of relativity while working eight-hour days at the Patent Office, arrived at the beginning of April and stayed until mid-May. It is clear that the two men liked each other; the first letter that Laub wrote to Einstein after his return to Würzburg is not only brimming with ideas on physics that they had discussed but also chatty, jocular, and relaxed in the manner of old friends. Laub not only asked to be remembered to Mileva, but also to the Bessos; clearly, he had been introduced into Einstein's inner circle. Indeed, this letter marked the beginning of a voluminous and regular correspondence that continued for many years, in which Einstein often used Laub as both a confidant and sounding board.

Seelig records that on first arriving at Einstein's home, Laub found him kneeling in front of the fireplace trying to light a fire to alleviate the freezing conditions in the room. Laub quickly fell into a routine where he would meet Einstein at the Patent Office, and they would go out for lunch. Einstein's hours of work were 8:00 a.m.–noon and 2:00–6:00 p.m., so that there was plenty of time for both eating and physics discussion over lunch. Laub met Einstein again as he left work, accompanying him home and usually to dinner in order to make the best use of his limited free time. On Sundays they went walking together, visiting local beauty spots, for example, St Peter's Island, a few kilometres from Bern, talking animatedly about problems of physics. A letter from Einstein to Mileva, who had taken Hans Albert to visit her family in Kac at Easter, described Laub's visit, their long walks, and meals together and characterised Laub as a very pleasant but also a very ambitious man. He was pleased to have a collaborator who could do some of the necessary calculations that he didn't have time for. The tone of this letter is rather affectionate, although he complained of being lonely and that the house was very dirty.

On one occasion, probably during this visit, Einstein and Laub went together to hear a performance of *Götterdämmerung* at Bern's Stadttheater, taking seats in the fleapit, which was all they could afford. After the scene depicting the death of Siegfried, Einstein whispered to his neighbour: 'God knows, Wagner is not my sort of music, but the way he has Siegfried recount his destiny is a wonderfully ingenious piece of composition!'

The immediate fruits of Laub's visit were joint publications, Einstein's first. They produced two such papers on the relativistic treatment of the electrical properties of moving materials, in response to a paper that Minkowski, Einstein's old Polytechnikum mathematics Professor, had published early that year. Minkowski had moved to Göttingen in 1902, partially in order to work with his close friend David Hilbert, with whom he had studied as an undergraduate in Königsberg. He had developed considerable interest in the work of Lorentz and Poincaré on the electrodynamics of moving bodies.

It was Einstein's dislike of what he saw as superfluous learnedness in Minkowski's treatment of relativity that led to his papers with Laub. Einstein's 1905 paper was restricted to the relativistic properties of simple electrically charged particles such as electrons. Minkowski's paper extended this treatment to real materials with atomic structures that reacted to the application of electric and magnetic fields. However, his treatment of these now seems unimportant in contrast to the methods he pioneered. He had in the previous year realised that the Lorentz transformations could be conveniently expressed as rotations in a particular definition of a four-dimensional space.[4] In his paper, he developed this idea and introduced the now universally used conventions of four-vectors, tensors, and matrix methods that became crucial to Einstein's own development of general relativity in subsequent years. However, at the time when Laub first made him familiar with these results, Einstein reacted strongly against them and set out with Laub to derive essentially the same results in a much more conventional and tedious way, involving calculating each component of the electric and magnetic fields separately. There is a delicious irony in the great iconoclast and revolutionary of physics proving to be such a conservative in mathematics. The marked contrast of Einstein's philosophy to that of Minkowski is clearly illustrated by the fact that the first Einstein–Laub paper quickly moves from mathematics to the discussion of a concrete physical example. The second paper, also completed and submitted for publication to *Annalen* during Laub's first visit, criticised a particular formula in Minkowski's paper and derived a different one. This discussion rumbled on in the pages of *Annalen* for several years; nowadays the apparent disagreement is understood as a reflection of a difference in the underlying definitions of the physics quantities used by Minkowski and by Einstein and Laub.

Although the lineaments of Minkowski's new treatment were contained in the 1908 paper, he reserved a lecture at the annual meeting of the German Society of Scientists and Physicians in Cologne in September 1908 for a much more forceful and indeed poetic definition:

> Gentlemen! The picture of space and time that I wish to develop for you has grown from the soil of experimental physics. Therein lies its strength. Its implications are radical. From this moment on, separate space and separate time must sink into the shadows and only a form of union of the two retain an independent reality.

For once, such a portentous announcement was justified: while perhaps not immediately, Minkowski's space-time structure and the methods he devised were rapidly accepted. Indeed, as modified by Sommerfeld in 1910 and subsequently by Max Laue in his textbook on relativity in 1911, they are still used to this day. Sadly, the author did not live even to see his words published. In January 1909 he died following an appendicitis operation.

Another paper presented at the Cologne meeting where Minkowski had spoken was by Alfred Bucherer, Professor at Bonn University. He had measured the velocity of electrons by deflecting them in electric and magnetic fields oriented perpendicular to each other. His results verified the relativistic formulae and contradicted the earlier results of Kaufmann. Not only the mathematical

[4] The space needs to be complex, that is, to allow numbers to have both 'real' (i.e. normal) numbers such as those learned at school, and 'imaginary' components. 'Imaginary' numbers are the normal numbers multiplied by i, defined as the square root of (-1), so that, for example, the square of the imaginary number $2i$ is -4.

interpretation of relativity but also the experimental physical investigation combined to produce a wide acceptance of relativity. Bucherer and Einstein corresponded several times over his results that autumn.

Mileva and Hans-Albert returned from Kac sometime in the early summer. Einstein corresponded with Laub and Paul Habicht about various improvements to the Maschinchen and prepared his next set of lectures on the theory of radiation. Rather than lecturing at the crack of dawn, as in the previous semester, these were given between 6:00 and 7:00 p.m. However, the audience hardly increased: Besso, Chavan, and Schenk were joined by Max Stern, the only true student. Stern became a friend; when Einstein left Bern a year later, he wrote to Stern telling him: 'I am always available to talk to you. If you have any questions, just come along and see me. You will never disturb me because I can put down my own work at any time and take it up again immediately after an interruption.' The audience at Einstein's lectures was occasionally supplemented by Maja, who described her attendance at one:

> Albert Einstein didn't really sympathise with the rest of the faculty, perhaps partially because of his introverted rather than extrovert character. His appearance when he went to work suggested not only that he wasn't interested in relations with society but was never conscious of what he was wearing, his wild hair blown by the wind and he was often unshaven with long stubble . . .
>
> [. . .]
>
> At that time there were a lot of often poorly dressed but very intelligent Slavs studying in Bern, mostly Russian Jews, who didn't make a good impression in the clean and orderly city of Bern. When the author, who was then studying in Bern, asked the Bedell in which lecture theatre her brother was lecturing he asked completely astonished 'What, the . . . Russian is your brother?' The dots stand for the somewhat unedifying adjective that he really used!

Although his audience may have been tiny and his fees non-existent, the preparation of the lectures at least had not been time wasted. They concentrated his mind once again on the problems of radiation that had inspired the photoelectric-effect paper in 1905. Catalysed by papers by Lorentz, Jeans, and Ritz that appeared in the autumn and winter of 1908, he again treated the problem of radiation from a statistical point of view, analysing the fluctuations in the radiation field to strive towards a theory that could explain the form of the black-body spectrum over its full range. He pointed out that Planck's radiation formula was inconsistent with Planck's own initial assumptions and obtained a formula for the implied fluctuations that contained two terms. The first was clearly what would be expected from a statistically independent collection of point-like objects, while the second was precisely what would be expected from energy carried by waves. Furthermore, the two terms were independent of each other, clearly implying that electromagnetic radiation could only be understood in a treatment in which both particle and wave properties existed simultaneously. For the moment he contented himself with opining that the nature of radiation must be different from what had previously been imagined. The maturation of his thoughts about this question over the next few months would lead him, in his first talk at a physics conference, to launch into another leap of the imagination that would have profound consequences for the future development of physics.

Einstein's audience when he talked on 'Electromagnetism and the Relativity Principle' to the Zurich Physical Society on Thursday 11 February was much larger than at his university lectures. The talk took place at the university's Physics Institute, where Alfred Kleiner, the sponsor of his doctoral thesis, was Director. Kleiner already had Einstein in mind for a position at his institute at least a year earlier, when he had mooted the organisation of such a lecture, assuming that

Einstein's Habilitation would take some considerable time to be approved. This idea was put on hold when the Habilitation was unexpectedly rapidly confirmed, in large part because Forster's expected opposition did not materialise. At some point in the summer semester of 1908, Kleiner visited Bern to sit in on one of Einstein's lectures. He was unimpressed. Einstein wrote to Laub in July 1908 that the prospect of a professorship had evaporated. He wrote again almost a year later to describe what had happened, since the rumour of Einstein's incompetence as a lecturer had reached Laub's ears. He admitted that he had not lectured well during Kleiner's visit, both because he was not well prepared and also because of nervousness at having Kleiner in the audience, which that day consisted of only one other person. Einstein told Laub that it was his idea to suggest to Kleiner that he should retrieve his reputation by lecturing at the Zurich Physical Society. This however is patently incorrect, since in his letter of February 1908 – before Einstein's Habilitation was granted – Kleiner had already made this suggestion. Einstein informed Laub that he had, uncharacteristically, lectured well at the Physical Society and that this had opened the door to the Zurich professorship.

Kleiner had discussed the professorship with Einstein on the day after his Zurich lecture, as Einstein wrote to his old friend Ehrat a few days later, also thanking him for his hospitality; he had stayed with the Ehrat family during his visit. He told Ehrat that he had passed the examination of his lecture to the Physical Society and that he hoped to accept an offer of a university position starting in the autumn, provided the salary was sufficient. A few days later he received a formal offer from the university and confirmed that he would be able to start teaching in the autumn semester. However, the formal appointment had to grind through the wheels of the Swiss bureaucracy, as well as satisfy Einstein's demand that his salary should at least match what he was receiving at the Patent Office. In April, Einstein wrote to Conrad Habicht that the appointment in Zurich was almost certain. The appointment, as Extraordinary Professor[5] of Theoretical Physics, with salary identical to that of his current position, became official on 8 May. In the letter to Laub cited above, he announced the appointment in typical Einsteinian fashion, saying that he was now an official member of the guild of whores!

Several other people were involved in Einstein's appointment as well as Kleiner. Friedrich Adler, Einstein's friend from his Polytechnikum days, was at the time a Privatdozent in Zurich. Kleiner had pushed Adler to get his Habilitation in 1906. He was teaching many hours a week, often to packed lecture theatres, a telling contrast to Einstein. His interests concentrated on the philosophical aspects of physics; he was a dedicated follower of Ernst Mach. Indeed, in 1908, Adler had published a tribute to Mach on the occasion of his seventieth birthday. This catalysed an attack by an obscure Russian revolutionary by the name of Vladimir Ilyich Ulyanov, then resident in Geneva. Later rather better known under his pseudonym of Lenin, he attacked Adler's article and all of Mach's philosophy as reactionary. Adler responded to this attack energetically, carrying on an intellectual struggle with the Bolsheviks that continued for many years.

It may have been a realisation that Adler's interest in physics was waning in favour of other interests, primarily philosophy but also politics, that made Kleiner begin to think that Adler might not be the support that he was looking for in core physics. Nevertheless, the fact that Adler's father was the founder of the Austrian Social Democratic Party gave him a constituency among some of the political supervisors of the university that added to the strength provided by his teaching

[5] The 'Extraordinary' referred to the fact that he was not paid on the same scale as full professors but received considerably less! Nowadays such an appointment would be referred to as an Associate Professorship.

experience. However, it seems that Kleiner told Adler in the summer of 1908 that he was not first choice for the post. Nevertheless, it is a testament to Adler's innate nobility of character and indeed unworldliness that in an interview with the Director of Education of the University of Zurich he praised Einstein so highly in contrast to his own abilities that he made Einstein's appointment in preference to himself practically certain. Furthermore, he wrote to the university authorities to the same effect.

Another who played a role in the Einstein appointment was his friend Heinrich Zangger, by then Dean of the university's Medical School. He had helped to convince the director of education of the canton of Zurich that Einstein was an excellent candidate and let this opinion be widely known among his colleagues at the university. However, it was Kleiner who played by far the most important role, not only convincing his colleagues that Einstein was a world-class physicist but also that his being a Jew should not be held against him. Anti-Semitism was endemic at that time and university faculty members were certainly not immune. Therefore, it is reassuring that Kleiner considered that it was not compatible with its dignity for the faculty to adopt anti-Semitism as a matter of policy. Kleiner is recorded to have subsequently remarked that getting Einstein to Zurich was his most important achievement in physics.

While this academic excitement was taking place, applied physics work continued on the Maschinchen. Einstein wrote to Laub in March 1909, reporting that he was excited to receive another prototype of the Maschinchen and was planning various tests of its sensitivity. Conrad Habicht's long-requested visit to Bern took place in April, during the Easter vacation, where he worked with his brother Paul and Einstein on the Maschinchen. In September, Einstein again wrote to Conrad. By far the more important purpose of the postcard was musical; Einstein requested information on the sources of a collection of delightful Gavottes that Conrad and their friend Gustav Kugler, the organist of Schaffhausen Cathedral, had played to him. Only as an afterthought, he enquired whether there was further progress with the Maschinchen. After considerable efforts by all parties, stretching over the next few years and struggling with inadequate equipment and construction techniques, the device appeared to work satisfactorily. However, presumably because of its complex construction and temperamental operation, it never gained commercial traction. Although several were built and even sold, Einstein gave up the idea of patenting it because of a lack of interest from manufacturers. Eventually Paul Habicht did manufacture the instrument himself via a company he founded in 1927, which also produced a variety of other electrical measurement devices, but the company was not a commercial success.

On 6 July 1909, Einstein wrote to the appropriate Cantonal official to resign his post at the Patent Office, effective from mid-October, when he would take up his Professorship in Zurich. Hot on the heels of achieving this first academic post, Einstein received his first academic honour. The University of Geneva, which had been founded by Jean Calvin, was celebrating its 350th anniversary and decided to do so in style, including the granting of no less than 110 honorary doctorates. The professor of physics, Charles-Eugène Guye, nominated some of the most distinguished scientists of the day, including Marie Curie and his friend Wilhelm Ostwald. Guye, who was a distinguished researcher in electricity, had taught Einstein at the Polytechnikum. He had become interested in aspects of relativity first from reading Lorentz and Abraham's papers; he devised experiments to distinguish between their theories on the properties of the electron. He was very impressed by Einstein's 1905 paper on relativity and over the next decade carried out a series of careful experiments on the movement of electrons in electric and magnetic fields that agreed with Einstein's predictions. He therefore included Einstein in his list of recommendations, with the result that an impressive letter arrived at the Patent Office whose contents were in a stylised calligraphic font that led Einstein to believe that they were in Latin. In fact, the invitation to travel to Geneva to receive an honorary degree was in French. In any case, Einstein did not bother to puzzle out the meaning and instead dropped it into the waste basket.

When Guye noticed that Einstein had not responded, he asked their mutual friend Chavan, who was a Geneva native, to intervene. Chavan, explaining the reason for the university's celebration and that Einstein was duty-bound to attend, was able to bundle him onto the train to Geneva directly after he had posted his letter of resignation from the Patent Office. According to Einstein's recollection more than forty years later, Chavan however omitted to tell him that he was to receive an honorary degree, so Einstein travelled in his normal work clothes and summer straw boater. On gathering at the hotel restaurant that evening, he met various acquaintances from Bern. When queried as to what his role was in the festivities, Einstein admitted that he had not the slightest idea. Some there were aware that he was on the list to receive an honorary degree. On being informed, Einstein, given his state of undress, proposed to absent himself from the festivities. He was told that this was impossible, and he gave way; no doubt he amused himself immensely by walking in the academic procession in shabby suit and boater. At least he did not have to doff his boater to the Rector, since Ostwald's recollection was that their names were simply read out. The subsequent celebratory dinner provided Einstein with further sources of typically cynical amusement:

> The festivities ended in the *Hotel National* with the most magnificent festive meal that I ever partook of in my entire life. This provoked me to remark to my neighbour, a Genevan patrician, 'Do you know what Calvin would have done if he were here?' When he said no and asked me what I thought, I replied 'He would have had us all burned at the stake for sinful gluttony.' The man did not speak another word to me all evening.

Having achieved his first academic position and his first academic honour, Einstein crowned a memorable year by attending his first scientific conference. In the previous summer, he had received a visit from Rudolf Ladenburg, then an assistant at the University of Breslau, working on the properties of gases. He had spent time as a student in Cambridge, as had his friend Max Born. Both had read Einstein's relativity paper and realised its importance, so that Ladenburg was keen to make Einstein's acquaintance. Planck wanted to encourage Einstein to attend the next Congress of the Society of German Scientific Researchers and Doctors in 1909, which was due to be held in Salzburg. Planck was on vacation near Interlaken and had hoped to visit Einstein; when he failed to do so, he asked Ladenburg to extend a verbal invitation, thinking that Einstein would find it harder to say no in person. Ladenburg was a charming and persuasive man, so that Einstein was not only convinced to go to Salzburg but the two also became friends – a relationship that endured throughout their lives. Ladenburg, who was also Jewish, eventually took up a position at Princeton in 1931, two years before Einstein arrived.

Einstein's movements over the summer of 1909 are unclear. At some point in August, the family went on a walking tour in the Engadin in southeast Switzerland. The family also stayed for some time with Conrad Habicht, probably at his house in Schiers but possibly at his parent's house in Schaffhausen. The postcard to Conrad Habicht early in September relating to duets and the Maschinchen, quoted above, was presumably a result of their making music together during this visit. Einstein arrived in Salzburg before the opening session of the conference, at which he witnessed a demonstration that must have interested him. This concerned Brownian motion and was given during Henry Siedentopf's lecture. Using new dark-field illumination optics developed by the Carl Zeiss company, of which he was director of the microscopy division, Siedentopf demonstrated the phenomenon during his lecture using a suspension of tiny particles of silver. In answer to suggestions that the apparatus could be used for quantitative measurements, Einstein opined that the difficulty would be in keeping the temperature constant.

Einstein's own talk in Salzburg was the first in the afternoon session of the physics section of the conference on Tuesday 21 September 1909. The session, which was crowded with more than 100 attendees, was chaired by Planck. He introduced Einstein's talk, entitled 'On the More Recent Changes which Our Views Concerning the Nature of Light Have Experienced', although Einstein later amended the title somewhat in the published version to 'On the Development of Our Ideas on the Nature and Constitution of Radiation.' Einstein began his talk with a trumpet call to battle that must have caused murmurs of astonishment among some of the most senior members of his audience. He quoted a textbook published as recently as 1902 that had said that the hypothesis of the aether was no longer a hypothesis but essentially an established fact. He then said that, on the contrary, the aether was now obsolete and that many of the observed properties of light could be explained much more easily by Newton's corpuscular theory than the wave theory. 'It is therefore my opinion that the next stage of the development of theoretical physics will bring us a theory of light that will be a sort of fusion between the wave and [particle] theories of light.' The object of his talk, he announced, was to show why such a conclusion was inescapable.

The first part of his lecture contained a masterly review of the experimental facts that had led to the theory of relativity, concentrating on the Fizeau and Michelson–Morley experiments, which had mandated the Lorentz–Fitzgerald contraction along the direction of motion. He showed that relativity could reconcile all the observations. He picked out one aspect of relativity, the equivalence of energy and mass ($E = mc^2$), to illustrate that the release of light corresponded to a decrease in the mass of the emitting body, clearly implying that light was an entity in its own right, as had been assumed by Newton, not a motion of another medium. As shortcomings of the wave theory, he adduced in particular the independence of the velocity of electrons in the photoelectric effect from the intensity of the incident light. If Lenard was at the meeting, he would have been incensed by Einstein's cavalier dismissal of his favourite 'triggering' hypothesis to explain this as having been pretty much abandoned.

Einstein continued by pointing out that a major shortcoming with wave theory was that, while an oscillating point charge produced an outgoing spherical wave, no point charge could absorb an incoming spherical wave, so that the entire theory was not reversible even in theory, in contrast to many processes in thermodynamics. He used the example of the production of secondary electrons from X-rays to show that the X-ray energy could not be in the form of a spherical wavefront. The crucial part of his argument was built upon the earlier 1909 paper (i.e. a thermodynamic analysis of the fluctuations of the radiation field in a cavity) showing that the derivation of the Planck law for black-body radiation was in fact incompatible with wave theory. According to Einstein, Planck's formula could not be wrong, because Rutherford and Geiger as well as Regener had recently confirmed Planck's value for the electronic charge. He went on to outline a 'toy' theory in which each photon had associated with it a field behaving like a plane wave whose amplitude decreased with distance from the source; according to Einstein's assertion, the interference between these fields could give rise to effects indistinguishable from the standard wave theory of light.

Whether Einstein's resumption of his seat caused a stunned silence or a cacophony of professors rising to their feet to have their say is not known. What is recorded is that Planck exercised his right as chair to lead the discussion. He was unconvinced by Einstein's revolutionary suggestion and expressed himself unwilling to give up Maxwell's equations, which seemed to be the inescapable cost of Einstein's ideas. He argued that Einstein's scenario was faulty and that the understanding of the mechanism involved in the emission of light was particularly uncertain. Further, he was convinced that Maxwell's equations could be maintained and that only the interaction of light with matter would need to be modified.[6] Stark joined in the discussion to support Einstein, saying that the ejection of electrons by X-rays from a metallic surface many metres from where the X-rays were produced could not be explained by the wave theory. Planck responded that X-rays were too little understood and counter-attacked by citing optical interference phenomena that would require a photon several metres in extent. Lorentz had raised this objection earlier, probably stimulated by his correspondence with Einstein. Einstein had the final word in the discussion, repeating his picture of a field associated with each photon that could mimic wave motion. Planck, as can be deduced from years of subsequent scepticism, was unconvinced.

Even if his talk had not convinced all his senior colleagues of the fact that radiation had a double nature, being both particle and wave (now known as wave–particle duality), many of the younger people looked back on hearing Einstein as one of the great experiences of their lives. Lise Meitner and Born both enthused about it. Wolfgang Pauli, one of the Young Turks of the coming quantum-mechanical revolution, although at the age of nine he was too young to attend in person, considered the talk a masterpiece. Certainly, the conference made a big impression on Einstein. For the first time he met the great names in physics, with many of whom he had corresponded over the last few years. These included Sommerfeld and Johannes Stark. In particular he finally met Planck; the two found each other congenial and their friendship was established on a firm, personal basis in addition to the long-standing admiration for each other's work in physics. He and Sommerfeld also struck up a strong admiration and friendship, despite Sommerfeld not being immune to the endemic anti-Semitism of the time. In December 1907, he had written to Lorentz and characterised Einstein's work as containing unhealthy dogmatism and partaking of the 'abstract-conceptual manner of the Semite'. However, Sommerfeld's prejudice seems to have been only theoretical, since he appears to have taken care of Einstein on the Thursday evening of the conference, when he suffered one of the periodic stomach upsets that had begun during his undernourished student days. In a letter to Edgar Meyer, who was about to leave his position as a Privatdozent in Zurich, sent shortly after he had arrived back in Bern, Einstein gave the strong impression of only slowly recovering from the intellectual 'high' of the meeting; he explicitly mentioned the plethora of happy memories he had brought back from Salzburg before going on to detail a calculation that the two had discussed during the meeting. A few weeks later, the Einstein household migrated to Zurich, where Einstein was now an Extraordinary Professor of Theoretical Physics. Only a handful of people had any inkling of quite how extraordinary he would prove to be.

[6] Eventually it turned out that both Einstein and Planck were correct. Light does propagate as discrete entities called photons, but the wave–particle duality that Einstein implicitly revealed to the world at Salzburg means that Maxwell's equations have indeed been preserved.

Chapter 8

'Oh what a tangled web . . .' (1909–1910)

Before shifting the scene to Zurich and Einstein's activities in his first academic position, it is necessary to consider an episode that spans the end of his time in Bern and the Zurich period. As discussed, Einstein's relationships with women were complex. Many of his relationships began through music. This was certainly true of those with Ernestina Marangoni and Marie Winteler. Marie's mother, Pauline, became at least as dear to him as his own mother. His subsequent relationship with his cousin Elsa also had a strong maternal component. He was certainly attracted to a pretty face, as evidenced by an anecdote of Einstein's being distracted from a boring explanation of psychoanalysis by staring at a particularly beautiful young student. He appreciated intellectual women, although in later life he regularly expressed misogynistic views of their true intellectual capabilities. It is generally agreed that Mileva was of rather plain appearance, as Einstein himself admitted when he first became involved with her. However, he had loved her intellectual courage and her ability to share his obsessive interests in physics. It was this latter bond that was clearly weakening towards the end of their stay in Bern; by the time of the move to Zurich, it had almost disappeared.

It was inevitable that, as Einstein established a role in academic life commensurate with his achievements and intellectual stature, his outlook and character began to change. In their student days, it had been he and Mileva against the world. They laughed at their professors' pomposity, rigidity, and lack of imagination. They were both indignant at Einstein's initial inability to break into the established academic hierarchy. However, those same academics were now beating a track to Einstein's door; he was speaking at major conferences on physics; and his papers were becoming accepted as of the first importance. He was perceptibly entering the establishment that he and Mileva had mocked from their high tower of isolated interdependence. He had left the tower and could not take Mileva with him, even had he wanted to. Whatever physical attractions she had held for Einstein had faded. No longer involved with Einstein's work in any major way, although she may well have still checked papers and derivations, her life became increasingly centred on Einstein himself and her child. Although her mother reported that Mileva's views were listened to with respect by the eminent physicists gathered in her house, other reports, starting from the earliest days of her marriage, tell in contrast of a silent brooding presence that observed, rather than contributed. A natural maternal partiality can explain the discrepancy. As Mileva withdrew, Einstein expanded. David Reichinstein observed him in the period just after he arrived in Zurich:

> Einstein's soul swiftly took wing. Increasing success supplied his soul with a feeling of superiority; he developed a generous wit. The sensation of insecurity, his feeling of not being equal to the struggle of life seemed to disappear gradually . . . the helplessness of his soul could already be seen to disappear during his first years in Zurich.

It is within the above context that the following three facts need to be interpreted. First, at some point substantially before September 1909, Einstein's relationship with Marie Winteler was rekindled and they had a passionate, at least from Einstein's side, affair, which lasted in some form until

Einstein. Brian Foster, Oxford University Press. © Brian Foster (2026). DOI: 10.1093/oso/9780198794875.003.0008

at least the summer of the following year. Secondly, someone else from Einstein's past re-entered his life. In May 1909, Anna Meyer-Schmid, known as Anneli to Einstein and other intimates, saw the announcement of Einstein's appointment as Extraordinary Professor in Zurich in a newspaper and sent him a letter of congratulation. They had met in 1899 during Einstein's vacation in the Hotel Paradies (see Chapter 3). Ten years later, she was a married woman living in Basel and there is no reason to believe that she had any reason for contacting Einstein other than pleasure in the success of an old acquaintance. He replied with a letter which, although brief, is surprising in its warmth. The third and final fact is that, sometime around October/November 1909, Mileva became pregnant.

These are the salient facts that must be interpreted by a biographer. This is made difficult by their apparent contradictory nature. Unfortunately, much of the relevant information has been destroyed. Helen Dukas, Einstein's devoted secretary, is thought, after his death, to have disposed of many documents that she considered not to reflect well on her late employer. Certainly, matters relating to this period would have been in this category. Many of the documents that became available from the Winteler family archive in 2015 had been ripped up by Marie and were only reconstructed after considerable effort; many others must have been completely destroyed. Indeed, Marie herself in her recollections in old age mentioned that she had burned them. Of the letters relevant to Anneli Meyer-Schmid, her reply to Einstein's suggestive letter was returned to her husband and is lost. The only possibility to make sense of the facts presented above is to identify what is most probable, noting that any set of conclusions is just one of several possibilities.

The evidence is that, of the three women involved in these events, the most important for Einstein was Marie Winteler. The letters revealed in 2015 were little short of astonishing to Einstein scholars. The previously prevailing judgement was that the affair of 1895–1897 in Aarau was a youthful infatuation that Einstein ended because he became bored with Marie's rather silly superficiality. He was apparently dazzled by the bright lights of Zurich and the attractions (mostly her intellectual companionship) of Mileva. The new evidence, revealing not only a new affair in 1909, but also renewed contact in 1899, gives the lie to this interpretation. Already in 1899, Einstein wrote to Marie inviting her to reopen communication and to put the trauma of the recent past to one side in favour of future enjoyment of each other's company (see Chapter 3). It is not known to what extent they may have corresponded in the decade between 1899 and 1909. Neither is it known how often they met. Since during this period Marie's employment was spasmodic and her place of residence fluctuated among her family, it seems rather likely that they would have been thrown together on occasion. This is particularly the case since Michele Besso, Einstein's closest friend, was Marie's brother-in-law and Einstein's sister, Maja, was in love with Marie's brother Paul, whom she married in 1910. Certainly, the murder of Pauline Winteler, Einstein's beloved substitute mother, by her son Jost Junior (see Chapter 7) in 1906, must have united them in mutual grief. Although there is no documentary evidence, it seems inconceivable that Einstein would not have attended Pauline's funeral. It is known that Marie was devastated by these terrible events, in which her mother died in her arms. She ceased to teach at junior schools in the region and instead in 1906 enrolled in a home economics course in Aarau, probably at the Ladies Academy that Maja had attended. In seems likely that at this time she was sacrificing her career as a teacher to look after her devastated father Jost in Rössligut.

Marie's own continued interest in Einstein over these intervening years is attested by the problems that Maja had when she arrived in Aarau in 1899 to attend the Ladies Academy. As an unwelcome reminder of Einstein, Maja felt she was an irritant to Marie; on several occasions this led to tensions between them. This is not the behaviour of someone who had put her involvement with Einstein behind her. Marie's account in old age that she had repeatedly refused Einstein's proposals of marriage during their first love affair implies that she must subsequently have had second thoughts. He blamed himself for her refusal of his proposals, apparently because his prospects of being able to support her were so poor. Towards the end of his period in Bern, it was clear that this difficulty had vanished and that he could indeed support a wife. His problem, and Marie's, was that he already had one.

The tone of the young Einstein's letters to Marie differs subtly from those to Mileva. Letters to Marie, both those from his youth and those from the period now under consideration, are devoid of the flippant, jocular tone of his letters to Mileva. There is no playfulness such as pet names, 'Dolly and Johonsel', but simply passion and anguish. It seems that Einstein never lost his youthful love of Marie but rather yearned for more than a decade to be reunited with her. His relationship with Mileva was indeed a 'rebound'. He certainly convinced himself that he loved Mileva and probably did for a while. However, it seems to have been a love engendered by circumstance rather than of the heart. One circumstance was his determination to defy the almost hysterical opposition of his mother to his marrying Mileva. Another was his dependence on a sympathetic voice and sounding board to support his investigations in physics. Once these circumstances vanished, his relationship with Mileva shrivelled away and he was thrown back into an enduring obsession with Marie, kept alive by infrequent meetings and possibly by correspondence that was subsequently destroyed. Einstein was not one to be deterred by considerations of conventional morality and he certainly had no respect for marriage as an institution. There are many quotes to substantiate this, including that marriage must have been invented by an unimaginative pig. His behaviour throughout his life shows that he had no compunction about carrying on affairs with women irrespective of their, or his, marital status. Thus, at some time well before September 1909, he and Marie took up where they left off more than a decade before and became lovers.

The approximate time frame for this consummation can be established from a letter that Einstein wrote to Marie on 15 September 1909. One thing strikes the reader immediately: Einstein uses the intimate '*Du*' form, something that, strangely enough, he had never done during their first relationship. He began by writing:

> I have written to you many times since you didn't want to appear at the window in Zurich when, despite Michele's command, I turned up. However, I wrote to Bern, in the belief that you had pitched your wigwam there for a while. So I could thereby console myself that perhaps Michele, for schoolmasterly reasons, had not given my letters to you. I am still living by the memory of the few hours of you that a miserly Fate granted me.

From this it can be deduced that their affair had reignited during the spring or summer of 1909 and that its existence was known to Besso, who obviously disapproved. Whether it was also known to his wife Anna, Marie's sister, and a close friend of Mileva, can only be conjectured but seems likely. Marie was at this time staying with her other sister, Rosa, in Oberwil bei Büren, about twenty kilometres north of Bern. Einstein went on to write:

> I escape my eternal longing for you only by intense work and contemplation. At least give me your reasons for running away from me as if I were a leper! My only happiness would be to see you again, or to get a letter from you. Don't you like me anymore or don't you feel justified in fulfilling my little requests? I feel like a dead man in this life full of duties and devoid of love and pleasure.
>
> I beg of you, let me hear something from you and I greet you with all my heart,
>
> Your,
>
> Albert'.

This letter leaves little doubt as to the intensity of Einstein's passion. He wrote to Besso in November and mentioned 'Mental equanimity lost because of M. not yet restored.' Before the discovery of the new letters, this 'M.' had been thought to refer to Mileva and the unpleasantness with regard to Anneli Meyer-Schmid (see below). However, 'M.' clearly refers to Marie. Letters from March, July, and August 1910 give further evidence for the all-consuming nature of Einstein's infatuation. In March, he chastised Marie for now doubting his sincerity, whereas in the previous year she had trusted him implicitly. He wrote that 'I think of you with the most heartfelt love in every free minute and am as unhappy as any person could be. A mistaken love and a mistaken life is how it always seems to me.' He wrote of their previous happy encounters in various spots north of Bern, presumably mostly before the Einstein family left in October 1909 for Zurich. The mental disturbance caused by these lovers' quarrels between him and Marie reflected itself in Einstein's behaviour and was evident to those close to him. In a letter to his mother in April 1910, he wrote: 'The bad mood that you detected[1] in me was nothing to do with you. Don't be concerned. When one is depressed or annoyed by something, it does no good to bang on about it to other people. One has to turn it over in one's own mind, however unsuccessfully, alone'. In July Einstein sent a postcard reproaching Marie for her silence. There seems to have been little contact between them in 1910, since Einstein's new job and the much greater distance between Zurich and Rosa Bandi's home where Marie was still staying made trysts much more difficult. Marie seems to have ceased writing. The silence was too much for Einstein; he sent a postcard and then a letter in July to try to elicit a reply from Marie. When a response finally came, it dashed whatever hopes he still harboured. His letter of 7 August is painful to read. It seems to be a reaction to a letter from Marie finally breaking off their affair. Einstein's letter begins with the sentence 'When I read your letter, it was as if I were looking on as my grave was being dug. The remnant of the happiness that I still retained is destroyed; only the bleak life of duty remains.' He recalled the 'moments of pure happiness [. . .] fifteen years ago and last year' and wished Marie all the best for her future. He asked her not to think of him if it was with hate and bitterness and assured her that he was not a betrayer.

It may be that Marie's last letter contained the news that she would become engaged to Albert Müller, the manager of a watch factory in Oberwil bei Büren. However, despite the air of finality of the letter from Einstein quoted above, given the obvious depth of his passion, it is possible that this was not the last time that the two met. The circumstantial evidence, although it is not much more than hearsay, can be found in a persistent rumour in the Einstein family that Marie bore a child to Einstein.

This rumour was mentioned to Einstein biographer Dennis Overbye by Aude Einstein, the first wife of Einstein's grandson Bernhard. The rumour apparently gained credence from the resemblance[2] of Marie's son, Paul Albert, to Einstein.

The possibility that Einstein fathered a child with Marie would have seemed completely bizarre before the discovery of the letters confirming their relationship in this period; now it seems at least possible. Does a sexual encounter between Marie and Einstein at least conform to the known facts about his whereabouts at the appropriate time? Chapter 9 gives details of Einstein's stay in Switzerland from around 3–17 October 1911. He stayed first in Luzern with Maja and her new

[1] The meeting to which this refers probably took place at the wedding of Maja Einstein to Paul Winteler, which took place in Zug on 23 March 1910.

[2] Readers may judge for themselves by comparing a photograph of a young Paul Müller reproduced in Rogger's book with one of the young Einstein.

husband Paul Winteler, and then in Zurich. The possibility of a liaison with Marie in this period cannot be ruled out. Paul Albert Müller was born on 8 August 1912. This is roughly 42 weeks after the latest time that Einstein could have met Marie. Only a few percent of babies are born at or later than 42 weeks after the first menstrual period before pregnancy. It is therefore logistically possible but unlikely that Einstein was the father of Paul Albert. Marie and her watch maker were married on 16 November 1911, by which date Marie would have suspected that she was pregnant if a liaison with Einstein had indeed taken place. Whether the marriage was hastily arranged or long planned is unknown, but the former seems likely since Besso did not mention it in his letter to Einstein written on 23 October. In a response on 26 December to a now-lost letter from Besso that contained the news of Marie's marriage, Einstein wrote: 'I welcome Marie's wedding wholeheartedly. With this, a dark spot in my life vanishes. Everything is now as it should be',

In a later letter to Besso, Einstein wrote that he was pleased that Marie was having 'a little boy (?), whose sort-of uncle I am', presumably referring to the fact that Maja Einstein was married to Marie's brother Paul. It does indicate Einstein's curiosity about the baby. The question mark next to 'boy' is a clear indication that he would like Besso to tell him the baby's sex and by inference that he was born safely. Besso would know because, as the husband of Marie's sister, he really was the baby's uncle. Einstein remained interested in Marie and her family. In January 1914, he wrote to her widowed sister Rosa Bandi, with whom she had often stayed, asking to be remembered to Marie and to her husband, Einstein's 'namesake and general substitute', Albert Müller. While 'namesake' is an appropriate epithet to describe Müller, 'general substitute' or 'general representative' (the German word '*Vertreter*' is ambiguous) seems strange indeed.

Whatever the truth about Einstein's relationship with Marie, for some time there had been difficulties in his relationship with Mileva (Figure 8.1). Somehow, she had become aware of the correspondence that had begun between Einstein and Anneli Meyer-Schmid mentioned earlier. Einstein was always careful to ensure that any mail that he did not wish Mileva to know about was addressed to him at the Physics Institute. For example, just before Christmas 1909, he had written to Conrad Habicht asking him for the name and publisher of a music book that he wished to give to Mileva as a Christmas present, presumably for her to practise on their newly acquired piano. He directed Habicht to reply to his university address, a clear indication that Mileva would open any mail that arrived at their apartment.

If Einstein's mother had noticed changes in his behaviour during the affair with Marie, it is certain that so too would Mileva. Her life and attention had increasingly narrowed down to Einstein and her son, so that, even had she not been of a jealous disposition, she could hardly have failed to notice that something had happened. Mileva knew Einstein's character well enough to be sure that he was seriously engaged with a woman. Perhaps Einstein carelessly brought home a letter from Anneli that Mileva found; more likely she visited him at his office and came across the letter there. Whatever the sequence of events, Mileva was incensed. She connected Einstein's change of demeanour with his renewed contact with Anneli. It seems that Mileva confronted Einstein directly and convinced him to return Anneli's second communication with a brief note saying that he didn't understand it. This apparently came to the attention of Anneli's husband, who wrote to Einstein, enquiring what was going on. Mileva, without consulting Einstein, replied to Anneli's husband, claiming that both she and Einstein were puzzled as to what had led Anneli to write a somewhat inappropriate letter. When Einstein discovered what had happened he was mortified and wrote an apologetic letter to Anneli's husband, blaming himself for the entire business and apologising for Mileva's letter, attributing it to her strong jealousy.

Figure 8.1 Mileva and Albert Einstein in 1912.

There can be little doubt that Einstein was deeply embarrassed by this farcical episode. It also seems highly unlikely that he would risk further contact with Anneli, especially by meeting her, after sending such an abject letter to her husband and risking Mileva's further unpredictable jealousy.

The final fact that needs to be assimilated in an understanding of these complex events is Mileva's falling pregnant, which probably occurred in October or November 1909, just after their move to Zurich. It is clear from the letters cited above that Einstein was at this period desperately in love with Marie but that any sexual relations between them had by then either ceased or become very infrequent. Einstein was a man with a strong sexual appetite and therefore he continued to have sexual relations with his wife. However, it seems extremely unlikely that he would have planned or agreed at this point to have another child with Mileva. The pregnancy was presumably therefore either an accident or deliberately engineered by Mileva in the hope of tying Einstein to her at a time when she could sense that she was in danger of losing him. Which of these possibilities is correct will never be known. Eduard Einstein, always known by the nickname Tete or Tetel, was born on 28 July 1910.

Can the various apparent inconsistencies between the above facts be reconciled? The explanation that seems most likely is the following. The affair with Marie began (again) in the spring of 1909 and quickly became serious, probably including sexual relations. As Einstein became more and more besotted with Marie, he began to discuss with her how they could spend their lives together. He would have had to convince her that he was willing to leave Mileva and his son and live with her, eventually marrying her. For a while she believed him, and they spent a happy few months, snatching trysts in the beauty spots on the northern outskirts of Bern, between their respective homes. Already in September, however, Marie was getting cold feet and by the following March she had become increasingly estranged from Einstein. What caused this estrangement is presumably that she began to doubt Einstein's assurances that he would leave Mileva. The crucial moment would be when, since it could hardly be concealed, he told her that Mileva was pregnant.

Marie would surely not have believed that Einstein would leave Mileva while she was pregnant with their second child. Although Einstein tried desperately until the summer of 1910 to convince her that somehow all could be arranged, in August, after Tete's birth, Marie decided that she could waste no more of her life on a man whom she loved but who seemed incapable of providing them with any sort of family life. Meyer had presumably been paying court to the very beautiful Marie for a while and provided a convenient escape from her destructive involvement with Einstein, whose desperate letter to her in August may never have been answered.

Proceeding further with this interpretation, the episode involving Anneli was simply a misapprehension on Mileva's part. Einstein began a harmless flirtation, something that throughout his life he was unable to resist, which Mileva happened to discover and to which she attributed Einstein's very apparent infatuation with another woman. Nothing happened between Anneli and Einstein except the exchange of letters detailed above. This latter conclusion is supported by letters that Einstein wrote to Anneli much later, in 1924, after she contacted him with a strange query about whether she should give details of their friendship to an American women's organisation in return for a substantial cheque. Einstein's response is heavy with sarcasm and puzzlement as to how a group of American women knew about their relationship but ends with nostalgia for their lost youth: 'It's good, by this chance to have heard again from Anneli, and that from this the beautiful days of our youth have emerged from the depths, a time when one didn't have grey hairs and was not the harassed big beast, from whom everyone wants something.' This does not sound like the response of someone who had fathered an illegitimate child with his correspondent.

While the explanation above seems the most likely, and the one that does least damage to Einstein's reputation, there are certainly other possibilities. One variant is that Einstein did in fact seduce Anneli sometime in May 1909 and that Mileva's jealousy of her was correctly aroused. After the exchange of letters culminating in Einstein's grovelling apology to Anneli's husband, in which Einstein protested that she had behaved with the utmost rectitude, their relationship ended. Either in parallel, or directly afterwards, Einstein's relationship with Marie blossomed. What speaks against this interpretation is the tortured passion that Einstein's letters to Marie reveals. Could he really have moved within weeks from seducing Anneli to desperate protestations of a love for Marie that had burned in his heart for a decade? His obvious prostration when their affair was finally ended does not allow any doubt that this love was genuine. It is also clear that Marie believed that he loved her. Since Einstein himself would surely not have told Marie anything about Anneli, how could she have been the cause of Marie and Einstein's estrangement? There seems to have been no connection between the two women except a distant one via Maja, who knew them both, although Anneli only very slightly.

Other information unearthed by Dennis Overbye, however, speaks for this variant interpretation. In his book *Einstein in Love: A Scientific Romance*, Overbye writes that Anneli had only one child, Erika, who was born on 10 March 1910, so was indeed conceived around late May/June 1909. Erika was brought up believing that she was Einstein's daughter. Erika's husband had been told that Einstein and Anneli had met in that May of 1909 and that Erika was the result. Erika's eldest son remembered that for many years she kept a framed photograph of Einstein on her bedside table.

Many years after Anneli's death, Erika wrote to Einstein ostensibly to ask the meaning of his letters to her parents that she had found among her mother's possessions. Perhaps she was hoping that he would admit to being her father; if so, she was disappointed. Einstein's reply explained away the episode as a product of Mileva's jealousy. His jibe that this was typical of a woman of such unusual ugliness demonstrates that his anger with his wife still burned even at a distance of more than forty years. The distant tone of Einstein's letters both to Anneli in 1926 and later in the 1930s, and to Erika, could be explained by the fact that he was ignorant of the possibility that he might be Erika's father.

Erika continued to seek evidence even after Einstein's death. Indeed, her daughter recalled that she was somewhat obsessive about finding such evidence. In 1976 she wrote to Helen Dukas, Einstein's former secretary and, along with Otto Nathan, his literary executor, enquiring whether there were any letters from Anneli among Einstein's papers. No reply has been found, and it is unlikely that she received one. Surely Anneli would not have told her only daughter that she was Einstein's illegitimate child unless she herself believed it? Either they had had sexual relations, even if only a single fleeting encounter, or the passing years and Einstein's worldwide fame awoke in Anneli what would now be called a 'groupie' mentality, whereby she eventually convinced herself that what she had greatly wished, and what Mileva had feared, had actually happened.

The last words in this controversial and unedifying tale should go to Marie. Her marriage to Müller was unhappy; by 1927 she had determined to end it. She wrote to Einstein, asking for his help in obtaining a position that would allow her to care properly for her two sons. It is sad to note that she used the formal '*Sie*' form of address, indicating their long estrangement. Einstein was touched by her plight, indicating to Besso that he would try to sell one of his manuscripts to raise money for her support. In fragile mental health towards the end of her life, Marie was committed to an institution in Waldau. One of the doctors there remarked in her case notes that a student named Albert Einstein had turned her head. In Waldau she wrote:

> Also later on I remained defiant [in refusing to marry Einstein], which caused me great pain because I had realised for a long time that it was the fault of one woman, who destroyed our love. However, from her guilt arose a calamity for her. But I realise, I am becoming personal.

Was she referring to Mileva, whose becoming pregnant destroyed Marie's chance of happiness with Einstein, and who reaped the calamities of Eduard's eventual insanity, her own desertion by Einstein, and her physical breakdown? Or was she referring to Anneli, whose affair with Einstein she had somehow discovered and whose 'calamity' was bearing Einstein's illegitimate and never-acknowledged child? The answer may never be known unless more undiscovered Einstein letters are found, or further light is thrown on the subject by DNA analysis.

Chapter 9

Extraordinary Professor at Zurich (1909–1911)

When the Einstein family arrived in Zurich on 14 October 1909, they moved into an apartment in a large house, Moussonstrasse 12. Einstein was gratified to find that he already knew one of his neighbours – Fritz Adler, his erstwhile 'rival' for the Zurich appointment. The two families quickly became intimate. Einstein and Adler would repair to the attic for smoky evenings of philosophy and physics. The families would climb up to the Einsteins' old haunts in the hills above the city on Sundays. Mileva was delighted to return to the spot where she and Einstein had met and first become intimate. Her initial hope was that the change of scene would restore their relationship to what it had been in that seemingly far-off happy time.

For the first time in his life since ceasing to be a student, Einstein was now free to concentrate on physics without any other distraction. However, he soon found that life as a professor did not necessarily mean he had more time for his research. The time that had been taken in adjudicating patents at Bern was instead substantially devoted to teaching and academic administration. Einstein's record as a teacher was somewhat mixed, as described in the saga of his appointment to the Zurich faculty. He could be obscure and at times rambling, mentally preoccupied by questions related to his current research; at other times he could be inspirational. His students found that the occasional lack of narrative clarity was more than offset by the glimpses of inspiration at the frontier of knowledge that Einstein's lectures afforded. His surviving lecture notes from the winter semester of 1909 show him launching straight into the introductory course on mechanics with a rather detailed written exposition, delivered in four lectures a week. He also taught thermodynamics and organised a weekly seminar. His notes on the kinetic theory of gases in the summer semester of 1910 are much sketchier than the introductory mechanics text – an indication of his much greater security in this area, which continued to be central to much of his research work. In the next semester, he produced another detailed written script, this time tackling electricity and magnetism. He took the opportunity when discussing how to measure small amounts of charge to explain the principle of the Maschinchen.

Accounts of Einstein's lectures from this time show both that he developed significantly as a lecturer and that he had a good relationship with his students. The fact that attendance at his classes was much higher than at Bern, averaging around 15, would certainly have increased his motivation. Hans Tanner, who subsequently became a teacher of physics and maths at Einstein's old employer, the Technikum at Winterthur, reminisced:

> I attended all Einstein's lectures in the years 1909–11 [. . .] We felt most at home in and after the Colloquia. These took place once a week between 8 and 10 p.m. Here it was up to us students to discuss a theme. Einstein never let slip how little we understood; often, at the end, he would ask 'Who is going to join me at the Café Terrasse?' There the discussions carried on. [A] typical experience: we were chatting until closing time in a café in Bellevueplatz. When they had closed the doors behind us, Einstein asked 'Does anyone want to come home with me? This morning I got a piece of work from Planck that must contain an error. We can read it together.' So, my friend Karl Gaule, who was later a Privatdozent at

Einstein. Brian Foster, Oxford University Press. © Brian Foster (2026). DOI: 10.1093/oso/9780198794875.003.0009

> the University of Danzig but sadly died young, and I climbed up the Zürichberg with our honoured teacher. In his flat he gave us Planck's paper to study. 'You have a look for the error,' he said, 'while I make you a cup of coffee.' We buckled to it with a will. After a quarter of an hour our host returned with steaming cups of coffee: 'So, have you found the mistake?' We demurred. 'You must be wrong, Professor. There is no mistake.' 'Yes there is. Because this and this imply this and this,' he said, demonstrating to us via a simple dimensional analysis. He was correct. He saw immediately the physics content behind the formula, which to us remained an abstract formula . . . I well remember how, in a few exquisitely crafted sentences, he explained to us that time was also relative. That there is no objective definition of simultaneity and that the time between two events depends on the movement of the observer. I think that was, and is, the greatness of Einstein as a scientist: to be able to approach a problem free from the weight of any tradition, not from any pleasure in criticism, but from his own personal imperative to understand everything and to clarify everything for himself. Thus he could throw out erroneous theories that had been accepted as true for centuries.

Tanner went on to do a PhD and has the almost unique distinction of having been Einstein's doctoral student. Although he had begun as an experimentalist, he switched to theory as he came more and more under Einstein's spell. Tanner was given a wide choice of thesis topics but was steered towards a particular one in the kinetic theory of gases. Having spent weeks trying to understand it without success, Einstein admitted that indeed Boltzmann was difficult to understand and perhaps Tanner should address a particular detail which Einstein then identified. Encouraged, Tanner made progress and shortly afterwards visited Einstein at home:

> He was sitting in his study in front of a pile of paper. He was writing with his right hand while holding his younger son in his left hand and answering his older son Albert, who was playing with building blocks, whenever he asked him something. With the words 'Just a minute, I'm almost finished,' he handed over the baby-sitting duties to me and carried on working. I could observe from this his powers of concentration.

Many years later, after Tanner was a well-established teacher in Winterthur, he met Einstein by chance at a conference. He grabbed Tanner's arm and introduced him to a senior colleague as his first and only Swiss graduate student.

Another person who worked with Einstein in this period was Ludwig Hopf. He came from a wealthy Jewish family of hop merchants from Nuremberg. He had studied at the Universities of Berlin and Munich; at the latter he was a student of Sommerfeld, who had introduced him to Einstein at the Salzburg conference in 1909. Since Hopf was a genial young man whose humorous outlook on life matched Einstein's own, the two immediately became friendly. Hopf was also an excellent pianist, a fact even more likely than his geniality to recommend him to Einstein. After his arrival in Zurich, Einstein felt the need of an assistant to help him with his research as his teaching duties took more and more of his time. Being only an associate professor, he was unable to obtain resources to hire an assistant; he seems to have approached Hopf, who had independent means, which meant he could take up Einstein's offer without salary. Having finished his PhD thesis with Sommerfeld on the onset of turbulence in fluid flow in July 1909, Hopf arrived in Zurich, most probably late in 1909 or early in 1910. By the summer of 1910, the two were hard at work, with Hopf having switched his interests to the quantum problems with which Einstein was at that time preoccupied.

David Reichinstein, a physical chemist with a deep interest in physics, was another attendee of Einstein's Zurich lectures and seminar. He and Einstein became close in later years until they fell out over Reichinstein's biography of Einstein, which he published in 1932 despite the subject's strenuous objections. It is certainly liberally scattered with inaccuracies, particularly about this period, and in some ways casts as much light on the author's character as Einstein's. Discursive, pretentious, and at times downright bizarre, it nevertheless does give one or two vivid vignettes of Einstein in Zurich. He painted a very similar picture of the Einstein household to that of Tanner, in particular Einstein's ability to work through domestic distractions. He related how Einstein was overcome by fumes from his faulty stove and was only rescued from suffocation by his friend Heinrich Zangger. He also recorded the bohemian lifestyle that Einstein still pursued. Tanner had remarked on Einstein's scruffy appearance at his lectures; Reichinstein relates an encounter at the theatre when Einstein, having come directly from his office, dined in the interval on two sandwiches that Mileva had brought with her. Indeed, Einstein and his wife were not only often at the theatre but attended almost every concert at Zurich's main concert hall, the Tonhalle. He would arrive with his clothes still covered in chalk dust and embarrass Mileva by greeting the higher echelons of Zurich society at the interval with his mouth full of sandwich.

Although Einstein was now a professor, his income was identical to what it had been at the Patent Office. Indeed, in many ways the family was worse off, since Einstein now had to maintain a public position, which required expenditure that was unnecessary for a patent clerk. In order to make ends meet, Mileva took in student lodgers.

The Christmas festivities of 1909 were not relaxing for Einstein. It seems that at least Conrad and possibly also Paul Habicht spent the Christmas period with the Einsteins. While Mileva practised her piano playing using Einstein's gift of sheet music (see Chapter 8) and Hans Albert played with his toys, the older boys played with theirs – the Maschinchen. Einstein found time to write a number of New Year's greetings, including to Michele Besso and Jakob Laub. He began the letter to Laub by apologising for the dilatoriness of his correspondence, something which became an almost constant refrain in his letters, especially to Besso, in subsequent years. His letter to Laub contains the excuse that he has been particularly busy with his lectures but that these were going well and giving him pleasure, particularly the easy relationship that he had established with his students. In March, he pleaded with the Habichts to spend the Easter vacation with them in order to finish drafting a paper on the Maschinchen.

At the beginning of March, Einstein had hosted an important guest – the distinguished physical chemist and subsequent Nobel laureate Walther Nernst, who interrupted his journey to Lausanne. Nernst and his colleagues in the physics department of the University of Berlin were carrying out a comprehensive series of measurements on the thermal behaviour of metals at low temperatures. Nernst was impressed by the fact that he had found that Einstein's predictions from his 1906 paper on the quantum behaviour of heat in solids fitted his own data well. He had also heard of relativity from his colleague Planck in Berlin and although he certainly didn't understand it, he was happy to interpret his respected colleague Planck's enthusiasm as greatly to Einstein's credit. His conversations with Einstein increased his standing further in Nernst's estimation, which would facilitate Einstein's subsequent move to Berlin. The arrival in Zurich of a scientific luminary such as Nernst specifically to talk to Einstein also greatly increased the latter's standing with his colleagues in the university.

Another arrival in April, but on a more permanent basis, was that of Ernst Zermelo, a distinguished mathematician who had been appointed as Professor of Mathematics in the University of Zurich. He had been an assistant of Planck's in Berlin in the late 1890s. In 1896–1897, Zermelo was involved in a famous controversy about the second law of thermodynamics with Ludwig Boltzmann. Although Zermelo subsequently moved his interests towards mathematics, he remained very knowledgeable about physics. By the time he arrived in Zurich he had established a considerable reputation. Einstein often consulted him on various mathematical problems. He also appreciated his deep interest in music; Zermelo published several papers on aspects of musical theory. In time the two became close friends; indeed, Reiser thought that Zermelo was the closest person to Einstein during his time in Zurich. Certainly, by the time Einstein left Zurich for Berlin in 1914, Zermelo and he were intimate. Zermelo was a fiery character, who was far from popular in the Zurich mathematics department. He had a serious case of tuberculosis and although he made an excellent recovery, his health was often precarious. It was the fact that this often prevented him from lecturing that gave his colleagues an excuse to rid themselves of this abrasive character. In April 1916, Zermelo resigned his chair and moved to the Black Forest in Germany.

Although it is unclear if either or both Habicht brothers spent time with the Einsteins in the Easter vacation, another family certainly did. Lucien Chavan, who had first studied with and then become Einstein's friend in Bern, spent a few days at the Einstein home in Zurich. He was accompanied by his wife, who had become friendly with Mileva. The two men spent some time discussing experiments with telephones, Chavan's professional preoccupation, and corresponded about this for the next few months. Einstein, as part of his duties as co-director with Professor Kleiner of the undergraduate experimental laboratory, was on the lookout for useful undergraduate experiments and acquired a few telephones for this purpose with Chavan's help. Einstein was clearly fond of Chavan and went out of his way to try to arrange a rendezvous in Bern when he was passing through on his way to give a talk on his photon work at a meeting of the Swiss Physical Society in Neuchâtel on 7 May 1910.

After a long hiatus in communication with his mother, Einstein wrote what in retrospect was a rather portentous letter to her just after the visit of the Chavans at the end of April. There are three reasons for its significance. One is the admission of his being distracted, which can now be interpreted as the result of his affair with Marie Winteler. The second is the mention of the fact that he had been approached to take up a full professorship by a great university, whose identity he was unable to divulge, but was in fact the German University of Prague. The third was a request to pass on his greetings to his uncle and aunt in Berlin, with whom his mother was staying, and their children, Paula and Elsa. This is the first mention of his cousin Elsa since they had played together in the courtyard of the Einstein compound in Munich. With his wife heavily pregnant and his affair with Marie still tormenting him with desire, Einstein re-established contact vicariously with the woman who would eventually become his second wife.

Music as always, was the most important thread in Einstein's life other than physics. In additional to the public concerts at the Tonhalle and elsewhere, private music-making was an almost incessant activity. The Einsteins' apartment at Moussonstrasse was a magnet for intelligent people and talented musicians. He renewed his acquaintance with Alfred Stern, Professor of History at the Polytechnikum, at whose house he had often been a guest in his student days. In a letter to Stern in May 1909 arranging a meeting, he joked that his violin playing was worse than ever. Their musical evenings were now enhanced by Stern's eighteen-year-old daughter Antonia, an excellent violinist who was studying at the Zurich Conservatory. A particularly important musical relationship was with his old mathematics teacher at the Polytechnikum, Adolf Hurwitz, whom Einstein called on more or less immediately after moving to Zurich. Hurwitz, who was an excellent pianist, liked nothing more than to arrange musical evenings in his home. He had already played with Einstein when he was a student, so he was delighted when he heard of his return to Zurich. He invited Einstein to his home and their Sunday music-making became a fixed point of their week. Hurwitz's eldest daughter, Lisbeth, was also a good violinist and her diary, sadly now destroyed, gave a graphic picture of these musical parties (see Figure 9.1). Einstein would arrive at the Hurwitzs' door punctually at 5 p.m. with the whole family and before knocking would shout that 'Einstein's here with the entire hen house!'

Figure 9.1 Einstein playing chamber music with Adolf Hurwitz and his daughter Lisbeth on the roof of their home in Zurich.

In addition to the Sunday afternoon gatherings, there were also many evenings of music-making. Lisbeth Hurwitz was sixteen when she first recalled playing with Einstein. Her impression of him was of a friendly but rather child-like man who was extremely musical. They played through a Bach concerto twice, presumably the Concerto for Two Violins in D minor, BWV 1043 (the 'Double' concerto).

Einstein's passion for making music meant that he delighted in playing chamber music in as many different combinations as possible. He regularly played piano trios with Henri Bas-Bulaneck, a pianist whom Einstein characterised as excellent if rather cold and undemonstrative, and the cellist Raphael Lewinowitsch. The latter, who was an engineer, also played with Einstein and Conrad Habicht.

On 28 July 1910, Mileva gave birth to a boy, called Eduard. Einstein wrote many notes proudly announcing the happy event to friends and family. The birth was not easy and neither mother nor son made a rapid recovery. Indeed, two weeks later, Einstein wrote that he was forced to remain at home and to give up his musical evening visits to look after them. Hans Albert had been sent to stay with friends in Winterthur and remained there for some time while his mother convalesced. While Hans Albert was a sturdy and straightforward boy, Eduard was a sickly child, a dreamer and introvert.

Shortly after Eduard's birth, Einstein was pleased to welcome his friend Sommerfeld to Zurich. This was the result of many letters that had passed between them over the preceding months relating to Einstein's major scientific preoccupation at that time – the quantum. Einstein's lecture at the Salzburg conference had excited many but had stunned others into disbelief and incredulity, ensuring that the controversy over the nature of radiation continued to rage unabated. Einstein struggled to reconcile the perplexing puzzle of his results that electromagnetic waves carried corpuscles of energy with the continuous wave theory of light embodied in Maxwell's equations. He was essentially alone in this conviction and was vigorously opposed by both Planck and Lorentz. His lecture to the May congress in Neuchâtel simply restated his belief that radiation was quantised. He motivated this with a statistical example showing that the distribution of energy in two connected volumes filled with radiation was identical to that of system of point-like particles. This could not be explained on the basis of the superposition of waves.

Since both Mileva and Eduard were still in poor health, Einstein gave up his usual holiday in the mountains and remained in Zurich for the summer. He tried to attract as many of his friends and colleagues to visit him as possible. He and Sommerfeld spent a few days together, working with Sommerfeld's protégé Hopf, now Einstein's collaborator. In addition to long discussions about the quantum nature of radiation, the three friends found the time for other activities. Sommerfeld wrote a note to his wife Johanna from the Apfelkammer (Oepfelchammer in the Zurich dialect), a famous drinking rendezvous in the old town, more or less unchanged today from its appearance in Einstein's day; every inch of wooden bench and wall is carved with patron's graffiti accumulated over the centuries. Sommerfeld's note contained light-hearted additions from his two companions to the effect that they were taking refreshment via a great many light quanta. The morning had not been entirely dominated by work either. Einstein did not let slip the chance to play with Sommerfeld, whom he knew to be a talented pianist; Sommerfeld reported that they had played Bach together.

As a young man in Göttingen, Sommerfeld had distinguished himself almost as much for his musical talents as his science. These came into full play in the Carnival season, which was marked in Göttingen, as in most of Catholic Germany, with prodigious and at times manic festivities. Music-making was the most important part of the parties that Sommerfeld attended. His performance of Chopin, Mozart, Beethoven, and Wagner arrangements as well as his accompaniment of *lieder* had excited widespread admiration. Sommerfeld's passion for music almost matched that of his host, Einstein, surely a factor in their becoming close friends.

The Sommerfeld visit catalysed the completion of two papers jointly authored by Einstein and Hopf. Their aim was to buttress Einstein's claim that radiation was quantised during its transmission, not just when it was absorbed. Their papers sought to refute suggestions that Rayleigh's law, describing the energy distribution of the radiation emitted from a black body, could be evaded in a wave theory of light. It was the violent disagreement of this law with experiment that had led Planck to postulate the quantum nature of the absorption of radiation. The papers were submitted to *Annalen der Physik* in August 1910. Although these publications reinforced Einstein's belief that a continuous-wave theory was untenable, they caused more controversy. Some of this came several years later from none other than Einstein's friend but often critic, Max Laue.

Shortly after Sommerfeld's departure, Einstein received a letter from the Austro–Hungarian Ministry of Education, enquiring if he would be willing to accept an appointment to a full professorship in Prague. This had been in the air for some months, Einstein having been first choice of the faculty for the position. However, the Ministry had other ideas and preferred to offer the chair to Gustav Jaumann, one of their own citizens and the faculty's second choice, who also had the advantage (from the Ministry's perspective) of not being Jewish. Jaumann however, apparently upset that Einstein had been the faculty's first preference, eventually declined so that the choice had reverted to Einstein.

The decision as to whether to accept the Prague appointment was a difficult one for Einstein. He, but particularly Mileva, enjoyed living in Zurich and they had just had an addition to the family. Einstein's students, having heard of the initial choice of the Prague faculty, got together under Tanner's direction to draft a letter to the university leaders calling on them to ensure that Einstein did not leave. The petition makes it apparent that Einstein's lecturing had greatly improved since his first faltering attempts, which had called his original appointment into question. The students particularly appreciated his informal style, friendliness, and responsiveness to their questions. The university authorities reacted to this letter with great rapidity, trying to ensure that

Einstein remained by increasing his annual salary from 4,500 to 5,500 francs and committing to reduce his teaching load to the norm for his position. However, all of these considerations had to be balanced by both financial and prestige questions. The salary associated with the Prague chair would be almost twice that of his just-enhanced salary. In addition, he would be a full professor, in charge of an institute, and have reduced teaching commitments. Another factor that probably influenced Einstein was his affair with Marie Winteler. The bleakness of his life after the loss of Marie, as well as the complications of his possible relationship with Anna Meyer-Schmid, may have made leaving Switzerland seem more attractive.

Einstein's next paper, which he completed in October, demonstrated his ability to multitask. Putting aside his struggles with a quantum theory of radiation, he became excited about a paper by eminent Polish theorist Marian Smoluchowski on opalescence, a phenomenon by which light scattered from objects in a gas or solid leads to interesting optical effects. The term derives from the opal gemstone, which exhibits the milky appearance typical of the effect. Einstein's interest in the problem followed directly from his work on Brownian motion. He analysed the problem near to the critical point at which a liquid becomes a gas, or vice versa, using his expertise in statistical mechanics and thermodynamics. He completed a full theory of the opalescence caused by fluctuations in the density of the scattering particles that Smoluchowski had only sketched in outline. The scattering of light, first discussed by Lord Rayleigh, from the fluctuations characterised in these papers is the reason why the sky appears blue.

A distraction from Einstein's own job concerns was the situation of his old collaborator Laub, whose relationship with his employer, Philipp Lenard, had deteriorated markedly. Einstein was trying to extricate Laub from a difficult situation by actively attempting to find him a position elsewhere. Einstein made some enquiries with an acquaintance in Chile; these eventually bore fruit, in that Laub left Lenard's employment a few months later. He moved to South America and spent most of his subsequent career in Argentina.

A very welcome communication arrived early in November 1910. This was a letter from Emil Fischer, Nobel Laureate and Professor of Chemistry in the University of Berlin, offering Einstein support for his research from an anonymous donor to the tune of 15,000 marks over three years. Although the money was intended to defray research expenses, such as book purchases, there were no formal restrictions on how it could be spent. The involvement of Planck, and in particular Nernst, whose work on specific heats using Einstein's theory was mentioned explicitly in the letter, can be seen in this offer. Since the annual sum offered corresponded to roughly his salary in Zurich, it would have allowed him to remain in Zurich at roughly the income he would receive were he to move to Prague. However, the money disbursed by Fischer was independent of Einstein's position, so would also move with him were he to decide to go to Prague. At any event, this handsome offer must have reduced the importance of financial considerations in Einstein's decision on the Prague chair. Einstein hastened to reply to Fischer and accepted the offer gratefully.

Einstein must have decided to go to Prague sometime after mid-November, since he reiterated his intention to stay in Zurich to the faculty, in response to the proffered salary increase, at around that time. He may have said this in some frustration at the lack of any news from Vienna, capital of the Habsburg empire, of which Prague was one of the most important cities; another reason could have been a belief that anti-Semitism might still frustrate the appointment. It does seem as if religion was indeed causing a problem. The aged Emperor Franz Joseph was of the view that no one who was entirely irreligious should be placed in a position of teaching the young. His consent was needed to confirm Einstein's appointment, but Einstein's last statement about his religion, when he received Swiss citizenship, stated that he had none. Apprised by his Prague colleagues of the problem, Einstein completed a questionnaire in which he stated his religion as Mosaisch, the

Austrian term for Judaism. After this there was a complete silence, although Einstein was surprised to read news of his move in the first December issue of *Physikalische Zeitschrift*. He remarked in a letter to Stern on 6 December that he had heard nothing formally about the appointment. At last, Carl Graf von Stürgh, Minister of Education,[1] wrote to Einstein on 15 December, confirming that his appointment was being sent to the emperor for his signature. He would take up the chair on 1 April 1911.

The turmoil of Einstein's emotional life and the confusion relating to his future employment that December made the solace of music even more necessary than usual. In the same letter in which Einstein informed Stern about the uncertainty surrounding his Prague move, he thanked him for sending a ticket to a concert at the Tonhalle. The programme included a performance by violinist Carl Flesch, who delighted Einstein with his playing of Joseph Joachim's Violin Concerto in the Hungarian Style, Op. 11. He also played an Aria by Tenaglia, an Allegro by Nardini, and an arrangement by himself of a study in octaves by Paganini. The pianist in these pieces was Fritz Niggli, brother of Einstein's old friend and musical partner, Julia; Fritz was a regular 'four-hands at piano' partner of Einstein's friend Adolf Hurwitz. The remaining works were Mozart's 'Haffner' Symphony, K385 and Brahms 'St Anthony' Variations, Op. 56. Einstein described Flesch to Stern as 'the magician, extracted wonderful things from his violin. He has a fascinating sense of rhythm and also wonderful tone production even in the most difficult passages.' The critic of the *Neue Zürcher Zeitung*, while also praising Flesch's playing of the Joachim Concerto, was somewhat less complimentary about the violin and piano pieces, hinting at some intonation problems, particularly in the Paganini piece.

The New Year celebrations of 1911 brought also, finally, the formal letter confirming Einstein's appointment to Prague, his duties, and his salary, including a generous sum of more than one quarter of his annual remuneration to cover moving and other expenses. It is clear that Mileva did her best to try to dissuade her husband from taking up the appointment. She loved Zurich, where she had many friends and perhaps as importantly, happy memories of her earlier life with Albert. She contemplated a move to Prague with dread. However, within a couple of days Einstein had made up his mind to accept the position, as he wrote to his friend Lucien Chavan on 17 January 1911, inviting him and his wife, who had just lost her father, to visit them before they left in March. The letter is very upbeat; Einstein remarked on the increasing acceptance of his theories, his successful current work and the enjoyment that he derived from his teaching duties. The evening before he had delivered a pellucidly clear account of relativity aimed at the layman to the Zurich Scientific Society. There was a long and vigorous discussion after the lecture. After Einstein had fielded all their questions, the members of the society expressed their regret at the imminent departure of their esteemed colleague to Prague.

Einstein had been invited to lecture at Leiden on Friday 10 February and wrote to Lorentz telling him how much he was looking forward to meeting him for the first time. Perhaps to make up for disappointing Mileva by the Prague move, he took her with him on the trip. Mileva's mother had arrived in Zurich and was entrusted with looking after the two children; their neighbour, Fritz Adler, was asked to offer support. The Einsteins stayed with Lorentz and his wife over the weekend before going on to visit Einstein's uncle Caesar Koch in Antwerp. By 15 February, they were back in Zurich. It is clear from Einstein's letter to Lorentz, thanking him for his hospitality, that the visit had been a great success, not only socially but also scientifically.

[1] In a bizarre coincidence, as Prime Minister, Stürgh would be shot and killed in 1916 by Einstein's friend, Friedrich Adler.

The remaining time in Zurich was hectic. Einstein not only had to wind up his own affairs but also to participate actively in the commission to select his own successor. In the past, he had tried to convince Adler that he ought to take the position, but by now Adler had followed his original inclination and, against the advice of his politician father, had moved from physics to politics. He first became a political journalism in Zurich and then, not long after Einstein's departure, returned to his native Vienna as one of four secretaries of the Austrian Socialist Party. Einstein's successor, although he remained for only a year, was also a subsequent Nobel Laureate, Peter Debye, then aged 27 and a protégé of Sommerfeld.

After staying with Conrad Habicht overnight at his parent's house in Schaffhausen on Saturday 1 April, the Einstein family continued on to Munich, where they arrived the following day. Einstein spent a delightful evening with Sommerfeld and Debye, meeting the latter for the first time and being very impressed. Indeed, the evening seems to have been unusually convivial. In general, Einstein did not drink alcohol, but this was not an invariable rule. An amusing postcard to Conrad Habicht exists on which in an extremely unsteady hand Einstein has written 'Sadly completely drunk both under the table. Your poor Steissbein and wife.' 'Steissbein'[2] was the nickname by which Einstein was known as chair of the Olympia Academy. No doubt Einstein imagined that Mileva was as drunk as he was. However, this seems unlikely as she has very primly written in tiny and very precise handwriting along the edge of the card 'A friendly greeting from your M. Einstein'. The following morning, the family left Germany and continued on to Prague.

[2] Literally 'coccyx'.

Chapter 10

Prague (1911–1912)

The Einstein family stayed initially in the Victoria Hotel in Prague. Although Einstein's contract began on 1 April 1911, he did not officially take up his duties as Director of the Institute for Theoretical Physics until 12 April, in good time for the start of the semester on 20 April. It is likely that his colleagues in Prague had helped him to find suitable accommodation, since already by 5 April he reported to his friend Lucien Chavan that they had moved into an apartment in the Smichov district of the city. The Einsteins' apartment had three bedrooms and was in a grandiose six-storey building that had been constructed the previous year. Their income was sufficiently high that for the first time they could afford to employ a maid. That and the presence of her mother must have given Mileva some breathing space to find her feet in her new environment. Einstein thought their apartment was beautiful, although two days later he wrote to Heinrich Zangger that it contained only their trunk and a push-chair for Eduard. Einstein's attempts to buy requisites to turn the empty apartment into a home immediately acquainted him with the unhealthy political situation in the city. When he was about to buy things from a shop run by Czech speakers, his predecessor wrote to him urgently recommending that he use a 'German' shop instead.

Prague in 1911 was indeed a place of contrasts, which Einstein was quick to notice. In his letter to Zangger, he remarked on the beauty of the representative buildings compared to the squalor and dirt of the average living quarters. The tension between the German-speaking and Czech-speaking inhabitants had also become immediately obvious. About one third of the population of Bohemia was ethnically German, the remainder being Slav; the struggle between these two groups had been going on for centuries. It had been brought to a head by the Austrian defeat by Prussia in the war of 1866, which led to the end of absolutist Habsburg rule and the foundation of the Dual Monarchy. Within this framework, a form of representative democracy was established, and Hungary attained a substantial degree of independence. In Bohemia, however, a constitution was promulgated from Vienna that gave very substantial electoral and other advantages to the German-speaking population, causing considerable unrest. While concessions were made to the Czech population in 1879, they failed to reduce the tension generated both by the subordination of the local Bohemian Diet to the parliament in Vienna and in particular by the status of the Czech language. One of the concessions, sanctioned by the emperor in 1882, was the splitting of the venerable Charles University in Prague into separate Czech- and German-language universities.

Anton Lampa had met Einstein at the Salzburg conference in 1909 and been highly impressed. Lampa, after studying and teaching at the University of Vienna, where he became a disciple of Ernst Mach, had been promoted to one of two chairs of physics in the German university at Prague in the year before Einstein arrived. He determined to bring him to Prague when his colleague in the Chair of Mathematical Physics retired in 1910. That he succeeded in doing so was a testament to his persistence, since there were many difficulties both inside the university and with the Austro–Hungarian government.

Lampa typified the linguistic tensions between the German and Czech populations, which were also reflected in the two universities, whose faculties were of similar size even though the Czech

Einstein. Brian Foster, Oxford University Press. © Brian Foster (2026). DOI: 10.1093/oso/9780198794875.003.0010

university had more than twice as many students. Lampa had been born in Budapest, the son of a railway engineer who moved to Bohemia. Although he had attended German-speaking schools, he was bilingual in German and Czech. Nevertheless, according to Frank,[1] he would return a purchased postcard if it had both German and Czech rubrics and ask for one solely in German, insisting that it was the postmaster's duty to keep a stock of cards with only German rubrics.

Einstein, a Swiss citizen (he was careful not to renounce this when he was required to become an Austro–Hungarian citizen in order to take up his chair) and a Jew with a Slavic wife, did not fit easily into this highly polarised society. Frank described his appearance as more that of an Italian violin virtuoso than a professor. As remarked earlier, his achievements in physics had given him a self-confidence lacking in his younger days. He no longer doubted the merit of his radical ideas. His abstraction from everyday matters in favour of physics and his general good humour meant that he treated everyone, from high functionaries to the janitor, with the same benign informality. At a reception in a fashionable hotel to welcome him to the university, Einstein arrived in his customary dishevelled attire. He was mistaken by the concierge for an electrician who had been called to repair a fault.

It was customary for new members of the Faculty of Philosophy to call on all their colleagues, some forty in total, at home. Einstein began these visits and charmed those he met, taking the opportunity to visit interesting buildings in their localities. After a while, however, he grew bored with the same staid conversations and terminated the visits, so that those who came towards the end of the faculty list or who lived in uninteresting parts of the city awaited his visit in vain. They naturally took offence and so, unintentionally, the newcomer alienated several of his colleagues.

The furnishings of the Prague flat included a piano on which Hans Albert, who began formal piano lessons in October, could practice. Having enough space, Einstein began to extend invitations to his friends to visit. In the first letter that he wrote from Prague to his friend Michele Besso, he pleaded for him to come and occupy the guest bedroom, telling him that Prague's beauty alone amply justified a long journey. Indeed, Einstein tried to convince him to move to Prague. Besso had left the Patent Office in 1909 and at this time was working for various companies near his parental home of Trieste, none of which positions he found particularly fulfilling. He might therefore have been tempted to join Einstein, who opined, tellingly about his own emotional state, that both of them would be less lonely. Fortunately, given Einstein's subsequent peregrinations, Besso resisted the temptation to move to Prague.[2]

In contrast to his dissatisfaction with aspects of his living conditions, including bed bugs, water that was hazardous unless boiled, the uncongenial class structure, and general dirtiness, Einstein was very satisfied with his new Institute. His teaching duties were considerably reduced compared to Zurich and although he thought that his Prague students were less capable and enthusiastic, he was happy to find one or two whom he considered competent. He had a well-stocked library, essential for a theoretical physicist of his encyclopaedic interests. His office on the third floor was both enormous and well-lit by four large windows, which looked out onto a large park surrounded by mature trees. From these windows, Einstein observed people taking constitutionals. He quickly noted that in the morning, only women were to be seen, while in the afternoon, only men. He then discovered that the grounds were attached to the Bohemian asylum for the insane and that the strollers were those inmates sufficiently well to be trusted outside the building. On Philipp

[1] There are several errors in Frank's account. In particular, Lampa was not a student at the German-speaking University of Prague but studied at the University of Vienna. It was there as a privatdozent that he met Mach, who had returned to Vienna in 1895 after 28 years in Prague.

[2] Besso was to return to Switzerland in 1915 and to the Patent Office in Bern in 1920.

Frank's first visit to Einstein, he pointed out the inmates to his visitor, remarking that the walkers represented that fraction of the insane yet to start working on quantum theory.

The arrival of the new professor had been widely heralded in the Prague German-language newspapers. For the first of many times in his life, he experienced the trappings of fame. His first public lecture was held in the context of the monthly meeting of the Federation of German Scientists and Physicians in Bohemia. He discussed the theory of relativity in the auditorium of his office building at 7pm on 24 May. According to a witness 'The entire Prague intelligentsia streamed into the largest lecture theatre in the institute . . . Einstein spoke clearly and animatedly . . . and at times with refreshing humour. Some of the audience were astonished that relativity theory was so simple.'

When Einstein arrived, the cultural and musical life of Prague was about to be reinvigorated. Always in the shadow of Vienna and often dependent on its artists, both its native sons Dvořák and Smetana were dead, while Janáček was at a nadir of his career and producing very few compositions. In any case, such 'native' composers were deprecated in the German-speaking cultural venues that Einstein was mostly constrained to attend. There were no less than three opera houses: the Estates theatre, where Mozart had conducted the first performances of *Don Giovanni* and *La Clemenza di Tito*; the Czech National Theatre; and its German-speaking counterpart, the Landestheater, of which Mahler was a previous musical director. Chamber music and orchestral venues included the Rudolfinum and the mammoth 3,500-seater hall in the Lucerna.

Verdi's *Rigoletto* was playing at the Landestheater when Einstein arrived, but musical Prague was waiting with considerable anticipation for the arrival of its new musical director, Alexander Zemlinsky. A distinguished composer who had turned to conducting in order to survive financially, his first concert, a performance of Beethoven's *Fidelio*, in September 1911, was a great success. He assisted his brother-in-law, Arnold Schoenberg, by organising guest conductor opportunities in Prague. One such appearance by Schoenberg is interesting in that his *Pelleas und Melisande* was inserted into a concert by Pablo Casals, whose star billing guaranteed a sell-out despite the novelty of Schoenberg's music. This concert in February 1912 is also notable in that Casals had hoped to play the Dvořák concerto but, because of the tensions between the two communities, instead played concertos by Saint-Saëns and Haydn. Zemlinsky also briefly engaged Schoenberg's student, Anton Webern, as an assistant conductor in 1911. During Einstein's stay in Prague, Zemlinsky presented Weber's *Der Freischütz*, Wagner's Ring cycle, *Tannhäuser*, and more congenial to Einstein, Mozart's *Don Giovanni* and *Die Zauberflöte*. In addition, Prague had venues in which light opera and operettas were presented, while cinemas were becoming increasingly popular.

Whilst the cultural events were enjoyable, physics was always paramount. Einstein's research direction was moving away from trying to construct a self-consistent theory of radiation based on the quantum, with which he had wrestled, fruitlessly, for so long. It is worthwhile to consider why Einstein had so little success in his attempts to derive a new quantum theory of radiation. As Stone points out, he was basically unlucky. He naturally assumed, as is the case in other physical laws, that the constant denoting the strength of a process would appear in the basic equations characterising it. For example, Newton's law of gravitation contains the gravitational constant that defines the strength of gravity. However, a modification of Maxwell's equations that explicitly includes Planck's constant, analogous to the gravitational constant in Newton's law, is not fruitful for two reasons. Firstly, Maxwell's equations successfully treat the average behaviour of light, which can indeed be described as a wave, averaged over many billions of quanta known as photons and insensitive to the individual properties of each. Secondly, because the photon has no mass and travels at the speed of light, its energy and momentum are directly proportional; although both are quantised in units of Planck's constant, its value does not enter the theory because it appears on both sides of the equation linking energy and momentum and therefore cancels. Faced with this unfortunate conspiracy that defeated his usually successful methods of developing his theories, Einstein finally admitted defeat. In his letter to Besso mentioned above, he wrote that he had given up wondering whether quanta really existed. Instead he was working his way through a variety of phenomena using the quantum idea. He was encouraged by the success of his theory of specific heats of materials which had been in most respects confirmed experimentally by Nernst and colleagues. In May he sent a paper to *Annalen der Physik* in which he built upon a paper published six months earlier. In these, he modified his previous theory of specific heats by taking into account, on a statistical basis, the thermally induced molecular motions in a solid.

Einstein's interest in certain aspects of relativity, dormant since his work with Laub, had been revived by Paul Ehrenfest, who was destined to play an important part in Einstein's life. Ehrenfest was particularly interested in the properties of rigid bodies.[3] In 1909, while working in St Petersburg with his Russian wife, who was also a physicist, he had published a paper pointing out what came to be known as the 'Ehrenfest paradox'. The problem concerned the Lorentz contraction in special relativity. According to this, the circumference of a rigid circular disk rotating about its central point contracts, and since the radius and circumference are proportional to each other, the radius must also contract. However, since the radius is perpendicular to the direction of motion, it ought to be unaffected by the Lorentz contraction. Since a rigid body by definition cannot deform, the contradiction is clear.[4] Through this work, Einstein renewed his acquaintance with Ehrenfest, whom he had first encountered in 1907. After correspondence with Ehrenfest, Einstein published a short refutation of a paper by Vladimir Varićak, who had proposed that Lorentz contraction was a psychological, rather than a physical, phenomenon.

Einstein made various statements in later life that 'he puzzled incessantly' about gravity in the period 1908–1911. However, is seems likely that Einstein's silence on this subject was because he had almost no time to think about it. He was instead wrestling with quantum phenomena. This changed in the early summer of 1911, in part catalysed by Ehrenfest and the interesting problems relating to rotating bodies, and thereby accelerating frames, in relativity. He had been all too aware in 1907 that the concluding section of his article on relativity in the *Yearbook of Radioactivity and Electronics*, in which he had discussed accelerating systems, left much to be desired. It was founded on an approximation very limited in applicability and although he was not yet able to go beyond that approximation, he had made enough progress by June to produce a paper.

Einstein began his new paper with two remarks. First, he was dissatisfied with the treatment of the 1907 paper. Secondly, he had also realised that the bending of light by a gravitational field, which he had dismissed as unmeasurable in 1907, could in fact be measured. The technique required the observation of stars whose position in the sky meant that their light passed near to the limb of the sun. In the new paper, Einstein restated the Equivalence Principle, pointing out that a homogenous gravitational field and a uniformly accelerated system are completely equivalent; just as for velocity in the original relativity theory, there is no preferred 'absolute' reference frame for acceleration. Such an assumption immediately explains the experimental

[3] 'Rigid bodies' in physics and mathematics are considered to be macroscopic objects which do not deform under external forces. A solid disk of steel rotating as discussed by Ehrenfest would typically be considered a rigid body. The paradoxes in relativity concerning such bodies are caused by the 'classical' definition being inadmissible, since it requires information, for example, stresses and strains, to move through the body instantaneously, rather than being limited by the speed of light.

[4] Einstein and Max Born had also realised the existence of this paradox in discussions at the Salzburg conference after Born's talk there.

fact that falling bodies in a vacuum fall at the same rate in a gravitational field irrespective of their material, since if, for example, gold fell more slowly than any other element, the 'gold standard' reference frame of acceleration could be preferred. He went on to show, as he had in 1907, that the original relativistic result that energy carried an inertial mass must also carry over to the mass associated with the gravitational field, or 'gravitational mass'. This is implied if the gravitational field is completely equivalent to a uniformly accelerated system. The same assumption implies that light from the sun should be shifted to lower wavelengths, analogous to the Doppler shift derived in the original relativity theory for constant velocities. So far, so very similar to the 1907 paper. However, in the penultimate section, Einstein realised that the results on the slowing down of clocks in a gravitational field that he had derived in 1907 and derived again in the current paper implied that the velocity of light also varied, in direct proportion to the gravitational field. In the final section, Einstein made a simple calculation of the size of the deflection to be expected when light from a distant star passed near to the sun and was detected on earth and made a plea to astronomers to attempt to measure it.[5]

A letter arrived at the end of August from Erwin Freundlich, an astronomer in Berlin, who had been alerted to Einstein's paper by Leo Pollack, a privatdozent at the Institute of Geophysics in Prague. Freundlich thought that he would be able to measure the effect that Einstein had predicted. In reply Einstein, who had consulted an astronomer at Prague, recommended some photographic plates taken in Hamburg as possibly useful. Thus began a correspondence that continued for several years and had significant consequences for Freundlich's career.

Einstein's renewed work on gravity was interrupted just as he was completing his paper. He was surprised to receive a letter from Brussels from a certain Ernest Solvay. Solvay was a rich industrialist who was fascinated by physics, particular the problem of gravity, which he had been working on for more than 30 years. He and Einstein may even have been acquainted, since both had been present at the 350th anniversary celebrations of the University of Geneva: Solvay had also received an honorary doctorate there for his contributions to industrial chemistry. His letter to Einstein contained an invitation to what he called an international scientific congress for the elucidation of some current questions concerning the molecular and kinetic theories.

The invitation from Solvay arose from a chance encounter with Walther Nernst in Brussels in the spring of 1910. Solvay mentioned that he would like to expose his theories to the leading physicists of the day, mentioning Planck, Lorentz, Poincaré, and Einstein by name. Nernst, who had highly sensitive antennae for the possibility of research funding, at once saw an opportunity and proposed to gratify Solvay's wish by organising an international meeting of the leading European physicists. Roughly a year later, Nernst's labours bore fruit in the invitation that had landed on Einstein's desk. In all, twenty physicists accepted invitations to the conference. Einstein was invited to present

[5] The bending of light by a massive object had been calculated by the Englishman John Michell and the Frenchman Pierre-Simon Laplace in 1784 and 1796, respectively. They had postulated the existence of what we today call black holes. Even more directly relevant, the bending of light from a distant star by the sun had been calculated, but not published, by Henry Cavendish in 1784, after reading Michell's paper. In 1805, the Berlin astronomer Johann Georg von Soldner published his result. He obtained the same answer as Einstein in his 1911 paper, which is not too surprising since Einstein's calculation in this paper (although not in the final formulation of general relativity, where the deviation is a factor of two larger) could be equivalently expressed as that of a particle (photon) with an inertial mass proportional to its energy attracted by a much more massive object.

an overview of 'Specific Heat and the Theory of Quanta'. He was thus forced, somewhat against his inclination, to return his attention to quantum matters because of the significant effort that preparing his contribution entailed. This work continued until the conference began at the end of October 1911.

There were additional distractions. The visitors Einstein had been so keen to invite began to arrive that summer. Besso and his family stayed with the Einsteins during a sweltering August. Heinrich Zangger, his friend and colleague from Zurich, visited in early September. Zangger was already engaged in political manoeuvrings in Zurich aimed at bringing Einstein back to his *alma mater*, although it was no longer the Polytechnikum but had that year become the Eidgenössische Technische Hochschule or ETH, with enhanced prestige and the ability to grant doctoral degrees. However, Zangger found his host very ill with a chest infection. Other visitors included Mileva's sister, who provided company during her husband's long absences that autumn.

The first period away, which lasted from around 23 September to around 16 October, was caused by Einstein's attendance at the 83rd meeting of the German Association of Scientific Researchers and Doctors in Karlsruhe. He did not present a paper but took part in several discussions after other presentations, for example that of his friend Sommerfeld on the quantum and the photoelectric effect. It was at this meeting that he first met Fritz Haber, the eminent chemist, who was giving the keynote address. Haber was about to move to Berlin as director of the Kaiser Wilhelm Institute for Physical Chemistry and Electrochemistry, in which position he would play an important part in Einstein's life for the next twenty years. Einstein's assistant Ludwig Hopf, who had accompanied him in the move from Zurich to Prague, also gave a talk in Karlsruhe. The two left Karlsruhe together and then went their separate ways at the Heidelberg railway junction. Hopf took up a position in Aachen, where he married and eventually returned to an area related to his original thesis topic, turbulence, by moving into the new field of aeronautical engineering.[6] Einstein acquired another assistant, Emil Nohel, who came from a rural Jewish village in Bohemia. It does not seem that he made much impact on Einstein's studies in physics, but he interested him in stories about life in his village. Here, the Jewish inhabitants used Czech in everyday life but on the Sabbath only German, as a sort of substitute for the Hebrew that they could no longer speak. The peculiar situation of the Jews in Bohemia, ground between the Germans and the Slavs, together with the ancient remnants of Jewish life to be found in the old city, may have reawakened Einstein's interest in his Jewish heritage.

Having taken leave of Hopf, Einstein travelled to nearby Heilbronn to visit his mother. Pauline Einstein had moved there to act as housekeeper for a merchant, Emil Oppenheimer. He then travelled to Luzern and stayed with his sister Maja and her husband Paul Winteler. He arrived in Zurich in time to give a series of lectures at the ETH that he had agreed to deliver some months previously in response to an invitation from his old friend Marcel Grossman. The six lectures on thermodynamics and relativity theory were delivered from 9–14 October to a group of middle-school teachers on a vacation course. Einstein wrote a letter to Mileva, now lost, probably from Heilbronn, to which he received an affectionate reply. She wistfully remarked that she would have liked to have attended the Karlsruhe meeting and seen 'all those fine people'. She asked Einstein

[6] Hopf became an authority on aerodynamics, largely through his experience with test flights in the First World War. As a Jew, he was forced out of his chair in 1934 and suffered terribly under the Nazis until he was finally able to leave Germany in March 1939 and to settle in Ireland as a professor in Trinity College Dublin. Tragically he died a few months later from thyroid failure.

tentatively whether she should indeed come to Zurich to meet him, as Einstein's letter had presumably suggested. She signed off touchingly with her old nickname, 'D' for Doxerl; whether she actually came to Zurich is uncertain.

When Einstein finally returned home, he found some complicated matters that had been building up over the summer were awaiting his attention. He had been in discussion since August with Willem Julius, professor of physics at the University of Utrecht, about the possibility of moving there. Julius had been impressed by the lecture that Einstein had given in Leiden during his visit to Lorentz. Einstein had replied to this letter saying that he was flattered but declining the potential offer since he had only just settled in Prague. Einstein also asked him in his reply about Julius's work on the shift of spectral lines from the sun, which however was ascribed to a much more conventional cause than gravity. In mid-September, Julius contacted him again, renewing the offer of a professorship after having discussed it within the Utrecht faculty. Einstein was tempted and asked a number of detailed questions about pensions, etc. He also wrote immediately to Zangger, who was by then deeply involved in trying to align the political constellations to engineer Einstein's return to the ETH. Einstein disingenuously noted in his letter to Zangger that the salary under offer was 6,000 gulden. He wondered whether ETH could make a quick decision and suggested that these matters as well as a discussion on his work on gravity could be had at their leisure at the Karlsruhe meeting, which Zangger also attended, and later in Zurich.

Hardly had Einstein arrived in Zurich from Luzern when Zangger wrote a long letter to a friend of his, Ludwig Forrer, who happened to be one of the seven members of the ruling Swiss Federal Council. In it he answered one by one the 'misconceptions' that Robert Gnehm, the President of ETH, had adduced against Einstein's appointment. Gnehm was mostly concerned about financial and space implications at a time of considerable upheaval at ETH. Zangger asserted that Einstein neither needed an assistant nor any laboratory space, that if necessary, he (Zangger) could accommodate him, that Einstein was enthusiastic to teach, and although not an easy lecturer was inspiring, comparable to Marie Curie and Paul Langevin. He offered to bring Einstein from Zurich to Bern to visit Forrer on Saturday 14 October. Gnehm in fact attended Einstein's lecture that morning, an indication that he was open to the possibility of Einstein returning. When Einstein got back to Prague, he found a letter from Julius awaiting him, reminding him that the faculty at Utrecht were waiting for his answer. He immediately wrote back to Julius, explaining that there were moves to bring him back to Zurich and asking for a few weeks' more grace. A few days later, Zangger forwarded a telegram from Forrer in which he indicated that he was working to create a position for Einstein. Einstein replied that he was gratified but also urged Zangger to greater speed because of the Utrecht position. This was where the situation was left when Einstein boarded the train to Brussels for the Solvay Conference.

Einstein had been dismissive of the value of the conference before the event in correspondence with friends, writing to Besso that he considered it a witches' Sabbath. However, he had also written to the driving spirit, Nernst, that he was looking forward to it. He arrived in the luxurious Hotel Metropole on 29 October. Sommerfeld commented that the Metropole was '. . . really stupendously swanky. We each have a private bath and W.C. in our rooms. I bathe each morning. We are the guests of Solvay even at all lunches, dinners, etc. No lunch of fewer than five courses. Crazy!' The splendour of his surroundings found Einstein, unsurprisingly, wanting. 'Einstein of course was not wearing a tuxedo yesterday at the dinner with the Solvay family. He doesn't own one.' There was a reception on the first evening and then eighteen of the leading physicists in Europe got down to serious business on the following morning. Twelve had been asked by Nernst to prepare summaries of particular topics.

Solvay opened the meeting and summarised his theories of gravitation, which were met with a polite, presumably embarrassed, silence. Lorentz, the chair, then opened the proceedings proper with a report on aspects of black-body radiation. Einstein's contribution, towards the end of the meeting, was nominally on the quantum theory of aspects of heat transport in materials but also covered a much wider sphere of the general applicability of the quantum idea. The discussion after his talk was animated but not particularly illuminating. Einstein was disappointed in particular by Planck, who still insisted doggedly that only emission and absorption of radiation, but not its propagation, were governed by the quantum hypothesis. He was also disappointed by Poincaré, who was antipathetic to the whole quantum idea. Einstein later characterised him to Zangger as negative and, despite his undoubted technical skills, with little real understanding of contemporary physics. Einstein did however remark to Zangger that he thought that Planck, in private, might finally be coming around to accepting Einstein's ideas that radiation too was quantised. Solvay himself closed the conference by castigating the participants that, despite the beautiful results discussed, they had made no progress in isolating the basic building blocks of the universe. He furthermore told them that their discussions hadn't changed at all the basic ideas that he had expounded at the beginning of the Congress. Despite the disappointment implied by his closing remarks, Solvay was impressed enough with the conference that subsequently, with the help of Lorentz, he founded an 'International Institute of Physics' which was charged primarily with convening further meetings following the Solvay Congress model.

Einstein's verdict was that he had learned little new from the Solvay conference. Indeed, he remarked that the proceedings with regard to quantum theory reminded him of lamentations over the ruins of Jerusalem. However, there were positive aspects related to individual contacts: reaffirming his relationship with Lorentz, Planck, and Sommerfeld and in establishing a number of new and important friendships, in particular with Marie Curie and the other French physicists, certainly Langevin and Perrin, but not perhaps Poincaré. Irrespective of their personal relations, Einstein's encounter with Poincaré paid dividends in the very positive letters of recommendation for the Zurich appointment that he, as well as Marie Curie, subsequently wrote on the request of Pierre Weiss, Professor of Physics at ETH. Just as the conference was ending, the liaison between Curie and Langevin was splashed all over the Paris papers by Langevin's jealous wife, who had stolen their love letters from the apartment they had rented for their trysts. Curie, as a widow, was free to enter a new relationship. In a misogynism typical of the time, this did not stop the Parisian press from attacking her virulently as a foreigner, which was at least accurate, and as a Jew, which was not.

On 7 November 1911, the unprecedented award of a second Nobel Prize to Madame Curie, this time in Chemistry, was announced. The publication of extracts from the stolen love letters in *l'Oeuvre* shortly afterwards caused a sensation that reached the ears of the Nobel Committee. The furore was not lessened when Langevin challenged the Editor of *l'Oeuvre* to a duel. Fortunately, when the protagonists met, the Editor, citing the good of French science, refused to discharge his pistol. Svante Arrhenius of the Nobel Committee wrote to Madame Curie suggesting that she might not wish to attend the Nobel Ceremony the following month. With great dignity Curie replied that she did not accept any connection between her science and her private life. She left for Stockholm and received her second Nobel Medal from the King of Sweden on 10 December without incident.

Unsurprisingly, Einstein took a very relaxed view on matters such as the Curie–Langevin affair. On his return to Prague he informed Zangger of his pleasure at meeting Curie, his impressions of her as an admirable character, and that she had promised to visit Prague with her daughters. On 23 November, in response to further calumnies in the French press, he wrote to her expressing his admiration for her spirit and strength of mind. This letter must have given her support when shortly afterwards she received Arrhenius's letter referred to above. These exchanges with Einstein established a friendship that was to endure until Curie's death in 1934.

Einstein's return trip included a visit to his Uncle Caesar in Antwerp and to Utrecht to discuss their offer of an appointment. However, a return to Zurich was now becoming more likely. On 15 November, Einstein finally wrote to Julius to decline the Utrecht offer. Some of his reasons for declining the position, given his letter on the very same day attempting to prod Zangger into further action on the ETH chair, were disingenuous to say the least. The next postal delivery brought a letter from his friend Grossmann, semi-officially offering him the chair at ETH. The following day, he wrote again to Julius, telling him about the offer from Zurich and that he was likely to accept it. Einstein asked him not to announce any offer to Debye, which Einstein had recommended, until Einstein had had time to close with ETH. Julius's reaction on receiving this letter, which Einstein had the grace to admit was 'strange', is not recorded. A few days later, Julius replied very politely, agreeing to Einstein's request. On 20 November, Einstein telegraphed and subsequently wrote to Zangger indicating his acceptance in principle of the position, subject to negotiations with Ghenm. The administrative wheels turned exceedingly quickly, no doubt lubricated by Zangger, who seems to have been a consummate political operator despite being under considerable personal and emotional strain at the time. Not surprisingly, he was an object of considerable irritation to many of the ETH professors, who couldn't understand why he, a professor at the university, was interesting himself so energetically in their affairs.

A letter from Gnehm to initiate negotiations was sent to Einstein on 8 December. The complications were not yet over, however; on the same day as Gnehm wrote, Lorentz enquired whether Einstein was really committed to Zurich. By then he had decided to give up his own chair and was hoping that Einstein would be his successor. Since he assumed that Lorentz was still trying to induce him to go to Utrecht and having received Gnehm's letter, Einstein replied that it would be highly embarrassing if he were not to accept the Zurich offer. Had he realised that Lorentz was about to offer him his own chair in Leiden, then his response might well have been different. However, by the time Lorentz wrote to him in February, explaining the situation and asking if he were interested in taking his chair, it was too late. Einstein travelled to Zurich, staying with Zangger, in order to meet Gnehm on the morning of 21 December. He expressed his satisfaction with the discussions to Zangger and began announcing to his friends that, subject only to approval by the ETH and Federal Councils, he would be returning to Zurich in the autumn of 1912. Paul Habicht, who had recently had a great success in demonstrating the Maschinchen in Berlin, wrote that if Einstein returned to Zurich then there would be again be a mensch (roughly translated as an honourable man) in Europe – an interesting indication of how the geographical location of Prague was perceived in Switzerland!

Einstein had in the autumn invited both the Habicht brothers to spend the Christmas holidays in Prague, but they couldn't make the trip. Instead, Einstein invited his old classmate at the Polytechnikum, Jacob Ehrat from Winterthur. On his return from his appointment with Gnehm, Einstein and Ehrat took the night train together to Prague on 22 December, where they spent the Christmas vacation. That Einstein could still be thoughtful and affectionate towards his wife is shown in a copy of the novellas of C. F. Meyer that he gave to Mileva inscribed 'For Christmas from Santa on the recommendation of your Babu'.[7] They probably returned to Switzerland with Ehrat, as Mileva wrote that they spent the New Year in St Moritz, surrounded by unearthly beauty. Just after

[7] 'Babu' was the pet name Mileva was using for Einstein at that time.

their return to Prague, Einstein's mother arrived for a visit. Einstein seemed to enjoy showing the sights of the city to both Ehrat and his mother and wrote that thereby he really got to know the architectural glories of Prague.

As Pauline Einstein's visit drew to a close at the end of January, Ehrenfest wrote to Einstein asking if he might visit him. Without a permanent position, Ehrenfest was travelling around Europe to see if he could find an opening. On 23 February he arrived at Prague station, where he was met by Einstein and Mileva. The three talked inconsequentially at a nearby coffee shop but as soon as Mileva left for home, they immediately began an intense discussion on statistical mechanics, which held a deep fascination for them both. They walked to Einstein's office, where the conversation had to be broken off when Einstein rushed off for his regular appointment to play string quartets. Ehrenfest was left in the company of Lampa, who had taught him and his contemporaries, Philipp Frank and Lise Meitner, at Vienna University. After dinner, Ehrenfest was escorted to the Einsteins' apartment. The Einsteins returned at around midnight, after which Ehrenfest and Einstein continued talking until 1:30 a.m.

The subsequent days of intense discussion forged a bond between the two that was cemented by their mutual love of music. Ehrenfest was a good pianist and by the third day of the visit he and Einstein were playing Brahms sonatas together. Ehrenfest also forged a strong relationship with Hans Albert, sitting next to him at meals and taking him on excursions. Hans Albert loved to listen to his father and their guest making music together. He beat time when they played Händel sonatas and sang when the adults played Brahms songs. As Einstein and Ehrenfest walked by the Moldau, they discussed Einstein's latest ideas on the problem of gravity, a discussion they carried on after dinner. Ehrenfest gave an excellent seminar on radiation theory at the university.

Ehrenfest's stay in Prague was extended since he had discovered that his passport had expired and considerable bureaucratic delay was incurred in getting it renewed. By the end of the visit, Einstein had decided that his new friend would be an ideal replacement for him in Prague. Finally, the time came for Ehrenfest to continue his peregrinations and visit Smoluchowski in Lemberg (modern Lviv in Ukraine). After a last session playing Brahms together, he and Einstein walked to the station, chatting about mostly music and the current state of Russia. Mileva met them at the station and reported that Ehrenfest's passport was still in transit; it only caught up with him later in his journey. Ehrenfest's extended visit had been a great success. He himself was enormously excited by the warmth of the relationship he had established with Einstein, as can be seen from the doubly underlined 'What will the future bring' that he wrote as the final entry in his diary that day.

Ehrenfest was an extremely gregarious and intense personality. His diaries and work notebooks reveal an organised, meticulous man, who numbered his physics thoughts, eventually reaching beyond 7000. Often the velocity of his thoughts exceeded his ability to record them, and he resorted to shorthand scribbling. He filled some of his workbooks from the back with lists of literature to read and music to explore. Friendship was very important to him. Meitner experienced this when they were students at Vienna, when she felt the necessity to keep him at a distance because of what she saw as his need to probe deeply into the feelings of all his acquaintances. With Einstein he found a soul mate, writing in his diary 'Yes, we will be friends. Was awfully happy.' Einstein in later years wrote 'Within a few hours we were true friends – as though our dreams and aspirations were meant for each other.'

Einstein had been appointed to the ETH chair by the Swiss Federal Council on 30 January 1912. Immediately he had requested to be released from his Prague appointment. With his future apparently settled, he returned to his work on gravitation. He had made considerable progress since the paper predicting the bending of light rays. The variability of the velocity of light in a gravitational field opened up the possibility of parameterising the gravitational potential via the speed of light. In the first quarter of 1912, Einstein produced two papers on gravity. In the first, submitted to *Annalen* on 24 February, he used the Equivalence Principle, that gravity is equivalent to acceleration, to derive an equation describing the relationship between the varying velocity of light and the mass density producing a static (not changing with time) gravitational field. This was a linear equation, in which the velocity of light, c, appears only as itself and not as some more complex function, such as c^2.

Even before this paper was published however, Einstein realised that it could not be correct. He wondered about withdrawing it, but in the end decided to add a footnote pointing out that the equation must be incorrect and then as soon as possible to write a supplementary paper explaining the error. This was submitted towards the end of March. It showed that the pattern of variation of the speed of light in a gravitational field must be more complex than linear. In the interim, Einstein had realised that the gravitational field itself contained energy. Since special relativity shows that energy and mass are equivalent, this energy must itself generate a gravitational field that feeds back on the original, so that the problem must be non-linear. This had the traumatic effect for Einstein of forcing him to abandon at least partially the basis of his approach to date. The Equivalence Principle could only be true locally, that is, for infinitesimally small distance scales. Attempting to solve the problem of gravitation was pushing Einstein further and further away from his comfort zone.

The note added in proof at the end of the second paper is very significant. In this note, Einstein remarked that the equations of motion of a point-like particle in a static gravitational field could be very economically expressed in a mathematical form known as a Hamiltonian (or, equivalently, a related quantity known as the Lagrangian). The fact that in nature events will always find the path expending the least energy,[8] known as the 'principle of least action', is equivalent to minimising this Hamiltonian quantity considering all possible outcomes, that is, finding the shortest distance between two points. In the 'flat' space explored in Euclid's *Elements* (written around 300 BC and the basis for the geometry learned at school), the shortest distance between two points is a straight line. This is indeed the result produced by minimising the Hamiltonian appropriate to a Euclidean space. The space considered by Einstein, however, in which the velocity of light varies from point to point, gives an answer different to a Euclidean straight line. This expression of the problem of gravity in terms of a Hamiltonian and an associated 'line element', the equivalent to the Euclidean straight line, was to have profound consequences in Einstein's eventual construction of general relativity. The 'Ehrenfest paradox' described above had been the first indication to him that Euclidean geometry might not be appropriate for accelerated reference frames; now he began to think of gravitation in a form that involved modifying the geometry of space and time.

In addition to taking the first steps towards general relativity, Einstein was also busy with reviewing his earlier creation. Sometime in the spring of 1911, Einstein had agreed to write a contribution on relativity to a handbook of radiation physics. He almost certainly began work on it early in 1912, perhaps even catalysed by Ehrenfest's visit. It was never published since Einstein worked on it until early in 1914; before the whole volume could be gathered together, the First World War supervened. The manuscript is a fascinating snapshot of how his views on relativity were maturing and the effect of his current work on gravity. Another seminal influence was the book on relativity

[8] Imagine arriving at the bottom of a mountain you wish to climb. As you look around and scan the possible routes, your brain will subconsciously apply a version of the principle of least action to attempt to find the route that will minimise the energy required to reach your goal.

that Max Laue had published in 1911, and which Einstein had warmly praised. Einstein began the review article by taking yet another route to special relativity, interestingly distinct both from his original 1905 paper and his 1907 yearbook article. The 1912 review begins with a detailed analysis of Lorentz's electron theory, surely brought to the forefront of his mind by their recent close interactions. Having established that Maxwell's equations and Lorentz's work imply the constancy of the velocity of light, Einstein showed that this is incompatible with either the relativity principle (that the laws of physics are identical in any frames in constant relative motion) or with the Galilean–Newtonian rule for the addition of velocities. With the aid once again of Fizeau's experiment, he demonstrated that it is the rule for addition of velocities that must be sacrificed. In support, he cited Willem de Sitter's astronomical observations of stellar binaries, which showed that the velocities of the component stars vary as they rotate around their centre of gravity and demonstrated that the velocity of light is independent of that of its source. Only at this point does Einstein go on to discuss how to decide on the simultaneity of events, clocks and the Lorentz transformation. The third section of the review is interesting since it is a beautifully clear mathematical exposition of four-vector and tensor algebra, the basis of the treatment of relativity that Minkowski had pioneered and Laue refined in his book. This section was probably written after his return to Zurich and his collaboration there with Grossman. This sort of mathematics was to be fundamental to the final formulation of general relativity.

The carousel of academic positions that resulted from Einstein's and Lorentz's moves continued in the spring of 1912. Einstein was involved in sending recommendations for candidates to replace Debye, who was moving from the University of Zurich to the chair in Utrecht that Einstein had turned down. Einstein was very positive about both Laue and Ehrenfest. Ehrenfest may certainly have been handicapped by being Jewish. Debye, who was on the commission to find his own replacement, wrote a markedly anti-Semitic comment in a letter to his former supervisor, Sommerfeld, who was also thinking of employing Ehrenfest as a Privatdozent, presumably on Einstein's suggestion. Eventually the choice for the Zurich position fell on Laue. Once Lorentz accepted that Einstein could not be obtained as his replacement, he wrote to Ehrenfest to explore his interest in the post. Ehrenfest received this letter on 3 May (Julian calendar) while on a cruise on the Volga with his family. He was overjoyed. By October 1912, he and his family had moved to Leiden. In December he gave his inaugural lecture, entitled, appropriately for Lorentz's successor, 'On the Crisis of the Light–Aether Hypothesis'.

Meanwhile, around 12 April, Einstein spent a week in Berlin for discussions with several colleagues, including Planck, Nernst, Haber, and Freundlich. He had been invited to stay with Emil Warburg, the doyen of German experimental physicists and President of the Physikalisch-Technische Reichsanstalt, the Imperial Physical Laboratory. He preferred, however, to stay with his uncle, Jakob Koch, in Charlottenburg. His outline schedule shows more cultural activities than physics meetings, including visits to a concert, an art exhibition in Salon Wertheim, the theatre, and the Ethnographic Museum. He also scheduled three afternoon meetings with Warburg. No doubt they discussed physics; Einstein later told Zangger that Warburg had tried to convince him to come to Berlin. However, the frequency of the meetings was probably related to the fact that Warburg was a fine pianist.

Of all these encounters, the most important for his future life was a meeting with other relatives living in Berlin. Einstein's maternal aunt Fanny and her husband Rudolf lived in Haberlandstrasse, a couple of kilometres south of the Kochs, with whom Einstein was staying. Hermann, Einstein's father, had been heavily in his cousin Rudolf's debt after the vain attempt to keep his business afloat. Their daughter, Einstein's cousin Elsa, divorced with two young children, lived above them. Her younger sister Paula Meier, a widower with a twelve-year-old son, Hans, seems to have also lived either with her parents or nearby. Einstein, rebounding from his affair with Marie Winteler and the news of her pregnancy, was desperate for new female companionship. He first began to court Paula, who had a reputation as a happy and carefree woman. Perhaps he was rebuffed; at any rate he immediately transferred his attentions to Elsa, who seemed to welcome them. They went on a trip to the Wannsee, a famous Berlin beauty spot. Einstein poured out many of his concerns and troubles to his receptive cousin. In particular, he talked about the state of his marriage and his relationship with his mother and sister. Almost immediately after his return to Prague, Elsa wrote with details of how they could correspond without her family's knowledge.

Einstein's reply to Elsa was quick and passionate. He told her that he had to have someone to love and, whether she liked it or not, she was that person. He concluded the letter with kisses, sexual innuendo, and underlining for emphasis the intimate '*Du*' form of address. He gave her an instruction to write to him, if she wanted to keep in contact, at the Physics Institute, Einstein's by now time-honoured method of keeping things secret from Mileva. He made a commitment to destroy her letters; fortunately for posterity, she kept his as treasured possessions. Elsa's reply, which Einstein destroyed as promised, must have voiced second thoughts, as he wrote again at the beginning of May that he was sad at her desire to break off their relationship, reiterating that he loved her. The two clearly suffered from indecision. In a further letter at the end of May replying to one from Elsa, who seems to have wanted after all to carry on the relationship, Einstein decided that it would be better to end it. Nevertheless, he was careful to promise to send her his work address in Zurich so that she could write and he ended by sending kisses; not perhaps a very convincing way to break off a relationship.

The start of the spring semester saw the arrival of Otto Stern in Prague. The future Nobel Laureate came from circumstances very similar to Einstein's previous collaborator, Ludwig Hopf. Both were Jewish and independently wealthy from their families' involvement in the grain business. Stern had been born in Upper Silesia and subsequently moved to Breslau. Having attended Sommerfeld's lectures in Munich, he returned to Breslau where he completed a doctorate in physical chemistry on the properties of solutions. Like Hopf, he could afford to come to Prague and work with Einstein without being appointed to a formal position. Stern's interests lay mostly in thermodynamics and not particularly with Einstein's current preoccupation with gravity and relativity, but since Einstein had so few people with whom he could talk physics, he was happy to discuss thermodynamics. Stern left a vivid picture of his first encounter with Einstein at the Prague Institute. He couldn't find anyone who looked as if he might be a professor but eventually came across a man without jacket or tie working at a desk. He was wearing what seemed to be a labourer's shirt with a large rip in it. This was Einstein; Stern found him very congenial.

As the spring turned into summer, Einstein continued to worry away at the gravitation problem that was dominating his work. Although he had found a solution to the gravitational field of stationary objects, he made little progress with the dynamic solution. He was clear that the equation that governed gravity must be nonlinear, that is, feed back upon itself, since the energy within the gravitational field must itself generate gravity. He realised that the concept of the velocity of light varying in a gravitational field was necessary but of insufficient richness to describe gravity; rather, the velocity of light must depend on both the position and direction in the gravitational field. This dependence on a direction, known mathematically as a 'vector', excluded his original 'scalar' field, which only depends on a number. His thoughts strayed back to his friend Ehrenfest's paradox of the rotating rigid disk, whose Lorentz contraction in the direction of motion meant that its circumference was no longer the 2π times its radius that Euclid demanded. His conviction grew that this obvious departure from Euclidean geometry for accelerated, and therefore by his Equivalence Principle also gravitational, phenomena required the invention of a new geometry of space and time.

As these thoughts were crystallising, Einstein was asked by his friend Zangger to contribute an article to a celebratory volume to mark the opening of the Institute for Forensic Medicine that Zangger had been trying for years to establish at the University of Zurich. Zangger's wish was that the volume should illustrate the breadth of the subjects that impinged on forensic science. He might have expected Einstein to contribute some area of physics related to medicine, such as the behaviour of solutions or colloidal suspensions in liquids, the area that had first brought them together. However, Einstein's concentration was completely on gravitation. It was only his deep gratitude to Zangger for his help in arranging the Zurich appointment, which he expressed on many occasions, that could have induced him to interrupt his labours to write an article. What he wrote is likely to have been simply a snapshot of his current preoccupations; it is incontrovertibly the most important physics result ever published in a journal devoted to pathology and government medicine!

The article questions whether a gravitational analogue of electromagnetic induction exists, in which an electric current induces a magnetic field, and vice versa. This was the next step that Einstein had set himself after 'solving' the problem of the static gravitational field, which was analogous to the electrostatic problem. He now set up a thought experiment in which a massive spherical shell surrounds a point mass at its centre. Classically, as Newton showed, there is no gravitational force exerted on such a particle anywhere inside the shell. However, if the external shell is accelerated, Einstein proved, using the gravitational equation derived in his recent papers, that the point mass experienced an 'induced' force. Qualitatively, this is expected since the Equivalence Principle means that the acceleration of the shell is equivalent to an additional gravitational field in the direction of the acceleration. By far the most important comment in this brief contribution to Zangger's celebratory volume was that this situation was relevant to earlier work by Mach. He had proposed replacing Newton's postulate of 'absolute space' with the notion that inertia was defined with respect to the 'fixed stars', an idealised far-off massive shell representing all the matter in the universe. Einstein's view was that his new result gave significant physical credence to Mach's philosophical construct. This would have an important bearing on the genesis of general relativity and its application to cosmology, the science of how the universe evolved and behaves.

June 1912 saw an abundance of good news arrive from old friends and colleagues. Einstein was surprised by letters from Hopf and Conrad Habicht. Both announced their engagements, to Einstein's astonishment in Habicht's case, since he had considered him a confirmed bachelor. Einstein invited Habicht and his fiancée to visit them as soon as they returned to Zurich in August. Habicht's brother Paul wrote to announce further progress with chasing down the sources of noise observed in the Maschinchen. Most momentous of all was an image from Laue, showing that X-rays produced interference patterns when scattered from a crystal. Einstein was overcome with admiration for Laue's work, which he described in a letter to Hopf as the most marvellous thing he had ever seen. He also remarked that it established scattering from individual molecules in the crystal. For this work, von Laue, as he became after his father was ennobled by the Kaiser in 1913, received the Nobel Prize in 1914.

Still absorbed in thinking about gravity, Einstein prepared to shake the dust, as well as the soot, of Prague from his heels. He had mixed feelings; although he had complained bitterly to friends about the unsanitary and dirty conditions and the political tensions, he had greatly appreciated some aspects of the city. In particular, the wonderful architecture and the atmospheric mediaeval buildings had made a deep impression on him. Clark, quoting Marianoff, said that Einstein remembered:

> ... 'the solemn sounds of the organ in Catholic cathedrals, the chorales in Protestant churches, the mournful Jewish melodies, the resonant Hussite hymns, folk music, and the works of Czech, Russian, and German composers.' This was the world in which he sought relaxation, moving in 'a sort of dream' while his mind concentrated on the work that mattered.

In addition to beauties of the urban environment, he had found good friends; in particular his musical activities had flourished. The floor beneath his offices housed the mathematical institute of Georg Pick, twenty years older than Einstein but like him a Jew and an enthusiastic and talented violinist. Unsurprisingly, the two immediately became friends, often going on walks together. Pick organised regular string quartet evenings which, as noted in the description of Ehrenfest's visit, were a top priority for Einstein, even higher than entertaining a guest. Pick also brought together other musicians for chamber music. Frank reported that Pick, who had in his youth been an assistant of Mach's, discussed differential geometry with Einstein and mentioned to him the work of Levi-Civita and Ricci in this area. This work was eventually to become important in the formulation of general relativity. However, these intellectual seeds fell on as yet stony ground, as the evidence shows that Einstein did not realise their importance until long afterwards.

Moritz Winternitz, who was Professor of Sanskrit, was also a violinist who attended Pick's musical gatherings. His sister-in-law, Ottilie Nagel, an elderly spinster, was a piano teacher who became a regular musical partner of Einstein. She maintained her disciplined pedagogical approach with him, so that he complained good-humouredly that it was like playing with a sergeant in the military. When she learned of his impending departure for Zurich, she extracted a promise from him that he would only recommend a replacement who was a good violinist. When Philipp Frank was introduced to her as Einstein's successor, Ottilie immediately suggested that they play together. When Frank informed her that he had never held a violin in his life, she complained bitterly that Einstein had let her down.

Another musical and philosophical group met at a salon in the Old Town Marketplace run by Berta Fanta, the wife of a rich pharmacist. Among the attendees was the philosopher Hugo Bergmann, who at that time was a librarian. He often accompanied Einstein on his walk back home from his institute. He later married Berta Fanta's daughter and eventually became Dean of the Hebrew University of Jerusalem. Johanna Fantova, who married Berta's son Otto, would become an important part of Einstein's life in his final years. Other attendees of the Fanta salon included the writers Max Brodt and, very occasionally, Franz Kafka.[9] The main attraction for Einstein was probably participation in the chamber music that Bergmann organised, but he also took part in discussions on the philosophy of Kant and Hume.

Even Mileva was not entirely negative about her stay in Prague. In a postscript to Einstein's letter to Besso on 26 December 1911, intended for his wife Anna, she wrote:

> I am pleased that our stay here probably seems to be coming to an end and that we will swap Prague for Zurich is a great joy. I will certainly take some happy memories of Prague away with me; we now know several nice people here and the unpleasant things I will try not to think about [. . .] We go often to the theatre and to concerts . . . Little Albert really enjoys going to school; he also loves his piano lessons, which he has been doing for two months now.

In February 1912 she wrote 'You can imagine how overjoyed I am about returning to Zurich. For the last few days, this happiness has made Prague seem beautiful. [. . .] We want now to go really intensively to the theatre and concerts, mostly the comic opera . . . '.

---

[9] Just before his death from tuberculosis, Kafka spent six months in Berlin in the winter of 1923–1924. Elsa Einstein recalled that she dreaded his visits because of his peculiar mannerisms.

On the other hand, when she was safely back in Zurich, she wrote to Helene Savić that:

> We are well and are all, big and small, very happy to have turned our backs on Prague. To me, in particular, because of the children, our stay there was so unpleasant. Hygienic conditions there are such that sometimes it is really hard for me with my boy. You know already that there is no drinking water there; milk is of just as doubtful quality; the air is always full of soot, there are no gardens, and there is very little free room, so that the children really have to live inside. Little Eduard, whom I took there as a healthy child, started gradually to weaken, lost his appetite and sleep, and in the end hardly ever ate, showed signs of rickets, and felt bad, and so did I. Albert also did not enjoy life there as much as in Switzerland, so I was very happy when it was finally decided to go back to Zurich.

At the end of July, the family packed their belongings, which were to go into storage until they found permanent accommodation, and then caught the train to Zurich. As they left the Austro-Hungarian Empire, Einstein's thoughts were concentrated on the problem of gravity. He had realised that a Euclidean space and time was manifestly incapable of supporting appropriate equations. He had remembered the lectures given by Carl Friedrich Geiser on Gaussian surfaces, one form of a non-Euclidean geometry, that he had attended as a student at the Polytechnikum. Could this be a solution to his problem? Einstein was travelling in the right direction since his fellow auditor at Geiser's lectures, Grossmann, was waiting in Zurich for his old friend to arrive. Together, they would take the next steps on the road to a theory of gravitation.

Chapter 11

Zurich again: General relativity begun (1912–1914)

The Einstein family launched themselves back into life in Zurich with gusto. Already by 10 August 1912, they had found somewhere to live. Their choice was a contrast to their previous abodes in the city. Rising slowly above the ETH and the university, the Zurichberg in their youth had been relatively undeveloped. This was gradually changing, and their attention was drawn by an advertisement for a newly constructed apartment block, the Hofburg, Hofstrasse 116, close to the Dolderpark. It was situated on the corner of the Hofstrasse and the Keltenstrasse, which led down the hill to the ETH Physics Department and Observatory in a short walk of around ten minutes. Mileva was apparently delighted with the imposing house, set in a still-rural environment, even though the house was surrounded by building materials, having been constructed the year before. They became the first tenants, renting a rather expensive six-room apartment on the ground floor. The children occupied one sunny room looking out onto the courtyard, shaded by trees, while the large room that opened onto the courtyard became the music room. That autumn, strains of Bach Solo Sonatas and the Double Concerto drifted into the courtyard. Einstein played the latter with his former music partner, Lisbeth Hurwitz, while the owner's daughter, Carola Gentner,[1] herself a reasonably competent violinist, listened entranced in the courtyard.

Einstein lost no time in re-establishing contact with his many Zurich acquaintances, in particular Heinrich Zangger, who had been so instrumental in engineering his return, as well as Ernst Zermelo. However, the person whom he particularly sought out was another who had strongly pushed for his appointment: his old friend and former classmate Marcel Grossmann. Immediately after establishing Mileva and the children in their new home, Einstein was plying Grossmann with questions about geometry, exclaiming 'Grossman, you have to help me, or I will go insane!' Grossman, who had been appointed as Professor of Descriptive Geometry at ETH in 1907, was now also the head of the Mathematics and Physics section of ETH. Despite this heavy administrative and teaching load, Grossmann was only too happy to respond to pleas for help with the mathematical techniques that by now Einstein was certain were required to bring his work on gravity to a successful conclusion.

Another old friend and colleague who was a vital sounding board in this period was Max Laue, who now held Einstein's old position at the university, which had been vacated by Peter Debye. He recalled the atmosphere of that time as Einstein wrestled with the gravity problem:

---

[1] The good relationship between the Gentners and Einstein did not last long. With a typical lack of regard for household details, Einstein thought he had agreed verbally with Herr Gentner that heating costs were included in the rent. Since this did not appear in the lease he signed and Herr Gentner's recollection differed, an acrimonious legal case ensued. The verdict was delivered just as Einstein left Zurich in February 1914. Einstein was found to be liable for the heating plus interest, legal costs, and fines.

Einstein. Brian Foster, Oxford University Press. © Brian Foster (2026). DOI: 10.1093/oso/9780198794875.003.0011

> After the colloquium [which Einstein ran], Einstein went for dinner to the *Kroenenhalle* (sic) restaurant with anyone who wanted to join him. At that time, General Relativity was at its beginnings, and I recall many disputes with Einstein over this subject. He was completely under the spell of these ideas and in the following years kept coming back to it, sometimes jumping to it from a completely unrelated subject.

By the summer of 1912, Einstein was convinced that the key to solving the problem of gravity was indeed to be found in a new description of the space-time geometry of the universe. His transition from the constancy of the velocity of light, in the special relativity of 1905, to the description of gravity with a variable velocity of light in his Prague work and the discovery that this too was insufficiently rich, led him inexorably in this direction. He had appreciated the importance of the definition of the shortest distance between two points as characterising geometries. This 'metric' was a straight line in the Euclidean geometry of space but much more complex in the geometries that he had concluded were necessary to solve the problem of gravity. In a special-relativistic four-dimensional geometry, for example, this distance represented the space-time separation between two events. In the most general four-dimensional case, the metric is represented by the sixteen components of a four-by-four matrix of numbers known as a 'tensor'. Einstein assumed[2] that the metric is symmetric (the element in the first row and second column equals that in the second row and the first column, etc.). This reduces the number of independent components to ten.

Einstein's work until now had shown him that the gravitational equations he sought had to depend only on the quantities that defined the metric of the appropriate mathematical spaces, or the rates of change of these metric components. Furthermore, if gravity were to reduce to the Newtonian theory as a special case, the rates of change involved should be no more than second order,[3] since Newton's equations are second order in the gravitational potential. Einstein asked Grossmann, who was not an expert in this particular branch of mathematics, to go to the library and consult the mathematical literature to search for an appropriate geometry in which this distance measure does not change as the coordinate system changes. The following day Grossmann returned with the answer: such a geometry did exist – Riemannian geometry. He told Einstein however that this could not be what he was seeking, since it was a non-linear theory. Einstein was able to reassure him that, because gravitational energy generated more gravity, this was indeed precisely what was required.

It can be assumed that Einstein received an intensive tutorial from Grossmann on the tensor algebra appropriate to manipulate objects in Riemannian geometry. He then entered a period that he characterised to his friend Sommerfeld as more taxing and harassing than any in his life to date, noting that, from this experience, he had conceived a great respect from the subtler methods of mathematics. This contrasts with his earlier opinion, as recounted by Otto Stern, that 'Any theory of physics that contains more mathematics than sine and cosine is a priori nonsense.' His growing expertise in the techniques of tensor algebra that Grossmann had revealed to him is evinced in the 'scratch' or 'Zurich' notebook that he used while in Zurich to record his thoughts on how to formulate the theory of gravity in this new geometrical form. This notebook has been extensively

---

[2] One thrust of Einstein's later work on unified theories of electromagnetism and gravity relaxed this condition, trying, ultimately unsuccessfully, to represent electromagnetism via the additional independent terms in the metric.

[3] 'Second order' means the rate of change of a rate of change with respect to coordinates. For example, velocity is the rate of change of position with time; acceleration is the rate of change of velocity. Velocity is thus first order, whereas acceleration is second order with respect to a particle's position.

analysed by scholars to try to divine Einstein's thought processes as he wrestled with the problem of gravity. Inspired by analogy with electromagnetism, Einstein rather quickly decided that the master equation must be of the form that the metric tensor, expressing the curvature of space under the influence of gravity, should be proportional to a tensor consisting of the sources of energy and momentum in the volume of interest. The latter is known as the 'stress-energy' or sometimes 'energy–momentum' tensor and is the analogue of the matter density that generates gravity in Newtonian theory.

Einstein approached the problem of constructing the correct equation with a two-pronged attack: one method was to investigate the mathematical structure of the rich and complex theory developed by Bernhard Riemann in the 1850s and enhanced by Elwin Bruno Christoffel in 1869 and Gregorio Ricci-Curbastro and Tullio Levi-Civita in 1901, whose 'absolute differential calculus' is now known, thanks to Einstein and Grossmann, as 'tensor calculus'; the other method was to work from known physical principles, such as energy–momentum conservation and the fact that, for weak gravitational fields, Newton's theory should emerge as a special case of a more general theory. As Einstein worked systematically through both approaches, he filled his notebook with derivations of various equations that were candidates to describe gravity.

Einstein kept his physics goals for his theory of gravitation at the forefront of his thinking in pursuing the physical branch of his attack. Although these goals changed somewhat in importance and emphasis as his work progressed, the desire to incorporate the principle of equivalence was always strong. Indeed, the result of the collaboration with Grossmann, the so-called Entwurf paper[4] of 1913, begins with the statement 'The theory described below is predicated on the conviction that the proportionality between the gravitational and inertial masses of bodies is an exact law of nature that must manifest itself as one of the foundations of theoretical physics.' Another guiding principle was that the equations of a correct theory of gravitation must be covariant, that is, have the same form irrespective of the frame in which they were expressed. Einstein believed, incorrectly as it turned out, that this is equivalent to a generalisation of the principle of relativity – not only are all inertial frames equivalent, the basic premise of special relativity, but all frames, inertial and non-inertial, are equivalent, thus removing the preferred status of inertial frames in physics, which had become no more acceptable to Einstein than the aether had been in 1905. Finally, at least in his initial work in the Zurich period, he wished to incorporate what he eventually labelled Mach's principle, that the property of inertia was a result of the gravitational effects of the surrounding massive universe. He had obtained indications of this before leaving Prague in his investigation of an accelerated massive shell, which was published as his contribution to the *Festschrift* for the opening of Zangger's institute.

In the subsequent 100 years, two of these three principles have not survived in the way Einstein formulated them. Indeed, as it is not an essential ingredient in the theory of gravitation, Einstein ceased to employ Mach's principle shortly after he formulated general relativity. However, it is necessary to establish what had always been a crucial goal for Einstein: the extension of the principle of relativity to arbitrary motion. As discussed later on, the covariance of general relativity was sacrificed in the Entwurf theory, although it was later recovered in the final theory. Both contemporary and subsequent authors have insisted that establishing covariance is merely a mathematical manipulation that is a necessary but not sufficient property for a theory of gravity. Indeed, contrary to Einstein's view when wrestling with gravitation, any theory, even Newtonian

[4] The paper's title is '*Entwurf einer Verallgemeinerten Relativitäts Theorie und einer Theorie der Gravitation*' ['Outline of a Generalised Theory of Relativity and a Theory of Gravitation'].

gravity, can be expressed in a covariant form. Einstein was misled here by his understandable tendency to build on his previous work. In the 1905 theory of special relativity, Lorentz covariance is both necessary *and* sufficient to establish the theory. It is important to remember, when considering the various missteps and blind alleys that Einstein took in the next few years, that the development of mathematical techniques gives subsequent authors insights not available to Einstein and Grossmann during their collaboration.

During the summer and autumn of 1912, Einstein and Grossmann were absorbed with attacking the gravitational problem by a variety of different routes, which can be traced by an analysis of Einstein's Zurich notebook. This analysis implies that Einstein's first approach was via the 'mathematical' branch of his dual strategy. Grossmann had pointed out to him that the Riemann tensor, which characterises the curvature of spaces in Riemannian geometry, was a promising place to look for generally covariant equations describing gravity. From it could be derived the components of the metric tensor, representing the potentials from which the gravitational field could be derived, analogously to the Newtonian theory. The Ricci tensor is a particularly promising tensor derived from the Riemann tensor by a mathematical operation known as 'contraction'. This is a type of projection onto a lower-dimensional space, similar to projecting a two-dimensional line in Euclidean space onto x and y axes. It transpires that the Ricci tensor was indeed the key that would eventually unlock the solution to the problem of gravitation. Unfortunately, with the key in their grasp, Einstein and Grossman turned it the wrong way in the lock, shook their heads and returned it to the shelf for three long years.

The reasons for this disastrous false path, which resulted in Einstein almost losing what became a race to solve gravity, are several and complicated. The two friends had found plausible equations of gravity by following the clues in the mathematical structure of Riemannian geometry. The next step was to ensure that these equations would lead to physically reasonable consequences. These included that they should satisfy the conservation of energy and momentum; that in the 'weak-field' limit, away from very massive objects, they should reproduce the results of Newtonian gravity; and, not as fundamental but of great importance to Einstein, that they should satisfy the generalised relativity principle, that is, that physics was invariant under arbitrary motion.

When trying the Ricci tensor as the basis for his gravitational metric, Einstein did not even reach evaluation of his third point, which, it transpired, was unattainable, even in his final theory; he already failed on the first two. The standard formulation of Newtonian theory is very far from being covariant, so in order to compare the equations of any new theory to it, the coordinate system used must correspond to an inertial frame, within which the Newtonian theory has its familiar form. Einstein chose what are now known as 'harmonic coordinates' to achieve this. However, his understanding of how such coordinate restrictions should be used was very different to that of today. Benefiting from a century of development in mathematics, it is now clear that such operations are what are called 'gauge transformations', which do not affect the physical predictions of the theory. A particular gauge transformation can be chosen arbitrarily to make whatever problem is being attacked more tractable. Einstein, in contrast, was under the impression that such operations were a one-off choice that, having been made, became intrinsic to the theory. Forced then to work with the truncated equations after he had restricted himself to harmonic coordinates, he could find no way to satisfy energy–momentum conservation and reduce to Newtonian gravity in the weak-field limit.

A further problem for the Ricci tensor solution was that Einstein was once again led astray by trying to build on his earlier work. For static gravitational fields, he thought that his new theory should reproduce his earlier results characterising gravity via a position-dependent velocity of light in a special relativistic, equivalently called 'flat' or 'Minkowskian', space-time. This earlier work had been conceived before he embraced the necessity for Riemannian geometry and necessarily therefore was based on a flat spacetime. Unfortunately, despite it seeming a reasonable expectation that in the weak-field limit the metric would reduce to a single potential, in the final theory there are non-zero off-diagonal terms in the tensor, which forbids such a reduction. However, these terms play no part in Newtonian equations of motion, so their presence or absence in the weak-field limit in irrelevant. The existence of these off-diagonal terms in the metric is equivalent to saying that static gravitational fields cannot in general be represented in flat space-time.

For these reasons, Einstein's attempts to derive his earlier results, starting from the Ricci tensor, failed. There seems to be some disagreement among Einstein scholars as to whether the failure just described also contributed to Einstein's decision to give up the 'mathematical' line of attack. In any case, he already had enough discouragement so that, probably as the autumn of 1912 turned into winter, he reluctantly gave up on the Ricci tensor. He also abandoned a variant approach suggested by Grossmann of splitting it into two, at the expense of reducing the covariance properties. This 'November tensor' was so named because it was used in the first of the stream of papers, starting in November 1915, which led to Einstein eventually solving general relativity. Back in 1912–1913, however, the November tensor caused Einstein difficulties similar to those from which the more general Ricci approach had suffered and which for the moment proved insuperable. Discouraged, Einstein turned to his 'physical' strategy to find a solution.

Einstein now searched for the correct equations by imposing properties such as energy-momentum conservation at the outset rather than as a subsequent test of candidate equations. He and Grossmann found they could conserve energy and momentum if they restricted the covariance properties of their candidate equation drastically. Although Einstein was initially reluctant to do this because of the implications on extending the principle of relativity, he resigned himself to it. After all, his initial goal had been to generalise the equivalence of inertial frames of special relativity to accelerated frames; achieving this limited goal required much less than the full covariance he had been subsequently striving for. In fact, by restricting the covariance to a group including at least linear coordinate transformations, he was able to generate equations that not only satisfied the conservation laws and reduced to the Newtonian limit, but also seemed to be, as he erroneously believed, unique. Sometime in late February or March 1913, he felt that it was time for him and Grossman to buckle down to writing a paper that would summarise their progress so far. By 14 May, they had finished.

The Entwurf paper appeared first as a booklet printed by Teubner in June 1913; however, it seems likely to have been submitted first to *Zeitschrift für Mathematik und Physik* and in parallel printed as a booklet to reduce journal publication delays. It consisted of two parts: a 'physics' part, authored by Einstein, and a 'mathematical' part, written by Grossmann. At the start of their collaboration, Grossmann had made it clear that he wished to concentrate on the mathematical aspects of their work and to take no responsibility, or credit, for the physics. His contribution was a detailed introduction to what, subsequent to this paper, became known as tensor calculus. In it, Grossmann set out a self-consistent, although somewhat clumsy and rapidly superseded, notation, as well as introducing several techniques going well beyond the work of previous authors. Various parts of this mathematical exposition feed back into Einstein's section. After the introduction already quoted, Einstein began with a recapitulation of his earlier work in both special relativity and gravitation, in particular the role of the variable velocity of light in a static gravitational field. Succeeding sections generalised to an arbitrary field, the introduction of a metric, the invariant length measure, and the construction of the stress–energy tensor. The crucial part of the paper is section 5, in which Einstein began by writing the tensor form of the analogy to Poisson's equation, which is the basis of Newtonian theory as well as electrostatics. After mistakenly ruling out a fully covariant solution, he applied energy–momentum conservation to set up the equations linking the gravitational field and the stress–energy tensor. These Entwurf equations were the basis of his efforts for the next two years. In section 6, Einstein used liberally the results from Grossmann's section to show that the formalism they had developed reproduced Maxwell's equations. The final section purported to show that a scalar theory of gravity was unacceptable. Einstein subsequently realised that the argument given was spurious and that a self-consistent scalar theory of gravity was in fact possible. Indeed, he spent some considerable time himself puzzling over such a theory, as will be discussed later.

While Einstein strained every sinew to solve the mysteries of gravity, his family, based in their comfortable apartment on the hill above the ETH, slipped back into a semblance of their old life in Zurich. Music, both concerts in the Tonhalle and private chamber music among their wide circle of friends, was (as always) a vital ingredient. Musical gatherings at the Hurwitz home were renewed with enthusiasm. Einstein appreciated the strides that Lisbeth Hurwitz had made as a musician since their last regular encounters. On 19 October, he and Lisbeth played Händel's Concerto for Two Violins[5] together. Mileva was only too aware, however, that life was not as it had been, even in the previous sojourn in Zurich, let alone in their student days. The extent to which Einstein was absorbed by his work can be measured quantitatively in the marked drop in his recorded correspondence in the first nine months after their return. Mileva wrote to her confidant Helene Savić in December 1912 that:

> My big Albert has become a famous physicist who is highly esteemed by the professionals enthusing about him. He is tirelessly working on his problems; one can say that he lives only for them. I must confess with a bit of shame that we are unimportant to him and take second place.

Mileva's depression was visible also to her friends, who knew that her deep love of music, to which she devoted more and more attention in the intervals between her domestic duties, was one way to lighten her mood. The Hurwitz family made a special effort to please her by repeatedly including pieces by Schumann, a particular favourite of hers, in their regular Friday-evening music-making. They did this both in November 1912 and throughout February 1913. Other composers were also explored, including Anton Rubinstein and Arcangelo Corelli. The whole Einstein family visited on Sunday afternoons; by now Hans Albert had made sufficient progress with his piano playing to take some part in the concerts. However, neither this nor Schumann could lift Mileva's mood, which was not improved by severe leg pains that hindered her from walking and were exacerbated by the Swiss winter chill. Einstein became irritated by her black moods and on several Friday evenings went alone to the Hurwitz house. It was rare for him to miss these evenings, although the pressure of completing the Entwurf calculations told at the end of February, when he could only send his violin for the use of Auguste Piccard, a physicist at that time completing his doctoral thesis under Pierre Weiss at ETH. Other colleagues and friends also habitually gathered at the welcoming Hurwitz household to take part in or listen to the music-making.

At least one thing had brightened Mileva's mood during her travails in the winter of 1912–1913 – the prospect of a trip to Paris with Einstein. He had been invited to attend the annual meeting of the French Physical Society and give a lecture on a rather minor piece of work that he had

---

[5] Presumably one of the Concerto Grossi arranged for two violins and piano, unless Lisbeth had misremembered the composer and the piece was Bach's 'Double' Concerto, one of Einstein's favourite pieces.

completed in Prague in 1912 before he became totally immersed in gravitation studies. This paper assumed that radiation was quantised but used classical thermodynamic methods to show that every absorbed photon caused a single molecule of a gas to disassociate if the system was in thermodynamic equilibrium. A subsequent supplement generalised the problem to a wider range of light wavelengths.

Shortly before they left Zurich for Paris on 24 March, Mileva wrote to their old friend Maurice Solovine, who was then living in Paris. She suggested that he move into their hotel in Paris during their stay, as they would feel lost without a guide to the great city. Mileva in particular hoped that Solovine could show her the sights while Einstein was absorbed by meetings with the French physicists. Einstein added a postscript saying how much he looked forward to seeing his old friend again and that he was rather nervous about his lecture, since he had agreed to give it in French. Helped out linguistically at times by Pierre Langevin, Einstein delivered the lecture to his own satisfaction on 27 March. He subsequently wrote to Elsa that he was positively embarrassed by the attentions that were paid to him. Their correspondence had also been greatly slowed by Einstein's absorption with general relativity but he still protested his desire to spend time with her, without his cross, as he labelled Mileva.

Both Einstein and Mileva seem to have greatly enjoyed visiting Paris and seeing Solovine again. Their social interactions with the French physicists, with whom Einstein had first become acquainted at the Solvay Conference, were also a success. He wrote an effusive thank-you note to Marie Curie, to which Mileva added a postscript, while Einstein conveyed Mileva's thanks along with his own to Jean Perrin, whose work on Brownian motion he had so admired. Mileva seems to have established a bond with Marie Curie, who had achieved a great scientific career against all the odds that had defeated Mileva. The two families made a tentative arrangement to meet for a walking holiday in Switzerland in the summer.

Einstein seems to have taken a little break from his arduous research work in the weeks after their return form Paris, the calm before the storm that he anticipated after the publication of the Entwurf paper. In May, he tried to arrange a musical evening with Conrad Habicht, sending him a part of a Handel piece he hoped to play with him – perhaps the Concerto Grosso that he had been playing with Lisbeth. Habicht's new wife as well as his brother Paul were also cordially invited. The message implies that Hans Albert would accompany them on the piano, indicating that he was now at a level when he could play the simple piano accompaniments typical for Baroque music. Indeed, after Einstein's return home each evening from ETH, Hans Albert regularly accompanied his father's violin playing on the piano. Although Einstein's relationship with Mileva was deteriorating markedly (to the extent that Hans Albert noticed that something was wrong), with the children he remained happy, playful, and affectionate. However, he could be stern and administer corporal punishment when they misbehaved.

The postcard to Habicht made no mention of the Maschinchen; Einstein's interest therein seems to have been another victim of his relativity obsession. Paul had made considerable strides in improving it for commercial exploitation but ultimately it was not a success on the market, presumably because it turned out to be too sensitive and unstable. Shortly after sending the postcard, Einstein spontaneously wrote to the head of the Winterthur school where he had taught in his youth, recommending Conrad for a vacancy there. Unfortunately, his informant, his friend Jürg Gasser, had mixed the brothers up; it was Paul who had applied for the position, to which he was appointed shortly thereafter. Another even older friend with whom Einstein was in contact at this time was Hans Wohlwend. Having spent several years in Karachi, where he had learned English, he now lived within walking distance of the Einstein residence, down the Zurichberg towards the centre of town. Somehow Einstein managed to find time to take English lessons from Wohlwend, who many years later recalled their attempts to pronounce 'such wonderful sentences as "there is a mill on the hill" with a finely rounded "L" at the end'. Finally, Jakob Ehrat was teaching in Winterthur and was often in contact with his old Polytechnikum classmates; indeed, he had until 1911 been assistant to Marcel Grossmann. Einstein tried to convince Ehrat to lodge with Rosa Bandi, Marie Winteler's sister, who was trying to support herself and her two young sons by opening a lodging house in Winterthur. Einstein had re-established contact with her on his return to Zurich; it seems likely that part of the attraction of visiting her was a vicarious contact with, and the likelihood of learning news of, Marie. Another link was that Michele Besso was supplementing his sister-in-law Rosa's income to provide for his two nephews. Ehrat resisted Einstein's match-making, however. He had both lodging and marriage plans of his own; the following year he married his landlady Margrit Schmid.

While Einstein had been absorbed in his gravitational work in the first half of the year, far away in Berlin machinations were underway which would result in his moving there the following year. While recuperating in the Engadin, Fritz Haber had occupied some time in wondering how to attract Einstein to Berlin, despite his refusal of a position from Emil Warburg a few months earlier. His first thoughts were to attach him to his own Institute of Physical Chemistry. However, after his return to Berlin and discussions with Max Planck, Walther Nernst, and others, the scheme gradually became modified to creating a new Kaiser Wilhelm Institute of Theoretical Physics. A vacancy in the Prussian Academy of Sciences, the premier scientific academy in continental Europe, had just arisen. In addition, one of the two research professorships of the Academy had been vacant for more than two years, since the death of Jacobus Henricus van't Hoff. This fortunate conjunction provided the foundation on which a scheme to attract Einstein could be built. Planck, Nernst, Haber, and Warburg proposed on 12 June that Einstein, despite his youth, should be elected to membership of the Academy. It is ironic to note that, despite his intention to convince his colleagues of Einstein's outstanding ability, Planck was unable to restrain himself from commenting that he had over-reached himself in his postulate that light propagation was quantised.

June 1913 saw Einstein's resumption of work on gravity, catalysed by visits from Besso and Paul Ehrenfest. Besso was the first to arrive; it seems likely that he spent the first half of June in Zurich. The two friends were soon buried in discussions on gravity. Besso left Zurich on 18 June, Ehrenfest having arrived earlier the same day; Besso had planned to stay longer but was called back to Trieste for a business meeting. As a result of this intended overlap, Ehrenfest and his wife Tatiana had booked lodgings in a boarding house rather than staying in Einstein's guest room. Perhaps this was just as well, as in the end the Ehrenfests stayed for almost a month. During this time, Ehrenfest saw Einstein virtually every day, indeed, he noted that there was only one day without Einstein; they divided their time between talking physics and making music, including pieces by Haydn, Brahms, and Bach, the latter underlined by Ehrenfest in his diary, as he was relatively unfamiliar with his works. Ehrenfest also renewed his acquaintance with Hans Albert. Ehrenfest and Einstein spent most of their physics time discussing gravity, as recalled by Max von Laue: 'The liveliest discussions happened in summer 1913, when the temperamental Ehrenfest was visiting Zurich. I can see even now a great crowd of physicists gathered around Einstein and Ehrenfest as they climbed the Zurichberg and Ehrenfest crying out "I have understood it!" in jubilation.'

Ehrenfest and Einstein would certainly have been discussing the work that Einstein had begun with Besso before Ehrenfest's arrival. This work is contained in a fascinating document, known as the Einstein–Besso manuscript, analysing the precession of the perihelion of the planet Mercury. Einstein had attempted to explain the discrepancy between the measured value and the Newtonian prediction in 1907, when he first began seriously to consider a relativistic approach to gravity (see Chapter 7). With Besso's aid, he had attempted to use the Entwurf theory to calculate the expected precession.

Perhaps the most striking thing about the Einstein–Besso manuscript is the evidence that Besso was working at almost the same level as Einstein. Subsequent scholarly work by the editors of the Einstein Papers Project and others can divide up the pages of the manuscript quite convincingly into those contributed by Besso and those by Einstein. The first and largest section was written jointly during the June visit. The next section was written by Einstein after Besso returned to Trieste and the final section was written solely by Besso, probably in 1914. It is clear that Besso not only had a solid understanding of tensor calculus, which Einstein presumably taught him in a crash course at the beginning of his visit, but also appreciated the subtler points, and indeed weaknesses, of the Entwurf theory. The first part of the manuscript, which they discussed together during the June visit, was mostly written by Einstein. In particular, he carried out most of the tensor manipulation and the difficult integrals that needed to be evaluated. Besso contributed some algebraic manipulation and obtained the expression for the precession they were calculating. The next section seems to have been written by Besso when Einstein was absent, as it recapitulates the whole calculation, as if Besso were reminding himself of what they had done. The two friends went on to consider the first-order correction to their original calculation and later to consider the sun in rotation, rather than their original, static, assumption. Besso covered page after page with complicated calculations, with the occasional annotation from Einstein when Besso made a slip. They went on to widen the investigation to other features of Mercury's orbit.

The Berlin plot thickened in July. Planck and his colleagues had taken soundings on the reaction of the Academy members to their proposal for Einstein's membership. Armed with the promise of a salary supplement from the generosity of rich industrialist Leopold Koppel, on 11 July Planck and Nernst took the night train to Zurich. Accompanied by their wives, they were ostensibly taking a short holiday break. In fact, they were on a mission to get Einstein's agreement to come to Berlin as a salaried member of the Prussian Academy of Sciences and subsequently the Directorship of a probable new Kaiser Wilhelm Institute. Einstein was fully aware that Mileva would hate to leave Zurich, particularly for Berlin, crowded as it was with members of the Einstein family. Their relationship had continued to deteriorate but perhaps enough regard was left to make him hesitate; he did not give an immediate answer after an interview with the Berlin dignitaries. They and their wives left the city for a day trip on 13 July. Einstein met their return train and, in a bizarre arrangement more suitable to a spy novel than *Zeitschrift für Physik*, they had agreed that if Einstein had decided to accept the position, he would wave a white handkerchief. Craning from the train windows, they espied Einstein giving the positive signal. The following night, they took the train back home to Berlin.

Einstein immediately wrote to Elsa with the joyous news that, by the spring of 1914, they would be neighbours in Berlin, subject to his election by the Academy. Putting his excitement to one side, on the following day Einstein returned to routine ETH faculty duties, writing an assessment of his assistant Otto Stern's Habilitation thesis. He had followed Einstein to Zurich from Prague and, after some resistance from the ETH authorities, had been appointed as Einstein's scientific associate in the winter semester of 1912–1913. Stern's thesis was accepted, and he became a Privatdozent in August. Meanwhile, on 24 July, the Prussian Academy of Sciences elected Einstein as its youngest member.

From approximately 4–10 August, the Einstein family joined the Curies on their long-planned tour in the Engadin region of Switzerland. They stayed initially in Flims, from where they probably took the postbus to Samaden, walking via St Moritz across the Maloja Pass and eventually across the Italian border to Lake Como. Eve Curie, in her biography of her mother, described the scene as the families (without Mileva, whose leg troubles precluded her from the strenuous activity the others were engaged on; she presumably took the postbus between their hotels) walked through the glorious scenery. The children took the lead, while the two scientists followed, Einstein engaged in a determined attempt to explain the Entwurf theory to Marie Curie. The youngsters were amused by Einstein's analogy of people in a free-falling elevator, an analogy no doubt sharpened by the sheer drops never far away from their path. The necessary intimacy of a shared holiday gave Einstein a deeper understanding of Marie Curie's character. Although he clearly had the greatest respect for her as a scientist, he found her cold and given to complaining. The latter may be related to the fact that she had only recently recovered from illness. These barbed observations are contained in a letter that he wrote to Elsa shortly after their return to Zurich. This letter contains the astonishingly naïve observation that he and Elsa would have a wonderful time together in Berlin without causing pain to Mileva, because Elsa could not take from Mileva what she did not possess: his love.

On returning from the Engadin, Einstein had sent off the text for the talk he had been invited to give at the meeting of the German Scientific and Medical Society in Vienna that September. He had begun to have second thoughts about the Entwurf theory, in particular the fact that he and Grossmann had had to give up on the idea of the general covariance of the equations. He expressed some of these worries to his mentor Lorentz in a letter sent shortly after his return. Perhaps his attempts to explain his ideas to Marie Curie had led to renewed concern, as is often the case in such situations. Only a day later, however, he was reassured. He had come up with an argument that convinced him that general covariance was excluded if the conservation laws were to be valid. He wrote again to Lorentz to explain that his reservations had proven groundless.

The discovery of an additional eighteen pages of the Einstein–Besso manuscript in 1998 makes it seem likely that Besso returned to Zurich for a few days in the second half of August 1913 to carry on working with Einstein. This is plausible, since Einstein had expected that Besso might return after his hurried departure in June. Einstein was certainly in Zurich in the second half of August as Lisbeth Hurwitz wrote in her diaries that on 22 and 23 August, she, her father, and Einstein were playing music together and that a photograph was taken of them on the Hurwitz's veranda (see Fig. 9.1). Grossmann and Mileva also attended and subsequently they all went for an excursion into the mountains. The fact that Besso is not mentioned as being present and that his handwritten date on the additional pages of the manuscript is 28 August gives a clear window of possibility for a second Besso visit to Zurich.

In the period between Besso's visits, Einstein had worried away at the problems they had been considering. Their work in deriving the metric appropriate to the sun in order to calculate the orbit of Mercury had led Besso to wonder whether their functional form was unique. In late June they discussed this again. Einstein had considered a situation in which a region of a space-time filled with matter surrounded a matter-free 'hole'. It could be shown that, in such a situation, the metric tensor was not unique. Besso, apparently recording a conversation that he had had with Einstein, wrote that he thought that this was irrelevant, provided that any physically measurable events were unchanged by this arbitrariness. This is a very understandable viewpoint to a modern physicist, since it corresponds to a similar situation in quantum-field theories, which permits arbitrary choices for the analogue to the metric, or 'gauge fixing' as it is known. It also corresponded to the situation in electromagnetism, in which the electrical potential is arbitrary since only its rate of change with distance, which corresponds to the electric field, produces measurable effects. However, for reasons that are difficult to reconstruct from the manuscript, Einstein convinced Besso that the situation was different for gravitation. Michel Janssen believes that this 'reflect[s] both a genuine difficulty that Einstein saw in Besso's proposal and a certain impatience with Besso's criticism of a new and promising way to justify the limited covariance of the Entwurf field equations' – indeed, so convinced was Einstein of this erroneous justification that he repeated the 'hole' argument in print no less than four times in the next eighteen months. It is richly ironic that, in January 1916, after Einstein had finally arrived at the correct equations of gravitation, he wrote to Besso explaining why the 'hole' argument had after all been incorrect. In his letter, Einstein more or less exactly repeated Besso's original point, that is, that only physically measurable events were significant in the theory!

Einstein seems to have retained their manuscript after Besso's return to Trieste. He kept it for some months, adding to it and presumably wrestling with the various numerical errors it contained to try to reproduce the observed value for the precession, before returning it to Besso early in 1914. In the letter that accompanied the manuscript, Einstein wrote that it would be a crying shame were Besso not to finish it off and publish it.[6] Einstein also remarked that it was a real pity that Besso had no access to any scientific journals in his backwater in Trieste. It is a testament to what a remarkable man Besso was, that he was capable of working at the forefront of physics, albeit strongly guided by Einstein, while employed in an insurance firm in a sleepy backwater of the Austro-Hungarian Empire, remote from any intellectual stimulation. This was one of the reasons why Einstein so cherished his friendship. The famous *New York Times* headline that some years later announced the confirmation of general relativity referred to the fact that only twelve wise men in the world could comprehend it. One of those putative twelve was surely Besso.

Einstein had by now become an oracle, to which both young and old physicists made pilgrimage. That summer saw an extended visit from a promising young theorist named Gunnar Nordström, who had a position in the University of Helsinki. He had for several years been working on a scalar theory of gravitation, an idea which Einstein had previously worked through and then discarded. Nordström's initial theory had indeed not proven tenable, but he had continued to work on it and by the time he arrived in Zurich to consult Einstein, he had a much more successful variant. Einstein was surprised to find that it did indeed seem to be self-consistent, and he spent considerable time working with Nordström that summer. Since Nordström's theory did not use the Equivalence Principle, Einstein did not find it congenial; however, it was much simpler than his theory. There were differences between Nordström's predictions and Einstein and Grossmann's. There was no bending of light around massive objects such as Einstein had predicted. Einstein was thereby reminded of the importance of experimental tests of such a phenomenon. He was therefore pleased to receive another letter from Erwin Freundlich in Berlin, who was busy trying to come up with experimental tests of Einstein's predictions. Freundlich was about to get married, and the couple intended to spend their honeymoon in Switzerland. Einstein invited them to visit him in September. They did so and indeed accompanied him to the Swiss Physical Society meeting in Frauenfeld, giving Käthe Freundlich her first taste of the idiosyncrasies and tribulations of being a physicist's wife.

In his talk to the Swiss Physical Society, Einstein gave a very clear exposition of the basics of the Entwurf theory, shored up by his recent (erroneous) 'discovery' that the conservation laws implied that general covariance was impossible. Directly after Einstein's talk, Grossmann summarised the mathematical part of their work.

On 10 September, the Einstein family left for a late summer vacation. They called in on Conrad Habicht in Schiers, who, like Freundlich, also had a new wife. They then continued on to

[6] Although Besso added substantially to the manuscript, he never completed it. It seems from his outline of the contents that his plan for the paper was much too ambitious. The manuscript lay unsuspected among his papers until long after both his and Einstein's death. The calculation for the perihelion in the Entwurf theory was carried out in December 1914 by a student of Lorentz, Johannes Droste, who obtained a value less than half of that observed.

the summer house of Mileva's parents in Kac. They stayed there in stifling heat for a little over a week. Einstein and Mileva were due to leave for Vienna, where Einstein was to give an invited talk on 21 September at the annual meeting of the German Scientific and Medical Society. That morning, Einstein's two sons were received into the Orthodox Church. Einstein could hardly have been unaware of this, since both his wife and in-laws would have considered it a major event. He does not appear to have attended the service, so it can safely be assumed that it was a matter of indifference to him.

The German Scientific and Medical Society Congress was a huge event, attended by thousands of scientists; 350 of them packed into the main auditorium of the University of Vienna's Physics Institute to hear Einstein's address on the morning of 23 September. After a general introduction to the subject and the basic physical principles that he had followed in his work, Einstein first discussed Nordström's scalar theory, taking the opportunity to explain the mistake he had made in ruling out such theories in the Entwurf paper. He went on to discuss his and Grossmann's work in a rather more detailed way than he had in the Frauenfeld talk. In the subsequent discussion session, Einstein was attacked on all sides. The discussion was sufficiently tendentious that it reached the newspapers. That afternoon the *Reichspost* reported that a heated argument had taken place and the following day the *Arbeiter-Zeitung*[7] had a two-page article on relativity theory. Einstein was rapidly becoming front-page news.[8]

As usual at such events, some of the most interesting discussions at the Vienna conference took place between lectures. Hungarian physicist and future Nobel Laureate Georg von Hevesy, who had attended Einstein's inaugural lecture as a student in Zurich in 1909, was at the meeting. He was working with Ernest Rutherford in Manchester and thereby familiar with the Rutherford nuclear atom. This had been established in 1911 by Rutherford and his students Hans Geiger and Ernest Marsden (the latter still an undergraduate at the time) by scattering alpha particles from a gold foil. The brilliant young Niels Bohr was also working in Rutherford's group. In 1913, Bohr postulated that electrons, despite being accelerated as they orbited the nucleus, did not lose energy by radiating light except in energy quanta defined by multiples of Planck's constant. His predictions for the spectra of photons emitted from atoms of these elements seemed to be in excellent agreement with spectroscopic data by Edward Charles Pickering and Ralph Fowler, Rutherford's son-in-law, subsequently a distinguished theoretical physicist. Later, more precise and easily interpretable data was added by Henry Moseley. However, the mechanism by which Bohr's ideas could be understood, in particular how and when atoms emitted photons, remained deeply mysterious. Von Hevesy told Einstein about Bohr's theory and the experimental data in Vienna. According to von Hevesy's account in letters to Bohr and Rutherford, Einstein's eyes opened wide in surprise and he said that, if correct, this was one of the greatest discoveries.

Einstein left Vienna for Berlin on 24 September, while Mileva returned to Novi Sad and the children. While Einstein no doubt had topics to discuss with his future colleagues in Berlin, the

---

[7] This was the organ of the Austrian Social Democratic Workers Party, of which Einstein's old friend Fritz Adler was by then the General Secretary. It would seem highly likely that Adler attended Einstein's lecture and had a hand in the newspaper coverage. Certainly, Frank reported that Einstein met Viktor Adler, Fritz's father and the founder of the party, on this trip.

[8] This was not the first time Einstein had made the headlines. A year earlier, Philipp Frank had been startled to read an account of special relativity in a Vienna newspaper entitled: 'The Minute in Danger. A Sensation in Mathematical Physics' by the physicist Ernst Lecher.

most important aspect of his trip for him was his reunion with Elsa. After leaving Berlin, he travelled on to visit his mother in Heilbronn, who subsequently joined him in visiting an aged and apparently senile uncle in Einstein's birthplace of Ulm. He was back in Zurich, accompanied by his mother, by 9 October, slightly late for the beginning of the final semester in which he would have any responsibility for undergraduate teaching.

Given the euphoric tone of Einstein's letter to Elsa immediately after his return to Zurich, it can be assumed that their meeting had gone well. That their relationship was developing is confirmed by a letter in mid-October, in which he reported that his domestic situation had deteriorated still further and that an icy silence reigned in the house. No doubt Mileva, with her highly attuned antennae, had detected the tell-tale signs of the brightening of Einstein's mood and associated it with the usual cause, a woman. Certainly, Einstein's mother seemed to be aware of the budding liaison, and given her endemic hostility to Mileva gave it her tacit approval. Pauline Einstein's presence in the house can only have further depressed Mileva's spirits. It certainly seems that the relation between Einstein and Mileva now began a rapid deterioration, so that by the end of 1913 they were almost totally estranged. In letters to Elsa from this period, Einstein refers to Mileva as 'an employee who I can't dismiss'. He continued that 'I have my own bedroom and avoid being alone with her. In this way I can survive our cohabitation very well.' He remarked that she was 'an unfriendly, humourless creature who gets nothing from life and whose very presence sucks life's pleasure from others.' He even implied that she had the 'evil eye'.

The 1913 Solvay conference was held 27 October–3 November. Einstein requested leave of absence for this week to allow himself and his colleague Pierre Weiss to attend. It can therefore be assumed that he left for Brussels on Friday 24 or Saturday 25 October. Einstein played a much less prominent part in this conference than in the first. He had explicitly asked to be excused from being a *rapporteur* on any of the topics of discussion. In any case, he was still completely absorbed by his work on gravitation. Despite the interest the topic held for Solvay himself, this was peripheral to the theme of the conference, which was the structure of matter. The Rutherford nuclear atom was hardly discussed. Although Bohr's theory was certainly known to Einstein from his discussions with von Hevesy (see above), and to the other prominent participants such as J. J. Thomson and Rutherford, it was not mentioned in the conference. Thomson in his report concentrated on his own highly complicated atomic model which envisaged electrons bound with positive charges at stable points inside the atom. Einstein himself, in common with most German physicists at that time, never evinced any particular interest in models for the structure of the atom, nor indeed, later in life, in models for other 'elementary' particles. The reason for this is presumably that all required aspects of quantum theory with which he felt uncomfortable. Other than having to toast their host, Solvay, at a banquet, he confined his only substantial remark at the conference to his old love: statistical mechanics.

On 7 November, Adriaan Fokker, a student of Lorentz, arrived in Zurich to work with Einstein. He had just completed his thesis, a substantial part of which he had carried out in collaboration with Max Planck. Fokker was an enthusiastic musician, who as a student had conducted various musical groups.[9] At first, they worked on various topics in radiation and quantum physics, but as Fokker reported to Lorenz, he felt sure that gravitation would somehow come into their

[9] Gradually Fokker's musical pursuits supplanted his physics; during the Second World War, he became obsessed by the musical theories of two other eminent scientists, Christian Huygens and Leonhard Euler, and explored various microtonal systems in many well-known compositions.

sights. Before long they were working on putting the scalar theory of Nordström, who had by then returned to Helsinki, into the tensor language of the Entwurf theory.

On 22 November, Gustav Roethe wrote on behalf of the Prussian Academy of Sciences to apprise Einstein of his election to the Academy with a special salary of 12,900 marks, removal expenses, a number of other benefits, and no teaching responsibilities. According to Lisbeth Hurwitz's diary, on the following day Einstein and Mileva attended a Hurwitz musical evening. After the playing was over, they all discussed the Berlin move. There was a generous counter-offer on the table from Ghnem, the Director of ETH, that would have both increased Einstein's salary and absolved him of any teaching duties. Einstein himself was not without doubts as to the wisdom of relocating to Berlin. He intensely disliked the Prussian mentality, whose pale reflection in Munich had driven him out as a schoolboy. However, there can hardly have been much doubt as to Einstein's final decision; both the prestige of the Berlin position and the prospect of renewing his relationship with Elsa were very strong attractions. Indeed, some months later, Einstein wrote to Zangger that Elsa was the main reason for his move. He accepted the offer on 7 December. The residual uncertainty and questions that nagged at the back of his mind about aspects of the Entwurf theory were probably behind his uncharacteristically insecure remark to another of his old ETH classmates, Louis Kollros. After a farewell dinner that Einstein's friends had organised at the Kronenhalle at the end of December, Kollros had accompanied him back home up the Zurichberg. On their way, Einstein had remarked that 'The gentlemen in Berlin are betting on me as if I were a prize hen. I am not sure that I can still lay eggs!'

Christmas brought little cheer to the Einstein household. Mileva was dreading the move to Berlin, particularly because Pauline Einstein was also planning to move there in the next few months to look after her brother Jacob, whose wife Julie was on her deathbed. Mileva's relationship with her mother-in-law could hardly get any worse. She was however particularly incensed when Pauline sent a postcard on 21 December addressed to Einstein alone and sent the children's Christmas presents to Maja for onward transmission, rather than to Mileva. The latter was so annoyed that she returned the presents accompanied by a letter saying that from henceforward neither she nor the children would have anything to do with Pauline. Music was, however, still a treasured escape for Mileva and Einstein. On 23 December, both attended another musical evening at the Hurwitz home. Directly after Christmas, Mileva left for Berlin to stay with the Habers and look for accommodation for the family, leaving Einstein to go to the Hurwitz house with Gasser, who came from Winterthur for more music-making just before New Year's Eve.

Mileva returned to Zurich in mid-January 1914, having found a suitable apartment in Dahlen, the area in Berlin in which Haber's Institute was situated and where Einstein would initially have office space. She had sparred with the various members of the Einstein family she had encountered, including Elsa, to whom she was rude and of whom she was suspicious. Einstein's response to this was to refuse to discuss members of his family with Mileva, although by this stage they were communicating as little as possible anyway. Around 20 January, Einstein congratulated Freundlich, who now was confident that, with Planck's help, he had been able to get the funding to equip the planned expedition to photograph the solar eclipse that summer in Russia. Although unconcerned about many aspects of the move to Berlin, including accommodation, Einstein was careful to ensure that suitable musical arrangements were in hand. He assured Freundlich, who was a cellist, that they would set up the string quartet they had discussed as soon as possible after his arrival.

That last winter in Zurich, with long periods of cold, grey, and gloomy weather, saw further work on gravitation, despite a brief period spent investigating a question of quantum physics. A great deal of the work on gravitation was done with Fokker, in whom Einstein found a congenial collaborator. The result was a very beautiful paper, in which they demonstrated that Nordström's scalar theory was a special case of the Entwurf equations in which the velocity of light is a constant.

Einstein also continued to work with Grossmann that winter. He produced several brief papers in response to criticisms of the Entwurf theory from Max Abraham and Gustav Mie, who attacked from differing, but equally vehement standpoints. Of the two, he had great respect for Abraham as a physicist, although he considered him to be misguided about gravitation. In contrast, he had little time for Mie's somewhat baroque theories. His attitude to these criticisms was combative; always ready for a good argument about physics, he made a musical analogy in a letter to his friend Zangger, quoting Figaro from the Mozart opera: 'Would the Count [Almaviva] like to do a little dance? Just say so! I will strike up.'

In the course of their work, Einstein and Grossmann discovered an unwarranted assumption that they had made about the covariance properties of the gravitational field tensor. When they relaxed this, they were delighted to find that the Entwurf theory had the maximum covariance properties consistent with the condition that the metric tensor was uniquely determined throughout all space by the distribution of matter and energy. Their discussion was largely contingent on the 'hole' argument mentioned above. Sadly, as also discussed earlier, such a condition is not required in nature. Their work led Einstein to even greater conviction that they were on the right track. Their final joint paper discussing these issues appeared after Einstein had already left Zurich for Berlin. The realisation that the Entwurf theory was in fact incorrect and the final push towards general relativity would only come more than a year later, after Einstein's collaboration with Grossmann had ended.

Mileva was fully occupied in preparing the move to Berlin and with the health of little Tete, which was once more causing concern. He was bedridden for several weeks; the doctors prescribed a healthier climate, so that Mileva and the boys left for Locarno at the end of March. Einstein had already left Zurich. He had completed his goodbyes to his scientific colleagues, giving his final paper to the Zurich Scientific Society on 9 February, on the theory of gravitation. He also said goodbye to his musical friends. Lisbeth Hurwitz's diary entry for 16 March read: 'This was the last time that Mr Einstein played at our house (Schumann Quintet, with myself as second violin, Mozart Quartet). Soon he will leave for Berlin.' Despite her favourite composer Schumann being on the programme, Mileva sat alone and silent throughout the evening. On 21 March, the day of his departure, Alfred Stern presented Einstein with a copy of his book inscribed 'To his dear friend Albert Einstein in remembrance of Alfred Stern.' Einstein completed his teaching duties at the end of the semester and immediately caught the train for Rotterdam, where he visited his Uncle Caesar overnight before travelling on to stay with the Ehrenfests in Leiden. Mileva and the children's stay in Locarno was a relief for Einstein, as it would give him a couple of weeks alone in Berlin. There he could renew his intimacy with Elsa without his wife's jealous, gimlet eye always fastened upon him.

Chapter 12

Berlin, war, marriage breakup, and general relativity completed (1914–1915)

Einstein's March 1914 visit with the Ehrenfests in Leiden was stimulating. His days were filled by music and physics. By the end of the trip, he proposed that Paul Ehrenfest and he should use the familiar 'Du' form together. Ehrenfest was overjoyed at this signal mark of confidence and affection. In those far-off days of linguistic formality,[1] the degree of intimacy that this entailed was significant and Einstein was sparing in granting it. Even such old and intimate friends as the Habicht brothers and Heinrich Zangger were never similarly honoured.

Einstein arrived in Berlin on Sunday 29 March 1914, where he stayed with his Uncle Rudolf. It can be assumed that he met Elsa very shortly thereafter. The rest of the week was taken up with establishing himself in his assigned room in Fritz Haber's Institute and visiting various colleagues in the city. Only on 2 April did he write to Mileva, who was still with the children in Locarno. In fact, the letter was written more to the children than to his wife. It was addressed 'Dear Mama' and signed 'Papa'. He gave brief news on his doings in Berlin and the situation with their new apartment, which was still being renovated. Their furniture was due to arrive the following week. Mileva and the children eventually arrived in Berlin at the end of the month.

The early correspondence from Berlin makes it clear quite how bowled over Einstein was by his new scientific base. Berlin was the centre of European science at the time and, as he was introduced to the scientific luminaries there, he found himself more and more distracted by the intellectual stimulation they catalysed. His obsession with gravity was swept away temporarily in this euphoria and he dabbled in a wide variety of issues, mostly quantum related but also photochemistry and several others. No doubt these more chemical topics were suggested to him in conversations by his fellow denizens in Haber's Institute. In a letter to Wilhelm Wien in June he remarked that Berlin

[1] Since this question of duzen, use of the intimate form of address in German, will come up again and again in the course of this book, it is worth clarifying its significance. This is particularly the case since even native German speakers younger than say 40 may be puzzled by the emphasis placed on it. In modern German, the distinction between the formal Sie and the intimate Du has almost disappeared; for example, it is common for teenage shop assistants to address venerable customers as Du, something that would have been utterly unthinkable in Einstein's time. Then, Du was strictly limited to family members, to and between children, animals (!), and intimate friends and lovers. To agree to duzen was a rather formal procedure, initiated by the elder or higher-ranking interlocutor and considered a major event. Some indication of the importance placed on it can be seen from a letter that one (future) Nobel Laureate, Erwin Schrödinger, by then successor to Max Planck as Professor in the University of Berlin, wrote to another, Einstein, in March 1930. Two paragraphs, more than half of the letter, are devoted to effusive thanks for Einstein addressing him as Du, even though Schrödinger thought it might have been just a slip of the pen.

Einstein. Brian Foster, Oxford University Press. © Brian Foster (2026). DOI: 10.1093/oso/9780198794875.003.0012

was extraordinarily inspiring. This whirl of intellectual and organisational activities was such that he was unable to play any music during these first weeks, an exceedingly rare occurrence in his life thus far.

That Einstein was by now a recognisable figure in the wider German-speaking world as well as in specialised scientific circles is evinced by the invitation he received to write a piece for the *Vossische Zeitung*. This was the German equivalent as a paper of record to *The Times* of London. The article was intended to introduce him to the German public on his arrival in Berlin. This very clear summary of his work on relativity appeared on 26 April. In the concluding paragraph he implied that he had already succeeded in carrying out a generalisation of relativity to accelerated motion and gravity that only awaited experimental confirmation by a measurement of the curvature of light in a gravitational field. He made a similar remark a couple of months later in the letter to Wien referred to above and in his inaugural lecture. Einstein delivered this on 2 July in the new rooms of the Prussian Academy of Sciences on Unter den Linden. This very brief and general address outlined the current state of theoretical physics, highlighting the quantum hypothesis and relativity. Einstein once again implied that he had solved the extension of relativity to accelerated motion and gravity and that it awaited experimental confirmation. The next months were to disabuse him of this conviction.

At least it seemed that the experimental test of the bending of light by massive bodies was going ahead. Erwin Freundlich was absorbed in preparations for his expedition to observe and photograph the solar eclipse, so that he had not yet been able to organise the string quartet that he had discussed with Einstein. Freundlich had concluded that a new expedition was necessary after examining most of the extant photographic plates. It turned out that they could not be used to look for the predicted effect since their exposure times were determined for other investigations. One such, ironically, was the search for intra-Mercurial planets that might explain the precession of Mercury. What was required to test Einstein's predicted effect was a much longer exposure and settings suitable to observe distant stars near the sun's limb during totality. The zone of totality was due to pass across Scandinavia, Russia, and parts of the Ottoman Empire on 21 August 1914. The longest duration of totality was in Russia, so this was chosen by Freundlich as the goal of the expedition. Einstein exerted himself to find funds, approaching Max Planck among others. He even at one stage offered to contribute himself. A combination of official and private funding sources eventually was found and planning for the expedition began in earnest.

Presumably Mileva and the children moved into the apartment that Mileva had found the previous winter, near to Haber's Institute, as soon as they arrived in Berlin. Einstein moved there sometime towards the end of May, since in mid-May, he was still sending letters from his uncle's apartment in Charlottenburg, where his mother Pauline had been diagnosed with stomach cancer. It was at this point that Hans Müsam, who was to become Einstein's closest friend in Berlin, entered his life. He treated his mother and other members of the family without charging, a gesture that Einstein still gratefully remembered more than thirty years later. Pauline was operated on, initially with success, early in July.

Ehrenfest had been bombarding Einstein with letters relating to the statistical behaviour of ensembles of resonators. He was desperate to discuss the topic in person, so he travelled to Berlin and stayed with Einstein and his family in Ehrenbergstrasse between 29 May and 4 June. In fact, they mostly discussed gravitation and Ehrenfest, as always, went out of his way to entertain and instruct Hans Albert, drawing two-dimensional men for him in his diary. Although Ehrenfest appeared not to notice anything amiss, Einstein's domestic arrangements did not endure long after his departure. Einstein began to absent himself from home for days at a time, without any indication of his whereabouts; presumably he was with Elsa. A memorial, written by Anna Besso to document a conversation that she had had with Mileva years after the event, records that Einstein attempted to install a lodger in the apartment without consulting Mileva. Her presumption was that this was an attempt to drive her out. At some point in mid-July, she reached breaking point. Rather than sitting with the children awaiting Einstein's rare and unannounced returns home, which inevitably resulted in shouting matches, she decamped from Ehrenbergstrasse to Haber's nearby villa on the invitation of his wife Clara, with whom Mileva had become friendly on her earlier house-hunting visit. They were thrown together by mutual exasperation over brilliant but unsatisfactory husbands. While Mileva's marriage ended in divorce, Clara's ended tragically with her suicide.

Haber found himself having to mediate between Einstein and Mileva. It rapidly became clear that this was an unenviable task. It seems that all the pent-up resentment of Mileva's behaviour in the matters of Anna Schmid and Marie Winteler, together with his strong attraction to Elsa, combined to make Einstein resolve to terminate his marriage. The only force pulling against this decision was the inevitable loss of his children, which was indeed a powerful deterrent. There can be no doubt that Einstein was extremely fond of both his sons. The prospect of losing them was enough to induce him to make one last attempt to live with Mileva. He felt it necessary to commit to paper his conditions for this around 17 July. It forms a sad indication of the level to which their relationship had sunk.

Conditions

A. You will take care:
   1) that my clothes and washing are kept in good order
   2) that I get three regular meals a day served *in my room*
   3) that my bedroom and workroom are kept in good order and especially that my desk is reserved for *my use only*

B. You give up all personal relations with me except where necessary for social occasions. In particular you will not expect
   1) that I will sit with you at home
   2) that I accompany you outside the house or on trips

C. You explicitly commit yourself to observing the following points in your relations with me:
   1) You expect no affection from me and will not reproach me for lack of it
   2) You are to cease any conversation with me when I request it
   3) You will leave my bedroom or workroom immediately without demur when I request it

D. You commit yourself not to demean me in the eyes of my children either by speech or action.[2]

As a sort of postscript to this 'contract', Einstein added a number of unpleasant remarks both to Mileva and for her to read out to Hans Albert, whom he accused of being lazy.

To the contemporary eye, this is a shocking document. Even attempting to consider the very different *mores* of more than a century ago, it still seems vindictive, petty, and degrading. Just

[2] Emphasis in italics is Einstein's.

how different attitudes were in those far-off days is indicated by the fact that Mileva wrote back almost immediately *accepting* Einstein's conditions. However, even this was not enough for Einstein. He replied insisting on confirmation that she really understood what she was agreeing to and explicitly pointing out that they were entering into a business relationship. After remonstrations from Haber, Einstein did soften this somewhat in the following letter to allow the possibility that their relations might improve with time. The tone of this sequence of letters does give the impression that the 'contract' was written while Einstein was in a rage, which gradually cooled. Eventually, Mileva thought better of her acceptance and instead they initiated the process of separation.

Einstein's old friend Michele Besso, who had been his confidant in the assignation with Marie Winteler, now arrived in Berlin, presumably at Einstein's request. On Friday 24 July Einstein went to Haber's villa where he, Mileva, and Haber drew up an agreement whereby the couple would separate, and Einstein would provide financial support for his family after their return to Zurich. Einstein agreed to provide a total of 5,600 marks per year from his salary of 12,900 marks. Mileva also registered her visceral dislike of Einstein's family, in particular Pauline and Elsa, by insisting that she should never have to give up her children to any of them. Einstein left to give a lecture on quantum theory to the German Physical Society and Haber agreed to get the document drawn up by lawyers. Early the following week, a meeting was set up at Haber's lawyers which was attended by Haber, Besso, and Mileva. Einstein was supposed to attend but did not. Presumably Besso agreed on his behalf and then took the document back to Einstein to be signed.

Meanwhile, Elsa had left the scene of conflict for a vacation in the Alps with her two daughters, probably on 25–26 July, just after the meeting in Haber's villa. Einstein wrote to her complaining about how difficult it was for him to give up his children, which he implied he was doing for her. Finally, on Wednesday 29 July, Mileva and the children arrived at the station to board the train for Zurich. Mileva had apparently made one last attempt to save her marriage by visiting Einstein, but he was now sure that separation and eventually divorce formed the only solution. He entrusted his family to Besso's care, who was travelling to Zurich with them. Standing side by side with Haber, he waved goodbye, tears streaming down his cheeks. Haber remained with him for the rest of the day. Einstein was inconsolable and remarked that he would not have made it through that terrible day without Haber's help and sympathy. He remarked to his confidant Zangger that, during this period, he awoke every morning and recollected the loss of his children as if it were a dagger thrust into his heart. As he cried himself to sleep on 29 July, he could console himself with the knowledge that Zurich was only a day's train journey away and that he would see Hans Albert and Tete often. However, a world crisis was brewing that would dwarf his personal troubles while at the same time greatly exacerbating them by cutting him off from the children he loved. As British Foreign Secretary Sir Edward Grey is said to have remarked, the lamps were going out all over Europe.

The Archduke Franz Ferdinand, heir to the throne of Austria–Hungary, was assassinated on 28 June 1914. The complex structure of alliances that had grown up in Europe since the turn of the century was both inflexible and unstable. When Austria presented Serbia, blamed for inciting the assassination, with an ultimatum, Russia, traditional protector of the Slav nations, stepped in to confront Austria. Germany stood by its ally Austria, while France fell in behind Russia. Once Russia began to mobilise its troops on 30 July, Germany and Austria followed suit the next day. Despite desperate diplomatic attempts to avert catastrophe, the mobilisation plans of both sides, once launched, were too inflexible to be stopped. Germany declared war on Russia on 1 August. The Germans declared war on France on 3 August and invaded Belgium on 4 August, which finally drew a prevaricating British government into hostilities. The First World War had begun.

The dismay with which Einstein received these events, coming almost simultaneously with his own personal tragedy, can only be imagined. Throughout his life he had had a deep aversion to militarism, particularly of the Prussian variety; it had been the main cause of his renunciation of his original citizenship and flight from Munich. The war put the generals even more firmly in charge of all aspects of life in Berlin. Einstein's beliefs also placed him in a tiny minority among his colleagues in the Prussian Academy and in the wider German scientific community. Almost to a man (there were very few women) they fell in behind the Kaiser, many with a full-blooded enthusiasm that surprises. His relationship with Haber was strained but survived the test, even though Haber immediately became a captain and by 1915 was deeply involved in the production and deployment of gas weapons on the battlefield. Nernst also became an officer and even Max Planck initially welcomed the war, in which his two eldest sons served. His old assistant, Ludwig Hopf, who had moved into the new field of aeronautics, worked on the development of military aircraft. Einstein's attitude is clearly demonstrated in a postcard he sent to Ehrenfest on 19 August:

> Unbelievable things have now begun in Europe's madness. In such times it is apparent to what sort of pathetic herd of cattle one belongs. I drift along in my peaceful contemplations, feeling only a mixture of pity and disgust. My family is also in Switzerland and will stay there. I remain alone and in constant tranquillity in my large apartment. My trusty astronomer Freundlich will experience Russian captivity rather than the solar eclipse. I fear for him.

Freundlich, with fellow astronomer Walther Zurhellen and an expert technician, had set off for the Crimea earlier in the summer, when the possibility of war had seemed distant. This was after all only the latest in a succession of Balkan crises over previous decades. Unfortunately for Freundlich, this one proved to be the real thing. He and his colleagues were arrested, together with their sophisticated cameras, which only increased the suspicion that they were spies rather than scientists. They were interned in Odessa for several weeks until an exchange with captured Russian officers could be arranged. Einstein knew of Freundlich's plight and may have intervened with colleagues influential with the military to try to obtain his release. Despite success in this regard, Freundlich's equipment was not returned.[3] Although Freundlich and colleagues[4] were released in September, it took several weeks for them to find a way back to Berlin through a Europe divided by the front lines of war. In fact, Freundlich's observation would have failed even had there been no war. An expedition from the Lick Observatory from still-neutral United States led by William Wallace Campbell, a correspondent of Freundlich's, was allowed to proceed by the Russian authorities but was foiled by cloudy skies.

The repulse of the German advance at the Battle of the Marne in September led to the realisation that, unlike the Franco–Prussian War of 1870, this conflict was not going to be won in a matter of weeks. Slowly, as the initial war of manoeuvre evolved into trench warfare, news of alleged German atrocities during the invasion of Belgium began to filter back to Germany through the heavily censored media. The general reaction was to characterise these reports as enemy propaganda. Many groups protested against these slurs on the German army and hence the German nation. Scientists and intellectuals were also scandalised. This led to the drawing up of a round-robin that was signed by almost all of Einstein's prominent colleagues, including Planck. Eventually ninety-three leaders of the world of science and the arts subscribed to the manifesto, which was addressed to the 'cultured world' and was prominently published in

[3] It eventually was recovered and returned to Berlin only in spring 1923.

[4] Zurhellen was killed in 1916 in the trenches in France.

German newspapers on 4 October. It denied the reports of atrocities, implying fabrications and claiming that such things were impossible from the nation that had produced Beethoven and Goethe. It promulgated the proposition, astonishing from today's perspective, that German culture and German militarism were identical. Einstein refused to sign this document. Since he had remained a Swiss citizen and had refused to take up Prussian citizenship on election to the Prussian Academy, this was not so serious. What was serious, however, was that he, along with only three others, signed a counter-manifesto produced by Elsa Einstein's old friend and physician Georg Nicolai that called for an end to the madness of war and the establishment of a European state. Needless to say, this document received no publicity whatever, although it was widely circulated among his colleagues, which certainly did not reduce Einstein's isolation.

As the cataclysm enveloped Europe, Mileva's own private tragedy intensified. That she still loved Einstein and hoped for a reconciliation is clear. She found herself now however cut off from him by the dislocation of war. Travelling between neutral Switzerland and Germany became fraught with difficulty. Their furniture, which Einstein had agreed to ship almost in its entirety except for a couple of beds and items originating from his own family, was delayed for weeks because trains were needed for military purposes and were travelling east–west; there was essentially no traffic north–south between Berlin and Zurich. Mileva and the boys stayed in a hotel until 7 September, when money problems precipitated a move to the Bahnhofstrasse home of Hans Wohlwend, Einstein's old friend and music partner. When Mileva wrote the same day to Einstein asking for more money, Einstein sent a savage response. He listed all the funds he had so far provided, including the furniture shipment, and stated that he couldn't send any more money since he had none himself until his next monthly pay cheque arrived. Although he committed himself to send 400 Swiss francs per month, this was substantially less than the 7,000 annually he had promised in their legal agreement. However, he did adduce various other payments he was making on her behalf in addition to the monthly allowance. He also accused her of turning the children against him, presumably on the basis that he had received no communication from them since they left Berlin. Indeed, as late as December of that year he was complaining that Mileva was presumably not passing on his letters to Hans Albert, since he had received no reply. It is more likely that Hans Albert was sufficiently upset with the separation that he could not bear to communicate with his father. It was inevitable that he would hear his mother's side of the story, although even Einstein's side would hardly have been less damaging. Finally, in his September letter, Einstein accused Mileva of turning his friends against him, presumably assuming that she would be complaining about him to Wohlwend.

Keeping his family supplied with money was, for Einstein, another problem that was exacerbated by the war. As the cost of the conflict mounted and the Allied naval blockade bit harder and harder, the value of the German mark fell with respect to the Swiss franc. Despite Einstein carrying out his obligation to send almost half of his salary to Zurich, it bought less and less. In the first year of the war, the drop was only around 10%, but by mid-1917, the German mark had almost halved in value, although there was a brief recovery in late 1917 after Russia had been knocked out of the war. Even in those first days of their return to Zurich, Mileva found it hard to make ends meet. She looked around desperately for employment, including teaching mathematics and piano, although the latter would have to await the arrival of the furniture from Berlin. After this eventually arrived, Mileva found in November an apartment in Voltastrasse in the district she knew so well, up the hill behind the ETH.

Just after the New Year of 1915, Einstein sent a reprint of his last paper on gravitation to Paolo Straneo, Professor of Mathematical Physics at the University of Turin. There was nothing unusual in that action, except that the receipt of the request had stirred a recollection of happier and carefree times in Einstein's mind. He reminded Straneo that they had met once before many years earlier, at the household of the Marangonis in Casteggio, where the teenage Einstein had frequently played the violin to the accompaniment of the captivating Ernestina.

In January, Einstein wrote to Hans Albert at their new address. Einstein too had relocated, from the large apartment with its unhappy memories of his family's brief occupation to a much smaller apartment near the city centre, Wittelsbacherstrasse 13. He had sent the children Christmas presents, including additions for Hans Albert's sled, in great demand on the Zurichberg during a particularly snowy winter in Switzerland. The letter congratulates his son on being in Zurich, a much pleasanter place for a boy than Berlin. He hoped that they could do some mountain walking together in the summer and that by then the war would be over. He also urged him to continue with his piano practice, a wish that he repeated in subsequent letters, telling him that if he could make beautiful music, he could give himself and others much pleasure. Einstein indicated that he had finally been able to return to making music himself. He told Hans Albert that he was playing in a benefit concert that evening on behalf of two impoverished artists. This was the first of many similar events throughout the rest of his life in which Einstein would play his beloved violin in public on behalf of good causes.

While his familiar world of family and peacetime civil society had been crumbling around him, two things had sustained Einstein: the comforting, undemanding affection of Elsa, and physics. In the six or so months after his family's return to Switzerland, gravitation was not the obsession it had been before or was to be shortly afterwards, although he embarked on a long and fruitful correspondence with Tullio Levi-Civita on various aspects of the tensor manipulation that he and Grossmann had published. Eventually he had to concede that one aspect of the proof of an important element of the theory was incomplete. However, nothing in this rather technical discussion shook his faith that the Entwurf theory was, if not the final answer, at least well on the way to it. This conviction allowed him to turn his attention to other matters. Somewhat to his surprise, he found himself, as he remarked to Besso, becoming an experimentalist in his old age.

The reason for Einstein's interest in the topic of the magnetic properties of materials is another reminder of his youthful activities, this time as a Patent Clerk. In an effort to replenish his bank account, depleted by his payments to Mileva, Einstein agreed to provide an expert opinion in a patent case brought by a German company against the Sperry Gyroscope Company of the US. Thinking of gyroscopes while writing this opinion presumably started Einstein speculating on their other roles in physics. The prevailing idea on why some materials were magnetic was that small gyroscope-like circulating currents, presumably electrons, aligned themselves so that each tiny current added to produce a net macroscopic effect. That there was a force between such electric currents had been known since André-Marie Ampère's work almost a century before.

Einstein, who was probably familiar with Maxwell's work on this subject, thought that there might be a way to test this idea by suspending a magnet or a ferromagnetic rod between the poles of an electromagnet and switching its polarity. If this was done at the appropriate frequency using alternating current, the natural resonant frequency of the suspended magnet could magnify the deflection due to the force on the electron loops sufficiently to make it observable, allowing the parameters of the tiny circulating electron currents to be determined. This idea might never have got beyond an idle thought on a Berlin tram had not a gifted experimentalist been in Einstein's social circle. Wander Johannes de Haas, then working as a researcher at the Physikalisch-Technische Reichsanstalt, was the son-in-law of Hendrik Lorentz. Einstein, through the Lorentz connection, was friendly with both him and his wife, who was also a good physicist. Presumably they talked about Einstein's idea casually and de Haas, keen to work with someone of Einstein's reputation, offered to set up the experiment at the Reichsanstalt.

Einstein and de Haas made rapid progress, aided by two experts from the Reichsanstalt, Wilhelm Jaeger and Ernst Orlich. By February they had obtained a positive result, which Einstein announced in a lecture to the German Physical Society. The resulting paper, entitled with no lack of ambition 'Experimental Proof of Ampère's Molecular Currents', appeared in April. The description of the theory and experimental setup is quite detailed. However, in contrast to the requirements of modern research papers, there is no discussion of the experimental uncertainties. The only mention is a bald statement that 'we have to assign an uncertainty of around 10% to our measurement'. In fact, the systematic errors of their determination were almost certainly much larger: a host of factors needed to be taken into consideration, including the effect of the earth's magnetic field, the alignment of the cylinder in the electromagnet, etc. Although all of Einstein's theoretical preconceptions expected the result they obtained experimentally, in fact the modern accepted value, due to factors unknown to Einstein and de Haas at the time, is twice as large. De Haas carried on the investigations after he returned to the Netherlands on 1 April to take up a position as a schoolteacher. He left Einstein to organise the move of his family's possessions, which turned out to be both difficult and expensive.[5] He also left Einstein to continue to try out ideas to improve the experiment on his own for several more months. They jointly published several more papers and comments over the succeeding months. Both Einstein and de Haas were honoured for this work by the award of the Freiherr von Baumgartner Prize of the Imperial Viennese Academy in June 1917. De Haas, who eventually was appointed to a research position, substantially owing to his collaboration with Einstein, continued to work in this area for the next ten years, long after Einstein had become absorbed back into theoretical pursuits. De Haas eventually accepted that the larger value obtained by many other experimenters was in fact correct, thereby implicitly admitting that his and Einstein's original result was flawed.

In the midst of the tragedy of war (and indeed related to it) came the personal tragedy of the suicide of Clara Haber. She was herself a practising research chemist, the first German woman to obtain a PhD in chemistry, until she married Haber. She subsequently fell into the expected role of housekeeper and companion, although she continued to work as an unrecognised assistant of her husband. She was however strongly opposed to Haber's research on chemical weapons, which resulted in the first release of poison gas in Belgium on 22 April 1915. Haber supervised the release, returning home at the end of the month. On 2 May, Clara used her husband's military-issue pistol to shoot herself. Her reasons are unknown, but it seems highly likely that they were related to her objections to her husband's war work. Einstein was unable to reciprocate the help that he had received from Haber when Mileva and the children returned to Zurich, since Haber travelled to the Eastern front to supervise the first release of poison gas there on the day after his wife's suicide. Mileva, who had warmed to Clara because of the obvious parallels in their lives, learned of the tragic news from Einstein, who in the tersest possible way mentioned it in the middle of a letter complaining about financial matters and her custody of the children.

It certainly seems that all was not well in Einstein's relations with his children. Despite having prepared and told several people that he intended to go on a walking tour in the Alps with Hans Albert, he had to change his plans when, sometime in early July, he received a postcard. In it,

[5] Einstein, having carried out a very similar task for Mileva a few months before, was becoming an expert in arranging movements of furniture. He wrote in jest to de Haas saying that if he ever wished to demonstrate his love for a woman, he should offer to arrange a house move for her in wartime.

very curtly, his son refused to accompany him. Clearly the separation of his parents continued to upset him greatly and, unsurprisingly, he blamed his father. Disappointed, Einstein decided to accompany Elsa and her children to Rügen from 15 July–5 August. This was his first vacation since adolescence not spent in the mountains. However, with one or two exceptions related to his children, he spent his future vacations by sea or lakeside. Hans Albert recalled that in later life even the sight of a mountain depressed his father. This complete change in Einstein's outlook may be ascribed to several factors. One was certainly a growing passion for sailing; he described himself a few years later to Besso as a sailing nut. There were painful associations that linked mountains with his first marriage and the separation from his sons. Most important were the restrictions on his ability to take strenuous walks imposed by his digestive complaints after 1918 and in particular his heart attack suffered while in the Alps in 1928, which he explicitly said precluded him from high altitudes (see Chapter 19).

After further exchanges of letters, Einstein finally managed to organise a trip to visit Hans Albert and Eduard in late August and September. This proved to be another nightmare of organisation. The German Empire, bureaucratic enough in peacetime, became impossibly so during the war. Einstein was singularly ill equipped to cope with the myriad forms and permits that he needed to obtain to travel to a neutral country. Indeed, an unauthorised trip to visit Göttingen to lecture on gravitation in June had already alerted the authorities to this subversive non-German citizen who seemed unaware that travel was no longer possible at the spontaneous whim of pre-war days. On 29 August, Einstein set out for Heilbronn, where his mother was once again living as housekeeper to Emil Oppenheimer. Only when Einstein arrived did it occur to him that his trip onwards to Switzerland would require documentation. When he visited the local police station, they reminded him that even travel inside Germany for a non-citizen required a permit that he did not possess. He mailed his passport to Elsa, who was charged with convincing the authorities to approve his trip post facto. At the end of the week, the passport was returned with the appropriate paperwork for him to continue his journey to Zurich, where he stayed in the Hotel Glockenhof. He managed to spend time with the children on two occasions, but Mileva was worried that he would turn them against her, so the visits were terminated. Einstein received support from Zangger and Besso, with whom he spent most of his time while in Zurich. He left to spend the weekend in Luzern with his sister Maja and Paul Winteler, returning to Zurich on Monday 13 September. He was unable to make any further visits to the children. He and Zangger did visit Romain Rolland, the outspoken pacifist who had left France for Vevey on Lake Geneva. However, Einstein does not seem to have met his Uncle Caesar, who had fled to Geneva from Antwerp, which had been besieged and taken by the Germans in October 1914.

Einstein's return trip was no less eventful than the outward leg. He found himself stranded on the Swiss–German border at Konstanz/Kreuzlingen, again for want of the appropriate stamp. This time it was Zangger who had to deal with the authorities at the German consulate in Zurich. Einstein was stranded in Kreuzlingen for three days. He could of course have returned to Zurich himself, a short train journey away, but he noted in a letter to Zangger that he would then have had to pay various social visits that he had managed to thus far avoid. He seems to have often expected his friends to sacrifice their time to suit his own convenience. Indeed, Zangger would spend a great deal of effort in the coming years in acting as an intermediary between Einstein on one hand and Mileva and the children on the other. Finally, Einstein received his passport, duly stamped, on 21 September and arrived back in Berlin on the evening of the following day.

The Berlin to which Einstein returned was beginning to feel the effects of the British naval blockade. By the autumn of 1915, German imports and exports had fallen by a factor of around two and food staples were beginning to disappear from the shelves. Philipp Frank described a visit to

Einstein during the war. He was invited to lunch but demurred, pointing out that in such times of shortage, no one could cope with extra guests. Einstein dismissed this with the remark that his uncle had more food than average so that he would be striking a blow for social equality were he to eat with them. On arriving at Einstein's uncle's apartment, Frank was introduced to Elsa. She remarked:

> I know very well what a talented physicist our Albertle is. These days one has to pick up food containers of all different sorts, and one never knows how to open them. They are often of some unknown foreign construction, rusty, dented, with the opening key lost – but there hasn't been one yet that our Albertle couldn't open!

Shortly after Einstein arrived back in Berlin, he began months of activity as strenuous and exhausting as any he had ever known. The centre of his attention swung back to gravity. It is not certain why Einstein suddenly realised that the Entwurf theory had to be discarded, or rather, as pointed out by Janssen and Renn, rebuilt from the foundations. After all, he had known for two years that the perihelion of Mercury was not correctly predicted. This was confirmed when Johannes Droste published his own calculation based on the Entwurf theory in December 1914. Einstein might have taken refuge in the possibility that some unknown body or bodies between Mercury and the sun caused the discrepancy. There is no doubt a great deal of truth in Janssen's view that Einstein tended to believe those results he wanted to believe and that he was adept at cooking up whatever arguments suited his needs, irrespective of whether they really held water. Indeed, his recent investigation with de Haas was a case in point.

Einstein and his trusty sounding board Besso would certainly have discussed their earlier work, including the Mercury calculations, during their meeting in Zurich. Perhaps this reawakened his doubts that rotating reference frames were properly treated in the Entwurf theory. Alternatively, the issue of rotating frames may have come up at a meeting with Erwin Freundlich, probably on 29 September, which was mostly related to Einstein's attempts to relieve Freundlich of his tedious routine duties at the Berlin observatory. There is some circumstantial evidence for this in a story that Freundlich told to the mathematician Walter Ledermann many years later. According to Ledermann, Freundlich claimed to have been the first to alert Einstein to the use of Riemannian geometry to describe gravity. While typical of the distortions and inaccuracies, amplified in the telling, that occur in stories about Einstein, there may be a grain of truth in this one. Perhaps the passage of years had magnified and simplified in Freundlich's memory his role in the development of the final theory of general relativity. It is also possible that he had mentioned Riemannian geometry to Einstein when they had first met but Einstein had immediately forgotten about it (see also Chapter 14 for another possible source of this recollection). As Felix Klein's PhD student in Göttingen, Freundlich would certainly have been more familiar with such concepts than Einstein. Whatever the catalyst was, the first indication that Einstein had realised that there was a major problem is a reference in a letter to Freundlich of 30 September. In this, in a tone of some excitement, he asked Freundlich to check a calculation showing that the Entwurf prediction for the gravitational field in a slowly rotating reference frame did not in fact reduce to the correct answer in the Newtonian limit.

The failure of the Entwurf formulation in rotating reference frames was the kiss of death as far as Einstein was concerned. Whereas he might have discounted the problem with the perihelion of Mercury as related to undiscovered astronomical bodies, rotation was central to his programme that all motion, accelerated as well as constant velocity, was relative. In his letter to Freundlich he pleaded that he really needed someone with a fresh eye to check the equations, which had become too familiar to him. He of course also worked feverishly himself. A few days later, he wrote to his regular correspondent Lorentz that he had discovered another error. In his major review of what he for the first time called 'general relativity', written at the end of 1914 for the Prussian Academy, he had proved that the Lagrangian he had used in his 1914 paper with Grossmann was unique.[6] He had now discovered that this was not the case, or rather that it was a consequence of the restrictions he had placed on the covariance of the gravitational equations that the rotation problem now impelled him to withdraw. Once it was clear that the Entwurf formulation was not forced on him, he began to reconsider all the possibilities that he had looked at in 1913 in the initial brainstorming with Grossmann.

At first, it seems likely that Einstein tried to remain inside the Entwurf framework, tinkering with some of the assumptions. It is also possible, as he himself implied in his paper describing his work, that he returned to the calculations of the 'Zurich Notebook' and simply put his trust in the mathematical beauty of the covariant formulation based on the Riemann tensor. Janssen and Renn make a persuasive, although not conclusive, case that it was the former; it is on their analysis that the following discussion is mostly based. More or less the minimum change that Einstein could make to the Entwurf framework was to modify the definition of the components of the gravitational field. He switched to a slightly more complicated definition, using the 'Christoffel symbols'. It turns out that these are a very natural choice since they play an important role in Riemannian geometry and define the shortest distance between two points. When he made this change, Einstein was struck by the fact that the resulting equations were essentially identical to those that he had worked out in the Zurich notebook from the Ricci tensor. This meant that, if he could find a way to return to this formulation, the much broader covariance that he had craved but given up regretfully in 1913 would once again be within his reach. The problem in 1913 had been that he was unable to prove that energy and momentum were conserved in the formulation derived from the Ricci tensor. Now he was armed with the Lagrangian formulation that he had developed in 1914, which made this task much easier. He established that energy-momentum was indeed conserved in a particular coordinate system, then carried out his usual procedure equating the new gravitational tensor, the 'November tensor' referred to in Chapter 11, to the tensor representing the total energy. This system of equations appeared highly promising to Einstein. Although its covariance was restricted to a class of transformations known as unimodular, this class was much broader than had been the case for the Entwurf system. Furthermore, Einstein convinced himself that rotational frames were also properly described. This was sufficient for him to rush to write a draft publication at the end of October.

To understand Einstein's haste, it is necessary to go back to the lecture course on relativity that he had given in Göttingen in June. Göttingen at that time was probably the premier university in the world for mathematical physics and mathematics. Although Klein was the doyen of this rarefied intellectual world, David Hilbert was by now a driving force. Einstein was very taken with his interactions with Hilbert during the week of his Göttingen lectures and flattered that Hilbert took such an interest in his exposition of general relativity. Indeed, Hilbert became sufficient interested that he began to try to resolve what he identified as problems with Einstein's Entwurf formulation. Both he, and others of his Göttingen mathematical colleagues, such as Klein, were privately scornful of Einstein's mathematical abilities. Hilbert famously remarked that 'Physics is really much too difficult for physicists.' He also said that 'Every boy in the streets of Göttingen understands more about four-dimensional geometry than Einstein.' Perhaps in retaliation, Einstein remarked

[6] Although this only follows after an additional assumption about the form of the Lagrangian that Einstein admitted he could not formally justify.

that 'The people in Göttingen sometimes strike me, not as if they want to help one formulate something clearly, but as if they want only to show us physicists how much brighter than us they are.' From this it can be seen that there was some considerable, mostly friendly, rivalry at work here.

Einstein learned sometime in the autumn that he was in a race with Hilbert to solve gravitation. Sommerfeld, an old friend and collaborator of Hilbert, gave Einstein the news that Hilbert had found an error in Einstein's 1914 review of the Entwurf theory and, presumably, that he was busily at work on the problem of gravity in general. This catalysed Einstein not only into a flurry of intellectual activity, but also feverish composition of lectures and papers. In November he lectured to the weekly meetings of the Prussian Academy on four successive occasions. This was not a grand and polished exposition spread over four well-planned lectures but rather a sequence of snapshots of his work in progress. As such, quite often he had to revise what he had described the previous week. His reason for this (at times unseemly) haste, other than the exuberance of his own velocity, was presumably to ensure that he had priority over Hilbert in publishing the final result.

Einstein's series of presentations to the Prussian Academy began on 4 November 1915. Here he presented the modifications he had made over the past three weeks that incorporated his insights relating to the Christoffel symbols and the happy return to the much broader covariance of basing his work on the Ricci tensor. He wrote, somewhat sheepishly, in the introduction to the published version of the paper that he had made a misjudgement in his previous presentation to the Academy. He had realised that his choice of Lagrangian was not only arbitrary, but wrong. He then went on to say that he had returned to his original conviction, that general covariance was the best guide to the correct equations. However, he then proceeded immediately to limit the covariance by restricting himself to unimodular transformations. He ended the introduction with the claim that anyone who really understood the solution he was about to outline could not fail but to be charmed by it because of the beauty of its mathematical foundation. As if forming a defensive redoubt, he deployed the big mathematical guns of Gauss, Riemann, Christoffel, Ricci, and his correspondent Levi-Civita.

The rest of this paper is a succinct exposition of Einstein's recent work: the replacement of his former ansatz for the components of the gravitation field, which he characterised as a fatal mistake, with the Christoffel symbols, which have additional symmetry compared to his original choice; the derivation of the new gravitational field equations in the Lagrangian form, which allowed a simple proof that energy and momentum were conserved; the demonstration that the weak-field case reduces to Newton's law, using a specific coordinate system that he used for this purpose only, a freedom that he had now realised existed and that he used here for the first time; finally, in a single paragraph, that the old bugbear, a slowly rotating coordinate frame, also left the new set of equations invariant. Janssen and Renn's view is that, having originally arrived at his new formulation by tinkering with the physical assumptions of the Entwurf theory, he found it simpler and more convincing to follow an exposition based on the 'mathematical' approach that he had followed in his original 1913 investigations. That the two approaches gave identical answers was deeply reassuring to Einstein. His paper was a tour de force. However, it was also, admittedly subtly, wrong.

Returning to his lonely Berlin apartment after his lecture to the Prussian Academy, Einstein still had an uneasy feeling that he hadn't yet quite arrived at the Holy Grail of the final theory of general relativity. The restriction to unimodular transformations seemed artificial. He was correct in this feeling but unfortunately his first attempt to circumvent it was unsuccessful. Once again harking back to his much-loved analogy to Maxwell's equations, he remembered that, if formulated for electromagnetism, his new equations would be generally covariant. He proposed therefore, following his old antagonist Gustav Mie,[7] that all matter was in fact electromagnetic in nature. He also assumed that gravitational forces were an important part of the make-up of matter, in which case his new equations would also probably be generally covariant. With this assumption, he perhaps also hoped to spike Hilbert's guns, since he knew that his rival had espoused Mie's ideas. Einstein referred to such a postulate in his paper as admittedly audacious. Pais considered it quite mad. The return for the audacious assumption was that now Einstein's equations were no longer restricted to unimodular transformations but, expressed in a suitable coordinate system, could be seen to be generally covariant. One week after his initial lecture to the Prussian Academy, Einstein delivered a second lecture in which these ideas were put forward. It appeared in print as an addendum to the original presentation.

Einstein wrote to Hilbert several times during this period, giving a snapshot of his thinking as his sequence of presentations to the Prussian Academy progressed. Hilbert wrote back, encouraging him to come to Göttingen to hear him present his own new ideas. Einstein declined however, citing a recurrence of his stomach complaints, the first concrete indication that the intense work of the past few weeks together with the poor wartime diet was having an effect on his constitution. He was undoubtedly alarmed by Hilbert's rather vague indications of his own progress in solving that subject 'too difficult for physicists', which Hilbert presented to the Mathematical Society of Göttingen on 16 November; he spoke about further developments at a meeting of the Royal Society in Göttingen on 20 November. He sent some sort of synopsis of his work to Einstein, who replied hurriedly on 18 November, going out of his way to imply that Hilbert's ideas, which Hilbert had claimed were completely different from Einstein's, were in fact 'identical, as far as I can see, to the ones that I have discovered in the past weeks and already reported to the Academy.' He went on to imply that anyone could find covariant equations from the Riemann tensor; the difficulty and what had delayed him for three years was to show that they reproduced Newton's laws in the appropriate limit. He no doubt considered it unlikely that Hilbert would concern himself with such mundane aspects of physics. He concluded his letter triumphantly with the news that, at his third lecture the same day, he would announce to the Academy that his calculations gave the correct answer for the advance in the perihelion of Mercury.

Einstein's third lecture marked the final moment of his emergence from the Slough of Despond in which he had toiled since 1907. Now he could visualise the full theory of gravitation. Armed with the confidence that the broad covariance of his new equations had given him, he turned to the familiar calculation that he had begun with Besso back in his Zurich days. Although, early the previous year, Einstein had returned their joint manuscript on the perihelion of Mercury, he either remembered the calculations very well or had likely kept a note of the key steps. This allowed him very quickly to recalculate the effect with his new formalism. When his calculation resulted in a value in excellent agreement with the measured value, in contrast with the much smaller effect that he and Besso had earlier calculated using the Entwurf theory, he was 'stunned by joyous excitement for days thereafter'. This could be no mistake. Not only were his soon-to-be-final equations covariant, but also they must be correct. Having once failed, with Besso, to derive the observed perihelion advance, he was convinced that this agreement must mean that he had finally arrived at the correct equations .

[7] Einstein didn't give Mie the satisfaction of a reference, referring instead to 'quite a lot' of people who held this view!

The second important result contained within Einstein's presentation on 18 November was that the bending of light was in fact twice as great as he had previously calculated.[8] As discussed in Chapter 10, Einstein had already predicted the bending of light by massive objects in 1907, as indeed had von Soldner in 1805. The 'extra' bending predicted by general relativity can be understood in a very simplified way: mass bends light, but in general relativity, mass in addition bends the space through which the light propagates. The lecture of 18 November relied heavily on the work that Besso and Einstein had carried out in 1913 and 1914, albeit with the wrong relativity equations. There is no doubt that Einstein had been the dominant partner in this enterprise, and that he was in a desperate rush to publish. It does however seem somewhat churlish that he did not give Besso an honourable mention with a footnote, as he had at the end of the special relativity paper of 1905. It would indeed have made a pleasing symmetry. With typical easy-going generosity, Besso didn't seem to mind. Einstein sent him a postcard on 17 November, announcing his success in the Mercury calculation, in attaining general covariance and his postulate that gravitation must have a role in the structure of matter. Had he waited a few more days before sending the card, he would not have included the third of these achievements.

One further lecture and one final step remained to put the cap on Einstein's crowning achievement. While redoing the Mercury calculations, he realised that his 'audacious' assumption that matter could in some unspecified way be reduced to electromagnetic interactions was in fact unnecessary. The addition of a simple term to the energy–momentum tensor could achieve the same effect. He had known of such a possibility back in his Zurich notebook but had discounted it since it did not give the results he had expected for weak static gravitational fields. His calculation of the Mercury perihelion advance involved such weak static fields which, in his final theory, he could see were demonstrably not spatially 'flat', as all his experience stretching back a decade had led him to expect. He was therefore now free to add the additional term, which, in the tensor world of general relativity, could be essentially considered as a constant offset that had no effect on the dynamical predictions of the theory, such as the bending of light or the perihelion calculation. It did moreover give a pleasing symmetry to the way in which energy–momentum entered the gravitational field and matter tensors in the equations. Energy–momentum conservation could now be plausibly inferred simply by inspection. Einstein was freed from his 'audacious' assumption. It no longer mattered whether matter might be constructed from some unspecified sort of electro-gravitational soup. On 25 November, he gave a final lecture to the Academy in which the 'Einstein equations' of general relativity as they are now understood were written down.

Einstein, as he remarked to Besso, was completely exhausted. It was however a happy exhaustion, buoyed up with the conviction that he had finally found the theory of gravity that he had sought for the previous decade. He lost little time, in between writing up his Academy lectures, in informing his usual correspondents. Besso,[9] Zangger, and Sommerfeld all received postcards or letters in November informing them of the happy news. Already, in the letter to Sommerfeld on 28 November, he implied that he got to his final equations by returning to a mathematical approach.

---

[8] Einstein's euphoria over the correct prediction of Mercury's perihelion in many ways explains his very relaxed attitude to the results of the eclipse expedition to measure the bending of light by the sun in 1919, to be discussed later. After all, as he implied in a letter to Sommerfeld, how could this be doubted since the Mercury calculation was correct?

[9] Besso and his family had recently moved from Gorizia to Krummenau in eastern Switzerland, where his father-in-law Jost Winteler lived in retirement. Besso moved to Zurich early in 1916. He had some months before, on Einstein's recommendation, met Zangger. The two became close friends, sharing much of the strain in being Einstein's plenipotentiaries with Mileva and the children.

The evidence to the contrary cited by Janssen and Renn implies that this was a rationalisation, but, if so, it was one that became settled in Einstein's mind and was to have a profound effect on his future work.

One correspondent who was not included in the flurry of Einstein letters was Hilbert. It is likely that Sommerfeld informed him of Einstein's work and there may well be missing letters between Hilbert and Einstein. There certainly seems to have been communication of some kind, although it may have been through third parties. Pais records a second-hand account that Hilbert had at some point in this period written an apology to Einstein, denying that he had poached any of Einstein's ideas because he had completely forgotten his lectures at Göttingen a few months previously. The first recorded letter that Einstein wrote to Hilbert after his final Academy lecture is dated 20 December, in response to Hilbert informing him that he had been elected a corresponding member of the Royal Academy in Göttingen. In it, he cleared the air by acknowledging that he had harboured some bitter feelings towards Hilbert – certainly he had complained about him in correspondence with Besso – but that he had succeeded in overcoming these feelings and hoped that they could continue a friendly relationship. Hilbert's reply is not known, but it can be assumed that he heartily concurred. Certainly, his nomination of Einstein as a member of the Royal Academy of Göttingen is evidence that he had no ill feeling; nor indeed had he any reason. As is now clear from an analysis of the proofs of the paper that he submitted on his gravitation theory, he had not derived the Einstein equations. Although Hilbert had certainly spotted some flaws in Einstein's work after his Göttingen lectures, he was led astray by others. In particular, he seems to have accepted the 'hole' argument (see Chapter 11), which led him to an artificial restriction that destroyed the covariance of his initial ansatz. Indeed, his version had more similarity to the Entwurf approach than it did to Einstein's final equations. Although the equations that were eventually published in Hilbert's paper of March 1916 had the correct form, this was added at the proof stage, presumably after he had had sight of Einstein's publications.

Hilbert's respect for Einstein can be gauged from the fact that the quotation given above about Göttingen children knowing more about differential geometry than Einstein is incomplete. Hilbert went on to say 'Yet, in spite of that, Einstein did the work, and not the mathematicians.' The magnitude of Einstein's achievement was succinctly summarised by his collaborator and friend, the ever-good-hearted Besso. He forwarded Einstein's postcard of 17 November to Zangger, writing: 'I enclose the historic card of Einstein, reporting the setting of the capstone of an epoch that began with Newton's apple.'

Chapter 13

Music and the war years in Berlin (1915–1917)

Einstein had finally achieved success in his quest to explain gravity. His personal life, however, continued to be full of trials and vexations. The immediate problem was what to do at Christmas, 1915. Michele Besso had suggested that Einstein come to Switzerland since Hans Albert was expected to visit Besso's son Vero in Krummenau, where Besso was staying at his father-in-law Jost Winteler's home. Einstein did not like this suggestion, as he interpreted it as Mileva not trusting him to be alone with the boys. Furthermore, he had been offended by a rare letter from Hans Albert at the end of November, which he characterised as nasty. Besso assured him that this was easily understandable in the circumstances and that things would be different if father and son spent some time together. Mileva also wrote to him in the same vein, so he relented and for a while the trip was back on, although the venue was now to be the Zugerberg, about 20 kilometres south of Zurich. Finally, however, Einstein cancelled his trip. He had heard that the German–Swiss border was almost completely closed and that several acquaintances had been turned back despite their papers being in order. On 18 December he sent Hans Albert a postcard telling him that he would come at Easter instead. He followed this with a letter on 23 December in which he told him that he had sent money for a Christmas present and indicating that he was so exhausted from his efforts on general relativity that he needed the Christmas holiday to rest and recover.

It can be assumed that Einstein was under continuous pressure from his new partner as well as his old. Although he was still living in his small bachelor apartment in Wittlesbacherstraße, he was a frequent visitor to Elsa's apartment on the fourth floor of Haberlandstraße 5 and took many of his evening meals there. It seems clear that Einstein and Elsa's more-then-cousinly relationship was widely enough known to cause scandal, certainly to her parents, whose apartment was on the floor below Elsa's. She herself clearly wanted to marry Einstein, to which Mileva was a severe impediment. Einstein, however, had at various times ruled out any idea of remarrying, most recently early in January 1916. Nevertheless, presumably in response to further sustained pressure, on 6 February Einstein wrote to Mileva suggesting that they convert their separation into a divorce. He suggested that this could be quick and based on the legal document which Haber had had drawn up in the summer of 1914. One month later, Einstein returned to the subject, deploying several new arguments to convince a clearly highly reluctant Mileva. The first was that his relationship with Elsa had to be normalised since it was harming the marriage chances of Ilse, Elsa's elder daughter. Bizarrely, he attempted to increase Mileva's sympathy, unlikely to have been aroused so far, by telling her that Ilse only had one eye. This certainly didn't seem to reduce her attractiveness to Einstein, as will be seen below, but neither did it influence Mileva. The second thread of Einstein's argument was straightforward bribery: he promised Mileva more money than had been agreed in the Haber document. Having completed his reasons for wanting a divorce, he proceeded to unravel the argument for the necessity of his marrying Elsa by insisting that he would never give

Einstein. Brian Foster, Oxford University Press. © Brian Foster (2026). DOI: 10.1093/oso/9780198794875.003.0013

up living alone, which he considered an indescribable blessing. Therefore, after the war was over, the children could safely visit him in Berlin.

The whole matter came to a head when Einstein made his postponed visit to Switzerland. He arrived in Zurich on 6 April and was delighted that his sons visited him in his hotel room on the following morning. They brought with them a letter from their mother, asking for a meeting with Einstein. According to Besso, her purpose in requesting the meeting was to assure herself that Albert really did himself want the divorce, rather than succumbing to pressure from others. Einstein refused to meet her. He seems to have developed an extraordinary aversion to seeing Mileva. Instead, he wrote back praising her care of the children, with whom he was immensely pleased, and pressing her once more about the divorce, which he suggested could most easily be achieved in Berlin.

On 9 April, Einstein and Hans Albert set off for an expedition to Seelisberg, perched above Lake Lucerne. They got along famously, but it snowed and, with no signs of improvement in the weather, they decided to cut their trip short and returned to Zurich by 15 April. On the following morning, Einstein showed Hans Albert some experiments in the physics department, probably of the university. Then things went downhill. His son asked Einstein to return home with him and see his mother. Einstein refused. He blamed Mileva for prompting Hans Albert, but it is equally likely that he was simply, in the unsophisticated way of a young boy, trying to bring his parents back together. Hans Albert angrily left his father. For the remainder of his stay in Switzerland, Einstein saw nothing more of his family and the discussions about divorce were discontinued. After a trip to Luzern to see his sister Maja and her husband Paul, Einstein returned to Berlin on 27 April.

The publication of the November 1915 papers on gravitation at last gave Einstein more leisure to enjoy the musical life of Berlin, which, despite the war, was still probably the most varied and exciting of any city in the world. The major expansion of the city after the foundation of the German Empire in 1870 had brought with it new concert halls, theatres and other places of public entertainment for the rapidly growing population. Only around 800,000 in 1870, it had reached more than 2 million people by the start of the war. To serve the musical appetites of this multitude were fourteen concert halls, with a total capacity of around 12,000, ranging from the Philharmonie, seat of the Berlin Philharmonic and its conductor Arthur Nikisch, with 2,200 seats, to the 191 of the Harmonium-Saal. In addition, there were three opera houses. The doyen was the Royal Court Opera, where Richard Strauss was Erster Hofkapellmeister between 1898 and 1908 and subsequently musical director until 1919. The Deutsche Oper, founded in 1912 in the Charlottenburg district in the west of the city, had by the start of the war become a significant competitor. The third house, the Theater Unter den Linden, which subsequently became the Metropole Theatre, specialised in light musical comedies and operetta.

Strauss was the dominant figure in the city's musical life, both as composer and conductor; he had long been at the cutting edge of German musical development. Indeed, Arnold Schoenberg, hardly a conservative, once remarked that he was never revolutionary and that the only revolutionary in their time was Strauss. For example, Strauss had an enormous success with the first Austrian performance of *Salome* in Graz in 1906, which he conducted. Daring musically, morally, and in production, the premier was attended by a stellar array of maestros: Gustav Mahler, Schoenberg, Giacomo Puccini, Alexander Zemlinsky, and six of Schoenberg's pupils, including Alban Berg. However, by the outbreak of the war, Strauss had become rather conservative both in his choice of repertoire to conduct and in his compositions. In this he was at least catering for what he considered the musical taste of the Berliners. This was certainly true of his employer, Kaiser Wilhelm, who did not at all care for Strauss's music but nevertheless considered that he was an ornament to his court; however, the Empress was scandalised by much of Strauss's output. The Kaiser, hearing the reports of the success of *Salome*, asked his employee why he hadn't written any marches. Strauss expressed ignorance of the genre. The following day, Strauss was again called to the palace where for three hours he was forced to listen to military bands marching up and down the courtyard. Thus encouraged, he composed two military marches, Op. 57, which, while certainly not among his most distinguished works, won decorations for their composer from an appreciative monarch in the form of the Crown Order, Third Class.

Musicians were no more immune to the epidemic of jingoistic patriotism at the outbreak of war than were scientists. In August 1914, Schoenberg wrote to Mahler's wife Alma 'Now we will throw [Bizet, Stravinsky & Ravel] these mediocre kitschmongers to slavery, and teach them to venerate the German spirit and to worship the German God.' On 20 August, on his score of the first act of *Die Frau ohne Schatten*, the opera whose composition occupied much of his attention during the war, Strauss wrote 'The day of victory at Saarburg. Hail to our brave troops. Hail to our great German fatherland.' He soon recovered from this euphoria. He refused to sign the manifesto of the German artists and intellectuals that Max Planck and almost all of his colleagues had signed. Strauss's attitude to the war was very self-centred: he resented the increasing personal discomforts and the effect on his family life. He had lost his entire life savings, which had been invested in Britain; they were first frozen and eventually seized. In order to rebuild his savings, he was forced to compose and conduct harder than ever, just when he had been looking forward to a less hectic schedule.

Although the various subscription concerts were advertised as usual in September 1914, the Berlin Philharmonic lost many of its members, who were called up as reservists. Any orchestra member who had to do war work was given a daily *ex gratia* payment of one mark from orchestra funds. On 12 September, there was an addition to the normal season – a benefit concert for the victims of the war. This was the first of many such events. They were dominated by one composer, Beethoven, whose compositions also became increasingly prevalent in the normal subscription concerts. All-Beethoven concerts became the norm rather than the exception. Popular violinist Franz von Vecsey played the Beethoven Concerto in one such programme on 22 October. Artur Schnabel and Carl Flesch played the complete Beethoven violin and piano sonatas in October and November. The *Eroica*, often publicised using the German translation of its nickname, *Die Heldensymphonie*, was a particular favourite. The 11 October edition of the *Berliner Tageblatt* advertised no less than three separate performances of the *Eroica* on 18 or 19 October. For example, on 18 October, Strauss conducted Weber's Overture to *Der Freischütz*, Haydn's Military Symphony, the '*Eroica*', and Wagner's *Kaisermarsch*.

The apotheosis of Beethoven was not limited to civilian Berlin: it was widely prevalent in the army. The notion was widely propagated that Beethoven, having witnessed the resurgence of the German nation after the initial defeats and humiliation of Prussia in the Napoleonic Wars, was a supporter of German military strength and greatness. The newspapers published an article from an artilleryman who wrote from the trenches that Beethoven's Fifth Symphony was the symphony of war. He went on to characterise each of the four movements: the first, with the insistent rhythm of the opening notes that in the next war came to embody victory, represented mobilisation, later episodes depicting the battles; the second movement, the love at home that had been left behind; the third the privations but also the gaiety of battle; the final movement celebrating victory. Beethoven was also used as a bludgeon with which to subdue the culture of non-Germans. In May 1915, the Berlin Philharmonic gave two concerts in occupied Brussels under Felix von Weingartner with Artur Schnabel as soloist. Every piece in both concerts was by a German composer, with Beethoven and Wagner being the most represented. Newspaper reports talked about the moral conquest of Belgium being accomplished not by troops alone but with the reinforcement of German music. His distaste for Beethoven, to which Einstein admitted in his later years and which was certainly not evident in his youth, may well have begun with the distortions and bombast of the war.

It is understandable that in the weeks after his triumph of November 1915, Einstein's own physics activities were limited to little more than explaining his path of discovery, and the various missteps, to his friends and colleagues. However, a few weeks after the Christmas holiday, he put pen to paper to summarise his new theory of general relativity in a way which had not been possible in the still-evolving thoughts documented by the written reports of his November lectures. The result appeared in March. In the 'Foundation of the General Theory of Relativity' Einstein puts aside all the confusion of the previous decade and gives a pellucid formulation of the whole programme of relativity, including why it must be generally covariant. In this paper he refers to the 1905 theory, for only the second time, as the 'Special Theory of Relativity'. This paper can be seen as a punctuation mark at the end of the most heroic endeavour of Einstein's career. However, it was more a semicolon than a full stop. Having set down the theory with exemplary clarity, his next task was to attempt to understand its implications for the universe around us. In this endeavour, Einstein's colleagues had already begun, with remarkable rapidity, to put their hands to his plough.

Acceptance of general relativity took place much more rapidly than had been the case for the special theory. In December and January, Hendrik Lorentz and Paul Ehrenfest were both busy working their way through Einstein's papers, although the former with more success than the latter. Ehrenfest struggled, firing off a series of letters to Einstein asking for explanations of various technical points. Ehrenfest was an extremely good physicist, but in Lorentz and Einstein he was trying to keep up with two of the greatest of all time. As Pais writes:

> Ehrenfest, aged 35, in Leiden, ten miles down the road, is also hard at work on relativity. His reply to Lorentz's letter shows a glimpse of the despair that would ultimately overwhelm him: 'Your remark "I have congratulated Einstein on his brilliant results" has a similar meaning for me as when one Freemason recognizes another by a secret sign'.

In February 1916, Einstein reported to Stern that Lorentz, Ehrenfest, and Hilbert had accepted the new theory; perhaps more importantly, so had Planck, ever the conservative in considering bold new concepts in physics. Another enthusiastic convert was Max Born, who had come to Berlin in the spring of 1915 at the invitation of Planck as an associate professor. He was almost immediately called up, initially as a radio operator, before being commissioned and put to work on sound-ranging, attempting to identify the position of artillery by using the sounds it produced. Since he was stationed in Berlin, he was able to continue his family life with his wife, Hedwig. He and Einstein had corresponded together for many years and soon became close friends. Since their apartments were very close, Born often visited at lunchtime. Einstein then began to pay evening visits to them. Hedwig became a great favourite of Einstein's, as evinced by their affectionate correspondence. Their friendship was aided by the fact that all three shared left-wing political views and by the fact that Born was a good pianist. Hedwig recalled Einstein's first visit, his warm manner and the careless way in which he removed his detachable collar and shirt cuffs before picking up his violin and launching into Haydn with her husband. During those dark days of the war, they spent many hours playing sonatas together. Since by now Einstein had acquired a piano for his small apartment, it is likely that they also made music during Born's lunchtime visits. Having finally solved gravitation, Einstein had been able to return to his passion for music-making and, as he told Hans Albert, by the start of 1916 he was once again playing the violin a great deal, as well as the piano. He used his time-honoured method for lubricating his social interactions by suggesting that his son practice the piano part for a violin sonata so that they could play together at their next meeting.

Another early recruit to the apostles of general relativity was Karl Schwarzschild, Director of the Astrophysical Observatory in Potsdam. Like Born, he too was serving in the army but unlike Born, Schwarzschild had volunteered at the outbreak of war and by the end of 1915 was a lieutenant on the staff of the commanding general of artillery of the Fifth Army at the Russian front. That he had some leisure time in between launching projectiles towards the Russians is evinced by a letter that Einstein received just before Christmas. In it, Schwarzschild remarked that he had been trying to understand Einstein's November papers and so had tried his hand at the general solution for the gravitational field of a point-like mass. In calculating the perihelion of Mercury, Einstein had been content to use a first-order approximation. Schwarzschild remarkably had obtained the full solution, which not only reproduced Einstein's answer (with a negligible correction) but also showed that it was unique. Einstein must have been flabbergasted; indeed, in his reply a few days after Christmas he admitted that he had thought that the general solution could not be so simple. He encouraged Schwarzschild to write up his work, which Einstein subsequently presented to the Prussian Academy.

Einstein and Schwarzschild continued to correspond. It was in a letter to him that Einstein first mentioned that he was investigating gravitational waves with the new gravitation equations. At the beginning of February, Schwarzschild wrote again with another surprising result. He had tried his hand at solving the next simplest problem to a point mass – that of an incompressible sphere of liquid. He commented to Einstein that, had he realised the trouble it would give him, he would never have begun. He found that as he compressed the same mass to smaller and smaller radii, there was a singularity, indicating that no smaller sphere could exist. This is now known as the Schwarzschild radius, which denotes the size of a black hole of a given mass. At the time however, it appeared a very puzzling result that neither Schwarzschild nor Einstein could understand. Sadly, Schwarzschild's amazing ability to produce world-class physics as the artillery thundered around the Russian front was brought to a tragic end. He contracted a rare autoimmune skin disease, to which he succumbed on 11 May. Einstein delivered a eulogy to the Prussian Academy in which among Schwarzschild's many achievements he highlighted that he was the first to solve the equations of general relativity exactly. He ended with the correct prediction that Schwarzschild's scientific work, to which all his energies had been devoted, would continue to bear fruit long after his demise.

At the same time as Schwarzschild was reporting curious aspects of general relativity, Einstein was also involved in a much less edifying discussion about the future of their mutual acquaintance, Erwin Freundlich. Others involved included Planck, Sommerfeld, and Hilbert. As discussed previously, Freundlich's duties at the Berlin Observatory were predominantly routine and dull and well below his intellectual capabilities. His heroic attempts to verify Einstein's theories, which had resulted in incarceration in Russia, had largely been carried out in his spare time or with a special dispensation. Freundlich was a prickly character, who invariably seemed to fall out with his superiors as well as colleagues and friends. Sommerfeld in particular had little time for him. Einstein himself often found Freundlich exasperating but felt obliged to try to get him a position more suitable for his talents, since he was one of very few astronomers interested in theoretical constructs such as general relativity. Indeed, he was shortly to produce a popular book about general relativity for which Einstein wrote a preface. Unfortunately, perhaps because of his proselytising novel ideas, his reputation was not high with his fellow astronomers. Even Schwarzschild, much more theoretically inclined than most, seems to have considered Freundlich to be a hopeless case. Thus, not even the influence of Einstein coupled with help from Hilbert, whom Einstein visited in Göttingen early in March to discuss placing Freundlich in the observatory there, was able to engineer an academic position for Freundlich. In the end, as will be seen below, the only way Einstein could get Freundlich a suitable post was to give him one himself.

The place among general relativists left vacant by the demise of Schwarzschild was taken up almost immediately by Willem de Sitter, another extremely talented astronomer, who wrote to Einstein in June with a suggestion for a coordinate system choice in which the solution of many particular situations became more transparent. Einstein was very excited by this suggestion, which he used to investigate the properties of gravitational waves. Unfortunately, in his lecture to the Prussian Academy and publication later that month, he made a calculational error that led him to the incorrect conclusion that there were three types of gravitational waves, only one of which transported energy and was therefore 'real'. Einstein thought that the two 'unphysical' types of waves could only be eliminated by specialising to a particular coordinate system. For the next two years, Einstein insisted to de Sitter and others that unimodular coordinates were 'special'. Einstein and de Sitter were involved in lengthy arguments in which the latter had no compunction in chiding Einstein for unwarranted assumptions such as that the universe had to be static, that is, although its components such as stars and gas were dynamic, the universe as a whole was on average unchanging. This assumption, to which Einstein clung tenaciously for many years, was subsequently to get him into further trouble.

Einstein's problems with gravitational waves were eventually solved when his old colleague Gunnar Nordström convinced him that his 1916 calculation was wrong.[1] In 1918, Einstein corrected the 1916 paper. Unfortunately, he made an error of a factor of two. Neither was this the last mistake that Einstein was to make about gravitational waves, as will be discussed later. The error in the 1918 paper was noticed and corrected in 1922 by Cambridge astronomer Arthur Eddington, who had been alerted by his friend de Sitter to Einstein's work in June 1916, just as de Sitter was beginning his correspondence with Einstein. Eddington was immediately fascinated. This was the first he had heard of Einstein's new theory since, because of the war, English physicists were cut off from German scientific publications. As will be seen later, Eddington's fascination bore fruit in that he was to play a major role in the eclipse observations that catapulted Einstein to world fame.

By the summer of 1916, the attention of all inhabitants of Berlin, even one as famously ascetic as Einstein, turned increasingly towards a single issue – food. The blockade of German ports by the British Grand Fleet was leading to shortages that were causing increasing problems to the authorities throughout Germany. An attempt by the German High Seas Fleet to bring the British Grand Fleet to battle and thereby loosen the grip of the blockade failed at Jutland at the end of May. Although the British lost more ships, they retained their superiority at sea and were not challenged again during the war. The German effort to bring Britain to its knees by unrestricted submarine warfare, which very nearly succeeded, could not help the Berlin housewife queuing for almost inedible 'war bread' and increasingly scarce potatoes. Indeed, by bringing the United States into the war in 1917, it made matters worse, as a squadron of US ships arrived in the North Sea to reinforce the blockade. The average calorific intake of a German male had by now fallen to under half that of his British counterpart. For the urban poor, and for women, the figures were lower still. Malnutrition was beginning to cause health problems, particularly in Berlin.

The foundation of a central office under the German War Ministry to control the distribution of food in May 1916 and the issuing of ration cards for staple foods was a recognition of fact, rather than a strategy. Unsurprisingly therefore, it failed to improve the situation. Early frosts ruined the potato crop, causing the infamous Turnip Winter of 1916–1917 that saw widespread famine. It

[1] A minor role in this was played by a future collaborator, Erwin Schrödinger, who worked briefly on general relativity at this time.

was not only Germany that suffered. Einstein's adopted country, Switzerland, surrounded by belligerents on all sides, was hard hit; it too introduced rationing by 1917. Nevertheless, the Swiss government was aware that the situation in Germany was much worse and that many Swiss citizens there were in dire need. It therefore supported Swiss in the countries at war with monthly food parcels. Einstein certainly benefited from this, as usual facilitated by the indefatigable Heinrich Zangger. Although as a relatively wealthy person, eating usually with Elsa's well-off parents, Einstein was not as deprived as many, nevertheless these months certainly had an effect on his health and must have contributed to his serious breakdown in the following year.

In Zurich, Mileva's health, never good, also began to fail. In July 1916, she was confined to her bed with a heart condition. Einstein was alerted to the problem by a letter from Besso, who hinted that he believed the condition had been brought on by the stress caused by Einstein's request for a divorce. Besso and his family had moved back to Zurich in late February. He began to teach patent law at ETH as a Privatdozent, receiving payment only via student fees. Despite his experience as a colleague of Einstein at the Bern Patent Office, he was too disorganised to make a success of such an undertaking and by 1919 had returned to the Patent Office. His presence near Einstein's family, however, was a boon for Einstein. He wrote to Besso that he had considered travelling to Zurich but had decided against it. His reasons for this were threefold. First, he refused to see Mileva under any circumstances and did not see how to avoid this if he went to Zurich. Secondly, he thought that his relationship with his sons was sufficiently bad at that time that they would not welcome his presence. Furthermore, he considered it not impossible that Mileva wasn't really ill at all and cautioned Besso to beware of her female wiles. Einstein did offer to come to Switzerland, although not to Zurich, to spend the summer vacation with his sons if Mileva was hospitalised.

Given Einstein's attitude, Besso and Zangger acted as Einstein's substitutes. In fact, Besso's presence, presumably because she associated him so closely with Einstein, seemed to upset Mileva. It was Zangger, therefore, who dealt with the situation, with the help of Mileva's friendly neighbour. As a medical doctor, Zangger could confirm that Mileva's illness was real. Indeed, Lisbeth Hurwitz was of the opinion that in mid-July she had almost died and that only sheer willpower not to leave the children to the care of Einstein's family, particularly Elsa, kept her going.

Despite his earlier offer, when Mileva was eventually hospitalised at the end of July, Einstein did not come to Switzerland. In a letter to Hans Albert, acknowledging that both he and Tete were now at home alone with only the maid to take care of them, he pleaded pressure of work for his remaining in Berlin. In fact, as he remarked on the same day in a letter to Zangger, he had no duties that particularly kept him in Berlin. Rather, he was worried, as before, that he would have to meet Mileva if he went to Zurich. He confided to Zangger that the reason this worried him was that he feared that Mileva would make him agree to relinquish access to his children, particularly should she die. This sudden fear of Mileva's persuasive powers seems bizarre, while the fact that he himself was relinquishing access to his children by refusing to go to Zurich did not seem to occur to him. In fact, Zangger thought that Einstein should indeed stay away, since he worried about the effect his presence might have on the sick woman. She had by now been diagnosed with 'brain tuberculosis' (i.e. a form of meningitis).

The children spent the school holidays with Mileva's old friend Helene Savić in Lausanne. Helene wrote to Einstein, who was very grateful for the details she gave him about the boys. In his reply on 8 September, Einstein regretted that his sons did not understand his separation from Mileva and that his relationship with them had deteriorated to such an extent that it would be better for them if he did not see them in future. He told Helene that his relationship with Mileva had irretrievably broken down and that they could never reconcile. He had every confidence that Mileva would bring the children up well and he was very pleased that she had now fully recovered. In fact, Mileva was far from fully recovered and her health was always to be precarious in future. Nevertheless, she was able to return home after being discharged from hospital on 25 October.

Einstein had written to Hans Albert that he was absorbed by work in the summer of 1916. In fact, he was in the middle of producing another deep insight into nature. This time he had returned to his old nemesis: the problem of the quantum. It isn't clear why he switched his attention from general relativity, although a clue may be found in a comment he made to Zangger a few months later in which he said that general relativity was in principle complete and as to what remains, what he could do didn't interest him and what interested him, he couldn't do! Of course, he was exaggerating and indeed much of his work in the next few years concerned the applications of general relativity. Nevertheless, psychologically he considered gravitation a solved problem and he was looking for new fields to conquer. He had for long been dissatisfied with Planck's derivation of the black-body radiation law, since it implicitly used two incompatible systems, quantum theories and classical electromagnetism. Einstein now saw a way to solve the problem, although the cost that he would eventually have to pay in sacrificing his own prejudices was a heavy one.

Einstein approached the problem in his time-honoured way – by applying statistical mechanics to a system in equilibrium. His starting point was that the Bohr system of the atom had proved a remarkable success in explaining the emission spectra of gases, which implied that atoms and molecules must be understood using quantum ideas. The system Einstein employed consisted of a large number of oscillators in thermal equilibrium with electromagnetic radiation. The structure of the oscillators was left open except that they had to have at least two different energy levels. The molecule can jump from the lower energy state to the higher by absorbing radiation of the appropriate frequency. It can go in the opposite direction by a similar process, in which the incident radiation stimulates the molecule to de-excite itself by a sort of resonance effect. This process is known as 'stimulated emission'. If there were no other factors, it is rather easy to see how the system would reach equilibrium but it would be essentially trivial. However, there is a third process, in which a molecule can fall to the lower energy state spontaneously, independent of the existence or strength of any radiation field. For obvious reasons, this is known as 'spontaneous emission'. Einstein assumed that the probability for this emission had the same mathematical form as that governing radioactive decay.[2] Although he was unable to calculate the probabilities of these effects because as yet there was no self-consistent quantum theory, he was able to calculate relative relationships between these probabilities, which later became known as the 'Einstein A and B coefficients'. By using the fact that the system was in equilibrium, he was able to deduce the form of the energy distribution as a function of temperature and to show that it reproduced Planck's formula for black-body radiation.

At the end of August, Einstein completed a second, more comprehensive paper on the quantum theory of radiation. In this, after repeating the considerations of the first paper, he showed that the emission of the photon in the processes he had discussed therein was 'directed', that is, not emitted as a spherical wave centred on the source but in a specific direction and carrying momentum.

[2] Stone considers this to be a remarkable leap of imagination. Indeed, it was. However, Einstein would probably have been aware that the gamma radiation from radioactive decay had been shown to be electromagnetic in nature by Ernest Rutherford and Edward Andrade in 1914, so that hypothesising a link between radioactive decay and the emission of electromagnetic radiation would not have seemed outlandish.

As he remarked in the conclusion of the paper, a weakness was that the time of emission of the photon and its direction were left to chance. In 1916, Einstein metaphorically shrugged his shoulders and muttered 'sufficient unto the day is the evil thereof'. He wrote in the paper that, despite the aforementioned weakness, he had full confidence in the reliability of the route that he had chosen, while, in letters to Besso, he triumphantly asserted that this work had essentially established the reality of light quanta. As time passed, however, he became increasingly unhappy about the 'dice-throwing' at the heart of quantum theory. One thing that he would have been both happy and indeed incredulous about is the subsequent importance of these papers. They form the foundation upon which that indispensable and ubiquitous tool of modern science, the laser, is based. Both in this applied sense therefore, as well as in their importance in the development of quantum mechanics itself, these papers could be considered as worthy of comparison with the general relativity papers of the previous year. This was truly Einstein at the height of his powers.

In September, Einstein finally managed to make the trip to Leiden that he had been struggling to arrange since January. A letter of invitation from Lorentz to lecture at the university smoothed the path to obtaining the necessary paperwork from the German authorities. He left Berlin on 27 September and remained in Holland until 12 October. He stayed with Ehrenfest, delighting in discussing physics with him and his wife, but particularly indulging in making music. His and Ehrenfest's passion for music was such that whenever they met, they played violin and piano duets together. Until this visit, Ehrenfest's favourite composer had been Beethoven. This changed after Einstein introduced him to the delights of playing Bach Chorales on the piano; as Ehrenfest noted in his diary, he opened up Bach to him. In a letter written after his departure, Einstein congratulated Ehrenfest on his luck in being able to play them for himself, rather than being in his own position of having to wait for someone else to play them to him. It can be concluded that Einstein was unable to play from piano sheet music, as indeed he himself often admitted. As for Ehrenfest, Einstein's advocacy seems to have resulted in his becoming addicted to playing Bach, to the extent that his work suffered.

During his stay in the Netherlands, Einstein was able to have long conversations with both Lorentz and particularly de Sitter, in which they explored the various solutions to general relativity about which they had been corresponding. This really marked the start of three years of intellectual wrestling that led them into weird and wonderful universes well beyond their initial imaginings. Although at least initially they talked at cross purposes because of confusion about coordinate systems, the central point of disagreement between them eventually centred around something that struck at the heart of Einstein's programme for understanding gravity: the relativity of all motion and in particular rotation. To achieve this, he was inevitably driven to Mach's[3] viewpoint that the origin of the inertia of a given object was its interaction with the totality of the 'fixed stars'. Already in their discussions in Leiden, de Sitter and Einstein had agreed that the unambiguous determination of the space-time metric in the Einstein field equations required *both* knowledge of the distribution of matter *and* assumptions about the behaviour of the universe at its edges ('boundary conditions'). Einstein was unhappy with this and sought first to circumvent the necessity for boundary conditions by ensuring that inertia vanished at infinity but was made non-zero at finite distances by 'Machian' distant masses. De Sitter pointed out that this idea was simply another version of Newton's absolute space, since the masses would have to be outside the visible universe and therefore could never be observed.

[3] Mach had died in February 1916; Einstein had written an appreciation of his work for the *Physikalische Zeitschrift* issue in April.

By 1917, Einstein had accepted de Sitter's view that boundary conditions must be specified in any solution of the field equations but with typical ingenuity got around the necessity for boundary conditions by abolishing the boundary! Einstein's new universe was spatially closed, so that, rather like an individual being confined to the surface of a sphere, infinitely long journeys can be taken without reaching an edge or boundary. Unfortunately, such a static universe is unstable since gravity will cause all the masses to collapse together. To get around this problem, Einstein introduced one of his most controversial hypotheses, that of a 'cosmological constant' whose function was to counter the attraction of gravity and stabilise the static universe. In puzzling over Einstein's idea, de Sitter discovered a solution of the field equations which contained only a cosmological constant and no matter – a 'vacuum solution'. This was anathema to Einstein, whose deepest instincts were that a solution without any matter to determine the metric was simply a return to Newton's absolute space.

Before leaving these issues for the moment, it is worth making two remarks. First, little did either Einstein or de Sitter anticipate that the cosmological constant would today be one of the hottest topics in particle physics and cosmology. Secondly, it is important to realise how limited the knowledge was that each had about the structure of the universe on which they could base their cosmological theories. A quote from Einstein to de Sitter illustrates this starkly: 'If coordinated Milky Way systems[4] really do exist (a view which, to the best of my knowledge, is not shared by all astronomers), there is no necessity for there to be any matter between them'.

Einstein's speculations on the nature of the universe were interrupted by a shocking event. On 21 October 1916, the Prime Minister of the Austro-Hungarian Empire, Count Stürgkh, was assassinated at his lunch table in a Vienna hotel. His assailant, Einstein's old friend from Zurich days, Fritz Adler, shot von Stürgkh three times, twice in the head, killing him instantly. Einstein must have heard about this the following day, since it was the lead story in the major Berlin newspapers. He was deeply affected by the news. As he wrote to Adler's wife Kathia a few months later, he considered Adler to be one of the most admirable and purest men he had ever known. He offered to help in any way he could. It can be presumed that she told Fritz of Einstein's letter since a couple of weeks later, Adler wrote to Einstein from his prison cell. In his enforced idleness, he had taken up physics again and hoped to publish a book discussing Mach's work. Although he had initially been overjoyed at Einstein's general relativity, he believed he had found an error in both Einstein and Mach's reasoning, which he hoped to have time to explain despite the fact that he expected to be sentenced to death at the end of his trial, which began on 18 May. Paper was rationed, so Einstein was spared a detailed account of Adler's theses, although not for long.

Einstein was moved by Adler's letter to try to do something to help. He wrote to his usual factotum, Zangger, asking him to organise a petition for clemency from the Zurich Physical Society. He offered Zangger help and support from Besso and another friend. In April, Einstein wrote to Adler offering to attend his trial as a character witness. In sending a substantial section of his book for perusal, Adler added that he appreciated Einstein's offer, which he said he would discuss with his counsel. Having Einstein as a witness to Adler's great qualities as a physicist and teacher was no part of their defence strategy and he was not called. Presumably the testimonial that was passed by the Zurich Physical Society was also not used. Having successfully fought off attempts to declare him insane by his estranged father, Viktor, his defence team (hoping to save his life), and the authorities (hoping to avoid a trial), Adler succeeded in his aim of making his trial a *cause célèbre*. It was headline news in most Viennese newspapers. For example, the *Arbeiter-Zeitung*,

---

[4] In modern parlance, "other galaxies".

the organ of the Socialist party of which Adler was a leading member, reported it in detail over its first eight pages; in particular, Adler's statement was covered verbatim over five full pages. In it, Adler reported that he had had to fight against his own defence counsel in order to make his statement. He contended that he could have committed no crime, since Austria was no longer a constitutional state with respect for the law. His affirmation that he killed von Stürgkh to protest against his dictatorial government and refusal to convene the Austrian parliament would have awoken sympathy in many readers. Nevertheless, Adler was convicted and sentenced to death. However, circumstances saved him from the martyrdom that he perhaps craved. The old Emperor Franz Joseph had died only a few weeks after Adler's attack. The new Emperor, his great-nephew Karl, was desperate to get Austria out of the war and probably sympathised more with Adler's attitudes than with those of his victim. The Emperor initially commuted Adler's sentence to eighteen years imprisonment in Stein an der Donau, northwest of Vienna; after representations from his father, Karl eventually pardoned him just before the end of the war in 1918.

Meanwhile Einstein had dutifully ploughed through Adler's book. He wrote to Besso that it made entirely worthless points at great length with enormous zeal. He agonised over how to reply to a man whom he considered would almost certainly soon be executed, perhaps within a few weeks. To his great credit, he worked through Adler's manuscript in detail, correcting and commenting as he went.[5] Adler replied in haste just before his trial began, wrote again, and then a silence of almost a year ensued. Einstein's heart must have sunk when he received another letter from Adler in July 1918 from his prison cell. Adler apologised for the long silence but said that Einstein had rarely been out of his thoughts because he had spent the whole year working on general relativity! The fruits of this labour eventually also arrived on Einstein's desk. At least, now that Adler was in no danger of imminent extinction, Einstein felt able to say what he really thought. He went through the second text again with enormous patience, taking two and a half days before he got to the stage where he was convinced that there was no point in reading the rest since it was built on the shaky foundations of earlier errors. His implied advice was that Adler should go no further with it; if it were published, he would make no comment on it. He ended his letter in a very friendly way, promising to send books of interest to him in prison. Typically, Adler[6] ignored Einstein's advice and published his books; that on Mach and his ideas in 1918, that on relativity in 1920.

Other than his musical activities with Born referred to above, there is little explicit evidence of Einstein's violin playing during the war years in Berlin. He played in string quartets in Berlin, as he had done in every city in which he had lived since leaving school. In May 1917, he reported to Hans Albert that he was active in a string quartet, although it is unclear whether Freundlich was the cellist, as had been planned in advance of Einstein's arrival in Berlin. One thing that is

---

[5] Frank records that Adler's defence lawyers made copies of this manuscript and circulated it to several psychiatrists and physicists in an effort to prove that he was insane. Frank wrote that he too received such a copy.

[6] After Adler's release from prison, he resisted calls to lead a Communist revolution in various parts of the Austrian Empire. He had always distrusted the Communists and, as discussed in Chapter 7, he and Lenin had exchanged diatribes for years. His father Viktor Adler, leader of the Austrian Socialists, had died on Armistice Day. He would have been pleased and proud that Fritz played a leading role in the establishment of the socialists in the new Austrian republic. Fritz then spent many years as a leader of the Socialist and Workers International in Brussels. He fled to the US after the invasion of France in 1940, having played a heroic role in helping his fellow Jews to escape Nazi Germany. He and Einstein corresponded quite regularly until 1920, but after that only very infrequently. They met occasionally during the interwar years. Adler died in 1960, having returned to the home of his and Einstein's youth, Zurich.

documented is his participation in musical evenings at Planck's house. In his youth, Planck had been undecided about whether he should be a professional musician or a physicist. Fortunately for science, he decided on the latter, no doubt realising that it is easier to be an amateur musician than an amateur physicist. He was an excellent pianist as well as having an exceptionally broad range of musical interests; for example, he wrote a book on harmonic systems in music, paralleling the interests of Einstein's former collaborator Adriaan Fokker. He also threw himself into musical performance, conducting choirs and hosting musical soirées at his home.

Lise Meitner left an account of one of the musical gatherings at Planck's home. After graduating in Vienna with classmate Ehrenfest, she had become only the second woman to gain a doctorate in physics. Arriving in Berlin in 1907, she had attended Planck's lectures unofficially, as women students were not permitted until 1909. She was initially forced to carry out her experiments in a room that Planck was nominally using at Humboldt University; she was only allowed to enter the building by the goods entrance. In 1918, she worked with Einstein on an experiment to test aspects of quantum theory, although Einstein found a flaw in his conception before they had progressed very far. She was subsequently to gain fame, although controversially not to receive a share of the Nobel Prize, as the co-discoverer with Otto Hahn of fission, the induced splitting of atomic nuclei by neutrons. Having trained as an X-ray operative and nurse to help the war effort, she had left Berlin in 1915 and served first with the Austrian army and then in Germany. In November 1916, she had just returned from these war duties to her work in the Kaiser Wilhelm Institute for Chemistry, when she attended one of Planck's musical events. She wrote about it to her collaborator Hahn, who was serving in the army, assisting in the deployment and study of the gas and chemical warfare innovations pioneered by their boss, Fritz Haber. The musicians that evening included Planck, who played the piano, and his favourite son, Erwin, who had recently been released from a French prisoner-of-war camp, who played the cello. Meitner wrote:

> Yesterday I was at the Plancks' house. Two heavenly piano trios by Schubert and Beethoven were performed. Einstein played the violin. When not playing, he delivered himself of some deliciously naïve and extraordinary views on politics and the war. That there can exist so learned a man who in this day and age never picks up a newspaper is indeed amazing.

In addition to his own music-making, Einstein certainly attended many concerts during the war years. The government believed that these were an important part of keeping up morale on the home front and as remarked earlier, also played a role in the glorification of German culture through its composers. It was in 1917 that Einstein is first recorded as hearing the playing of the great German violinist, Adolf Busch, who was to become an admired friend. Busch, then in his mid-twenties, was at that time making a name for himself in Berlin. Resident in Vienna, he had been born in Siegen in Westphalia. He shared Einstein's conservative tastes in music, being at that time involved in something of a feud with Schoenberg in Vienna. On 5 May 1917, Einstein attended the opening concert of the Deutsche Brahms-Gesellschaft Festival with the Berlin Philharmonic conducted by Artur Nikisch. The concert began with the Tragic Overture, Op. 81 and ended with the First Symphony. The highlight for Einstein was Busch playing the Brahms Concerto, which he subsequently described as the greatest musical experience of his life. Later that year, on 1 December, Busch played the Brahms Double Concerto with cellist Paul Grümmer with the Berlin Philharmonic under Max Fiedler.

Although Mileva had returned home in October 1916, her health remained poor, and she was confined to bed. In mid-November she had a relapse, which Besso blamed on her being upset that Hans Albert would not show her a letter he had received, presumably from his father. From this it can be deduced that in addition to her physical complaints, Mileva's emotional state was also fragile. Einstein had in September accepted that he would not bring up the question of divorce again. He remarked that he had learned how to resist tears from his relatives.

The question of how to help Mileva care for the children was a recurring theme in Einstein, Besso, and Zangger's correspondence. One casualty of the situation was Hans Albert's piano playing. Mileva herself was too ill to give him lessons. Einstein in his letters[7] to Hans Albert was clearly concerned that his son should continue to play music, which had been such an important element of his own life. He need not have worried. Firstly, Hans Albert seems to have been motivated. Einstein was pleased to hear that he had been playing Mozart sonatas to entertain Mileva on her return from hospital and asked him which ones he was playing. Hans Albert replied with a sketch of the opening bars of K331 in A major, implying that he could not only play the theme, which is a beginner's piece, but also the variations, which are not. Einstein in his reply recommended him to work hard at Mozart sonatas, since with them began Einstein's own passion for music. Secondly, the faithful Besso stepped in to ensure that Hans Albert carried on with piano lessons. In March 1917, for example, he took him along for lessons with Henri Bas-Bulenek, with whom Einstein and the cellist Raphael Lewinowitsch had played piano trios during Einstein's time in Zurich. Einstein recalled that Bas-Bulenek's playing was intelligent but rather cold.

Somehow, with the aid of her maid, Mileva managed to care for the children until the end of April, when she had a relapse. On 2 May 1917, she was once again hospitalised. After a month of care, she returned home but was able to do nothing more than lie in bed on the balcony, from whence she could see the peaks of the Alps and breathe the fresh air. She had to return to hospital again twice during the summer of 1917.

It is typical of the tangled web of Einstein's female attachments that, at exactly the time when he was trying to cope with the illness of his wife, one of his old flames arrived at his sister Maja's door. Marie Winteler stayed with her and Paul for eight days at the beginning of May 1917. Marie had still not recovered from her affair with Einstein in 1909–1910. She poured out her heart to Maja, who was relieved to report to her father-in-law, Jost Winteler, that at least Marie seemed no longer to bear a grudge against Einstein.

Once Mileva had returned to hospital in May, Einstein was again faced with the question of what to do to ensure that his sons were cared for. Tete was a very sickly child and spent much of this period also undergoing medical treatment, which at least meant he was normally being looked after. His poor health caused the postponement of his entry to school. Einstein blamed Tete's frail health on inheritance from Mileva and reproached himself for having conceived him. When Mileva's health had worsened at the end of April, Hans Albert had been taken into Zangger's home. Einstein considered bringing Hans Albert to Berlin and teaching him himself, although he knew that Mileva was highly unlikely to agree to this. Another possibility was lodging him at a school, such as that where Einstein's former pupil, Hans Tanner, taught, or that attended by Vero Besso. His favoured solution however was to lodge Hans Albert in Luzern with his sister Maja and Paul, who was also Besso's brother-in-law. He asked Besso to approach them, since he did not wish to pressure them into accepting by asking them himself. This at least is what he told Besso, but he knew that Maja would be sure to agree unless Paul forbade it. Einstein probably thought that Besso and his wife Anna, Paul Winteler's sister, would have more chance of convincing him than Einstein would. Einstein was adamant that Mileva's

[7] An amusing incident occurred in this correspondence towards the end of 1917. Einstein had set his son a trigonometric problem. Hans Albert had posted his answer to Berlin. The military censor had opened it, scanned the contents and then corrected Hans Albert's answer.

wishes with regard to Hans Albert should be disregarded if they were not in the best interests of his son.

In the end, none of these schemes came to fruition. Zangger would have counselled that removing one of her sons from her care would be likely to have a very serious effect on Mileva's already fragile mental state. Somehow, Zangger and Besso managed to cope until the end of the summer of 1917. Mileva had earlier written to ask her mother to come to Zurich and look after her, but her mother was too ill. She was also prostrate with concern over Milos, Mileva's brother, who had been missing at the Russian front for many months. Mileva's younger sister Zorka, who had been unable to come when Mileva first fell ill, probably because of the difficulty and length of time required to get wartime travel authorisations, came to Zurich instead. She arrived at the end of August. She impressed Lisbeth Hurwitz as being solid and reliable, although her movements were impaired by a limp even more pronounced than Mileva's. Zorka stayed with Mileva, bearing the brunt of caring for the children, until her behaviour began to become more and more eccentric. By February 1918, she had to be placed in a mental institution, leaving Mileva once again ill and without assistance.

As if Mileva's illness were not enough, in the same letter of 3 March 1917, in which he thanked Besso for arranging Hans Albert's piano lessons, Einstein indicated that he also was sickening. These were the first symptoms of a digestive illness that was to prostrate him for almost a year towards the end of 1917. His sister Maja had always blamed his subsequent health problems on his poverty as a student and the resulting irregular and insufficiently nutritious diet. The chronic shortage of food, although alleviated by parcels from Switzerland facilitated by Zangger and by provisions from Elsa's extended family in agricultural south Germany, must have been the catalyst for a virulent relapse in this underlying condition. The Turnip Winter of 1916–1917 caused a crisis of health all over the capital. Initially, Einstein was diagnosed with gallstones. The diagnosis then shifted to a stomach complaint and then finally an ulcer. Soon his stomach became so sensitive that the only food he could tolerate was a sort of porridge or rice pudding cooked for him by Elsa.

A vivid picture of Einstein in Berlin at this time is painted by Rudolf Humm, a young man from Einstein's former school district of the Aargau in Switzerland. In 1917, he was studying in Göttingen. In May he arrived in Berlin in order to attend Einstein's lectures. In his diary he noted that he had visited Einstein's apartment and found him in stocking-feet. When Einstein asked him his business, he said he wanted to be inducted into the community of physicists:

> Then I asked him for three of his papers. He searched for one of them for a long time without success, wondering where it could be and complaining about the disorder and his forgetfulness. We went from room to room and then stood at a loss in front of the bookcases. He had a rather austere apartment and didn't seem to have a maid. I stayed shamelessly for one and a half hours before accompanying him to the tobacconist and then to the door of a friend of his. At first he was reticent, but then he warmed to me and never stopped talking. He is wonderful for students, once one understands how to set him off by posing questions and swapping ideas; then he carries on by himself. I was astonished by the clarity and penetration of his thought. He is almost never in doubt and where he is, the doubts are very specific. He admitted to being a poor calculator: he worked more intuitively and seemed to distrust what we get up to in Göttingen; he never thought in such formal terms. The richness of his imagination was anchored firmly in the real world. He remarked that he imagined gravitational waves using an analogy of an elastic body, squeezing his fingers as if he had a rubber ball in his hand. He is very cautious and totally a physicist. He doesn't immediately rush off into abstractions like we do in Göttingen, which he explains by the fact that he has had to divest himself very slowly of his prejudices. Thus, he couldn't see at the outset how there could be general invariance but had to arrive at this realisation by a step by step approach. In hindsight it seemed very plausible but he himself had to overcome

> a distaste for the quantities that he needed to use – the curvature tensors – because they seemed much too non-intuitive . . .
>
> [. . .]
>
> To construct a theory of the universe purely from imagination seemed attractive and could have useful results. However, history teaches that such endeavours usually fail. The only theory built on pure thought that corresponds with reality is the kinetic theory of gases. The diversity inherent in the tensor formulation is much too great for one to know which one should choose in order to describe electromagnetism. The experimental data is much too limited to serve as a useful guide. Certainly, great and specific imaginative powers are necessary but he rejects all attempts to describe the universe by the power of pure thought. It would be presumptuous in the extreme to build a complete theory of the universe when there are so many phenomena about which we can have only the most limited conception. Thus, he had been trying for the last decade to account for why the electron doesn't fall apart but has had to give up.

Assuming that Humm reported correctly, it is indeed a pity that in his later career, Einstein did not take his own advice.

By June, Einstein felt well enough to set off to visit his children in Zurich. He left on 29 June, travelling first to Frankfurt to lecture and then moving on to Heilbronn to visit his mother, Pauline, who was also ill. He arrived in Zurich around 7 July, staying at first in a hotel and subsequently with Zangger, in order that the later could observe his state of health more closely. It was now that Zangger became convinced that he had an ulcer. Anna Besso devised an extremely efficacious treatment: causing him to lie down with hot-water bottles on his abdomen. After staying in Zurich only a few days, Einstein left with Hans Albert for Arosa, where they visited Tete, who had been hospitalised there to benefit from the healthy air. Einstein and Hans Albert spent some days in Arosa, walking in the mountains and rowing together on the nearby lake. Subsequently, Einstein stayed for several weeks with Maja and Paul in Luzern, having decided that he could achieve the rest and recuperation his doctors prescribed much more cheaply there than at the spa resort they had recommended. They were joined there by Pauline Einstein. Hans Albert also accompanied him to Luzern but was called back to Zurich on 31 July by Mileva, who was always nervous when her children were staying with Einstein's family. Maja took good care of Einstein, preparing his gruel and continuing the heat treatment methods for his stomach prescribed by Zangger. Feeling significantly better, Einstein left Luzern in the last week of August, travelling via Zurich, where he remained for a couple of days, staying with the Bessos. At some point in his trip he visited Kathia Adler, who had moved her family to Unter-Aegeri in the canton of Zug. He also met up with his old schoolfriend and music partner, Jakob Ehrat, in Zurich. He spent an evening with Conrad Habicht before, with some difficulty, he crossed the Swiss–German border. He proceeded to Sigmaringen to visit a friend, Father Brandhuber, who was a priest but also a member of the Prussian parliament. Einstein's health had recovered enough for him to take several strenuous walks with Brandhuber, whose opinions he found surprisingly congruent with his own.

By the middle of September 1917, Einstein was back in Berlin; not however to his old ascetic bachelor quarters, where his lease was about to end, but to an apartment opposite to that of Elsa. It had become vacant, and Elsa had snapped it up for Einstein. Although nominally, for appearance's sake, in different apartments, in practice they were now living together. The bohemian Einstein, perpetually trumpeting his determination never to give up his freedom, was slowly but surely subdued, although never fully subsumed, into the bourgeois way of life that he had for so long despised.

Chapter 14

The war, the empire, and Einstein's marriage end in chaos; Remarriage and the eclipse (1917–1919)

At the end of 1916, Einstein embarked on a project for which he had often said he considered himself very unsuited: the popularisation of his theories of relativity. He was probably encouraged in this endeavour by the book by Erwin Freundlich to which he contributed a preface, as well as the earlier success of his friend Max von Laue's book on special relativity. He completed the first draft of his book in December 1916 and it was published in the spring of 1917. Especially after Einstein's rise to world fame, the book understandably became very popular and went through many printings, revisions, and translations into the major languages. It deserves this popularity.

The original German version of the book was entitled *On the Special and General Theories of Relativity – A Generally Accessible Account*. It forms an invaluable introduction to these crowning achievements of Einstein's life. Written in the first person, as if an avuncular teacher were sitting by the reader's side, it begins with a recollection of his own childhood: the wonders of Euclid. These he imagines will have been the common experience of most who pick up the book. By a clear exposition of reference frames, measuring rods, and the motion of objects in moving trains, Einstein establishes the basic tenets of special relativity. He does not shy away from using formulae and simple algebra, writing down the Lorentz transformations explicitly. There are of course sections that seem strange to the modern reader. In discussing the motion of electrons he wrote:

> In the theoretical treatment of these electrons, we are faced with the difficulty that electrodynamic theory of itself is unable to give an account of their nature. For since electrical masses of one sign repel each other, the negative electrical masses constituting the electron would necessarily be scattered under the influence of their mutual repulsions, unless there are forces of another kind operating between them, the nature of which has hitherto remained obscure to us [to which is added the footnote] The general theory of relativity renders it likely that the electrical masses of an electron are held together by gravitational forces.

Einstein elaborated on the above idea in a talk to the Prussian Academy that appeared in its *Proceedings* in April 1919. It is a strange production, in which Einstein seems more intent on providing some more fundamental basis for the cosmological constant than understanding the structure of elementary particles. He complained that the cosmological constant was a particularly grave blemish on the beauty of the theory. Having linked the gravitation tensor with the electromagnetic field tensor in a rather artificial way, he obtained the cosmological constant as a consequence rather than it having to be inserted into the Einstein equations by hand. The modified gravitational field equations produced a 'negative pressure' that Einstein asserted could stabilise the mutual electrical repulsion of the electron's constituent particles. However, there is no quantitative estimate of the size or properties of the particles that would be so produced. This was not indeed possible, since

Einstein. Brian Foster, Oxford University Press. © Brian Foster (2026). DOI: 10.1093/oso/9780198794875.003.0014

he admitted at the end of the paper that the theory was not fully determined and any spherically symmetric distribution of electric charge would be allowed.

Although Einstein had good reasons to think that the electron was composite and hence was right to point out that an additional force was required to overcome the electrostatic forces between charged particles, it is difficult to see how he could have thought that gravitational forces, many orders of magnitude weaker, could solve the problem.[1] Einstein wrote to Hermann Weyl, the brilliant theoretician then in Zurich, early in 1917 that:

> The question of whether the electron should be treated as a singular point, whether true singularities are at all admissible in the physical description, is generally of the greatest interest. Maxwell's electrodynamics has a finite radius of the electron in order to explain its non-zero inertia but finite energy.

The question Einstein posed to Weyl is indeed one that still puzzles today's physicists and is an area of intense research in for example superstring theory. The idea that gravity could bind a composite electron represented Einstein's first tentative steps towards a unified theory of gravitation and electromagnetism, an endeavour to which his final years would be devoted. However, as Pais remarks, '[As] is not unusual for him in his later years, a thought comes, is mentioned in print, and then vanishes without a trace.' In fact, Einstein did return to this idea briefly in 1927, but in a paper on a technical aspect of the mathematics he had used. He tersely noted in a footnote that 'Subsequent research unfortunately demonstrated to me that such an approach would not lead to a satisfactory theory of the electron.'

Another idea that was suggested to Einstein around the time of his lecture to the Prussian academy described above turned out to be much more important in the future development of physics. In May 1919, Theodor Kaluza wrote to Einstein with an idea to unify gravity and electromagnetism. Einstein's excitement at this idea, to embed the unified theory in five dimensions rather than four, is tangible in his reply. He told Kaluza that although he had often thought of expressing the electromagnetic field in terms of the Christoffel symbols that played a central role in general relativity, it had never occurred to him to extend the number of dimensions in which they operated.[2] He admitted that he found the idea extremely interesting and pointed out a number of very promising properties that this theory seemed to have. He cautioned however that Kaluza should avoid the mistakes for which he chastised Weyl, with whom he had been having a running battle by letter for several months about the latter's unified theory. Einstein insisted that it was essential to confront the theory with real physical phenomena. He carried on working and thinking about Kaluza's idea, exchanging two further letters about the details until, a few weeks later, he came up with what he, and apparently also Kaluza, saw as a major problem. The influence of the gravitational field on the motion of an electron in Kaluza's theory was many orders of magnitude greater than could be physically admissible. Until this was solved, Einstein did not feel that he could present Kaluza's work to the Prussian Academy as he had originally intended. Working out this problem at least to the level where Kaluza felt he could publish the work led to a delay until 1921. Even then, Kaluza in a footnote admitted that he hadn't solved the problem. His subsequent

[1] In fact, modern theories contain a 'strong' force, carried by the 'gluon', that overcomes the electrostatic forces inside composite particles, such as the proton. As for the electron, it seems to be point-like and have no structure, at least to the precision of today's most powerful accelerators.

[2] This was not the first time that such an extension to more than four dimensions had been suggested. Einstein's old collaborator Gunnar Nordström had had the idea back in 1914, although before Einstein's development of general relativity, he had been constrained to implement it on a flat space-time metric.

work with Oskar Klein (no relation to Einstein's correspondent the eminent mathematician Felix Klein from Göttingen), who in 1926 added a quantum interpretation to Kaluza's classical ideas, developed an area that is still an active field of research. Indeed, in many respects Kaluza–Klein theory is the precursor of string theory.

Returning to Einstein's popular book on relativity, it necessarily becomes more difficult as it progresses towards general relativity. Today, whereas all undergraduates in physics will be exposed to special relativity, often in their first year of study, general relativity normally remains an option for the final year. In expounding Minkowski space as a prerequisite, Einstein uses imaginary numbers with essentially no explanation, which would certainly be a step too far for a popular book today. The centrepiece of his explanation of the principles of general relativity is the equality between gravitational and inertial mass. There are times where he uses that well-worn phrase of the textbook author, 'It can easily be shown that . . .' for example, in discussing that light no longer moves in straight lines when viewed from an accelerated frame. Nevertheless, his explanation of why the geometry of general relativity is not the Euclidean geometry with which he began his exposition is masterly. After mention of the experimental evidence for general relativity, the correct prediction for the perihelion of Mercury, Einstein mentioned the two other experimental consequences, the bending of light by massive objects and the redshift of spectral lines in a gravitational field. With typical Einsteinian flourish, he ended the book with the statement that he had no doubt that these two phenomena would also eventually be observed.

Although the original edition ended with the experimental consequences of general relativity, the third edition in 1918 added another section on the implications of general relativity for cosmology that can usefully be mentioned here. Einstein illustrated the concept of a finite universe without limits by considering the analogy of beings limited to the surface of a sphere. He introduced the concept of elliptical space, with which he had in fact himself only become familiar after he had completed the first draft of the original book. It was Freundlich, who had studied differential geometry in Göttingen with Klein before switching to astronomy, who alerted him to this possibility in February 1917, perhaps after Einstein had shown him his draft.[3] The additional section concludes by remarking that general relativity implies that a static, homogenous universe must be spherical or elliptical. Subsequent editions of the book as the decades advanced added appendix after appendix in which these conclusions were elaborated, modified, and moulded to the dynamic universe that observational astronomy gradually revealed.

While Einstein had been vacationing in Switzerland in the summer of 1917, back in Berlin the bureaucratic wheels necessary to fulfil a long-standing promise to him were slowly turning. Part of the original offer of his appointment to Berlin had been the establishment of a Kaiser Wilhelm Institute for theoretical physics. These plans had been derailed by the war and for three years nothing more had been said. However, the situation changed when wealthy Berlin industrialist Franz Stock offered a donation of 500,000 marks. This, together with matching funds from the Kaiser Wilhelm Society and another donation from the Koppel Foundation, provided an annual budget of 75,000 marks, guaranteed for ten years. This allowed the foundation of the Kaiser Wilhelm Institute for Physics Research. On 9 July 1917, the proposal for the Institute, of which Einstein was to be the first Director, with an honorarium of 5,000 marks, was agreed by the Senate of the Kaiser Wilhelm Society and announced in the *Vossische Zeitung*. On 1 October, the Institute was formally established. In order to save money, the Institute's headquarters was to be Einstein's

[3] This may have been the source of the recollection recorded by Ledermann and mentioned in Chapter 12 that Freundlich had introduced Einstein to Riemannian geometry.

apartment in Haberlandstrasse. Its terms of reference were to offer grants to support first-class research work across both experimental and theoretical physics, to be carried out in institutes and universities throughout Germany. In December, Einstein requested permission to appoint a part-time secretary for 3½ days a week from 1 January 1918. Einstein already had a candidate in mind: Elsa's daughter, Ilse. This would produce a significant perturbation on Einstein's family life.

Another appointment that Einstein was able to make finally fulfilled what he felt were his obligations to Freundlich. Although Einstein had limited freedom of action, since all approved projects had to be agreed by a Board of Directors, he was able to ensure that Freundlich would be relieved of his tedious duties at the Berlin observatory. In December, Freundlich discussed with Einstein the terms of an appointment to a three-year position, during which he would devote his energies to the experimental investigation of general relativistic effects of light deflection and red shift. Although it was intended that the position would begin on 1 January 1918, there were the usual difficulties with finding a laboratory willing to host Freundlich's work. That this was problematic even though Freundlich's salary was paid by Einstein's Institute is a telling testimonial to Freundlich's low standing among his fellow astronomers. Eventually, the observatory at Potsdam agreed to host him and Einstein received approval to make the appointment on 21 January 1918.

Although Einstein's health was significantly improved after his summer break, it gradually deteriorated again as the combination of the pressures of the war regime with its food shortages, the academic politics that the Kaiser Wilhelm Institute forced on him, and his own research, told on his constitution. By Christmas Eve 1917, he was once again prostrate. He sent Christmas greetings to Hans Albert, remarking that he had to remain in bed for four to six weeks. In January he wrote to Michele Besso, who was about to move to Rome to catalogue his uncle's library.[4] In his letter, Einstein mentioned that he was intending to stay in bed for at least the next four weeks. This was in any case more comfortable because of the freezing temperatures in his apartment resulting from the shortage of fuel. He went on to complain bitterly that his resources were not able to cope with the continued cost of Tete's treatment at the sanatorium in Arosa. Only these financial worries and the discomfort from his ulcer can partially excuse the accusatory and petulant tone of his remarks to Besso and Heinrich Zangger, who were exerting themselves enormously on his behalf over the care of his wife and children. He continued his complaints about his two friends in a letter to Hans Albert at the end of January, in which he accused Besso and Zangger of wasting his sons' inheritance on expensive treatments and pampering for Tete. He put his hope in Hans Albert soon being sufficiently mature than they could arrange their affairs themselves without needing the help of strangers!

It is clear that Hans Albert, at the age of 13, had developed a mind of his own. In January of 1918 he wrote twice, favouring his father with his own prescription of a patent medicine to help his health problems and giving further information on progress with his piano playing. He was now receiving lessons at the Music School of the Zurich Conservatory and had graduated to Beethoven Sonatas. He played the 'Moonlight', although he admitted that he couldn't yet manage at full speed. By March he had graduated to Chopin waltzes. The second letter, which he wrote in reply to that of his father mentioned in the previous paragraph, is remarkable. It clearly reveals his resentment at his father's treatment of his mother and his absence from their lives. Although his hurt feelings

[4] Besso seemed to imagine that this would be a permanent occupation, but his wife Anna came to a different conclusion and remained in Zurich.

had previously broken out in his relatively infrequent letters to his father, Hans Albert's protective instincts towards his chronically sick younger brother seem to have been strongly aroused by Einstein's writing that Tete was being pampered in his mountain sanatorium. Hans Albert lashed out at his father's ignorance of his sons' situations. The hurt he felt is clear in his words:

> You are completely unable to judge what is the best thing for Tete, because you have no idea of what he has gone through. Herr Zangger is the best judge, since he has looked after Tete continuously. We can be really grateful that he takes such good care of us. What would have happened to us when Mummy became ill had he not been here. Think about that! Something should really be done about *his* health. You don't know anything about us, or we about you and you have no notion what we need or is necessary for us; I know nothing about you; I know much more about Herr Zangger, or any of the people around here, than I do about you and that is really a pity.
>
> I am now playing the fifth Beethoven Sonata (C Minor), which I am really enjoying.

The bathos in this last sentence makes the preceding paragraph even more affecting. Einstein, unsurprisingly, let this letter remain unanswered until at least the middle of March. His own illness and the helplessness he felt at being so far from his children expressed themselves not only in his exasperation with his friends who were trying to assist him but also in numerous schemes for looking after his sons, with which he bombarded Zangger, Maja, and others. They all involved taking the children away from Zurich and lodging them with one or other of Einstein's relatives or with him and Elsa; they all foundered on Mileva's implacable opposition and understandable desire to keep the children with her, even if at times short-term arrangements were required when she had to be admitted to hospital.

Einstein's health gradually improved with continued bed rest under the ministrations of his faithful nurse Elsa and a staple diet of rice pudding. By March he had gained weight, reaching 60 kg, which shows quite how emaciated he had become during his illness. During his stay in Lzern the previous summer, he had reported his weight as 72 kg. Nevertheless, he still had to be very careful to avoid any exertion. In April 1918, David Hilbert, in transit through Berlin, called at his apartment but found Einstein out. In his subsequent letter apologising for his absence, he told Hilbert that he had attended the weekly meeting of the Academy for only the second time in more than three months. He apologised that he would be unable to attend Max Planck's lectures at Göttingen because of his ulcer, mentioning that he had recently suffered a gastric attack brought on by playing the violin for just one hour. It must have been particularly trying for Einstein that his illness also affected his beloved violin playing. Indeed, he wrote to Hans Albert that he had been forbidden to play the violin since he became ill the previous December, although he had obviously been unable to resist the occasional truancy. The state of his health also caused him to change his plans to spend several weeks of the summer vacation in Switzerland with his sons. This came as a major disappointment to them both. Instead he decided to accompany Elsa, Ilse, and Margot to the Baltic coast. It was not until the late summer of 1918, after this two month stay at Ahrenshoop, that he considered himself to have regained sufficient health to resume a fairly normal lifestyle.

As Einstein's health had declined, so had Mileva's. Although she had spent some time at home in the second half of 1917, she complained of back pains and kept to her bed. Zangger wrote to Einstein in December that she was doing reasonably well but his diagnosis of an inflammation of the spinal cord that had spread to the vertebrae was disturbing. His letter was complete with a graphic diagram of the spinal cord illustrating the problem. The only cure was Nature's: continuous bed rest. Shortly after Mileva's letter, her sister Zorka, who helped around the home, broke down and had to be placed in an asylum. This meant that Zangger had once again to step into the breach, since Besso was mostly in Rome during this period. Zangger continued to keep an eye on Tete in the sanatorium in Arosa. At the end of January, he wrote to Einstein that he thought that his son would be able to return home in February and begin school in the spring. Once again, Zangger took Hans Albert into his own home and cared for him. During all this, he had also to cope with having been elected Dean of the Medical Faculty, which he dreaded, and with Einstein's frequent carping about the extravagant expense to which he considered that he and Besso were subjecting him. Without Zorka to help her, Zangger insisted to Einstein that the best, and furthermore the cheapest, solution was to have Mileva admitted to hospital again. Worst and most tragic of all for Zangger, during these months his daughter Gertrud fell ill with a fever, developed pneumonia in March and died.

Einstein's enforced bed rest allowed him no escape from the pressure exerted by his faithful nurse Elsa to regularise their relationship. At the end of January 1918, despite Zangger's strong opposition to upsetting Mileva again, Einstein once more wrote to her on the question of divorce. He clearly hoped that bribery would work where previous appeals had failed. Einstein's ability to be more generous resulted from the defeat of Russia in the war. There was a ceasefire on the Eastern Front in December 1917 followed by the Treaty of Brest-Litovsk with the new Soviet government in March 1918. This led to a burst of confidence that Germany would eventually win the war and a concomitant recovery in the exchange rate of the German mark. Einstein made various proposals to Mileva for a generous financial settlement that included the reversion of the substantial funds from the Nobel Prize, if and when he received it. This was the carrot; the stick was a threat never to give anything more than the current sum if Mileva refused the divorce.

Mileva's reply to Einstein's letter, rather like that of her son mentioned above, wrenches the heartstrings:

> Exactly two years ago, you pushed me over the brink with letters like this, which I still haven't got over and now you think that, because, as you say, I don't have a fever or attacks, I am having it too easy so you send me another. I can't answer you today; I really have to get advice from my lawyer; my duty to the children doesn't allow me to do anything else. I already wanted to do this previously but I never felt well enough so I kept putting it off and then I became ill. Also, Besso told me that you had given up the idea and no longer considered it. Dear Albert, why do you persecute me interminably? I would never have thought it possible that a man could treat a woman, who had dedicated her love and her youth to him and borne his children, as badly as you have me. You cannot imagine what I have suffered in the last two years, not to mention what happened previously. And now when I haven't even properly recovered, the whole thing starts again. I really haven't deserved this from you. You also could have spared me the endless hard and unfeeling complaints about Tete; you knew that he was ill and something had to be done to stabilise him, particularly as I am sick and cannot care for and provide for him. And how can you characterise Tete's treatment as pampering? You shouldn't have said that for Zangger's sake. He has taken so much trouble for us, never sparing his time or money to help us; he has done things that were really your duty, because he cares for you and you should acknowledge that and not wound him. It's a great misfortune that we spent more money this year than planned; if you had let our health deteriorate, you could have rejoiced over saving a couple of thousand Francs. You should also realise that everything is so expensive today – some things have trebled in price; how am I supposed to cope? I don't understand you. Do you really want to take Elsa's husband's place. From her you will learn how hard it is to survive – do you really want to put your wife and children into the same position. Remember that for me it is worse because I have essentially no capital and that after my illness it will take a considerable time before I can think of earning anything. And your health hasn't improved since the summer, Albert, and you have no idea how much I worry about that. And if your illness should end badly, and one has to consider the possibility, do you want to take

away from me in the case of a divorce the only secure thing we have, the pension and leave me to the mercy or otherwise of your second wife and whether she would pay it to me, so that I would have to fight with those people for the rest of my life. I beg you to consider all these factors; it is impossible for you not to realise that it is completely inappropriate that you should conclude your letter with a threat.

I hope that your health soon improves and beg you to do everything you can to ensure this, which really is the most important thing.

Miza.

Another reply from Mileva to Einstein's letter also exists, presumably later than that quoted above. This is much more conciliatory and was drafted for her by Zangger with input from Besso. In informing Besso of the developments, Zangger expressed an unusual irritation with Einstein's behaviour. He asked Anna Besso to show the draft to Mileva, amend it as she thought fit, and then show it to Mileva's lawyer friend, Dr Zürcher. The letter stated that Mileva found it difficult to concentrate on such matters during her illness, that she didn't really understand why Einstein wanted to do this but didn't want to stand in the way of his happiness. Things would be easier to arrange after the war was over.

In Besso's absence, Anna was drawn further into the Einstein family rescue operation. She agreed to do an audit of Mileva's finances, which she reported in detail to Einstein. He wrote back to her in flattering terms and expressed his gratitude for her help, suggesting that she take over the financial management and that he would send his remittances for Mileva and the children to her instead. He then made the mistake of trying to recruit her to his side on the question of divorce. He once more attempted to enlist sympathy by drawing the tragic picture of Elsa's two daughters, whose future marriages and happiness he opined were being blighted by his and their mother's irregular life together. This enraged Anna, who, having already heard from Zangger and her husband of Einstein's complaints about them, replied in a blistering letter.

It is possible from Anna's letter to divine a lifetime of frustration with Einstein's selfish behaviour, starting with his treatment of her sister Marie in his schooldays, continuing with the rekindling of their affair while he was in Zurich, and his exploitation of her husband's good nature to always be at his beck and call. She refused to take the responsibility for Mileva's finances and poured scorn on his picture of Ilse and Margot's plight. She considered that Elsa had only herself to blame for the situation by throwing herself at Einstein as soon as she had the opportunity. Being aware of Einstein's frequent assertions in correspondence with her husband that he had no intention of remarrying and recalling that he had made the same assertion to her in Zurich the previous summer, she also taxed him with his own inconsistency. Unsurprisingly, Einstein was taken aback by Anna's frankness and took grave exception to the list of home truths that she had delivered to him. He wrote to Mileva that she should not look to Anna for further advice and that he was cutting off all communication with her. It was of course only a few months since he had benefited from Anna's hospitality and nursing, which he had praised to Zangger, while staying in Zurich. This altercation clearly made a major impression on Einstein, as he was still remarking to Besso on Anna's (as he considered it) impertinent letter four months later.

By April 1918, Einstein could report to Zangger that Mileva and he were communicating about divorce arrangements in a perfectly amicable way. He admitted that he had done much damage with his previous approaches and also apologised for his unjustified criticism of Zangger and Besso over the previous few months. The most outrageous of these had caused Zangger to take the unusual step of typing his furious response, to ensure that there would be no ambiguity from his usual crabbed and often indecipherable handwriting. In this letter, he compared Einstein's conviction that cheaper accommodation could be found for Tete in Arosa to a hypothetical situation in which he, Zangger, sent someone to Berlin to tell Einstein that he did not require such an expensive electrometer to do his experiments.

In further efforts to persuade her to grant him a divorce, Einstein wrote to Mileva with an offer to deposit 40,000 marks in securities in Switzerland, to be used if he did not receive the Nobel Prize or to be offset against it if he did. The receipt of several valuable scientific prizes in the previous few months allowed Einstein to make this generous offer, which also included a further 20,000 marks to be deposited in a Berlin bank.

It appears that, during the months in which he was impressing on both the Bessos and Mileva that he ought to get married, he was increasingly uncertain as to who should be the bride. Einstein's move to Haberlandstrasse brought him not only into intimate contact with Elsa but also her two young daughters. Both were teenagers but, as Ilse left her teens, Einstein began to be attracted to her in ways beyond that of an unofficial step-father. This attraction increased in January 1918, when they were thrust together even more. Einstein had hired Ilse as his part-time secretary in his position as Director of the Kaiser Wilhelm Institute. At some time in the next few months, the relationship became sufficiently intimate for Einstein to discuss marriage – not to Elsa, but to her daughter, Ilse.

That Einstein's proposal to Ilse was actually made, bizarre as it seems, is attested by an anxious and rather pathetic letter that Ilse wrote to Georg Nicolai, an old friend of her family, and through them an acquaintance of Einstein. Nicolai, who was a few years older than Einstein, was a pacifist and an ardent campaigner against the war. He was a doctor by profession and had been called up into the army as a medical officer but then was broken to the ranks and subsequently court-martialled for insubordination and sent to a provincial backwater. In the early summer of 1918, he deserted. Arriving surreptitiously in Berlin, he hid in the Einstein household until 20 June, when he managed to steal an aeroplane and fly to Denmark.

Overbye states that Ilse was in love with Nicolai. If this were true, there must be a suspicion that Ilse made up the whole episode of Einstein's proposing to her in order to excite Nicolai's jealousy. However, there are several indications that Einstein's attraction to her was not a figment of her imagination, although it cannot be ruled out that she may have embellished the facts for her own purposes. For example, back in April of 1916, Einstein had referred to missing the company of Elsa and the 'little minxes'. In May of 1918, he wrote a very flirtatious letter to Ilse, thanking her effusively for a letter and drawing that she had sent to him while visiting Nicolai in his place of detention. Finally, the detailed nature of Ilse's letter to Nicolai and the authentic voice of Einstein that can be detected therein makes it hard to believe that Ilse could have entirely manufactured the episode.

The letter written on 22 May asks for Nicolai's advice as to what Ilse should do with respect to the dilemma she was facing. She said that the question of whether Einstein should marry her or her mother had arisen. Einstein had told her that he loved her very much. She loved him in return, but apparently in a different way, since she felt not the least physical attraction towards him. However, she was aware that he most certainly lusted after her, since he had told her so himself as well as admitting that he found it difficult to control himself. Unsurprisingly, she considered him as a father rather than a lover. She was torn between wishing to share her life with him and the conviction that the love between a husband and wife should be qualitatively different to what she felt for him. Finally, she was concerned about the effect on her mother, who was superficially unconcerned and allowing Ilse to make up her own mind, but who Ilse suspected, surely correctly, would be devastated at losing the husband she had been manoeuvring to get for so many years. There would also be significant social fall-out, of particular concern to her grandparents. She asked Nicolai what outcome he thought would be most conducive to the happiness of all three members of this love triangle. It would have been interesting to have had sight of Nicolai's reply, but it has not been preserved. It is indeed lucky that anything is known about this extraordinary incident, since Ilse's letter began with a command to Nicolai to destroy it after he read it. He did not obey; perhaps the reason why no answer from him exists is that he couldn't decide how to answer or simply delayed until he could talk to Ilse herself after his arrival in Berlin.

Whatever Nicolai's counsel, nothing more came of a marriage between Ilse and Einstein. In early June, Einstein talked jokingly of his little harem in Haberlandstrasse to Max Born, who could not have guessed that this description contained more than a little truth. However, by the time Einstein, Elsa, and the girls had departed for their Baltic holiday at the end of June, Einstein was once more writing to Besso rejecting his advice not to marry Elsa. This letter is particularly interesting, since, in it, Einstein told Besso that he would have remained faithful to Mileva if she had not made their married life so unbearable. Concluding the letter on less controversial ground, Einstein informed Besso that contemplating the mysteries of the quantum while the Baltic waves washed quietly on the shore had convinced him that radiation was certainly quantised.

Some correspondence that Ilse would have dealt with as she agonised over her future was from Gustav Bucky, who was Director of the X-ray Department at the Pharmacology Institute of the University of Berlin. In May 1918, he requested funding for X-ray research from the Kaiser Wilhelm Institute. Einstein declined and referred him to the Physikalisch-Technische Reichsanstalt. This brief correspondence has significance in that it is the first recorded between Einstein and Bucky, who, as a fellow Jew fleeing from the Nazis, would eventually become one of those closest to Einstein during his stay in the US.

Meanwhile, Zangger in Zurich continued to work for what he considered Einstein's best interests. As the war news worsened, the situation in Berlin became increasingly desperate. The spring offensive on the Western Front masterminded by General Erich Ludendorff faltered and had obviously failed by the summer of 1918. Zangger therefore hoped that he could attract Einstein to leave the ruins of Wilhelmine Germany and return to Zurich. He began to manoeuvre to create an appointment for Einstein jointly between the university and ETH. The head of physics at the university, Edgar Meyer, wrote to Einstein enthusiastically offering to sweep aside all obstacles to such an appointment. He affirmed that teaching duties could be reduced to whatever Einstein wished and essentially offered to write him a blank cheque. It is interesting to note that Meyer, in naming the other physicists with whom Einstein would be able to work, places Besso at the head of the list! Einstein was certainly tempted but for several reasons could not accept. One was the undoubted continued attraction of the scientific milieu in Berlin and that he did not wish to appear ungrateful to Planck and all his other colleagues who had been so kind to him. Another was probably the complications that Elsa and family moving to Zurich would entail. However, Einstein saw an opportunity to have the best of both worlds. He proposed to both Zangger and Meyer that he should come for a six-week period twice a year to give a concentrated series of lectures. To Zangger he remarked that he would bring Elsa along, who cooked for him. Discussions continued throughout the summer. Meyer was disappointed that Einstein would not move completely back to Zurich, but nevertheless grateful for any time that he could spend there. Gradually the idea of a joint appointment faded, and the university alone took the initiative. In October, the faculty agreed to the proposition. On 4 November, as the mutineers of the High Seas Fleet seized Kiel and the Imperial throne tottered, Einstein wrote to Meyer that he would be in Zurich to give his lectures on 1 February 1919.

The collapse of the German Empire accelerated. On 9 November, the Kaiser agreed to abdicate and crossed the border into the Netherlands. He was to live until 1941, dying in the conviction that Hitler had succeeded in destroying France where he had failed. Prince Max von Baden, the

Imperial Chancellor who had insisted that the Kaiser resign both the Imperial and the Prussian crowns, was forced himself to resign in favour of Friedrich Ebert, leader of the Socialist Party. The armistice was signed and at 11 a.m. on 11 November, the guns at last fell silent. Not only the sailors in Kiel were in revolt; the army and the workers, who were engaged in a general strike, also played an important part in the disorder that racked the nascent German Republic. Einstein was elated both by the end of the war and the fall of the militarists and the monarchy, although in later life he defended the Kaiser personally as being well intentioned but badly advised. The disorder affected all the Imperial institutions, including the University of Berlin, where the Rector and associates were relieved of their duties by a revolutionary council of students and workers and thrown into prison. Einstein, whose stock with the revolutionaries as a notorious anti-war intellectual and pacifist was high, tried to use his influence to free the academics. Born described the events in which he accompanied Einstein on a tram through the chaotic streets of Berlin to the Reichstag building where the students were meeting. When Einstein declined to approve of their behaviour, they passed him on to the new Chancellor, Ebert, a few streets away in the Wilhelmstrasse. No doubt having many more pressing matters requiring his attention, Ebert scribbled a note to the appropriate minister and eventually the Rector and colleagues were released.

Amidst the chaos of the immediate post-war period, the legal wheels ground imperturbably forward in the matter of Einstein and Mileva's divorce. On 6 June 1918, Einstein had signed the divorce agreement and financial settlement. The fact that he was in Berlin while the proceedings were taking place in a Swiss court led to almost farcical complications, as documents were passed backwards and forwards, arriving too late for Einstein to provide official replies. This carried on through October and November but finally on 23 December, Einstein could make a sworn deposition at the Berlin municipal court admitting his adultery with Elsa since the summer of 1914.

Even as the killing fields of the Western Front were abandoned, another deadly killer was reaching behind the lines to decimate both civilian and military personnel alike. The influenza pandemic of 1918 had its earliest recorded origins in late 1917, but it was not until a more virulent strain, often known as Spanish flu, struck in the summer of 1918 that the situation became really serious. By the time it died out in 1920, around a third of the world's population had become infected; 5% of the world's population died. Although Einstein escaped, many of his friends and colleagues fell ill. Zangger, for example, was doubly hit; not only did he himself contract the disease, but as Dean of Medicine, he was also responsible for dealing with the pandemic in the University Hospital in Zurich. He had developed pneumonia, a common effect of the disease that made it particularly deadly, and had had to recuperate in Davos. Besso, good hearted as ever, had accompanied him, despite his wife Anne being only slowly convalescent from the disease. Back on duty in Zurich in November, Zangger described to Einstein the chaos, with private vehicles being requisitioned as ambulances and his phone constantly ringing.

Chaos of a different type continued in Berlin, as the various communist and socialist factions tried to gain the upper hand. Einstein was rather active on the socialist side, speaking on 13 November to the Neues Vaterland organisation and involving himself with various left-wing groupings over the next few months. By December, Einstein could write to Besso in a hopeful tone that the military religion had vanished and, poignantly in the light of future events, that he thought it would never return. The food shortages became even more acute, always an eventuality likely to increase revolutionary ardour. The soldiers returning from the front were arriving back in Berlin by December. National elections were scheduled for 19 January 1919.

Just after the New Year celebrations, Einstein decided to change his travel plans and left for Zurich earlier than originally intended, arriving there with Elsa sometime before 9 January.

The reason for this change in plans is unclear but may well be related to the political situation in Berlin. Already in November, Ilse, who was a convinced revolutionary, had had to shelter in the streets during an exchange of fire. Chancellor Ebert had called loyal troops into the city to suppress the mutinying sailors. On Christmas Eve, an attempt at a military solution led to a two-hour bombardment of the Royal Stables building on Palace Square but the troops were forced to withdraw when crowds of Berliners intervened. Fierce fighting broke out from 5–15 January, in which the Army suppressed the communist revolutionaries, locally known as 'Spartacists', leaving over a thousand casualties. On 15 January, Einstein sent a postcard to the Borns, contrasting the sunshine and abundant chocolate in Zuoz, where he and Elsa were staying, with the bullets flying in Berlin.

The musical life of Berlin was certainly disrupted but not stopped by the revolutionary violence. Even as early as November 1918, performances had to be cancelled as there was chaos on the railways and soloists could not reach the city. By December, the situation had deteriorated further, and January saw extensive disruption of planned concerts. For example, on 10 January, artillery was being deployed and mines laid in front of the Beethovensaal. Marching troops meant that public transport could not reach the concert hall. However, the audience, showing a great thirst for their musical entertainment, could approach the hall from another direction. The members of the Berlin Philharmonic also arrived, although conductor Fritz Busch did not. Nevertheless, the concert took place. By mid-February, the fighting had died down, but the temperature had fallen so that, given the desperate shortage of fuel, the audiences shivered through the entertainment. By April 1919, things had improved somewhat. Georg Schünemann, distinguished critic of the *Deutsche Allgemeine Zeitung*, wrote on 9 April that 'While outside both foreign and domestic politics was controversial, musical life went along as usual. The concert-goer wrapped himself in his own world and, at the first sound of orchestra or choir, forgot all the worries that their well-meaning fellow citizens had secretly entrusted to them.'

Returning to Zurich from his vacation in Zuoz, Einstein began his lectures on the theory of relativity at the University of Zurich on 20 January. They were extremely popular, with an audience reaching 150. Einstein was able to see his sons often during this period. Now that his divorce was imminent, he appears to have overcome his previous revulsion at seeing Mileva. He is recorded as spending considerable time at Mileva's apartment, where he often played the violin accompanied by Hans Albert[5] at the piano. During the musical evenings in Mileva's apartment, Einstein apparently sat for a portrait for a well-known artist then living in Zurich, Dora Hauth-Trachsler. It can be safely assumed that Elsa was not invited to these evenings. His lectures complete, Einstein and Elsa returned to Berlin sometime between 19–23 February.

During his stay in Zurich, on 14 February 1919 Einstein and Mileva's divorce was granted, although he had to wait until 10 April for a copy of the decree to be delivered to him in Berlin. Part of the judgment was that Einstein was forbidden to remarry for two years. Presumably this was only enforceable in Switzerland, since on 2 June, Einstein and Elsa were married in a quiet ceremony in a Berlin registry office, finally settling the question of which member of the family Einstein would wed. It is remarkable that, despite the fact that Einstein had been notionally living opposite to, but in practice living in, Elsa's apartment for many months, he gave his address

[5] Trbuhović-Gjurić's narrative is problematic for this period. Not only does she attribute a heart attack to Einstein that is unrecorded elsewhere, but also she states that Tete accompanied Einstein on the piano. However, this is contradicted by Mileva who, in a letter to Einstein, explicitly states that Tete did not begin to learn the piano until several months later (see below).

on the marriage certificate as a lodging house on Uhlandstrasse in the Wilmersdorf district of the city, where the ceremony took place. It can only be assumed that someone, probably Elsa's old-fashioned parents, thought that this pretence was necessary to protect her reputation.

A few days before Einstein's second marriage, an event had occurred that would have a far greater impact on his life than even this union. Two teams of English astronomers had set up telescopes in Brazil and the island of Principe, off the west coast of Africa. The thing these widely different localities had in common was that both were on the path of totality of the solar eclipse of 29 May 1919. Einstein remained confident of the correctness of his theory of gravitation both because of its economy and beauty but also because of the successful prediction of the orbit of Mercury. Nevertheless, he was anxious that both of the remaining two phenomena that distinguished general relativity from Newtonian gravity should be experimentally tested, namely, the redshift of light in a gravitational field and the bending of light around a massive object such as the sun. Despite the best efforts of his colleague Freundlich and several others, it took years, including periods in which it seemed likely that Einstein's prediction had been ruled out, before eventually the redshift was verified. Both phenomena were extremely difficult to detect; it was the eclipse measurement that was to prove decisive.

The interest among British astronomers in Einstein's theory had been catalysed by the communications from Willem de Sitter to Arthur Eddington referred to in Chapter 13 and by de Sitter's subsequent papers elucidating the subject. Eddington in particular was excited by the beauty of Einstein's theory and determined to make measurements to test it. He also saw an eclipse expedition as a means to reaffirm the international nature of science that had been badly damaged by the war. The next suitable total solar eclipse would have a region of totality that began in South America, swung across the Atlantic, and crossed into Africa. Eddington himself led the expedition to Africa, fixing on the small island of Principe as likely to produce the most suitable observing conditions. His colleagues Charles Davidson and Andrew Crommelin travelled to Sobral in Brazil. Both took with them a variety of telescopes and photographic plates with which to record the spectacular event. Freundlich was forced to concentrate his efforts on trying to observe the redshift. The economic condition of Germany precluded funding an expedition to distant parts, so that Freundlich could only be a spectator as others tried to succeed where he had failed in that so different world that had been Russia in the summer of 1914.

The Brazilian expedition soon discovered that its optical equipment had serious deficiencies, so that their most useful observations were made with a small telescope that had been brought as a back-up. Eddington had an easier time, but many difficulties still beset him. On 4 June, he telegrammed to Sir James Dyson, the Astronomer Royal, that the weather had been cloudy but that useful plates had been exposed. One day later, Crommelin reported to Dyson that their weather conditions had been clear. On 16 June, Einstein wrote to his mother in Luzern that the initial reports from the expeditions were promising. As she battled with terminal cancer, Pauline Einstein was staying with her daughter Maja. She became perhaps more excited than her son about the expeditions, fretting that clouds might obscure the stars over Brazil. Einstein told Zangger that the dying woman was only interested in the eclipse because of her vanity.

Einstein made a second visit to Zurich to lecture, spending all of July and the beginning of August there. He admitted that he didn't really have anything new to say about relativity and only came to oblige the Zurich government. His other reasons were of course to see his sons and to visit his dying mother in Luzern. In order to spend as much time with her as possible, he concentrated his lectures around the weekend, spending the middle of the week in Luzern. He managed to find time to play music with acquaintances in Zurich but much more frequently with Hans Albert in Mileva's apartment, where Einstein lived while Mileva was away taking a cure with Tete. It was just after Tete's return to Zurich that he also began to learn to play the piano. He turned out to be extremely musical and made rapid progress on the instrument; by November, he was asking his father to send him the 'Notebooks for Anna Magdalena Bach' that J. S. Bach had presented to his second wife.

This period while staying in Mileva's apartment was a delight to Einstein, as he spent significant time with Hans Albert and began to appreciate his character. He also took him sailing on the Zurich lake and the Bodensee. On his way back to Berlin, Einstein stayed for one night with his old friend Conrad Habicht. He missed meeting Fritz Haber, who arrived in Zurich on vacation just after Einstein left. Both Haber and Planck were very concerned that Einstein might leave Berlin to return permanently to Zurich. Einstein was able to reassure Haber that he would remain in Berlin. Haber tried to ensure this by arranging, with Planck's help, a significant increase in Einstein's salary. Einstein was back in Berlin on 15 August.

Although Eddington arrived back in England on 14 July, the Brazilian expedition remained for another month in order to take photographs of the same region of the sky to be compared to those of the eclipse. The analysis of the photographic plates turned out to be difficult and time consuming. The first tentative reports came at the British Association meeting in September. Hearing from a Dutch astronomer who attended this meeting, Lorentz telegrammed Einstein on 22 September that the preliminary results were somewhere in between the Newtonian value and the general-relativistic prediction. Einstein lost no time in informing his ailing mother of the good news. In October, he felt confident enough to send a very brief notification to the journal *Naturwissenschaften* that claimed the expedition had confirmed his prediction on the basis of Lorentz's telegram. Various messages of congratulation began to arrive from across Europe. In a visit to Leiden from mid-October until the start of November, Einstein received the congratulations of his Dutch colleagues but in particular enjoyed many evenings of music-making with Paul Ehrenfest, with whom he stayed. The two also played music with the daughters of Ehrenfest's colleague Willem Julius during a visit to the Julius home. The musical aspects of the visit seem to have inspired Ehrenfest's daughters to emulate Einstein by playing the violin. He was charged with purchasing instruments for them, which he commissioned from a local Berlin luthier, probably Julius Levin, for whom he had great respect. He also commissioned a violin for himself.

In England, Eddington and Dyson prepared the ground for their formal announcement of the eclipse result. Dyson called a joint meeting of the Royal Society and the Royal Astronomical Society at the Royal Society's rooms in Burlington House on 6 November. The results from Brazil and Principe straddled the Einstein prediction. The Brazilian results, taken with the small back-up telescope, had the smaller estimated error. This measurement was larger than but compatible with Einstein's prediction with a reasonable probability; however, it excluded the Newtonian value with a high degree of certainty. Since the average of the Brazilian and Principe results was in excellent agreement with Einstein, the journal *Nature* on 13 November opined that, combined with the prediction of the perihelion motion of Mercury, these results could be considered to establish general relativity.

The reaction, first of the British press but subsequently that of the world's, was extraordinary. Sponsel makes a compelling case that this was no accident but rather the result of a carefully laid plan by Dyson and in particular Eddington. As remarked above, Eddington was the first British astronomer to take up general relativity. Mathematically inclined, he was greatly taken by the beauty of Einstein's mathematics and came rapidly to believe that it must be correct. He quickly latched on to the bending of light as a crucial test and, with Dyson, was the driving force behind the eclipse expeditions. He and Dyson were canny enough to know that it is not enough to make a measurement and then explain why it is important. Rather, by concentrating on three possibilities, all of which were exciting in their own way, they built up expectations of the eclipse result in advance among colleagues, interested scientists, including people such as Gustav Mie in Germany, and the press. This was certainly helped by the fact that Eddington was a superb populariser of science. The trichotomy, as Sponsel calls it, set up by Eddington was:

a) No deflection of light near to the Sun's limb is observed, in which case nothing about light and gravity is understood and all theories are wrong.
b) The deflection agrees with the Newtonian value, for a British audience having the virtue of thus disposing of the theories of perfidious foreigners.
c) The deflection agrees with the Einstein prediction, in which case 'REVOLUTION IN SCIENCE'.

As the famous headline in *The Times* of London of 7 November 1919 eventually put it: 'New Theory of the Universe. Newtonian Ideas Overthrown'.[6]

In fact, there were several other possible explanations for any particular deflection result but discussing them would have watered down the impact of the experiment. It is unlikely there would have been newspaper headlines saying 'Bending of Light by Refraction in the Solar Atmosphere Observed.' By carefully defining the terms of discussion, Eddington, Dyson, and colleagues focused the debate and raised the stakes. So convinced was Eddington of the fundamental correctness of general relativity, that in his data analysis he found reasons to discard some results that favoured the Newtonian deflection value. Since, in retrospect, it is clear that he ended up with the correct answer, it would be churlish to do anything other than accept Eddington's reasons for this selection.

The news of the London meeting spread rapidly. In particular, *The New York Times* published its famous headline on 10 November 1919:

> LIGHTS ALL ASKEW IN THE HEAVENS;
> Men of Science More or Less Agog Over Results of Eclipse Observations.
>
> EINSTEIN THEORY TRIUMPHS
> Stars Not Where They Seemed or Were Calculated to be, but Nobody Need Worry.
>
> A BOOK FOR 12 WISE MEN
> No More in All the World Could Comprehend It, Said Einstein When His Daring Publishers Accepted It.

The bright lights of scientific fame were briefly turned from Einstein to Max Planck, Johannes Stark, and Fritz Haber when they were awarded Nobel Prizes in November 1919. However, the spotlight swung inevitably back to Einstein. On 28 November, he published a piece in *The Times* of London describing his theories of relativity. It was entitled 'Einstein On His Theory. Time, Space and Gravitation. The Newtonian System.' In it, Einstein went out of his way to emphasise his gratitude to English science and to use the eclipse results and the triumph of relativity to make his contribution to healing the fissures in European science caused by the war. In typical fashion, he could not resist ending his beautiful article with a joke, which was picked up and repeated immediately in many circles:

---

[6] In somewhat different words, this is how Eddington himself described the importance the eclipse results would have at a lecture at the Royal Institution in February 1918.

> By an application of the theory of relativity to the taste of readers, today in Germany I am called a German man of science, and in England I am represented as a Swiss Jew. If I come to be regarded as a *bête noire*, the descriptions will be reversed, and I shall become a Swiss Jew for the Germans and a German man of science for the English.

He was inundated with requests from other newspapers to write articles, which he systematically declined, due to lack of time. Offers of very substantial funding for relativity research arrived from the German government.

Within a couple of weeks of the Royal Society announcement, Elsa was complaining bitterly to Ehrenfest that it was raining letters requesting autographs, that reporters and photographers were besieging the flat, and that everyone wanted Einstein's relativity book translated into their own language. The front cover of the 14 December 1919 edition of the *Berliner Illustrierte Zeitung* carried an illustration of Einstein, eyes downwards, looking pensive if not indeed depressed. The caption reads 'A new great of world history, Albert Einstein, whose researches have effected a complete revolution in our view of nature and whose insights are equal to those of Copernicus, Kepler and Newton.' Einstein's apotheosis into Great Man of Science was well underway. His life would never be the same again.

Chapter 15

Fame: Life and music in Berlin, and first visit to the US (1919–1922)

Einstein's life for the next decade and a half was centred on his apartment in Haberlandstrasse in Berlin. Here were not only his living quarters but also the nominal headquarters of the Kaiser Wilhelm Institute of which he was the head. His secretary, his step-daughter Ilse, also lived on the premises. It was to this address that his daily tormentor, the postman, delivered a growing flood of letters from around the world.

Haberlandstrasse was in the 'Bavarian quarter' in the west of the city, an area of solid respectability inhabited by civil servants, business managers and intellectuals. Many of these were Jewish. The Einsteins' neighbours included the writers Artur Landsberger and Else Lasker-Schüler, her brother-in-law, the famous chess Grand Master Emanuel Lasker, the drama critic Alfred Kerr, and the Rabbi Leo Baeck, probably the best-known theologian of Reform Judaism in Europe. Einstein became friendly with many of them, in particular Lasker[1] and his sister-in-law. The high density of Jewish inhabitants gave this district the nickname 'Jewish Switzerland'.

The Einstein apartment was on the fourth floor and had originally been that of Elsa's parents. In response to increasing infirmity, they migrated to the first floor, leaving Elsa and Albert in possession of the eyrie, which was serviced by an elevator. The apartment originally contained seven rooms: a library, living room or salon, dining room, and four bedrooms together with the usual ancillary rooms. Einstein's and Elsa's bedrooms, at least by the mid-1920s, were at diametrically opposite poles of the house. The salon was the centre of family life and contained an imposing grand piano, as well as the heavy dark furniture of Biedermeier style that Elsa had acquired. The dining room was the largest room in the house and separated from the salon by sliding doors. In 1922, the loft space was converted to form an office suite for Einstein, consisting of his office, a library, and another room. Einstein's desk was under a small window looking over the eaves and flanked by engravings of Faraday, Maxwell, and Schopenhauer. Planning permission was not obtained for this extension; Einstein had to pull many strings to have it retrospectively granted when this omission was discovered by the authorities in 1927.

Einstein's future son-in-law, Dimitri Marianoff, described how the salon appeared in the latter part of Einstein's residence:

[1] Einstein and Lasker, who first met in 1918, often went on long walks together, enlivened by animated discussion. The two friends had radically different views on politics and also relativity. Lasker had studied with David Hilbert and had a doctorate in mathematics. Einstein listened as Lasker expounded his version of relativity in which the velocity of light was infinite in a perfect vacuum with zero residual matter. Since Lasker had cunningly picked a postulate that could not be falsified experimentally, Einstein contented himself with an occasional attempt to interrupt the torrent of Lasker's exposition with an unsuccessful plea to consider the unnaturalness of such a theory.

Einstein. Brian Foster, Oxford University Press. © Brian Foster (2026). DOI: 10.1093/oso/9780198794875.003.0015

> It was large and spacious, furnished partly in the Empire style. It was comfortable, old-fashioned and unpretentious. The wallpaper, a dark green in colour, was badly in need of replacement. On the round table in the centre was a white embroidered linen tablecloth with a good Hamburg crocheted-lace edge. In a wall cabinet rested some very fine pieces of antique porcelain. A heavy silver frame containing a Russian ikon sat on a small table under a painting of a child by an old master. In a corner of the room stood a large, commonplace, open cabinet with shelves and shelves of medals . . . from universities, colleges, foundations and societies [. . .] In the opposite corner of the room was a grand piano with four or five violins under it and a printed copy of a Mozart original manuscript on top of it.

The family shared the apartment with various animals, presumably cared for by Ilse and Margot. There were two cats named Samson and Delilah and a green and blue cockatoo that Margot had taught to talk. Einstein on his way to his work in the penthouse would stop and listen to the bird.

The diarist and diplomat Harry Graf Kessler described a dinner in March 1922 in Haberlandstrasse. It was attended mostly by Einstein's Jewish friends, many of whom Graf Kessler considered to be rolling in money. Physics was represented by Emil Warburg, just about to step down as head of the Physikalisch-Technische Reichsanstalt, wealth by Einstein's benefactor Leopold Koppel and his friends and fellow musicians Franz von Mendelssohn and his wife. Although Graf Kessler characterised the Einsteins as rather naïve and unsophisticated hosts and the dinner as 'industrial' in the quantity and presumably the quality of the food, he thought the apartment attractive. Philipp Frank, a frequent visitor from Prague, thought that Einstein looked uncomfortable in the midst of this bourgeois comfort. Another guest thought that Einstein appeared as if he had found his way into the apartment by mistake and remained because he didn't know the way out. His way out was, of course, his work. As often with the arrival of middle age, this period marks the beginning of a gradual withdrawal from the fiery emotional attachment to both people and causes that had characterised Einstein's life until now. This does not mean that he was no longer attracted to pretty women – far from it – nor that he took up no new enthusiasms – his work for Zionism and in particular the Hebrew University of Jerusalem dates from this period. However, he recognised this gradual mellowing by two remarks that he made to his closest old friends. To Heinrich Zangger he wrote 'I have learnt to recognise the mutability of all human relations and learnt how to isolate myself from the heat and the cold, so that an equable temperature is pretty much guaranteed.' He informed Michele Besso that 'Here, everyone is close to me only to a certain limit, so life goes on almost without friction; this I have learned in life.'

The Haberlandstrasse apartment was also the home base from which, particularly in the first part of this period, an extensive series of travels to the far corners of the world began. It seems as if Einstein compensated for the drudgery of many aspects of his world fame by exploiting it to see parts of the world that held a particular interest for him, at the same time often earning substantial amounts of badly needed hard currency.

In 1919, however, foreign travel lay in the future. By the winter, it was clear that Einstein's sister Maja could no longer cope with their mother's illness. Pauline's abdominal cancer was advancing steadily and causing her excruciating pain. After staying with Maja for several months, during which Maja could not leave the house for more than two hours at a time, in August Pauline Einstein was placed in a clinic in Luzern. However, the cancer was too advanced for any successful medical intervention and in October Pauline returned to Maja's house. In despair, Maja wrote to Einstein. She believed his presence could help by raising the spirits of his mother, who was unaware of the terminal nature of her condition. Einstein was willing but had great difficulty in finding accommodation. He had originally thought that it would be possible to put his mother in what were probably his old rooms across the corridor in Haberlandstrasse. After great efforts, in which he tried to bring influence from various contacts in the government and city hall to bear, he eventually succeeded. Maja believed that 'steadily advancing anti-Semitism' in Berlin was a contributory cause to these delays.[2]

As Christmas neared, Maja's husband Paul, who was employed by the Swiss railway, organised a special carriage to convey the invalid, her doctor, a nurse, and Maja to Berlin. They left on 26 December, and on their arrival two days later, Pauline was moved into the room in their building that Einstein had acquired. Looked after by the nurse and Maja, Pauline's hold on life was tenacious, despite often being in agony. One doctor called in to ease her final days was Janos Plesch, who was building up a successful career in Berlin as physician to the rich and famous. An extraordinary person, Plesch was to spend a great deal of time with Einstein in years to come. Marianoff remarked that in the Einstein household it was almost traditional that any doctor in attendance on one of its members was admitted to the family circle, as Einstein's liking for the company of doctors was marked. However, none of them were able to do much for Pauline, whose sufferings finally ended on 20 February 1920.

Einstein's grief at his mother's death belied his words to Zangger and Besso about his emotional isolation. As he remarked in another letter to Zangger 'My mother died a week ago after dreadful suffering. We are all completely exhausted just from witnessing it; one really feels the bond of blood deep in one's bones . . . I am standing here as if in front of a wall because I can't picture the future at all.' His numbness apparently also gave way to tears.

Another relationship that contradicted Einstein's conception of himself as emotionally isolated was that with his friend Paul Ehrenfest, who was a desperately needy personality, subject to black depressions that eventually led to his suicide. Like a moth, he was drawn to the flames of Einstein and Niels Bohr, with both of whom he established devoted and indeed passionate friendships that could only exacerbate his deep sense of inferiority as a physicist. He wrote extremely long, rambling letters to Einstein, full of strange symbols, arrows, afterthoughts, and postscripts. They simultaneously exasperated and entertained Einstein, but he could at least read them easily, in contrast to his other indefatigable correspondent, Zangger, whose handwriting was frequently indecipherable. In one letter from this period, Ehrenfest included the following poem from an astronomy journal drawn to his attention by his astronomer colleague Einar Herzsprung:

> We thought that space was straight and Euclid true
> God said, 'Let Einstein be' and all was skew.[3]

---

[2] Einstein problems with accommodation continued in May 1921, while he was visiting the US. He received a letter from the Berlin District Housing Office stating that, due to the Berlin housing crisis and the fact that four persons were living in seven rooms, two rooms must be vacated for persons to be appointed by the Office. They must be given a share of the kitchen, bathroom, and toilet. Einstein, or rather Ilse on his behalf, replied that this was impossible since Einstein was a university professor entitled to have a separate room, and director of the Kaiser Wilhelm Institute, which was co-located in their private apartment. Einstein told Ilse to solicit supporting letters from the rector of the University of Berlin and State Minister Friedrich Schmidt-Ott. The string-pulling worked again; on 2 June the Housing Office wrote back to say it was reversing its decision.

[3] A variation on Alexander Pope's famous couplet on Newton:
Nature and Nature's laws lay hid in night:
God said, 'Let Newton be!' and all was light.

As a result of the reparations demands of the Versailles Treaty, Germany's economic weakness was increasing rapidly. A related phenomenon was the growth of anti-Semitism, marked among many other indications by the foundation of the precursor of the Nazi party in 1919. These developments emboldened several foreign universities to attempt to attract Einstein to leave Berlin. Leiden was one, first by the offer of a highly lucrative full professorship. When Einstein declined this, giving as his reason loyalty to Planck and his other colleagues, Ehrenfest hit upon the idea of a 'Special Professorship'. The momentum for this proposal was greatly accelerated by Einstein's visit to Leiden in October, described in Chapter 14. All the Leiden faculty were thrilled by its success and Ehrenfest received enthusiastic support when he touted the idea of the Special Professorship, which today would be called a Visiting Professorship. He proposed the idea to Einstein, who assented with joy. With the support of the highly influential Dutch Nobel Laureates Lorentz and Heike Kammerlingh Onnes, the proposal was agreed by the faculty. Lorentz was able to write to Einstein on 21 December, offering him a three-year position with an annual stipend of 2,000 gulden with duties involving no more than a couple of two-week visits to Leiden to interact with the physics faculty and students as he saw fit.

Typically, Einstein didn't realise that Lorentz was waiting for his written acceptance of the above offer, but he realised his mistake and wrote to accept in mid-January 1920. Lorentz thereafter moved quickly, the resolution passed the University Council unanimously and was sent to the government for approval. A long silence ensued. What was actually happening is bizarre and was hidden from even the most influential Dutch academics. The government was very worried about the contagion of revolution spreading from the new and chronically unstable Weimar Republic. In what seems more like an extract from a movie script, a certain Carl Einstein who cohabited with an exotic German countess had been involved in a communist uprising in Brussels in November 1918. The Foreign Ministry became suspicious that the German professor and the revolutionary architect were one and the same. Only after the President of the university and the Head of its Charitable Trust were called to a meeting at the ministry was the confusion resolved. This whole process dragged on to the extent that Einstein left Berlin on 6 May for his inaugural lecture before his appointment had been confirmed by Queen Wilhelmina. Indeed, it took until 21 September for the royal assent finally to be granted, requiring the postponement of his lecture. Instead, he gave an informal lecture on 'Space and Time in Modern Physics'. When the inaugural lecture was eventually given on 25 October 1920, with splendid pomp much to Einstein's discomfort, it was a great success.

The case of mistaken identity was not the only mishap on Einstein's trip to Leiden. After his stay the previous autumn, Einstein had agreed to have two violins made by a Berlin luthier for Ehrenfest's daughters. In addition, he had agreed to obtain a grand piano for Ehrenfest from the Hupfeld company, based in Dresden and Leipzig. While the latter was safely delivered to Leiden, Einstein decided to take one of the violins with him. The authorities clearly thought he was in the musical export business and trying to evade customs duty, so that the violin was impounded at the border. This aroused Elsa's scorn for her husband's naivety in declaring it, rather than pretending it was his own violin and then quietly leaving it with Ehrenfest! It was only after significant efforts that she eventually managed to get the violin sent on to Leiden, but it arrived badly damaged.

Einstein loved Leiden and even more or less fulfilled the requirements of his professorship by visiting regularly. Ehrenfest's house became a second home, filled with music and laughter (see Figure 15.1). Ehrenfest's wife Tatiana was not only an excellent physicist in her own right, who was an equal partner with her husband in their renowned 'Encyclopaedia' article on thermodynamics, she also designed the house they built on the corner of Witte Rozenstraat and Jan van Goyenkade in the southwest suburbs of Leiden. The substantial square house with ample gardens on three sides has a basement, two main floors, and attic rooms. One of the rooms was large enough for

Figure 15.1 Paul Ehrenfest at the piano, with Paul Jr on Einstein's knee, in 1920.

Ehrenfest to hold his renowned seminars, at which almost all the great physicists of the period gave talks and where Einstein's notes on a blackboard can still be seen. Seelig described Einstein's stays there: 'In the dining room there was always a little table for [Einstein] with a spread of milk, bread, cheese, biscuits and fruit ready. He exclaimed contentedly, "What more could anyone want than these things, a violin, a bed, a table and a chair?"'

After his trip in May 1920, he wrote to Ehrenfest that they must stay in close contact because they did each other good and their friendship meant that each felt less alienated from this world. Einstein also involved Ehrenfest in an activity that was growing in importance for him. While having ceased to be a practising Jew in his childhood, Einstein had always been sympathetic to other Jews suffering hardship. For example, he had often intervened on behalf of deserving Jewish students who were suffering discrimination when trying to matriculate. The growing anti-Semitism that followed Germany's defeat in the war made him even more sensitive to the plight of Jews both in Germany and elsewhere. Despite his previous antipathy to the ideas of Zionism, which he had expressed for example during his Prague stay, he began to support the movement. The area that excited his particular interest was the foundation of a Hebrew University of Jerusalem. His growing fame made the Zionist leaders, in particular Chaim Weizmann, who was a chemist by training, anxious to recruit him to their cause. Einstein was increasingly asked to involve himself in Zionist activities, for which he struggled to find time. Ehrenfest, a Jew, had himself suffered discrimination while working in Russia with Tatiana. To marry, they were both forced to renounce their religions, since Tatiana was a Russian Orthodox Christian and mixed marriages were forbidden. Einstein suggested to the Zionist leaders that Ehrenfest could help to organise a congress in Geneva on the Jewish University. In fact, the conference never took place but Ehrenfest's sympathy for this activity of Einstein's was never in doubt.

Just before Einstein's trip to Leiden, he finally met Niels Bohr, who gave a lecture in Berlin on 27 April 1920. During his visit, he and Einstein spent several hours discussing Bohr's quantum theory of the atom. On his return to Copenhagen, he sent the Einsteins a consignment of butter, very scarce at that time in Berlin. In his letter of thanks, Einstein referred to the pleasure that Bohr's visit had given him and that he was working his way through Bohr's papers. Reporting to Ehrenfest on their encounter, Einstein wrote that Bohr was like a highly sensitive child who wanders about this world in a kind of trance. They met again when Einstein and Ilse travelled to Kristiania, as Oslo was then called, to lecture to a Norwegian student society on 15, 17, and 18 June. On the return journey, Einstein and Ilse stayed in Copenhagen from 24–28 June. Einstein gave a lecture on 'Gravitation and Geometry' at the Royal Danish Astronomical Society. They met again in December 1920, when Bohr passed through Berlin to rendezvous with his wife. Einstein formed a great admiration and affection for Bohr that survived for decades despite their deep differences of opinion on quantum phenomena. Many years later Einstein wrote:

> That this insecure and contradictory foundation was sufficient to enable a man of Bohr's unique instinct and tact to discover the major laws of the spectral lines and of the electron shells of the atoms together with their significance for chemistry, appeared to me like a miracle – and appears to me as a miracle even today. This is the highest form of musicality in the sphere of thought.

Einstein's packed schedule meant that he had no proper vacation in 1920; in particular, his plans to meet with his sons were complicated by Hans Albert spending a great deal of the summer in Geneva improving his French. Einstein eventually invited them to meet him in Germany in October, since Switzerland was cripplingly expensive for a German. However, his summer was not without incident.

The background was that, in February 1920, there had been a furore about 'unauthorised persons', characterised as predominantly Jewish by the right-wing press, attending Einstein's university lectures on relativity. There had been a disturbance at the following day's lecture that Einstein considered to have had a clear anti-Semitic element. A few months later, right-wing journalist Paul Weyland launched a stinging attack on both relativity and Einstein personally. On 24 August, Weyland's Association of German Natural Scientists for the Preservation of Pure Science held its inaugural meeting in the Philharmonic Hall, bedecked with the swastika banners that had been a symbol of far-right politics in Germany since the previous century. Einstein had attended many concerts there and he attended this event, too, although certainly with less enjoyment. The meeting had been organised by Weyland and a Berlin physicist, Ernst Gehrcke. As the vituperation of the speakers increased, Einstein seemed imperturbable. That this was not the case, however, was apparent when, after some internal struggle, he was unable to resist penning a robust repost for the Berliner Tageblatt entitled 'My answer to the Incorporated Anti-Relativists'.

Both the event itself and Einstein's response unleashed a deluge of correspondence. Many friends and others, disgusted with the attack, wrote to Einstein in sympathy and support. However, his closest friends, the majority of whom were Jews themselves, were dismayed by his article. Ehrenfest and Lorentz regretted that Einstein had responded; Max Born's wife Hedwig, a favourite of Einstein, was particularly outspoken in deprecating the article. Ehrenfest considered it would have come much better from one of his colleagues such as Planck. Einstein replied defensively that the article was his 'sacrifice on the altar of stupidity'. It was certainly unfortunate that Einstein criticised Nobel Laureate Philipp Lenard in the article since, although Lenard entirely agreed with Weyland and Gehrcke, he had not attended or publicly supported the Philharmonic Hall meeting. Lenard was incensed by Einstein's article; it also had the unfortunate effect of prolonging the controversy, which otherwise would probably have died away at least temporarily, since only one of the series of anti-relativity lectures announced by Weyland actually took place. Instead, the 'showdown' with his critics that Einstein had requested was organised at the annual meeting of the German Society for Natural Science and Medicine.

The 'Relativity Showdown' between Einstein and Lenard took place in Bad Neuheim on 23 September, attended by more than 600 people. The two-hour session began with the reading of contributed papers related to gravity from Weyl, Mie, von Laue, and Grebe before the general discussion began. The newspapers had a field day, with the *Vossische Zeitung* taking up half of its front page under an enormous headline 'Battle around Einstein'. Lenard and Einstein jousted together according to the newspaper report rather like medieval savants arguing over theology, each taking and receiving intellectual blows. Subsequently several others joined in, including Max Born. It was no surprise that there was no Damascene conversion on either side, and neither was particularly judged to have got the better of the argument. In a perceptive comment on the situation, the *Vossische Zeitung* reporter wrote that a physics textbook written in 1892 began by stating that the goal of physics was to describe all the phenomena in nature by reduction to mechanical motion. Einstein's opponents continued to hold to this view, while Einstein and colleagues had discarded it and moved on.

Although all the reports of the Bad Neuheim meeting give the impression of a civilised encounter, this may have been somewhat sanitised. Max Born commented on the 'distress' that Einstein had had to endure at Bad Neuheim. Certainly, it left an extremely bad taste in Einstein's mouth, to the extent that he tended to avoid these annual meetings in the future. He felt that he had lost his composure during the discussion and blamed himself for not exercising more control. Some flavour of the encounter can be found in Lenard's diaries,[4] where he recorded the events at some length. He recounted that Nernst had attempted to reconcile the two after the debate but that as Nernst had drawn him towards a waiting Einstein, he had broken away and dashed to the cloakroom. While Lenard queued to retrieve his coat and umbrella, Einstein followed him and attempted to repair their relationship. However, Lenard refused, complaining that Einstein had publicly libelled him and there was nothing further to say. On the last evening of the conference, Planck approached Lenard. Planck had co-written an article for the newspapers in which Einstein was quoted as expressing regret for associating Lenard with Weyland and Gehrcke in his own earlier newspaper article. No doubt attempting to be conciliatory, Planck remarked that he too was actually downright anti-Semitic. Lenard's response in his diary was to imagine that, in that case, he must regret having ever brought Einstein to Germany!

The debate at Bad Neuheim certainly had a bad effect on both Einstein and Elsa, who subsequently became ill. On her recovery, they continued their trip together to Stuttgart, where Einstein gave a talk, and then went to visit their extended family in Hechingen. On 6 October, Einstein met his sons from the Zurich train and took them to stay for a few days with his friendly Catholic priest in Benzingen. Here Einstein reported that he had managed to borrow a violin and played in a performance of Haydn's 'Creation' oratorio. Einstein, Hans Albert, and Tete walked along the picturesque headwaters of the Danube before climbing a volcanic hill with many fortified ruins, the Hohentwil near Singen. After stopping in Konstanz and buying Hans Albert some music, he despatched the boys to meet their mother, who was taking a cure nearby. Einstein returned briefly to Benzingen before leaving for Leiden, where Ehrenfest had arranged a small conference to discuss the microscopic theory of magnets.

[4] Anyone wishing to gauge the depth of Lenard's anti-Semitic bile needs only to read a few entries of his diary, as published by Schirrmacher, around this date.

It was during Einstein's stay in Leiden that the next crisis occurred. At a literary event in 1916, Einstein had met Jewish journalist and author Alexander Moszkowski,[5] who had established a significant reputation for his humorous writing. He and Einstein quickly developed a friendship. Moszkowski was significantly older than Einstein and in considerable financial difficulties, having lost his savings after the war. With Einstein being the man of the moment, Moszkowski saw an opportunity to use their relationship to repair his finances. When he asked Einstein's permission to publish some of their conversations in a book form, Einstein, without giving the idea much thought, gave his permission. The book was due to be published that October and the publisher's notices to booksellers about the book had been distributed. They were seen by Max Born and subsequently other friends. Coming hard on the heels of the Philharmonic Hall meeting controversy, this was the last straw for the Borns. They were concerned that a book written by a Jew lauding another Jew and his theories would stoke up more anti-Semitism and were convinced that Einstein's personal reputation would be harmed by the perception that he had encouraged the book and was a publicity seeker. At first Einstein laughed it all off in his customary way as an amusing irrelevance but eventually he wrote a registered letter to Moszkowski, asking him to withdraw the book. It had however gone too far for that, as the print run had finished and it would have been unconscionably expensive for Moszkowski to withdraw it at this stage. To give him credit, he clearly valued his friendship with Einstein greatly, as evinced by letters, in particular from Moszkowski's wife, about his frantic efforts to prevent publication. He was pained by Einstein's attitude. When Moszkowski's efforts to prevent publication failed, he offered to insert a phrase in the introduction that the responsibility for the contents were his alone. This had to satisfy Einstein and his friends, and the book finally appeared in 1921. While it attracted some criticism from anti-Semitic elements, and Einstein himself received many letters of complaint about particular aspects, even Born subsequently agreed that it was in fact rather harmless.

In fact, Moszkowski came out of the controversy surrounding his book rather better than Einstein's friends, in particular Hedwig Born. Hedwig had been a favourite of Einstein's while they still lived in Berlin and was accustomed to speaking to him with great frankness. She certainly exercised this liberty on this occasion. Her letter of 7 October is full of invective against Moszkowski, to whose humorous writings she took visceral exception. She had apparently also heard negative things from her father, who had studied with Moszkowski in Heidelberg. Her husband was almost as outspoken, giving Einstein a list of steps that he should take if his initial request to Moszkowski was unsuccessful and asking for power of attorney to allow him to travel to Berlin, or indeed anywhere, to sort it out. In Born's defence, he was sensitive to this issue since he had been criticised for publishing a photograph and a brief biography of Einstein in his popular book on relativity. When it became clear that Moszkowski's book was going to be published, Hedwig wrote a seven-page letter to Elsa, to whom she had also been close. This was a reply to a letter from Elsa suggesting that she should correspond with her, rather than bothering Einstein. This must have infuriated Hedwig, as her reply can only be described as hysterical. She accused Elsa of allowing Moszkowski access to Einstein, of being exceptionally susceptible to flattery, and that she encouraged or was even the source of the publicity that followed her husband. She quoted Einstein as remarking that Elsa was a megalomaniac, with which assessment Hedwig concurred. These accusations must have stung since, except perhaps for the last, they were substantially true. This extraordinary missive roused Einstein to what must have been a singularly sarcastic letter, sadly not extant, which served

[5] Moszkowski's younger brother, Moritz, was a distinguished pianist and composer, then living in Paris.

to silence Hedwig. Although Einstein's relationship with Max Born seems not to have been permanently damaged by this episode, his previous intimacy with Hedwig seems never to have been completely re-established.

The Moszkowsi book is a source of some valuable information on the Einstein of the early 1920s. Based as it was on recent conversations, it is likely to be reasonably accurate in quoting what Einstein actually said. Some opinions are disturbing although not untypical of their time. Einstein is quoted as considering that women function at a much lower level than men: 'It is conceivable that Nature may have created a sex without brains.' This is remarkable piece of prejudice from someone who was a friend of both Marie Curie and Tatiana Ehrenfest, and indeed Hedwig Born. Einstein mentioned the importance literature had for him, stating that Dostoyevsky ranked higher than Gauss. He made the point that, had Dostoyevsky not existed, there would have been no *Brothers Karamazov*, but had Gauss not been born, someone else at around the same epoch would have come to the same conclusions. He was thoroughly enjoying *Brothers Karamazov*, a birthday present from his sister Maja, in March 1920. Its fascination still held him seventeen years later, when he told the English scientist and administrator C. P. Snow that it was his favourite book. In answer to a question from Moszkowski about Lord Rutherford and the structure of the atom, Einstein illustrated the difficulty of making reliable predictions in physics: 'We need to gain clarity on how the nucleus is composed of positive and negative charges, and it is my conviction,' he concluded, 'that there is no further layer of substructure.' This latter conversation is likely to have been very recent, as Einstein had published an article on 25 July 1920 in which he commented on Rutherford's 1919 discovery of splitting the atom in the context of a discussion on new sources of energy. This was very topical since, under the terms of the Versailles Treaty, most German coal was exported to France, which left a dangerous shortage for the coming winter. In the article, Einstein does a better job as a seer in clearly indicating the possibility of nuclear chain reactions, which is also mentioned in Moszkowski's book. Einstein realised that this could not lead to an energy source because of the electrostatic repulsion of the charged fissile fragments as they approached the nucleus. He was unaware of the electrically neutral constituent of the nucleus, the neutron, which was not discovered until 1932. This was to change everything and dominate the final part of Einstein's life.

On the whole, rather like David Reichinstein's book mentioned in Chapter 9, Moszkowski's book is at least as much about himself as Einstein, as evinced by a long rambling section on the occult. Indeed, much of the book consists of Moszkowski displaying both his erudition and credulity by expanding on various ideas that Einstein had mentioned. The following section in the English translation is worth quoting at length because of the insight it gives into Einstein's thoughts on music at that time:

> He is confessedly a classicist, and a sincere devotee of the revelations of Bach, Haydn, and Mozart. What fascinates and enraptures him above all is that which is directed inwards, which is contemplative and erected on a religious basis. The simple masterful flow in musical development and invention is all-important for him. The architectonic structure that we marvel at in Bach, the Gothic tendency towards heavenly heights, perhaps calls up in him sensations that emanate from his hidden wealth of constructive mathematical ideas . . . To mention only the main features, then, neither Beethoven as a composer of symphonies, nor Richard Wagner, denote the pinnacles of music for [Einstein]; he could live without the Ninth Symphony, but not without Beethoven's ensemble music. The number of composers and compositions which are not a necessity of life for him is very considerable. It includes the majority of romanticists, the erotically inclined school of Chopin and Schumann, which revels in sensation, and as already mentioned, the neo-German dramatic composers. He has much objective admiration for them, yet he does not conceal the fact that he also feels lively opposition in the gamut of his sensations.

He regards the properly modern productions as interesting phenomena, and has various degrees of disapproval for them, extending to complete aversion. It costs him an effort to hear an opera of Wagner, and when he has done so, he returns home bearing with him the *leitmotiv* of Meister Eckhard: 'The lust of creatures is intermingled with bitterness.' In general, he seems to take up approximately the point of view of Rossini. Wagner gives him wonderful moments, followed, however, by periods of acute emotional distress

[...]

Einstein also occupies himself in an active sense with music, and has developed into a very fair violinist, without claiming higher degrees of achievement. Among other things I once heard him play the violin part of a Brahms Sonata, and his performance approached concert standard. He draws a beautiful tone, infuses expression into his rendering, and knows how to overcome the technical difficulties. Among the supreme artists of his instrument who have exerted a personal influence on him, Joachim assumes the first place. Einstein still speaks with great enthusiasm of Joachim's performance of Beethoven's Tenth Sonata and of Bach's Chaconne. He himself plays the latter piece, for which the purity and accuracy of his double and multiple stopping fits him. Whoever chooses the right moment – this good fortune has not yet befallen me – may overhear Einstein at his pianistic studies. As he confessed to me, improvisation on the piano is a necessity of his life. Every journey that takes him away from the instrument for some time excites a home-sickness for his piano, and when he returns he longingly caresses the keys to ease himself of the burden of the tone experiences that have mounted up in him, giving them utterance in improvisations.

The regular run of concerts in which displays of bravura play an important part finds little favour with him; above all, he is not a worshipper of the orchestral conductor, whom he regards only as an interpreter and not as a virtuoso on the orchestral instrument. He expressed this idea in unmistakable words: 'The conductor should keep himself in the background.' I believe that his dearest wish would be to breathe in the tones without a personal or material medium, merely out of the air or out of space. Furthermore, I believe that there is an unfathomable connection between his musical instinct and his nature as a research scientist. For the ear, as we know from Mach, is the true organ that enables us to experience space, and thus things may occur within the ear of the investigator of space that may have a different significance from that of music which is representable in tones. I strongly doubt whether traces of compositional form occur in Einstein's tone-monologues, but perhaps they contain examples of an art for which the aesthetics of a distant future may find a name.

Einstein's musical activities in the 1920s in Berlin were as important to him as ever, although they had to be squeezed into his hectic travel and work schedule. His quartet with Erwin Freundlich is not heard of again, probably because, as will be discussed later, they became increasingly estranged. He is known to have played in a quartet with the distinguished architect and cellist Alexander Baerwald in the early part of the decade. Einstein told his sister Maja in 1926 that he was unable to be part of a regular string quartet, as he had been throughout his life until then. This was not due to lack of motivation but simply because he could not fit regular meetings into his schedule. Although he continued to play in ad hoc groups with amateurs like himself, increasingly he began to exploit his fame, many aspects of which he found distasteful, in a pleasant and positive way. He loved to attract world-class musicians to play with him. Few professional soloists enjoy playing with amateurs, or indeed playing at all in social settings, but the pull of fame was sufficient to attract them to Haberlandstrasse. Elsa seems to have played a part in at least organising the logistics of these musical encounters. In 1920, she arranged for a piano quartet session with two young Jewish prodigies, Andreas and Joseph Weissgerber. The pianist was probably Eugen D'Albert. Andreas was a pupil of the famous soloist and pedagogue Jeno Hubay. Andreas, who presumably played the viola in these quartets, was twenty at the time. He was making a big impression on the Berlin musical scene as a violin virtuoso. In 1921, he recorded Pablo de Sarasate's *Zigeunerweisen*. He went on to record several other works, including Beethoven's 'Spring' Sonata with D'Albert in 1923. His eighteen-year-old brother was a fine cellist, who together with Andreas and the great Claudio Arrau made up a popular piano trio. The brothers emigrated to Israel in 1933, where, together with Bronislaw Huberman, they are credited as the founders of what eventually became the Israel Philharmonic Orchestra. Andreas died of a heart attack at the age of only 41.

The teenaged Boris Schwarz, who was born in St Petersburg, was another Jewish prodigy violinist with whom Einstein played at this time. Boris and his father Joseph, a respected concert pianist, were invited to Haberlandstrasse on the strength of a recommendation. Einstein was so moved by Boris's rendition of the Bruch G-minor Concerto that he exclaimed during one passage that it was clear that the boy loved the violin. After the concerto was complete, Einstein grabbed his own violin and played Vivaldi with them, the first of many musical sessions that the three enjoyed. Their meeting occurred before 1922, since in the spring of that year, Einstein recommended Schwarz to Heinrich Goldschmidt, who had been his host on his visit to Norway in 1920.

Einstein regularly played violin and piano sonatas with one of the twentieth-century's leading interpreters of Bach and Mozart, Edwin Fischer. Born in Basel in 1886, he was resident in Berlin from 1918 and had a very successful career as a soloist and recitalist. He subsequently became a distinguished conductor, firstly of the Lübeck Musikverein and then of his own chamber orchestra. In the latter role he was one of the early pioneers of historically accurate performance. In 1919, Fischer married Eleanora von Mendelssohn, scion of the branch of Moses Mendelssohn's family that had, at least initially, remained in the Jewish faith. Moses's great-grandson Franz had been ennobled for services to the Prussian state through the Mendelssohn bank. Franz's granddaughter Eleanora was a famous beauty who had an initially brilliant career on the stage. The von Mendelssohn family was large; Eleanora's uncle was Franz (the younger) von Mendelssohn, head of the family bank who was the financial supervisor of the Kaiser Wilhelm Gesellschaft of which Einstein was a director; as such, he regularly countersigned Einstein's requisitions. This von Mendelssohn was highly influential and a pillar of the Weimar regime. He was also, like Einstein, a passionate amateur violinist and was a close friend of the great Joseph Joachim. He and Joachim often played chamber music together and indeed appeared in public benefit concerts together, a testament to von Mendelssohn's abilities as a violinist. Einstein and Franz von Mendelssohn were friends; they played together in the palatial Landhaus Mendelssohn, on what was known as 'Millionaires Row' near Herthasee Lake in Grunewald in the western suburbs. Franz's books carried a bookplate that pictured Felix Mendelssohn-Bartholdy, Joachim, and a Cupid-like figure playing the violin.

There was more than one member of the Mendelssohn family with whom Einstein played chamber music. Francesco von Mendelssohn, the younger brother of Eleanora, was a cellist. It seems likely that Einstein and Francesco became musical partners through Francesco's brother-in-law Fischer, although Elsa Einstein also had extensive social contacts in the Berlin Jewish community that might have extended to the heights of the von Mendelssohns. Both branches of the family were fantastically rich. Francesco lived in his family's villa near to Landhaus Mendelssohn at Koenigsallee 16 in Grunewald. It was furnished by Robert von Mendelssohn, who died in 1917, with Gobelin tapestries, El Grecos, and Rembrandts. Robert was an excellent cellist who, with his brother Franz, put together an astonishing collection of musical instruments, including no less than seven by Stradivari. One of these was the *Piatti* of 1720, one of the greatest of Stradivari cellos. Francesco, whose teachers included Pablo Casals, became an excellent professional cellist and played the *Piatti*, although he had to tussle for its possession with another cellist, Gaspar Cassado. Cassado, a few years older than Francesco, became his mother's lover; the Contessa, as she was known, was an ardent Fascist and decamped with Cassado to Italy directly after Robert's death.

Francesco and Eleanora were left to their own devices as rich teenagers in their enormous house. It is perhaps not surprising that both became notorious even in the hedonistic Gomorrah that was Berlin at that time. Francesco in particular was a flamboyant homosexual whose sartorial eccentricities were legendary. Einstein, with his indifference to appearances, may not even have noticed the idiosyncrasies of the young man as they played piano trios in the splendour of the Grunewald house as well as in the much less opulent salon of the Einstein apartment.

Another cellist who played with Einstein in Berlin was Eva Heinitz. She became somewhat of a specialist on early music and on the viola da gamba but was also a fine cellist. Many years later she recalled:

> A pianist asked me if I would like to play a Mozart trio with the famous Einstein. Who would say no? So we went to Einstein's apartment and played the Mozart B-flat Major Piano Trio. Einstein played the violin with a very soft tone, even when the music required more. His playing was perfectly correct, but totally uninteresting. But what a fantastic face! The face of the famous Albert Einstein is something I could never forget. It was like a landscape, not quite human, unforgettable, the face of one of the greatest minds in history.

The delights of music-making were a relief from the financial worries crowding in on Einstein. The decline of the value of the German mark and the necessity to support his family in Switzerland were constant concerns. The mark's fall was accelerating as it became clearer that Germany was being beggared by the economic terms of the Treaty of Versailles. By December 1920 Einstein had a two-pronged approach to this crisis. The first was to renew his attempts to convince Mileva and his sons to leave Zurich and settle in Germany. The second was to increase as much as he could his earnings in 'hard' currency, which at this point meant anything but German marks. On 15 December, Einstein wrote a letter reiterating arguments made over several years, urging his two sons to take his advice and move to Darmstadt. He accused them, although he really meant Hans Albert, of fanatical obsession with staying in Zurich and unnatural resistance to his proposals. He had admitted that Hans Albert would lose a year in school progression if they made this move. He does not seem to have considered the alternative of leaving Germany to take up one of the many offers of employment that flooded into his mailbox from around the world, thereby receiving his salary in the hard currency that he so desperately required. Irrespective of his devotion to his Berlin colleagues, which was certainly real, it seems likely that Elsa's attachment both to her aged parents and to her comfortable life in Berlin made such a move impossible. Einstein was disappointed to receive Hans Albert's letter turning down the proposal. He understood Einstein's dilemma but said they would prefer to take their chances in Zurich. They would look around to see what opportunities they had to earn money to support themselves at least partially. Lisbeth Hurwitz's diary documented the exceptionally frugal manner of the Einstein family's mode of living. She noted that Mileva made all the family's clothes herself, reusing old material wherever possible. To sugar what he knew would be an unpleasant pill for his father to swallow, Hans Albert concluded his second letter of refusal to leave Zurich with musical matters. He had taken up the double bass and was playing it in the school orchestra, of which he subsequently became president.

The second prong of Einstein's scheme to address his financial worries was to earn hard currency. His father-in-law had transferred more than 100,000 marks in stocks to Einstein and Elsa at the end of 1918 but the dividends from these were presumably in marks. The award of the Professorship at Leiden was a very welcome addition to his finances, which were also beginning to be swelled by royalties from his books. For example, by November 1921, Einstein was receiving royalties in excess of twice his annual salary, itself the maximum that a Prussian professor could receive. However, these remittances were in rapidly devaluing German marks. At the end of October1920,

Einstein was invited to give some lectures at the University of Wisconsin. Several other US universities had extended similar invitations, specifically Columbia University in New York. It occurred to him that since the US was by then the richest country in the world and the dollar exceedingly strong, perhaps this could be a source of hard currency. He began to think of a lecture tour in the US. During his visit to Leiden, he may have mentioned this to Ehrenfest, who was deeply involved in Einstein's financial affairs, having agreed to act as a 'fence' for his earnings in the Netherlands. Other remittances from non-German sources were, if possible, routed to the account that Ehrenfest controlled in Leiden. On orders from Einstein, Ehrenfest disbursed the money, usually to Mileva in Zurich. Another person who received funds clandestinely for Einstein was his distant cousin, Kuno Kocherthaler, who lived in Madrid. All this was in complete violation of the strict capital and foreign-exchange controls for a German resident such as Einstein. Thus, in order to avoid incriminating evidence should the financial authorities examine Einstein's correspondence, Ehrenfest had to write to him about his financial affairs in scientific code. A typical example is 'As I already informed you, we now managed to raise the concentration of the Au ions up to 6.96×10^{-3}, thanks to mixing a local preparation of 1.00×10^{-3} with one from Methu of 1.08×10^{-3}.' The numbers should be multiplied by a million to get sums in guelders; 'Methu' refers to royalties from the Methuen publishing house[6].

Ehrenfest entered into the plans for a lecture tour with enthusiasm, showering Einstein with advice on how he should conduct negotiations and, knowing his vagueness in such matters, begging him not to enter into any agreements without asking his friends. He suggested that he should request $29,000 for an extensive lecture tour, to be given in German. This was a staggering sum, given that the annual salary of the Head of the Princeton Physics Department at that time was $5,000. When Einstein requested $15,000 from Wisconsin, the authorities were so astonished that they wondered whether a mistake had been made in typing and a spurious zero inserted! Needless to say, Einstein's terms could not be met and by Christmas 1920 the whole project had languished. A few months later it was revived by a completely different plan for a trip to the US, in the service of Zionism.

Meanwhile, on 6 January 1921, Einstein set off on a journey to Prague, Vienna, and Dresden. He travelled on the first leg with an acquaintance, Otto Fanta, whom he had met at his mother Berta's salon in Prague. Fanta had studied chemistry at Berlin University and then become a teacher in Prague. He had collaborated with Einstein's old acquaintance Georg Nicholai on the script for a film to popularise general relativity, of which Einstein was aware. When it was released in the following year, Einstein objected to its advertisement as the 'Einstein film', giving the impression it was about him rather than his theory, which was certainly more likely to improve the box office takings. In 1925, Fanta married Johanna Bobatsch, who, as Johanna Fanta, was to play an important role in Einstein's final years.

[6] Such precautions were necessary. Paul Winteler in a letter to Einstein in August 1921 had mentioned some funds deposited in Lucerne from Einstein's investments with Swiss companies. In October, Einstein received a letter from the German authorities querying this, having been tipped off by the Postal Supervisory Authority, who were clearly intercepting and copying suspicious private mail. Presumably they suspected that Einstein was evading taxes. Einstein's defence was flimsy to say the least. He wrote back that he was receiving the monies for a third party to whom they couldn't be sent directly for political reasons. Astonishingly, this fanciful depiction of Mileva as some sort of Swiss *femme fatale* seems to have satisfied the authorities for the time being. However, unsurprisingly, this wasn't the end of the matter. In November 1923, Einstein received a very stiff letter from the local tax office requiring him to present himself there to explain a variety of irregularities, mostly related to his disposal of his Nobel Prize monies (see Chapter 17) but including a detailed account of all his earnings from writing etc.

When Einstein arrived in Prague, he was met by Philipp Frank, his successor as Professor of Physics there. As Einstein stepped off the train, clutching his violin case, Frank thought that, despite his rise to world fame as the second Newton, he still looked like an itinerant violin virtuoso. Rather than living in a hotel, with the attendant crowds of admirers, he stayed with Frank, whose domestic arrangements were somewhat ad hoc. Having just married, the housing shortage in the city compelled Frank to live in his office, which had previously been Einstein's; Hania, Frank's wife, rented a tiny bedroom nearby. In the morning, Einstein and Frank explored the coffee shops of the city, each with its own specialised clientele, for example, Jews, socialists, scientists, all familiar to Einstein from his earlier sojourn. They returned to Frank's room for lunch. While Frank and Einstein were talking, Frank's wife, new to the role of a housekeeper, cooked the liver they had bought on their morning stroll on a Bunsen burner in a corner of the office. In a story that belies the usual perception of Einstein as someone who paid no attention to what he was eating, Einstein noticed that Frau Frank was boiling their lunch. He immediately told her that this wouldn't do and that liver had to be fried, although it must be admitted that the reasons for his intervention were scientific rather than culinary.

That evening Einstein gave a lecture in a dangerously overcrowded hall at the Urania Society. Subsequently a select number retired to a smaller room for various speeches, including one by Friedrich Adler, who read a poem in which Poetry greeted Science. Einstein replied, saying, self-deprecating as usual, that they had been praising him far beyond his deserts and therefore he proposed to reply with music. A musical end to the evening had been planned in any case, so that a pianist was in attendance, with whom Einstein played a movement of a Mozart sonata. The article went on to say that 'Einstein showed himself, to the great surprise of the audience, to be technically much superior to the average amateur, and to be an accomplished violinist and an expressive musician.' Reinhold Fürth, a young physicist there that evening thought that 'he played quite well, but not outstandingly'. On the following morning, Einstein played with his former string quartet colleagues, including his old friend Georg Pick. That evening he returned to the Urania for an open discussion on relativity with Oskar Kraus, a professor of philosophy who was an opponent of relativity.

On the following day, 9 January, Frank and his wife had intended to accompany Einstein quietly to the railway station. However, it had become known that Einstein was sleeping on the foldaway bed in Frank's office, so that the physics department was besieged with enthusiastic admirers. One young man, brandishing a thick wad of papers on which he said he had proven how to extract unlimited energy from the atom using relativity, insisted on seeing Einstein, who finally agreed to an interview. Since time was pressing, Einstein cut the meeting short with the assurance that a longer discussion would not help since the foolishness of his idea was patently obvious. As Frank remarked, twenty-four years later, the atomic bomb would be dropped on Hiroshima.

The Franks and Einstein finally made their escape but were unable to shake off one particularly persistent, and distinguished, personage. Baron von Ehrenfels, the founder of Gestalt psychology, attached himself to the party. In the fifteen minutes it took them to reach the station, the Baron attempted to convince Einstein of the existence of an absolute space and time. Einstein was probably glad to board the train for Vienna.

In Vienna, Einstein stayed with his friend Felix Ehrenhaft's family. He gave three more lectures, one in the large hall of the Philharmonic, which held 2000, and tickets for which were fought over frantically. In the end the audience was 3000, one of whom was an eighteen-year-old Karl Popper, later to become a distinguished philosopher of science. Popper was the nephew of the well-known engineer and social scientist Jozef Popper-Lynkeus, a close friend of the late Ernst Mach. Einstein visited Popper-Lynkeus, who was a house-bound invalid, during this stay. Popper-Lynkeus, who had published many of the tenets of Zionism, in particular the necessity for a Jewish homeland, before Theodore Herzl, died a few months after Einstein's visit. Einstein went on to Dresden, where he gave a talk to students at the Technical University of Dresden on 17 January, returning to Berlin a few days later.

Shortly after this trip, Einstein wrote to his good friend Adolf Busch. Perhaps the most distinguished German violinist after the death of Joachim in 1907, Busch was also, like Einstein, a convinced pacifist. Einstein had a high regard both for Busch as a person and for his playing, saying that he found it impressive, although his strict rhythms were somewhat inflexible and lacking in colour; he blamed this on the Berlin *milieu*. Busch and his seventeen-year-old protege Rudolf Serkin[7] had given their first recital in Berlin the previous evening and had sent Einstein a ticket. Unfortunately, Einstein had a previous engagement to play at a celebration for an eighty-year-old friend and sent his profound apologies and deep regret that he had been unable to attend. He told Busch that 'his daughter' Margot had attended and been enthralled.

February 1921 was an eventful month. Einstein spent a significant fraction of it in the company of that extraordinary character Graf Kessler. A homosexual like Francesco von Mendelssohn, he was much more discreet and therefore admitted to the highest echelons of first Imperial and then Weimar Germany. He moved easily between the worlds of art, diplomacy, and espionage and led a life of astonishing richness of incident. After the war and despite his aristocratic military background, he had become a convinced pacifist, which led to his becoming friendly with Einstein. On 4 February, they both attended the German premier of Richard Strauss's ballet *Josephslegende*, which had been originally premiered in Paris in June 1914, just before the outbreak of the war. Graf Kessler, together with Strauss's usual librettist, Hugo von Hofmannsthal, had created the ballet in June 1912 after a memorable evening in a Paris restaurant. Sergei Diaghilev asked for suggestions for a piece on a biblical theme for the Ballets Russes. Jean Cocteau suggested David dancing before the Ark of the Covenant, while Nijinsky, Reynaldo Hahn and Marcel Proust had other suggestions. Graf Kessler described the scene as the discussion raged:

> Bowls with monstrous strawberries stood on the table, glasses of champagne and liquors which glittered in all colours; the Aga Khan, the richest Muslim Prince of India, sat... in an oriental costume completely covered with fabulous genuine pearls, and even larger rubies and emeralds [...] A belated pair danced the tango.

Although Graf Kessler's memory seems to have conflated several separate events, the description gives a good idea of the circles in which he moved. Kessler's diary entry proclaims that the Berlin performance had been an enormous, almost unprecedented success and had been attended by the luminaries of Berlin society. *Josephslegende* has not however stood the test of time; indeed, it has been said to mark the first of Strauss's many subsequent failures with the musical public, as his artistic creativity began to diminish.

Ten days after the ballet, Einstein and Graf Kessler travelled to Amsterdam as emissaries of the German pacifist organisations to contact the International Trade Union Congress (ITUC) about Germany's situation with respect to the Treaty of Versailles. Einstein's unfamiliarity with sleeping cars amused Graf Kessler. On the long journey they talked about physics and he asked Einstein why, since the atom, with electrons orbiting a central nucleus, seemed to be analogous to the solar

[7] Serkin had made his Berlin debut a few days before in Bach's Fifth Brandenburg Concerto under Busch's direction. His keyboard playing was received so warmly that Busch suggested that he should play an encore. Serkin had nothing prepared so asked for suggestions. Busch joked that he should play Bach's 'Goldberg' Variations. After the conclusion forty-five minutes later, only three people were left in the audience: Busch himself, Artur Schnabel, and Alfred Einstein. Had Albert not been on his trip to Prague and Vienna, he would surely have also stayed, like his distant cousin, until the end.

system, general relativity could not be used to describe it. He posed the intriguing question of whether size was the last absolute. Einstein agreed that the necessity to apply completely different laws of physics depending on the size of the object being described was deeply puzzling.[8] After their rather inconclusive meeting with the ITUC the following day, they went to the Rijksmuseum and looked at Rembrandt's *The Night Watch*. At first Einstein was disappointed but eventually he was overcome with admiration, telling Graf Kessler that such an ability to see magic and wonder in the everyday can only also be found in Dostoyevsky.

Einstein's growing interest in Zionism was mentioned earlier. In February 1921, he was contacted by Chaim Weizmann via the Berlin Zionist leader, Kurt Blumenfeld, to see whether he would be willing to accompany him on a fund-raising trip to the USA. A particular goal was to raise funds for the library of the planned Hebrew University of Jerusalem. After some consideration, Einstein decided to go, provided that Elsa could accompany him and that they should have separate rooms throughout the trip, ostensibly so that Einstein could work undisturbed. Elsa may also have made this stipulation since she ascribed their Haberlandstrasse sleeping quarters being as far apart as possible to Einstein's stentorian snoring. Einstein had hesitated over the decision because of his planned attendance at the Solvay Congress in Brussels, with which the US trip overlapped; he was the only "German" scientist invited. His Berlin colleagues, particularly Haber, pressed him hard to attend as a sort of Trojan Horse to rehabilitate German scientists with their former Allied Power colleagues. Despite this pressure, he decided to accompany Weizmann, although aware that his fame would be exploited for the Zionist cause. The recent discussions on lectures in the US no doubt played a part in his eventual acceptance. However, he reduced his requested honorarium at Princeton to whatever sum the university thought appropriate.

On 21 March, the Einsteins entrained for Rotterdam, where they met Ehrenfest, who had travelled the short distance from Leiden. On the following day they boarded ship, which sailed on to Southampton, where Weizmann and the other Zionists boarded. Weizmann had received telegrams from his American colleagues advising him that Einstein should be left behind because his demands for money from US universities the previous year had been leaked and caused outrage. Weizmann refused and ordered the US Zionists to repair the damage. The two scientists now met for the first time and developed a friendship that, despite sometimes stormy disagreements, lasted for life. In addition to enjoying Weizmann's company, he also relished that of Mrs Weizmann, who found him 'young, gay and flirtatious'. During the Atlantic crossing, Einstein participated in a concert in which he played some of his beloved Mozart sonatas.

When the Zionist party arrived four hours late into New York on 2 April, they received a tumultuous welcome. Undeterred by their long wait, a crowd of several thousand, predominantly Jewish, supporters of Zionism remained to welcome their leaders. After transferring into a small vessel,

[8] In considering this question Einstein did not apply the lessons he had learnt in his own career about 'horses for courses'. Special relativity is indistinguishable from Newtonian mechanics for almost all macroscopic problems except for speeds approaching that of light. General relativity is indistinguishable from special relativity in the absence of gravity, or equivalently, acceleration. What was required was a theory of the atom that was indistinguishable from Newtonian mechanics except at distances approaching or smaller than atomic size. Bohr had made giant steps towards such a theory and, building on his work, Heisenberg, Schrödinger, and others were in a few years to produce quantum mechanics. As will be seen, however, Einstein never accepted it. Since both relativity and quantum mechanics merge into Newtonian mechanics in the appropriate limit, links between all these theories should be possible. However, it must be admitted that, a century later, a theory that unifies quantum mechanics and general relativity remains elusive.

which was filled with journalists, to take them to the pier, Einstein had to field their questions. This strained his vestigial English; although Elsa had a reasonable command of the language, a translator was provided for the interviews. A movie of Einstein walking on deck was also made. The press coverage was both extensive and laudatory. For example, *The New York Times* front-page report entitled 'PROF. EINSTEIN HERE, EXPLAINS RELATIVITY', mentioned his name in virtually every paragraph. The opening paragraph of the article is worth quoting:

> A man in a faded grey raincoat and a flopping black felt hat that nearly concealed the gray hair that straggled over his ears stood on the boat deck of the steamship Rotterdam yesterday, timidly facing a battery of cameramen. In one hand he clutched a shiny briar pipe and with the other clung to a precious violin. He looked like an artist – a musician. He was.

The article also quoted Elsa as saying 'Whenever he became weary in the midst of his work he went to the piano or picked up his violin and rested his mind with music. He improvises . . . He really is an excellent musician.'

Excellent musician he may have been, and world-famous physicist he certainly was, but for the next few weeks he did little else than let himself be paraded 'like a prize ox' to draw rich donors to the cause. He and Weizmann attended lunches and dinners as well as a major meeting at the Metropolitan Opera House, in between holding court at their rooms in the Commodore Hotel, next to Grand Central Station. On 8 April, both Einstein and Weizmann were awarded the freedom of the city and the freedom of the state of New York. On 12 April, nearly 20,000 people attended a rally at the Armoury. On 14 and 15 April Einstein managed to escape from his Zionist duties to give two brief lectures, first on general relativity at City College of New York and then special relativity at Columbia University, followed by four further lectures at City College on 18–21 April. On 25 April, he and Elsa were released from the daily grind of Zionist fund-raising and journeyed to Washington D.C., where they and a delegation from the National Academy of Sciences were received by President Warren Harding. On the following day, Einstein spoke at the National Academy in Washington, briefly at a scientific session in the morning and later that evening at a dinner. He also met with US Supreme Court Justice Louis Brandeis, probably the most distinguished Jew in the country. Brandeis and Weizmann were locked in a bitter power struggle[9] about the direction of American Jewry. Weizmann had warned Einstein against Brandeis, but they liked each other immensely and became friends.

On 2 May, the Einsteins arrived in Chicago, where they met prominent physicists including Nobel Laureate Robert Millikan. Einstein lectured on the special and general theories of relativity in the University of Chicago's Mandel Hall on 3–5 May. On 3 May, ' . . . hundreds without tickets who had hoped for a glimpse of the celebrated physicist were turned away in disappointment'. Einstein was 'received with a tremendous burst of applause lasting for several minutes as he came up on the stage'. During this visit, he was shown around the city by German-born lawyer Siegmund Zeisler. He seems to have had an enjoyable evening of music-making, playing the Bach Double Concerto with Zeisler's violinist son, Ernest, who was studying mathematics at Chicago University. On their way back to the East Coast, the Einsteins visited the Yerkes Observatory in Wisconsin,

[9] Brandeis supported Judge Julian Mack, the president of the Zionist Organization of America, whose opposition to Weizmann had fractured Jewish opinion in the US. Whereas most of the rank-and-file members supported Weizmann, the executive, dominated by Mack and Brandeis, had a majority against. These disagreements rumbled on throughout the visit and forced Weizmann to go over the head of Mack and proclaim Keren Hayesod to provide the money to establish the Jewish homeland in Palestine. This he did on 17 April 1921, during their stay in New York.

often called the 'birthplace of modern astrophysics', founded in 1892 by George Ellery Hale with a donation from Charles Yerkes. On 9 May, Einstein received an honorary degree from Princeton.

While at Princeton, Einstein gave five lectures on relativity which were eventually to form one of his most famous, and successful, publications, the 'Meaning of Relativity' which has never been out of print since the first edition of 1923. This masterly summary of the theory of relativity is quite technical, really a physicist's introduction to relativity. It is an adapted version of the three 'technical' lectures that he gave, after the initial two 'popular' ones. The first chapter of the book, after a somewhat philosophical introduction to space and time, launches directly into the mathematical definition of tensors, ending with writing Maxwell's equations in a rather complicated tensor form. The second chapter covers the special theory, highlighting the contradictions that it solved with respect to the aether. It derives the Lorentz transformation and introduces the Minkowski interpretation of space-time. Einstein then worked through a variety of applications to material bodies, ending with perfect fluids. The third chapter deals with setting up the mathematical tools necessary for the general theory. In introducing the need to move to a non-Euclidean geometry, Einstein uses again the Ehrenfest paradox about the area of a rotating disk. Where appropriate, he replaced his original treatment with simpler ones derived from recent developments in tensor algebra. The final chapter continues the discussion of the general theory, showing that in the limit of small velocities and low gravitational fields, general relativity reduces to Newtonian gravitation. Einstein then discussed some of the consequences of general relativity, including the bending of light by the sun and the precession of the perihelion of Mercury. In the latter he again used results obtained after the 1915 papers, in particular the Schwarzschild solution described in Chapter 13, to simplify the derivation. He continued by discussing the applicability of Mach's ideas of the relativity of inertia and the beauty of a closed static universe, ideas that while certainly beautiful did not survive much beyond the book's publication and were greatly revised in future editions. It is interesting that Einstein makes no mention of the cosmological constant in the book, although he had done so in the actual lectures in Princeton.

After his welcome diversion into physics, Einstein rejoined the Zionist party in Boston on 17 May and on the following day he visited Harvard. After more Zionist meetings in New York and then Cleveland, Ohio, he and Elsa probably visited Niagara Falls on the return trip to New York. On 27 May a reunion with one of his oldest friends took place. Max Talmey, who had fed the schoolboy Einstein with reading material and ideas, had moved to New York to work in the Mount Sinai Hospital in 1895. Some years after their happy reunion in 1921, Talmey wrote a book on relativity in which he left a picture of Einstein the musician:

> The following remarks characterizing expressively the influence of music on him are attributed to Professor and Mrs. Einstein (*Kiel. Zeit. fur Kunst und Wissensch., October 1920*): 'In my life the instinctive prevision of the artist has played no small part. This explains also my great love for music. I never had proper musical instruction. But the piano and violin are my most faithful companions. Here I feel my recreation in all strenuous work. It is to Bach and Mozart that I always retreat. Mozart's serene melodies full of celestial harmony have fascinated me.' To this Mrs. Einstein added: 'He then forgets the world around him, indulges in reveries while ex temporizing *(sic)* on the instrument, and abides in the Elysian fields.'

Also at the end of May, the Einsteins were dinner guests of Gano Dunn and his wealthy wife Julia at their New York apartment in Washington Square. Dunn had accompanied Einstein for most of his US tour, including his visit to The White House. Many of the city's richest and most influential people were at the dinner, as well as a sprinkling of musicians, including the pianists Harold Bauer and Ernest Schelling. The highlight of the evening, certainly for Einstein and also for his hostess, was after dinner when he and Bauer played a Mozart sonata. Julia Dunn wrote:

> No living man plays Mozart the way Bauer does and added to this there was a peculiar tenderness which anticipated and filled out and waited for all the places where the violinist followed his own fancy, rather than that of a received model. He played musically, reverently, enthusiastically and with a tremendous beauty of form and accent which everyone felt – Bauer most of all, as he said afterwards. It was the moment of consecration for all the achievements, imperfect as they are, which have made people think this a beautiful room and this a sympathetic place.

On 30 May, Einstein and Elsa embarked on the *RMS Celtic*, bound for Liverpool, where they arrived on 8 June. Freundlich was on hand to greet them and to act as scientific interpreter; since his mother was English, his own English was excellent. The Vice-Chancellor of the university entertained his guests to lunch. On the following day, Einstein travelled to Manchester. He first spoke to the Jewish Students Society, which had collected more than 2,000 books for the library of the Hebrew University of Jerusalem. Afterwards, Einstein was awarded an honorary degree by the University of Manchester and gave the Adamson Lecture to an audience of 1000 people. Although it was a 'popular' lecture, he spoke in German. He was cheered as he left the building.

The Einsteins next travelled to London, where they stayed with Lord Haldane, who had been a pillar of the Liberal Party, reforming the British Army as Secretary for War and serving as Lord Chancellor in Herbert Henry Asquith's government until hounded from office for his supposed German sympathies in 1915. He was making a slow political transition from the Liberals to the fledgling Labour Party when Einstein arrived. A year later, Haldane would serve in Ramsay MacDonald's first Labour government, achieving the unique distinction of having been Lord Chancellor for both the Liberals and Labour. He was also distinguished by being both a politician and an intellectual. He was a Fellow of the Royal Society[10] and wrote extensively about philosophy. Two months before Einstein's visit, Haldane completed his *The Reign of Relativity*, a broad survey of the philosophical concept of the relativity of experiences in which Einstein's theories form only one of twenty chapters, of which the last three deal with humanity's relationship with God. Nevertheless, the book shows that he was one of the few famous people to meet Einstein who actually understood his scientific work. Arthur Eddington remarked that Haldane understood Einstein's work better than any other British philosopher or many who used its mathematical apparatus to calculate physical phenomena, although he admitted that this was not much of a testimonial! Part of the motive for Haldane's invitation had been that he was aware that Einstein's visit would be good publicity for *The Reign of Relativity*. Indeed, it ran to three editions in four months, the last of which had several revisions resulting from their conversations in Haldane's elegant house in Queen Anne's Gate.

On the morning of 10 June, Haldane and Einstein attended a meeting of the Royal Astronomical Society chaired by its new President, Eddington. That evening, Einstein was guest of honour at a glittering dinner attended by the Archbishop of Canterbury, as well as scientific luminaries such as J. J. Thomson, Sir Charles Sherrington, President of the Royal Society, Frank Dyson, and Eddington. Haldane's own distinguished family was represented by his brother, a physiologist, and their nephew Graeme.[11] Prime Minister David Lloyd George had been expected to attend but had not

---

[10] In May, Einstein had been elected as a Foreign Member of the Royal Society

[11] Graeme Haldane was a research student of Lord Rutherford's at the Cavendish Laboratory in Cambridge. He subsequently became an electrical engineer and helped to found the UK National Electricity Grid.

been well and sent his apologies. After dinner there was a reception, in which another cross section of British high society participated, including George Bernard Shaw, Dean of St Paul's Cathedral William Inge, distinguished Jewish Socialist and Zionist Professor Harold Laski, commander of the ill-fated Salonika Expedition in the First World War General Sir Ian Hamilton, and Haldane's publisher John Murray. No doubt Einstein, having little English, had the distinct impression of once more being exhibited like a prize bull. At least Haldane, who had been a student at Göttingen, spoke fluent German, and Freundlich was no doubt hovering at Einstein's elbow to assist the flow of conversation.

The following day, which was a Saturday, allowed the Einsteins some relaxation. A photographer arrived to take various studies of Einstein, Haldane, and of them both sitting together. On Sunday, they lunched at the Rothschilds, where other guests included Lords Crewe and Rayleigh[12]. That evening they dined at the London house of the well-known society hostess, Lady Homer, wife of Sir John Homer. After dinner Einstein received a reward for his patience in allowing himself to be lionised by high society: an evening of music. According to *The Jewish Chronicle*, he played 'selections from the works of Bach, Mozart, and Schumann. Lord Haldane said that Professor Einstein played with a feeling and an insight which one did not often find in the best professors of music. "He is a master of the violin," he said, "as well as a master of mathematics."'

On the following morning, 13 June, Haldane escorted Einstein to Westminster Abbey, where he paid tribute at the monument to Sir Isaac Newton. In the evening, he gave a lecture, 'The Development and Present Position of the Theory of Relativity', at King's College on the Strand. Haldane presided, introducing Einstein by saying:

> [We give] a British welcome to a man of genius. The highest knowledge is a possession of which the world at large [is] proud, and genius [knows] no frontier. [We are] grateful to Germany for having given us the genius of Einstein, just as Germany has been grateful to us for haying given to the world the genius of Newton. [This] morning [I was] touched to observe that [my] distinguished guest left [my] house to gaze on the tomb of Newton in Westminster Abbey. What Newton was to the eighteenth century, Einstein is to the twentieth. They [have] made vast revolutions in the outlook of things, and yet there [has] been a continuity in their work which lifted it to the highest level of human thought. Einstein [has] given [us] a new conception of the universe.

After a dinner attended by the senior physicists and the teaching staff of King's College, Einstein departed for Haldane's home.

On 14 June, Lindemann, who had taken up the position of Professor of Experimental Philosophy at Oxford in 1919, arrived to motor the Einsteins up to Oxford, where they spent a day *incognito* looking around the city. Einstein was deeply impressed by Oxford, which he found redolent with tradition. On 17 June they departed for Berlin.

Einstein seems to have greatly enjoyed his visit to England, despite the 'prize ox' aspects. He wrote both to Haldane's aged mother, whom he had not met but knew vicariously through her devoted son, and to Haldane himself, expressing deep gratitude for their hospitality. He wrote to Ehrenfest that he was particularly pleased with the English and that as long as they had a leading position, all would be well.

---

[12] Clark's description of this lunch is very confused. He states that Rayleigh and Einstein had last met at the Solvay Conference a decade before. In fact, John William Strutt, Nobel Laureate and 3rd Lord Rayleigh, had been invited to the Solvay Conference of 1911 but was unable to attend. In any case, he had died in 1919, two years before this lunch. The Rothschilds' guest was Robert John Strutt, FRS, 4th Lord Rayleigh, who although also a physicist was not in the same league as his redoubtable father.

If Einstein hoped for a quiet period in which to get back to work, he was to be disappointed. On 29 June he gave an interview to a young female journalist from the *Nieuwe Rotterdamsche Courant*, in which he seems to have relaxed into a lively discussion of his experiences on his visit to the United States. In this interview, he made various disparaging comments on American men, calling them 'toy dogs for their wives'. He also characterised the US as a cultural wasteland outside the largest cities in which men were interested only in work. He went on to make several admiring comments about other aspects of American life and science, as well as remarking that Prohibition, which came into force just after he left, was not supported by the majority but was likely to persist. His remarks appeared more or less verbatim in the newspaper and were quickly taken up in the US, where, unsurprisingly, the positive ones were ignored while the negative ones caused outrage. The Zionist organisation was dismayed and made urgent representations to Einstein. He denied any responsibility, characterising his remarks as a private conversation. He attempted to repair the damage by giving another interview in July in which he tried to draw the sting from his comments, although he didn't deny them, saying that he had been quoted out of context. This was no doubt a lesson to him that there is no such thing as a private conversation with a journalist unless it is specifically designated as 'off the record'.

It is startling to realise the extent of Einstein's fame as revealed by the adulation he experienced during his trip to the US. This is reinforced by the furore over the *Courant* article, which elicited no less than five editorials in *The New York Times*, the country's premier newspaper of record, castigating his ingratitude for the hospitality shown to him. Missner ascribes Einstein's astonishing fame in the US to a number of coincident factors. The first was a real interest in his scientific work, which happened to chime in with a national mood of self-improvement. Newspaper reports of space curving, stars shifting and four dimensions intrigued the general public. 'Einstein has destroyed space and time' was the grandiloquent headline in many newspapers. The craze for the paranormal led many casual readers to associate the 'fourth dimension' with spiritualism. The very name of the theory, 'relativity', led many people to think that it must be related to the relativity of all knowledge, and with good reason, since Lord Haldane's book *The Reign of Relativity* made precisely this point. The famous *The New York Times* headline that only twelve wise men could understand the theory gave it a cachet that encouraged people to believe that if only they studied it carefully, they too could join this exclusive club. The other side of this coin was hostility to a theory imported from Europe, which played to a wave of xenophobia and concern about Bolshevik political revolution that was sweeping the country around the time that the eclipse results catapulted Einstein into the headlines. Thus, Einstein's theory could not be ignored because it too was revolutionary and might be a communist plot. All of these factors gave Einstein's work a penetration into the popular mind and culture unequalled by anything since Darwin's theory of evolution sixty years previously.

A similar conclusion was drawn by Lord Rutherford in December 1933. Subrahmanyan Chandrasekhar, already a distinguished astrophysicist and newly elected Prize Fellow of Trinity College, Cambridge, recorded a conversation around the fire after a dinner with Rutherford, Eddington, and others. Rutherford's response to the question of why he was not as famous as Einstein was to say that Eddington was responsible. He continued:

> The war had just ended; and the complacency of the Victorian and the Edwardian times had been shattered. The people felt that all their values and all their ideals had lost their bearings. Now, suddenly, they learnt that an astronomical prediction by a German scientist had been confirmed by expeditions to Brazil and West Africa and, indeed, prepared for already during the war, by British astronomers. Astronomy had always appealed to public imagination; and an astronomical discovery, transcending

> worldly strife, struck a responsive chord. The meeting of the Royal Society, at which the results of the British expeditions were reported, was headlined in all the British papers: and the typhoon of publicity crossed the Atlantic. From that point on, the American press played Einstein to the maximum.

The discussion of relativity theory had somewhat died down by the time that Einstein arrived in New York in 1921, just at exactly the right time to rekindle the public excitement. What then occurred was that the reporters waiting for his disembarkation from the SS *Rotterdam* mistook the cheering crowd of thousands as having come to greet Einstein. In fact, they were almost exclusively Jewish Zionists who had come to greet Weizmann. Einstein's own warm and eccentric personality meant that this initial misapprehension was quickly overtaken by a real effect. Soon even the Zionists gave Einstein at least equal prominence as that given to Weizmann; he was cheered to the echo whenever he appeared in public. Having been launched into the stratosphere of fame, Einstein continued to orbit there for the rest of his life.

Chapter 16

The Nobel Prize and travels to France, Japan, Palestine, and Spain (1920–1923)

Einstein's relationship with his two sons continued to be problematic. He habitually forgot their birthdays. The fact that Hans Albert mentioned this with a light-hearted tolerance in an interview many years later shows that it still rankled. He wrote to his father on 14 May 1920, mentioning casually but deliberately that it was his sixteenth birthday. Einstein did remember his eighteenth birthday, at least to some extent. He wrote on 5 May 1922 that he wished him a happy birthday 'contrary to my habit . . . because I happen to be thinking about it.' It is difficult to understand why he almost deliberately ignored these occasions. Birthdays were an irrelevance for him, although he did often remember his sister Maja's; he admitted this was the only one whose date he had firmly in his mind. He must have been aware that birthdays mattered to children, especially sons of whom he was unarguably very fond but who saw him extremely infrequently. Hans Albert was reduced to buying Moszkowski's book to try to discover more about his father, who told him to swap it for something else because it wasn't worth reading. Early in 1921, Hans Albert replied to his father to thank him for a photograph he had sent by writing 'It is great that photography has been invented, since at least we can see you that way.' While no doubt meant humorously, an element of sarcastic reproach is clear. Hans Albert wrote to Einstein in early December 1921, saying that they were worried because they had received no letter for such a long time. Tete wrote a letter to his father on New Years Eve 1921 that contains the following heart-breaking wish: 'It would be great if you could celebrate Christmas with us once. That has never happened for as long as I can remember.'

Einstein's attempts to induce Mileva and his sons to leave Zurich for Germany were a source of continued tension, although the pressure henceforward gradually reduced. This was probably related to two things. First, he eventually realised that Mileva and the children were determined not to move whatever compelling logical reasons he adduced. Secondly, the financial pressure on him to find the Swiss francs he needed to support them gradually lessened as the royalties from his various books in foreign translation (funnelled through Ehrenfest and others) and royalties from foreign companies (funnelled through his brother-in-law Paul Winteler) began to become substantial. In the summer of 1921, however, he still found the cost of a vacation in Switzerland far too high, so he arranged with Mileva for the boys to travel to Germany. However, Berlin, tainted by the presence of Elsa and the other members of the Einstein clan whom she detested, was still a step too far. Instead the children took the train to the Baltic coast, where they met Einstein on 18 July. They lodged in Wustrow, a tiny village on the Baltic coast north of Rostock, adjoining Einstein's former vacation destination of Ahrenshoop. The west of the village faced onto the Baltic, the east onto an enclosed body of water known as the Saaler Boden. Here Einstein and his sons enjoyed the sea air and sailed on a hired yacht. Einstein was clearly thrilled by this chance to spend time alone with his sons; they all particularly enjoyed the sailing trips.

Elsa's two daughters, Ilse and Margot, were also on vacation in Ahrenshoop. Although living so close by, Einstein only visited them once, briefly. He explained this as necessary since his sons

Einstein. Brian Foster, Oxford University Press. © Brian Foster (2026). DOI: 10.1093/oso/9780198794875.003.0016

were jealous and resented any time he spent away from them, which seems plausible. However, to his wife, he told a different story: that he had had to rush away because it was getting dark and they had no light in their lodgings. Both Elsa and her daughters suspected that something else was going on. The source of these suspicions can be traced to the fact that Einstein, Hans Albert, and Tete were being looked after in Wustrow by Anna, who was the maid in their Berlin apartment. She had recently been dismissed for reasons which seem to have been related to her being an attractive young woman. Einstein believed that she had been 'frozen out by his womenfolk.' She was working out her notice and would leave on 1 September. When Elsa hired another, apparently also attractive, maid in October, Einstein joked 'Another beautiful maid! Poor old me. You will have to get a veil for her when I come home.' From this it can be deduced that Einstein's penchant for amorous escapades with pretty young women showed no signs of waning.

Rather than returning directly to Berlin from the Baltic, Einstein and his sons travelled east to Kiel, arriving on 10 August. They stayed with Hermann Anschütz-Kaempfe. He was originally an art historian who had turned his hand to inventing. He and Einstein had met in 1914 when Einstein delivered an expert opinion in a patent case relating to a gyroscopic compass that Anschütz-Kaempfe aimed to produce in his Kiel factory. They became friends and Einstein spent considerable time advising him on improvements to his compass design. Einstein, at that time a convinced pacifist, would have had second thoughts had he realised that Anschütz-Kaempfe's device, put into production in 1927, would be the basis for navigation for German warships and U-boats in the Second World War. While his boys enjoyed themselves in the palatial grounds of Anschütz-Kaempfe's Kiel house and became favourites of both his and his wife's, Einstein worked with him and his associates on design improvements for the compass. In addition, Einstein had been invited to bring his violin with him; Frau Anschütz-Kaempfe had begun to play hers again and, as she remarked, filling her husband's ears with her scratching. Einstein returned to Berlin around 17 August; his sons joined Mileva, who was staying nearby on the island of Föhr off the western coast of Schleswig-Holstein.

At the end of August 1921, Erwin Freundlich wrote to Einstein about various internal matters at the Potsdam Observatory. Freundlich had joined the staff at Potsdam in 1920, after leaving the position funded by Einstein's Kaiser Wilhelm Institute. As remarked previously, Freundlich was a prickly character who was highly unpopular with his astronomer colleagues. The new Director of the Potsdam Observatory was Hans Ludendorff, the younger brother of General Erich Ludendorff, architect of the final stages of the German war effort. Hans Ludendorff had been waging a scientific war with Freundlich for several years over Freundlich's attempts to find the expected relativistic red shift of light. It was presumably in a continuation of their feud that Freundlich told Einstein that Ludendorff was proposing to make three questionable appointments to the Observatory's senior staff. Prompted by this, Einstein wrote to the Ministry questioning these appointments. In his letter he mentioned two staff at the Observatory who had alerted him to this; one was clearly Freundlich, the other is unknown. The Ministry replied, requesting the names of the 'whistle blowers'. Any remaining correspondence has not been preserved, but the three contentious appointments went ahead. Einstein was clearly embarrassed by his intervention, to the extent that, after a conversation with Ludendorff some months later, he asked the Ministry to behave as if his letter had never been sent. This is likely to have accentuated his frustration with Freundlich, which had gradually overwhelmed his long-standing friendly feelings for him as an almost lone voice among German professional astronomers in favour of relativity. He had latterly expressed a low opinion of Freundlich as a scientist to a number of his close colleagues.

Matters came to a head just before Christmas 1921. Einstein had handed over to Freundlich the manuscript for his 1916 exposition of the principles of general relativity. Freundlich thought that he had given it to him and later claimed that he was intending to sell it for the benefit of the Einstein Donation Fund, funding the building of the Einstein Telescope at Potsdam. The acerbity of Einstein's comments over this issue leads to the possibility that he believed that Freundlich was trying to sell it abroad for his own benefit. Einstein claimed that he had asked Freundlich to return it already in the summer and he now peremptorily demanded its return. According to mutual friend and editor of the journal *Naturwissenschaft* Arnold Berliner, Freundlich was too upset by Einstein's letter to reply. Berliner tried to mediate but Einstein was obdurate. He replied to Berliner, who had also tried to mollify him on the telephone, that he considered Freundlich to be a liar and wanted nothing more to do with him except unavoidable scientific interactions. On Christmas Day 1921, Einstein wrote to Ludendorff resigning as a Trustee of the Einstein Fund. Thereafter, Einstein and Freundlich's relations remained, at least for the next decade, extremely frosty. Einstein did retrieve the manuscript, which was forwarded to the Hebrew University of Jerusalem library. In return, in 1925 the university donated a sum to the fund for the Einstein Telescope, which was received, no doubt with a wry smile, by Freundlich.

During their vacation on the Baltic, Einstein had suggested to Hans Albert that he might like to accompany him to Italy in October, where Einstein had agreed to give a series of lectures. The original idea was that Einstein would call at Zurich to meet his son and then they would proceed together to Verona. However, Einstein discovered that it was cheaper to travel through Austria rather than Switzerland. After pleas both from Mileva and Hans Albert, he agreed to return via Zurich. This pleased Tete, who had been practising a Mozart sonata to play for his father. Einstein arranged to meet Hans Albert at Innsbruck on 16 October. They travelled on to Bologna and thence to visit his sister Maja and her husband Paul, who had moved to Florence at the end of 1920. The emotional strain of having Maja's dying mother in the house combined with problems at his workplace, where he had been passed over for promotion, caused Paul to lapse into a depression. For a while their marriage was under great strain. Paul decided to retire and, having no children or particular ties to Luzern, they decided to move to Italy, where Paul's Swiss pension would stretch much further. They went first to Florence, then shortly before Einstein's arrival to a two-room rented apartment in Fiesole, in the hills just to the north. The location of the apartment, Via Verdi, was indeed apt as Maja quickly befriended several of the skilled musicians in the nearby fourteenth-century monastery of San Francesco. One in particular, Father Caramelli, who conducted the local choir, accompanied Einstein during his visit. They played in the monastic church, while Einstein also played, no doubt Bach solo sonatas, in the open air by the light of the moon behind the cloisters. The stone on which he sat to play is still pointed out to tourists.

After this musical interlude in Fiesole, Einstein toured Florence, which he had not visited before, with Hans Albert, Maja, Paul, and Michele Besso's sister Bice. He wrote to his friend as they sat outside the Palazzo Vecchio that they were sitting in Florence together, as if it were perfectly normal.

On 21 October, Einstein and Hans Albert returned to Bologna, where they probably stayed with Federigo Enriques, professor of geometry. On leaving, Einstein left a note for Enriques's daughter Adriane saying 'Study, and generally the pursuit of truth and beauty, is an area in which we can remain children all our lives.' On the following day, Einstein delivered the first of three lectures, in which he resurrected his youthful command of Italian. The Italian newspapers complimented his delivery, although a translator was on hand in case a word did not come immediately to mind. His old friend Tullio Levi-Civita, with whom he had corresponded intensively during the genesis of general relativity, attended the lectures. On 27 October, Einstein and son went on to Padova, where he also lectured, this time attended by another great mathematician. This was Levi-Civita's doctoral adviser, Gregorio Ricci-Curbastro, famed for the Ricci tensor and Ricci curvature in tensor calculus. Einstein and Hans Albert then left Italy for Zurich, arriving on 29 October. During his stay, Einstein spent as much time as possible playing music with his sons, no doubt also listening and applauding Tete's playing of Mozart piano sonatas. Lisbeth Hurwitz also played, both at Mileva's apartment and at her mother's, where they had previously so often made music. Lisbeth remarked on Einstein's increased weight but also that his violin playing was as wonderful as ever. On 2 November he left Zurich for Leiden to visit Paul Ehrenfest, another music partner. From Leiden he wrote to Ilse to commiserate with her on the breaking-off of her engagement to a journalist, who a year earlier had visited their apartment to interview her step-father. Einstein returned to Berlin on 27 November, stopping to give a lecture in Dusseldorf the day before.

Einstein's work in physics at this time had undergone one of his periodic oscillations back to quantum issues and the photon. Einstein had puzzled for years over the contradictions that had encrusted the concept of the photon. How could it behave like a wave in optical experiments and in Maxwell's equations but like a particle when it ejected an electron from the surface of a metal? He devised an experiment that he believed would be able to disentangle these two seemingly contradictory properties. His conviction was that the Doppler effect, where the frequency of radiation emitted by a source depends on its speed, was purely a wave effect and would not affect the particle behaviour of the photon. Since he was convinced that emission of energy from an individual moving atom must involve the particle-like aspect of the photon, he wanted to use a prism to deflect light produced from atoms moving with different velocities, which would separate out different wavelengths just as a prism produces a rainbow-like image from a narrow beam of sunlight. He calculated that classical wave theory would produce a different effect from a quantum photon theory. Although prepared to do the experiment himself, he turned to his colleagues at the Phvsikalisch-Technische Reichsanstalt, as he had with his experiments on magnetic materials with Wander de Haas. This time Hans Geiger, sometime scientific assistant of Lord Rutherford, and Walther Bothe, a young physicist recently returned from a prisoner-of-war camp in Russia, were eager to help their distinguished colleague.

A letter from Hendrik Lorentz in November 1921 discussed Einstein's experiment. He paraphrased what he understood of Einstein's view that light has two aspects: a 'guiding' radiation that carries approximately zero energy and 'shows the way' to the photon, and a quantum particle that accompanies the guiding part. He ranged far and wide over the possible ways of testing this hypothesis, for example pointing out that it would have difficulty accounting for the angular momentum of a photon, which Arnold Sommerfeld had shown to be necessary in a recent publication. However, right at the start of his letter, by simple considerations, he convinced himself that both normal light and Einstein's postulated atomic radiation would behave identically. By December 1921, Geiger and Bothe reported to Einstein that they had confirmed his predictions for photons generated from the moving atoms. Einstein was delighted and reported the results to the Academy. However, he had made what he later described as a monumental error in the classical prediction, which turned out to be identical to that for the quantum. Both Ehrenfest and Max von Laue tried to convince Einstein that he was wrong almost as soon as he told them about the experiment. He refused to believe them for several months, insisting to Ehrenfest that he had checked his calculations. Once he realised his mistake, he published a correction, which showed that his theory and classical wave calculation in fact gave identical results. As he remarked sheepishly to Max Born, death was the only certain way to avoid screwing things up.

Einstein's work in the area of general relativity in this period was quite low key, limited to not much more than a change of heart over the paper that Theodor Kaluza had sent to him in the spring of 1919. Kaluza proposed adding extra dimensions to space-time in order to unify electromagnetism and gravity. Einstein wrote to him in October, just before he left for his Italian and Dutch trip, offering to present his paper to the Prussian Academy at the end of November. He told him that he preferred his ideas to those of Hermann Weyl. It seems likely that this postcard was catalysed by a meeting with Einstein's mathematical collaborator, Jakob Grommer from Göttingen. They had been working together on areas relating to Weyl's theories and it may be that, in discussion, Einstein remembered Kaluza's work. A couple of months after presenting Kaluza's ideas to the Academy, Einstein and Grommer published a paper[1] showing that a spherically symmetric solution in Kaluza's theory was impossible. The then-current conception of what an electron was had been postulated by Einstein's old sparring partner, Gustav Mie, in 1912 and 1913. He characterised a charged particle as spherically symmetric, with a high electric-field intensity in a finite volume and appropriate equations of motion. Einstein and Grommer's findings therefore excluded such charged particles from being described in Kaluza's theory.

In the late summer of 1921, Einstein had received a letter from Jun Ishiwara, professor of physics at Tohoku University in Sendai in north Japan. He had studied in Germany and attended the colloquium series that Einstein had organised when a professor in Zurich. He conveyed an invitation from Sanehiko Yamamoto, proprietor of a Japanese progressive newspaper, to give a lecture tour of Japan. The honoraria indicated were substantial, which awoke Einstein's interest in obtaining more hard currency as well as chiming in with a long-standing desire to see more of Asia and its peoples. The origin of the invitation was apparently that Yamamoto had originally asked Bertrand Russell, the famous British philosopher, to come to Japan and to give him the names of the three most significant persons alive so that he could also invite them. Russell replied with only two names: Lenin and Einstein. Since Lenin was somewhat preoccupied with dismantling Tsarist Russia and constructing the Soviet Union, the invitation was extended to Einstein. The negotiations dragged on for several months, much to Einstein's annoyance, but were eventually satisfactorily concluded in January 1922. In March, Einstein moved the trip back by one month, now proposing to leave for Japan in October.

David Hilbert celebrated his sixtieth birthday on 22 January 1922. His colleagues in Göttingen were arranging events to mark the occasion. Einstein had been invited by Max Born, by then also a professor in Göttingen, to attend the festivities. When he indicated that he would like to come, Richard Courant contacted him with a view to participating in a musical tribute. The plan was to play the first movement of Beethoven's 'Piano Quartet in E-flat major', which is an arrangement by the composer of the wind quintet Op. 16. It seems that Hilbert was particularly fond of this piece. Einstein was invited to play with Born as pianist, Courant's wife, and an unnamed cellist. However, on 18 January Einstein decided that he could not afford the time and sent Hilbert written congratulations instead.

In February, composer and teacher Juliusz Wolfsohn wrote to ask if he could dedicate his next series of compositions based on old Jewish folk melodies to Einstein. Wolfsohn was a celebrated pianist who was particularly noted for his Chopin interpretations. His compositions were very popular at the time, although now completely forgotten. Einstein in his reply was appreciative of Wolfsohn's work in keeping Jewish traditions alive and accepted the dedication with pleasure.

[1] The paper was published in *Scripta Universitatis atque Bibliothecae Hierosolymitanarum. Mathematica et Physica* a new journal published under the auspices of the Hebrew University of Jerusalem library; Einstein had agreed to be editor. The journal was unusual in having parallel Hebrew and German text; Grommer produced the Hebrew version. The guiding hand in this venture was Immanuel Velikovsky. In 1950 he gained great notoriety from the publication of his best-selling book *Worlds in Collision*, which clearly demonstrated that he had not learned much astronomy from his interaction with Einstein! He moved to Princeton in 1952 and became a frequent visitor at Einstein's house in Mercer Street.

In an interesting aside that throws light on his lack of competence as a pianist, he admitted that he has not been able to play Wolfsohn's music as he was unable to read piano music from sight.[2]

On 18 February, Paul Langevin wrote to Einstein to renew the invitation to lecture in Paris that had been cancelled in the autumn of 1914. There was a highly political side to this invitation, given the extremely bad relations between France and Germany at this time. That this estrangement extended very much into the scientific sphere was demonstrated by the lack of any invitations to German scientists to attend the Solvay Conference that Einstein had skipped in order to make his trip to the USA. These considerations were uppermost in Einstein's mind when he declined, with clear regret, Langevin's invitation to give the Michonis Lectures. However, he changed his mind after a conversation with Walther Rathenau, the Foreign Minister of the Weimar Republic, a friend and fellow Jew. Rathenau convinced him that it would be extremely valuable to reduce tensions between the two countries if he made a successful visit to France. In early March, he wrote to Langevin announcing that he had changed his mind and that he would after all travel to Paris in the first half of April.

Langevin was delighted with Einstein's acceptance and wrote to him recommending that he arrive at the end of March. He put a small apartment near the university that belonged to a colleague at Einstein's disposal. Here he could live quietly and undisturbed. Although Elsa was also cordially invited, Einstein decided that it would be easier to live a quiet life devoted to discussions with his colleagues if he came alone. He made sure that his old friend Maurice Solovine was given the address of the apartment on condition that he kept it entirely secret. He was met at the French border on the evening of 28 March by Langevin and Charles Nordmann, an astronomer and writer who had done as much as anyone to popularise relativity in France. On arrival in Paris, they avoided the waiting reporters by crossing to a little-used side door of the Gare du Nord.

Einstein gave his first lecture at the Collège de France on 31 March 1922 to a select group that was entirely French and predominantly eminent scientists. Langevin and a detachment of gendarmes were stationed at the narrow wicket of the imposing gates to the campus. They held back the enormous press of people trying to gain entry and ensured that only those invited were allowed in. Marie Curie, Léon Brillouin, Paul Painlevé, the philosopher Henri Bergson, and others were there to hear Einstein's remarks. As in Bologna, he delivered them in the local language, this time prompted when necessary by Langevin. Nordmann characterised his French as fluent, although with a slowness that was not without its charm. Einstein spoke about the basic principles of general relativity, along the same lines as he had in Princeton. He returned to participate in more informal discussions on 3, 5, and 7 April. On 6 April, he attended a meeting of the French Philosophical Society and participated in discussions on relativity. Others who participated included Langevin, Painlevé, Lévy, Perrin, Jean Becquerel, and Bergson. One forum that he did not attend was the grandest of all, the Academy. Anti-German feeling was crystallised by thirty members, who announced that they would leave the room as soon as Einstein entered. Hearing of this, Einstein solved the problem by declining to attend.

Einstein had been insistent on keeping formal events to a minimum and enjoyed the quiet lunches he had in his apartment with Langevin and a small group of friends, although he did not

[2] The German of Einstein's letter is somewhat ambiguous. Having never had formal piano lessons, it can be surmised that Einstein's problem with reading piano scores was unfamiliarity with the bass clef, which as a violinist he would never have used. Although Einstein often played the piano, there is no recorded instance of him playing from sheet music, rather than abstract improvisation.

escape entirely – he had to attend a dinner at the École Polytechnique and a reception at the home of Becquerel, who had succeeded his father Henri, the discoverer of radioactivity, as Professor of Physics at the Natural History Museum.[3] Around 100 people attended this reception, a subset of whom left the salon and went through to the library for an intimate discussion with Einstein, who held forth holding a glass of champagne. He went on to the Theatre-Français, where he attended a performance of Molière's *L'Avare* and Pierre de Marivaux's *Jeu de l'Amour et du Hasard*. He was observed to laugh heartily during the performance. Earlier in his stay he had enjoyed two operas, Alfred Bruneau's *L'Attaque du moulin* at the Opéra Comique, and Jacques Offenbach's *Damnation de Faust* at Concert Colonne.

Einstein left Paris early in the morning of 9 April. He had expressed a wish to Nordmann to visit the battlefields of the war. Nordmann drove Einstein, Langevin, and Solovine north towards St Quentin and Reims. Einstein was horrified by the scenes of devastation that met his gaze. He was still shaken when Nordmann dropped him at the station to take the train to Cologne, from whence he went to Kiel. His friend Anschütz-Kaempfe had asked him to appear at a hearing there of a particularly vexatious patent case. He returned to Berlin on the evening of 14 April. At the next meeting of the Prussian Academy, his colleagues demonstrated their disapproval of his consorting with the enemy, as they saw it, by ostentatiously leaving him surrounded by empty seats.

As remarked earlier, Rathenau had encouraged Einstein to make his Paris trip. He and Einstein had become friendly after meeting, probably early in 1917, at a reception given by Moszkowski. They encountered each other often at similar events in Berlin, such as a gathering at Moszkowski's home in October 1918 when Einstein also talked until midnight with his friend Emanuel Lasker, the chess Grand Master. Einstein's friendship with both Rathenau and Weizmann meant that he could easily bring them together informally to discuss Zionism's international agenda. He arranged a meeting just before leaving for Paris in March 1922, at the end of which Weizmann expressed a wish to meet Rathenau again soon in London.

The steady growth of right-wing violence against Jews and socialists – there were 300 assassinations between 1918 and 1923 – made Einstein sufficiently afraid for his friend Rathenau that he visited him shortly after his return to Paris with a dual purpose. One was to brief him on his contacts while in France; the other was to try to persuade Rathenau to step away from official life. This was desirable both for his own safety and to reduce the danger to other Jews from right-wing groups. They were enraged by Rathenau's prominence as a Jewish capitalist in the Weimar government. This second point was strongly argued by Einstein's companion and Zionist leader Kurt Blumenfeld. Einstein interjected occasional points about Rathenau's personal safety. Both were wasting their time. Rathenau, who had just transformed Germany's relationship with Bolshevik Russia, believed that he was the right man for the job and that he was fulfilling his destiny by working for the benefit of Germany. His ferocious work ethic and enormous talents were being fully employed. Einstein later remarked to Blumenfeld that Rathenau would take on the job of being Pope if it were offered to him, and would probably do it rather well.

In the mid-morning of 24 June, 1922, Rathenau climbed into his luxurious open-topped limousine, which his chauffeur drove along his invariable route to the Foreign Ministry. Rathenau had been warned by the police that such a regular procedure placed him in great danger, but he ignored the warning. A car full of right-wing fanatics, who had familiarised themselves with his routine, pulled behind him and when the chauffeur slowed at an intersection, drew abreast.

[3] Remarkably, his grandfather and great-grandfather had also held the same chair.

A twenty-five-year-old former naval officer armed with an automatic pistol pumped five bullets into Rathenau, blowing away part of his jaw. An accomplice followed this up with a hand grenade.

Einstein was devastated, if not surprised, by the assassination. He wrote to Rathenau's mother, expressing his sympathy and his own personal pain at the loss of such an exceptional figure. Her reply, explaining that his letter was one of the very few that she felt she must answer because she knew how much her son had valued Einstein, is deeply moving. He attended both the funeral and the memorial service and published a brief tribute in the *Neue Rundschau*. As the shock wore off, he began to consider the implications of the assassination on his own safety. He received letters warning him that he was a likely future target and decided that he should keep a low profile. At the start of July, while staying in Kiel with Elsa, he decided to cancel his planned appearance at the annual meeting of the German Scientific and Medical Association in Leipzig, arranging for Max von Laue to take his place. He also asked von Laue to take over from him as acting Director of the Kaiser Wilhelm Institute, exploiting the situation and his impending departure for Asia to rid himself of what had become an increasingly uncongenial task. In addition, he stepped down from the International Committee on Intellectual Cooperation, which he had been convinced (with considerable difficulty) to join. A secondary factor here was his disgust with the League of Nations over its failure to mitigate the disastrous economic effects of the Versailles Reparations on Germany. However, a few weeks later, after energetic lobbying, he changed his mind but sent his apologies for the first meeting of the committee in August, citing pressure of work and preparation for his Asia trip.

It may have been as part of Einstein's plan to keep a low profile in Berlin that he told Max Planck that he would be spending the summer in Zurich with his sons. He wrote to his friend Solovine that he was officially absent from Berlin, but actually present. In fact, Hans Albert and Tete came to him in Berlin in mid-July. They did not stay in Haberlandstrasse, which would no doubt have been a concession too far for Mileva; rather they stayed in what Einstein referred to jokingly as his 'Spandau castle', a cabin on a kleingarten (allotment) that Elsa had rented for him. Einstein commuted each day from his apartment and he and the boys spent their time in making music and sailing on one of the myriad lakes in the suburbs of western Berlin. The boys had both been excited about their vacation, particularly playing music with their father. Although there was no piano in the log cabin, Einstein hoped they would be able to find one nearby, or they could travel to nearby Charlottenburg and borrow that of his friend Dr Moritz Katzenstein.

The friendship between Einstein and Katzenstein was one of the strongest he made in Berlin. Born in 1872 to a prosperous Jewish merchant family in Fulda, Katzenstein trained as a doctor and then became a highly innovative and respected surgeon. He also held academic positions at Friedrich-Wilhelms-Universität (now Humboldt University) in Berlin; his Habilitation was in 1911, he became an Extraordinary Professor in 1913, and full professor in 1920. It is unclear precisely when they met but the first occurrence of Katzenstein in Einstein's correspondence is in the summer of 1919, when Einstein mentioned frequently sailing with him in a letter to his mother Pauline. They were on intimate terms in 1920, when Einstein presented a signed portrait of himself to Helene, Katzenstein's highly cultured wife. By then Katzenstein was head of one of the two clinics of the state hospital in Friedrichshain. He was very much an innovator in surgical procedures, with a real enthusiasm for scientific research. It was no doubt Katzenstein's delight in research that brought him and Einstein together, a relationship strengthened by their enthusiasm for sailing and Katzenstein's possession of a fine and elegant yacht.

During the visit of Einstein's sons, Elsa was conspicuous by her absence from Spandau, presumably another condition of Mileva's. That the antagonism between Mileva and Elsa had been communicated to her sons is evident by a conflict between Hans Albert and his father. It appears as if Hans Albert came across a photograph of Einstein with his second family and reacted badly to it, causing his father considerable anger, which he still remembered the following year.

Hans Albert's musical skills had been exercised in Zurich by accompanying Professor von Gonzenbach, a violinist, as well as their old friend Lisbeth Hurwitz. He had played the Beethoven Violin Concerto with von Gonzenbach as well as sonatas by Bach and Tartini. Clearly von Gonzenbach's playing was not up to Einstein's standard, as Hans Albert wrote to Einstein that he looked forward to hearing Tartini's 'Devil's Trill' Sonata played nicely for once. He had also played piano four-hands pieces with his mother. Tete also wanted to play some Corelli with his father. Einstein was hoping that, with Elsa not present, he could convince their former maid Anna, with whom he was clearly still in contact, to look after them once again. Einstein thoroughly enjoyed his time with his sons. However, he had to disappoint Hans Albert in one request. Both he and his mother had written to Einstein just before their vacation, almost begging that Hans Albert should accompany him to Japan. This was quite a step for Mileva to take, since she must have been aware that Elsa would also be going. However, Hans Albert was left behind, presumably because he would have missed a substantial part of the first two semesters of the engineering course that he was about to start at the ETH.

The period between the end of Einstein's summer vacation with his sons and his departure for Japan was marked by three scientific papers that span the full gamut of his interests. The first, written jointly with Paul Ehrenfest, was begun during Einstein's visit to Leiden in May. It treated one of the most important experiments in quantum physics, that of Otto Stern and Walther Gerlach. This used silver atoms moving in a spatially varying magnetic field to demonstrate that angular momentum is also quantised, as Sommerfeld, extending Bohr's theory, had earlier postulated. Einstein and Ehrenfest tried, not terribly successfully, to elucidate the mechanism whereby the silver atoms align themselves so that half have their angular momentum pointing along the direction of the magnetic field whereas the other half point in the opposite direction. After exploring various options, the two friends metaphorically shrugged their shoulders at the difficulties of any sensible interpretation of quantum phenomena.

The second paper, completed in August 1922, was written in partial fulfilment of Einstein's agreement with his Japanese sponsors and was eventually published in December in the Japanese journal *Kaizo*. In 'On the Current Crisis of Theoretical Physics', Einstein looked back on the developments in physics since he became an active researcher. He noted the completion of a programme of 'classical' physics, including his beloved statistical mechanics, culminating in general relativity. This was all based on continuous smooth functions in a continuous space; in contrast, quantum physics was based on discontinuous quanta that somehow miraculously, in the limit of large distances and numbers of atoms, coincided with the classical predictions. Yet, mathematically, it seemed unnatural or even impossible to construct a quantum theory based on differential equations that intrinsically must operate on continuous functions. In the final paragraph, Einstein looked forward to the construction of a new mathematical language, presumably algebraic, that could accommodate the discrete nature of the quantum world. Three years later, the young Werner Heisenberg, who was disappointed that his hero Einstein did not attend the Leipzig conference, would take the first steps in the direction that Einstein had indicated.

The third paper was a brief comment on another publication by Franz Selety, which pivots back towards Einstein's cosmological interests. The comment itself is important only in that it indicates a clear shift in Einstein's attitude towards Mach's principle, which had played such an important part in the genesis of general relativity. Einstein seems to have finally internalised the work of Willem de Sitter and others that showed that there were solutions of the Einstein Equations that contained no masses. It was thus untenable to hold to the idea that gravity and inertia were consequences of interactions with the totality of the 'fixed stars', as Mach had contended. In this note, it is clear that Einstein had now moved to a position in which he used Mach's principle as a selection criterion. Hence his preference for a static, bounded universe in which Mach's principle does indeed hold.

In September, Tatiana Ehrenfest and her daughter visited Berlin for a few days. This happened to coincide with the solo recital debut of Einstein's friend Rudolf Serkin in the Singakademie on 13 September. Einstein, Ilse, and Margot accompanied the Ehrenfests to the concert. Among other pieces, Serkin gave the first performance of the C minor piano sonata by his mentor, duo partner, and eventual father-in-law, Adolf Busch. Busch was as well-known a composer as he was as a violinist during his lifetime, although his music has subsequently vanished from the concert hall. Another visitor in September was Einstein's old friend from Zurich, Heinrich Zangger, who was presumably in Berlin for a scientific meeting.

Around 1 October 1922, the Einsteins left Berlin to travel first to Zurich and Bern, where they met many old friends, in particular Michele Besso. On 6 October they left for Marseilles, where they were to board the Japanese ship *Kitano Maru*, which was to be their home for the next six weeks. The passengers were mostly English, on their way to various intermediate colonial ports, and Japanese. Einstein found their company generally congenial, although not their habit of photographing him whenever he appeared on deck. However, it was the lack of company that he most enjoyed. The expectation of solitude had indeed been one of his major incentives to agree to the trip. Away from the distractions of Berlin, he savoured the isolation and got down to serious contemplation as well as reading of books, including philosophy and medicine. A book by Kretschmer on mental illness caused him a rare moment of self-analysis. He wrote in his travel diary that 'I felt myself as if probed by forceps. Hypersensitivity changed into indifference. In my youth, inward-looking and alienated from the external. A pane of glass between myself and others. Groundless mistrust. A world constructed from paper. Distancing from pleasures.' However, most of his time was spent as usual on physics rather than recreational reading. He was working towards a unified theory of gravity and electromagnetism.

As the *Kitano Maru* sailed on, they passed the volcano Stromboli on their way through the Straits of Messina to the Suez Canal. The heat became extremely uncomfortable at night as they sailed across the Arabian Sea and Einstein suffered with diarrhoea and haemorrhoids to the extent that he consulted a fellow passenger, a Japanese doctor. The Emperor of Japan's birthday was on 30 October and was energetically celebrated by the Japanese passengers and crew. After a festive dinner, an impromptu concert introduced Einstein to Japanese music. He was unimpressed, characterising the singing of one young man as like a cat whose tail had been trodden on. The performer accompanied himself on a Japanese instrument that Einstein described as guitar-like, with a short fingerboard, from which with wild gestures he occasionally produced a note that seemed completely divorced from what he was singing.

They stopped in Colombo in what was then Ceylon (modern-day Sri Lanka). Einstein felt exploitative when being pulled through the streets in a rickshaw. They next called at Singapore, where Einstein had promised Weizmann to try to extract funds for the Hebrew University from rich Jewish inhabitant Sir Menasseh Meyer. Weizmann had alerted the sizeable Jewish community to Einstein's arrival and arranged an exhausting programme of visits and social events. These culminated in a dinner at Meyer's palatial house that was so long and sumptuous that Einstein felt sick and had to leave the table. Although he wasn't sure at the time if his importunities had pierced Meyer's protective mental armour, a donation to the university was eventually forthcoming.

Several days later, the *Kitano Maru* arrived in Hong Kong. Saturated by his efforts in Singapore, he avoided as many events with the Jewish community as he could, although two Jewish businessmen drove the Einsteins around. He liked Hong Kong, with its many sea inlets and wooded hills dropping down to them. He also appreciated the relatively cool air when they took the funicular railway up to Victoria Peak.

On 13 November, the ship arrived at Shanghai, where the German community took centre stage. The German consul met the Einsteins and accompanied them to a typical Shanghai meal in which a seemingly endless procession of communal dishes appeared on the table. Einstein compared Shanghai unfavourably in terms of cleanliness with Hong Kong.

Sometime during the day, Einstein was informed that he had won the 1921 Nobel Prize in Physics, which had been carried over until 1922. His citation[4] read 'For his services to Theoretical Physics, and especially for his discovery of the law of the photoelectric effect.'

The 1922 award was made simultaneously to Niels Bohr, 'For his services in the investigation of the structure of atoms and of the radiation emanating from them.' Each was very pleased for the other, as well as that they had received the award at the same time. However, despite the fact that the large sum of money in hard currency that accompanied the award would secure the future of Mileva and his sons in Zurich, as previously agreed in his divorce settlement, Einstein did not consider the event of sufficient importance to note it in his travel diary.

When the Einsteins arrived in Kobe on 17 November, they were met by a large welcoming crowd. His sponsors, for whom the visit had a clear commercial rationale, had been sure to drum up interest both in Einstein and his theories before his arrival. As well as the German consul, there was a large group of journalists, excited not only by his arrival but also the news of the Nobel Prize. More welcome members of the reception party were a small number of Japanese who had studied in Germany, some with Einstein himself. In particular, Jun Ishiwara was an almost constant presence at Einstein's side during the visit. He had been forced to resign his physics chair in 1921 because of an affair with a married woman and was not scientifically very active. Instead, he had turned to popularisation of science and also poetry. The party moved on to Kyoto, where the city, people and countryside, particularly the sun setting over Mount Fuji, made a very positive impression. On the following day, they travelled to Tokyo.

It was in Tokyo that the visit really got under way. There were mobbed with reporters and thousands of well-wishers at the railway station, finally extricating themselves to travel to a reception. When they arrived at their hotel, they found masses of floral tributes awaiting them. The following day Einstein gave two lectures, one lasting 150 minutes, the other two hours, in the main auditorium of the University of Tokyo. At least half the time was taken up by Ishiwara's sentence-by-sentence translation. The German ambassador reported that thousands of Japanese people had attended the lectures, at a significant cost per person. On the following day Einstein attended a luncheon given by the Academy of Sciences at the University Botanical Gardens. This was followed by a meal with their host, Yamamoto-san, the head of the Kaizōsha publishing company, and colleagues. They went on to a theatre in which the dramatic action was accompanied by three male singers and a Japanese orchestra. Einstein was once more unimpressed by the 'twittering' music and the lack of any discernible musical structure.

The following day marked the ceremonial high-point of the visit. The annual chrysanthemum festival was celebrated in the Imperial Palace Garden in the heart of the city. Attendance required morning dress, which naturally Einstein did not possess. No doubt the German embassy, who were

[4] The fascinating story of how the Royal Swedish Academy arrived at this form of words, the role of anti-Semitism and scientific prejudice that ensured there was no mention of relativity, and how Carl Wilhelm Oseen steered a brilliant political operation to ensure awards to both Bohr and Einstein, to the latter by emphasising the photoelectric effect 'law', is told by Friedman.

firmly in charge of this part of the visit, provided it. The Taisho Emperor, as he became after his death in 1926, had not been seen in public for many years due to mental illness, so the ceremony was presided over by the Empress and the Crown Prince and Regent, Hirohito. While 3000 people attended, only select groups of dignitaries formed semicircles to be presented to the Empress. Einstein, in the diplomatic group, was stared at rather more than the Empress, who conversed with him graciously, if briefly, in French. In the ensuing buffet, Einstein was introduced to what seemed to him to be the entire gathering. He was at least able to appreciate the beauty of the palace gardens and the spectacular chrysanthemums.

The next few days were spent in Tokyo in less formal ways. The Einsteins were shown various aspects of life in the city, for example being invited to dine at Mr Yamamoto's, to Einstein's eyes very small, house. After a sumptuous lunch, he attended a concert at a music conservatory which once more left him unmoved and uncomprehending. He was much more appreciative of Japanese art, which he considered reflected the elegance and calm of the Japanese people. On the following evening, 24 November, he gave two more mammoth lectures, ending at 10 p.m., to sold-out audiences. He continued to have discussions with Japanese physicists at the university for the next few days, interspersed with excursions to a variety of Japanese cultural events. One evening featured entertainment by geishas, who entranced him with their elegance and sophistication. On another afternoon he was greatly impressed by a traditional tea ceremony. On 27 November, he was finally able to hear music to his taste when he took his violin out of its case after dinner in a private house, playing pieces by Gluck, Max Hauser, and Bach. A Japanese violinist played a piece by Henryk Wieniawski.

On 30 November, the Einsteins listened to musicians of the Imperial Court play some of the ancient music that was heavily influenced by the Chinese. The musicians played the flute, other wind instruments that used a reed, and various Japanese traditional stringed instruments. This seemed to be much more to Einstein's taste, as in his travel diary he characterised it as reminding him of a Chorale, with wonderful tone colours. This musical enjoyment was bracketed by events involving tens of thousands of students, which he found very tiring. Nevertheless, he was on duty as a musician again after a glittering dinner at the German Embassy attended by the diplomatic corps. However, he considered that he had played very badly due to fatigue and lack of any practice.

The climax of Einstein's musical endeavours in Tokyo took place on 1 December after yet another gala dinner, this time at the Imperial Hotel, attended by around 300 people. A newspaper article in the *Kahoku Shimpo* was entitled 'Dr Einstein's Exquisite Performance.' It reported that he gave a brief speech:

> Since I visited Japan, I have had many experiences every day which have given me much food for thought. Above all, Japanese-style paintings are most impressive for their refined brush works and elegant atmosphere, far beyond western-style paintings. And I also found Japanese people are very friendly, honest and open-minded in their character. I am really pleased to be able to address you here tonight.

He went on to perform Beethoven's 'Kreutzer' Sonata with a piano professor, who was either American or British, from the Tokyo Ongaku Gakko, now the Tokyo University of the Arts. According to the newspaper 'His performance was greeted with loud cheers from the guests. "We cannot fully understand the theory of relativity, but we can surely appreciate the splendid violinist." Besides physics and his theory, that night he devoted himself to music with his eyes shining like the evening star.' The characterisation of Einstein's performance as 'exquisite' should perhaps be taken with a grain of salt. So polite a people as the Japanese, describing an honoured guest, would be unlikely to be critical, particularly as it is doubtful if many of the audience were familiar with the piece. Nevertheless, it is remarkable that Einstein felt able to perform a work as long and technically challenging as the 'Kreutzer' Sonata in public, since he certainly had almost no time for practice and is highly unlikely to have been able to rehearse with the pianist.

On the afternoon of 2 December, Einstein left Elsa in Tokyo and took the train for the eight-hour trip to Sendai, where the usual enormous crowd waited to greet him. He then returned southwards, calling at the famous beauty spot of Nikko, where he and his colleagues climbed to the lake at 1300 metres, descending in a snowstorm. On the following day they toured the temples, including the grave of Tokugawa Ieyasu, founder of the Tokugawa Shogunate that ruled Japan from 1603 until the Meiji Restoration in 1868. He returned briefly to Tokyo on 6 December and picked up Elsa for the next stage of their itinerary to Nagoya and then Kyoto, where they visited the Imperial Palace. German biochemist colleague and pianist Leonor Michaelis arrived and Einstein lost no time in playing music with him that afternoon before another public talk in the evening. A further two talks were given the following evening in Osaka. While Einstein noted that all of this was not too tiring since his hosts were so accommodating and polite, Elsa's nerves seem to have been frayed by his neglect of her. The two had the latest in a series of arguments when Einstein arrived back from his lecture. On 10 December, another two public talks took place in Kyoto. The following two days were devoted to sight-seeing in Kyoto and Kobe. On 14 December, Einstein gave a talk to university students on the origins of relativity, which has tantalised scholars for years. As discussed in Chapter 6, this was never written up and only a Japanese verbatim transcript exists, whose translation is littered with ambiguity. After more sightseeing, the party travelled to Nara on 17 December. More sight-seeing brought them eventually to Fukuoka on the southern island of Kyushu and further public talks. On Christmas Day, Einstein was exhausted by the time he got to Moji, but always rejuvenated by playing the violin, was able to delight a party of children with Christmas tunes. Einstein managed one more violin concert in Moji, interspersed with what seems to have been an early form of karaoke, in which each of the businessmen present sang something. On 29 December, Einstein and Elsa embarked on the *Haruna Maru*, the brand-new sister ship of the *Kitano Maru*, and waved goodbye with regret to a land that Einstein had come to greatly appreciate – except, perhaps, for its music.

In fact, Einstein subsequently set down his thoughts on Japanese music in an article that he wrote while he was staying in Nikko and which was published in *Kaizo* magazine in January 1923:

> Japanese music, which has developed partially or even completely independently of ours, was of particular interest for me. Only when one first hears completely different art does one approach the ideal of being able to separate the conventional from the essentials that derive directly from the human condition. The differences between Japanese and our music are really fundamental. Whereas in our European music, harmony and architectural structure is common and seems to be essential, they are absent in Japanese music. On the other hand, both use the thirteen-note subdivision of the octave. Japanese music seems to me to be a sort of impressionistic painting of immediate subliminal feelings. Indeed, it seems that to attain the artistic effect, the absolute pitch is not so decisive. It gives me the impression of being much more to do with the stylised representation of the human voice as well as those sounds of nature that bring to the mind feelings of emotion, such as bird song or the breaking of the waves on the shore. This impression is strengthened by the important role of percussive instruments, which have no particular pitch but are particularly suitable for rhythmic characterisation. The extremely subtle rhythms are for me the most interesting aspect of Japanese music. I am completely aware that the intimate subtleties of this type of music are a closed book to me. In addition, long experience is required to differentiate the banal from the personal style of the artist; the relationship between the spoken word and the song, which is so important in so many pieces of Japanese music, completely escapes me. The artistic approach of the Japanese spirit seems to me to be typified by the use of the soft

wooden flute rather than the much harsher metallic one. This is also a reflection of the characteristic preference for the charming and delicate that shows itself both in Japanese painting and in the design of everyday utensils. I particularly enjoyed the music that was played during a stage play or a mime dance, for example a *Noh* play. What in my view inhibits Japanese music from developing into a true art-form is the lack of any formal structure or architecture.

On New Year's Eve 1922, the ship arrived in Shanghai, where the Einsteins stayed with a certain Mr Gatton, whom Einstein considered a snob but whose hospitality was greatly enhanced by the provision of a good piano. After a forgettable New Year's Eve party and similarly uncongenial New Year's reception the following day, Shanghai and its Chinese inhabitants again made a bad impression. He discovered that the Chinese were fond of playing European music no matter how incongruous to the occasion, provided that there were plenty of trumpets. In Hong Kong, Einstein again enjoyed an expedition up to Victoria Peak, the last part of which he completed on foot. He greatly appreciated the solitude of the next few days aboard ship, thinking incessantly about a unified theory. Before they arrived in Colombo, he believed, not for the last time in his career, that he had solved the problem. He began to draft a paper that he asked Planck to present to the Prussian Academy in February during his absence; it was published in the *Proceedings* in March 1923.

The approach Einstein took in his paper was to exploit the developments in the mathematical theory of tensors that both Levi-Civita and Weyl had recently developed, as well as what was known as the 'affine connection' or transformation, of which Arthur Eddington was a strong proponent. He used these techniques to develop a unification that he believed overcame the weaknesses that he had identified in similar previous attempts by Weyl and Eddington. Weyl's approach he had long characterised as unphysical; that of Eddington he found uncongenial because it resulted in forty arbitrary functions of the space-time coordinates. The crux of his paper was the construction of a physically motivated mechanism to determine these quantities. This was based on the properties of the energy of the system, as characterised by a Hamiltonian function, at its boundaries. He chose a suitable Hamiltonian with some minimal assumptions. By removing the assumption that the metric tensor must be symmetric, which he had made in his 1915 papers, he identified the symmetric part with gravity and the antisymmetric part with electromagnetism. He then proceeded to show two things. First, he demonstrated that in the limit that there were no electromagnetic fields, the new equations reduced to those of general relativity. Secondly, he showed that in the limit of infinitesimally small electric currents, at least a part of Maxwell's equations could be recovered. Einstein concluded by remarking that these equations had the freedom to allow the possibility that positive and negative charge had different properties.

It took Einstein only a few weeks to realise that he had made an error and that in fact the solutions for electromagnetism implied that masses of the positive and negative charges were identical. The irony here is that he was unaware that this was, in fact, a desirable feature. Einstein knew that the particles carrying positive charge, which we now know to be protons, behaved very differently to negatively charged electrons. He assumed that this was some inherent asymmetry in electromagnetism, whereas today it is understood to be simply a reflection of the fact that the proton is a large, heavy composite of three quarks, two with 2/3 of the electron's charge and the other with −1/3, whereas the electron is structureless and elementary. This paper, and the following short correction that he presented to the Academy in April 1923 stand as exemplars of the impossible task that Einstein had chosen for himself and that dominated the remainder of his career. It was not only that two other fundamental interactions of nature that require unification, that is, the weak and strong forces, had yet to be discovered. Crucially, he showed no interest in the increasing indications in his lifetime from experiments on nuclei and cosmic rays that such radically different forces must exist. This presumably stemmed from his antipathy towards the quantum theory that was manifestly required to understand these phenomena.

On 10 January 1923, the *Haruna Maru* arrived in Singapore, where the Einsteins were once more entertained by Menasseh Meyer, whom Einstein labelled the plutocratic Jewish 'Croesus'. Until this point, they had retraced their outwards voyage in reverse, but now they sailed up the western coast of Malaysia, stopping at Malacca and Penang. After this short diversion, they arrived in Colombo on 19 January. On one of their excursions they were passengers in a rickshaw pulled by a man who had been an elephant keeper at Hamburg's Hagenbeck Zoo, of which he retained fond memories. He was so pleased by their tip that he returned before their final departure with a bunch of bananas for their journey.

On 1 February, the Einsteins arrived in Port Said, where they disembarked for their journey to Palestine. They travelled overland through the desert and then up to Jerusalem, which they reached a day later. They stayed with British High Commissioner Sir Herbert Samuel in his magnificent residence, which had formerly belonged to the Kaiser and was situated on the Mount of Olives overlooking the city. Samuel had been Home Secretary under Asquith but had resigned after Lloyd George took over as Prime Minister in 1917. As a practising Jew, his appointment to Palestine in 1920 had caused considerable controversy. He showed the Einsteins to the Damascene Gate of the Old City. Einstein was taken by other colleagues to the Haram al-Sharif (Temple Mount), where he was not particularly impressed with the Dome of the Rock nor with the Al-Aqsa Mosque. Descending, he viewed the scene of men rapidly bowing in front of the Wailing Wall with great sadness, characterising his fellow Jews as a people who had a past but no future. Later after lunch he met a figure from his Prague days, Hugo Bergmann, who had moved to Jerusalem and was trying to set up the library at the Hebrew University for which Einstein had been so energetically raising funds.

On the following day, Einstein and Elsa were driven to the ruins of Jericho, up the Jordan Valley, and then back to Jerusalem. Einstein was deeply moved by the imposing scenery. On 5 February, they visited two Jewish settlements west of Jerusalem. In the evening, Einstein played with an unnamed solider in the Samuels' residence, which he remarked was an extended session because he was suffering from musical deprivation. At a reception[5] that night at the home of Menachem Ussishkin, one of the leading Zionists, he met Palestine's attorney-general Norman Bentwich and his wife Helen, who promised to assuage his thirst for making music. The following night, he and Elsa came to dinner and afterwards Einstein played in a Mozart string quartet with Bentwich playing viola and his two sisters the remaining instruments. Both Bentwich and his wife recorded that Einstein looked very happy during the performance and Helen thought that he played extraordinarily well. The Einsteins subsequently visited the Christian holy sites before leaving on the 8 February for Tel Aviv.

Einstein was very impressed with the modern Jewish city that he found when exploring Tel Aviv. After he became the first person to be honoured with the freedom of the city, he and Elsa went on first to Jaffa and then to Haifa, where they climbed Mount Carmel. Afterwards, they participated in an evening sing-song of psalms and Jewish folk melodies. On 11 February they went to Nazareth and Lake Tiberias, visiting the first kibbutz. On their final day, they returned to Jerusalem, where Einstein gave a popular lecture on relativity in German to a packed auditorium. They left early

[5] Although Einstein's travel diary does not mention this reception, Bentwich does.

the following morning, but Elsa felt very ill by the time they arrived in Port Said. On 15 February, 1923, Elsa recovered somewhat and on the following day they embarked on the liner *Ormuz* for Toulon.

On either 21 or 22 February, Einstein and Elsa arrived in Toulon, where they left the great majority of their luggage. They took the train to Barcelona, managing to lose their passports at the border and having to have temporary ones issued by the Swiss Consulate. On their arrival, they proceeded to a modest hotel. The proprietor, recognising Einstein from his picture in the newspaper, rushed to his room to explain that he was expected at the Ritz. He found Einstein sitting on his bed, playing his violin. Einstein insisted that he was a modest citizen who chose hotels with this in mind. He gave three lectures on relativity in French, followed by a more popular lecture on the philosophical implications of relativity. On 27 February, they attended a civic reception at City Hall. They also managed some sight-seeing: the Poblet Monastery and the Roman basilica at Tarrassa. The following day Einstein attended a performance of the Catalan national dance, the sardana. He characterised it as a work of art closely related to sport and treasured the musical recordings of the dances with which he was presented.

On 1 March 1923, the Einsteins travelled on to Madrid. They were met by an old acquaintance of Einstein's, Blas Cabrera, as well as the German Ambassador and other worthies. Their cousins Lina and Kuno Kocherthaler, who had acted as recipients for some of Einstein's hard currency deposits, were also there. Another three relativity lectures were given, interspersed with banquets and receptions. On 4 March, Einstein was inducted into the Academy of Sciences. King Alfonso XIII presented Einstein with his certificate of membership. At a reception that afternoon, Einstein was observed to enjoy himself in the midst of a company of admiring society females. He also enjoyed playing a duet with well-known violinist and director of the Madrid Conservatory Antonio Fernandez Bordas, who had been a pupil of the legendary virtuoso Pablo de Sarasate.

On 6 March, the Einsteins managed to give their solicitous hosts the slip and spent the day in Toledo with the Kocherthalers, Kuno's wife Maria, their son Julio, and an art historian. Einstein noted in his travel diary that strolling through the city and viewing some wonderful paintings made this one of the most enjoyable days of his life. On the following day it was back to duty, although a rather elevated one, as they had an audience with the King and Queen Mother. After receiving an honorary doctorate at the university, he gave his final lecture on the philosophical implications of relativity on 8 March. He was rewarded with another evening of chamber music with Bordas. They were then free to spend several days looking at the dazzling art collections of the Escorial and the Prado, where Einstein particularly enjoyed pictures by Velazquez, El Greco, Goya, Raphael, and Fra Angelico. On 12 March, they travelled to Zaragossa for two lectures on relativity in French; a third, scheduled for Bilbao, was cancelled. A final musical event occurred on 13 March, when the pianist and composer Emil von Sauer happened to be in the city. He agreed to play with Einstein after a banquet organised by the German consul. Afterwards, the Einsteins attended an operetta *La Viejecita* at the Theatro Principal. After lunch with von Sauer the following day, which was Einstein's forty-fourth birthday, they set off on the long journey back to Berlin via Zurich.

Chapter 17

Family and female complications require the solace of music (1923–1928)

Even before Einstein arrived home from his expedition in March 1923, he resigned from the League of Nations Committee for Intellectual Cooperation. He had been outraged by the occupation of the Ruhr by French and Belgian troops on 11 January 1923 in order to extract the reparations imposed as a consequence of the Treaty of Versailles. He considered that the League's failure to remonstrate effectively about what he considered an arbitrary use of force was inconsistent with his own pacifist views and with his solidarity with his German colleagues. Pierre Comert, the Director of the Information section of the League and secretary of the committee, who had spent considerable effort in dissuading Einstein from his previous resignation in the summer of 1922, was outraged. While his friends on the committee, such as Marie Curie, were deeply disappointed and felt snubbed, his enemies charged that his actions were proof of his untrustworthiness. At least the election of Lorentz as his replacement was a source of some satisfaction.

This was not, however, the end of Einstein's involvement with the League. He had rapidly begun to have second thoughts, so that when Marie Curie wrote to him in January 1924, gently placing before him both her regard for him personally and the potential for good that the Committee's work promised, she was pushing at a partially open door. An exploratory letter from distinguished Oxford classicist Gilbert Murray, his ex-colleague on the committee, allowed him indirectly to send a message to the League authorities in his reply that he would welcome being re-elected to the committee. In June 1924, he was happy to accept re-election. He attended his first meeting in Geneva in July, which made a positive impression on him. Einstein attended a performance of Jacques Offenbach's opera *Orphée aux enfers* in Geneva. Although he did not find subsequent committee meetings so positive, often complaining of his boredom, he did attend assiduously for the first couple of years. The meetings alternated between Geneva and Paris, the Geneva venues being convenient for trips to his family in Zurich. There is no doubt that the advent of Lorentz as Chair in 1925 also encouraged him to participate, and he was placed on several subcommittees, including the one to found an International Meteorological Institute.

As time passed however, Einstein became less and less involved with the League. This was due to a combination of his own inclination, a serious illness in 1928, and manoeuvrings by a colleague, Hugo Krüss. Krüss was a high-ranking civil servant who became Director of the Prussian State Library in 1925. He was well connected in many departments of the government, including the Foreign Ministry. Einstein was not considered to be an ideal representative of German academia by these bodies. Krüss had ingratiated himself with Einstein when they both served on the Subcommittee of Sciences and Bibliography; they had something in common as Krüss had a PhD in Physics from the University of Jena. He began to stand in for Einstein at meetings of the League committee and its subcommittees, so that gradually and unofficially he took over Einstein's functions. The arrival of Alfredo Rocco, the Italian fascist minister, as a member of the committee, seemed to be the last straw for Einstein. The two had clashed furiously in 1926, when Rocco was appointed to the board of the Committee's Institute in Paris. Einstein handed in his resignation at

Einstein. Brian Foster, Oxford University Press. © Brian Foster (2026). DOI: 10.1093/oso/9780198794875.003.0017

the August 1930 meeting and formally stood down in 1931, when he considered that his mandate had in any case expired and by which time his influence in a German government moving increasingly to the right had long been extinguished. Krüss[1] formally succeeded Einstein at the July 1933 meeting of the committee.

The occupation of the Ruhr that had precipitated Einstein's resignation from his League role had its roots in the catastrophic weakness of the German economy. The galloping inflation meant that reparations payments in marks were worthless; only the actual products of German industry or its mineral wealth had value. Both were centred in or near the Ruhr. Not only were these thereby lost to the German economy, but the government also encouraged the Ruhr workers to lay down their tools in passive resistance to the occupiers. For this to succeed, the workers needed to be financially supported, which forced the government to print yet more paper money. It was not long before drastic consequences ensued.

On 26 March 1923, directly after his return from Spain, Einstein reported to the Ministry and requested that his salary as director of the Kaiser Wilhelm Institute should once more be paid. The Ministry agreed and informed him that from 1 April, he would be paid 854,310 marks monthly! He also responded to a request to various scientific bodies by the *Vossisches Zeitung* newspaper, asking about their status. The results were published in a special supplement of the Easter Sunday edition on 1 April. Einstein pointed out that the real-terms funding of his institute had depreciated to 1/20 of that available before the war.

By the summer of 1923, hyperinflation had set in. Ordinary purchases required wheelbarrow loads of paper currency. A student at Freiburg University is recorded as ordering a cup of coffee, ascertaining the 5,000 mark cost from the menu. Having enjoyed it, he ordered another but was bemused when the final bill was for 14,000 marks. He was told that if he needed to save money and wanted two cups of coffee, he should order them both at the same time. As autumn drew on, the situation became critical. On 7–8 November, riots began in Berlin because of the price of bread. These quickly developed an anti-Semitic element. Einstein, still nervous from the assassination of Walther Rathenau and aware of specific threats against his life, fled to the safe haven of Leiden. He remained there until just before Christmas. It was probably during this stay that he played solo Bach at the house of Heike Kamerlingh Onnes, a performance that remained long in the memory of Kamerlingh Onnes's wife.

The government was finally able to stabilise the hyperinflation by introducing a new currency, tied to, but not convertible into, gold. The issuance of the Rentenmark was controlled and strictly limited by the Finance Ministry rather than the Central Bank. It was introduced in October and replaced the worthless old paper currency in mid-November 1923. These measures quickly tamed the hyperinflation. The episode however had many catastrophic consequences that echo down the generations into the current German psyche. Anyone with savings deposits in banks who was unable to move the money abroad or buy valuables, lost everything. Einstein fortunately had several stashes of hard currency and a variety of foreign income, but he and his extended family were also hard hit. The economic hardship gave oxygen to extremists of both left and right. As Einstein was fleeing the riots in Berlin, on 8–9 November 1923, Adolf Hitler,

---

[1] Krüss supported Hitler and became a member of the Nazi party. He was responsible for the looting of many books from libraries in the occupied areas of western Europe, in particular France. He organised the evacuation of millions of books from German libraries during the Second World War, saving most of them from destruction. He returned to a devastated Berlin in April 1945, where he shot himself in the ruins of his library.

General Erich Ludendorff, and other Nazis tried and failed to seize power in the state of Bavaria in the Munich 'Bierhalle putsch'. Hitler's subsequent trial, imprisonment, during which he dictated *Mein Kampf* to Rudolph Hess, and early release enormously increased his profile inside Germany.

Another source of foreign currency available to Einstein was the Nobel Prize. With this, as specified by his divorce agreement, Einstein expected finally to place Mileva and his sons on a firm financial footing. Having been unable to attend the presentation ceremony and receive the medal and diploma because of his Asian trip, Einstein was invited to deliver his Nobel Lecture at the Annual Meeting of the Nordic Association of Natural Scientists. After some sailing with his friend Anschütz-Kaempfe in Kiel and a brief stay with Niels Bohr in Copenhagen, he travelled to Gothenburg. On 11 July 1923, he regaled the packed audience with a lecture that emphasised the first part of the citation, his services to theoretical physics, rather than the second, the photoelectric effect. His talk was entitled 'Fundamental Ideas and Problems of the Theory of Relativity.' He led his audience through the postulates of relativity, the special and then the general theory. Only at the very end did he turn to the problems then engaging him, the new methods introduced using the affine connection and the unification of electromagnetism and gravity. The last substantive point that he made was one that in a single sentence crystallises a thread he would follow for the rest of his life: 'In particular, in my opinion a field theory can only be satisfactory if it contains singularity-free solutions for the elementary electrical bodies.' This conviction sharply divides him from the thrust, shortly to germinate in the minds of Werner Heisenberg and Paul Dirac that produced the relativistic quantum field theories of modern physics.

The Nobel Prize also led to a clarification of Einstein's citizenship status. The prestige of the award led to some undignified squabbling among two contenders for his allegiance, that of his birth, Germany, and his adopted homeland, Switzerland. Einstein originally asked that the Nobel Medal and Diploma should be sent to the Swiss Embassy in Berlin. This caused problems with the German authorities, so that the Nobel Committee cut the Gordian knot by despatching the materials to the Swedish Embassy. In correspondence with the German authorities in February 1924, Einstein agreed to accept the interpretation that he became *de jure* a German citizen when he accepted appointment to the Prussian Academy in 1914. His insistence at the time that he retain his Swiss citizenship and the numerous occasions during the war years in which he had stated that he was not a German citizen were quietly overlooked.

Although for many years in his youth he had had no passport, Einstein now embarked on a quest for several. In particular, he wanted a diplomatic passport to obviate the onerous bureaucracy relating to foreign travel. This was demonstrated when, in October 1924, he was refused re-entry to Germany under his Swiss passport because the authorities chose to doubt that he was resident there. The Swiss government were initially reluctant to provide the diplomatic passport, noting his dual nationality, although they agreed to approve one were he to fail to convince the German government to provide it. In February 1925, Einstein finally obtained a German diplomatic passport, which he retained until he returned it to Adolf Hitler's government after he left Germany in 1933.

Although the Nobel Prize money was intended to provide financial security for Mileva and the children, it also caused more strife between them and Einstein. The divorce agreement of 1919 mandated that the interest should be transferred to Mileva without condition but that she could not dispose of the capital without Einstein's consent. It also specified that the capital should be deposited in a Swiss bank account. However, in a letter in May 1923, Einstein informed Mileva that 45,000 Swiss francs would be transferred to his business friend, Albert Karr who would pass it on to her, which she wanted to use to purchase a house in Zurich; the balance, roughly 133,000 Swiss francs, would be invested in dollar bonds issued by several governments. Einstein was concerned that the inflationary catastrophe playing out in Germany might have ramifications in other European countries, including Switzerland. He took advice from one of his financier friends, who advised that the US was the most stable economy.

Unfortunately, Einstein had addressed the letter to Mileva incorrectly and it was returned to Berlin and then forwarded to Leiden, where he had been staying. Since he left before it arrived, it had to be re-addressed to Zurich by Ehrenfest. On receipt of the greatly delayed communication, Mileva panicked and sent Einstein a telegram asking him to wait before depositing the funds in the US. Her subsequent letter very clearly stated that she preferred the money to be deposited under her name in Switzerland, as specified by the divorce decree. Einstein's reaction was brief and to the point. He angrily reproached Mileva for ingratitude, mistrust, and bad behaviour and, ignoring her request, informed her that he was going to proceed with the US deposit under her name but under his supervision.

Distraught by Einstein's abusive communication, Mileva turned to her old friend and adviser Emil Zürcher, who wrote to Einstein that he was legally in the wrong and that Mileva's agreement was essential for any arrangements for the capital. Mileva also consulted her old intermediary from her brief Berlin interlude in 1914, Fritz Haber, who wrote a very formal letter to Einstein asking to discuss the matter with him. After a flurry of letters and interviews among all the parties, it was eventually decided that the US investments would proceed. However, in early June 1923, Hans Albert wrote to his father, remarking that they had no idea where he was and complaining that he hadn't written to them for a very long time. He exhorted him to tell them when the money intended for the house purchase would arrive. This was followed shortly afterwards by a second letter, in which Hans Albert berated his father for his incompetence in telling Mileva that she would receive 45,000 francs whereas Albert Karr had informed them that he had only received 40,000. In fact, 45,000 francs were transferred, but Einstein had confused Karr by mentioning a smaller sum of 40,000. Clearly Mileva thought that she was being robbed and had confronted Karr, as evinced by a justifiably annoyed letter that he wrote to Einstein, asking what on earth was going on.

The saga of the Nobel Prize money and the house purchase marked the beginning of a period of tension with Einstein's elder son that was to persist for many years. Hans Albert had grown into a young man whose emotional makeup was in many respects uncannily similar to that of his father. He was stubborn, knew his own mind and did not hesitate to let others know it too, was very fond of his mother, with whom he nevertheless had a tense relationship, and often had a blind spot about what might hurt or offend. In the autumn of 1922, he had begun studies for an engineering degree at ETH, against the opposition of his father, both to the subject and the location. He lived at home and, again like his father, music was a major interest. He was a member of the ETH student orchestra, where he played the double bass. His mother wrote to her old friend Helene Savić around this time that:

> [Hans] Albert is hard at work with his studies; he has already passed two exams with excellent results, and there remains just one in the next one and a half to two years. He plays his piano very well, and he likes to play with other instrumentalists, so a lot of music is being played here, which gives me great pleasure.

Einstein received a letter from his old friend and collaborator, Marcel Grossman, who informed him that Hans Albert had done well in his course on descriptive geometry.

Hans Albert resented the semi-detached nature of Einstein's relationship with his family, the infrequency of his father's communications, his even rarer visits and the often vague and changeable proposals for joint vacations. These frustrations burst to the surface in the letter about the transfer of the Nobel Prize money referred to above. It must be said that his letter is both rude and impertinent; it deeply angered his father, even more so because many of the accusations were justified. Hans Albert was protective of his mother and saw the anguish which his father's behaviour often inflicted on her; he taxed Einstein with completely failing to understand the effect he was having on her. Einstein replied in kind with a letter that was fortunately intercepted by Mileva before it reached their son. Mileva wrote to Einstein begging him to write a friendlier letter to Hans Albert and explaining how upset they all were by Einstein's decision to lodge the remainder of the Nobel Prize money in the US.

Einstein was exceedingly angry with his son for several weeks and rather than writing a friendlier letter, as Mileva had suggested, cut off all communication with him. His friend Ehrenfest, to whom he had confided the situation, tried to persuade him that he was being childish, without success. In June, Einstein wrote to Tete, telling him that because of Hans Albert's behaviour, he was not going to take him on holiday. Instead he invited Tete to come alone, either to Lautrach, west of Munich, where Anschütz-Kaempfe had fitted up an old castle as a meeting centre for academics, or to Kiel, again as a guest of Anschütz-Kaempfe at accommodation provided at his factory. He sent this letter poste restante to Zurich in order to evade interception by Mileva. Tete replied immediately, clearly in a quandary, torn between wanting to see his father and loyalty to his brother, who, he informed his father, was distraught. In mid-July, Einstein wrote once again to Tete. He urged him to come alone to Lautrach and, bypassing Hans Albert, forwarded a letter for Mileva from the American bankers related to the Nobel Prize money. The following day he wrote to Tete again, advising him to convert the notorious 45,000 francs into dollar bonds as the Swiss franc was sinking. Tete was twelve years old at this time.

Anschütz-Kaempfe, whom Einstein had informed of Hans Albert's letter, intervened in this unedifying saga. He was very fond of both father and son; he wrote to Hans Albert and offered to try to reconcile him with his father. Hans Albert and his mother were very grateful, and both wrote to the industrialist, expressing their bemusement at Einstein's anger. Hans Albert went to see Anschütz-Kaempfe in Lautrach at the end of July 1923. As a result of this meeting, Hans Albert wrote a grovelling letter of apology to his father. This seems to have mollified Einstein. He picked up Tete at Stuttgart railway station; they stayed overnight in Einstein's birthplace, Ulm, paying 570,000 marks for their hotel room. They arrived in Lautrach on 8 August for a ten-day stay. Hans Albert had stayed on after his meeting with Anschütz-Kaempfe. He and his father were reconciled and spent most of the time playing music together; Einstein remarked that his violin playing was greatly improved by their sessions. Hans Albert joined his father again in September at Anschütz-Kaempfe's Kiel base, where they played more music and sailed together. In a letter to Elsa from Kiel, Einstein reported that Hans Albert was well disposed towards Elsa and daughters (there had been some unpleasantness the previous summer, which had also greatly annoyed Einstein). However, he opined, Hans Albert was totally insensitive to human feelings.

Zangger, who had also been drawn into the controversy over the Nobel Prize money, reported that Hans Albert was learning a lot about the realities of life from the process of the house hunting and purchase. They found a large house, Huttenstrasse 62, with four stories, each containing a sizeable apartment. The plan was for Mileva and the boys to occupy one and to let out the others to generate income for the family. Einstein visited Zurich in July 1924 and pronounced himself satisfied with the house, which was near to the ETH and had a good view of both the lake and the mountains.

In June 1924, Einstein wrote to Hans Albert that 'I am pleased, dear Albert, that you are having so much pleasure with your subject. Anschütz would be pleased if you became his successor;[2] but far be it from me to try to influence you. Everyone must take these most important decisions in life for themselves.' As usual, Einstein gave his son good advice; unfortunately, he himself failed to take it. For Hans Albert was soon to embark on the most important decision of his life – marriage – and Einstein strove with all his might to influence him against it.

At some time in the first half of 1925, Hans Albert began a relationship with a neighbour from their previous apartment on Gloriastrasse, Frieda Knecht. They had known each other since 1914, when Mileva and the children moved to Gloriastrasse, which Frieda's father owned. Their first recorded meeting was in the summer of 1918, when Hans Albert was fourteen. He and Frieda, who was a violinist, played violin and piano music together. As his father could testify, this often leads to a deeper involvement than just music-making. Frieda obtained a doctorate in German Literature at the University of Zurich. She became a German teacher and speech therapist in Zurich. Einstein's sister Maja met her several years later and found her 'much better than the impression given me by my brother and his ex-wife. She is almost ten years older than [Hans Albert], rather ugly and misshapen, but intelligent, modest and tactful.' This relatively positive view was not one shared by either Mileva or Einstein when the relationship first came to their notice in the summer of 1925.

When both Hans Albert and Tete arrived in Berlin in July 1925, once again music-making with their father was their main occupation. Einstein had to attend his League of Nations committee meeting in Geneva at the end of July, leaving the boys to travel to Kiel, where Hans Albert was to work in Anschütz-Kaempfe's factory as an intern. On his way back from Geneva, Einstein stopped off in Zurich to see Mileva. She gave him the latest details on what she saw as Hans Albert's infatuation with Frieda. After Tete returned home at the end of the school vacation in mid-August, Hans Albert and Einstein remained in Kiel, playing music together in the evenings after Hans Albert had finished working in the factory. Einstein got the impression that there was as yet no commitment either on his son's or Frieda's side. He wrote to Mileva that Hans Albert was inexperienced with women and did not understand how to get on with them. He invited Hans Albert to come to Berlin, where they could play Bach Inventions together and where he hoped to convince his son that the woman he had chosen would be a disaster. However, Hans Albert declined the invitation.

The parallels between the uproar surrounding Einstein and Mileva's marriage and that of Hans Albert and Frieda's are uncanny. The similarities even ran to a concern with bringing 'bad blood' into the family. Einstein became increasingly convinced that Frieda was of inferior genetic material, for no other reasons that can be discerned apart from her diminutive stature and the fact that her mother had had some sort of mental breakdown. He consulted Zangger, who, in April of 1926 tried to convince Hans Albert that madness on both sides of the family[3] was a dangerous recipe for future progeny. Zangger spent considerable time and effort in talking separately to both Hans Albert and Frieda. Einstein also enlisted Anschütz-Kaempfe to try to convince his young apprentice. Anschütz-Kaempfe went to the extent of sending for Frieda, interviewing her and, no doubt predisposed by Einstein's opinion and seeming to share his obsession with heredity, concluded that she was indeed a degenerate psychopathic dwarf. Influenced by Einstein's opinion of Hans

[2] Anschütz-Kaempfe, who was childless, had talked about leaving his factory to Hans Albert in his will, but the subject was eventually quietly dropped.

[3] Mileva's sister Zorka had been institutionalised in Switzerland in 1918. In the summer of 1920, since her parents were no longer able to afford the fees, Mileva organised her return to her parents' home in Novi Sad, where, in 1938, Zorka drank herself to death, alone except for her fifty cats.

Albert's lack of experience in the world, Anschütz-Kaempfe attempted to give him a large cheque so that he could buy some decent clothes and see more of the world. Hans Albert, to his credit, refused to accept the proffered money.

The similarity with Einstein and Mileva's own youthful experiences extended to Mileva's attitude to the marriage, which was as negative as Pauline Einstein's had been twenty-five years earlier. Mileva was also concerned about the age difference and the unpromising heredity; also like Pauline, she succeeded only in alienating her son. She brought in Frieda's father, who also had concerns. Bizarrely, their mutual hostility to Hans Albert's marriage plans brought Einstein and Mileva closer together. At one point in their exchange of letters, Einstein even claimed to be 'faithful' to her! He now often stayed in her apartment when in Zurich. As his ex-wife drew closer, his son drifted away. Einstein was convinced that Frieda was carrying out a campaign to ensnare his son and was bombarding him with gifts and letters. He tried cutting off relations with him for a while but when they were re-established, Hans Albert was still obdurately determined on marriage. Hans Albert and Frieda's adopted daughter Evelyn later remarked 'Albert had such a hell of a time with his parents over his own marriage, you would think that he would have had the sense not to interfere with his son's. But no. When my father went to marry my mother there was explosion after explosion.'

Throughout all the parental pressure, Hans Albert quietly continued his studies, graduating in 1926. He visited Einstein in Berlin early in 1927, who offered help in finding him a job and gave him some money. Finding Hans Albert still intent on marriage to Frieda, Einstein grudgingly accepted the situation but enjoined Hans Albert not to have any children. His son simply did not reply and neither did he accept his father's offer to find him a job. Instead he accepted a position as engineer at a steel fabrication firm in Dortmund, where he started work in February 1927. Neither does he seem to have reacted to Einstein's outburst in February 1927, in which he forbade Hans Albert from bringing Frieda to Berlin, which he said he simply could not tolerate. Hans Albert had by now presumably learned that if he ignored his father's outbursts, he would eventually come around. Hans Albert and Frieda married on 17 May 1927 in Zurich; it is not clear if Mileva attended the ceremony, but Einstein did not.

It took a considerable time for Hans Albert's parents to accept the marriage. At Christmas, 1927, Zangger had met Hans Albert in the street and was so shocked by his appearance and the amount of weight he had lost that he wrote to Einstein, asking him to take an interest in his son's well-being. Einstein replied early in January 1928:

> Dear Zangger, It is good of you to write to me about [Hans] Albert's condition. However, the lad cannot be helped. He has thrown himself away on this completely worthless and useless old bag who allows him to go to pot, doesn't cook for him and exerts such a baleful influence on him. If he remains with her, he is lost. He can only get out of it himself – if that were at all possible. Anyway, as long as he remains under her influence, I am not going to concern myself about him. I fought hard enough, while there was still a point.

In 1928, Mileva wrote that:

> [Hans] Albert used to be such a healthy and robust boy, a paragon of a healthy young man who for many years was never sick. And now, what becomes of him during this half a year! His wife does not know how to look after him, she thinks only of herself.

Hans Albert eventually recovered his health. The couple had their first child, Bernhard Caesar, in 1930, followed by two others, who sadly both died young. However, despite these tragedies, their marriage was successful, lasting until Frieda died in 1958.

The weight that Einstein in particular, but also Zangger, Anschütz-Kaempfe, and Mileva, put on heredity with respect to Hans Albert's marriage seems somewhat strange to modern eyes. However, it is important to remember the enormous influence that eugenic ideas had in this period. Switzerland had introduced eugenic marriage laws in 1907, only four years after Einstein and Mileva had married there. In 1905, the Society for Race Hygiene had been formed in Berlin. Right up to the Nazi period, eugenics was not only a powerful force but also a mainstream one. Its ideas were promoted by socialists and fascists alike. It is in this context that Einstein's repugnance towards his son's marriage and his forceful insistence that if Hans Albert and Frieda married, they must not have children, must be understood.

As Hans Albert drifted out of Einstein's life in his efforts to start his own family, Tete came into sharper focus for his father. What that focus revealed in 1923 was a gifted but somewhat idiosyncratic child. His father was delighted by his love of music, remarking that it was the best of companions, smoothing the path through life. Tete played a great deal of Bach, through which he grew to love the music of that other master of the fugue, Reger. He also loved Debussy. With regard to music to play with his father, he found the Haydn violin sonatas dull, although Einstein disagreed. He recommended that his son play both Haydn and Scarlatti on the spinet. As his musical studies progressed, they began to take him well beyond Einstein's beloved Mozart and Bach. Heinrich Zangger wrote to Einstein about Tete's playing during a visit to the Zangger household: 'How beautifully the lad plays Beethoven; he pulls you into the music, which he clearly enjoys while remaining modest and working hard to improve. Einstein, despite everything, you have been lucky!' By the autumn of 1926, Tete had developed a passion for Arnold Schoenberg's compositions. One of his letters to Einstein contains a long disquisition on Schoenberg's musical theories, which he identified as being the major musical revolution of the century. He explained the 12-tones system used in the Wind Quintet, Op. 26, which had been presented to an astonished musical world only two years earlier. While he knew[4] his father disliked Schoenberg's music, he put this down to his lack of professional expertise!

Tete was also very fond of competitions, mostly but not exclusively literary, run by local shops or businesses. He won several prizes for his short stories, a watch in a drawing competition, a penknife in a shooting competition, and ten bananas for his poetry. This penchant for writing often humorous poetry was something he had inherited from his father. He showed Tete's verses to his old friend Moszkowski (by now completely forgiven for his indiscreet book), who produced such things professionally; Moszkowski approved. Among these relatively normal pastimes for a teenager, other, stranger characteristics began to emerge. His letters to his father took on an anachronistic and formal tone and, ominously in the light of subsequent developments, talked about feeling as if there were multiple personalities inside his head. He lacked confidence and motivation, finding school very dull but grateful that it passed time that he would otherwise just fritter away. He wrote to his father questioning the value of science and indeed of all intellectual endeavour, which in his opinion was worthless. This elicited, highly unusually, an immediate response from Einstein, who defended intellectual activity and told Tete that his attitude was rooted in fear of his own failure. To Mileva, Einstein wrote that some of Tete's writing was brilliant but characterised

[4] Tete's belief that his father disliked Schoenberg's compositions is likely to have been based on remarks made after a rehearsal of Verklärte Nacht, Op. 6, that Einstein attended in April 1924. Although Tete thought that it was the learned twelve-tone technique that Einstein disliked, it is difficult to think of two more different pieces than Op. 6 and Op. 26. The former is the apotheosis of high romanticism, although both this and twelve-tone music were equally objectionable to Einstein.

creative work on literary matters, as a primary occupation, as absurd. When, in 1928, Tete published some verses in the left-wing journal, *Das Forum*, entitled 'Kleiner Katechismus', his father warned him that he must be aware that his surname rather than the quality of his work might be the decisive factor in gaining publication.

Tete did seem to enjoy the very limited time he could spend with his father. The various tensions between Einstein and Hans Albert referred to earlier meant that he and his father spent more time alone together. He did not enjoy sailing, unlike Hans Albert, for whom it became a life-long passion; Tete's preference was for walking holidays, bringing Einstein back to the mountains from which he had long been absent. In July 1926, they undertook two separate walking trips, punctuated by Einstein's attendance at a League of Nations Committee meeting in Geneva. On their way, they called in on Michele and Anna Besso in Bern, while in Geneva Einstein visited his old friend Lucien Chavan, with whom Tete stayed. Their first tour was in the Uri Alps, near Bern. After leaving Geneva for the second trip, they travelled up the Rhone Valley into the Valais, visiting Crans Montana and then crossing over the Rawil Pass to Lenk. Einstein expressed surprise that he had coped so well and enjoyed it so much. Their plans to meet up with Besso were abandoned when his nephew, Benvenuto Bandi, son of Rosa Winteler-Bandi, committed suicide. It was Benvenuto's father who had been shot dead by Jost Winteler Junior in the terrible tragedy in Aarau in 1906.

Einstein clearly relished his son's company and pressed him to come to Berlin. Tete spent the Easter vacation of 1927 in Berlin and then in Rheinsberg, seventy-five kilometres to the northwest. Much music was made together; they seem to have particularly enjoyed a concerto by Giuseppe Tartini. Einstein called in to Zurich on his way to one of his Geneva committee meetings in early July 1927, where he reported to Hans Albert that he was playing Haydn, Mozart, and Schumann and enjoying the 'unrivalled' pianism of Tete. After his Geneva meetings were over, Besso and Einstein spent three days together, initially at Visp, before Einstein departed for Zuoz, where Margot Einstein was convalescing. Tete joined him for a walking holiday of around twelve days. Although Einstein was concerned about both the frailty of Tete's health and his tendency to withdraw into a private world, he hoped that time and maturity would improve matters. Alas, his hopes were not to be realised.

Even Einstein's sister, Maja, gave him cause for concern in the 1920s. She had been very ill with a septic abscess in 1922, to the extent that, in those days before antibiotics, her life was thought to be in danger. She recovered but her husband, Paul Winteler, was regularly ill, both physically and psychologically. They had bought a charming house in Sesto-Fiorentino, about six kilometres outside Florence. The house, given the name Samos by their friends, was surrounded by 0.7 ha (almost 2 acres) of land. Although their finances were very limited, they had obtained a mortgage for 10,000 Swiss francs from the Lucerne Cantonal Bank that was guaranteed by some stocks deposited by Maja's friend Sophie Schnyder. Paul had a small pension but no other income except from his paintings, which only sold to friends and family. Maja hoped that they could live mostly from produce from their land and by taking in paying guests. Neither of these things worked out as they had hoped and by the end of 1924 they were falling into debt. Schnyder became nervous about losing her funds and Maja, desperate, asked her brother for help. He had already contributed 1,000 Swiss francs for Maja's medical expenses in 1922, but he responded that he would donate the required 6,750 francs, rather than the loan for which Maja had asked. Einstein's opinion of Maja and Pauli's financial responsibility was such that he probably concluded that they would never be able to pay off a loan anyway. Both were overjoyed when the money was transferred from Einstein's 'secret' account in the Netherlands where his foreign earnings were deposited under the control of Paul Ehrenfest. Indeed, Maja wrote that when the news arrived, she had played a 'Te Deum' of a Bach Toccata and Fugue in Einstein's honour. Such musical offerings would often be required to mark Einstein's regular contributions for repairs and the like over the next decade.

It was perhaps the relationship with Elsa that gave Einstein most cause for concern in the 1920s. The halcyon days of passionate love letters written before he moved to Berlin had long since departed. They had quickly settled into a cosy, at least for Elsa, domesticity in which love subsided quite rapidly to affection, then tolerance with occasional downright exasperation. They were very different; Elsa, while certainly not unintelligent, particularly in assessing personality, was utterly uninterested in Einstein's work. However, she was passionately interested in things that bored him – namely Berlin society. She also insisted on her rights as Frau Professor and bitterly resented being sidelined, as she often considered had been the case in their long Asian trip in 1922–1923. Just as with his sons, he almost never remembered her birthday. When not bickering in Haberlandstrasse, they were exchanging misunderstanding via brief and irregular, on Einstein's part, postcards and occasionally letters. In communications at the end of 1925, she twice rebelled against what she considered his peremptory instructions. He was astonished when she chided him for writing what she thought were angry letters. She loved to travel but Einstein almost never invited her to accompany him on his trips to Paris and Leiden in this period. In Paris, he preferred to spend his spare time with his scientific friends, playing music, for example as part of a string quartet with the distinguished mathematician Jacques Hadamard. At least his and Elsa's excursions to the concert halls and theatres of Berlin gave both pleasure, although Einstein increasingly attended musical performances without Elsa as the decade progressed.

One major problem between Einstein and Elsa was her inability to hide her jealousy of the succession of lovers with whom Einstein became entangled. The flirtation with his step-daughter Ilse and the maids in the Berlin apartment has already been remarked on. Whenever he travelled without Elsa, she was convinced that he was seducing other women, even the wives of his friends. Despite Einstein's vehement denials when directly accused of this in October 1926, he certainly subsequently developed a passionate relationship with the woman in question, Margarete Lebach. The extent of Elsa's watchfulness can also be gauged from a narrative by Esther Salaman, a Russian student who attached herself to Einstein between 1922 and 1925. Despite telling Salaman that very few women are creative, he interested himself in her career. He once remarked to her:

> I like Paris, but I don't want to go there, or anywhere. I should not like to learn a new language; I don't like new food or new clothes. I'm not with people much, and I'm not a family man. I want my peace. I want to know how God created this world. I'm not interested in this-or-that phenomenon, in the spectrum of this-or-that element. I want to know His thoughts, the rest are details.

As an attractive woman in her early twenties, Salaman found no difficulty in getting Einstein's attention and often visited his apartment. She reported that events such as a woman refusing to leave the front door of the apartment block because she claimed 'the Professor' had jilted her were not unusual. In the summer of 1925, she was in the Einstein apartment when he asked her to go sailing with him. Elsa at once appeared in the corridor. To Einstein's perhaps feigned amusement, she scotched the invitation by remarking on Salaman's weak eyesight. Salaman, sizing up the situation, said that also she could not swim. Nothing more was said about sailing.

In June 1923, someone entered the Einsteins' life who would cause Elsa much more heartache than Esther Salaman. Betty Neumann was a 23-year-old niece of Dr Hans Mühsam, who was among Einstein's closest friends in Berlin. They had met when Mühsam had treated Einstein's mother for cancer. Ilse Einstein had been acting as her step-father's secretary since the foundation of the Kaiser Wilhelm Institute, but her health was always frail. It happened that Betty was staying with her uncle when Ilse fell quite seriously ill. No doubt Einstein mentioned to Mühsam that he was at a loss without a secretary and Mühsam suggested Betty, who had some secretarial qualifications. She acted officially as Einstein's sole secretary for some six weeks, deferring her return to her home in Graz in Austria until August. She left a charming account of her time with Einstein, in which she describes interrupting him playing the soprano part of the Mozart song Das Veilchen, K476, on his violin. This setting of a poem by Goethe tells of a violet whose only desire is to be pressed to the breast of a beautiful young shepherdess but instead, unnoticed, is crushed under her foot.

Other than the obvious symbolism of the choice of Mozart piece, Betty's account of life with Einstein is demure and uninteresting. In fact, soon after she began work, Einstein fell deeply in love with her. He spent a considerable part of his train journey to Schloss Lautrach for his vacation with his sons (see above) wondering how he could maintain contact with her now that she had returned to Graz. He wrote to her as soon as he arrived at Lautrach, proposing that she should join his family and that he would look after her like a daughter. He already used the intimate *Du* form of address. He wrote several more times during his stay and on the journey to Leiden, in which he begged her to write to him, direct to Haberlandstrasse, where he implied that he would take steps to ensure that her letters did not fall into the wrong hands. Elsa, no doubt prompted by her husband, arrived at Mühsam's residence early in October to suggest a part-time arrangement for Betty, since the convalescing Ilse could manage no more than four hours work a day. Betty returned to Berlin early in November, no doubt to take up this offer. However, Einstein almost immediately fled to Leiden to escape the bread riots. As soon as he arrived, no doubt inflamed by a brief meeting with Betty organised just before his flight, Einstein wrote to her and proposed to take a job with a large salary at New York's Columbia University if she would come with them. The salary would allow him to support her without any problems. Betty refused this offer of a love triangle with Einstein and Elsa. In the New Year of 1924, Einstein was pleased to hear that Betty's engagement to a man even more her senior than Einstein had fallen through. He began to meet her clandestinely, at least sometimes at Mühsam's residence, apologising to her for using the *Du* form in front of others.

At some point in January 1924, matters came to a head. Unable to convince a besotted Einstein to give Betty up, Elsa was more or less forced to accept the lesser of two evils. She dreaded the humiliation of her society friends hearing about Einstein and Betty's clandestine meetings, so she agreed that they could meet in Haberlandstrasse on two afternoons a week. Einstein announced this concession to Betty with glee, instructing her to ensure that she was nice to Elsa when they met. She complied, bringing Elsa a bunch of flowers. The arrangement, which Einstein referred to as a 'clover leaf' (i.e. triangle), continued until Einstein's visit to Kiel in May, when Betty returned to Graz for a six-week visit to her mother. The secretarial situation was complicated by Ilse's marriage to Rudolf Kayser on 16 April 1924. For some time, specifically in August, Betty did again act as Einstein's secretary, although Ilse was on duty, presumably part time, in September. By October of 1924, the strain was telling on all the parties. Einstein reluctantly agreed to bring his relationship with Betty to an end. In doing so, he not only lost Betty, who was deeply hurt by his action, but was also estranged from Hans Mühsam. Einstein and Mühsam were only fully reconciled in 1942 after both had left Germany. On 28 October, Einstein wrote to Betty saying that they must stop seeing each other in order that she could find someone her own age and settle down and have a family. He signed off with a familiar sentiment: 'I am just an old guy who has to accept the facts as they are and look in the stars for what is forbidden to me on earth.' Presumably subconsciously, he used almost the same words as he had in separating from Marie Winteler in their youth in Aarau.

Einstein's problems with secretaries continued. Ilse seems to have helped out, part time, until the start of 1928. From roughly July 1924, Siegfried Jacoby, who also acted as Berlin correspondent for several Jewish newspapers, acted as secretary. It is likely that Elsa played a major role in ensuring the appointment of a male for this position. However, he resigned in January 1927, having been admonished by Elsa for opening some of Einstein's letters without authorisation. A student, Edwin Sieradz, spent a few months in the position late in 1927. After he left, Elsa began to search again, spurred on by Einstein falling seriously ill with a heart complaint. She heard from a friend, also born in Hechingen, whose sister, Helen Dukas, was looking for a secretarial position. Dukas was appointed after first an interview with Elsa on 13 April 1928 and then the following day with Einstein himself, who was confined to bed at the time. She served as faithful secretary, gate keeper, and eventually general factotum thenceforth until Einstein's death.

Einstein could be one of the kindest of men to persons who did not approach him too closely. He supported Carl Stampe, a poor musician who originally taught Margot Einstein music, for three years, including letting him occupy his own bedroom while he was away and then an anteroom for three months. Indeed, he had considerable difficulty in getting Stampe to leave when the room was required to store their coal! He consistently supported poor Jewish students, often from Russia, who applied to him for help when they arrived in Berlin. An impoverished doctor, who had not charged for his services when he treated Elsa's mother, had a daughter with tuberculosis. She went to Switzerland for treatment. Einstein quietly got her treatment bill reduced at his own expense by asking Zangger to intervene.

Although he was so remiss in remembering the birthdays of his family, he did write to congratulate friends. On 15 July 1923, he wrote a loving and respectful letter to Lorentz on his seventieth birthday. November 1926 saw the eightieth birthday of his old friend Alfred Stern, whose hospitality and musical evenings had been so important to him as a young student in Zurich. Reminded by another music partner from this period, Julie Ansbacher, Einstein wrote a warm and affectionate letter to Stern, who reciprocated, fondly remembering their times together and informing Einstein that a family trio had serenaded him on his birthday with music by Bach.

The decade of the 1920s seems to mark a gradual diminution in Einstein's contact with the closest of his friends, Besso and Zangger. This can be partially attributed to Einstein's manifest failings as a correspondent and partially to fewer emergencies in Mileva's household in Zurich, into which either or both of the friends were usually drawn. In September 1924, Einstein attended a conference in Vienna, followed by the annual conference for German scientists and medical doctors in Innsbruck. At the end of the month, he visited Zurich briefly; there was as usual much musical activity and he also took Tete to a performance of Shaw's *Pygmalion*, with which he was familiar from a German translation that Ilse's husband, Rudolf Kayser, had given him. He had been introduced to Shaw at Lord Haldane's reception in 1921 and they became friends. At the start of October, Einstein travelled to Luzern to give a talk at the annual gathering of the Swiss Association of Scientists. Besso, his wife Anna, Zangger, and Marcel Grossmann also attended the meeting. They all appear to have stayed in the same hotel and to have greatly enjoyed their reunion, sending postcards to mutual acquaintances such as Einstein's former mentor, Jost Winteler and Einstein's friend, Lucien Chavan.

The frequency of letters drops off markedly after the Luzern reunion. Zangger often seemed unsure of where Einstein was. Zangger was seriously ill for several months in early 1925. Following his recovery, he was active in the Hans Albert marriage saga. He spent much of the autumn of 1926 in Berlin, where he was contemplating taking up a senior position, and spent several evenings with Einstein as well as attending the university Physics Colloquium. Much of their conversations in this period involved their mutual friend, Besso, who had developed serious psychological issues, probably depression. Besso's son Vero thought that his father was in danger of losing his position at the Patent Office, since rather than dealing with the expected average workload of 200 patent applications, Besso was only processing fifty. This had come to a head when another employee, who had been threatened with dismissal for inadequate performance, had pointed out that he had dealt with more applications than Besso. Einstein considered the problem to be that Besso was such a fount of knowledge that he spent too much time answering questions from his colleagues. In addition, he was such a perfectionist that he had difficulty finishing his own work. Zangger asked Einstein to write this down is such a way that he could use it with his various influential contacts in the Swiss government. As usual, Zangger's influence was brought to bear expertly and in the correct places to ensure that Besso retained his position.

Zangger was also involved in the sad case of another of Einstein's old friends and collaborators, Marcel Grossman. His health had begun to deteriorate in 1924. After a visit in 1925, Einstein described his symptoms to Zangger. After he was forced to ask for several teaching dispensations, Grossmann eventually consulted Zangger himself, who diagnosed multiple sclerosis. Grossmann took early retirement in 1927 and died in 1936.

After the Luzern meeting in October 1924, Einstein made his regular visit to Leiden, stopping off in Antwerp to congratulate his Uncle Caesar on his seventieth birthday. As usual, he enjoyed his time with the Ehrenfests and their effervescent children. The eldest, Tatiana, by then twenty, had become a very gifted mathematician. Two years later, he reported to Elsa that she was exceptionally talented; he had asked her to solve some of the maths that he was working on, and she had done so within a day. He remarked that if only she were a man, something important might come of it. The requirement to visit Leiden once or twice a year, on top of his work for the League of Nations and his stays in Kiel, was becoming too much for Einstein. In November 1926, a month after returning from Leiden, he wrote to Ehrenfest asking for his help to be relieved of his professorship and the obligations it entailed. Although the request must have been bitter gall to Ehrenfest, who worshipped Einstein and loved his visits, he replied in the most accommodating way that the university would be only too happy to continue his appointment without any requirement for his physical presence. Einstein was pleased, informing Ehrenfest that he should use his salary after deducting his expenses to fund a deserving student. In fact, Einstein did subsequently visit Leiden from time to time, his last stay being in August 1932. The fountain pen that he had used to write all his general relativity papers until 1921 can still be seen in the Boerhaave Museum in Leiden.

Max Born continued to be a regular correspondent of Einstein's, as, less frequently, was his wife Hedwig. In early 1926, Born asked Einstein's help to get one of Hedwig's plays staged. He passed it to his son-in-law, Kayser, who had influence with theatres such as the Berlin's Volksbühne, and subsequently to another impresario, Leopold Jessner. In giving his opinion to Hedwig, while generally praising her work, he could not resist one of his typically misogynistic remarks that, as a woman, she was incapable of giving her characters any depth. In early 1927, he got involved with the messy divorce of his fellow popular idol, Charlie Chaplin, by giving a quote supportive of the great comedian at the request of an American newspaper. As he had in 1922, he disparaged American society as dominated by women.

Another characteristic of the 1920's was a renewed involvement with and fascination by practical matters. Einstein spent a great deal of time in Kiel with his friend Hermann Anschütz-Kaempfe. Indeed, he spent so much time there, advising the engineers in the factory on aspects of the gyroscopic compass, that Anschütz-Kaempfe refurbished an apartment for him. Situated very close to Kiel's harbour, where a boat was moored for his use, it came with a cook and domestic help, not to mention a baby grand piano. Here, Einstein could not only help practically in the factory but also contemplate his theories without the usual interruptions in Berlin. His work in Kiel resulted in a number of patents. He brought Hans-Albert and Tete to his apartment for several vacations, borrowing Anschütz-Kaempfe's sailing boat on countless occasions. Once the compass went into production in 1925, however, Einstein's reason to visit weakened. His last recorded visit was in October 1926, when he agreed the financial compensation due to him on future sales with Anschütz-Kaempfe's company. No further correspondence between them is extant after November 1926, although Einstein mentions him as doing a great deal of business in a letter to Hans Albert in July 1927; Anschütz-Kaempfe may have reduced his involvement with the company and moved to his other home in Munich. He sold the company to Zeiss in 1930 and died in Munich in 1931.

Einstein was involved in a long and detailed correspondence on another practical matter with Auguste Piccard, Professor of Physics at the Free University of Brussels. They had met at the Luzern conference in the autumn of 1924. Einstein had proposed that a slight difference in the electric charges of the proton and electron might explain the Earth's magnetic field by inducing a motion of the charges in the core. Piccard proposed various experiments to test the equality of the electron and proton charges over the next year, explaining his ideas to reduce systematic errors in detailed letters. Einstein replied in matching detail. Although Piccard did not detect any difference in the charges, he did succeed in increasing the sensitivity of his experiment by an order of magnitude. He went on to test the measurements of Dayton Miller, who claimed in the early part of the decade to have detected evidence for the aether. Although Miller stuck stubbornly to his guns, the many negative results contradicting his experiment led to it being eventually discounted. Only in the 1980s did reanalysis of his results reveal that he had made a number of errors in his analysis.

Einstein began a fruitful collaboration with Leo Szilard, a Hungarian national who had originally studied engineering but switched to physics. He completed his PhD at the University of Berlin in 1923. His thesis, on thermodynamic fluctuations, a subject still close to Einstein's heart, had brought him to Einstein's attention. Szilard was extremely inventive and by 1926 the two were applying their thermodynamic expertise to a pumpless refrigerator. Refrigerators were just starting to be introduced to the home but usually required large and cumbersome pumps. As they progressed with their design, Szilard ensured that they protected their intellectual property with patents, which Einstein helped to draw up. They were particularly active in 1927 and sold various of their patents to companies, which earned them a substantial sum, although their various designs had little commercial impact.

By 1924, Einstein's memories of the exhaustion and irritations of his trip to Japan, Palestine and Spain in 1922–1923 had faded, while the attractions of a long sea voyage had remained. He had been invited to visit South America on several occasions; early in 1925, he accepted an invitation from the University of Buenos Aires catalysed by a group of rich Jewish Argentinian families who also part-funded the trip, including first-class travel and a $4,000 honorarium. Quickly, the tour was extended by funding from an Argentinian Jewish charity and several neighbouring countries. A memory of their conflicts during the Japanese trip, as well as the fall-out from his recent affair with Betty Neumann, may have induced Einstein not to invite Elsa. His step-daughter Margot was to have accompanied him, but her participation was cancelled at the last moment due to illness. It was therefore Einstein alone who travelled to Hamburg on 4 March 1925 to embark on the SS *Cap Polonio* the following day. On the evening of his arrival in Hamburg, he played Mozart on a borrowed violin with his friends the Robinows.

As Einstein had hoped, he enjoyed the sea voyage greatly. He met a number of interesting passengers, including the well-known author Else Jerusalem, whom he nicknamed the 'Panther Lady.'

After spending a great deal of time together on the voyage and after their arrival, she seems to have decided that he was paying her insufficient attention and boycotted the final parts of his stay. After a brief stop in Lisbon on 11 March, the ship ploughed across the Atlantic. Einstein greatly enjoyed the solitude of his cabin, in which he thought about physics and read Chaucer's *Canterbury Tales*. He emerged only at mealtimes, with the inevitable exception of music-making. He wandered into the family room and entertained the children and their mothers with his piano improvisations. Every day he played string quartets by Mozart and Schubert in his cabin with a Mr. Holländer, Mrs. Ohnesorg, and the first violinist of the ship's orchestra. Although the temperatures were punishing and the sweat dripped from him, he was in heaven. During a concert in which the passengers took turns to entertain each other, Einstein and his quartet played Mozart's Eine Kleine Nachtmusik, K525. Einstein, presumably with a pianist, then played the Beethoven Romance in F Major, Op. 50.

All too soon the voyage ended. On 22 March they docked in Rio de Janeiro before proceeding to Montevideo and thence to Buenos Aires on 25 March. Einstein's host was Bruno Wassermann, a rich German merchant. His stay revolved around a number of lectures, which he sometimes delivered in French, some for the general public but most for the scientific community. These were interspersed with political and tourist activities, which included an audience with the President, as well as Einstein's first trip on an aeroplane, a seaplane which flew him around Buenos Aires. On 1 April, he was guest of honour at a celebration attended by 4,000 people, which was arranged by the Jewish community to celebrate the formal foundation in Palestine of the Hebrew University of Jerusalem. Although this was a joyous occasion for him, he soon found that founding a university was much easier than making it work, as will be discussed in Chapter 18. Two old friends awaited him in Argentina. One was his cousin Robert Koch, his schoolmate and rival for the affections of Marie Winteler in Aarau. The other was Georg Nicolai, veteran of anti-war causes in Germany and confidant (and perhaps lover) of his step-daughter Ilse. Nicolai had become Professor of Physiology at the University of Cordoba, probably leaving Germany to avoid taking responsibility for an illegitimate son he had fathered with a friend of Ilse's. The local German community mostly boycotted events during Einstein's visit, considering him to be an unpatriotic pacifist. The farewell dinner on 22 April was a lively occasion, ending with singing and guitar playing. After a considerable delay during which a violin was located, the guest of honour obliged with pieces by Schumann, Mozart, and Beethoven.

On 24 April Einstein arrived in Montevideo, which apparently had a superior musical life to Buenos Aires, as he lost no time in going to the opera. On the following evening, he attended a performance of Verdi's *La Traviata*; on 26 April, much less to his liking, it was Wagner's *Lohengrin*. He certainly greatly preferred Montevideo to Buenos Aires, which he compared to 'New York, attenuated by the south', where only money and power counted for anything. His pleasure at his reception in Uruguay was demonstrated by his donating his lecture honorarium to set up a prize for a deserving student. After three days aboard a highly unsanitary ship, the *Valdivia*, he arrived in Rio de Janeiro on 4 May. The scientific community in Brazil at that time was very small; his lectures, again given in French, were therefore addressed to the general public and took place in overcrowded and stuffy theatres. Having fulfilled his final commitment, he enjoyed a day's sightseeing, including taking the funicular to the top of the Corcovado, not yet crowned by the statue of Christ the Redeemer. On 12 May, exhausted, he boarded the *Cap Norte* for the voyage back to Hamburg.

Einstein had taken a great deal of pleasure in the exotic flora and fauna of South America, although he had mostly seen it in controlled environments such as botanical gardens. He was given many presents during the trip, some of which he brought back and handed on. One was a collection of butterflies that he thought his sons might like. He offered Mileva a collection of cacti that the director of the botanical gardens of Rio de Janeiro had given to him. He suggested in his letter that she would be welcome to accompany Tete to Berlin for the summer holidays and stay with him and Elsa in Berlin. Although Tete did indeed spend the summer of 1925 in Berlin, Mileva did not. The typically tactless way he ended his letter of invitation would not have helped. He made an analogy to Gottfried Keller's novella *Der Landvogt von Greifensee*, in which the protagonist invites the loves of his life to spend time with him. Mileva would not have been amused when Einstein commented that to emulate Keller's story he would also have to invite Marie Winteler!

In October 1925, Mileva, perhaps emboldened by Einstein's friendlier tone, had suggested she might write a memoir for publication. This brought an end to their friendlier relations and caused a predictable explosion, in which Einstein reproached her for never being grateful for all the things he had done for her, including the cacti. He opined that nonentities should not write books about those who have done something. She should keep quiet but if not, he would grin and bear it as he had already had enough rubbish written about him by his friends. It was only because he cared about her that he wrote so candidly, he said, adding that it was not just children who needed a slap every now and again but also adults and particularly women.

The rigours of the South American trip were enough to convince Einstein to spend more time in Berlin. This gave him more leisure for music-making. In December 1924, he bought a new Blüthner piano. He complained to Maja that he was unable to take part in a regular string quartet group, as he had for almost all his adult life, because he did not have sufficient time. However, he certainly played with many ad hoc groups, often with distinguished professional musicians. These musical soirées were one reason to install a better piano in Haberlandstrasse, but they also took place in other venues, in particular the palatial house of the von Mendelssohns.

There were other venues where amateur music-making merged into professional concerts. One such was the Russian Embassy, which in the 1920s had become a cultural centre for the left-leaning cultural intelligentsia. One of these concerts, in 1927, was immortalised both by Lydia Pasternak-Slater and her father, the painter Leonid Pasternak (see Figure 17.1). The artist and his wife, the concert pianist Rosa Kaufman-Pasternak, and their daughters had moved to Berlin in 1921, while her brother, the author and future Nobel Laureate Boris Pasternak, had remained in Russia:

> One time my mother was playing the grand piano when someone asked her if she wouldn't accompany Einstein who was also present. But Einstein objected. He said: 'I really wouldn't dare to come forward, after such an artist!' But my mother talked him into it. My father made a sketch. And that's how the fine page with the violin playing Einstein came to be.

The Pasternaks subsequently continued to make music with Einstein in private gatherings. The cellist Raya Garbousova made her Berlin debut in 1926 with the pianist Michael Taube, who introduced her to Einstein. They played chamber music together, after which her verdict was that Einstein wasn't very good, having very odd vibrato and playing out of tune.

In addition to Leonid Pasternak, Einstein befriended several other artists. He had some talent in drawing and his ability to lose himself in deep contemplation would have made him an excellent sitter. One friend who took advantage of this to produce several works depicting Einstein in the 1920s was Max Liebermann. Many years later, Einstein remarked that a particularly atmospheric painting from 1921/1922, now in the possession of the Royal Society, looked more like Liebermann than him. Emil Orlick was another artist who produced several sketches of Einstein in this period, lithographs of which Einstein would sign and distribute to particularly favoured

Figure 17.1 Einstein playing the violin at the Russian Embassy in Berlin, sketched by Boris Pasternak in 1927.

correspondents or friends. Born in Prague, Orlick moved to Berlin in 1905, where he made many portraits of prominent musicians, including Strauss, Bruckner, Schoenberg, Zemlinsky, and the violinist Bronislaw Huberman.

Another attendee at almost any Berlin cultural event of any significance was Harry, Graf Kessler. His diaries hold a mirror to the decadent Weimar Berlin as immortalised by writers such as W. H. Auden and Christopher Isherwood, whose escapades made Einstein's serial philandering seem like a poetry reading in comparison. Einstein is often mentioned in Kessler's diaries. On 12 February 1926, Kessler organised a dinner party that Einstein and Elsa attended, she no doubt with much more eagerness than he. Perhaps making a silent protest, Einstein was wearing boots with his dinner jacket. Nothing would induce him to wear his 'Pour le Mérite' decoration, although Elsa remarked that she had finally persuaded him to pick up his Royal Society and Royal Astronomical Society gold medals from the British Embassy. He didn't however bother to open the parcel.

Nevertheless, he had written to the Secretary of the Royal Society on the Award of the Copley Medal in 1925:

> The man who has discovered an idea which allows us to penetrate, to whatever slight degree, a little more deeply the eternal mystery of nature, has been allotted a great share of grace. If, in addition, he experiences the best help, sympathy and recognition of his time, he attains almost more happiness than one man can bear.

In June 1927, both Kessler and Einstein attended a dinner at the home of the publisher and drama producer Samuel Fischer, also attended by dramatist and novelist Gerhard Hauptmann. The latter engaged in an argument with Einstein about astrology, which Einstein dismissed with scorn. The theatre critic Alfred Kerr subsequently steered the conversation to theology, asking, sneeringly, whether it could be true that Einstein was religious. Einstein replied:

> Certainly, you could call it that. When one tries to penetrate the secrets of nature with our limited means one finds, behind all the relationships that are discoverable, something remains, subtle, intangible and inexplicable. Awe in front of this power, beyond anything that we can grasp, is my religion. To this extent, I am, in fact, religious.

Einstein was often invited to professional concerts by their organisers or the performing artists. In March 1924, the pianist Paul Schramm invited Einstein and his family to his two Berlin concerts. After Einstein had attended the first and presumably talked to his host after the event, Schramm reassured him that he would not have to listen to any Liszt in the second concert! In November 1924, Einstein was sent tickets for a performance of Beethoven's *Missa Solemnis*. In thanking Siegfried Ochs, he admitted that, although he agreed that the piece was inspired, he found Mozart and Bach more to his taste.

Einstein was also frequently invited to perform in public. Of all these invitations, few could have been as bizarre as the one that arrived in the summer of 1926. Kurt Singer, a Jewish neurologist, who was also music critic for the magazine *Vorwärts* and conductor of the Berlin Doctor's Choir and Orchestra, invited him to play one of the two violin parts in the first movement of Brahms's String Sextet, Op. 18. The occasion was the opening of the first International Congress on Sexual Research. Einstein's response, unduly modest on both counts, is delicious: 'Dear Sir, unfortunately I feel that neither my sexual nor my musical abilities are sufficient to allow me to accept your kind offer.'

Invitations to concerts that Einstein did not refuse in this period were extended by Toni Mendel. She had met Einstein in the early 1920s, when she and her husband Albert were members of the same pacifist organisations. Her husband, a rich clothing manufacturer, died in 1922, leaving the attractive Toni ample means to enjoy herself. She would arrive at Haberlandstrasse in her imposing chauffeur-driven car and whisk Einstein off to their concert. She accompanied Einstein to musical performances much more frequently than did Elsa. Although Toni provided both tickets and transportation, Einstein did require some money, which was always in Elsa's care. Blazing arguments sometimes resulted. Although Toni did her best to ingratiate herself with Elsa by gifts of chocolates, she was received only with cool politeness. Since Einstein often spent the night at the Mendel mansion in Wannsee, where he was known to play the piano loudly until the early hours, Elsa's tolerance is commendable.

The concert landscape in Berlin in the 1920s was still dominated by the three great German 'Bs' – Bach, Beethoven, and Brahms – but it was also a centre of new music. Perhaps the most famous premiere of the decade was *Dreigroschenoper* (derived from John Gay's eighteenth-century *Beggar's Opera*) by Berthold Brecht and Kurt Weil in the Theater am Schiffbauerdamm on 31 August 1928.

Although the first half of the first act was received in silence, the *Kanonensong* was a major hit, paving the way for one of the most enthusiastic receptions in the history of the theatre. By the beginning of the following year, more than twenty theatres in the German-speaking countries were staging the opera. Einstein had met Brecht but had been put off by his difficult character and his lurid reputation, as relayed by Einstein's son-in-law Kayser. Einstein was not in Berlin for the premiere of *Dreigroschenoper* but attended one of the later performances, persuaded by Margot. He did not enjoy it; in particular he disliked Weil's music. Brecht himself had been interested in relativity since the sensation of the 1919 eclipse results. In 1931, he attended a lecture given by Einstein to an audience of 2,000 under the auspices of the Marxist Workers' School organisation in Berlin. His knowledge of Einstein's work strongly influenced his play *Life of Galileo*, which he wrote in 1938. In contrast to *Dreigroschenoper*, this play gave Einstein great pleasure when he read it, as he told the author in a letter in 1939.

Of the great names of contemporary composition, Richard Strauss had departed Berlin for Vienna, founding the Salzburg Festival in 1919. However, Arnold Schoenberg moved to Berlin in 1926 to take up the directorship of the master class in composition at the Prussian Academy of the Arts that had been left vacant by the death of Ferruccio Busoni in 1924. Schoenberg's pupil, Alban Berg, was overjoyed when the young conductor, Erich Kleiber, heard a piano reduction of his opera *Wozzeck*, which he had been trying to place since its completion in 1922. Kleiber resolved to give its first performance in Berlin, where he had been appointed Director of the State Opera in 1923 at the age of 33. After a record-breaking number of rehearsals, the first performance took place on 14 December 1925. The reviews on the following day in the Berlin newspapers were generally rather positive; although admitting that the work was extremely difficult, several complimented the composer on his work.[5] Einstein would certainly have been invited to the first-night reception by the Austrian ambassador to honour an Austrian composer, as was the cream of Berlin society. The Demann String Quartet entertained the attendees with Schubert. Einstein's friend Erich Kleiber would also have invited him to the performance. Whether he would have attended, given that he would have hated the opera, is another matter, but he was spared the choice since he was in Leiden at the time.

Arnold Schoenberg certainly did attend his pupil's first night. He had written a curious letter to Einstein earlier that year, complaining of not having been put first in a list of German composers produced by Hugo Leichtentritt, which he seemed to ascribe to anti-Semitism. He admitted however that he had only written to Einstein to establish contact so that he could arrange a meeting to discuss Zionism and the establishment of a Jewish state. He hoped for a positive response, but it seems that Einstein did not reply.

[5] A myth has grown up that the first performance saw a riot among the audience. This is reported in *Before the Deluge* by Otto Friedrich, which records fist fights, derisive laughter, boos, and riots. A similar myth grew up around the first performance of Stravinsky's *Rite of Spring* in Paris twelve years earlier. Certainly, there was audible disturbance from some sections of the audience for both works and some critics were vituperative. However, Theodor Adorno, a pupil of Berg recalled that: 'After the Berlin premiere of *Wozzeck* and the dinner at Töpfers where [Berg] was fêted and, like an embarrassed adolescent, scarcely able to respond, I was with him until late into the night, literally consoling him over his success. That a work conceived like *Wozzeck*'s apparitions in the field, a work satisfying Berg's own standards, could please a first-night audience, was incomprehensible to him and struck him as an argument against the opera.' The successful run enjoyed by the opera in Berlin and the fact that it was subsequently taken up by other companies adds weight to Adorno's account.

One contemporary Jewish composer who did impress Einstein was Ernest Bloch. While in Paris in March 1926, he attended a concert by the Croatian cellist Mirko Bjelinski-Waisz in which a piece by Bloch was played. Einstein wrote to Bloch, presumably thanking him for his compositions, which often had a distinctly Jewish inspiration.

In parallel with the radical innovations of Schoenberg, Berg, and Hindemith, the last of whom became a Professor at the Berlin Hochschule der Musik in 1927, another reaction to Romanticism was taking place. At the forefront of a move towards more historically informed performances were Einstein's friends the pianists Artur Schnabel and Rudolf Serkin and Schnabel's regular duo partner, violinist Carl Flesch. Schnabel and Flesch had a long-standing trio with the cellist Hugo Becker, who in 1930 was replaced by Gregor Piatigorsky. Piatigorsky had slept on a park bench in Berlin for some time after his escape from Russia via Poland. Slowly making a name for himself in Berlin, he had made a big impression on Schnabel when they played Schoenberg's *Pierrot Lunaire* together in 1924. Shortly afterwards Piatigorsky became principal cellist of the Berlin Philharmonic. He was the obvious choice to replace Becker, who had also been his teacher, when Schnabel and Flesch decided to revive their trio. Piatigorsky and Einstein became friends..

Piatigorsky's autobiography contains a story from this period:

> Scientists' perception of matter and time can be . . . puzzling. Albert Einstein, whose ardent love of music often brought him into contact with musicians, once asked me to dinner. As I arrived at his apartment at Haberlandstrasse in Berlin he asked me, 'Did you ever see a Japanese violin?' I said, 'No.' 'A Japanese student at the university made one. Would you like to see it?' I said, 'Yes.' 'That's fine. Just wait a minute.' Einstein disappeared. Alone in the apartment and hungry, I waited almost two hours for his return. At last he rushed in, out of breath. Not taking his coat off, he handed me the violin. The coarsely constructed violin was unattractive. While I thought of what to tell him, I felt that his interest in the violin and my opinion of it suddenly disappeared. Relieved, I suggested having dinner at a good restaurant nearby. He ate heartily, spoke of music, and asked me questions, and I, of course, did not attempt to touch his field. Before saying *auf wiedersehen* he asked how I had liked his playing the violin the week before at the home of friends. I remembered too well, but hesitated to answer. 'How did I play?' he repeated. I said, 'Eh, relatively well!'[6]

Fritz Kreisler was another of the great musical names living in Berlin during this period. He had kept a pied à terre apartment in the Kurfürstendamm area of the city since 1913, but in 1924 he moved into a house almost as palatial as that of the von Mendelssohns and quite close to them in Grunewald. It was surrounded by vast swathes of lawn and included a formal garden. Kreisler and his formidable wife Harriet,[7] who organised every detail of his life, were ferried around the city in an enormous Lincoln Cabriolet. Kreisler was also a friend of Einstein and, like Piatigorsky, had

---

[6] The last two sentences in this quotation form an Einstein anecdote to which many names are attached. It is probably apocryphal in the sense that those who recount it would have said it, had they thought of it at the time rather than when writing their memoirs. Even more names are attached to variations on 'Albert, can't you count?' in exasperation at some supposed lapse in Einstein's ensemble playing. Famous names associated with this quotation include Schnabel, Arthur Rubinstein, and Fritz Kreisler.

[7] It is probably Harriet (Plesch's reference is to 'the wife of one of our friends, a great musician') of whom Janos Plesch recalled Einstein saying that she was a creature that he could kill in cold blood. Not only did Harriet regulate every detail of Kreisler's life, but she was also anti-Semitic and maintained that Kreisler, who had received Catholic baptism while studying in France as a youth, was not Jewish in origin. To the statement that Kreisler had not a drop of Jewish blood in his veins, Plesch repeated the remark of a cynical colleague, that Kreisler must in that case have been much more anaemic than he looked.

a fund of stories about him. One he recalled in an interview reported in the New York *Musical Observer*:

K: Nature requires a balance. I love the violin, but I am fascinated by many other things – most of all by books, chiefly incunabula. I was more flattered, not long ago, when a bookseller's paper called me 'the erudite incunabulist' than I have been by many a critical eulogy of my playing.

I: Your friend. Professor Einstein . . . has a similar 'balance', hasn't he, in his violin playing?

K: Yes, he has, and he plays the violin very well, indeed [Kreisler laughs]. But let me tell you something funny about that. Not long ago Dr. Einstein played the violin in aid of a charity benefit at a very small German village. The local newspaper reporter had no idea who the violinist was. When he asked, afterward, he was told: 'Why, that is the world-renowned Einstein.' The reporter, knowing nothing of the author of the theory of relativity, and thinking that he had heard one of the greatest violinists in the world, got out his encyclopaedias, rolled up his sleeves, and described Einstein's playing as one of the greatest marvels in the history of music. 'The other geniuses of the violin,' he wrote 'would have paled last night had they been there to hear the playing of the greatest master of them all.' Einstein was delighted. He treasures the clipping. I think he is prouder of it than he is of all his achievements in the realms of physics, mathematics, and celestial mechanics. He is a delightful and lovely man – a great man.

The story that Piatigorsky told concerning the Japanese violin is probably a somewhat garbled recollection of a visit that Einstein received in November 1926 from two of the sons of Masakichi Suzuki. They had brought with them from Nagoya in Japan three violins that their father had made. Einstein and the Suzukis blind tested the violins and ascertained that they sounded better than either of Einstein's own violins. At the urging of his visitors, Einstein kept the best one and wrote to thank Suzuki Senior for the gift, which he said had given him more pleasure than any from a person not known to him. Indeed, it became for a while the favourite of the several violins that he owned. One of the two sons who brought the violins was Shin'ichi Suzuki, subsequently the famous violin pedagogue and founder of the Suzuki Method. For the rest of his life he treasured the lithograph of the Orlick sketch that Einstein had given him, dedicated 'To Mr Shinichi Suzuki, with very friendly remembrance, November 1926'. Like many who met Einstein, Suzuki's relationship with him grew in his memory, eventually to Einstein being his guardian and taking him to many concerts during his years in Berlin. The letter to his father and the autographed sketch, as well as perhaps Piatigorsky's story, are however the only extant evidence of their relationship.

Chapter 18

Bose–Einstein and quantum mechanics (1924–1927)

Gradually, between his Far Eastern journey and the trip to South America, Einstein sloughed off almost all of his tedious duties as head of his Kaiser Wilhelm Institute. His deputy, Max von Laue, did the job efficiently and with the minimum of interaction with Einstein. Von Laue's brilliant book on special relativity, still a classic, had been the first to bring the subject to the non-specialist, but he had been sharply critical of some of Einstein's ideas and was not in the least intimidated by him. Einstein grew to respect von Laue greatly, granting him the rare honour of 'Duzen'.

Passing on his administrative duties allowed Einstein more time for his own research. He was to write a new chapter in the subject that was his first love, statistical mechanics. The seminal idea came from Satyenda Nath Bose, who had won prizes as a student in several Indian institutions before he became a lecturer in an institute of the University of Calcutta in 1916. He was already interested in relativity and had translated Einstein's 1916 general relativity paper into English. In 1921, he moved to the new University of Dhaka, where he set up major parts of a physics department. He also had a heavy lecturing load, including aspects of thermodynamics. Without easy access to textbooks, he relied on describing phenomena with self-consistent explanations with which he himself felt comfortable. Planck's law of radiation, in which Planck had first postulated an indivisible quantum of radiation, caused Bose a great deal of trouble, since he found neither Planck's original derivation nor subsequent attempts self-consistent. The same feeling concerned others, including Planck himself, Einstein, and Peter Debye, the latter of whom had presented in 1912 a self-consistent derivation, but which relied on classical electromagnetic theory, onto which the quantum idea was bolted. Einstein's own new derivation in his 1916 paper still had to rely on assumptions, even if plausible (and indeed correct) ones.

Bose concentrated on deriving the Planck law from the purely statistical considerations that Boltzmann, drawing on earlier work by Maxwell, had used in the previous century to derive the laws of thermodynamics. To do so, he treated the photons inside a hot cavity as a gas. Since the properties of a gas are fully defined by the momenta and positions of its constituent particles, Bose realised that for a particular volume of cavity, the number of states could be calculated by implementing the Planck quantum postulate as defining 'cells' of momentum that were proportional to Planck's constant cubed, as Planck himself had pointed out in 1906. The power of three is determined because there are three orthogonal spatial dimensions in which the momentum vector can point. The quantum hypothesis forbids subdivision of such cells. From this basis, Bose was able to derive the constant factors in front of the energy dependence of Planck's law.[1] As he testified at the

[1] In fact, he had to insert an additional factor of two by hand, which in his paper he attributed to the two polarisation states of the photon. Since the whole thrust of his paper was to avoid assumptions of the wave nature of light, which he found a weakness in Debye's 1910 paper on the derivation of the Planck law, he must have found this assumption uncongenial, since how can a particle have a polarisation? The factor of two actually comes from the two possible directions of the photon's intrinsic spin. However, the concept

Einstein. Brian Foster, Oxford University Press. © Brian Foster (2026). DOI: 10.1093/oso/9780198794875.003.0018

end of a long and distinguished career in scientific administration in India, he then thought that he was home and dry. He applied Boltzmann's counting ideas to the contents of the momentum cells and obtained the Planck law. However, in applying Boltzmann's statistics, he had implicitly assumed that the objects in the cells were identical and indistinguishable. In contrast, Boltzmann assumed that all entities were at least in principle distinguishable. Had he followed Boltzmann's prescription exactly, Bose would have obtained the incorrect Wien law. In his subsequent recollections, he admitted that he had not realised that he was doing anything novel; in fact, he was opening up a new branch of physics.

For a very simple case, it is easy to see why two distinguishable particles have different behaviours from two that are indistinguishable. Imagine that each of two distinguishable particles, which can therefore be labelled 'A' and 'B', can populate two large boxes. There are four different possibilities: Box 1 has A *and* B; Box 2 has A *and* B; Box 1 has A, Box 2 has B; and Box 1 has B, Box 2 has A. This is the situation assumed in classical Maxwell–Boltzmann statistics. If the particles are indistinguishable, that is, both are labelled 'A', there are only three possibilities: Box 1 has AA, Box 2 has AA, or each box has one A. This corresponds to Bose's assumption. If the boxes are considered to be energy states, with one higher than the other, then under the Maxwell–Boltzmann assumption the higher energy state is occupied 3/4 of the time, whereas for Bose's assumption of indistinguishable particles, it decreases to 2/3. The more particles are considered, the larger becomes the tendency for them to 'bunch together' in lower-energy 'boxes'. It is this tendency that contributes to the reduction in probability for higher photon energies in black-body radiation, characteristic of the Planck law and agreeing with the data.[2]

In March 1924, Bose composed a brief paper outlining these ideas and sent it to the destination of choice for Indian physicists, the *Philosophical Magazine* in London. It is unclear whether his paper was rejected or simply that the review took a long time. In any case, Bose became anxious that it would be rejected and decided to try a different route. He wrote to Einstein, excusing himself by asserting that, although he had never met Einstein, he considered him to be his teacher as he was the teacher of all physicists. Bose asked for his help in translating his paper into German so that it could be published in the *Proceedings of the Prussian Academy*. As soon as Einstein read the paper, he realised that it was important and should be published, although it took some time before he realised quite how important it was. Nevertheless, he immediately saw how Bose's ideas could be applied more generally. Specifically, he decided that it must be the key to understanding the behaviour of quantum gases. For decades, the translation of quantum ideas to the behaviour of ideal gases had been delayed by an inability to reproduce some of the basic results of thermodynamics. In particular, the ground-state energies of macroscopic quantum gases were much too low. In a paper that he read to the Academy on 10 July 1924, Einstein applied the Bose method to an ideal gas consisting of single atoms, which was more complicated than the photon case because the number of atoms was fixed. Einstein was able to recover the classical relationships between pressure and the kinetic energy of the atoms and show that the quantum behaviour would

of spin was only first proposed for the electron by Compton in 1921 and subsequently by Pauli in 1925; it was discovered by Uhlenbeck and Goudsmit in 1926.

[2] For completeness, there is a third possibility: each box can only accommodate one particle. Then there is only one configuration for two indistinguishable particles, with each box containing one 'A'. If the particles are distinguishable, then there are two possibilities, Box 1 contains 'A' and Box 2 contains 'B'; and vice versa. The behaviour of indistinguishable particles with half-integer-spin quantum numbers, known as fermions, is described by 'Fermi–Dirac' statistics; systems of such particles will tend to have higher energies than either of the previous possibilities.

manifest itself only at very low temperatures. At such low temperatures, he could demonstrate that entropy would tend to vanish, that is, the gas would become a highly ordered state much more like a liquid or solid. This had been observed experimentally but not previously understood theoretically. This behaviour (known as Nernst's theorem and later in various forms as the third law of thermodynamics) can be qualitatively understood by returning to the box analogy above; more and more identical particles 'bunching' together into the same 'box' clearly leads to a highly uniform system. Indeed, it is this coherent bunching of countless photons that allows them to appear as the macroscopic waves that formed the accepted optical theory for centuries.

Over the next year, Einstein continued to contemplate the puzzling implications of Bose's idea. Conversations with various colleagues, in particular with Paul Ehrenfest, an expert in statistical mechanics and a sceptic about Bose's method, only confirmed his belief that this was a fundamental advance. Bose himself arrived in Berlin in October 1925 after a one-year stay in Paris. He asked to see Einstein immediately. Apparently, Einstein's first question was about how he arrived at the method discussed in the paper that Einstein had translated. Although Einstein introduced him to many of his famous colleagues and proposed two areas on which he could work, Bose achieved little during his stay in Berlin. He returned to India in the summer of 1926, if not enriched by physics achievement, then certainly by his observation of a period in which the quantum world was once again revolutionised by the establishment of quantum mechanics.

At the end of 1925, Einstein wrote a second paper in which he explicitly drew attention to the assumption of indistinguishable particles as at the root of the new statistics initiated by Bose. He went on to point out the implication that as gases cooled, the tendency for atoms to populate the lowest energy state would increase, leading to a situation analogous to that of a liquid-gas phase transition, even though there was no direct force attracting the atoms together. In some sense, the lack of any physical principle *forbidding* the tendency of the atoms to 'flock together' mimics a force of mutual attraction. The possibility to form what is now called a 'Bose–Einstein condensate', which can be considered a new state of matter, was more or less ignored for many years and only experimentally confirmed in 1995 by Eric Cornell, Carl Wieman, and Wolfgang Ketterle, leading to the award to them of the 2001 Nobel Prize in Physics.

Another section of Einstein's paper dealt with the fluctuations in energy of two volumes separated by a membrane through which only molecules near to a particular energy could pass. This was grist to Einstein's thermodynamic mill and just the latest insight he had gleaned by decades of examining such fluctuations for different systems. After straightforward calculations, he found that the formula for the squared average of the fluctuations contained two terms, the first the 'classical' term for independent molecules, together with a second that resembled an 'interference' term between the molecules, similar to the pattern formed on the far side of a barrier containing two holes struck by a water wave. However, only waves that are coherent, in this case in the same volume and with identical frequency, can interfere. Einstein drew the remarkable conclusion that each gas molecule must have an associated wave with a frequency proportional to its energy. The fact that this idea had been prompted by the work of a young French student, Louis, Prince de Broglie,[3] does not make it any the less remarkable or courageous. Just as he had pronounced decades before that light waves were particles, he now seized on a symmetry that greatly pleased him – that radiation and matter were merely different manifestations of a single entity: waves had

[3] De Broglie was the younger brother of Maurice, Duc de Broglie, who was a distinguished experimental physicist known to Einstein. Maurice had been one of the secretaries, along with Lindemann, at the first Solvay Conference.

particle properties and particles could be waves. Although this 'wave–particle duality', which has a central role in quantum phenomena, is a synthesis of the work of many people, a case can be made that it should be mostly credited to Einstein, whose insight 'at both ends' of the problem was decisive.

Einstein had been alerted to the work of de Broglie by his old friend Langevin, who was de Broglie's doctoral supervisor. Langevin apparently thought that although de Broglie's work was absurd, so was Bohr's concept of preferred orbits in atoms, so that even the maddest theory shouldn't be rejected out of hand, as it might contain a grain of truth. Langevin sent Einstein a copy of the thesis in the summer of 1924, presumably to be reassured that it wasn't completely mad in advance of de Broglie's oral examination. Einstein read it by December and considered it a remarkable document, writing to Langevin that 'De Broglie's work has made a great impression on me. He has lifted a corner of the great veil [of Nature]'; he also wrote in similarly glowing terms to Lorentz. The importance with which Einstein viewed de Broglie's work can also be deduced from the enthusiastic references he made to it in his second paper on quantum ideal gases discussed above. Einstein warmed to de Broglie's thesis because it also gave a rather straightforward geometrical explanation for the hitherto mysterious Bohr–Sommerfeld selection rules for the interactions of atoms and photons. De Broglie postulated that every particle had associated with it a wave whose frequency was proportional to its momentum. The preferred Bohr orbits in atoms corresponded to those in which the wave associated with the particle could interfere constructively with itself, that is, could form a standing wave, which only occurs for integer multiples of the frequency around the closed orbit.

The work of de Broglie was a major influence on Erwin Schrödinger and the beginnings of what would become quantum mechanics. Before that story can be told however, it is necessary to deal with a bombshell dropped by Niels Bohr early in 1924. Back in 1922, the US physics community had been excited by a report from Arthur Holly Compton, who had been experimenting with scattering X-rays. His data indicated very clearly that the process being observed was a photon striking a bound electron in an atom and losing energy in the process, thereby reducing the photon's frequency. This was impossible to understand in a wave theory interpretation but natural if photons and electrons were scattering as 'real' particles. Einstein thought this result sufficiently important to explain it in an article for the general public in the *Berliner Tageblatt* of 20 April 1924. Bohr however greatly disliked Einstein's photon hypothesis and was firmly in the Planck camp, retaining wave transmission of electromagnetic energy and quantising only the absorption and emission. Compton's results caused great problems for Bohr's efforts to retain a classical wave theory, to the extent that he was forced to give up energy conservation in individual interactions, only retaining it on average over many. This 'BKS theory', named after Bohr and his two collaborators, Hendrik Kramers and John Slater, sparked intense interest in the German physics community. On 28 May, Einstein gave a colloquium on the subject in Berlin, in which he clearly showed his scepticism.

Walther Bothe read the BKS paper and immediately discussed it with his colleague Hans Geiger at Berlin's Physikalisch-Technische Reichsanstalt. They agreed that it should be possible to test the BKS theory experimentally by seeing whether the X-ray photon and the electron it scattered from could be detected simultaneously. On 7 June 1924, possibly stimulated by discussions after Einstein's colloquium, they submitted a brief note to *Zeitschrift für Physik* in which they outlined an experimental test of the BKS theory. The difficulty of the experiment was to measure coincidences between the two particles with sufficient resolution that real coincidences would significantly exceed random ones. It is similar to an observer seeing a pool table fully set up, leaving the room and then returning 30 minutes later. All the balls are in the pockets, but the resolution of the observation is too poor to distinguish between two hypotheses: one that a master player cleared the table at one visit and the other that two complete beginners took hundreds of shots to fluke all the balls into the pockets. An observer who had remained in the room and watched, with a resolution of a fraction of a second, would have no difficulty in distinguishing these hypotheses. The temporal resolution of an experiment in physics in those days was limited by the accuracy with which an observer could press the button of a stopwatch, typically around 0.05 s. Bothe and Geiger calculated that this was at least an order of magnitude too crude, so they devised an ingenious if cumbersome method. The X-ray scattering from the electron was detected in one 'needle counter', a precursor of the Geiger–Müller tube still in use today, while the recoiling electron was detected in another. The light pulses from electrometers connected to the counters were focused onto a film 1.5 cm wide and 75 m long, moving at around 20 cm/sec. The higher the speed the better the resolution but the more film would be needed. In fact, processing the film was the limiting factor; Bothe remarked that during the experiment the amount of film drying in the laboratory gave it the appearance of a laundry. The experiment took many months to complete, The result, published in the spring of 1925, showed that the photon and electron were indeed coincident; the probability of the BKS theory's predictions being correct was only slightly better than one in a million.

Einstein was elated by the confirmation that light was indeed composed of photons. However, another cloud on his quantum horizon was gathering over Göttingen, where Max Born had been joined in the winter of 1922–1923 by a brilliant student, Werner Heisenberg. Heisenberg was a protégé of Arnold Sommerfeld in Munich but spent some months in Göttingen while Sommerfeld was visiting the USA. Just how bright and self-confident Heisenberg was can be judged by the following anecdote. Sommerfeld had previously brought Heisenberg with him to Göttingen in the summer of 1922 to attend lectures given by Niels Bohr. Heisenberg was then a twenty-one-year-old in his fourth semester of study. After a long and complex lecture, Heisenberg put up his hand and challenged Born about an aspect of the Stark effect in hydrogen. Bohr deflected the question but was clearly worried by it, so after the lecture, he approached the callow student and they went for a walk together up the Hainberg, a small hill to the east of the city. Heisenberg characterised this discussion, ranging across physics and philosophy, as one that changed the course of his life. Bohr invited him to come to Copenhagen the following spring.

Given Heisenberg's precocity, it was unsurprising that a few months later, Born characterised him in a letter to Einstein as just as bright as Wolfgang Pauli.[4] Born also thought that Heisenberg was much nicer than the famously sarcastic Pauli,[5] who Born admitted daunted him and whom he could never convince to do anything he was asked. He also remarked that Heisenberg was a very good pianist. Indeed, on his twenty-first birthday, Heisenberg was invited to Born's home where they played together for the first time. They played Mozart and Beethoven concertos on two pianos, with one playing the solo part and the other the piano reduction of the orchestral

---

[4] Pauli had just left Göttingen for a year in Copenhagen with Niels Bohr, which he followed by accepting a position at Hamburg University.

[5] Pauli's reputation for outspokenness began early. Around 1920, when still a research student of Sommerfeld's in Munich, he attended a crowded colloquium. In the discussion afterwards, the visiting Einstein made a remark. Pauli stood up at the back of the room and commented 'You know, what Mr Einstein said is not so stupid.'

score. Heisenberg thought the Beethoven was unbelievably beautiful. After completing his doctorate with Sommerfeld on aspects of turbulent flow in liquids, Heisenberg returned to Göttingen to work with Born as a Privatdozent in 1924.

After two decades of wandering in the quantum wilderness, in which seemingly arbitrary rules had to be conjured up to agree with the experimental evidence from for example the observation of atomic spectra, a pressure was building that new approaches were needed. Einstein himself also felt this. In September 1924, he gave an interview to University of London mathematician and philosopher Thomas Greenwood. In it he conjectured that the standard differential-equation approach to quantum mechanics could not work and that a new type of function would be required. The same idea was occurring to Heisenberg. He first met Einstein in Göttingen just after his return in 1924. During a brief visit by Einstein, he and Heisenberg took a walk together, during which they discussed quantum problems. According to Heisenberg, Einstein 'had a hundred objections' to the BKS theory that Heisenberg had enthusiastically pressed on him.

Only briefly downcast by Einstein's lack of enthusiasm, Heisenberg remained convinced that a new approach was required, even though it turned out not to be the BKS theory. His conviction was reinforced by his spending the next few months in highly complicated calculations of quantum effects in atomic spectroscopy. Discarding these in disgust, he spent six months in Copenhagen from September 1924, working with Bohr on problems of atomic physics. In these months, he greatly enlarged his perspective on physics, as well as enjoying many musical sessions, accompanying the cellist Kramers on the piano. On his return to Göttingen in the spring of 1925, Heisenberg decided that it was best to concentrate on quantities that could be measured rather than abstractions such as the path of electron orbits around nuclei. His inspiration for this came from Einstein and special relativity, with its emphasis on measurement, as he intimated to colleagues on a cycling tour. However, his attempts to apply this philosophy to the hydrogen atom foundered on complicated mathematics. He retreated to the simplest non-trivial system in physics, a one-dimensional harmonic oscillator, such as a weight attached to the end of a spring constrained to move only vertically. When pulled downwards, the weight oscillates around its rest position with a known mathematical form until its energy is dissipated by friction as heat. Planck had fallen back on such systems when trying to understand black-body radiation at the turn of the century. Before he could get down to work on this, Heisenberg suffered a debilitating attack of hay fever. He asked Bohr's permission to flee to the wind blown rocky island of Heligoland in the North Sea on 7 June, where he found not only relief from pollen but more importantly few distractions to contemplation. After a few days he had worked out how to write down a table of quantities that could reproduce the results of the old quantum theory. After a night of continuous work, he proved that his methods conserved energy. He began to draft a paper, returning to Göttingen on 19 June.

Heisenberg first communicated his ideas to Pauli, who was highly excited, and then to Born and their young colleague Jordan. In July 1925, Heisenberg published his harmonic oscillator calculations in a paper entitled 'On the Reinterpretation of the Quantum Theoretical Kinematic and Mechanical Relationships'. Born and Jordan realised that the tables of numbers that Heisenberg had drawn up and then laboriously worked out how to multiply were in fact mathematical entities known as infinite-dimensional matrices.[6] In December 1925, they published 'Towards Quantum Mechanics', in which they outlined the necessary elements of the theory of matrices, in particular that matrix multiplication is non-commutative, that is, a matrix A multiplied by a matrix B does not equal B multiplied by A. They then applied Heisenberg's ideas, together with Bohr's allowed

[6] Matrices are simply ordered sets of numbers that obey particular rules of arithmetic.

orbits, to atomic transitions, writing down for the first time a key result that the difference between A multiplied by B and B multiplied by A for particular combinations of variables was proportional to Planck's constant. This is the key to the application of the new quantum mechanics. They reproduced Heisenberg's results for the harmonic oscillator with their new techniques and then used them to re-express Maxwell's equations for electromagnetism in the new language.[7] The capstone of the arch was placed in August 1926, when all three authors published a paper that extended the treatment to many more systems, deriving results on atomic selection rules and the intensities of atomic spectral transitions, including a key atomic property known as the Zeeman effect. They also applied the new theory to the thermodynamics of cavity radiation, linking as they did so to the work of Debye, Bose, and Einstein described above, and returning full circle to Planck's original inspiration.

While Heisenberg's original paper had been very difficult to understand and had puzzled most of its readers, the two papers discussed above awoke enormous interest. Born was quick to communicate what was being developed, and on 15 July he wrote to Einstein and announced that Heisenberg's new paper, which Born had just sent to *Zeitschrift für Physik*, while somewhat mystical, was certainly both correct and profound. Jordan wrote to Einstein in October about the paper that he, Born, and Heisenberg were preparing. Part of it related to Einstein's beloved thermodynamic energy fluctuations and the derivation of Bose–Einstein statistics; Jordan no doubt felt the need to run their ideas past the greatest of experts in this area, as well as the one likely to be most critical. Indeed, Einstein was unconvinced. In November he wrote to Ehrenfest that 'Heisenberg has laid a great quantum egg. They believe it in Göttingen (I don't).' Jordan showed Einstein's reply to his October letter to Heisenberg, now back in Göttingen. Heisenberg wrote to Einstein on 16 November, responding to some of the criticisms in the letter to Jordan. He went out of his way to appease his hero Einstein, in the final paragraph almost apologising for the whole approach and its various inadequacies but sticking to his basic premise that only quantities that could actually be measured should appear in the theory.

On 2 December 1925, Einstein arrived in Leiden for the celebrations of the fiftieth anniversary of Lorentz's doctorate. He stayed there for almost two weeks, attending the day of celebrations in Leiden and the banquet in Haarlem. During the Leiden celebrations, he delivered a congratulatory message from the Prussian Academy. The genesis of this had caused Einstein considerable irritation since the Academy had insisted on removing sections that praised Lorentz for his efforts to bring European science back together after the war. Clearly even in 1925 there was still sensitivity about the exclusion of German scientists by the rest of Europe. Bohr also attended the Lorentz festivities, staying with Ehrenfest; Einstein and he spent considerable time discussing quantum issues.

---

[7] Heisenberg is not an author of this paper because Born and Jordan had been working on a mammoth paper on atomic spectra for some considerable time before Heisenberg left for Heligoland. Once Heisenberg returned and passed on his results to Born, he and Jordan immediately began to rewrite their paper on this new basis. Heisenberg could not participate because he no sooner arrived in Göttingen then he left again, first visiting Ehrenfest in Leiden and then travelling to Cambridge to see Ralph Fowler, Rutherford's son-in-law, whom he had befriended in Copenhagen. The peripatetic physicist was just as much a feature of life in the 1920s as 100 years later. Fowler asked for a copy of the proofs of the paper, which Heisenberg sent him on his return to Göttingen. Fowler passed them to his research student, Paul Dirac, who studied the paper over the weekend; by Monday he had established the train of thought that was to lead to his equation, the prediction of antimatter, and for this golden generation, the by-now almost obligatory Nobel Prize.

The year 1926 was probably the most exciting ever to have been a physicist. Whereas the discovery of general relativity had electrified the general population, physicists were less enthusiastic. Once the initial admiration had dissipated, Einstein's theory became rather a backwater studied by a tiny community; it remained so for the best part of the next half-century. The study of the atom was entirely different. It was central to physics and all physicists were intensely interested in developments of quantum theory. In particular the spring of 1926 saw a bewildering number of advances. Bose, who was still in Berlin but increasingly a spectator of the great events rather than an active participant, left a graphic picture of the intellectual ferment in a letter to a friend in May 1926:

> Everybody (every physicist) seems to be quite excited in Berlin, about the way things have been going on with Physics, first on the 28th last Heisenberg spoke in the Colloquium about his theory, then, in the last Colloquium, there was a long lecture on the recent hypothesis of the spinning electron (perhaps you have heard of it). Everybody is quite bewildered, and there is going to be very soon a discussion of Schrödinger's papers. Einstein seems quite excited about it; the other day, coming from the Colloquium, we suddenly found him, jumping into the same compartment, where we were, and forthwith he began to talk excitedly about the things we have just heard. He has to admit that it seems a tremendous thing, consider [sic] the lot of things which these new theories correlate and explain, but he is very much troubled by the unreasonableness of it all. We were all silent, but he talked almost all the time; unconscious of the interest and wonder that he is exciting in the mind of the other passengers.

The Heisenberg lecture that Bose mentions took place on 28 April 1926. For two hours, Heisenberg explained quantum mechanics to many of the greatest physicists in the world. When the flurry of questions after the lecture had died down, Einstein collared Heisenberg and asked him to walk home with him. Once they had arrived in Einstein's tower-room study in Haberlandstrasse, Einstein asked Heisenberg to explain the philosophy of his matrix mechanics, and in particular why he insisted on using only observable quantities. Heisenberg, having been inspired by special relativity, thought that he was on safe ground with its creator. He was therefore dismayed when Einstein queried whether he seriously believed that none but observable magnitudes must go into a physical theory. When Heisenberg defended himself by saying that Einstein had done exactly the same in denying the existence of absolute time and that only clock readings were admissible, Einstein responded that he might well have done that, but it was nonetheless nonsense and that indeed it was the theory that told us what could be measured. Heisenberg was taken aback and before long they were discussing Mach's philosophy, which Einstein, in the wake of general relativity, had by now essentially repudiated. Einstein went on to ask whether the new quantum mechanics supported the concept of a photon being emitted in the de-excitation of atoms. Heisenberg had to admit that he wasn't yet sure. He asserted however that it must be left to experiment to determine if the new quantum mechanics was correct. He was convinced that the simplicity and beauty of the theory, as he saw it, was a strong aspect in its favour. Einstein certainly thought it was neither simple nor beautiful, but he was willing to concede that experiment must have the last word.

Göttingen was not the only centre of the ongoing quantum revolution. Bose in his letter quoted above mentioned 'Schrödinger's papers'. Independently of Heisenberg and much more inspired by de Broglie's work, Erwin Schrödinger developed his own distinct quantum theory, which became known as 'wave mechanics'. Schrödinger was born in Austria in 1887 and received his doctorate from the University of Vienna. After military service, during which he found time to produce two papers on general relativity, he spent periods in Jena, Stuttgart, and Breslau before taking up

Einstein's old chair at the University of Zurich in 1921. His discussions with Einstein about Bose–Einstein statistics led him to increasingly deep absorption in the problems of quantum theory. In March 1926, Schrödinger published his first paper[8] on wave mechanics. Einstein's reaction was to turn to Schrödinger as a saviour to lead him out of the Slough of Despond of matrix mechanics into which he felt the 'Göttingen trio' were dragging him. He wrote to Ehrenfest that Born and Heisenberg were wrong after all and that Schrödinger's idea that de Broglie waves accompanied particles such as electrons was much clearer and on a better physical foundation. However, he was concerned that the waves were not propagating in space and time, like electromagnetic waves did, but rather in an entity's 'configuration space', a mathematical description of the totality of its independent parameters, such as position and momentum. It was entirely unclear what the physical meaning of such waves was, but using the mathematics he had developed, Schrödinger was able to derive what became known as the 'Schrödinger equation' and to use it to predict the energy levels of the hydrogen atom. In essence, what Schrödinger did was to work out the equations implied by the result of de Broglie that his matter waves could give electron states in the atom that did not radiate energy. Despite his worries about the interpretation of the Schrödinger waves, Einstein wrote to the author to congratulate him.

Heisenberg, whose parental home was in Munich, heard that Schrödinger was giving a lecture there in the early summer of 1926. Although impressed by the ease with which Schrödinger's equation allowed the calculation of the hydrogen atom's energy levels, Heisenberg strongly disagreed with Schrödinger's interpretation of the characteristics of the matter waves. When he questioned Schrödinger after the lecture, he was more or less shouted down by Wien, the hugely distinguished head of experimental physics. Discouraged, he told Bohr of his experience, who decided to invited Schrödinger to visit Copenhagen in September. Schrödinger shared Einstein's, and Wien's, distaste for the sudden 'quantum jumps' that matrix mechanics implied took place in transitions between atomic states. His interpretation of wave mechanics involved a gradual transition as the matter waves associated with the initial and final states interfered. Bohr and Heisenberg countered that this was incompatible with the distinct quanta of energy required by Planck more than twenty years previously to explain black-body radiation. No meeting of minds took place, but Schrödinger left with much food for thought.

Later in 1926, Born, who according to Heisenberg had himself almost discovered the Schrödinger picture of quantum phenomena, worked out that the two theories gave equivalent answers by considering the scattering of electrons from an object. The key was to interpret Schrödinger's waves as being related to the probability of a particular occurrence. Suitably constructed so that the probability of any phenomenon could not exceed one, known as normalisation, the square of Schrödinger's wave function gives the probability for a particular process. When interpreted in this way, Schrödinger's theory coincided perfectly with the predictions of the Göttingen matrix mechanics. Schrödinger himself published a paper in the spring of 1926 showing that the wave and particle pictures were mathematically equivalent.[9]

Heisenberg had been appointed to a university position in Copenhagen as Bohr's assistant in May 1926. They discussed the interpretation of quantum mechanics for months, usually late at night, so that by the end of the year, both were exhausted. In February 1927, Bohr decided that he

[8] The paper was said to have been drafted while Schrödinger, whose marital and sexual life made Einstein's seem a model of rectitude, was on a skiing vacation with one of his many mistresses.

[9] In fact, Schrödinger's paper did not fully show the equivalence and it was left to John von Neumann to publish the definitive proof in his book *Mathematical Foundations of Quantum Mechanics* in 1932.

needed a holiday and departed to ski in Norway, leaving Heisenberg alone. Once more the solitude led to great advances. One evening after midnight, Heisenberg suddenly recalled Einstein's view after his Berlin colloquium that it was the theory that decided what could be observed. As he walked through Faelled Park, he realised that the macroscopic phenomena that can be observed, such as the track that an electron makes in a cloud chamber, is not in fact the trajectory of the electron but rather the result of its interaction with other objects, such as the water molecules around which the visible droplets of water condense. These collisions change the path of the electron; therefore they disturb the properties of the object being examined. Such a perturbation limits the accuracy with which the trajectory can be reconstructed, since it can only be an average over a multitude of microscopic collisions. When he returned to his office, Heisenberg was able to put these thoughts into mathematical form. A more exact localisation in space requires more and more interactions with the particle, which in turn mean that its direction and speed become more and more uncertain.

On 23 February 1927, Heisenberg wrote a fourteen-page letter to his friend Pauli in which he set forth his ideas in a series of numbered points. He began by asking what was meant by the 'position' and 'orbit' of an electron. His first point contained the by-now classic illustration of quantum uncertainty, which became known as the 'gamma-ray microscope'. If the position of an electron is to be determined with a given precision, then it must be observed with light of wavelength than can resolve to that accuracy. Classical optics implies that the shorter the wavelength, the better the resolution, but shorter wavelengths implies higher-energy photons so that when the photon hits the electron under investigation, its momentum is changed. This means that it is impossible to determine the position and the momentum of a particle simultaneously to arbitrary precision. Pauli's reply to Heisenberg's thoughts was positive. A few days later Heisenberg sent him the draft of his paper 'On the Observable Content of Quantum Theoretical Kinematics and Mechanics', whose content was very similar to his earlier letter. Both included the proof that the uncertainty principle (that the product of the uncertainty on the position of a quantum object and its momentum must exceed Planck's constant) was a consequence of Heisenberg's earlier discovery in quantum mechanics that the momentum multiplied by position was not equal to position multiplied by momentum.

In fact, Heisenberg's thought experiment of the 'gamma-ray microscope' contains a clear contradiction. In discussing the effect of light on an electron, Heisenberg was implicitly using both the wave and particle picture of quantum objects. When Bohr returned to the Institute, he immediately felt uneasy about this aspect of the paper, which Heisenberg had submitted to *Zeitschrift für Physik* towards the end of March. Bohr was certain that his idea of complementarity was more central than Heisenberg's uncertainty principle. The key to their disagreement was that Bohr thought that complementarity meant that it was necessary to consider both wave and particle interpretations simultaneously to solve a quantum problem. Heisenberg on the other hand was convinced that these were simply alternative perspectives on the same underlying physics and that everything could be solved within one, preferably his matrix mechanics, approach. After a period of considerable conflict, which Heisenberg admitted at one point caused him to burst into tears,[10] Heisenberg admitted that Bohr's criticism of his microscope analogy was valid and that in a full

[10] The uncertainty principle paper illustrates not only Heisenberg's symbiotic relationship with Pauli but also his desire for the approval of the two people he most admired in physics: Bohr and Einstein. It is important to remember that Heisenberg, although at the forefront of quantum physics, was still only twenty-six years old. He knew that Einstein did not approve of quantum mechanics after their conversation in Berlin

treatment of it, both wave and particle aspects needed to be considered. Nevertheless, Bohr agreed that Heisenberg should add a postscript to the proofs that gave credit to Bohr's point of view without changing the thrust of the paper at all.[11]

Bohr was able to summarise his views on complementarity at a conference on the shore of Lake Como to celebrate the centenary of Alessandro Volta in September 1927. Rising after a rather obscure Bohr lecture,[12] Heisenberg humbly acknowledged his indebtedness to Bohr for his efforts to elucidate the uncertainty principle. Now Heisenberg was once again in public harmony with one of his two physics idols, although in private they still disagreed about what actually happened inside the atom. Indeed, it is likely that Bohr's lecture was particularly difficult to follow because he was at pains to obscure the real differences between Heisenberg's interpretation and his own. However, neither a public nor private meeting of minds with the other hero, Einstein, would prove possible. Einstein had presented a paper to the Prussian Academy on 5 May that attempted to show that exact particle trajectories were compatible with Schrödinger's Equation, but several weeks later he pulled the written version during typesetting after discussions with Walter Bothe and Jakob Grommer revealed flaws in his reasoning. Heisenberg had already heard from Born and Jordan that Einstein was convinced that it was possible to know the track of particles with a precision that violated the uncertainty principle; he wrote to Einstein later in May asking for more information but he does not seem to have received a reply.

A showdown was on the horizon, as the leading figures of world physics gathered in Brussels for the Fifth Solvay Conference, which took place from 24–29 October 1927. Lorentz, the chairman, had originally asked Einstein to report on quantum statistics at the conference, but as the date drew closer, and after the debacle with his Prussian Academy paper, Einstein began to regret his acceptance. On 17 June he wrote to Lorentz that he didn't feel able to give the talk as he had been unable to keep up with the latest developments in quantum theory, whose statistical foundation he anyway did not accept.

The participants stayed at the Hotel Britannique, but the sessions were held in the Institute of Physics that Solvay had founded, which had premises in the Free University of Brussels Institute of Physiology in the Parc Léopold. The photograph of the participants was taken outside this building and shows twenty-eight men and one woman, Marie Curie (see Figure 18.1).

---

referred to above, and that the core of his opposition was to the very indeterminacy that this paper propounded. He went out of his way to mention Einstein, who appears in the fifth line of the paper and then in two successive footnotes. The latter of these could be interpreted as an attempt to co-opt Einstein by citing his work using Bose–Einstein statistics in the quantum theory of gases as some sort of precursor to the uncertainty principle. Such a link is, to say the least, tenuous. In any case, he guaranteed Einstein's opposition by the closing line of the paper: '... thus, via quantum mechanics, the law of causality is definitively shown to be invalid.'

[11] This episode reveals the limits of oral recollections of principals many years after the event. Heisenberg in his interview with Kuhn in 1963 was convinced that this addendum was added before the paper was sent to the journal. However, the existence of proofs that do not contain the addendum makes it clear that it was added after the paper was submitted. These proofs were given to Linus Pauling, at that time a Guggenheim Fellow visiting the great quantum physicists in Europe, by Max Born in June 1927.

[12] Much later, Einstein encapsulated the difficulty in understanding Bohr by writing 'When I called Bohr's train of thought musical, what I meant was that the logical approach could only be accessed purely intuitively, as we experience it in music. For truly creative work, it always has to be like this.'

Figure 18.1 Perhaps the largest brain power ever gathered together at one place. Delegates at the 1927 Solvay Conference. Seated in front row, left to right: Irving Langmuir, Max Planck, Marie Curie, Hendrick Lorentz, Einstein, Paul Langevin, Charles-Eugène Guye, Charles T. R. Wilson, Owen Richardson. Middle row, left to right: Paul Debye, Martin Knudsen, William Lawrence Bragg, Hendrick Kramers, Paul Dirac, Arthur Holly Compton, Louis de Broglie, Max Born, Niels Bohr. Back row left to right: August Picard, Émile Henriot, Paul Ehrenfest, Édouard Hersen, Théophile de Donder, Erwin Schrödinger, Jules-Émile Vershaffelt, Wolfgang Pauli, Werner Heisenberg, Ralph Fowler, Léon Brillouin.

The statistics for Nobel Prizes were somewhat more gender favourable, since of the twenty-eight men, sixteen had won or would win the Nobel Prize, whereas Marie Curie had two of her own.

The conference got under way on the morning of Monday 24 October with two sessions. There was a reception on Tuesday morning by the Free University followed by an afternoon session. Wednesday had two full sessions. Heisenberg attended the opera that evening. On Thursday, the participants travelled to Paris to attend the centenary celebrations for the great French expert on waves and optics, Augustin-Jean Fresnel. The sessions resumed in Brussels on Friday afternoon and terminated with a discussion session on Saturday morning. After the conference concluded, the King and Queen of the Belgians hosted a lunch, at which Einstein met Queen Elisabeth for the first time. She was a Wittelsbach, a scion of the royal house of Bavaria and a fellow violinist, who became a life-long friend and chamber-music companion of the physicist. After the closing dinner hosted by Armand Solvay, son of the deceased founder Ernest, Einstein paid a flying visit to his favourite uncle, Caesar Koch, in Liège.

Although every session of the conference was fascinating, it was perhaps the conversations outside the sessions, and the final discussion session on Saturday, that are most interesting. The presentation by de Broglie was given on Tuesday afternoon. It was followed by a lively discussion, in which one of the contributions, from Pauli, called into question the mathematical basis of what de Broglie had presented. Pauli's manner of questioning was typically robust. The contribution of Heisenberg and Born, who both talked, splitting the time between them, took place on Wednesday morning. Although the discussion session directly after the Heisenberg/Born presentation was brief, those outside the formal proceedings were anything but. Both Heisenberg and Ehrenfest left descriptions of these animated conversations. Ehrenfest was squarely on the side of Bohr and had little sympathy for what he saw as Einstein's obstinacy. Indeed, he chided him that his attitude to quantum mechanics was similar to those who had opposed relativity. At one point during Compton's report to the conference, he noticed Einstein chuckling. He passed him a note saying 'Don't laugh: there is a special department of Purgatory for 'Professors of Quantum Theory' in which they are lectured to on classical physics for ten hours per day.' Einstein replied 'I am only laughing at their naivety. We will see who will be laughing in a few years time.' Ehrenfest told his Leiden students that every morning Einstein would arrive at the breakfast table with a new thought experiment that he claimed invalidated the uncertainty principle. Heisenberg also recalled the morning challenge. Einstein would pose the question, he, Heisenberg and Bohr would discuss it on the way to the formal sessions, the 'Copenhageners' would discuss it over lunch and by supper time, Bohr would present Einstein with the answer. The following morning, the whole process began again. The standard interpretation is that Einstein was defeated in these discussions and what soon became the 'Copenhagen interpretation' of quantum mechanics triumphed. Certainly, this was Heisenberg's narrative, as he wrote to his parents just before leaving Brussels. However, he was hardly a dispassionate observer and a careful examination of what was actually written down at the time, rather than recollected many years afterwards, tells a somewhat different story.[13]

In the discussion session on the final morning of the conference, Einstein outlined another thought experiment. An electron passes through a slit sufficiently narrow that its de Broglie wave properties cause it to behave analogously to light, so that it diffracts into a circular wave front. In front of the slit, a hemispherical photographic film sensitive to the passage of an electron is placed far away from the slit so that the circular wave emanating from it would arrive at all points on its surface simultaneously. If the wave function is a complete description of the electron, as the Copenhageners maintained, and the square of the wavefunction is proportional to the probability of the electron's being detected, why cannot two or more signals be detected simultaneously on the plate, since the wave function is non-zero everywhere on the film? Since this does not happen in experiments, then there are only two possibilities: either quantum mechanics is non-local, that is, some sort of action-at-a-distance must tell the other possible electron signals that the wave function has already 'collapsed', that is, given a signal at a particular point; or, quantum mechanics is incomplete, and in addition to the wave function, some 'guiding wave' or similar phenomenon must also accompany the wave function, which determines the position of the electron signal on the film.

In the discussion that followed Einstein's intervention, no answer was given, except a rather vague statement from Dirac that quantum mechanics and the relativity principle were incompatible. Not only is this vague, but also it is somewhat surprising since in the following year the speaker

[13] The following account follows that of Guido Bacciagaluppi and Antony Valentini.

was to produce a relativistic formulation of quantum mechanics. It seems that none of the other participants really understood the point that Einstein was trying to make, and which he continued to make for the rest of his life, most famously in the Einstein–Podolsky–Rosen paper (see Chapter 27). The reports of the other participants, mostly written long after the event, recall only Einstein trying to circumvent the uncertainty principle. In fact, as demonstrated in the proceedings of the conference written essentially contemporaneously, one of his concerns was a much deeper one. This concern is one that, to this day, physicists and philosophers continue to grapple with and which has generated several entirely new major fields of study, including the philosophy of quantum mechanics, quantum information theory, quantum entanglement etc. Although Ehrenfest said at the time that he was ashamed of Einstein, perhaps he, and the Copenhageners, should have listened to him more carefully. It may be that they were all tired of hearing Einstein's once and future mantra, 'God does not play at dice,'[14] to which Heisenberg recalls Bohr replying 'But still, it cannot be for us to tell God, how he is to run the world.'

The year 1927 not only saw a revolution in physics but a revolution, in the sense of a circular movement, of physicists. Within a few months of each other at the end of 1926, both full professors in Leipzig died. On 1 October, Max Planck would retire from his chair at what is now the Humboldt University of Berlin at the statutory age of 67, after almost forty years in post. At the same time, several other senior physicists were due to retire. In the summer of 1927, Peter Debye accepted one of the chairs at Leipzig, leaving a vacancy at ETH, Zurich. The appointment of Erwin Schrödinger to succeed Planck left a vacancy at the University of Zurich. Although Sommerfeld was the major power broker, Zangger made sure that Einstein was involved in both of the Zurich vacancies, where he strongly recommended the famously astringent Pauli for the ETH position, despite the opposition of the faculty there. It is notable that Einstein's ranking of the various names that Zangger suggested shows both the extent of his familiarity with the contemporary physics scene and the excellence of his judgement of ability. By the end of this game of musical academic chairs, almost all the Young Turks of quantum mechanics were established in prestigious full professorships. Heisenberg joined Debye at Leipzig, Pauli went to ETH, Gregor Wentzel to the University of Zurich, and Oskar Klein replaced Heisenberg as Bohr's assistant in Copenhagen. Now all of equal academic status, Heisenberg, Bohr, and Pauli began to use the intimate *Du* pronoun of address with each other.[15]

Before leaving Einstein's activities in quantum mechanics in this period, it is necessary to mention an unfortunate and rather unsavoury episode involving the experimental physicist Emil Rupp, who had been a student of Philipp Lenard, Einstein's old opponent, at Heidelberg. Rupp obtained a PhD in 1920 and then in 1926 submitted his Habilitation. This thesis was on the subject of the light emitted by 'canal rays'. These are atoms or ions (atoms with missing electrons) accelerated in a partially evacuated cylinder between two metallic contacts at a high voltage difference. Rupp's Habilitation was published and caused quite a stir. It was brought to Einstein's attention, and he was offered a position as an assistant to Robert Pohl at Göttingen. However, others expressed doubts in print about Rupp's results, which to many seemed too good to be true.

---

[14] The first time that Einstein used this form of words was in a letter to Max Born of 4 December 1926, in which he wrote: 'Quantum mechanics very much demands attention. But an inner voice tells me that it is not the genuine article. The theory offers a lot, but it hardly brings us closer to the Old One. At any rate, I am convinced that *he* does not play dice' (Einstein's italics).

[15] Bohr normally wrote to Pauli in Danish, which Pauli had learned when working with Bohr, but as it happens the Danish intimate pronoun is also 'Du'.

Einstein had an experiment in mind, for which Rupp seemed to have precisely the right expertise. Einstein published his idea, which was aimed at distinguishing the abrupt emission of a photon from the classical continuous emission of a light ray, in the prestigious journal *Die Naturwissenschaften* in March 1926. When he saw Rupp's results, he contacted him to see if he would be interested in carrying out his experiment. The basic setup was very simple. The atoms in the canal rays emit light which can be focussed by a lens onto a grid of alternating wires and holes. Einstein thought that if the photon hypothesis were correct, then the presence of the grid would be irrelevant. If the classical theory were correct, then the interference fringes of the light as they were observed behind the grid would be more visible in some places than others, as parts of the light would be blocked by the metallic grid. Rupp set to work enthusiastically, spurred on by a blizzard of communications between Einstein and himself as he reported back his progress almost on a daily basis. Alarm bells might have started to ring for Einstein when he would query some disposition of Rupp's apparatus that could not have worked, to which Rupp would reply that he had made a mistake in describing it. It also turned out that the results in Rupp's Habilitation thesis could only have been obtained had he introduced a rotatable mirror, which he had not mentioned in the text, and even worse, that his observations supported the classical theory! Einstein seems to have assumed that Rupp must have had the mirror and 'unconsciously' rotated it, and anyway he had changed his opinion about his original proposal and now expected that the classical result would be observed. He told Rupp this before he reported his next set of results – which confirmed the classical result. However, they did so in a way that made no theoretical sense. When Einstein pointed out the various problems, Rupp two days later sent a revised set of data that answered Einstein's points in every respect, although he claimed to have taken the data before receiving Einstein's letter. Einstein continued to find further inconsistencies, met each time with admissions by Rupp to unfortunate errors in transcription.

The exchange of letters and admission of errors went on for most of May 1926. It might have been expected that Einstein would by now have become suspicious, but in fact when Rupp finally came up with a self-consistent set of measurements and data points on 3 June, Einstein was overjoyed and wrote that the results represented a convincing confirmation of the theory. Einstein and Rupp published linked articles on the results and their theoretical interpretation in the Proceedings of the Prussian Academy. The reprints were bound together with a cover page displaying both titles. Rupp's results were quoted in succeeding years by figures no less eminent than von Laue and Heisenberg. Einstein and Rupp were still exchanging letters, though at a much-reduced frequency, in 1930.

However, Nemesis was at hand. In 1930, a student in the Munich group had tried to repeat Rupp's experiment, with null result. Indeed, the results showed that Rupp's result could not have been obtained with the apparatus he had used. After an almost unbelievable farrago of chicanery on Rupp's part, many other suspicious results surfaced and eventually he was dismissed from his position with the AEG firm – actually he was dismissed twice, since he was rehired on the basis of another set of fraudulent results – and in 1935 the German Physical Society went so far as to publish a notice that no references were to be made to any of Rupp's papers and no further papers from him would be accepted for publication.

Jeroen van Dongen, on whose researches the above account is based, believes that Einstein's behaviour can be understood by a combination of several factors. Firstly, he had no reason initially to believe that Rupp was a fraud. Secondly, he tended to stop looking once he had obtained the result he expected, as indeed was earlier remarked in his experiments with Wander de Haas on the electron's magnetic moment. Finally, he now looked more and more to mathematics for inspiration rather than experimental physics, so that he did not feel particularly motivated to ensure Rupp's results were correct. Whatever the reasons, Einstein's reputation suffered scarcely a blip; he was, after all, guilty of nothing more than some doubtful judgement. Rupp, on the other hand, was destroyed as a research scientist. However, remarkably, Rupp ended his life as a respected Institute Director of Graphical Technology in the German Democratic Republic and a Professor in what is now the Technical University of Chemnitz. He died in Leipzig in 1979. Apparently the 'ground zero' of 1945 had wiped the slate clean.

In addition to his research, Einstein continued to explain science to the general public and to maintain his interest in philosophy. In March 1924, he wrote a beautiful account of the role of geometry in physics for the prestigious literary and cultural quarterly *Die Neue Rundschau*. It appeared in the January 1925 edition and represents one of the clearest accounts even written of the basis of geometry and its role in physics. It is particularly strong, unsurprisingly given its author, in the interplay of mathematics and physics in this development. In September 1924, he wrote a review of Alfred Elsbach's *Kant und Einstein* for the *Deutsche Literaturzeitung*, a well-respected literary journal founded in 1880 and relaunched with this volume, casting its net wide over literature and science; following Einstein's review was one by Max von Laue of a book on the Bohr atom. Einstein's article is a long and extremely learned review in which he demonstrated his considerable philosophical acumen. Also more philosophical in nature than physics was Einstein's lecture at the annual gathering of the Swiss Society of Scientific Researchers in 1924, in Luzern. Entitled 'On the Aether', the lecture points out that, although special relativity killed the 'luminiferous aether', the gravitational field of general relativity was in some sense an aether. With a confidence entirely justified by events, Einstein opined that physicists would always want to retain the fundamental aspects of general relativity.

In 1927, the UK commemorated the bicentenary of Sir Isaac Newton's death with a variety of festivities. To one of these, Einstein sent a communication whose central message was a plea to re-establish the causality that both Newton and Einstein cherished, and which was under attack from quantum mechanics. In March 1927, he wrote two tributes to his great predecessor in the form of articles entitled 'Newton's Mechanics and its Influence on the Development of Theoretical Physics' and 'On the Two-Hundredth Anniversary of the Death of Isaac Newton', in which with wonderful clarity he extolled Newton's insight. Knowing Kepler's laws and the value of the acceleration due to gravity on the earth's surface, and armed with only a rough idea of the distance from the earth to the moon, he was able to construct the universal inverse-square law of gravity that had stood for two centuries and was still accurate enough for all but the most extreme situations. Einstein stood in awe of this astonishing achievement.

Einstein's work in other areas of physics in this period was not extensive and mostly directed towards unifying gravity and electromagnetism. In a contribution to the *Festschrift* for the 1925 Lorentz celebrations, he failed to describe both electron and a positively charged particle with the same mass in a unified field theory. This led him to the conclusion that there was no further point in trying to unify gravity and electromagnetism. If only he had taken his own advice, he would have spared himself twenty-five years of nugatory toil. His difficulties with electricity continued in answer to a letter from George Rainich in April 1926. Here Einstein gave up the possibility of treating electrons as point-like singularities and posited that electrons in motion could only be considered by treating them as continuous distributions of charge.

In 1926, Einstein again became interested in the Kaluza theory, which aimed to unify the electromagnetic and gravitational interactions by expanding the number of spatial dimensions. This was catalysed by Ehrenfest, who had met Oskar Klein on a visit to Bohr in Copenhagen. Klein wrote to Einstein in September 1926 and sent him an offprint of a paper in which he set out a theory rather similar but distinct to that of Theodor Kaluza. Early in 1927, Einstein presented two papers to the

Prussian Academy on developments of Kaluza's theory. In the written version of the second, he admitted that there was nothing new in his lectures, since it was all contained in Klein's paper. The obvious conclusion would have been that he had not read the paper Klein had sent him. However, he had written to Ehrenfest a few days after it arrived, remarking that he had found it impressive. Something similar happened with respect to Einstein's paper with Jakob Grommer, which was published in the *Proceedings of the Prussian Academy* in February 1927. Hermann Weyl wrote to Einstein to point out that he had already set out the main ideas of Einstein and Grommer's paper in the 1923 fifth edition of his famous book *Space, Time and Matter*.[16] These events give the distinct impression that Einstein was becoming semi-detached from the physics literature, reading without remembering much of what he read, and operating in a way increasingly independent of mainstream thought and publications.

Having gratefully relinquished day-to-day management of the Kaiser Wilhelm Society Institute of Physics to von Laue, Einstein found himself increasingly, and acrimoniously, involved with another sort of academic politics. His involvement with Zionism in the 1920s had somewhat diminished, although he was still in close touch with Zionist leader Chaim Weizmann. Indeed, he was often used by the German government as an informal go-between with Weizmann. Sometimes, Einstein was pleased to find that music and Jewish causes combined. Early in 1927, he accepted the honorary presidency of a Committee in Jerusalem to promote Jewish music in Palestine, instigated by David Schor of the Moscow Academy of Music, who was the driving force to establish schools of music in Palestine. In December 1927, he played in a Beethoven piano trio with his friends Francesco von Mendelssohn and the celebrated Chopin interpreter Bruno Eisner at a sold-out concert for the benefit of the Fund for Aged Jews in Berlin (Figure 18.2); also performing was the famous contralto Sigrid Onegin. However, most of his energy for Jewish causes was poured into his favourite – the foundation and development of the Hebrew University of Jerusalem.

In January, 1924, Einstein accepted the invitation to join the board of Governors of the University. Judah Magnes, who had been appointed Chancellor of the university, was a trained Rabbi, born in San Francisco in 1877, who had emigrated to Palestine in 1922. Together with Weizmann, he could lay claim to the original suggestion to the Eleventh Zionist Congress in 1913 to found the university. He was influential with rich American supporters of the Hebrew University and had corresponded with Einstein during his New York visit in 1921. Magnes had suggested organising a meeting with New York intellectuals interested in the Hebrew University. When Einstein endorsed the idea provided that it could also be used to raise funds, Magnes demurred with a rather brief and frosty letter. Thus, their relationship did not begin auspiciously and was to deteriorate further.

As soon as he arrived in Palestine, Magnes joined the University's Jerusalem Committee and quickly became the local driving force for its establishment. He was an undistinguished scholar but a highly capable academic administrator and politician, which made him the diametric opposite of Einstein. The fact that both were committed pacifists did nothing to stop an all-out battle for the direction of the Hebrew University breaking out between them.

The Hebrew University was formally dedicated on 1 April 1925, while Einstein was in Argentina. It was after the second meeting of the Board of Governors, held in Munich from 23–24 September 1925 and chaired by Einstein, that the conflict between Einstein and Magnes began in earnest.

[16] Weyl wrote a paper developing these ideas in 1929 (see Chapter 19), which is widely credited as marking the foundation of modern gauge theories, such as the standard model of particle physics.

Figure 18.2 Einstein rehearsing piano trios, probably in December 1927. From left to right: Francesco von Mendelssohn, Bruno Eisner, Einstein.

The root of this conflict lay in the very different conceptions that Einstein and Magnes had of their roles. For Einstein, Magnes was a scientific bureaucrat of the type that he was familiar with in Berlin, suitable perhaps to manage but only in agreement with a strategy set by the eminent scientists of the Board of Governors. As far as Magnes was concerned, a committee could only advise on a general direction and should leave the management of the new organisation to those who understood and could influence the actual situation on the ground in Jerusalem. Since each had elements of right on his side, the stage was set for a bitter struggle.

Magnes fired the first shot and, to give him credit, it left Einstein reeling with its sheer effrontery. The Board of Governors Meeting, having been held in Europe, was overwhelmingly attended by Europeans, with the influential Americans, who had the financial resources, relying on the minutes to understand what had been discussed and agreed. Realising the importance of this and wishing to exploit his great influence with his fellow Americans, Magnes decided to issue a revised version of the minutes that had already been circulated. Although the great majority of the text was identical, there were many crucial alternations to decisions on the role and organisation of the Board of Governors and the other university organs. For example, in Magnes's version, the administrative seat of the Board was not defined, whereas in the original version it was to be in London; according to Magnes, it was to meet only once a year rather than twice. Einstein's outrage and astonishment is evident from the letter that he sent to Magnes on 29 December 1925.

Magnes' response to Einstein's letter was insouciant. He claimed that he had agreed with Leo Kohn, the secretary of the Board of Governors, that he would edit and then circulate the minutes.

He insisted that Solomon Ginzberg, based in Palestine, was in fact the 'official' secretary of the Board and had had a contract as such since April 1925. In any case, he proposed to compare the two versions of the minutes and send a detailed commentary around for discussion. A few days later, Magnes further muddied the waters by writing to Einstein proposing a wholesale reconstruction of the Board, which would then be able to decide on which version of the minutes it preferred. Einstein, irate and outmanoeuvred, thrashed around in correspondence with Kohn and with Weizmann, who had been elected President of the Board of Governors *in absentia* at the Munich meeting. In March 1926, Einstein lost patience with Magnes's manoeuvres and refused to communicate with him further, saying that he would leave the decision on the minutes to the next Board meeting.

Weizmann was torn between his understanding that an energetic administrator like Magnes was essential, particularly since he was adept at raising funds from rich Americans, and his sympathy, as a distinguished academic himself, with Einstein's view that Magnes was grossly incompetent in academic matters. He visited Jerusalem in spring 1926 and expressed his worries in a letter to Einstein. He wrote that Magnes was leading the Hebrew University into strange directions, the only scientists so far appointed were of low quality, and Magnes's dictatorship must be broken. However, Weizmann saw no immediate way to get rid of him, since the university desperately needed funds only likely to be forthcoming from rich Americans. Of these, by far the most influential in the affairs of the university was Felix Warburg, scion of the Hamburg banking family and President of the American Friends of the Hebrew University. He had known Magnes for around twenty years and had defended him in earlier controversies. With Warburg,[17] Magnes's position was unassailable. Before long, Einstein was threatening to withdraw from all involvement with the university. This was certainly heavy artillery, as both Weizmann and Magnes were aware of the prestige that Einstein brought to the fledgling academic entity and were desperate to retain him. However, this threat soon became overused, and its potency diminished. Magnes came to Geneva in July 1926 to try to mend his fences with Einstein, who however refused to discuss the minutes dispute, although he did discuss what he saw as the future of the university. In the summer of 1927, Kohn wrote to Einstein saying that Magnes was essentially ignoring the Board of Governors and going in his own direction, making bizarre academic appointments. Einstein once again wrote thunderingly to Kohn that if the Board accepted Magnes's recommendations, he would not only resign from any involvement with the university but also make it publicly known that he considered it utterly compromised as a serious academic institution.

In additional to the personal antipathy, which was clearly strong, there were deeper reasons for the struggle. Einstein and Weizmann wished the Hebrew University to develop along European, and specifically German and British, lines, respectively, with an academic head equivalent to a British Vice-Chancellor, rather than a US model favoured by Magnes.[18] There was also tension as to whether or not the new university should be purely a research organisation or should begin

[17] Although generally exasperated with rich Americans, Einstein had considerable respect for Warburg and maintained cordial relations with him in spite of his support for Magnes. However, he did consider Warburg a bad musician. A violinist, Warburg owned three Stradivari violins.

[18] In UK universities, the Vice-Chancellor is the head, to whom an administrative officer is answerable. The Chancellor, nominally superior, is in practice a purely honorific post. In US universities, the head is more likely not to be an academic, but a public figure much more involved with fund raising, to whom an academic leader answers. Einstein's model is set out in some detail in a letter he sent to the Board of Governors in advance of their meeting in June 1928, which he was unable to attend because of ill health.

to teach undergraduates. In addition, Magnes and Weizmann had significant political differences. Magnes was an early advocate for what would now roughly approximate to a 'two-state' solution for the future of Palestine, which was anathema to Weizmann. This complex conflict, in which Einstein was continually outmanoeuvred, to his exasperation and disillusionment, would rumble on for years to come.

Chapter 19

Illness and recuperation; Fiftieth birthday and unified theories (1927–1929)

At the beginning of June 1927, an article appeared in the *Neue Freie Presse* of Vienna, in which the interviewer asked Einstein a broader than usual range of questions covering the arts and politics. When asked about the importance of cinema,[1] Einstein replied: I believe in moving pictures because they offer the great mass of humanity – those that lead one-dimensional, monotonous lives – colour and an illustration of greater beauty in life.' In response to a question about Expressionism, Einstein said that:

> The Expressionistic movement in modern art and literature demonstrates a pronounced change in the psychology of the world since the war. It is caused by the desolation of these sad times; humanity has turned inwards on itself. Expressionism is a symbol of this and represents a search for comfort from the sorrows of everyday life.

On the influence of music, he said:

> Through music, one can understand the subtleties of thoughts that cannot be expressed in words. Music is the crystallisation of experience. Life surrounds people just as one is surrounded by water when one jumps into the sea. People cannot understand life because it presses in on them too closely. Music lifts us above the surface so that we are able to gain a clearer perspective.

Finally, asked about pacifism, he stated 'I am a pacifist. I believe that unless Europe becomes completely pacifist, it must destroy itself.' Sadly, political developments in Germany were eventually to change his mind on this central tenet of his beliefs.

During the summer of 1927, Einstein made the acquaintance of Catherine and Alexandre Barjansky. Catherine was a sculptor who worked mostly in wax that was subsequently reproduced in ceramic; Alexandre was a virtuoso cellist. They had moved to Berlin so that Catherine could organise an exhibition of her work. She was hoping to make busts of famous Berlin personalities while she was there and of course Einstein was at the top of her list. Shortly after her exhibition opened, she received a letter from Margot, who as a sculptress herself, had presumably visited the exhibition. She asked whether Catherine would be willing to model her father. She eagerly accepted and began sittings. Einstein soon discovered that her husband was a distinguished cellist:

[1] Cinema was at that time undergoing revolutionary changes, both with the development of various colour processes, such as Technicolour, and the advent of sound. The first true 'talkie', *The Jazz Singer*, appeared in October of 1927 but around ten movies where the sound was recorded separately and played synchronously with the film had already appeared. *The Ten Commandments* was released in 1923, with substantial parts in Technicolour, although the great majority of movies continued to be made in black and white until the 1930s.

Einstein. Brian Foster, Oxford University Press. © Brian Foster (2026). DOI: 10.1093/oso/9780198794875.003.0019

> [Einstein] adored music, and we had many musical evenings with him, as we lived only five minutes' distance from the Einstein apartment. He often played with my husband, either at his home or at ours. He played patiently, adoringly, but his taste was ultraconservative, limited to classical music. Once, I remember, he was offered a magnificent rare violin, which he refused, saying such an instrument should belong only to a great musician. Playing was his favorite recreation.

In July, Einstein received an invitation, conveyed by his friend Lord Haldane, to give the prestigious Rhodes Lectures at Oxford. At first, he refused but then came under various pressures, including from the German government, to accept. The government's motive was that his acceptance might lead to the restoration of the right of Germany to nominate Rhodes scholars, which had been suspended after the First World War. These prestigious awards had been instituted in the will of Cecil Rhodes, paying for students across the world to study in Oxford. Einstein's initial refusal had however led to an invitation being extended to the US educator Abraham Flexner, whose Rhodes lectures, published as *Universities: American, English, German* in 1930, led to the formation of the Princeton Institute of Advanced Study. Both this institution and Flexner himself would subsequently play important parts in Einstein's life. Einstein proposed to come to Oxford in 1929 but it was not until 1931 that he eventually delivered the lectures, one year after German Rhodes scholarships had been reinstated.

On 9 July 1927, Einstein arrived in Zurich, where he met Tete, before travelling on to Geneva on 11 July to attend meetings of the International Committee on Intellectual Cooperation (ICIC) Subcommittee on Bibliography. There was a gap of one week before his next set of meetings, which he chose to spend alone in Leukerbad, a spa town above the Rhone Valley beyond Sion. Here he walked and worked, studiously ignoring the rest of the guests, before returning to Geneva on 20 July to attend the ninth session of the ICIC. Before he left again, he wrote to Michele Besso, urging him to meet him in Geneva or, failing that, in Visp so that they could spend a few days together. Besso met him in Visp, from whence they travelled to Saint-Maurice and Berisal, a village above Brig on the road to the Simplon Pass, before Einstein left for Zuoz on 1 August. Margot Einstein was convalescing there from an abdominal complaint. On the following day, they were joined by Tete. They spent a week there, discussing Nietzsche and Tete's impressions of Italy, where he had spent three weeks with Mileva. Then he and his father hiked to Maloja and over the Septimer Pass. During this trip, Einstein was informed of the death of Alfred Karr, his old Zurich friend and financial adviser. Einstein returned to Berlin by the middle of August.

The remainder of 1927 was relatively quiet. Einstein participated in the Solvay Conference in Brussels in October, as discussed in the previous chapter. He also began to correspond with two men who were to play an important part in his life. Alfred Einstein, the distinguished musicologist and critic, had moved to Berlin from Munich in the late summer to take up an appointment as music critic of the *Berliner Tageblatt*. He moved into an apartment about 400 metres away from Einstein's. Given the existence of two A. Einsteins in the same suburb of Berlin, it is unsurprising that occasionally each would open letters meant for the other. They quickly became friends and indeed over the next few years became close. Einstein often addressed him as 'Cousin'. Alfred sent congratulations for Einstein's fiftieth birthday, to which Einstein replied saying how glad he was to have met him and to have him close by. It is unsurprising that they were drawn to each other, given their shared interest in music and the similarity of their childhoods. Both were brought up by non-observant Jewish bourgeois families in Munich. In an autobiographical sketch, Alfred reminisced about his early life in Munich: 'When I reached my ninth or tenth year of age my mother decided that my sister and I ought to take music lessons. My sister studied the piano and I the violin.' With minor alterations, the same passage could have described either Einstein.

Einstein's other correspondent was Leopold Infeld, whom Einstein had first met in 1920 as a Polish student desperate to study at the University of Berlin. In 1927, he was teaching physics at a Jewish secondary school for girls in Warsaw and had sent Einstein a manuscript in the autumn of 1927 on unification of gravity and electromagnetism, although Einstein had to be alerted by a colleague to look for it among the pile of manuscripts he had received. Einstein began to correspond with Infeld, who worked closely with him in Princeton in the years immediately before the Second World War. Together they wrote the famous book *The Evolution of Physics*.

At the beginning of January 1928, Einstein's step-daughter Ilse Kayser travelled to Celerina in the Engadin, where she was treated for what seems to have been a nervous breakdown. She had had a laparotomy the previous December to investigate her chronic stomach complaints and had developed an anaphylactic shock after the operation. The faithful Zangger was consulted and made all the arrangements. Ilse's husband Rudi travelled with her to Switzerland, where they were joined by Margot. In mid-February, Zangger wrote that Ilse was making progress but that her mother must be kept away from her. The consulting doctor was concerned that Ilse wanted to return to Berlin, and that this would destroy her recovery, which would require another four months of treatment. Although the details are unclear, it seems as if the doctors were convinced that Elsa's domineering nature was a major contributory cause of Ilse's psychological disorder.

On 4 February 1928, Hendrik Lorentz died after a short illness. Einstein was distraught at the loss of someone who had been a father figure to him in physics. He revered him both for his science, in particular his theory of the electron, a foundation upon which Einstein constructed special relativity, and for the sweetness and generosity of his nature. Einstein hurried to Leiden by train, arriving on 7 February and taking part in the funeral obsequies two days later. On the following day, he delivered a eulogy in the Main Auditorium of the University of Leiden. This gives ample testimony to his deep respect for and admiration of Lorentz. It ends with a plea to carry on Lorentz's work on the international collaboration of intellectuals. On his return to Berlin, he wrote to Ilse and Margot in Celerina that 'I was in Holland to say farewell to Lorentz. One becomes more solitary to prepare oneself better for the final departure.'

In the same letter, Einstein had indicated that he and Elsa were undecided on whether to visit them in the Engadin. On the one hand, Einstein knew that the doctors were adamantly opposed to his wife's presence. On the other, he was under pressure from Elsa, who was concerned about Ilse's condition; he was also aware that they both needed a holiday. Einstein had remarked on this in a letter to his sister Maja, admitting that both he and Elsa had gained weight and gone completely grey. In any case, he had already accepted an invitation to speak at the newly founded Davos Lecture Courses. Scheduled for mid-March, they were to be held in the valley adjacent to that containing Celerina. A few days later they decided to take a vacation and leave for Zuoz, about 15 kilometres away from Ilsa's clinic. Einstein's friend Wilhelm Meinhardt, chairman of the Osram company, owned a villa there that he placed at their disposal.

Reading of Einstein's appearance at Davos in the newspaper, his old school friend and musical partner Hans Byland invited the Einsteins to visit him and his family. They lived in Chur, conveniently on the route from Zurich to Zuoz. On 23 February, Byland and his family were waiting on the platform to greet their guests, who disembarked from a third-class carriage. Byland's son described the scene: 'My father strode up smiling to Professor Einstein, who at first looked puzzled. Then he unceremoniously let the two suitcases he was carrying fall onto the platform and embraced his old friend.' Once at home it did not take the two long to resume their old pastimes. Byland produced an edition of Biber's 'Mystery Sonatas,' which had only been rediscovered twenty years before. Einstein immediately unpacked his violin, saying that they must play together once more. They did, Einstein playing from sight. This was no mean feat, as not only do the Sonatas contain passages with dense and rapid notes, but also they require *scordatura*, in which the normal tuning of the open violin strings is changed. Even experienced professionals can find this disorienting. Byland's son and daughter, who were twins, seem also to have sung for their distinguished visitors before they departed for Zuoz.

Einstein and Elsa enjoyed a week or so of rest and relaxation, with Elsa visiting Ilse only briefly on a couple of occasions. Early in March, Einstein was called away to give evidence in a patent case between the AEG and Siemens companies. Acting as an expert in patent cases was something Einstein did quite often in this period, for an appropriate fee. He travelled to Leipzig and returned late the same night. There was no taxi available, so he toiled up the steep slope to the Meinhardt villa through the snow. He was apparently carrying a heavy suitcase and the effort strained his heart.[2] There was no immediate effect and for a while he carried on as normal. A few days later, Einstein and Elsa made the short journey to the Grand Hotel Kurhaus in Davos, where Einstein was due to give a keynote lecture at the festive opening of the first Davos Lecture Course. This was designed to allow the many tubercular guests, who were sent to Davos for its invigorating air, to continue or to restart their education.

In the afternoon of 18 March, Einstein gave a wonderful lecture in the great hall of the Grand Hotel to an audience of 700 people. He talked in simple terms for the general public, ranging widely across theoretical physics, *On the Basic Concepts of Physics and Their Most Recent Transformations*. In his talk, Einstein ranged from the Greek philosophers to Newton and how he had established a framework of space and time that stood for centuries. The development of the concept of the field as embodied in the electric and magnetic fields of Faraday and Maxwell led to the slow unravelling of Newton's universe and the establishment of the theory of relativity, first special and subsequently general. Einstein pointed out that only the strict causality of Newton then remained and that was now challenged by the apparently statistical nature of the theory of the atom and its constituents, quantum mechanics. Although he concluded with the hope that causality could be restored through appropriate developments in field theory, he did admit that this was not the view of most of his younger colleagues who were at the forefront of the new developments.

Not content with regaling the Davos residents with his first love, physics, Einstein went on to display his prowess with his other passion, music. On 24 March, he played pieces by Bach, Mozart, and Beethoven with pianist Dora Brunner and cellist Walfried Liebscher to raise money for the foundation that was running the lectures. A sketch of Einstein playing at this event was made by Emmerich Haas. Einstein's passion for playing music is demonstrated by his fulfilling this engagement despite evidence that his heart complaint was worsening. Elsa wrote to Zangger two days before the concert telling him that 'this gigantic, robust man' was nevertheless suffering with a heart problem. Zangger immediately wrote to Einstein to try to understand what the problem was. Einstein and Elsa were due to visit the recently widowed Luise Karr-Krüsi in Zurich. Elsa

[2] The only source for this story is the autobiography of Janos Plesch. This book is an astonishing read; the density of name dropping is prodigious, with the author's friends ranging from Popes – in the plural – the top echelons of the Catholic Church, almost every contemporary composer and virtuoso musician from Toscanini to Kreisler, and leading figures from literature and the arts, including Marlene Dietrich and Oskar Kokoschka. It stretches credulity to believe all of Plesch's stories. Indeed, Einstein's good friend Konrad Wachsmann, the architect of his summer house, characterised Plesch's book as 'imaginative and glamorised'. In this story, it seems bizarre that Einstein, who famously on long journeys often never opened suitcases packed by his wife, should be carrying a heavy suitcase for a day trip to Leipzig. Perhaps he was transporting many files relating to the patent dispute but Carl Seelig gives a somewhat more plausible story, also involving suitcases. In this version, Einstein refused to let the old Hotel Kurhaus porter carry his and Elsa's baggage, doing so himself with almost fatal results. Whatever happened, it brought his illness to a head after a period of several months in which he had felt unwell.

told Zangger that they were due to arrive there on 24 March, presumably directly after Einstein's concert. It may well be that the exertion of playing caused his heart problem to worsen, since they seem instead to have stayed in Davos until at least 26 March. Once they arrived in Zurich, Einstein, presumably on Zangger's medical advice, remained in bed for a considerable time. However, he felt well enough to unpack his violin and play Mozart and Bach with Luise Karr-Krüsi. On 31 March, he wrote to Hans Albert that he had a distended heart, high blood pressure, and an irregularity with his pulse. He believed that he would need to stay in bed for months to recover. Despite this, and protests from his younger son and from Mileva against his travelling in his weakened condition, Einstein decided that he needed to go home and by 11 April, he and Elsa were back in Berlin.

The next few weeks saw Einstein submitting to the examination of several of his doctor friends. Zangger, who had initially treated him in Zurich, saw Einstein again after travelling to Berlin in mid-April for discussion about his possible relocation there, which never came to fruition.[3] Zangger prescribed eating plenty of fruit and taking short walks provided the weather was not too hot. However, Einstein seems to have remained in bed, although on 20 April, Elsa wrote to Ehrenfest that her husband was feeling considerably better. Only by 24 May was he able to get out of bed, although he felt himself to be very rickety.

Janos Plesch was the principal attending physician after Zangger's departure, much to the disgust of Rudolf Ehrmann, another of Einstein's medical acquaintances. Ehrmann considered Plesch to be a quack, although Einstein thought that this was simply jealousy over Plesch's wealth and connections. By the early summer, Plesch finalised his diagnosis as inflammation of the pericardium together with an accumulation of liquid secretions. His treatment was bed rest and a strict salt- and nitrogen-free diet, including diuretic measures. Einstein also stopped drinking coffee; from now on he consumed the caffeine-free Kaffee HAG, which he prepared himself, although by 1931 he allowed himself the occasional cup of freshly roasted coffee from beans brought back from his first trip to California. Einstein seems to have become disillusioned with his medical friends during his illness and after leaving Berlin to recuperate later in the summer, was treated by two new doctors. However, he stayed in contact with Plesch by letter.

It is somewhere around this period in 1928 that a middle-aged Russian called Dimitri Marianoff entered Einstein's life. Marianoff was born in 1889 in Ukraine and went to school in Kiev. For a while he studied at the Sorbonne in Paris. He returned to Russia and became an editor of several art magazines. In 1917, he became involved in the nascent Russian movie industry and by 1918–1919 was director of the largest movie company in Moscow. In the 1920s he appeared in Berlin as a production assistant for Phoebus movies. By 1929–1930 he was leader of the department of the Russian Trade Delegation responsible for cultural movies. During the 1920s, Marianoff, charming and gregarious, was a constant presence in Berlin social life.

In his book, *Einstein: An Intimate Study of a Great Man*, co-written with journalist Palma Wayne and published in 1944, Marianoff stated that he first met Einstein in 1926 after a performance of Alexei Granovsky's Moscow Theatre Company in the *Theatre des Westerns*. Marianoff had been engaged as Granovsky's head of public relations. He was approached in the foyer after the performance by Isaac Steinberg, who had been a minister in Kerensky's government

[3] The process had dragged on for many months. Zangger, despite several concessions from the Berlin authorities, had formally refused the offer of a chair at the University of Berlin in the autumn of 1928. New negotiations had begun early in 1929 to offer Zangger an independent institute. It was from these that Zangger withdrew in March 1929 after he too developed serious heart disease.

after the first Russian revolution that had overthrown the Tsar. Steinberg had enthused about the performance and said that he had attended with Einstein and his family. Einstein had said he wanted to know more about Russian theatre and invited Marianoff to visit his home the following day.

Almost nothing in the previous paragraph, derived from Marianoff and Wayne's book, is correct; it is worth devoting some effort to clarifying this as it establishes the unreliability in particular of the dates given in the book. Granovsky's company played in the *Theatre des Westerns* from 11 April–16 May 1928. The performances were in Yiddish. Steinberg was in fact Minister of Justice in Lenin's first coalition cabinet, not Kerensky's. Einstein was confined to bed by his heart problem for the entire period of the Granovsky performances. It is therefore highly unlikely that he could have attended any of them. It is possible that Elsa attended with Steinberg, who was an acquaintance of Einstein's, and conveyed a message to Marianoff but there is no evidence for this. It is also possible that Marianoff was confusing the Granovsky tour with that of Habima, the world's first Hebrew-speaking acting company, which did take place in 1926 and which Einstein did attend. However, the first (indirect) mention of Marianoff in any of Einstein's correspondence is in a letter from Tete on 28 October 1928, followed by a letter from Einstein to his sister Maja in January 1929, when he alludes to an 'actual, real-life friend' of Margot Einstein's. As a suitor of the almost pathologically shy Margot, Marianoff would not have remained unmentioned in family correspondence for almost three years. This is confirmed by a remark by Elsa quoted in *The New York Times* report on their wedding, which states that Marianoff had entered Margot's life several months before. If any further evidence of the factual mistakes in the book were needed, only four pages later than the description dissected above, Marianoff gets the date of his marriage to Margot Einstein wrong by a year – he quotes 1929, rather than the correct date of 29 November 1930.

However he obtained his introduction to Einstein's house, there is no reason to disbelieve Marianoff's qualitative descriptions of life there. His impressions of Haberlandstrasse were described in Chapter 15. The purple prose with which the book is replete can be gauged by the description of Einstein's entry to the room on Marianoff's first visit, presumably gilded by his publicist co-author, which characterises Einstein as the man whose visit to an emperor or a nation results in a national holiday.[4] In his first visit to Einstein, most probably in 1928, they discussed Russian theatre, about which Marianoff was an expert and Einstein had an interest, and Einstein's enthusiasm for Leo Tolstoy. Marianoff then fascinated both Einstein and Margot by his expertise in graphology. The accuracy of his characterisations of various writing placed in front of him was such that it led to a further communication by Margot, who responded to his charm and sophistication with an intimacy that over the next two years led to marriage.

Marianoff's encounter with Einstein may not have been entirely innocent. His connections with the Russian state through its trade delegation, and his relationship with Soviet Commissar for Culture and Education Anatoly Lunatscharski, whose literary secretary he had once been, led to persistent subsequent rumours that he was a Russian spy. It could indeed have been his link to Lunatscharski that brought him into Einstein's circle. Einstein and Lunatscharski had met and become friendly at the Russian Scientific Researchers Week held in Berlin in July 1927. Almost

[4] It was this sort of exaggerated journalistic intrusion into his private life as much as the factual inaccuracy of the book that incensed Einstein, so that he refused to sanction its publication and broke off relations with his by-then former son-in-law in 1944.

inevitably, Lunatscharski, who seems to have spent a great deal of time in Berlin, was also a close friend of Plesch, who sympathised with the Soviet system and spent a considerable period in the Soviet Union. Plesch's autobiography demonstrates familiarity with the activities of OGPU (the forerunner of the KGB). The Berlin police were carefully monitoring the activities of this organisation. A report in August 1932 mentions Marianoff as one of the two closest assistants of Arthur Normann, the chief Soviet spy in Germany. Their suspicion was that Marianoff was running a spying operation from Einstein's house in Haberlandstrasse, taking advantage of Einstein and Elsa's long absences during 1930–1933. The *volte face* that Einstein executed with regard to the Stalinist show trials of 1930, which he initially condemned and then one year later revised his opinion, may well have been a result of Marianoff's influence.

While Marianoff was beginning his relationship with Margot Einstein, her stepfather was slowly convalescing. Hans Albert visited him, probably in early May. He was no doubt concerned about his father's condition, but he also attended an interview for a job at a railway company in Berlin that his father had sounded out for him. He decided not to take the job, which was outside his expertise, but Elsa suggested he exploit the offer to improve his situation in his job in Dortmund, which he did.

Einstein agreed to join a committee to foster international cultural cooperation, which was the brainchild of Antonina Vallentin. A remarkable woman, Vallentin was born to a Jewish family in Lemberg (now Lviv in Ukraine), then a Polish-speaking part of the Austro–Hungarian Empire. She left to study at universities in the west and during the war married a Jewish businessman. However, the marriage did not long survive. Vallentin became a journalist, being a Berlin correspondent of *The Manchester Guardian* and then editor of a magazine on international affairs, *Nord und Sud*. The influential salon she ran was well known for its liberal and internationalist slant. She became a close but informal adviser of Gustav Stresemann, Foreign Minister of the Weimar Republic for six years until his death in office in 1929. Indeed, he referred to her as his energetic second-in-command. Vallentin became a close friend of Elsa Einstein, although Konrad Wachsmann, architect of Einstein's summer house in Caputh thought that Einstein disliked her. In addition to Einstein, Vallentin invited others to join her committee: the former and future Prime Minster of the UK Ramsay MacDonald; the founder of the post-Versailles German Army General Hans von Seeckt; playwright, author, and Nobel Laureate Gerhard Hauptmann; and author H. G. Wells, who had met Leo Szilard in London and been informed by him of Einstein's activities. Wells was keen to meet Einstein, although Szilard thought that their having no common language would be problematic.

With the help of his new secretary Helen Dukas, Einstein began to address his backlog of correspondence. The editors of the illustrated weekly magazine *Reclams Universum* of Leipzig received a shorter than expected reply to their request for an overdue contribution: 'I have this to say about Bach's works: listen, play, love, honour – and keep your trap shut!' He was obviously quite pleased with this epigram as he used it again, replacing Bach with Schubert, a few months later. By the end of May, Einstein was out of bed and on his feet again but saw no virtue in advertising the fact too widely, as he was able to work without interruption. On 10 June, Hans Albert and his wife Frieda arrived. This was Frieda's first meeting with Einstein. Given his previous attitude to his daughter-in-law as expressed in his letters, which had ranged from the downright abusive to the merely rude, the visit went surprisingly well. Einstein discovered that Frieda in person was much more acceptable than she had seemed on paper. He promised to provide them with a piano when they were settled in a suitable apartment. The families parted in harmony around 15 June 1928.

Einstein's friend Toni Mendel suggested that the Einstein family should accompany her to a villa she proposed to rent for the summer in Scharbeutz, northeast of Lubeck on the Baltic coast.

Plesch opposed this, since the medical consensus was that damp and humid climates were not good for cardiac problem such as Einstein's. The patient however had other ideas; no doubt being surrounded by a 'hareem' of women looking after him was an attractive prospect. The families left for the Baltic on 5 July; initially Toni, Elsa, and Margot accompanied Einstein. They were joined by Ilse and Rudi a week or so later, Ilse having substantially recovered from the nervous condition that had required her treatment in the Engadin. Einstein had a very relaxing time for the next three months; he was photographed lounging on a deckchair in an elaborate Chinese dressing gown. While other members of the family came and left, Elsa stayed. It cannot have been easy for her living cheek-by-jowl with Toni Mendel, whose relationship with Einstein was, to say the least, ambiguous. Margot left for Hiddensee towards the end of August; Maja took her place and stayed for around two weeks – the longest period she had seen her brother for many years. Indeed, she had not seen him at all for seven years. They went sailing together on the Baltic, despite his doctors having forbidden such exertions, which indeed prostrated Einstein for some days afterwards. Toni returned to Berlin just after Maja left. It seems that Tete joined Einstein and Elsa around then and that they returned together to Berlin, Einstein reluctantly, on 2 October.

Family members were not the only visitors to the Baltic retreat. Chaim Weizmann and Leo Kohn arrived on 21 July. They discussed the status of their attempts to displace Judah Magnes from the predominant position in the affairs of the Hebrew University of Jerusalem into which he had manoeuvred himself and to which, limpet-like, he continued to cling. Unable to attend the Board of Governors meeting in May because of his health, Einstein had sent a detailed memorandum advocating among other things the appointment of a position modelled on the Vice-Chancellor of a UK university who would take over academic affairs from Magnes, the Chancellor. Magnes, whose power base lay in the support of the rich American donors, particularly Felix Warburg, continued to provide a master class in academic politics. Recognising that the pressure from Einstein, supported by Weizmann, was becoming dangerous, he gave way, only to frustrate them by calling into question the qualifications of their preferred candidate and then offering the position without authority from the governing council to someone he felt would do his bidding. Einstein had earlier threatened to withdraw from all involvement with the university and to publicise his reasons. Frustrated and outmanoeuvred, he wrote to Weizmann in June resigning from his position on the Board of Governors but indicating that he would not publicise his decision. Weizmann and Kohn certainly tried to dissuade him from this step. Weizmann, who was exhausted and sickening, had floated the idea of resigning his leadership of the Zionist organisation and displacing Magnes himself. Einstein however was highly sceptical that Weizmann was likely to take this step. For the next few months, others would also attempt to convince Einstein to remain involved with the Hebrew University but at least for the time being, he withdrew.

The other visitors to Einstein's place of convalescence were two doctors who seem to have been vacationing in Scharbeutz. Emanuel Libman was a professor at New York's Columbia University, who seems to have first treated Elsa, who had also developed a heart condition. He and another doctor, Leopold Lichtwitz, who was the Director of the hospital in Altona in Hamburg, also examined Einstein. Soon both patients were singing Lichtwitz's praises. By September, Einstein was talking daily walks lasting 45 minutes. Elsa noted that Lichtwitz was now the only physician treating Einstein, the others merely continuing as friends. She is presumably referring particularly to Plesch, whom Einstein nevertheless kept informed of his condition.

The news of Einstein's illness had prompted the major newspapers to review their obituary files. In October 1928, *The Observer*, the venerable British Sunday newspaper, commissioned an appreciation, which, for obvious reasons, was never published. Since Bertrand Russell had written the hugely influential *ABC of Relativity* in 1925, he was an obvious choice. It is worth quoting the end of the article as an example of the views of a distinguished mathematical philosopher on Einstein's place in science:

> Contemporary estimates are notoriously fallible, but most contemporaries would be inclined to accord to Einstein the highest degree of eminence. Indeed, his intellectual calibre would by many be judged as of the same with that of Galileo or Newton. In an age when physics has produced a large number of great men and a bewildering variety of new facts and theories, Einstein remains supreme in the breadth and depth and comprehensiveness of his constructions.

The isolation and peace of Scharbeutz had catalysed further work towards the unification of gravity and electromagnetism. Already in his enforced bed rest in Berlin, Einstein had developed the idea of 'distant parallelism' that was to be his preoccupation for the next few years. Indeed, he had asked Max Planck to present a paper entitled 'Riemannian Geometry Retaining the Concept of Distant Parallelism' to the Prussian Academy of Science. The basic idea of distant parallelism is to introduce an additional four-dimensional field, known as a 'tetrad', at every point in a Riemannian geometry similar to that of general relativity. Such extra degrees of freedom allow directions to be defined as parallel far from their points of origin, which can never be implemented on a simple spherical surface. Even if two directions are parallel to each other on a closed curved surface at two given points, the lines defined by these directions must meet. Considering the specific case of a spherical surface, they will meet, at the latest, at the poles. Einstein himself, in a popular article written for *The New York Times*, of which more later, explained how his new construction could redefine parallelism. As an illustration, he used three-dimensional Euclidean space with two parallel lines intersecting a third line roughly at 90 degrees. A fourth line, in the same plane and parallel to the third, must also be cut by the two original lines. However, in his new Riemannian geometry, the fourth line could cut the first line but not necessarily the second. The distance by which the lines missed each other was proportional to the 'torsion' of the theory, which came from the extra tetrad degrees of freedom, allowing parallel directions to be defined on curved spaces.

Einstein became greatly excited as he began to explore this new (to him) mathematics. The question was whether it had the flexibility to contain within it the by-now accepted tenets of general relativity, while at the same time reproducing Maxwell's equations and thereby electromagnetism? Unfortunately, the theory was so complex that the answer to neither question was quickly forthcoming. As he struggled to extract the implications of the new equations, he worked with several collaborators. The first was his former assistant, Jakob Grommer, who took up a position in the Belorussian State University in Minsk in November 1928. Another was Chaim Herman Müntz, a highly competent Polish mathematician eking out a living in Berlin by private tutoring, who contacted Einstein in 1927 and began to work with him on his new ideas for a unified theory.

Perhaps less than a collaborator but more than a student was the roughly twenty-year-old Curt Schwarz, who had attended Einstein's lectures in Berlin in 1927. It did not take much correspondence before Einstein realised that Schwarz was brilliant and he began to invite him to his home on a regular basis. Schwarz was not only a prodigy, but also brash and self-confident to a degree. His letters contain biting criticism of Werner Heisenberg and ingratiating praise of Elsa as well as scatological comments worthy of Mozart. Mixed with these juvenilia are remarks showing great perception. For example, Schwarz castigates Einstein for not spending more time examining the work of others and tells him that Schrödinger's formalism is a complete description of atomic interactions. Einstein was sufficiently impressed to recommend Schwarz for a research stipend, despite the young man's brashness apparently extending to an unsuccessful attempt to seduce Margot Einstein. Sadly, Schwarz died early in 1929, a potentially brilliant life tragically unfulfilled.

With the help of Max von Laue, Einstein arranged a stipend for Cornelius Lanczos to spend a year in Berlin working with him on distant parallelism. Lanczos had sent his thesis, a highly original document that reformulates electromagnetism and considers electrons to be singularities in its solutions, to Einstein in December 1919. He asked for permission to dedicate it to him, which Einstein granted, having glanced through the thesis and been impressed. In 1924, Lanczos, having by then moved from Budapest to become a Privatdozent at the University of Frankfurt am Main, had produced a solution of the Einstein equations for cylindrically symmetric dust particles. Lanczos's compatriot Szilard, who was still busy working on refrigerator designs with Einstein, recommended him to Einstein as an assistant. However, Lanczos had his own ideas and research programme, distinct although related to that of Einstein, so that their collaboration was much less close than for example that between Einstein and Müntz. Einstein also found that he had to meet Lanczos's stipend from his own pocket initially, as he did not really begin the application process until October 1928. Lanczos nevertheless arrived early in November, having arranged for his Frankfurt duties to be covered by the young future Nobel Laureate Hans Bethe.

Lanczos seems to have developed in later life the tendency, also apparent in many others, to exaggerate his intimacy with Einstein. In the introduction to his book *The Einstein Decade* he states that they discussed their innermost thoughts, which Einstein did not discuss with anyone else. Although they met occasionally, and corresponded frequently, many years later when both were resident in the US, this statement seems difficult to believe, particularly since the documentary evidence from this period hints that Einstein was more irritated by Lanczos than intimate with him. Lanczos goes on to state that:

> Einstein had a strong musical inclination, and music gave him much needed relaxation after intense intellectual work. During his trip to Japan in the early twenties a precious violin with an exceptionally beautiful tone was presented to him. Unfortunately, moving it to the humid and changeable climate of Germany produced a crack in the body of the violin; this was repaired, but the old sweet tone never returned. In addition, the many occupations of the Berlin years made any regular practising on the violin impossible, with the result that his bow technique suffered and his tone became scratchy. He preferred to improvise on the beautiful grand piano that one of his admirers had given him.

Many things here are incorrect. For example, Einstein's dislike of methodical practice but almost compulsive violin playing at every opportunity during his Berlin years has been already documented in this book. Neither was he presented with a violin in Japan. Presumably Lanczos is thinking of the violin he received from the Suzuki brothers during their visit to Berlin mentioned in Chapter 17. The passage is included because of the remarks about Einstein's 'scratchy tone' and bow technique, which represent one of the few personal observations on Einstein's playing ability in this period. However, it is difficult to see when Lanczos would have had much opportunity to hear Einstein play, certainly in public, as he was convalescent for much of their year together. He may instead have remembered Einstein's playing at Princeton, when age had certainly begun to tell.

Although Einstein's confinement to bed meant that he was unable to play the violin, he became engaged in a theoretical debate on the subject with Ferdinand Müller, a luthier and dealer from Solingen, who had sent Einstein a manuscript in November 1928 in which he claimed to have discovered various 'secrets' of violin construction. Einstein was unconvinced by most of them. Müller persevered and presented Einstein with one of his violins, which, in July 1930, he lent to Max Born, one of whose children was a violinist. Various instruments were then exchanged among the three, one at least of which Einstein found excellent. However, Einstein seems to have brought the correspondence to an end by the spring of 1931.

The scientific collaborations outlined earlier, particularly with Müntz, enabled Einstein to make substantial progress with his new ideas in that autumn and winter. He began, once again, to become excited that he was on the verge of unifying electromagnetism and gravity. His excitement must have spilled out in an interview he gave on 3 November 1928 to an American journalist, ostensibly on the proposal to build a new 200-inch telescope in California. This appeared on the front page of *The New York Times* on the following day:

> 'EINSTEIN ON VERGE OF GREAT DISCOVERY; RESENTS INTRUSION
> ASSERTS NEW THEORY WILL STARTLE THE WORLD FAR MORE THAN RELATIVITY DID
> HE IS SILENT ON ITS NATURE
> HOLDS 200" TELESCOPE CAN'T GIVE MORE PROOF OF RELATIVITY THAN OTHERS HAVE DONE
>
> [...]
>
> Berlin Nov. 3rd – 'Not the eye but the spirit furnishes the proof of theories – and that errs most of the time.' In these words, Dr Albert Einstein, scientist and violinist, answered a question on whether the 200-inch telescope to be built in California could afford visual proof of his relativity theory.
>
> Dr Einstein later said that he was near a discovery of far greater moment than relativity, but declined to explain its nature for the present.
>
> Dr Einstein has just returned to Berlin . . . Near Lubeck the scholar hid from the public . . . For the first part of his sojourn he amused himself in a sailing boat. When his physician discovered this pastime he forbade his patient even that.
>
> From then on Dr Einstein sat on a sunny beach and appeased his desire to work by playing his violin to the waves. Now he has so far recovered as to be able to put into print some new theories . . .
>
> [...]
>
> Dr Einstein is still forbidden to work more than a few hours daily, but he remains most of the time in his little room softly playing his favourite tunes while his active mind wanders into the realms he has created with his theory of relativity.

The theme of music and research also appeared in comments Einstein made in response to a round-robin from Paul Plaut, a Berlin psychiatrist doing a study on creative personalities. When asked if playing music influenced him in other areas, he responded that 'Music does not *directly affect* research work, but both are fed from the same source of desire and nourish each other by the release produced through them.'

Einstein found that he yearned for the solitude and quiet he had been able to find by the sea in Scharbeutz and which was so lacking in the bustle of the Haberlandstrasse apartment. He found a haven courtesy of his friend Plesch, who had recently acquired a magnificent country estate in Gatow, on the bank of the Havel River on the eastern outskirts of Berlin. Built by rich industrialist Otto Lemm, the rectangular central mansion sits in extensive parkland, then known as Gut Lemm. Subsidiary buildings included a domed pavilion that abutted the river and servants' quarters at the entrance from the main road. Einstein was able to occupy part of these quarters, an apartment intended for the chauffeur, where, as he wrote to Besso, he lived like a hermit, cooking for himself.

In the solitude provided by Gut Lemm, Einstein read George Bernard Shaw's new book on socialism, which delighted him and which he subsequently sent to Tete and his sister Maja. Another book he read was *The Good Soldier Svejk* (or Schwejk as it was transliterated into German) by Jaroslav Hasek. It is unsurprising that Schwejk's bumbling passage through the Austro-Hungarian army with its mixture of humour, iconoclasm, and mockery of the military and militarism was greatly enjoyed by Einstein. As well as playing the violin, Einstein also had the use of an organ in the music pavilion, on which he would improvise, sometimes for hours.

Plesch recounted that, at weekends, large numbers of people would gather in river craft to listen, although none knew that it was Einstein playing. He first stayed in Gatow for two weeks in November 1928 and again 20 December 1928–5 January 1929, except for the period immediately around Christmas Day, when he met Ehrenfest in Haberlandstrasse.

It was in Gut Lemm that Einstein finished the draft for his paper on unified field theories, which he intended to ask Planck to present at the first meeting of the Academy of Sciences in January 1929. He wrote to Tete that with this paper, he considered that he had effectively completed his life's work. A few weeks later, and somewhat in contradiction, he wrote to Louis de Broglie that, although the paper contained no mention of quantum physics, he believed that it would provide a sounder basis than the current quantum mechanics on which to treat such questions in the future.

Einstein had been asked at the end of 1928 to contribute to a Festschrift for the seventieth birthday and retirement of Aurel Stodola, Professor of Mechanical Engineering at ETH who had taught Besso and had met Einstein in 1909, when Einstein was a new professor at the University of Zurich. His contribution was a *tour d'horizon* entitled 'The Current State of Field Theory'. In speaking of general relativity he wrote:

> The success of this attempt to derive subtle laws of nature in a purely theoretical manner from a conviction of the formal simplicity of reality encourages us to proceed along this speculative path, whose dangers must be kept vividly in mind by all those who dare to tread along it. [. . .] After these initial results, I have hardly any doubts that the combination of the Riemannian metric with the postulate of the existence of distant parallelism yields the natural representation of the physical properties of space in the framework of field theory.

On 10 January 1929, Planck presented Einstein's paper 'On Unified Field Theory' to the Prussian Academy of Sciences. It built on the structure he had outlined in two previous papers read to the Academy. In the first two sections, Einstein set up the extremely complex mathematical framework he needed. In the next section, he indicated how the equations of motion could be derived in the framework of distant parallelism by adding a small perturbation to the key tensor, manipulating the resulting equations and then examining them in the limit that this perturbation tended to zero. By doing this, he could show that the framework produced both the Maxwell and Einstein equations, describing electromagnetism and gravity, respectively. However, he admitted that this was only an approximate solution and that further work would be required to show if this formalism did indeed reproduce all the successes of electromagnetic theory and general relativity.

The repercussions of Einstein's highly abstruse exposition, which makes the great general relativity papers of 1915 look like child's play, were extraordinary. Firstly, it contained mistakes. Both Müntz and Lanczos informed Einstein of errors, which Einstein acknowledged in a brief addendum added in print to the end of the paper. He also drafted a fuller correction which he used to produce a new paper some months later. Secondly, it produced a sensation in the world's press and Einstein was inundated with requests to explain what his advance meant. Einstein reacted with a mild dose of schizophrenia. It had only been a few weeks before that he had told *The New York Times* that he was near a discovery of far more importance than relativity. Now, however, he was quoted as saying that an elucidation clear to the layman was not possible and that he had made a purely mathematical extension of the general theory of relativity. There was nothing to be excited about and he couldn't understand why a newspaper should take an interest in it:

> 'There will be only a few mathematicians who will be inclined to read it and, although I never did make that statement which was ascribed to me that only eleven people in the world could understand relativity, I really don't believe that there will be more than a handful of people who will take the trouble to follow its argument.'

> Although it will take about a fortnight before the paper will be issued, it is understood that copies are being made available for fellow scientists in various countries, including Prof. A. A. Michelson of the University of Chicago, Professors Millikan and Russell and Jean [sic] of England.
>
> It was explained that the extension of Prof. Einstein's relativity theory, embodied in his new paper, would lead to the mathematical conclusion that matter cannot be formed without electricity or that electricity is the original source of all existence.

The press closer to home were also interested. Hans Reichenbach was recognised as a leading philosopher of science, with a long-standing interest in relativity. He and Einstein had been on good terms with each other. Reichenbach published an article in *Vossische Zeitung* on 25 January that gave a general description of Einstein's ideas as they had been explained to him by their author. Einstein was incensed by the article, not because of its accuracy but because he considered it an unpardonable solecism that Reichenbach had published in advance of the paper's appearance in the proceedings of the Academy. He wrote to the editor of *Vossische Zeitung* to that effect. The editor replied defending his actions and passed Einstein's letter on to Reichenbach, who was deeply hurt that Einstein should have not written directly to him if he thought he had cause for complaint. He considered Einstein's reproaches particularly unfair because he had long defended him and relativity against the attacks of others. Einstein's reply was unrepentantly belligerent, although he ended his letter by proffering the olive branch that everyone was fallible. Reichenbach grasped it, although still obviously distressed by the episode.

Despite his anger over Reichenbach's explanatory piece of journalism, indeed possibly catalysed by it, Einstein seems to have changed his mind yet again by the end of January. Despite having said that an elucidation clear to the layman was not possible, he tried to do precisely that by writing an article for *The New York Times* and *The Times* of London, explaining the context of his January paper and the concept of distant parallelism. It was probably this article and the interest that it generated that led to the Selfridges department store in London pasting a copy of Einstein's paper from the Academy proceedings in one of its windows. Arthur Eddington reported that crowds gathered outside the store to read it.

Another problem that arose from the paper echoes the criticism that the brilliant young Schwarz had levelled at Einstein: that he did not read enough of the literature relevant to his interests. The appearance of the two preparatory papers in June 1928 had already elicited a note from Roland Weitzenböck, Professor of Mathematics at the University of Amsterdam, who pointed out that the subject of distant parallelism had been treated by him in 1921 and by a long list of authors subsequently, including Padova's Giuseppe Vitali and Princeton professor Luther Eisenhart, to whom Einstein wrote in apology. To Eisenhart he blamed his illness for his not being aware of others' work in the area. Perhaps even more embarrassing was the fact that Weitzenböck, in his contribution to the Academy proceedings designed to correct Einstein's omissions, had himself ignored the seminal contributions of eminent French mathematician Elie Cartan. This omission was probably not accidental given the bad blood that continued to fester between German and French scholars. Cartan in May 1929 pointed out to Einstein that, not only had he published in this area for many years, but he had explained a simple example of distant parallelism to Einstein verbally during Einstein's visit to Paris in 1922. In evident embarrassment, Einstein replied that he hadn't understood what Cartan had told him in Paris and certainly didn't think at that time that such ideas had any application to physics. He suggested that to set the record straight, Cartan should write an article setting out the full history of the mathematical treatment of distant parallelism. This could be associated with an article from Einstein giving an overview of his own work. These articles were published early in 1930.

The somewhat awkward episode between Einstein and Cartan did not interrupt their cordial personal relations. These were renewed when they met in Paris in November 1929 on the happy occasion of Einstein receiving the honorary degree from the Sorbonne that had had to be postponed by his illness. The only negative aspect to an otherwise pleasant and successful week in Paris related to Einstein's election as a Foreign Member of the Academy of Science. Einstein was shocked to discover that his old friend and highly distinguished colleague, Paul Langevin, had not been elected to the Academy's ranks. In these circumstances, Einstein considered that his own election would be both embarrassing and personally painful for him. He therefore indicated that he could not at that time accept, despite his election having been announced in several newspapers. Putting this academic politics to one side, Einstein concentrated on physics, giving two lectures, with a substantial component related to distant parallelism, at the Institut Henri Poincaré. Cartan and he continued to work on this subject in 1930.

In his letter in which he reported the Selfridges incident, Eddington told Einstein that he was unimpressed by his paper, but that if calculations to the next approximation were successful, he might change his mind. Eddington's doubts were shared by most of their other colleagues. The reservations of both Müntz and Lanczos have already been mentioned. Einstein told Besso that he was certain that, if Besso didn't stick his tongue out at it, then he would be a hypocrite. It was, however, that great iconoclast Wolfgang Pauli who was, unsurprisingly, the most outspoken in criticising Einstein's ideas. He attacked Einstein on all fronts: directly in a letter, indirectly to colleagues and publicly in print. It is worth quoting from these various sources at length, not only because they represent, if somewhat exaggeratedly, the opinions of the vast majority of physicists but also because they are often amusing:

> Dear Mr Einstein, I thank you greatly for sending me the corrections to your latest contribution to *Mathematischen Annalen*, containing such a clear and beautiful overview of the mathematical properties of a continuum with the Reimann metric and distant parallelism. I want to tell you what I and a large part of the younger generation of physicists think about the physical aspects of this subject.
>
> Contrary to what I said to you early this year, there is no longer any argument from the standpoint of quantum theory that would favour distant parallelism. This is because Weyl and Fock have been able to show that the Dirac theory of electrons can be carried over into the relativistic theory of gravitation . . . You will no doubt reply that, for the moment, you don't want to hear anything more about quantum theory. I know that, but I deeply regret it. However, I have to tell you further, that it seems to me that the derivation of your field equations . . . is unconvincing and that even the simplest consequences of them bear hardly any resemblance to the widely known and well established facts of physics. The first thing to be deprecated is that even to the first order, Maxwell's equations only appear in differential form. Secondly, no integral of the total energy or momentum of the field exists and it doesn't seem that one can construct its energy-momentum tensor. Moreover, what has happened to the perihelion advance of Mercury and the bending of light around the Sun? They seem to have been lost in your general dismantling of general relativity. I am clinging on tightly to this beautiful theory, even if you are betraying it!
>
> With your remark that you are a long way from being able to prove the physical validity of the equations, you take the words right out of the mouths of your critical colleagues, so to speak. It is only left to them to congratulate you (or should I rather say, to commiserate with you) on having defected to the pure mathematicians. I am not so naïve as to believe that any criticism from others will change your mind. However, I am willing to bet you anything you like that within one year from now you will have given up completely on distant parallelism, just as you did with affine connections. And I don't want to spur you to contradict this letter by continuing it further and thereby delaying the natural demise of the theory of distant parallelism.

This is strong stuff, and it does Einstein great credit that, rather than sending an angry rejoinder, he replied in a detailed and patient way and asked Pauli to look again at the theory with a fresh pair of eyes. He might have been more perturbed had he seen what Pauli wrote to Jordan:

> It seems Einstein has been spouting terrible rubbish about a new distant parallelism. He wants to argue that the mere fact that his equations have not the slightest resemblance to the Maxwellian theory implies that they must have something to do with quantum theory. With this sort of nonsense he can only impress American journalists, not even American physicists, let alone European ones . . .

or to Ehrenfest: 'By the way, I now no longer believe a single syllable of distant parallelism; the dear Lord seems to have completely abandoned Einstein.'

Hermann Weyl also received a typical Pauli communication. Weyl, in laying the foundations of what is now known as gauge theory in an extraordinary paper in 1929, had undermined Einstein's unification by convincingly postulating that electric charge was intimately associated with the matter waves of quantum mechanics, rather than gravity. This viewpoint, developed over the next decades into the standard model of particle physics, is the one that currently reigns supreme. Pauli wrote to Weyl:

> And here I have to do justice to your work in Physics. When you constructed your earlier theory[5] [of gravitation and electromagnetism] it was pure mathematics and Einstein could rightly criticise and complain. Now the hour of your revenge is at hand; now Einstein has shot himself in the foot with distant parallelism, which also is pure mathematics and has nothing to do with physics, and you can complain!

Pauli reserved his most trenchant criticism for the scientific literature. In a review of a popular exposition of Einstein's distant parallelism theory by Lanczos, Pauli wrote in 1932 that:

> It is indeed a bold statement of the editorial staff to include a review of a new field theory of Einstein under the label '*Results* of the Exact Sciences.' In recent times his unfailing ingenuity and dogged energy in chasing a particular goal has given us on average about one such theory a year – wherein it is psychologically interesting that each theory is for a while typically characterised by the author as the 'definitive solution'. It is thereby possible at the appearance of another work on this subject to be tempted to proclaim, in an adaption of the familiar historical cry, 'The new field theory of Einstein is dead, long live Einstein's new field theory'. (Although one should not hide that some factors indicate that the basic idea of Einstein's newest theory may turn out to have a longer lifetime than previous unified theories).

Pauli's somewhat conciliatory final parenthesis turned out to be too optimistic. Instead, he won his bet as, by 1931, Einstein had ceased to work on distant parallelism. It is possible to picture his sheepish grin as he wrote ruefully to Pauli on 22 January 1932: 'You were right after all, you rascal!' Unlike several of his other methods to achieve unification, he never returned to this one. Tilman Sauer draws an interesting parallel with the situation of the Entwurf theory that preceded the final version of general relativity (see Chapter 11): it was not so much that anything showed conclusively that distant parallelism was the wrong strategy, more that progress became very slow and painfully technical, and that other approaches more congenial to Einstein's philosophy began to seem more promising. This was catalysed by the departure of Einstein's co-workers – Lanczos to return to his duties in Frankfurt in the autumn of 1929, Müntz to Leningrad where he had obtained a chair and Grommer to Minsk. It is interesting to note that the old Einstein, fearlessly launching himself into great unexplored vistas of science with only his physical intuition to guide

[5] See Chapters 14 and 16 for Einstein's attitude to Weyl's unified theory.

him, had now developed into a rather tentative mathematician, dependent on a professional to guide him around calculational obstacles. He found the right companion on this arid journey in the Austrian Walther Mayer, an expert in differential geometry, who began to work with him early in 1930. Although they initially produced two papers using distant parallelism, it wasn't long before they decisively changed direction, back towards a variation of the extra dimensions of the Kaluza–Klein approach.

By the time of the presentation by Planck of the January 1929 distant parallelism paper, Einstein was starting to feel well enough to rejoin, slowly and carefully, the scientific community of Berlin. By the end of that month, he arranged to meet von Laue at a meeting of the Academy. By March, he was once again accepting invitations to address public meetings and to be present at committees. In between, he was pleased to receive a visit from Harriet Cohen, a strikingly beautiful British pianist. She became best known for her playing of contemporary music, particularly that composed by compatriots such as Arnold Bax, with whom she had a long-lasting romantic relationship,[6] and Vaughan Williams, who dedicated his piano concerto to her. However, she was also an enthusiast for Bach's keyboard music. Einstein's cousin Alfred was greatly impressed by her and may well have suggested the visit. Unsurprisingly, it was Bach, rather than Bax, that she played for Einstein.

Habitually dismissive of birthdays as he was, there was one that Einstein simply could not ignore. On 14 March 1929, he turned fifty, an occasion that he feared would give rise to an enormous worldwide ballyhoo. His strategy was to become invisible, thereby avoiding events such as the unveiling of a bust by Kurt Isenstein at Einstein Tower in Potsdam on his birthday. Images of himself always made him uneasy; this one he described to his sister Maja as exaggerated. He had at hand the ideal hiding place, Plesch's country estate at Gatow; he arrived there on 12 March, staying for almost a week. While the telegrams and greeting cards accumulated by the sack-full in Haberlandstrasse, he cooked for himself in the chaffeur's apartment. Only on the day of his birthday did he relent sufficiently to allow his immediate family to join him for a meal. Elsa, Ilse and Rudi, and Margot brought stuffed pike and trimmings with them. According to the enterprising reporter who tracked him down, Einstein was wearing plain trousers, a brown sweater, and slippers. The family departed at 5.00 p.m., leaving Einstein to resume his reclusive existence.

The sacks of birthday greetings had all to be answered, a job to which Einstein wearily applied himself when he returned to Haberlandstrasse. Most received a copy of a verse of his own composition produced on a duplicator. The most favoured also received a personal message, either on the back of the card or in a separate letter. The senders ranged from the mightiest to the unknown and humble. Both the German Chancellor and Social Democratic leader Hermann Müller and the US President Herbert Hoover sent messages, although, unsurprisingly, there is no record of the reactionary President of Germany, Paul von Hindenburg, having done so.

Among the host of scientists who paid tribute to Einstein was Sigmund Freud. He and Einstein had met briefly in 1927; earlier in 1929, Einstein had written to Freud asking him to support a Jewish charitable cause. Tete's interests meant that Freud's name frequently came up in the correspondence of father and son. Two years later, in 1931, Einstein and Freud were to initiate a famous series of letters on the causes and possibilities of avoiding war. Their exchange of letters around Einstein's birthday, when they were known to each other predominantly through their scientific work, which each found mutually incomprehensible, is also striking. In a brief message on 13 March, Freud had congratulated Einstein on his past and future happiness. Einstein's

[6] Her love-life was legendary; her other lovers included D. H. Lawrence, Vaughan Williams, Arnold Bennett, H. G. Wells and two British Prime Ministers: Lloyd George and Ramsay MacDonald.

reply would no doubt have been grist to Freud's psychiatric mill. It queries why Freud particularly emphasises Einstein's happiness, since he hardly knows him well enough. Some weeks later, Freud sent a long response, equally psychologically revealing, in which he begged Einstein not to reply and to destroy after reading. He gave his reason for considering Einstein to be happy: his envy that Einstein's subject was the physical sciences, where only trained individuals could have a valid opinion. In contrast, everyone felt free to pronounce on his subject of psychology. He ended with the touching sentence: 'Think of me as one of the very many that are glad that Albert Einstein lived and worked in their time.'

Another interesting psychological insight is given by Einstein's reply to birthday greetings from Nahum Sokolow, Chair of the Zionist Executive in London:

> I have on the whole had more good fortune than I deserve and in what I have done gratified more my own desires than any ethical considerations, indeed I have often neglected personal duties for the sake of my work to a degree that anyone else would have been condemned for.

Weizmann also sent a telegram of congratulation from London. His health had finally given way after years of stress and he had only just emerged from a month's stay in a sanatorium in North Wales.

Letters arrived that were redolent with memories of childhood and youth. Otto Neustätter, friend of Einstein and Ernestina Marangoni (see Chapter 1) in Pavia, reminded him of the starry skies under which they had hiked in northern Italy and reflected that Einstein had now penetrated their secrets. Hans Byland's ebullient congratulations also contained a note of sadness in noting the death of their old teacher and mentor, Jost Winteler, a few weeks earlier. He mentioned that music was never silent in the Byland household and that, during that winter, they had enjoyed attending concerts by the Klingler Quartet and Bach by that other famous violinist named Albert: Schweitzer! Einstein in his reply to Marie Barthelts about their playing together when he was a young man (see Chapter 4) wrote that he had become an ailing, grey-haired man, but his happy spirit and pleasure in making music had remained. Besso sent a brief but eloquent letter; Einstein replied that nothing in later life approached the friendships of youth.

Celebratory events around the world were covered in newspapers. *The Jewish Chronicle* reported on an event held in New York, appropriately at the Metropolitan Opera, on 16 April 1929:

> The speeches delivered at the Einstein celebration meeting in New York, and the messages which were sent, are to be bound in book form and presented to Professor Einstein during the Zurich-Zionist Congress. Professor Einstein sent a message to American Jewry in connection with the event, referring to the gifts to the Jewish National Fund in his name, in which he said: 'The biggest national fortune of a people is its land, and therefore of all the presents I have received the dearest to me are the contributions which will help to build a home for the Jewish people.'

In a similar vein, Einstein greatly appreciated the activities of both the Jewish National Fund and German Zionists, who organised the purchase of land for an Einstein Forest to be planted in Palestine. By the time of his birthday, 2,000 trees had already been planted at the site just northwest of Jerusalem. He received with great joy a basket containing a selection of all the varieties of fruit grown in Palestine and some Palestinian honey.

The reporter who had hunted down Einstein in Gut Lemm had noted that the birthday presents on a sidetable in the Gatow apartment included 'a sketch of a sailing yacht friends are having built for him and the plan of a wooden summer cottage'. The sailing boat, which certainly gave him more pleasure than any of the other gifts he received, was generously provided by three Jewish financiers. One, Henry Goldman, was based in the US and was a founder of Goldman–Sachs; the other two, Siegfried Bieber and Otto Jeidels, were based in Germany. The *Tümmler* (Porpoise) was seven meters long and had a small cabin into which four persons could squeeze but which contained berths for only two. It was constructed in Kiel and was due to be finished in May. It was in use at the Einsteins' rented house in Potsdam, which abutted on the Templiner See, by August 1929, but had to wait to be tied up at its intended home, the second of the gifts mentioned above.

For several years, the Einstein family had discussed the idea of purchasing a holiday home. Initial suggestions had been mostly in the Alps, but the discussions remained desultory: Einstein blamed his wife and step-daughters for being unable to agree on anything. Plesch claimed that he had suggested to Berlin Mayor Gustav Böß that the city should provide a summer house as a gift for Einstein. Plesch wrote that he had first had to explain who Einstein was, which, if true, would have implied that Böß was extraordinarily uninformed. Initially, Böß suggested that Einstein should occupy a house owned by the city in Neu-Cladow. Elsa was taken aback when she visited the house to discover that it was occupied. It transpired that the residents had a clear legal right to remain, irrespective of the city's wishes. The city authorities then proposed to find a plot of land on which a house could be built.

At this point, a young architect called Konrad Wachsmann enters the story. He noticed in a newspaper that the city was considering gifting a house to its most famous scientist and decided that his expertise in wooden construction would be ideal. Like most in the city, Wachsmann knew in very general terms about Einstein's work. He was better informed than most, since Einstein's friend and neighbour, chess grandmaster Emanuel Lasker and he were part of a group of intellectuals who met in the Romanischen Café. Emboldened by Lasker's characterisation of Einstein as a modest and kindly man, he decided to visit Haberlandstrasse and offer his services. Einstein was away, but Elsa opened the door and after considerable misgivings, was persuaded to take the unknown young man seriously when she saw that he had brought an impressive car and a chauffeur to drive them to the proposed site. The city had offered a plot in Gatow, but Wachsmann pointed out that it was near a motorboat club and would be very busy and noisy in summer. Wachsmann took the matter in hand and convinced the relevant authorities to look for other plots but none in its portfolio was suitable. He then offered to find a site that the city could purchase. While busy with this search, political developments in the city council made it less and less likely that such a purchase would eventually be approved. The right-wing faction of the council objected to the idea of presenting anything to a prominent left-wing Jew. By May of 1929, Einstein wrote to Böß asking him to forget about any contribution from the city.

For a while, Einstein thought about abandoning the entire idea of a summer house in the environs of Berlin but, early in April, he took Wachsmann sailing near Caputh, borrowing Adolf Stern's yacht. Shortly afterwards, impressed by his conversation with the architect, Einstein decided to proceed, using his savings to fund the land purchase and construction. It was personal contact that led to the final site. Adolf and Elspeth Stern owned a large plot in Caputh and agreed to sell part of it to the Einsteins. Although not bordering the lake, the plot was beautifully situated with a fine view and was only a ten-minute walk from the jetty where Einstein could tie up the *Tümmler*. Einstein wanted a wooden structure, which was Wachsmann's speciality. He finalised the plans and once Einstein had signed off on them, applied for building permission on 12 May 1929. It was granted on 21 June with a speed that perhaps reflects residual embarrassment by the city authorities. Building went on apace over the summer, with the Einsteins keeping an eye on progress from a rented house nearby on the Templiner See. Being wood, mostly Oregon pine and Galician fir, the outer shell of the building went up very quickly, in around two weeks. By September, construction had progressed sufficiently that the family could take possession, although finishing work would continue until the following spring.

The Caputh summer house, whose address was Waldstrasse 7, was unpretentious. It consisted of an extensive cellar, the living quarters and an upper floor, half of which was taken up by a large balcony area with a wonderful view towards the lake and the Havel River. There was a spacious living room, with a monumental marble fireplace and large french windows that opened up to give access to another sitting area, sheltered by the overhang of the first-floor balcony. The land fell sharply away from this terrace, forming a garden area, containing a small tree that sheltered the balcony and, at least in spring, rather regimented flowers in rows running away from the house. Einstein's working and sleeping room and Elsa's bedroom were also on the ground floor. Margot's room and a guest room, mostly used by Ilse and her husband Rudi Kayser, were on the first floor, together with a room for the housekeeper, Herta Waldow. There was no piano,[7] although Einstein's violin was a constant companion. Neither was there a telephone; urgent calls could be made to a neighbour, whose coded trumpet call indicated for whom any call was intended. Helen Dukas remained based in Berlin, from whence she came out to Caputh a couple of times per week to deal with Einstein's correspondence. From henceforward, Einstein's home, where he could work away from distractions, play the violin, and sail, was Caputh. Haberlandstrasse, for him and to a lesser extent Elsa, became merely a *pied-à-terre*.

Not everyone had participated in the worldwide wave of admiration for Einstein's half century. There was no rejoicing at recognition for 'Jewish physics' among the cohorts gathering around the increasingly powerful Hitler. The anti-relativists led by Lenard and Gehrcke, quiet in recent years, were emboldened by the rise of fascist and anti-Semitic movements. A new source of anti-relativism appeared from the Roman Catholic Church. Archbishop of Boston Cardinal O'Connell made a startling intervention in a speech on 7 April 1929. He declared that Einstein's 'so-called theories' seemed

> . . . nothing short of an attempt at muddying the waters so that without perceiving the drift innocent students are led away into a realm of speculative thought, the sole basis of which [...] is to produce universal doubt about God and His creation . . . In a word, the outcome of this doubt and befogged speculation about time and space is a cloak beneath which lies the ghastly apparition of atheism.

These strong words were backed up by even stronger ones when a few days later the Cardinal announced that he had gathered new facts – that Einstein was not a true scientist and that relativity was 'false in its construction, plagiaristic in its main statement and atheistic in its tendencies'.

---

[7] Natalia Saz described a visit to Caputh in her memoires. She was stage manager for Klemperer in his staging of Verdi's *Falstaff* in Berlin in 1931 and was director of the Moscow Theatre for Children. Her main claim to fame was to have commissioned Sergei Prokofiev's *Peter and the Wolf*, collaborating closely with him and narrating the first performance. Einstein came to the first performance of *Falstaff*. Sats was then invited to Caputh, a visit that she described at great length, including gardening, weeding beds of strawberries with Einstein, and then a long session of playing Tchaikovsky pieces on the piano with him. Her assertion that there were any strawberry beds, or indeed any crop other than herbs, at Caputh was flatly contradicted by Herta, the Einsteins' maid, when explicitly asked about them by Herneck in his book *Einstein Privat*. In particular, she insisted that there was no piano in Caputh, a statement backed up by Herneck's conversations with Einstein's neighbours. Perhaps, writing many years later, Sats conflated a musical encounter in Haberlandstrasse with a separate visit to Caputh, As to the strawberry episode, a photograph of Einstein in the Caputh garden shows a substantial bed of some crop, possibly strawberries, but anyway certainly in contradiction to Herta's recollection.

The catalyst for Cardinal O'Connell's didacticism about physics, extraordinary from someone with no background in it, was presumably the extensive coverage in the US press about Einstein's birthday. This may have awakened anti-Semitism, where the Cardinal had a far from spotless record. The plagiarism charge is one often repeated by the anti-Semitic Ernst Gehrcke and his circle. Another factor was the Cardinal's ultra-conservatism. The spirit that condemned Galileo was alive and well on the east coast of the US. Quite what O'Connor made of Georges Lemaître,[8] Catholic priest, pupil of Arthur Eddington, and the first to use general relativity to propose an expanding universe in a paper published in 1927, is fortunately not on record. Since eventually both Popes Pius XII and John XXIII endorsed Lemaître's work, and the latter promoted him to the rank of monsignor, it can be concluded that the Church and relativity were by then reconciled.

Einstein ignored O'Connell's attack but responded to a related question from New York's Rabbi Herbert Goldstein, who, having read the Cardinal's reported remarks, sent a telegram asking Einstein to state whether or not he believed in God. Einstein replied that he believed in the God of Spinoza, who was manifest in the orderly harmony of creation, and not in a God who interests himself in the destiny and affairs of humanity.

With the birthday celebrations finally behind him, Einstein attended a concert at the Philharmonie that he, and indeed the musical world, would long remember. On 12 April 1929, Bruno Walter conducted the Berlin Philharmonic in a programme that would be inconceivable today. A chubby little boy, twelve years old, played the Bach E major Violin Concerto, followed by the Beethoven and finally the Brahms concertos. Yehudi Menuhin had been born in New York but moved to San Francisco at an early age, where he became a pupil of Louis Persinger, himself a pupil of Eugene Ysaye. Even in a Berlin inured to the appearance of violin prodigies, who had thrilled in previous decades to the young Bronislaw Huberman and a thirteen-year-old Jascha Heifetz, Menuhin seemed something special. Certainly Einstein thought so. Walter, watching him during the cadenzas, remembered that Einstein was in the front row, surprised and enthusiastic. After the concert, Einstein was trapped backstage for thirty minutes amidst the immense throng waiting to congratulate the boy, 'his hands high above his head where he had placed them to clap his approval, but he finally fought his way to the artists' room, embraced Yehudi, and exclaimed "Now I know there is a God in heaven."' Subsequently he said to the *Musical Courier* that:

> The talent of the boy is the greatest I have ever observed. The spiritual conception of everything he plays, whether by Bach or Brahms, plus the technical perfection with which he masters a large violin with his little, plump fingers, reminds me of my sensations 40 years ago when I heard the great Joachim play for the first time.

---

[8] Lemaître had first meet Einstein in the interstices of the 1927 Solvay Conference. Lemaître recalled that: 'Walking through the avenues of the Leopold park, he spoke to me about an article regarding the expansion of the universe, passed almost unnoticed, which I had written the previous year and a friend had him read. After some favourable technical observations, he concluded by saying that from the point of view of physics it seemed to him absolutely abominable. In an attempt to prolong the conversation, Auguste Piccard, who accompanied him, invited me to get into the taxi with Einstein, who needed to pay a visit to his laboratory at the University of Brussels. In the taxi I spoke about the recessional velocity of [galaxies] and I had the impression that Einstein was not very informed about astronomical phenomena.' Einstein also recalled the meeting, with a certain *schadenfreude*, as he had been able to inform Lemaître that Alexander Friedmann had published the same ideas a couple of years previously, a fact of which Lemaître had been unaware.

Five years later, Elsa confirmed the latter comment, writing to Menuhin that 'When we left the concert hall, my husband remarked to me that listening to you had made the biggest impression on him since he had heard Joachim play.'

Plesch also attended the concert with his friend Fritz Kreisler: 'Asked afterwards to give his impressions, Kreisler declared: "I feel rather sorry for the boy. He has missed all the joys of mastering his art. The rest of us have all had to fight hard for what mastery we have attained; for Menuhin mastery has been a gift from heaven"'. In fact, Kreisler need not have worried. Long after his triumphs as a prodigy, Menuhin became unhappy with his technique and wrestled much harder than Kreisler had[9] to regain mastery of his instrument.

Just after Menuhin's triumph, Einstein achieved a milestone of his own. For the first time since his return from Scharbeutz, he left Berlin and its suburbs. The destination of this first journey is interesting. He spent the period 17–22 April with Grete and Willy Lebach in Düsseldorf, afterwards calling on Hans Albert in nearby Dortmund and returning to Berlin on 25 April. He had first stayed with the Lebachs at the Düsseldorf annual meeting of the German Society of Scientists and Doctors in 1926. An intimacy had clearly developed between Einstein and Grete, as evinced by letters Einstein wrote to her during his stay in Gut Lemm around Christmas 1928. She obviously made a great fuss over Einstein, as he tactlessly reported to Elsa in his letters.

Satisfied that travel was now practicable for him, Einstein left Berlin again the next month to attend the Scientific Committee of the Solvay Institute in order to plan the next Solvay conference due in 1930. Since the death of Lorenz, his old friend Pierre Langevin had taken over the chair. While there, Einstein received an invitation from his friend Queen Elisabeth to visit her at the royal residence of Laeken. Her husband, King Albert, was away in Switzerland. An accomplished Alpinist, he was climbing in the Berner Oberland. In his absence, Queen Elisabeth was able to monopolise her guest. He arrived late, having walked through the park from the station. Although the queen had sent a car, Einstein had not expected one and walked past it, while the chauffeur had only looked for passengers disembarking from first class, not the third class that Einstein habitually used. Einstein and his royal host immediately got down to important matters. The queen was a violinist as enthusiastic, and of similar standard, to Einstein. She had a collection of great Italian instruments. Einstein unpacked his humble Suzuki and they played duos by Viotti. After tea under the chestnuts, the queen walked with Einstein through the famous Laeken greenhouses, after which they dined together. He returned to Berlin, where he was pleased to receive two photographs of his visit from Her Majesty. He replied, looking forward to their next musical collaboration, when he hoped that their group would be strengthened. Unfortunately, his still-precarious health precluded him from accepting a renewed invitation in November of 1929, when the queen heard that he was travelling to Paris to receive an honorary degree. However, he would take advantage of several other opportunities to make music with the queen in the next four years.

---

[9] Kreisler was famous for a similar reluctance to practice as his friend Einstein. Kreisler's wife Harriet was quoted, tongue in cheek, by Bishop Sheen in his funeral eulogy for Kreisler as having maintained 'If only Fritz had practised, he would have been a great musician.'

Chapter 20

Life and Music at Caputh; Journeys to Solvay Conference, England, New York and California (1929–1930)

The completion of the Caputh summer house in the autumn of 1929 marked a qualitative change in Einstein's lifestyle. Nevertheless, he still spent some time in Haberlandstrasse. An insight into his life in Berlin can be gleaned from the testimony of his two sons-in-law. Dimitri Marianoff, who moved in after his marriage to Margot Einstein in November 1930, described Einstein's daily routine, which began every morning, both in Berlin and in Caputh, with a hot bath:

> His day usually began at nine o'clock. He would appear in the door of the living room in a long, white nightgown down to his heels, such as Germans wear, or maybe in slacks with a pair of old slippers on his bare feet. Once or twice we caught him on his way to the academy in these slacks and slippers. He had forgotten to dress.
>
> Sometimes he shaved before going to the academy, sometimes after. He never uses a shaving brush or cream, but plain, everyday soap that he rubs in with his hand.
>
> He would go to his piano. This was really a private rite, and he did not like his family to listen. If he saw us he would stop. The room would resound with harmonic melodies, complicated cadenzas, pastoral tunes, bits of Mozart, Brahms, Bach (before whom he prostrates himself), Schubert and Beethoven.
>
> Sometimes I would come to the door and listen to a soft improvisation without plan, without structure, just some thoughts. I often saw him remove his hands from the keys, the face locked in profound concentration, with a rapt look.
>
> Perhaps this is not a bad place to stop and speak of his music. It is not the hobby that the world has been led to believe. Music is an integral part of Einstein's being. The playing of the piano or the violin is only a mode of expressing the continuity of his thinking in a mechanical form.
>
> First, it started the day for him in a right mood. Some mornings he would take his violin and for an hour or more he would practise only the monotonous notes of the music scales – up and down, up and down. It was plain that his thinking gained by this kind of preparation.
>
> [...]
>
> Music most certainly is of the emotions. Einstein, whose intellect is of polar purity and is pitched on an exceedingly cold strain, is not an emotional man. He reasons always from logic, never from that heady element that lashes the arteries and makes the heart pound.
>
> Poets love music – it inspires them and stirs the imagination; writers listen for the same reason, and composers also. But in Einstein's case it is in no way an inspiration for his work. He does not go to his music to help him create. He goes to it to help him to be calm. It is entirely a relaxation, a habitation to which his mind may go for mental rest, for sanctuary, from a culvert from the storm, just that.
>
> No-one who has seen Einstein playing Bach will ever forget the tranquil absorption of his face. It is the same accent one finds in that of Michelangelo's Jeremiah – a lofty human dignity, an expression of deep, utter goodness; and the respect in your heart for what is there is so great that you are on your knees before it.

Einstein. Brian Foster, Oxford University Press. © Brian Foster (2026). DOI: 10.1093/oso/9780198794875.003.0020

After his breakfast, always a simple meal of coffee, toast and eggs, he was invariably joined by his secretary, and the morning mail brought into the dining room to be sorted. [. . .] in winter he would wear an old brown leather jacket[1] which his wife had presented to him many years ago. This he wore day in and day out. He never changed his dress for guests or visitors [. . .] On days when it was cold and windy he wore a fine gray sweater of English wool which Elsa had given him on his birthday.

Einstein's other son-in-law, Rudi Kayser, confirms the above picture of Einstein's daily life:

He considers Shakespeare's mighty work, as great in its characterisation as in its poetic form, the high point of the world's literature. He is, furthermore, an admirer of the at present unfashionable Schiller. Drama appeals to him, since it is closest to human life. Lyric and epic forms are foreign to him. He is fond of the deep self-confessors of Russia, like Tolstoi and Dostoevski, and also of the sceptical laughter of Anatole France and Bernhard Shaw. Gerhard Hauptmann is for him the most profoundly affecting of contemporary German poets . . .

[. . .]

Association and conversation between Hauptmann and Einstein are, therefore, a source of joy for both [. . .] Einstein and [George Bernard] Shaw feel a similar mutual understanding: in this instance, both men are united by their sceptical humour and their social attitude.

But Einstein's greatest love is music, especially classical music. Here profundity and significance of experience is joined with beautiful form, and such a union, to Einstein, means the greatest human blessing. The human will to live as it is expressed hourly in the great and small things has been raised by means of music to an absolute force, which in turn absorbs every experience and dissolves it into a transcendent, beautiful reality. The school of German music from Bach to Beethoven and Mozart best manifests for Einstein the essence of music. It does not follow that he is dogmatic and despises other musical personalities and tendencies. He loves old Italian music, he also loves the German romanticists, but the peak of musical achievement is, in his opinion, this triple constellation . . .

[. . .]

He himself is a proficient violinist. Hardly a day passes that he does not pick up his violin and, usually accompanied by distinguished pianists, plays sonatas and concertos. He is also fond of chamber music and joins prominent professional musicians in trios and quartets. At such times he is completely carried away into the realm of music; he combines so much reverie and forgetfulness of the world with unfailing technique that he is lost to all else. He gladly consents to play at charity concerts, though two considerations make him hesitate. In the first place, he does not like to compete publicly with professional artists, since in his artistic modesty he considers himself only an amateur; and, in the second place, he fears he might hurt professional musicians, who secure all the advantages of a public appearance by participating in charity concerts.

[. . .]

Very recently, when Einstein again placed himself and his instrument at the service of charity, a wit remarked that Fritz Kreisler would have to deliver lectures on physics, since Einstein was so busy with concerts.

One such charity concert was documented by Einstein's friend, the distinguished Austrian pianist Bruno Eisner. Their concert in December 1927 with Francesco von Mendelssohn was already mentioned in Chapter 18. On 3 November, 1929, Eisner and Einstein again shared the

[1] A similar jacket, bought to replace this one, presumably lost with Einstein's other possessions in Berlin in 1933, was sold by Christie's in 2016 for £110,500. Apparently, it still carried the odour of Einstein's pipe.

stage, this time with Francesco's uncle Franz von Mendelssohn. The concert took place in Franz's palatial house in Grunewald, as recalled by Eisner:

> Shortly after my first return from Palestine we arranged a charity concert in the house of von Mendelssohn for some Palestinian cause. The programme contained among other things two Bach arias for Alto, piano and obligato violin. My wife sang the arias much to Einstein's admiration, with good reason in my opinion. He played the violin part. However, at the first rehearsal at Mendelssohn's house it became apparent that the part was much too difficult for him. It was extremely amusing to see how the noble, good-hearted, warm Franz von Mendelssohn, who was an excellent violinist, was childishly pleased by Einstein's discomfiture. Later that evening my wife undertook the delicate mission to convince E. to give up with the white lie that she hadn't sung the arias in public for a long time, that a lot more rehearsals would be necessary and she would sing some Schubert lieder instead. The next morning at 7.30am (the concert was that day at 5pm) Frau Einstein rang: 'Albertle (she was Schwaebish) has been practising since 6am. You should come for a rehearsal at 10am.' We made our way there through snow and ice to the great annoyance of my wife. However, when she was in front of Einstein she said 'Perhaps I have held up some earth-shattering discovery. Your time is so precious.' Einstein replied 'Don't pretend to me. I know I played very badly. Lets [sic] try it again. The more one has to do, the more time one has.' And indeed it went decently. Before we went into the performance he touched her affectionately on the shoulder and said 'You all conned me but it got me back into shape.'

As well as the pieces mentioned above, Olga Eisner sang four Schubert lieder, Eisner, Einstein and von Mendelssohn played the Bach Double Concerto and, joined by cellist Frau Peters-Moorsheim and violist Max Baldner, they played the Schumann Quintet Op. 44.

Eisner elsewhere in his memoirs describes other impressions of Einstein at this time:

> One of his passions was sailing. First music, then sailing and the beloved pipe .... He was an excellent sailor ... He was also very skilful with his hands. One night after dinner as we sat around the room E. was playing with an empty chair. Suddenly he stood it on its front legs and it stayed there. Probably this trick was a combination of his dexterity with his knowledge of the laws of balance. I tried to do this trick for 3 decades without success. I never found anyone else who could do it either.
>
> [...]
>
> One has also to talk about the negative aspects. He was a person like any other and one shouldn't idolise him by leaving out the 'other.' [He could be] very unpleasant and aggressive. This was another facet of his truthfulness and his inability to compromise .... I was often the witness of such outbreaks and towards the end once the recipient ... He could be very emotional.
>
> No, he was not vain with one exception – his violin playing. And even about that he had no reason to be. I think that his violin was almost as important to him as his science. He was intrinsically musical. If you were playing chamber music with him and he lost his place, there was no need to stop. He jumped back in at the right moment. His playing was very run-of-the-mill, colourless, not very black or white but more grey. But his phrasing was faultless.

Another assessment of Einstein's playing at this time came from Plesch:

> He was not a very good technical performer, but I don't know anyone who exceeded him in fervour and sensibility. He would practise very zealously for his beloved chamber-music evenings, but it was on this field that he felt the gap between desire and performance most deeply.
>
> His hand was not the characteristic one of the great artist. It was rather long and yet fleshy with pointed fingers; quite different from the bony fingers of Wagner of those of Franz Liszt [...] or from the hands of Kreisler, D'Albert, Orlik, Slevogt, Schnabel and other great artists I have known. With those pointed fingers of his, Einstein could produce a fine enough tone, but he lacked technical virtuosity, and, in particular, a fluent technique of bowing. He is well aware of his shortcomings in this respect

> and they often make him comically furious, particularly when he has to negotiate a technically difficult passage. As I have said, he is not a jealous or an envious man, but in such moments he does envy the great performers from the bottom of his heart. However, his own difficulties usually end up in a burst of mortified laughter.

Other opinions of his playing were also far from complimentary, one being that he had a bow stroke like a lumberjack.

Einstein gave two interviews for the American press in mid-1929. One, authored by the artist Samuel Woolf, who drew a striking portrait of Einstein during the interview, was published in *The New York Times* on 18 August. With reference to his violin playing, Einstein is quoted as saying 'I get a vast amount of pleasure from it. When I am too tired to read or work I find that music rests me. Much has been said of my playing but I am sure that I am the one who gets the most amusement from it.' The second was a wide-ranging interview to George Alexander Viereck that was published in October 1929 in *The Saturday Evening Post*.[2] In addition to a brave attempt to make distant parallelism and the fourth dimension understandable to his readers, Viereck also gave his impression of other aspects of Einstein's life:

> Professor Einstein looks more like a musician than like a mathematician. 'If,' he confessed to me, with a smile that was half wistful, half apologetic, 'I were not a physicist, I would probably be a musician. I often think in music. I live my daydreams in music. I see my life in terms of music.'
>
> 'Perhaps,' I remarked, 'if you had chosen to become a musician you would outshine Richard Strauss and Schoenberg. Perhaps you would have given us the music of the spheres or a fourth-dimensional music.'
>
> Einstein gazed dreamily – was it into the far corner of the room, or was it into space – that space which his investigations have robbed of infinity! 'I cannot tell,' he replied, 'if I would have done any creative work of importance in music, but I do know that I get most joy in life out of my violin.' As a matter of fact, Einstein's taste in music is severely classical. Even Wagner is to him no unalloyed feast of the ears. He adores Mozart and Bach. He even prefers their work to the architectural music of Beethoven.

Einstein's view of Beethoven was always somewhat ambivalent. In November 1929, he wrote to the author of two books on the composer's life, Hector Laisné, who had been recommended by Einstein's old friend Solovine. Einstein told the author that he had greatly enjoyed his account of Beethoven's feelings and the effect his compositions had on his auditors, which he felt had a clear message for their own time with its threats to culture.

The Viereck article quoted above is also interesting for the insight it gives into Einstein's insistence on determinism as the foundation for his world view. He concludes the interview by saying:

> Everything is determined, the beginning as well as the end, by forces over which we have no control. It is determined for the insect as well as for the star. Human beings, vegetables or cosmic dust, we all dance to a mysterious tune, intoned in the distance by an invisible player.

---

[2] *The Saturday Evening Post* is a distinguished American illustrated weekly, founded in 1821, which traces its origins, somewhat indirectly, back to Benjamin Franklin. Viereck himself was an interesting figure. Half-American, half-German, the German half reputedly descended illegitimately from Kaiser Wilhelm I, Viereck had supported the German cause in America during the First World War. He subsequently became a strong supporter of the Nazis, his activities on their behalf earning him a prison term in the US during the Second World War. In between, he wrote a book popularising science and a number of pen-portraits of prominent figures, of which this is one.

Such statements allow his problems with quantum mechanics to be put in context.

The year 1929 saw the award of a number of distinctions to Einstein, including the honorary degree of the Sorbonne mentioned in Chapter 19. One that gave him particular pleasure was the inaugural Max Planck Medal of the German Physical Society, which was presented to him by Planck himself in a ceremony in a crowded auditorium in the Technical University of Berlin on 28 June. He also took part in a ceremony to honour another eminent scientist. On the evening of 21 October, he travelled from Caputh to Berlin to broadcast congratulations to Thomas Edison on the fiftieth anniversary of his invention of the incandescent lamp. He paid tribute to the electrical industry thereby founded, although he made no mention of its role in his own childhood and youth, when his father and uncle tried and failed to exploit Edison's invention. Unfortunately, the combination of Einstein's German and problems with the transmission meant that the speech was unintelligible on the other side of the Atlantic.

Einstein's relationship with his younger son Tete absorbed an increasing amount of his attention during this period. Tete had enjoyed his stay with Einstein in Berlin over Easter 1929. His experience at his Zurich school as he approached his final exams that summer seems to have been a good one. His letters to his father show typical invention in poetry and aphorisms and a questing intelligence that delighted his father. He passed his final examination with a score sufficient to ensure his admission to one of the most competitive faculties in the University of Zurich, that of medicine. Einstein stayed with Mileva and Tete in Zurich from approximately 10–15 August while attending the meeting of the Jewish Agency Council, held directly after the Sixteenth Zionist Congress had met. Einstein received a tremendous reception when he addressed the Council, which was intended to broaden participation in Jewish affairs and Palestine from Jews who did not consider themselves Zionists. In his speech, Einstein lauded the work of Chaim Weizmann and his colleagues, typically taking a position as almost, but not quite, a Zionist. He was guest of honour at a banquet held by the American Jewish Physicians Committee for the Hebrew University of Jerusalem. His schedule was extremely packed, to the extent that he did not even have time to meet Heinrich Zangger, but he and Tete must have spent some time together. By 19 August, Einstein was back in Caputh.

In September 1929, Tete and a group of his school friends descended on Maja in Florence, overflowing her accommodation but enjoying themselves greatly. Einstein replied to Maja's report of the visit with delight that she had enjoyed Tete's company, whom he characterised as a fine character, full of life. Immediately on his return from Italy, Tete started his first year of medical study. His intention was to specialise in psychiatry; he had developed a fascination for Freud, whose picture hung over his bed. He had signed up for nine hours a day of lectures and related activities. That summer he drew on a postcard his characterisation of the mind of someone who had just finished his school certificate. His drawing contains a bewildering mixture of chemistry, Latin, Greek, literature, and physics terms, suggesting a mind bursting with incoherent knowledge that threatened to overwhelm it. The shock of independent study at university, together with the punishing lecture schedule to which he had committed himself, proved too much. By mid-December, he was causing concern to both his mother and to Einstein's trusty factotum, Zangger.[3]

[3] The progression of Tete's mental problems can most easily and coherently be followed by the exchange of letters between Einstein and Zangger between 1929 and 1933. Other sources are at times inconsistent and contradictory. For example, Trbuhović-Gjurić links a suicidal episode of Tete's to late November 1930, but this does not seem likely given other evidence of his mental state at the time; her source for this appears to be Marianoff's book, which has already been seen to be highly unreliable in dating.

Zangger, who had supervised much of Tete's treatment for his childhood physical ailments, wrote to Einstein on 19 December 1929, saying that Tete seemed depressed, distressed, and unable to concentrate. He blamed the brutally hard work required by his medical course and thought that it might be sensible for him to suspend his studies and to spend some time in the mountains in peace and quiet. A few days later, after consultation with various colleagues, he wrote to Einstein again. Although Tete was considerably better now that the Christmas break had begun, Zangger still recommended recuperation in the mountains and told Einstein that it would do the boy good if his father could join him. His final words, that the Einsteins were not an easy brood, seem heartfelt.

Einstein replied to Zangger immediately, agreeing that it would be better if Tete were studying something less enervating than medicine, which had probably given him mental indigestion. However, rather than answering Zangger's query about sending Tete to the mountains of the Engadine, he launched into one of his obsessive worries: that the mental illness in Mileva's family was manifesting itself in their son. If so, there was nothing that could be done. His concern for Tete is however evident, believing that it would be good for him to come to Berlin as he could talk to Einstein more easily than to his repressed mother, who Einstein was surprised had not written to him. Shortly before Christmas, Einstein wrote to Tete, urging him to visit him in Berlin where they could talk his problems through. Once the New Year festivities were over, Einstein wrote to Zangger again, complaining that the medical authorities (by then Prof Dr Maier, the director of the Burghölzli sanatorium in Zurich) were keeping his son away from him. Einstein was spending two weeks alone in Fürstenberg an der Havel, an isolated area of north Germany. He thought that it would have done Tete good to have joined him. He also opposed the idea of Tete suspending his studies, on the basis that he would see this as evidence of his own incapacity. Ten days later, Zangger replied from the Engadine, where he had accompanied Tete and who he reported was feeling much better. After Tete's daily mountain walk, they had discussed various points of chemistry, in particular electronic orbits, with which Zangger felt Einstein could help Tete at their next meeting. This was a subject that Tete, never at home in science, found particularly difficult; he regularly skipped his chemistry lectures. Zangger thought that, provided he took regular breaks and visited Maier for regular check-ups, Tete might after all be able to cope with his medical course.

For some time, Tete did indeed seem able to cope with his work. He visited his father in Berlin at the end of the semester in early March 1930. Einstein wrote to Zangger that Tete's reaction was not dissimilar to his own; he also had been in a sorry state during his first year of physics studies and that it took him a year after graduation before he could think again.[4] Their confidence that Tete had recovered was justified when he was able to pass his end-of-year examinations in the summer. He visited Einstein again, this time in Caputh in the autumn of 1930; Einstein was delighted with him, calling him a real and sympathetic comrade. They played music together in early November 1930, when Einstein was in Zurich to receive an honorary degree from ETH at its 75th anniversary celebrations.

The economic dislocation of the Great Depression would have important effects on both Tete and in particular Mileva. The effects of the US stock market crash on 29 October 1929 rippled through the rest of the world over the next few years. Einstein's extended family lost very significant amounts of money. In the summer of 1930, the bonds which Einstein

---

[4] This statement, repeated in his autobiographical notes published in 1947, is another example of Einstein's memory leading him astray, as discussed in Chapter 4.

had insisted that Mileva invest in US securities came to their term and required reinvestment. Fortunately, they seem to have held their value but Einstein's understandable nervousness about the financial markets influenced him to recommend to Mileva that they should be withdrawn from the US and invested in property in Switzerland. By August 1930, Mileva found a suitable house in her neighbourhood and purchased it, intending to rent it out, followed rapidly by a third house, so that the remaining funds in the US were severely depleted. This led to a major drop in the revenue that had formed an important part of Mileva's income, which would have been compensated by the rent from the new buildings. Unfortunately, however, the depression had by then hit Switzerland. The number of people able to pay the high rents that properties on the Zurichberg commanded diminished greatly, leaving Mileva's properties either vacant or occupied by tenants who fell into rent arrears. By the summer of 1932, the situation had become serious, and Einstein called in professional advice, which Mileva did not act on. The rest of the US securities had been used up by 1936 on building repairs, while the income from the two new properties did not even cover the mortgage repayments. With Einstein by then no longer in Europe, Mileva's financial affairs degenerated into a chaos that neither she nor her ex-husband was able to resolve.

An interesting vignette of Einstein at the beginning of the 1930s is provided by a student, Wolfgang Yourgrau.[5] In the late twenties, Yourgrau attended Einstein's occasional lectures at Berlin University, which he characterised as unimpressive, with the lecturer speaking quietly with a marked Swabian accent, pausing for long periods before continuing with a new approach to a problem. His writing often wandered obscurely across the blackboard. Yourgrau was involved in a serious train accident on his way back from visiting his father in Brussels and spent months in rehabilitation in hospital. In a gesture of remarkable kindliness, he was visited twice weekly by both Einstein and Elsa. Einstein kept his spirits up by talking of physics while Elsa brought him food. At the end of each visit, Einstein would slip some money into an envelope and leave it on his bedside table, which embarrassed Yourgrau, both of whose divorced parents were wealthy. Yourgrau recovered, graduated, and became an assistant to Erwin Schrödinger, by then at the University of Berlin, whom he idolised.

Yourgrau was also an accomplished cellist. He, and his similarly talented but eccentric mother, shared a music stand in a chamber-music ensemble conducted by the renowned Michael Taube, who subsequently played an important role with Einstein and Bronislaw Huberman in the foundation of the Palestine Symphony Orchestra. Yougrau recounted that, while they were playing one of Händel's Concerti Grossi, his mother caught sight of Einstein in the audience. In a loud stage-whisper she asked him who that man with the unkempt hair and collar covered in dandruff was? Mortified, he tried both to explain who it was and keep her quiet. He failed and, feigning a headache at the interval, crept away before the second half of the concert.

Yourgrau dreaded the occasions when he had to play in a quartet with Einstein. The quartet met on Friday evenings at a venue very close to the Haberlandstrasse apartment. It normally consisted of Einstein, Yourgrau's uncle, an old lady, and a cellist, whom Yourgrau replaced when indisposed. According to Yourgrau:

---

[5] Yourgrau's fascinating life began with a childhood oscillating between Berlin and Brussels, subsequently ranging from brawling with Nazi storm-troopers, escaping to Poland, journalism in Jerusalem, service with the US Office of Strategic Services during the war, physics research in South Africa and finally a chair in physics at the University of Denver, USA.

> It is true that the honoured master was extremely musical; he had a great love for Mozart and a slightly lesser for Bach. Unfortunately, his love of music and wonderful knowledge of the subject did not translate into ability as a violinist. I was greatly relieved when the cellist, for whom I had stood in, reclaimed his seat.

In Yourgrau's view, the cellist was the only competent member of the quartet.

Einstein began 1930 with the already mentioned solitary stay at Fürstenberg an der Havel. He returned to Berlin by 21 January when he attended the farewell dinner of American Ambassador Jacob Schurman, at the Hotel Kaiserhof.[6] He appeared in a concert in aid of the poor Jewish community at the synagogue in Oranienburger Strasse on 29 January. The concert was predominantly choral and starred the famous tenor Herman Jadlowker, who by then had retired from the stage to become the cantor of the synagogue in Riga. He was accompanied by the Oranienburg synagogue choir, reinforced for the occasion. Einstein played together with another violinist, Professor Alfred Lewandowski, the son of Louis Lewandowski, perhaps the most famous composer of Jewish liturgical music, whose 'Hallelujah' ended the concert. Einstein and Lewandowski played two pieces, accompanied by Arthur Zepke on the organ: the Trio Sonata in B-flat major, Op. 2 No. 3, HWV 388 by Händel and the slow movement of the Bach Double Concerto. A newspaper reported that Einstein's prominence in other areas in no way overshadowed his great abilities as a violinist.

After Tete's stay in Berlin during March, Einstein set out for Leiden for the first time for several years on 1 April. He stayed with Paul Ehrenfest for about a week. This also turned out to be Einstein's last formal visit to the university, although he did make several more brief stays in the city. Dima Marianoff, at the Hague on other business, joined Einstein at Ehrenfest's house.[7] Marianoff was struck by the chaotically untidy and spartan nature of the rooms, which neither Ehrenfest nor his daughter nor Einstein seemed to notice. On their return from a walk beyond the city limits, Marianoff and Einstein stopped at an inn and ordered sandwiches and coffee. Einstein realised that he had to pay:

> This was the first time I ever saw Albert pay for anything. He had little use for money. It seemed to bother him. When he had to use buses or take railroads, the exact fare was provided for him. so as to make it less difficult . . . . He managed to find a change purse with an old-fashioned clasp, looked into it with a puzzled expression for a moment, then with meticulous care searched for the right amount, and with an air of much relief, laid it on the table.

They discussed the rise of the Nazis. Einstein was inclined to downplay their importance, trusting in the innate good sense of the German people. His final words to Marianoff, who had pointed to the large number of workers joining the Nazi party, were 'I don't think this can be so bad as you imagine – no, I don't think so.'

Einstein's scepticism about the importance of the Nazi's seemed reasonable, despite the collapse of the coalition government led by the German Socialist Party (SPD) only a few days before. After all, Hitler had only five years previously still been in prison after the Munich Beer Hall Putsch.

[6] This is interesting only in that a photograph of Einstein and the US Consul General shows that Einstein is clearly wearing formal dress of white tie and tails, as would be expected for such an occasion at the time. The Einsteins' maid, Herta, interviewed many years later, claimed that he did not own a 'Frack', as a tailcoat is known in German and would not have borrowed one. Einstein also claimed not to have owned one. Many other photographs exist in which he also appears to be wearing a 'Frack', for example Fig. 21.4. Much as Einstein disliked wearing formal clothes, when they were expected, he obliged.

[7] Marianoff implies that this visit was sometime in 1931 but 1930 seems more probable.

The Nazis had twelve seats in the Reichstag from a total of 491 and had been supported by fewer than one million voters. However, at the elections of September 1930, the Nazis won 107 seats, overhauling the Communists to become the second largest party, supported by more than six million voters.

Immediately after his return from Leiden, both Einstein and Elsa wrote to the pianist Harriet Cohen[8] on 11 April 1930 with apologies for not attending her concert in Berlin. They were moving out to Caputh on 14 April, from whence it was impossible to return by public transport after the concert. Although his friend Grete Lebach was by then living in Berlin, they must have retained their house in Düsseldorf, since Einstein was in that city on 17 April. In some sense the 'official' opening of the Caputh house was on 4 May, when Max von Laue wrote the first entry in the visitors' book, donated by Einstein's neighbours, the Sterns. In keeping with the instructions that Einstein inscribed at the top of the first page, von Laue wrote in verse. Indeed, he also wrote the last entry, not quite three years later. This was appropriate, not only because von Laue was, with Ehrenfest, the closest of his physicist friends, but also because he was a frequent visitor to Caputh. He often came in the company of Planck. They would discuss physics and Academy of Science business with Einstein on long walks through the surrounding woods. Einstein characterised his relationship with Planck to von Laue in the following terms: 'Planck loves you like a son, for me he has at best the feeling of a reptile collector who has got a rare but difficult-to-classify specimen in his collection. He stands bemused in front of it, unsure of whether the thing bites.'

The relaxed atmosphere at Caputh was only interrupted by tension between Einstein and Elsa over his lady friends. Toni Mendel was a frequent visitor, but Elsa seems to have come to terms with Einstein and Toni's relationship and to have found Toni relatively congenial. Grete Lebach was a different matter. Even before the work on building the house had been completed, in October 1929, Einstein invited Lebach to go sailing with him. The Einsteins' maid recalled that whenever Lebach was expected, which was frequently, Elsa would leave and spend the entire day in Berlin, ostensibly to shop. On one occasion, Konrad Wachsmann recounted that Einstein's colleague Mayer collected a rather daring bathing suit from the *Tümmler* for washing. Assuming that it belonged to Margot, he handed it to Elsa. In fact, the owner was probably Lebach, and the aftermath was a prolonged shouting match between Einstein and Elsa. However, such scenes were brief interruptions in the rural idyll. A dog from a nearby house was a constant visitor, often accompanying Einstein on his long walks. A cat was also a great favourite with the female members of the household; neighbours would be paid to look after it during the winter when the Einsteins were not in residence.

Einstein had accepted the offer of an honorary degree from the University of Cambridge, which was to be conferred on 5 June 1930. One reason for his acceptance was the chance to meet Arthur Eddington, with whom he stayed during this visit. Although Eddington had been his strongest supporter in the UK and his measurement of stellar deflections at the 1919 eclipse had established general relativity, the two had met only a handful of times, most recently in 1926 at a conference in Berlin. Einstein had regretfully declined a series of invitations from Eddington to visit Cambridge in the 1920s, so he was pleased to finally be able to make the journey, even though it entailed the sort of social events and academic flummery that he disliked. By a happy chance, Einstein's visit took place just after the announcement of the King's Birthday Honours List, in which Eddington became Sir Arthur.

[8] Another link with Einstein was that Cohen had also been invited to give a concert at the International Congress for Sex Research (see Chapter 17). Unlike Einstein, she accepted and played at the second congress in London in August 1930.

After the degree ceremony, at which Planck was also honoured, the honorary graduands and distinguished guests were entertained at a garden party in the grounds of Magdalene College, whose Master was the current Vice-Chancellor. A Fellow, Dennis Babbage, recalled giving Einstein an ice cream during the event. Einstein would have enjoyed this rather relaxed event more than the next – a formal dinner at Trinity, the grandest of Cambridge colleges.

The dinner that evening was even more elaborate than usual for honorary graduands, as it also marked the installation of Stanley Baldwin, Prime Minister until the previous year, as Chancellor of the university. Baldwin had been a student at Trinity (where he had obtained an ignominious Third Class degree in history), so it was Trinity's responsibility to hold the dinner. The Master of Trinity, who presided over the dinner, was Sir J. J. Thomson. Attendees included Lord Rutherford, Henry Tizard, Eddington, Sir Joseph Lamor, and Sir James Jeans, but only Jeans and Thomson were on the high table with Einstein, although he was placed far away from either. The scene was described by the Rouse Ball Professor of Mathematics and Fellow of Trinity, John Littlewood:

> Bertrand Russell and I each met Einstein once. I was put next to him at the dinner at Trinity to Baldwin when he became Chancellor, on the dubious ground that I could speak German – Einstein then spoke no English. He was very good at speaking simply, and we managed. We talked mostly about music . . . Opposite me was the American Ambassador [C. G. Dawes], pale with anxiety about his speech (which turned out to be very odd). On his left [HRH] the Duke of Gloucester, and on his right, a sporting peer [Lord Queenborough]. These were all honorary graduands, created by Baldwin. The peer got considerably drunk, and heckled Einstein across the table – Einstein could not understand a word. The peer[9] boasted afterwards of having stumped Einstein on Relativity.
>
> All Bertrand Russell said about their meeting (apart from how wonderful Einstein looked, and his wonderful eyes) was that Einstein told him a dirty story. I said that Einstein was well known for consummate tact in adapting himself to his company.

After dinner, Einstein saw a familiar face, his former student, Esther Salaman, who had gone to study in Cambridge on his recommendation, and then married and had a child:

> The men wined and dined for two hours, and only then came up to the Master's Lodge where Lady Thomson and her daughter were entertaining the ladies. J. J. Thomson was walking round with Einstein, telling him about the famous portraits. Einstein looked his usual calm self. Francis Cornford asked me to come with him, and we approached J. J. 'Master, here is a pupil of Einstein.' J. J. stepped aside. Einstein looked very pleased to see me and I introduced my husband. Einstein said he had heard from his friends about us. It was a happy meeting. Then Eddington approached him, and Einstein's face lit up. We left them together.

On the following day, Einstein travelled to the University of Nottingham. He had originally been invited to attend the opening of the university by Professor Henry Brose, the head of the Physics Department. It had been founded as an outstation of University College London in 1927. Brose had worked on relativity for many years and had translated several of Einstein's articles. Einstein arrived very late, since on the journey he had made a stop at Newton's birthplace at Woolsthorpe. He lectured in German to a packed hall. An eye-witness account describes him in the following terms:

---

[9] Queenborough was a hugely reactionary Conservative politician who was placed next to the American ambassador because he had spent considerable time in the US. Queenborough was a supporter of Franco and Hitler and rabidly anti-communist, which presumably explains his hostility to Einstein, with his well-known socialist principles. Littlewood's use of the adjective 'sporting' is a reference to Queensborough's prowess at yachting: he was Commodore of the Royal Thames Yacht Club.

> Einstein was a remarkably serene and august person, with a halo of slightly greying hair. He stood and looked around himself, with a slightly lost air of benevolence and bewilderment, mixed. He had a benign and remote calm. I thought he looked with this wonderful hair, like a mixture between an amiable lion and one of those calm looking pandas that we see in zoos.

By 12 June, Einstein was back in Caputh. Frederick Lindemann was interfacing with the Oxford authorities in organising the Rhodes Lectures in Oxford, which Einstein had agreed to deliver finally in the following year. Replying to him, Einstein admitted that he felt unwell. He had recovered by 14 July, when Rabindranath Tagore, Nobel Laureate in Literature, visited Caputh. They had met first in May 1926, when Tagore had delivered a lecture at the Berlin Philharmonic Hall. After a government reception the following morning, Einstein had invited Tagore to take tea in Haberlandstrasse.

The visit in July 1930 made quite an impression on Tagore. He described Einstein's shock of white hair, burning eyes, and warm manner when he met him at the start of the track that led through the woods to his house. Their conversations both on the terrace at Caputh, and subsequently at another meeting at Toni Mendel's palatial estate, ranged widely over philosophy, religion, and music.[10] In the first meeting, they spent considerable time debating whether the concept of truth was independent of the existence of humanity. Einstein, unsurprisingly, held that it was, Tagore, that it was not. They discussed the question associated with Bishop Berkeley, in this case as to whether a table on the balcony exists when there is no one to observe it. Tagore concluded by exploring the possibility of consciousness freed of space and only existing in time and that his religion was the reconciliation of the Super-personal Man, the Universal human spirit, in his own person, which had been the subject of the Hibbert lectures he had just delivered in Oxford.

The second discussion took place on 19 August 1930, either in Toni Mendel's house or in the villa that her son-in-law (and nephew) Bruno Mendel had just constructed in the grounds. Mendel, who was a medical doctor, was a friend of Tagore. At this meeting, Einstein and Tagore mostly discussed music, although they began by discussing the German youth movement, one of whose centres, near Koblenz, Tagore had just visited. It is poignant to read Tagore's enthusiasm for this movement, given its distortion only a few years later into the Hitler Youth. Moving on to causality, Tagore likened the small freedom for personal initiative that he believed must exist to Indian musical conventions. In response to Einstein's questioning, Tagore explained the various forms of Indian music and the role of improvisation within it. They continued:

EINSTEIN: European music is not older than 400 years; it was only from then that music, polyphonous music, has been introduced.

TAGORE: Has melody suffered in your music by this imposition of harmony?

EINSTEIN: Sometimes it does suffer very much. Sometimes polyphonous music swallows up the melody altogether.

TAGORE: Melody and harmony are like lines and colours in pictures. A simple linear picture may be completely beautiful – the introduction of colour may make it vague and insignificant. Yet

[10] Dima Marianoff recorded the conversations and published a version in *The New York Times* the following month, which Einstein vetted, although he later opined that the language difficulty between the two principals was such that the article should never have been published. Another version was published by a member of Tagore's entourage, which is subtly different in emphasis and content. The source used here is a subsequent version that reconciles the two printed versions and adds other relevant documents of Tagore's.

colour may by combination with lines create great pictures so long as it does not smother and destroy their value.

EINSTEIN: It is a beautiful comparison; the line is also much older than the colour. It seems that the structure of your melody is much richer than here; Japanese music too seems to be so.

TAGORE: It is so difficult to analyze the effect of Eastern and Western music on our minds. I feel deeply moved by the Western music – I feel that it is great, that it is vast in its structure and grand in its composition. Our own music touches me more deeply by its fundamental lyrical appeal. European music is epic in character, it has a broad background and is gothic in its structure.

EINSTEIN: Yes, yes, this is very true.

TAGORE: But this music is immense – I can never forget that how much I was affected by its power once when a Hungarian lady played on her violin some pieces of music, both classical and modern.

EINSTEIN: When did you first hear European music?

TAGORE: At 17 I first came to know it intimately when I first came to Europe, but even before that time I had heard European music in our own household. I had heard the music of Chopin and others at an early age.

EINSTEIN: There is a question which we Europeans cannot properly answer, we are so used to our own music. We want to know whether our music is a conventional or a fundamental human feeling, whether to feel consonance and dissonance is natural or is it a convention which we accept.

TAGORE: Somehow Piano confounds me. Violin pleases me much more.

Einstein and Tagore met for a third time in late September in Haberlandstrasse, after Tagore's return from Moscow. He had been accompanied by a group that included Margot Einstein, who was fascinated by Tagore and his thought.[11] Their final meeting in 1930 would be in New York in December, which Einstein visited on his way to California Institute of Technology (Caltech) in Pasadena.

On the day following Tagore's first visit, 15 July, Einstein attended a lunch at the house of Hugo Simons in honour of Aristide Maillol, the French artist and sculptor. Einstein had a family interest in sculpting through Margot's activities. Also attending were Max Liebermann, Renée Sintenis, a German sculptor, Julius Meier-Graefe the novelist and art critic, and his wife Anna Marie Epstein, an artist. Harry Graf Kessler was also there. He recorded in his diaries that he pointed out Einstein to Maillol when he entered the room. Maillot commented that he had a good head and wondered whether he was a poet.

On 23 July, Einstein was in Geneva to attend a meeting of his League of Nations committee, now chaired by Gilbert Murray, an Oxford classicist, following Lorentz's death. He travelled on to Zurich to visit Tete, staying with him and Mileva, at the end of July. Einstein brought with him one of his violins, which he deposited with Mileva in Zurich for his use when in the city. Despite having

[11] Marianoff also accompanied Tagore as an interpreter but in typical Marianoff fashion places the trip in December 1930, not the correct date of September. The Einsteins' housekeeper, Herta, mentioned that Marianoff and Margot visited Moscow soon after they were married, in December. Since he describes the ice, snow, and their fur coats, it may be that he and Margot visited Russia twice, once with Tagore and subsequently for a honeymoon.

several violins, he only had one bow, so Tete was requested to find one for him to use during their music sessions together. Einstein returned to Caputh on 4 August, the day after his sister Maja arrived. Immediately after his arrival, Einstein took Maja sailing on the *Tümmler*. She wrote to a friend that her brother and she were often on the water alone, which she greatly enjoyed. They sometimes talked a lot, sometimes little, and enjoyed each other's company. When Maja left to return to Florence, Elsa loaded her down with gifts of clothes and household items.

Einstein's summer of 1930 was a busy one. The Seventh German Radio and Audio Exhibition took place in the Charlottenburg district of Berlin. Einstein opened it on 22 August with a speech in which, fighting against a gusting wind and the noise of passing trams, he paid tribute to the pioneers who had made the discoveries necessary to allow radio to become ubiquitous in the industrialised world. He also joked that it was a scandal that those who listened to the radio had no more idea of the scientific ingenuity behind it than cows had of the botany of the plants they were eating. He also pointed out the unique potential that the radio had to further international reconciliation. Naturally his speech was broadcast on the radio and parts were recorded.

Norman Bentwich, who had been one of Einstein's hosts in Jerusalem in 1923, visited Caputh in September. He wanted to discuss the situation in Palestine. Einstein told him that he would no longer remain associated with the Zionist movement unless it began substantive talks with the Arab community to find an acceptable *modus vivendi*. Bentwich also noted that the Jewish community in Berlin had become seriously alarmed by the rise of the Nazi party following its gains in the recent Reichstag elections and that Einstein himself was anxious.

Tete also visited Caputh in September. Since there was no piano in the house, it is unclear how much music they played, but Einstein would have arranged for them to use a neighbour's piano, as he had on former occasions. He certainly took Tete sailing. He stayed until Einstein left for the Sixth Solvay Conference in Brussels, which began on 20 October.

This Solvay Conference was the first to be held after the death of Lorentz, its chair from the inauguration in 1911. The new chair, Paul Langevin, began proceedings by reading a telegram that had been sent to King Albert, expressing the participants' respectful greetings. He continued with a tribute to Lorentz and his work as chair and thanked his colleagues for their trust in selecting him as his successor. A telegram of greeting was sent to Lorentz's widow, as well as one to the widow of Ernst Solvay's son, Armand, who had taken over from his father as sponsor of the Solvay Institute. Arnold Sommerfeld, who had not attended since 1913, gave the first talk on magnetism and spectroscopy. The overall topic of the conference was magnetism, an area in which Einstein had done little work since his collaboration with Wander de Hass (who also attended the conference) in 1915 and his joint paper with Ehrenfest (who was not a delegate to the conference) on the Stern–Gerlach experiment in 1922. Indeed, Einstein made very little contribution to the conference proper, not giving a talk and only contributing two brief comments to the recorded discussion.

Although not particularly active in the conference sessions, Einstein, as in 1927, was at the centre of the informal discussions. These mostly took place in the Club de la Fondation Universitaire, where the delegates were staying. Several other prominent physicists, while not formally delegates, were attracted to Brussels and into orbits around the stellar participants. These included de Broglie, Ehrenfest, the prominent theoretician from the University of Louvain Charles Manneback, and Leon Rosenfeld. It was the latter who recalled the discussions between Einstein and Bohr. It seems as if Einstein had been thinking for some time before the conference of a thought experiment that would reveal the shortcomings of quantum mechanics. He considered a box that contained a source of photons and had a shutter that could be opened by a clock within the box for a time short enough to allow a single photon to escape. By weighing the box before and after the shutter opened and using the relativistic relationship between mass and energy, the energy of the photon could be deduced. Since Einstein assumed that the mass of the box could be weighed with arbitrary precision and since the uncertainty in the time of emission of the photon was determined by the time the shutter was opened, then the product of the uncertainties in time and energy could be made arbitrarily small, in contradiction to Heisenberg's uncertainty principle.

Rosenfeld observed Einstein enter the club smiling broadly after he had posed his question to Bohr. Bohr, on the other hand, looked like a dog who had received a thrashing, with hanging head. He insisted that Einstein's thought experiment must be impossible. Ehrenfest took a famous photograph of the two at this conference, deep in discussion of the problem. The following morning, Bohr emerged after a presumably sleepless night. Beaming, he told Einstein that he had forgotten about his own construct, general relativity. Bohr's whole concept of quantum phenomena intimately involved the process of measurement so that he had to consider the problem as a whole, including how the box could be weighed. He considered it to be suspended from a spring balance with a pointer on a scale whose position indicated the weight. When the photon left the box, energy was lost and therefore mass reduced, so the box moved slightly upwards. General relativity predicts that a clock in a gravitation field ticks at a rate dependent on that field. Bohr calculated that the uncertainty caused by the movement of the box in the gravitational field was sufficient to restore the Heisenberg relations between the energy and time of the photon. The generally accepted outcome of this duel is that, discomfited, Einstein never again tried to find a way around the Heisenberg uncertainty principle. In fact, this may never have been his intention in the first place. In July 1931, Ehrenfest wrote to Bohr that Einstein had long ceased trying to disprove the uncertainty relation, as will be discussed in Chapter 21.

With hindsight, it can be seen that both Einstein's formulation of the box paradox and Bohr's refutation of it are oversimplified. When the photon passes through a narrow slit, wave–particle duality implies that it has a range of frequencies and hence energies. Although the sum of the energies of the box and photon are fixed by energy conservation, neither is individually fixed until one is determined. Bohr mixed uncertainties derived from quantum and classical effects in an unjustifiable way when discussing the general relativistic corrections. Indeed, general relativity is unnecessary; the concept that energy and mass are equivalent is enough to sustain Bohr's argument. It is now clear that the 'Einstein box' can only be understood using quantum entanglement. This concept would not even be named, let alone fully understood, for another five years.

The Solvay conference ended with the by now traditional invitation to dinner at the royal palace in Laeken on 26 October. Beforehand, the participants of the conference gathered for a photograph in the gardens of the palace. At the dinner, Einstein was seated next to Queen Elisabeth, who was flanked on the other side by Langevin. Marie Curie sat next to King Albert. Others attending included Bohr, Enrico Fermi, and de Haas and his wife, Lorentz's daughter.

Einstein left Brussels on the following day to travel to England for another dinner, at the Savoy Hotel on 28 October under the auspices of the Joint British Committee of the Societies ORT–OZE. ORT promoted technical aid and agriculture among the Jews in Eastern Europe, while OZE concerned itself with their health. Before the main event, Einstein met Weizmann, who could not attend the dinner that night for what he called political reasons. Weizmann was anxious to discuss the simmering tension between Arabs and Jews in Palestine; Weizmann and Einstein disagreed on possible solutions.[12]

---

[12] Weizmann thought that peaceful coexistence with the Arabs was not possible until they recognised the principle of the Jewish National Home, whereas Einstein had given a number of interviews emphasising

Also on 28 October, Einstein had lunch at the house of George Bernard Shaw, attended by his old acquaintance Sir Herbert Samuel. Although Shaw and Einstein had met only once, ten years before during Einstein's previous visit to England, Shaw was well versed in Einstein's work and indeed had developed something of a passion for tensor algebra. The reason for this was his friend and biographer, the mathematician Archibald Henderson, who had worked with Einstein in Berlin in 1924. Already in the same year, Shaw in the preface to *St Joan* had mentioned Einstein as a representative of human genius. Shaw's dramatic works contain twenty mentions of Einstein in total. Nor was the admiration one-sided. Einstein once said: 'There is not a superfluous word in Shaw's prose, just as there is not a superfluous note in Mozart's music. The one in the medium of language, the other in the medium of melody, expresses perfectly with almost superhuman precision, the message of his art and soul.' He and Einstein greatly enjoyed each other's company; Einstein appreciated Shaw's razor-sharp wit. After their lunch, Samuel introduced Einstein into the public gallery of the House of Commons to watch it in session.

The dinner at the Savoy was presided over by Lord Rothschild. Einstein was the guest of honour and George Bernard Shaw the principal speaker. Others attending were Astronomer Royal Sir Frank Dyson, H. G. Wells, and distinguished philosopher Samuel Alexander. Both Einstein's and Shaw's speeches were broadcast; that of Shaw was also filmed, although only short excerpts of the film have survived. Einstein would surely have been embarrassed by Shaw's laudatory, and extremely long, speech:

> . . . there is an order of men who [. . .] are makers of universes on earth. They are very rare. I go back, two thousand five hundred years, and how many can I count in that period? [. . .] Pythagoras, Ptolemy, Aristotle, Copernicus, Kepler, Galileo, Newton, Einstein [. . .] And I must – even amongst those eight men I must even make a distinction. I have called them makers of universes, but some of them were only repairers of universes. Only three of them made universes. Ptolemy made a universe which lasted fourteen hundred years. Newton also made a universe, which has lasted three hundred years. Einstein has made a universe, and I can't tell you how long that will last. [. . .] I rejoice at the new universe to which he has introduced us. I rejoice at the fact that he has destroyed all the old sermons, all the old absolutes, all the old cut-and-dried conceptions even of time and space, which were so discouraging because they seemed all solid, that you never could get any further. I want to get further always. I want more and more problems. And our visitor has raised endless and wonderful problems and has begun solving them. [. . .] I now give you: Health and length of days to the greatest of our contemporaries: Einstein.

Einstein responded in German, translated into English directly after his speech. His remarks were much briefer than Shaw's. He began by talking about the aim of the evening, to improve the lot of the Jews of Eastern Europe, praising the work of both the ORT and OZE societies. He continued:

> It rejoices me to see before me Bernard Shaw and H. G. Wells, for whose conception of life I have a special feeling of sympathy. [. . .] I, personally, thank you for the unforgettable words which you have addressed to my mythical namesake, who has made my life so burdensome, who, in spite of his awkwardness and respectable dimension, is, after all, a very harmless fellow. But to you, however, I say

---

the importance of finding a peaceful solution by finding a path to co-operation without preconditions. Weizmann, as well as making it clear to Einstein in correspondence that he disagreed with him, was privately very critical of Einstein's stance, labelling its supporters 'extreme pacifists'. It is ironic that one of the leaders of this group was none other than Einstein's old sparring partner, Judah Magnes.

> that the being and fate of our people depend less upon external factors than that we remain true to our moral traditions, which have carried us through the centuries in spite of the heavy storms which broke in upon us. In the service of life, sacrifice becomes a Grace.

The evening was a fundraising success, with more than £4500 being pledged for the ORT–OZE.

Einstein spent the night with the Samuels in Porchester Terrace before travelling back to Brussels. Although there had been no opportunity during the Solvay Conference to play music with Queen Elisabeth, she had extracted a promise from Einstein to do so on his return from England. He arrived late in the evening of 29 October and found a quiet hotel near the station. On the following day he went to the royal palace at Laeken at 3 p.m. Einstein conversed with the King and Queen for an hour and then he and the Queen were joined by two others to play trios and quartets for two hours. After the music, Einstein was invited to stay for dinner. The scene can be reconstructed from a similar meal described by Catherine Barjansky, who began to tutor the Queen in modelling in the mid-1920s:

> Luncheon was served in her drawing room; a table already set was rolled in on wheels by two footmen; and another table was brought on which silver-covered dishes stood on electric heaters, and coffee was served in a blue thermos bottle. The footmen left, and we ate alone, waiting on each other.

Einstein was also struck by the fact he and the royal couple ate alone and served themselves, as well as by the simplicity of the meal: spinach, fried eggs, and potatoes. At some point in the visit, they strolled into the wonderful park surrounding the palace, where the King took a photograph of the Queen and Einstein (see Fig. 22.1), who wrote to Elsa that he greatly enjoyed his visit and appreciated the fine characters of his royal hosts.

From Brussels, Einstein travelled to Zurich and then quickly on to Florence to visit his sister Maja, where he arrived on 1 November. He wrote asking Michele Besso to come to Zurich, where Einstein was due to return to receive his honorary doctorate at the 75th anniversary celebration of ETH Zurich on 5 November. After the ceremony he returned to Berlin. On 9 November, the *Frankfurter Zeitung* published an article by Einstein to mark the 300th anniversary of the death of Johannes Kepler. In the article he described very clearly the sheer grind that Kepler went through to arrive at his results, thinking of hypotheses, ruling them out, and starting again until finally he found that the planetary orbits were ellipses with the sun at one focus. He concluded with the observation that Kepler's life's work showed that knowledge could only be gained not from simple observation but by the confrontation of observations with ideas.

Einstein published another newspaper article on the same day that centred around one of his other heroes, Baruch Spinoza. In the magazine of *The New York Times*, he analysed the causes of religion in primitive peoples and how they developed from fear of the unknown into a moral dimension. He charted how this could lead men of genius to conceive of a non-anthropomorphic God, in what he called a cosmic religion. This he believed was the driving force behind the work of Kepler and Newton and inspired many working to understand the basic principles of the universe today. Although he did not explicitly mention himself, he clearly implied that he was among these seekers after truth.

The reasons for Einstein's decision to publish his article in *The New York Times* at this juncture are unclear. The article may have been a belated riposte to the attack of Cardinal O'Connell of the previous year; it certainly expands on his telegrammatic answer to Rabbi Goldstein. That it was considered an important piece by *The New York Times* is borne out by the following full-page article, which explains and contextualises Einstein's references to Democritus, St Francis of

Assisi, and Spinoza. Einstein's article caused a storm of controversy among Christian groups of all denominations in the US, as he must surely have expected. If he had really wanted to make a quiet and private visit to the US the following month, concentrating on his scientific work, then this was the worst possible preparation. On the other hand, when published in German in the *Berliner Tageblatt* on 11 November, the article excited little interest. Another literary event took place in the US just before Einstein's arrival. Under the pseudonym Anton Reiser, Einstein's son-in-law Rudolf Kayser published his book *Albert Einstein: A Biographical Portrait.*[13] As the first 'authorised' biography of Einstein, in the sense that Einstein wrote its preface, it also generated considerable interest in the forthcoming Einstein visit.

Arnold Schoenberg, who was at that time living in Berlin, wrote to Einstein on 1 November asking him to sign a letter of support for Adolf Loos, a Viennese architect and friend of the composer. Loos and his supporters were convinced that Loos had never had proper recognition for his work; the letter Einstein was asked to sign, which had also been sent to Thomas Mann, after lavishly praising Loos's achievements, called for the establishment of a 'Loos school' where his acolytes could learn from the master. This somewhat bizarre request, particularly since Loos had been convicted of paedophilia in 1928, a fact not mentioned by Schoenberg, was politely refused by Einstein on 11 November. He had clearly never heard of Loos and hardly knew Schoenberg. Mann also refused to sign, although his brother Heinrich eventually did, together with James Joyce and Schoenberg himself. Nothing came of the initiative, however; Loos died in 1933.

On 29 November, Dima Marianoff and Margot Einstein were married in the local registry office. Marianoff recalled Einstein had found him and Margot together in her room shortly before and suggested that since they were clearly fond of each other, they should get married and do so before he and Elsa left for the US. Another reason for this sudden decision might have been a hint of scandal attached to their recent trip together to Moscow with Tagore. One week later, photographs show Einstein, Dima, and Margot outside the registry office (see Figure 20.1). The Associated Press article in *The New York Times* seems mostly fictional, reporting that Einstein only agreed to come if he could be back at his desk, in his 'soundproof room' [sic], in thirty minutes. Elsa was quoted as saying that Marianoff had arrived at their house several months ago and, from that moment, Margot's heart was lost; she said that they both valued their new son-in-law highly. *The Jewish Chronicle* took Einstein to task, as a supposed Zionist, for sanctioning a mixed marriage that took place on the Sabbath and for appearing particularly happy about it. There seems to have been no special celebration of the event and the principals returned to Haberlandstrasse for lunch.

One reason for Elsa not attending her daughter's wedding could be that she was preoccupied by packing for their trip to the US on the following day. Various attempts to attract Einstein to Caltech had taken place over the previous few years. Paul Epstein, who had begun his career in Moscow and then studied under Sommerfeld, had met Einstein in 1919 after he moved to Zurich. He had been a professor at Caltech since 1921. Already in December 1924, he had written to Einstein, urging him to accept Millikan's invitation to visit them. Among Epstein's inducements, which included the fact that Lorentz was also expected to be there in 1926, was that a good string quartet could be assembled and pianists for duos were available. Ehrenfest, who had visited in

[13] Since in the preface Einstein characterised Reiser as knowing him 'rather intimately . . . in bedroom slippers', the author's true identity was not difficult to guess. Einstein's antipathy to biographies, beginning with Moszkowski's ten years previously, resulted in his giving permission for the book to be published only in English, not in German.

Figure 20.1 Einstein and his step-daughter Margot with her betrothed Dimitri (Dima) Marianoff outside a Berlin Registry Office where they were to be married on 29 November 1930.

1924, had greatly enjoyed his stay. Nevertheless, despite repeated urgings, it took Einstein until 1930 to finally accede.

On the evening of 30 November, the Einsteins were waved off at the station by the newly minted Dr and Mrs Marianoff, Ilse and Rudi Kayser, and several others. They visited Einstein's favourite uncle, Caesar Koch, who now lived in Liège. After a family lunch, they caught the train to Antwerp. On the following day, having been joined by Einstein's assistant Walther Meyer and Helen Dukas, they boarded the SS *Belgenland*, bound for Los Angeles via New York and the Panama Canal. Also travelling on the *Belgenland* with the Einstein party was Estella Katzenellenbogen, a rich divorcee from Berlin who was a friend of both Einstein and Elsa. She had rather a similar relationship with Einstein as had Toni Mendel. She often provided her luxurious car for Einstein's use and accompanied him to concerts and the theatre. Like Toni, she probably also encountered Einstein's amorous advances.

The *Belgenland* was the pride of the Red Star Line's fleet. At 27,000 tons and more than 200 metres long, she was advertised as the largest ship to circle the globe. Einstein and Elsa had a splendid first-class suite, with two bedrooms and a shared lounge. Meyer and Dukas were of course accommodated in lower class cabins, third and tourist, respectively. However, the contentment that Einstein always felt on board ship was disturbed by importunate telegrams from reporters seeking interviews, particularly about his religious views. It did not take Einstein long to look for musical partners. He encountered Hendrik van Loon, a Dutch historian, journalist, humorist, and prolific author, who was travelling with his second wife, Eliza Helen, known as Jimmie. A man of heroic proportions, Einstein called him 'the fat Hollander'. By the fourth day of the voyage, when the waves had attained a size that confined Helen Dukas[14] below deck, they had already agreed to play violin duos together. Van Loon wrote to his son:

> This is a nice ship, 27,000 crew and 36 passengers. Einstein is on board. I shall try and catch him in the can one day and say, Now, lieber Meister, I know that you are in a hurry but how about these parallel lines . . . Do they really meet?

The two liked each other immediately. Both had brought along violins.[15] They entertained themselves and the little clutch of out of season passengers with evening concerts.

On the approach to New York on 11 December, boats full of reporters put out from Long Island to interview Einstein. They bombarded him with questions, which had to be translated into German and back into English by his 'fat Hollander', who had studied at Harvard and Cornell Universities. He had also translated a statement from Einstein that was distributed to the press. Finally, the *Belgenland* docked at the Hudson River pier and another group of reporters repeated the process, this time marshalled and translated by a combination of Elsa and the German Consul, Paul Schwarz. *The New York Times* reported that the questions included a request to define the fourth dimension in one word and discuss the virtues of his violin.

The *Belgenland* remained the Einsteins' base for the next four hectic days in New York. After the reporters had left the ship, Einstein broadcast two messages to the American people, for which $1000 was paid to his charitable fund. One of these was introduced by van Loon. Before they could have lunch, Einstein gave a final interview to the Jewish Telegraph Agency. After lunch, Schwarz entertained the Einsteins in his apartment and drove them around part of the city, including Chinatown, before they were given dinner by Einstein's old friend, Leonor Michaelis. Just as they had at their previous meeting in Nagoya in 1922, they played violin and piano sonatas together, on this occasion Bach and Brahms. After philosophical discussion around the dinner table with other guests, Einstein and Elsa returned to the ship.

The following day began with Zionist activities. Felix Warburg, his old sparring partner on the question of the governance of the Hebrew University of Jerusalem, and another banker, Bernard Kahn, met him in advance of the arrival of a delegation representing many Zionist organisations. The latter presented him with a 'Golden Book' detailing the many who had donated to Palestinian causes to celebrate his fiftieth birthday. Einstein was irritated by a speech by Menachem Ussishkin, in which he said that Einstein 'belonged to us'. The Einsteins had lunch with Adolph Ochs, the publisher of *The New York Times* and another influential Jewish New Yorker, and then toured the newspaper building. Ben Bennenson, a wealthy real estate developer and Mrs Michaelis drove the Einsteins to inspect Einstein's effigy in the 'Arch of Scientists' on the West Portal of the Riverside Church. They were met by its pastor, Dr Fosdick, and John D. Rockefeller Jr, who escorted the party to view the arch. George Viereck, who had interviewed Einstein a few months earlier, was also present. Einstein was the only savant depicted who was still alive, leading him to remark that he had better be careful of his future behaviour. That evening they and their party, including Estella Katzenellenbogen, attended a performance of *Carmen* at the Metropolitan Opera. Einstein had to acknowledge the applause of the audience by standing up in the General Manager's box

---

[14] She did not emerge until four days later, when, according to Einstein, she looked like a corpse on a brief vacation.

[15] In fact, at least according to the *Los Angeles Times* of 1 January 1931, Einstein had not brought his own violin with him, probably concerned about the effect of the changes of climate on their long voyage. He borrowed a violin, perhaps from a member of the ship's orchestra.

in the *entr'acte* and waving his handkerchief. He was escorted backstage after the event by Giulio Gatti-Casazza, the General Manager, to talk to the singers, including Maria Jeritza, who played Carmen, and the tenor Giovanni Martinelli. The Einsteins, evading waiting photographers, took a taxi back to the 'steamer,' as the *Belgenland* was designated in Elsa's eccentric and basic English.

At 11 a.m. on 13 December, Einstein was formally greeted at City Hall, a ceremony that had had to be postponed from the previous day. The Aldermanic Chamber was packed and thousands more cheered outside when parts of the ceremony were repeated for the cameras. Speeches were given by the President of Columbia University and New York Mayor Jimmy Walker; the former spoke learnedly at length, the latter Einstein found amusing. Einstein replied very briefly in German. The ceremony lasted for thirty minutes and then the party was taken to view the Egyptian Gallery and the Rembrandts at the Metropolitan Museum. A lunch reception was hosted by Felix Warburg, followed by a brief concert of lieder. In the afternoon, Einstein was taken to the NBC Studios, where he broadcast an address to Zionist youth on the current developments in Palestine, where the British authorities had recently made decisions that angered the Jewish settlers. Controversially inside the Jewish community, Einstein argued for cultivating good relations with the Arab population, which among many other advantages, would mean that the British authorities would not have to arbitrate disputes between the communities. After the address, Einstein returned to the ship, where Catherine and Alexandre Barjansky, who had moved to New York earlier that year, visited. After a very pleasant evening at the home of the Morris family, they went to a Hanukkah celebration attended by 18,000 people. Einstein received a jubilant reception. He gave a short speech, after which a singer from the Metropolitan Opera in a wonderful gold dress sang one of the great arias from *Tosca*. At the conclusion she sank to her knees before Einstein, who was greatly embarrassed. He was subsequently moved by a wonderful Hassidic song from a famous Kantor and by a young violinist who played Palestinian folk songs. The party did not return to the *Belgenland* until midnight.

The following day began with breakfast with Frau Michaelis and her daughter, who probably drove them to their appointments. They visited Abraham Flexner to discuss the institute he was trying to found at Princeton. This became the Institute for Advanced Study, of which Einstein was to become a faculty member. They went on to the suite of the great violinist Fritz Kreisler at the Hotel Savoy–Plaza, where Kreisler's wife Harriet appeared to have just got out of bed. Einstein and she did not get on, as evinced by the epithet he gave her, 'Megaera', the 'jealous one' of the Greek Furies. Harriet was notorious for the close eye she kept on all of Kreisler's activities. During his conversation with Einstein, Kreisler let slip the fact, much to Einstein's surprise, that he himself had composed the 'classical pieces' attributed to many different composers that he often played in his recitals. Kreisler did not admit this in public until 1935. Their final visit of the morning was to Tagore, who had arrived in the US after his trip to Russia. They were escorted there by Henry Morgenthau, former Ambassador to Turkey, Consul Schwarz, and their wives. Tagore had been ill and was due to return to India on the same day as Einstein set sail for California. Einstein was rather scornful of the way in which he thought Tagore had fallen for Soviet propaganda, particularly with regard to education, which Tagore thought was a model that India could follow.

After saying farewell to Tagore, the party were entertained to lunch by Schwarz. They went on to a matinee concert[16] starting at 2:30 p.m. at the Metropolitan Opera, in which Arturo Toscanini conducted a programme of Ermanno Wolf-Ferrari's Overture to *The Secret of Suzanne*, Beethoven's

[16] At this time, the New York Philharmonic Orchestra, or as it was then called, the Orchestra of the Philharmonic-Symphony Society of New York, played the same programme three times. The normal pattern was two successive evenings followed by a matinee, usually after a gap of a day. This was necessary to cope with the demand from subscribers.

Pastoral Symphony, the premiere of Kodaly's *Dances of Marosszek*, and Bach's Passacaglia and Fugue in C minor, BWV 582, arranged for orchestra by Resphigi. During the interval, Einstein, who had greatly enjoyed the Beethoven, was introduced to Toscanini, initiating their friendship. A review of the concert in *The New York Times* mentioned Einstein's presence solely to opine that relativity was as relevant to music as to science. The restrained and classical reading of Toscanini was contrasted with the histrionics of the more romantic interpretations. Toscanini's interpretation was much more to Einstein's taste.

Following the concert, John D. Rockefeller, Jr gave a reception in advance of Einstein's three final engagements of the day. He and Einstein discussed the scholarships provided by the Rockefeller Foundation, the criteria for which Einstein considered much too strict. Rockefeller agreed to examine any case that Einstein brought to his attention.[17] After the reception, Einstein briefly addressed two gatherings, one of Jewish doctors, before he left for the New Historical Society to give an evening lecture. Here he delivered the most controversial speech of his visit, in which he emphasised his belief in pacifism and made a passionate case for pacifists to refuse to do military service, not only in the event of war, but also in those countries where it was demanded in peacetime. If only 2% of those conscripted refused to obey, there would not be enough prison space to confine them, and the coercive power of the state would crumble. This was strong stuff, and when reported in *The New York Times* the following day gave rise both to strong support, particularly among the young, for whom '2%' became a slogan sported on lapel badges, and to polemical refutation from conservative elements. It certainly proved too strong for the original translator, who broke down after Einstein's first sentence and had to be replaced by a volunteer. Helen Keller, the celebrated deaf and blind educator, author, and activist, addressed the audience after Einstein had finished. One of her methods of communicating was Tadoma, in which she touched her interlocutor's throat, jaw, nose, and lips. Michelmore's summary of Einstein's New York experience was 'Fritz Kreisler and Einstein posed with their violins. Toscanini and Einstein discussed Mozart. John D. Rockefeller, Jr and Einstein argued economics. Helen Keller put her fingers on Einstein's lips and the two chatted for twenty minutes.' Einstein found his discussion with Keller very moving; Helen Dukas reported that he had tears in his eyes.

Monday 15 December was, in comparison to Sunday, a day of rest. The Einsteins were picked up by Arthur Fleming, a wealthy contributor to Caltech who, earlier in 1930, had visited Einstein in Berlin to urge him to come to California. He drove them to the house of his son-in-law Lloyd Wilton-Smith at Lloyd Neck on Long Island. It was a very cold day and Long Island sparkled with frost. Built four years previously on the seven-acre plot as a wedding present for Fleming's daughter, the house, known as Kenjockety, was designed by Bertrum Goodhue, who was also responsible for many of the buildings at Caltech. Einstein could relax in the beautiful house surrounded by trees and looking over Long Island Sound. They were shown the magnificent library and one of its treasures, a page from the Gutenberg Bible. Einstein found the luxury unbelievable, especially the indoor tennis court in a huge building. In the late afternoon, Fleming drove them back to the *Belgenland*, which sailed at midnight. The impact of Einstein's visit can be judged by the fact that on every day, except the last, there was a substantial front-page article in *The New York Times*. Einstein was exhausted and felt only a sense of great liberation as he watched the lights of New York recede into the night.

[17] This exchange was catalysed by Einstein's correspondence, beginning in 1929, with a brilliant Polish mathematical physicist, Myron Mathisson, who so impressed him that he exerted himself firstly to get him to submit a PhD thesis and then to obtain financial support. His difficulties with the bureaucracy of the Rockefeller Foundation exasperated him.

The weather warmed as they approached Cuba, the next port of call, where they spent two days. Einstein enjoyed the people, who seemed happy despite their grinding poverty, their lot somewhat alleviated by the plentiful bananas. He had the usual round of official visits and entertainments, including meeting the corrupt president Gerardo Machado. They reached the Panama Canal on 23 December, when during a brief stopover Einstein was once again whisked away by the German Embassy to a round of meetings, including with President Florencio H. Arosemena, who turned out to have been a student in Zurich, allowing him and Einstein to swap reminiscences. Through the canal and into the Pacific, the weather was still hot and humid. Einstein regularly passed the time by playing table tennis with Helen Dukas. On Christmas Eve there was a concert, during which Einstein played a Beethoven sonata. The orchestra then accompanied him in a lullaby by Goddard followed by 'Ombra mai fu' from Händel's *Xerxes*. These were received with thunderous applause. His colleagues, the musicians of the orchestra, were among the few to whom Einstein was sorry to say goodbye when the ship docked in San Diego early in the morning of 31 December 1930.

Chapter 21

Caltech and Chaplin; First visit to Oxford (1931)

In a mirror image of his arrival in New York, Einstein awoke on 31 December 1930 to find a large delegation of reporters had been shipped out to the *Belgenland* as it approached San Diego. Without taking his habitual bath or breakfasting, he fielded the usual inane questions with Elsa acting as translator. The sound of children's singing allowed him to break off the press conference and repair on deck, where he beamed at the assembled crowds and received the girls, dressed in white sailor's blouses, each of whom presented him with a Californian poinsettia, eventually filling his outstretched arms. He managed to escape to his cabin and then snatched some breakfast before he and his party disembarked through the press of the enormous crowd. They were driven around some of the San Diego beauty spots before arriving at a reception in Balboa Park, where Einstein was formally received by the mayor and presented with gifts from the Jewish community. After a private lunch at the Grant Hotel (not to the German taste as it was mostly spicy Mexican specialities), he addressed a few words into a microphone, probably broadcast on the local radio station. Paul Epstein, who had been instrumental in bringing Einstein to Pasadena with his promises of fine musical activity at Caltech, arrived for lunch. Arthur Fleming, who had crossed the country after waving goodbye to Einstein at New York, then drove the party up to Pasadena, where they stayed in Fleming's large Swiss-style chalet on 'Millionaires Row', South Orange Grove Avenue. Einstein was struck by the garden, with its typical Californian complement of palm trees.[1]

On New Year's Day, the annual Pasadena Tournament of Roses Festival took place. Einstein was driven with a police escort to a bank overlooking the procession route, where Fleming had organised seats out of public view in the manager's office. The relatively cool weather reduced the crowds somewhat, so that the usual number of serious injuries did not materialise. On the following day, Einstein visited the Physics department of Caltech, where he had discussions with Epstein and others. He also held a press conference, responding patiently to many mostly sensible and even penetrating questions related to physics. On the following morning he discussed cosmology with Richard Tolman before he and Elsa moved into their accommodation, which Einstein characterised as beautiful. It contained a grand piano and all domestic conveniences, including a refrigerator. An old friend of Elsa's from Hechingen, Barbara Siebert, who lived nearby, helped

[1] This account of Einstein's stay is taken predominantly from his travel diary, that of Helen Dukas and reports in the *Los Angeles Times*. Some idea of the interest generated by his visit can be gauged from the fact that his name appeared in a headline, usually on the front page, on 46 of the 59 days of his stay in Pasadena. He was also mentioned on all but one of the seven days after his departure. The newspaper even made fun of itself for the intensity of its coverage. The front-page cartoon on 3 January showed Einstein buried under multiple editions, each with a different headline such as 'Einstein lands', 'Einstein says', 'Einstein this', and 'Einstein that'. It seems almost nothing was too trivial to be recorded in the newspaper. For example, the headline on the front page of the 4 January edition was 'Einstein moves to Bungalow'.

Einstein. Brian Foster, Oxford University Press. © Brian Foster (2026). DOI: 10.1093/oso/9780198794875.003.0021

with setting up the domestic arrangements. Einstein soon became enchanted with the sunny California weather, the luxuriant vegetation and the friendliness and contentment of the inhabitants. He noted that the area was dominated by long straight roads filled with automobiles and was astonished that he had observed one for sale for $25.

A serious hole in Einstein's life was filled on 4 January, when he received a wonderful Guarneri violin on loan from Wurlitzer, a firm with an enormous showroom of musical instruments in downtown Los Angeles. Although Wurlitzer was more noted for pianos and particularly organs than violins, Einstein shrewdly observed that they would make a fortune from the publicity. He received not only a famous violin but also a visit from a famous violinist. Lili Schober had been born in Chicago but studied with Joseph Joachim in Berlin, where she met and fell in love with one of the foremost soloists of his day, Sasha Petschnikoff, whose Berlin debut in 1895 had caused a sensation; he was spoken of as the successor to Joachim. Lili and Petschnikoff married and had three children. They often appeared together on the concert stage, commissioning several new works for two violins. Sasha was however effectively a bigamist, living for many years a double life with a mistress whom he had installed in her own establishment. Lili eventually discovered the deception, separated from Petschnikoff, and, after hair-raising adventures during the First World War as the citizen of a belligerent country (Sasha Petschnikoff was Russian; after their separation she reclaimed her American citizenship), managed to escape from Germany with her children to her native US. In 1919, she settled in a house at the foot of the Cahuenga Pass on Highland Avenue in the Hollywood Hills. A popular picnic place in uninhabited scrubland nearby had, by the time Einstein arrived, become the iconic concrete shell of the Hollywood Bowl. Having retired after a successful concert career, Lili Petschnikoff acted as host to many of the famous musicians who came to perform at the Bowl, among them Einstein's friend Bruno Walter. It was however through another mutual friend, Amy Israel, wife of Berthold Israel who was the proprietor of the famous Berlin department store, Kaufhaus Nathan Israel, that Einstein was introduced to Lili. Einstein characterised Lili, whose pure white hair was almost as arresting as his own, as talkative. She not only provided a regular venue for making music, but also ferried him around in her car, which invariably contained her aged and eccentric mother and aunt in the back seat. She also often visited the Einsteins in their house for supper. Elsa accompanied Einstein to the Petschnikoff house for the weekly chamber-music sessions, where she would sit quietly in the corner repairing items of clothing. Einstein and Lili played works for two violins and piano, but she also encouraged Einstein to play Mozart violin and piano sonatas, which he would select at random. Lili considered that his fingers were fleet and his tone good. She and Elsa became close friends, exchanging letters until Elsa's death.

Lili was also a friend of the author Upton Sinclair, whose outspoken socialist views were unusual, to say the least, in Pasadena. Einstein had read some of his books and enjoyed them, writing to Sinclair that his wicked tongue was one of the joys of his life. In 1930, Sinclair had sent Einstein a complimentary copy of his book on paranormal phenomena, *Mental Radio*. He wrote to Sinclair with the surprising offer to write a foreword for a German edition, which Sinclair joyously accepted. In November 1930, he had heard of Einstein's intended stay at Caltech and wrote to offer his garden as a quiet refuge. Sinclair was also a violinist and suggested that Einstein bring his violin so that they could play duets together, provided that Einstein did not mind him playing occasionally out of tune. Einstein visited Sinclair's house, a few kilometres east of his own, on 9 January. He found himself in agreement with Sinclair's views, particularly on Russia, and agreed to be guest of honour at some small dinners for Sinclair's friends. He even attended a séance organised by Sinclair's wife, whose 'psychic powers' had been the basis for *Mental Radio*. Despite the attendance of a famous medium, nothing paranormal occurred. Sinclair also arranged for Einstein to view a film about Mexico by Eisenstein. Early in February, Sinclair published a question-and-answer exchange between him and Einstein in a New York socialist journal. Sinclair and Einstein enjoyed each other's company and corresponded for many years after Einstein ceased to visit California.

Probably either Petschnikoff or Sinclair recommended that Einstein visit the nearby Zoellner Music School. This had been founded by Joseph Zoellner, violist in the eponymous quartet, the other chairs of which were occupied by his children. Although they were no longer performing in public as a quartet, they played Beethoven and Mozart with Einstein in January.

Another musician whose acquaintance Einstein made during his stay in Pasadena was Artur Rodzinski, at that time conductor of the Los Angeles Philharmonic. Einstein had been impressed by Rodzinski's performance of Brahms' Fourth Symphony in a programme that also included Beethoven's *Leonore* Overture and pieces by Debussy and Ravel. He and his wife met the Einsteins on 12 January, when, according to the *Los Angeles Times*, Mrs Rodzinski accompanied Einstein's violin playing of Bach and Mozart on the piano. That evening another, rather less distinguished musician visited for supper. Emil Hilb had been born in Hechingen in 1890 and was a composer, conductor and arranger. Hilb was on the payroll of Paramount Pictures, where he composed the scores for several films and acted as a sort of musical director. He ingratiated himself with Elsa and Helen Dukas by bringing with him a variety of German delicacies, a commissarial role that would continue for many years.

Hilb also brought with him an idea that initially appealed to Einstein. He proposed to make a film of Einstein playing two pieces of his choice, either with piano or orchestral accompaniment. In return, Einstein would receive a cheque for $10,000, which he could use for distribution to the various Palestinian charities that he supported. There were stipulations in Hilb's proposal, including that the recording could be used for the film only and not transferred to a record for individual sale. Elsa wrote to Hilb, to whom she had taken a great liking, immediately on their arrival back in Berlin, enthusiastic about the $10,000 for charity. However, always reluctant to be recorded playing the violin, Einstein soon had second thoughts. Hilb's initial proposal subsequently morphed into a more ambitious one, to make a series of films for educational use on the great composers, which Dima Marianoff was to produce and direct in Europe. Einstein was to be involved only as a member of an advisory board, which nevertheless would give very valuable credibility to Hilb's project. Although Marianoff replied to a subsequent letter from Hilb in August 1931 with an enthusiastic acceptance, once again the proposal eventually ran into the sand. As Hilb remarked to Elsa, the times were not exactly propitious for drumming up money for endeavours of this kind.

It was Lili who organised a visit to Universal Studios, probably suggested by Elsa, who had a keen interest in the movies. Lili's son Sergei was an assistant director at Universal. Lili was also friendly with the producer and subsequently famous agent, Paul Kohner. He suggested that Einstein should be invited to visit the studios for a private screening of the anti-war film *All Quiet on the Western Front*, which had been banned in Germany. The visit was arranged to coincide with the birthday of Carl Laemmle, the head of Universal. Laemmle happened also to be a Swabian, born a few miles from Einstein's birthplace. Although Laemmle's English was much better than Einstein's, both spoke throughout their lives with pronounced Swabian accents.[2]

Einstein's party for the Universal visit included not only Mayer and Helen Dukas, but also Lili Petschnikoff and Estella Katzenellenbogen (see Figure 21.1). The latter, who was staying in

[2] A film of Einstein, Elsa, and Laemmle talking in German during this visit is available at https://www.youtube.com/watch?v=F9C_1uaaioM.

Figure 21.1 Einstein and party visit Universal Studios in Hollywood on 17 January 1931. From left to right: Paul Kohner, Albert Einstein, Helen Dukas (face obscured), Estella Katzenellenbogen, Elsa Einstein, Sergey (Sasha) Petschnikoff, Lili Petschnikoff, Walther Mayer, and Sigmund Moos.

a nearby hotel, had intended to remain with the Einsteins for the whole trip but shortly after this visit she received word that her daughter was ill and she rushed back to Europe. The screening was interrupted by the arrival of the actress Mary Pickford, whose fame had to be explained to Einstein. After the screening, which made a great impression on Einstein, the party was given a lavish lunch, complete with the sort of laudatory speeches that he detested. At the conclusion, he responded 'I thank you for all the lovely things you have said. If I believed them, I would be crazy and since I'm not crazy, I don't believe them!'

In his autobiography, Charlie Chaplin recounts that Laemmle telephoned to ask him to attend the lunch and meet Einstein. While Einstein toured the sets, Elsa suggested that they would be delighted to spend some time with Chaplin in a relaxed atmosphere, so visits to Chaplin's film studios followed by dinner at his luxurious mansion in Beverly Hills were arranged[3] for 14 January. After a striking performance of Japanese dancing performed in a replica Japanese temple set up in the gardens, they dined. While one of the guests confused Einstein with questions about the fourth dimension and spiritualism, Elsa told Chaplin a story about Einstein coming in for breakfast in Haberlandstrasse, preoccupied with a wonderful idea. After drinking his coffee, he went to the piano and in between playing, would stop to jot down ideas. After half an hour he vanished upstairs with orders not to be disturbed. Two weeks later, he came down, looking pale, and laid down

[3] Note that this version is contradicted by Helen Dukas, who maintained that Chaplin did not attend the lunch but responded to a relayed request from Elsa with the invitation to dinner.

two sheets of paper. Chaplin recalled Elsa as concluding that this was his theory of relativity.[4] Thenceforward, Chaplin and Einstein were firm friends.

It is unsurprising that Einstein and Chaplin found themselves in sympathy. The fact that they were arguably the two most famous men in the world gave them a unique bond and perspective on the vagaries of humanity. Chaplin's left-wing views, anchored in the desperate squalor of his upbringing in Victorian London and subsequently catalysed by their mutual friend Sinclair, were similar to Einstein's. They also shared a delight in music. Chaplin had taught himself to play the violin, piano, and cello, although the violin was his passion. Extremely unusually, he played both the violin and cello the 'wrong way around,' presumably because, being naturally left-handed, he assumed that it would be preferable for him to hold the bow in his left hand.[5] From the age of sixteen, he practised for several hours a day, picking up instruction whenever he could from members of theatre orchestras.

Not all of Einstein's busy social schedule was as enjoyable as dining with Chaplin. On the following evening, Robert Millikan, whose right-wing views made both Chaplin and Sinclair *persona non grata* at Caltech, ensured that Einstein played his part in the never-ending round of fund raising. He was guest of honour at a reception for 350 people at the Athenaeum, the Caltech Faculty Club and Residence. All were desperate to shake Einstein's hand. He, along with seven other distinguished scientists, including Millikan himself, Albert Michelson, and Edwin Hubble, gave short talks that were broadcast on the radio. This was the first time that Einstein had met Michelson; he paid tribute to Michelson's work in determining the velocity of light, which he credited with paving the way for the theory of relativity.

Einstein settled into a routine of evening music-making and daytime visits to the physics department, where he both lectured on his unified theory and listened to talks from the faculty as well as the astronomers from the Mount Wilson Observatory. For example, on 23 January, Hubble lectured on his observation of many 'island universes', today known as galaxies, and the implications for cosmology. On the following day, the Einsteins were driven to Palm Springs for a long weekend by their old friend Samuel Untermyer, a millionaire Jewish lawyer whose main practice was in New York. The Einsteins were guests at his home, The Willows. Einstein loved the desert climate and the hot sun, which would have been particularly welcome since the weather in Los Angeles had been unusually wet. He greatly enjoyed the visit, which included a trip to the famous El Mirador Hotel. Here Einstein and Elsa wandered around the gardens with their impressive display of cacti. They were driven out to see a local beauty spot, Palm Canyon, where a stream allowed palm trees to grow in the middle of the desert. The cacti of King C. Gillette, the razor-blade magnate, were also viewed with interest. They explored date and citrus groves, where they were informed that dates

[4] Either Elsa's, or more probably Chaplin's, memory is at fault here. This story resembles one that Elsa told on their arrival in New York in 1921 (see Chapter 15) and on several occasions elsewhere, but it relates to Einstein's general working habits, not to any particular discovery and certainly not to general relativity. Chaplin's grasp of dates and details is even more tenuous that that of Marianoff; he places this dinner in 1926.

[5] Although at first glance the violin may look symmetrical, it has many internal features that are designed for the normal method of playing, with the bow held in the right hand and the left hand stopping the string. Many alterations, usually not particularly successful in terms of tone production, are required to allow a violin to be bowed by the left hand. This is usually therefore a last resort when, for example, a player has some left-hand deficiency, such as a missing digit, that means the notes must be stopped with the right hand. It is a sign of Chaplin's strength of character that he persisted, even though his various ephemeral teachers would have begun by trying to persuade him to play in the normal way.

had two sexes and were grown in different plantations. Einstein was entranced by the beauties of the desert sunrises and sunsets.

Two weeks earlier, Einstein had been examined and pronounced fit to be allowed to ascend the 1500 metres to the Mount Wilson Observatory. This shows that his heart condition of previous years was now no longer considered to be a problem. On 29 January, he and Mayer were driven up to the Observatory and spent the afternoon and evening both seeing the telescopes and equipment and discussing the latest observations with Charles St John, whose work on the sun had inspired Einstein to do some thinking about solar physics. He observed Jupiter and its moons through the new 100-inch telescope.

On the following day, Einstein and his party were invited by Chaplin to attend the premiere of his movie *City Lights*. They dined at his house and then were driven to the movie theatre (see Figure 21.2). Chaplin was nervous about the movie's reception, at least in part because, long after Hollywood had converted to 'talkies', Chaplin insisted on making it as a silent, although it had a recorded musical soundtrack that he had composed.[6] He need not have worried; the film was an enormous success and after the lights went up, Chaplin was surprised and no doubt gratified to note that Einstein had tears in his eyes at the sentimental ending. They may have turned into tears of laughter after Einstein had a mishap with the lobster in aspic served at the post-performance reception, where the Einstein party stayed until the small hours. Chaplin himself immediately dashed off to New York to supervise the opening of *City Lights* there, having accepted an invitation to visit the Einsteins in Berlin on his impending world tour.

The Einstein party had two weekends of sight-seeing, the first in Santa Barbara on 7–8 February. On the following Sunday, they visited the Highland Springs and Cherry Valley beauty spots in Beaumont. Suitably rested, on 16 February Einstein made an appearance at a large banquet organised in his honour by the Jewish community of southern California. It acted as a fund-raiser for the Einstein Forest in Palestine. The organisers were careful to organise a musical interlude, in which local violin virtuoso Calmon Luboviski, accompanied by pianist Claire Melonio, played Bloch's *Baal Shem* and a piece by Saint-Saëns. Einstein's speech emphasised the importance of the Zionist ideal, financial support for the Jews in Palestine, the necessity to live in peace with the Palestinian Arabs and praised the work of Chaim Weizmann.

Another musician who intruded into Einstein's life at this point was Alla Moszkowski. She claimed to be the daughter of the composer Moritz Moszkowski, brother of Einstein's old friend and biographer, Alexander. She was a fine pianist who played at a musical evening in Beverly Hills organised by the sister of the philanthropist Julius Rosenwald, with whom Einstein was slightly acquainted. Moszkowski, who also claimed to be the widow of Count de Ribadeo, tried to convince Einstein to provide financial support for a music school. The normally trusting Einstein became suspicious and presumably subsequently checked with Alexander, who would have informed him that Moritz's only daughter had died some years previously. The affair did not end with Einstein's refusing to support the *soi-disant* Countess; some months later she wrote to Einstein in Berlin threatening him with legal action unless he withdrew his allegation that she was a fraud. The

[6] Chaplin was a gifted composer, who had before *City Lights* published several compositions, mostly songs. He went on to compose the soundtracks for all of his films. He could not read music, so others produced the musical notation and details of orchestration, but the ideas were his and he carefully supervised the final product. His prowess as a composer is lost in the shadow of his greatness as a comic actor but several of his songs, of which perhaps the most famous is *Smile*, have become staples of the repertoire. The only Oscar that he won in open competition was for best original score of *Limelight*, on its re-release in 1973.

Figure 21.2 Einstein and Charlie Chaplin outside the Los Angeles Theatre at the world premiere of *City Lights* on 2 February 1931.

view of Moszkowski's relatives was that Alla was a former pupil of the composer who had become deranged.

On 19 February, Einstein was driven out in the evening to the Irvine Ranch, then a featureless expanse of bean fields, now the city of Irvine. Since 1929, Michelson had been working there on a new determination of the velocity of light in a vacuum. He had constructed a mile-long straight steel tube of 36-inches diameter. In photographs of the time, this looks very similar to the LIGO instruments constructed 60 years later, which were used to discover gravitational waves and subsequently test general relativity with exquisite precision. Michelson's measurements were just beginning when Einstein visited, although they were suspended that day because the sun had been hot and the pipe was still cooling down! Einstein looked around the measurement hut and spent some time in discussion with Michelson before returning to Pasadena. Sadly, Michelson was by then mortally ill; he died three months later but the measurements were completed and published by his collaborators in 1935.

As the Einsteins' departure date at the end of the month neared, the density of social and political engagements increased even further. The Jewish community organised a concert and reception at Einstein's request on 20 February, in order to give the poorer members of the Jewish community who could not afford to attend a banquet a chance to meet him. On the previous day, Governor of California James Rolph arrived at the Einsteins' house, complete with frock coat and top hat, to pay his respects. The city of Los Angeles honoured Einstein with a ceremony on the steps of City Hall in the afternoon of 23 February. Einstein's agreement to attend the annual banquet of the Los Angeles Chamber of Commerce that evening as a guest of honour would have been reluctant. He was no doubt encouraged to attend by Millikan, who accompanied him to the event. Millikan may well have applied a sweetener to the pill; on the previous day, he took Einstein to Long Beach where they went sailing. On 24 February, a farewell lunch thrown by the city of Pasadena attracted more than 500 attendees. On the penultimate day of his visit, Einstein opened an Astronomy Study Centre in the presence of 5,000 enthusiastic local school pupils.

Just before his departure, Einstein wrote two articles. One was for the *Los Angeles Times* and was published on the front page of the Culture section of the Sunday edition on 1 March. In it, he stated his views on the social and anthropological origin of religions and the age-old conflict between them and science. He asserted that in fact the greatest scientists, such as Kepler and Newton, had been inspired by a 'cosmic religion', the conviction that there was a guiding hand behind the splendours of the universe which they had devoted their lives to understanding.

The second article was on the subject of quantum mechanics. Einstein with his Caltech colleagues Tolman and Boris Podolsky submitted to *Physical Review* a letter entitled 'Knowledge of Past and Future in Quantum Mechanics'. It is a variation on the thought experiment with the box on a spring balance enclosing a radioactive source that had for a while stumped Bohr at the Solvay Congress a few months earlier. Two particles going through two slits were now involved, one of which travelled straight to a detector, the other of which was reflected back to the box after a long journey. The long journey allows time for the box to be weighed and therefore, in the authors' view, for the time of arrival and energy of the second particle to be predicted with arbitrary accuracy. It would be fair to say that this brief note throws little further light on the deep mysteries of quantum mechanics. However, it does indicate that they were still very much in Einstein's mind and is the first occasion on which his thought experiments involved correlations between two particles, which was to prove a very fruitful area for future work.

At 1 p.m. on 27 February, Einstein and party boarded the luxurious special Pullman coach that would take them to New York, via the Grand Canyon and Chicago. On the following day, the party arrived at the famous Hotel El Tovar overlooking the Grand Canyon. Einstein agreed to don a Native American headdress, a treat for the waiting newspaper photographers, and was inducted as a chief of the Hopi tribe with the name 'Great Relative.' On the following day they were driven roughly 190 miles southeast to the petrified forest, now the Petrified Forest National Park, in freezingly cold weather. They returned to their train, passing through Albuquerque and Kansas City before arriving in Chicago. Einstein was visited briefly by architect Frank Lloyd Wright and his daughter before he emerged onto the rear platform of his carriage. Here he read a prepared speech in which, with Millikan's influence removed, he returned to his pacifist theme, once again urging refusal of military service. The train departed directly afterwards, depositing Einstein and party in New York in the early morning of 4 March.

The stay in New York was hectic. After a brief photo opportunity, Einstein breakfasted with Dr George Crile, a distinguished medical researcher, who enthused him with his account of having created synthetic cells from organic matter through which an electric current was passed. Einstein is recorded in the *Los Angeles Times* as believing that Crile had made enormous progress in making inanimate material live, an opinion not borne out by subsequent research. Equally unfruitful would have been a lecture Einstein attended entitled 'Fourth Dimensional Music'. He then energised a delegation of pacifists before attending a fund-raising dinner given by the American

Palestinian campaign. He would certainly have been relieved to board the SS *Deutschland*, which sailed for Southampton at midnight on 4 March 1931.

The Atlantic crossing was very stormy, as Einstein informed his sister Maja. Even he, an excellent sailor, felt the need to lie on the deck at times. Nevertheless, nothing could stop him from making music. On 8 March, the 175th birthday of Mozart was celebrated with a concert in which Einstein played violin in a Mozart Piano Trio (see Figure 21.3). On 14 March, they arrived in Cuxhaven, returning to Berlin by 16 March.

Hardly had the Einsteins removed their coats when the telephone in Haberlandstrasse rang. It was Chaplin, who was on a mammoth world tour to promote *City Lights*. He had arrived in Berlin from England roughly a week before to a tumultuous welcome from the citizens of Berlin and was due to leave that night on a sleeper train. The entire Einstein family was assembled when Chaplin arrived at the Haberlandstrasse apartment for dinner, bringing with him a companion, Sir Philip Sassoon. Chaplin and Sassoon had become very friendly during Chaplin's previous visit to Europe a decade earlier. According to Dima Marianoff, Chaplin was on tremendous form, sitting at the head of the table in Einstein's usual place and holding forth on a variety of subjects. Sassoon on the other hand remained rather silent. The ravages of the Great Depression that was now in full swing in the US and the UK had clearly made a profound impression on Chaplin, forming the seeds of another of his masterpieces, *Modern Times*. He considered that he had the solution to the depression and held forth on the Gold Standard and economics. Einstein nodded in agreement and proposed that all the world's money should be collected at one point and burned. This greatly amused both of them but had a very different effect on Sassoon, who looked more and more uncomfortable until he made a rather abrupt, although courteous, exit. Chaplin explained

Figure 21.3 Einstein playing a Mozart piano trio aboard the SS *Deutschland* on 8 March 1931.

that his discomfort and subsequent departure were precipitated by his being one of the five richest men in Britain. The Einstein–Chaplin socialistic consensus and in particular the prospect of all his wealth being consumed in the flames was, unsurprisingly, not a pleasant one. Chaplin went on to give hilarious impersonations of the British ruling class, including a dour Ramsay MacDonald, the sparkling wit of David Lloyd George, one of MacDonald's predecessors, and the proceedings he had witnessed in the Houses of Lords and Commons. He inspected Einstein's workroom, which, gratingly for Einstein, he entered rather as if it were a cathedral. When he reluctantly left to catch his train, he noticed on the piano some photographs of Einstein that Elsa kept handy for guests seeking souvenirs. When he asked for one, Einstein autographed it 'To Charles Chaplin – the great economist – from Albert Einstein'.

It was not long before Einstein set off on his travels again, this time to Oxford. He restarted his erratic diary for the month or so he spent in Berlin before departing, which provides an unique snapshot of his Berlin activities in the 1930s. In particular it sheds light on the intensity of his musical life. On 12 April, he played one of the Brahms string quartets and a Mozart divertimento. All of the participants other than Einstein were professionals; the violist was Erna Schulz, who was a pupil of Hubay and Joachim and had been part of the pioneering all-female Wietrowetz Quartet. Although she formally left in 1912, she continued to play as a substitute until its dissolution in 1917. After dinner on the following day, Einstein played Haydn Piano Trios with Erich Mendelsohn, the architect of the 'Einstein Tower', and Erwin Freundlich, who was a close friend of Mendelsohn's. This is the first indication that the decade-long estrangement between Einstein and Freundlich had begun to heal. Indeed, on 9 April Einstein and Max von Laue had been involved in once again trying to bring about a reconciliation between Freundlich and his perennial sparring partner and superior, Director Hans Ludendorff.

Later on 13 April, after his music-making with Freundlich and Mendelsohn, Einstein had an interesting idea related to cosmology. This had been germinating since his conversations with Edwin Hubble and Tolman in Caltech. Already in a lecture in February he had announced that, given Hubble's new data that all galaxies appeared to be receding, his original concept of a static universe was no longer tenable. On 16 April, three days after this idea had occurred to him, he submitted a paper. In this, he set the cosmological constant to zero, a relief to him since he had long considered its introduction a mistake; its purpose being to permit a static universe, it was now redundant. Indeed, he pointed out that such a static solution of his equations was in fact unstable and therefore, irrespective of Hubble's observations, invalid. He examined solutions to his equations assuming that the universe is expanding and discovered that the universe would approach a maximum and then begin once again to contract. After obtaining estimates for the age of the universe and its average density, whose basis is difficult to discover from what is written, he concluded that the expanding universe implied by Hubble's results was more natural in general relativity than was a static universe.

Having submitted his paper, that evening Einstein went to see what he considered a powerful performance of the play *Minna von Barnheim* by Lessing. On 18 April he played Haydn and Mozart Trios. The following day, a Sunday, was particularly busy. In the afternoon, he played Beethoven quartets and trios, received visitors in Haberlandstrasse and then in the evening went to the theatre with Estella Katzenellenbogen. She had invited him to see *Der blauer Boll* by Ernst Barlach at the Schauspielhaus. On the following Tuesday evening, Einstein entertained a gathering of friends including the Lebachs by playing violin duets with Fräulein Hermann.

On 23 April, Einstein attended a meeting of the Academy, where Haber lectured on catalysis. Afterwards he met with Max Planck and then attended a concert at Berthold Israel's palatial house.

He went on to visit his old friend Alexander Moszkowski; they probably discussed the case of Alla 'Moszkowski' mentioned earlier. Alexander, although mentally alert, was failing physically; he died three years later. The following Sunday saw another quartet gathering, this time including Einstein's cousin Robert, who was visiting Berlin from his home in Italy. They played pieces by Haydn and Schubert. On 27 April, Einstein visited a piano shop and selected a reconditioned Blüthner grand piano as a present for Maja's birthday. When it arrived in Florence in the summer, Maja was overcome with joy at its beautiful tone[7].

Einstein's journey to Oxford began on the evening of 29 April 1931, when he took the train to Hamburg with the banker Max Warburg, with whom he discussed Palestinian issues, particularly concerning the Technikum at Haifa, which later became Technion, the Israel Institute of Technology. After an overnight stay in Hamburg, he boarded the *Albert Ballin* for the voyage, whose brevity he regretted, to Southampton. There he was met by Lindemann, who came on board to escort him to dry land and his waiting Rolls-Royce. They stopped off at Winchester so that Lindemann could show Einstein the cathedral, which Einstein thought wonderful, and Winchester College, one of the oldest of the English public schools. That evening they arrived at Christ Church College, of which Lindemann was a Student.[8]

Einstein's impressions of Oxford were not initially favourable. His rooms, which he characterised as a 'small castle', had been vacated by Robert Dundas, who was then travelling in India. No sooner had he deposited his valise than he had to change into his dinner jacket, don a black gown, and join what he characterised as a club dinner. Seated at High Table and surveying the 500 or so undergraduates, he thought the proceedings both bizarre and boring. In his diary, he also remarked on the all-male nature of the college and how hideous life would be without women. This latter state, at least, he was in no danger of experiencing. A particularly importunate mistress of his, Ethel Michanowski,[9] had followed him to Oxford and was showering him with expensive presents. In a letter to her in which he admonished her to desist, he characterised Christ Church as a place in which sinful affluence reigned.

Despite his initial impressions, Einstein soon grew to appreciate certain aspects of Oxford life. He took pleasure from the English countryside and the tasteful and venerable architecture of Oxford. He liked the reticence with which colleagues passed in the quad, not audibly greeting but acknowledging each other's presence. By Sunday, 3 May, he was beginning to enjoy the conversation at the evening meal. On the following day, his friend and fellow Student of Christ Church, Gilbert Murray, Regius Professor of Greek, invited him to his country house, Yatscombe Hall, at Boars Hill overlooking Oxford. Murray and Einstein were long-standing colleagues

---

[7] After Maja fled Italy and joined Einstein in Princeton in 1938, the piano was lent to her friend the artist Hans-Joachim Staude. It is now in the library of the Astrophysical Observatory of Arcetri, near Florence.

[8] English 'public schools' are in fact private and charge substantial tuition fees; neither are Students of Christ Church students, but rather Fellows of the college.

[9] Their relationship began in the late 1920s and continued until Einstein left Germany in 1933. When Elsa discovered that her friend Michanowski was in Oxford, she wrote angrily to Einstein, who was unfazed. He replied that Michanowski had done what she enjoyed, which did not harm anyone, while refraining from doing what she did not enjoy, and which would have hurt others. Because of the former she followed him to Oxford, while because of the latter she did not tell Elsa. This he characterised as impeccable behaviour according to the best Jewish–Christian morality. Unsurprisingly, Elsa was not mollified by this communication and wrote further irate letters, also catalysing Estella Katzenellenbogen and surprisingly, Helen Dukas, to write to Einstein in protest. He was reduced to writing to Margot Einstein, asking her to intercede with her mother. He asserted that of all his women, he was only really attached to Grete Lebach.

on the League of Nations Committee on Intellectual Cooperation, which Murray chaired. The knowledge of current affairs possessed by the women who attended Murray's party impressed Einstein.

On 5 May, Einstein dined in Trinity College and afterwards listened to a presentation on novae and the subsequent formation of white-dwarf stars by Edward Milne. Milne was Rouse Ball Professor of Mathematics and for many years had been engaged in the study of stellar atmospheres, which had brought him into conflict with Einstein's friend Eddington. Einstein and Milne met for several more discussions during his visit, including the following night when Milne dined at Christ Church. Following dinner, they were joined in Einstein's room by Lindemann to discuss cosmology. After Einstein's visit, and perhaps catalysed by it, Milne began to elaborate what he called 'kinetic relativity', an alternative to general relativity.

Einstein spent the day of 8 May quietly but that evening was invited to a concert at a favourite Oxford chamber music venue, the White Barn, home of the Pearces just outside the city on Boars Hill. This was given by Einstein's old friends Adolf Busch and Rudolf Serkin, who played Beethoven Sonatas and a Schubert Rondo. Erna Schulz, the violist with whom Einstein had played quartets in Berlin a few weeks before, was also at the concert. She often had engagements in London and also taught there. She accompanied Einstein back to Christ Church.

On the following day, Einstein gave the first of his Rhodes lectures, for which he received the substantial honorarium of £900. It took place in the large Milner lecture theatre of Rhodes House, which was packed with an audience of 500. The lectures on the theory of relativity were delivered in German. That, and the difficulty of the topic, meant that Lindemann had taken the precaution of printing 1000 copies of an English translation of Einstein's summary of all three lectures, which was the basis for almost all the reporting that appeared in the major newspapers. Einstein took no prisoners, launching immediately into the mathematical detail of general relativity and admitting as he closed that the major problem was that he had not yet succeeded in incorporating electromagnetism into the same framework. The second lecture took place a week later, to an unsurprisingly much reduced audience. Entitled 'The Cosmological Problem', Einstein expounded his latest thinking on the expanding universe and his paper on the subject submitted a few weeks previously. The blackboard on which he wrote elements of his lecture was preserved, much to Einstein's annoyance as smacking of hagiography, in Oxford's Museum of the History of Science, where it can still be seen. The final lecture was even more challenging than the first, as Einstein gave a frank and mathematics-laden account of his struggles to achieve a unified field theory. The Dean of Christ Church, who had devoted his life to piecing together the original text of the Vulgate, the Latin version of the Bible, was unable to follow Einstein's thread. Much to the lecturer's amusement, he slept soundly through the entire proceedings.

After his first lecture, Einstein accompanied Lindemann to visit his father in his country estate down the Thames. On 11 May he was delighted to make music at Gunfield, the home of Margaret Deneke, which was in Norham Gardens just north of the city centre and very close to Lady Margaret Hall. The large music room at Gunfield was another well-known venue for chamber music in Oxford. Deneke was a professionally trained pianist who had played for Clara Schumann. She had devoted much energy to women's education, in particular to Lady Margaret Hall, a women's college popularly known as 'LMH', of which her sister Helena was a former bursar and German tutor. Margaret Deneke (known as Marga) was a tireless fund-raiser for LMH in particular but also for St Anne's College in Oxford; she undertook several trips to the USA, one of which resulted in a donation sufficiently large to build the LMH main building, including the dining hall, known as the Deneke Building in her honour. Not only was she very active in Oxford musical activities, but also she was a respected musicologist and most importantly for Einstein, spoke flawless German.[10] She had met Einstein at the Busch-Serkin concert and invited him to dine and play music thereafter.

Marga Deneke left an account of this first of many musical evenings with Einstein:

> In 1931, 32 and 33 Professor Albert Einstein came to Christ Church. We knew he was an amateur violinist and we invited him to play quartets with artists. Marie Soldat[11] was to lead, Arthur Williams (Klinger quartet) cello and Erna Schulz (Wietrowetz quartet) viola. He had heard them all at concerts in Vienna or Berlin and accepted the offer with alacrity.
>
> Only Erna Schulz had played with him before; we others stood in expectation. In he came with short quick steps; he had a big head with a very lofty forehead, a pale face, a shock of grey, untidy hair.
>
> Talk at dinner flowed on with ease. He was rather particularly pleased with our report of Donald Tovey's answer of enquiry into the moral teaching of Wagner's operas: 'Abstain from unknown drinks.' He threw back his head and laughed heartily; his large mouth displayed many gold-stopped teeth. ... He turned to mother with an engaging smile. 'You have provided a delightful meal; now the enjoyable part of our evening is over and we must get down to work on our instruments. Shall we play Mozart? Mozart is my first love – the supremest of the supreme – for playing the great Beethoven I must make something of an effort, but playing Mozart is the most marvellous experience in the world.'

The players started with a Mozart; under Marie Soldat's rich tone on her Guarnerius 'del Gesu' the Professor's borrowed violin sounded starved and raucous, but his rhythm was impeccable. Before passing on to a Haydn there was a pause for a chat. Marie Soldat commented on the Professor's long violin fingers, tapering usefully at the tips. The Professor said 'Yes, I never practise; I have never practised and my playing is that sort of playing . . .'.

At the end of the evening, Einstein drew the line at playing one of the Beethoven Op. 18 quartets, saying that if they played all that they could not remain sober. A few days later, the Denekes and the quartet returned to the site of the Busch–Serkin concert on the invitation of Mrs Pearce for more food and music-making.

Another musical evening at Gunfield let to an incident that illustrated Einstein's familiarity with the classical repertoire. The eminent musicologist and composer Donald Tovey came to stay with the Denekes. He had arranged a Haydn Piano Trio so as to give more work to the cello. When Marga Deneke, Marie Soldat, and Arthur Williams launched into the arrangement, Einstein was intrigued by this unfamiliar piece, which sounded Mozartian. He jumped up from the sofa and inspected the music, remarking that he could not believe that there was a Mozart Piano Trio with which he was unfamiliar. Another guest at Gunfield was Ernest Walker, like his close friend Tovey both an academic musicologist, a composer and a pianist. Walker had been Musical Director of

[10] Deneke's parents were native Germans who had emigrated to London, where her father was a banker. While living in London, they became friendly with the confusingly almost homophonic Benecke family. Paul Benecke, who owned the 'Benecke' Stradivarius violin of 1694, was Felix Mendelssohn-Bartholdy's grandson. Marga and Paul re-established their friendship when the Denekes moved from London to Oxford and Paul became a Fellow of Magdalen College, Oxford. On Paul's death in 1944, he left the family papers to Marga, who transferred them to the Bodleian Library, where the 'M. Deneke Mendelssohn Collection' forms one of the world's greatest collections of Mendelssohniana.

[11] This was musical royalty indeed. Marie Soldat-Roeger was in the first rank of virtuoso violinists at the turn of the century, among females only rivalled by Gabriele Wietrowetz, Leonora Jackson, and Maud Powell. All were pupils of Joachim; Soldat had also been a close friend of Johannes Brahms. Although Soldat had for some time concentrated on teaching, she was still a formidable violinist and chamber musician.

Balliol College, in particular organising its famous concert series, for twenty-five years until retiring in 1925 to concentrate on composition. He had played before Queen Victoria and accompanied Joseph Joachim and Pablo Casals, drawing compliments from both. Walker lived a few hundred metres away from Gunfield and was a regular guest at dinners and musical evenings there. Marga Deneke chose a piece of Walker's, the *Fantasia* String Quartet of 1923, to give Einstein as a parting gift. It is preserved in his collection of music and is inscribed 'To Professor Albert Einstein, in friendly recollection of May 1931 and Oxford and quartet evenings in Gunfield, from M. Deneke'.

Einstein began to fall into a pleasant routine, punctuated by his Rhodes lectures, of meals and discussions with eminent scientists, meetings with students and regular music-making. He had been elected an honorary member of the university musical club. The Vice President, Neville Coghill, had written to Lindemann in April, offering to get three members of a string quartet together that Einstein could lead. It seems likely that Einstein would have availed himself of this opportunity, although there is no concrete evidence of his having done so. He acquired a taste for vintage port and counted his blessings when he was seated near the fireplace in the draughty college hall. Although he found the formality of many of these encounters, particularly the necessity of practically living in evening dress, irritating, they did nevertheless also give him the opportunity for much private, and some public, amusement. On 15 May, the evening before his second lecture, the Warden of Rhodes House, Sir Francis Wylie, invited him to a celebratory dinner. Lady Wylie, knowing that Einstein's English was vestigial, thought that a non-verbal entertainment might be preferable and thus asked Deneke to arrange it. On doing the rounds of the guests with his host, Einstein was relieved to discover Deneke and Soldat and enquired whether they had brought instruments with them. Marga reported that:

> After dinner we were established around the piano; the Trustees saw their lion disporting himself with Bach's double concerto and Handel and Purcell Sonatas. He tucked the violin under his chin with a will, and tuned loud and long. Then he chose the piece he wanted and made suggestions: 'No repeats in the Adagio please and the Allegro not too fast, there are tricky bits for me.' We started obediently and he threw himself whole-heartedly into the music. He made no effort at all to discover what the Trustees might want to hear. Unashamedly we played for our own enjoyment, without the slightest pretence about performing. The trustees smoked in silence, witnessing their guest of honour having a happy evening. I doubt if they listened.

Einstein's diary records a less expurgated version, in which the other guests beat a hasty retreat once the music began.

On 17 May, shrugging off an unpleasant attack of lumbago, he dined in New College with J. S. Townsend, the Wykeham Professor of Physics, who had been a student of J. J. Thomson in Cambridge. After dinner Townsend escorted Einstein to hear a fine performance of the Mozart *Requiem* in the college Chapel. Four days later, Townsend showed Einstein around his laboratory. Einstein subsequently went to the River Isis (the Oxford name for the Thames) to watch the opening day of rowing in Eights Week, the highlight of the Oxford rowing season. Next, he attended a lecture in the University Church entitled 'The Ancient Music of the Coptic Church' by Professor Ernest Newlandsmith, a Carmelite Friar. The lecturer began by playing a Hebrew lament on his violin in memory of his recently deceased collaborator. He illustrated his lecture at the piano with excerpts from Coptic songs and liturgical music that he had noted down from performances by musicians in Egypt. These had been preserved orally since the time of the early Christian fathers. Indeed, Newlandsmith's thesis was that they substantially preserved the music of the pharaohs. Although the press coverage of the lecture seems to have been very positive, in his travel diary Einstein was unusually outspoken in his criticism, characterising Newlandsmith as a picturesque swindler. He thought his harmonisation of the Coptic melodies unbelievably sentimental and took exception to his enormous girth, red face, blue cassock and apparently inebriate delivery. Such unusual ire was almost certainly aroused by Newlandsmith's anecdote of the power of music to convert a 'Jewess' to Christianity without any further words being exchanged. Einstein's busy day ended with a gala dinner in his honour at Christ Church, in which, once again uncomfortable in evening dress, he was placed between two women whom he characterised as dragons, neither of whom spoke a word of German. At least he enjoyed his dessert and port.

On 22 May, Einstein visited Ruskin College, which was not part of the university, whose ethos of facilitating the attendance of adults at educational courses was one close to his own heart; he had himself given several courses to workers in Berlin. In the evening he was once again joined by Fraus Soldat and Schulz at the home of the Pearces for another musical evening. The Mozart and Haydn quartets included a young local cellist and finally they played a Mozart quintet with the addition of another local violist. Absorbed in the music, Einstein did not get back to Christ Church until midnight, when he would no doubt have had some difficulty in gaining admittance, since the College was usually locked up well before then.

The following morning, Saturday 23 May, saw Einstein pacing through the Divinity Schools wearing the gown of Doctor of Science, with which degree he was invested by the Vice Chancellor at a specially arranged Congregation. Marga Deneke, who was there, thought that he had not understood the Latin oration but noticed his face light up when he heard Mercury mentioned. In fact, Einstein's knowledge of Latin was greatly superior to the Public Orator's grasp of science, since Dr A. H. Poynton conflated the perihelion of Mercury and the discovery of light deflection at the 1919 eclipse. Rapidly returning to familiar ground, Poynton spent most of the rest of his speech on the philosophy of Euclid, Heraclitus, Epicurus, and Plato. Einstein caught enough of Poynton's meaning to realise his scientific confusion and his inapposite meandering into Greek philosophy. Having received his degree, Einstein continued on to Rhodes House, where he delivered the final of his three Rhodes lectures. After a well-earned afternoon snooze in his rooms, he joined Lindemann for dinner at Christ Church. A busy day ended with a discussion with pacifist students, who impressed Einstein in comparison with their German counterparts.

The following day, a rainy and unpleasant Whit Sunday, was for Einstein a day of rest from physics but full immersion in music. In the afternoon he visited the Pearces and played a Dvorak String Quartet with his usual colleagues, followed by a session of violin and piano pieces, probably played with Ernest Walker. Pieces included Mozart's Violin Sonata 'No. 4,' by which Einstein probably meant that in C major, K303. Afterwards he was joined presumably by Marie Soldat in his beloved Bach 'Double' Violin Concerto. After viewing Mrs Pearce's fairy-tale garden full of spring flowers, he returned to his rooms. Skipping dinner, he instead lay in front of the fire and recovered from his musical exertions, before joining a group of philosophers for a round-table discussion on space and time. He contrasted this enjoyable experience with that of his confrontation with Phillipp Lenard in Bad Nauheim back in 1920, again finding Germany wanting in comparison with England.

After a visit to a country garden with Lindemann on Monday 25 May, Einstein again rosined his bow for more Haydn and Mozart string quartets, this time at the Denekes. The evening was somewhat marred by squabbling between Soldat and Schulz, causing Einstein metaphorically to shake his head. Perhaps fortunately for the sake of Soldat and Schulz's continuing friendship, this turned out to be the last musical encounter of his visit. His remaining time was spent in a tour of the Ashmolean Museum and discussions with anti-war campaigners about his speeches in the US, which revived his enthusiasm for the cause. On his last full day in Oxford, he held discussions in the beautiful gardens of St John's College with a group of philosophy students. He and Lindemann dined in Christ Church for the final time after they had observed the boisterous ending of the "Eights Week" rowing on the river. On the following day, Lindemann's Rolls-Royce made the return journey to Southampton where Einstein boarded the *Hamburg* for the homeward voyage, not to the eponymous port but the slightly closer Cuxhaven. Despite being delayed by fog, Einstein as always enjoyed the voyage, particularly since he had a luxurious cabin to which he could retire to escape the gawping of the other passengers. Despite the constant requirement for his dinner jacket, he had greatly enjoyed many aspects of Oxford, particularly the music-making and the scholarly atmosphere. Oxford had also enjoyed him, as evinced by the prompt despatch of a letter from Lindemann at the end of June, offering him a Research Studentship at Christ Church at a salary of £400 per year for five years. He would be required only to be present for around a month during one of the three terms. Einstein accepted with alacrity, pleased as he said to be able to retain contact both with Oxford and Lindemann personally.

Once back in Berlin, Einstein almost immediately set out for Caputh, where he spent the entire summer and autumn. It did not take long before he was aboard his beloved *Tümmler,* accompanied, unusually, by Dima Marianoff, who did not often leave Haberlandstrasse, despite his wife's fondness for Caputh. Einstein's luthier of choice, Julius Levin, had sent him a violin to try out. Levin had known Einstein for several years and was particularly friendly with Elsa. He had taken one of Einstein's violins and by progressively thinning the wood of the top and bottom plates, in Einstein's opinion greatly improved the tone. He was enthusiastic about Levin's latest offering, a violin of his own construction. He wrote to Levin in July that he was positively intoxicated and bewitched by its tone. It seemed to play itself. He compared it favourably to the Guarnerius that he had had on loan in Pasadena and told Levin that he considered the instrument to be only on loan to him until Levin eventually disposed of it. It may be this instrument that Elsa was trying to get Levin to return to Einstein, presumably after some adjustments, in November of 1932 before they left for the US.

Paul Ehrenfest made his way to Caputh early in July 1931, at the start of a summer expedition that took him through France, Switzerland, and Italy, mostly in the company of his former student Gerhard Diecke, with whom he had also extensively toured Canada and the US during the previous summer. He had not seen his friend Einstein for many months, and they immediately buried themselves in a discussion about quantum mechanics. Afterwards Ehrenfest reported the discussions to Bohr, explaining to him that Einstein had never intended to question the Heisenberg uncertainty principle with his 'Einstein's box' thought experiment, but rather to focus on the broader incompleteness of quantum mechanics. Although Bohr almost certainly received Ehrenfest's letter,[12] it seems not to have lodged in his memory since both a lecture he gave at Bristol three months later and his 1949 account of his and Einstein's exchanges still interprets 'Einstein's box' as designed to question the uncertainty principle. However, the thought experiment itself remained fixed in his memory; he returned to it throughout his life to illustrate many of the surprising and mysterious features of quantum mechanics.

Also waiting for Einstein on his return to Berlin was correspondence with Freundlich, who after two previous failures because of poor visibility, finally in 1929 had mounted a successful expedition to view a solar eclipse. He travelled to Sumatra to view the eclipse on 9 May with an ingenious double-armed camera that simultaneously photographed both the stars near to the sun's limb and

---

[12] With typical concern for his friends, Ehrenfest addressed his letter to Bohr's wife with strict instructions that he was to be shown it only if not too exhausted. He also emphasised that there was absolutely no need to think that it required a reply.

a reference field. Over the next year he analysed the data and by June 1931 was ready to publish results that showed a deflection of light by the sun substantially in excess of the predictions of general relativity. On 12 June, Freundlich gave a talk on his results at a German Physical Society meeting, at which Arnold Sommerfeld was also presented with the Planck Medal. Subsequently, Freundlich exchanged several letters with Einstein in which he attempted to defend himself from Einstein's criticisms, which centred around the statistical compatibility of the four plates that Freundlich had measured. Einstein's criticisms seem well founded and Freundlich's result to have significantly too small an uncertainty. By now Freundlich, having been Einstein's sole defender among the astronomical community in 1915, had become himself increasingly sceptical about Einstein's theory and in succeeding years put forward his own, now forgotten, alternative theory.[13] He was no doubt encouraged in this by Einstein's own wavering as to whether the general relativistic prediction should be modified by the unified theories on which he and Mayer were still doggedly working.

By now, Einstein had given up on the concept of distant parallelism as a road to a unified theory of gravity and electromagnetism. Never downhearted, he had cast around for an alternative route and recalled Theodor Kaluza's idea of adding a fifth dimension to space-time that had intrigued him a decade earlier. One of Einstein's problems with this idea was the difficulty of accounting for the fact that the fifth dimension was so different from the others that it had not been observed. Now he and Mayer came up with the idea of grafting Kaluza's idea onto a four-dimensional space-time, so that every point in it had an additional five-dimensional space associated with it. In two papers, the first completed in the summer of 1931 and published in the *Proceedings of the Prussian Academy* in October, they worked out the appropriate differential calculus of these new spaces and formulated general relativity and Maxwell's Equations therein. At first it looked promising: electromagnetism seemed to be uniquely identified with a particular antisymmetical tensor and gravitation seemed to fit in a natural way.

Enthusiastically, Einstein began to trumpet the new theory in public. He had committed to give a lecture in Vienna[14] on 14 October in the main auditorium of the University of Vienna. He stayed with his colleague Ehrenhaft, as he had in his previous visit ten years earlier. The subject of his lecture was 'The Current Status of Relativity Theory', in which he set out his and Mayer's new results. Back in Berlin by 17 October, Einstein allowed his report, provided to a funding institution that had supported his stay in Caltech, to be published in *Science*. He described in layperson's terms the gist of the new theory, concluding that 'This theory does not yet contain the conclusions of the quantum theory. It furnishes, however, clues to a natural development, from which we may anticipate further results in this direction. In any event, the results thus far obtained represent a definitive advance in knowledge of the structure of physical space'. Sadly, none of these statements turned out to be correct and by the following year, this line of attack had followed distant parallelism into the waste-paper basket.

---

[13] Freundlich remained at Potsdam, constantly feuding with Ludendorff and others, until 1933. Since his father was Jewish, he was dismissed when the Nazis took power. He went first to Istanbul, then to Prague in 1937, from whence he was forced to flee after the German occupation in 1939. He then arrived in St Andrews in Scotland, where he was a neighbour of Max Born in Edinburgh. He eventually became Napier Professor of Astronomy, retiring to his native Wiesbaden and dying there in 1964.

[14] The Viennese authorities, concerned about the effect of Nazi anti-Semitic propaganda, which had led to riots against foreign and Jewish students in the university in June, 1931, insisted that Einstein should be accompanied by a bodyguard when travelling in the city.

The Great Depression was by now hitting Germany with full, disastrous, force. Einstein's own family network, formerly composed of either the comfortably off or the downright rich, now flocked to Haberlandstrasse begging Einstein for help. He and Elsa, sometimes together, sometimes independently, helped financially to the extent that even Einstein, never normally concerned with money, began to wonder whether they in turn would be reduced to penury. Despite his earlier good-natured mockery of Chaplin's attempts in this direction, he began to study what solutions could be found to get the German economy working again, explaining his ideas to a bemused Elsa. In the meantime, the solid German middle class that had been the foundation of the Weimar republic became progressively both impoverished and radicalised. The opening to subvert the republic for which Hitler had been positioning the Nazi party began to gape invitingly wide. Unfazed by the growing chaos from Nazi thugs in the streets, Einstein continued to explain his work to the public. On 4 October, he gave a talk entitled 'Why We Do Physics' in the main hall of the Berlin Planetarium, which was packed to capacity. He explained that physicists were generally spurred on in their work not by a desire to increase human comforts or to advance technology but because they sought a better understanding of the nature of the universe. Given the background, his message may not have been quite what his audience wanted to hear, although apparently the acoustics were so bad and Einstein's delivery so idiosyncratic that many of them would have been unable to make out his words anyway.

As the summer of 1931 turned to autumn in idyllic Caputh, Einstein was increasingly preoccupied with questions of pacifism and disarmament. His sights were focused on the International Disarmament Conference that had been called by the League of Nations to meet in Geneva early in 1932. He had the opportunity to meet British Prime Minister Ramsay MacDonald at a reception in his honour given by the German Foreign Minister on 28 July. MacDonald had visited Berlin primarily to discuss the desperate financial situation of Germany, but the Disarmament Conference was also on the agenda. MacDonald, flanked on one side by Planck and on the other by Einstein, was no doubt given the benefit of Einstein's views on both these issues (see Figure 21.4).[15]

The origins of the 1932 Disarmament Conference are complex. One factor was rising pressure from supra-national organisations, some of which had Einstein's support, that influenced some of the international delegations to the League. Another was the relative success of the London Naval Treaty of 1930 that gave credibility to the prospect of nation states voluntarily limiting aspects of their armed forces. Both the Treaty of Versailles and the League of Nations Charter had enshrined disarmament as a major pillar for which the signatories should strive. Einstein produced a flurry of letters and articles in September 1931 that built on his activities during his stay in the US. Central to his views was the necessity for some sort of world authority that, in contrast to the League, was not in thrall to nation states but possessed the means and determination to enforce a code of behaviour upon each of them through the concerted action of the rest. Only then could disarmament be rationally and safely implemented. This was the thrust of an article he wrote for the Dutch publication *Nieuw Europa* on 17 September. In an article in the US weekly *The Nation* published on 23 September, he concentrated more on the single question of disarmament, characterising the upcoming conference as of decisive importance. He called on all individuals to exert themselves

[15] MacDonald may not have been terribly attentive to Einstein's prognostications given his own concerns; only a few weeks later his Labour government fell after the party split irreparably over Chancellor of the Exchequer Philip Snowden's swingeing tax increases. From the wreckage, MacDonald salvaged his position as a figurehead Prime Minister in a National Government in which the Conservative Party of Stanley Baldwin had a great preponderance.

Figure 21.4 Einstein gives British Prime Minister MacDonald some advice at a German government reception on 28 July 1931. From left to right: Max Planck, Ramsay MacDonald, Einstein, Hermann Dietrich, Hermann Schmitz, and Julius Curtius.

to influence the participants of the conference to ensure a successful outcome. In early October, in a message to the German section of the War Resistors International Movement, he returned to the necessity for an organisation that could effectively promote international justice as a prerequisite for disarmament. The news of the Japanese invasion of Manchuria in September brought added urgency to this message.

Einstein had intended to return to Caltech for the winter but became involved in a protracted negotiation with Fleming, who acted semi-autonomously from the Caltech authorities, about the terms of his return. The question of permanent positions for both Einstein and Meyer, who by now was in his mid-forties, was also involved. A variety of stipends for Einstein's visit were proposed, all more generous than Millikan eventually offered when he and his wife visited Einstein in Caputh in October. Exasperated, Einstein informed Millikan and Fleming that he would instead remain in Europe, head south, and enjoy sun and relaxation. Somehow, the situation was retrieved, probably by Fleming, who generously offered to make his palatial apartment in the Caltech faculty club available to the Einsteins during their stay. Einstein relented and signed the contract with Caltech in November.

With his return to Pasadena for the winter now finalised, Einstein wrote an article for *The New York Times Magazine*. It was published on 22 November under the arresting image of an Achilles-like figure driving the masses to war. Einstein likened the holding of a disarmament conference without guaranteed security to a city council legislating on the length and sharpness of each citizen's dagger; he characterised not providing for effective law enforcement as merely leaving the weak to the mercy of the strong. He fastened on the Kellogg Pact of 1928, which had formally outlawed war, and proposed to make this a reality by the signatories, which included Germany, agreeing never to compel their citizens to undertake warlike action against their own moral convictions. Sadly, Einstein's suggestion would have had little impact in 1939, since he did not yet foresee a Nazi state in which those moral convictions would have been perverted to unimaginable evil.

By the autumn of 1931, Heinrich Zangger was optimistic that Einstein's son Tete was coping with life and his course work. That summer, Tete had written to his father regretting that they were unable to go on a walking holiday. The letter is full of Freud's theories and psychoanalysis and how men's character traits were the product of competition with fathers. He proudly announced that his literary career had begun, as the *Neue Schweizer Rundschau* had accepted twenty-six of his aphorisms for publication. These appeared in the December 1931 issue.[16] He also recommended that Einstein play J. E. Bach's D major Sonata, which Tete thought was a wonderful piece.

However, it seems likely that, even in this period, Tete continued to experience mental disturbance not apparent in his irregular contacts with Zangger. By that autumn, Prof Maier at the Burghölzli clinic indicated that Tete was an outpatient. In the letter mentioned above, Tete had informed his father that he had postponed his second-year examination because he had been unable to work. He blamed this on the complications of his love life. He was in some sort of relationship with an older, possibly married, student in his medical course. He had intended to visit Maja in Florence during the summer of 1931 but wrote to her that he preferred to stay in Zurich to pursue his love interest. In the Christmas vacation he enjoyed skiing with some friends. It was probably just after this that he had a traumatic sexual experience that resulted in a descent into schizophrenia. The Burghölzli doctors were of the opinion that this resulted from the prevalence of the disorder in Mileva's family; indeed they also considered Mileva to have a schizoid personality. As the winter wore on, Tete isolated himself in the Zurich apartment for months on end, playing Busoni arrangements of Bach, Mozart piano concertos, and Schubert sonatas. It is likely that an episode in which he tried to hurl himself from a window in their apartment, and hung pornographic pictures on his bedroom walls, much to his mother's distress, dates from this winter of 1931/32.

The colder weather eventually forced the reluctant Einstein away from the Caputh house back to Haberlandstrasse. On 11 November, he attended a dinner given by von Laue in honour of Millikan, who was still travelling around Europe. Three other Nobel Laureates attended, Walther Nernst, Planck, and von Laue himself. On 14 November, Einstein left for a trip to Leiden, which was conveniently on the way to Antwerp and the ship to California. He was pleased to see Ehrenfest again, whose personal life was becoming increasingly troubled by fits of severe depression. His wife spent long periods in Russia and returned infrequently to Leiden, where Ehrenfest's daughter Tania lived with him in his chaotic household. Tania was a highly competent mathematician; she received her PhD from Leiden one month after Einstein arrived and often aided him in his complicated mathematical endeavours towards unifying gravity and electromagnetism. Together they explored a new unified theory in five dimensions published by Oswald Veblen and Banesh Hoffman in 1930.

---

[16] Tete seems to have forgotten his aphorisms published in 1928, and his father's admonition that publication could be a result of whose son he was, rather than literary merit. His father had however praised his latest work and commented on the drafts in some detail in letters starting in the early summer of 1930. Some of the aphorisms in Tete's new collection give insights into his psychological state (e.g. 'Nothing is worse for someone than to meet a person, in comparison to whom all their striving and existence is worthless' and 'A wound on the body heals with a scar; a wound in the soul heals with deep knowledge').

Einstein also enjoyed visits to his friends in Leiden, in particular Adriaan Fokker, who entertained him and the Ehrenfests to dinner. He was also given a guided tour of the Natural History Museum and visited the Dutch parliament in The Hague.

Immediately after arriving in Leiden, Einstein gave a lecture on relativity in the small lecture hall in Ehrenfest's department. This was probably very similar to the one he had given in Vienna one month previously describing his progress in unification. On the following day, he talked about quantum mechanics. This was recalled, surprisingly vaguely, by Hendrik Casimir in his autobiography. Casimir was a student of Ehrenfest who had just received his PhD degree. He went on to become a leading theoretical physicist. Einstein repeated and expanded his 'Einstein's box' thought experiment that he had discussed with Ehrenfest in July in Caputh. He had also presented this in a university colloquium chaired by von Laue in Berlin on 4 November, in which he mentioned his work with Tolman and Podolsky. Casimir was convinced, as was Ehrenfest, that the thrust of Einstein's argument was not to question the validity of Heisenberg's uncertainty principle but to highlight some of the bizarre properties of quantum mechanics, such as 'action at a distance', with the implication that it was an incomplete theory. As had been requested by Ehrenfest, Casimir stood up at the end of Einstein's lecture to begin the discussion. He outlined what was already the conventional wisdom of the 'Copenhagen interpretation' propagated by Bohr. Einstein replied that 'I know this business is free of contradictions, yet in my view it contains a certain unpalatability.'

Einstein left Leiden for Rotterdam, arriving on 30 November. Elsa had by then joined him and together they boarded the MS *Portland* for the journey to California, sailing on 2 December. They had selected this ship since, unusually for a passenger liner, it did not stop first at New York. For Einstein this had the advantage of more uninterrupted days at sea, one of his greatest pleasures; for Elsa, it avoided the stress and publicity always associated with Einstein's visits to New York. They also discovered that the *Portland* had the virtue of both very comfortable accommodation and very few other passengers. As usual, Einstein and Elsa had separate sleeping cabins. The Atlantic crossing was stormy. Always a good sailor, Einstein mused on his future as the storm raged. The offer from Oxford and the drawn-out negotiations with Caltech, combined with the ever-darkening outlook for politics in Germany, probably catalysed a decision that the time had come to abandon his position in Berlin and to become as he put it, a migratory bird. Although he changed his mind shortly thereafter, the time was rapidly approaching when his hand would be forced by events outside his control.

Chapter 22

Return to Caltech and Oxford; Tete's breakdown; Nazis impinge on the Caputh idyl (1932)

After experiencing tropical, indeed sweltering, conditions passing through the Panama Canal, and a fierce storm off the coast of California, the Einsteins arrived in the Port of Los Angeles in Wilmington late on 30 December 1931. On the following morning, having forbidden any extravagant ceremony such as had greeted his previous year's arrival in San Diego, Einstein held a press conference on board ship at which Richard Tolman translated. Tolman then transported Einstein and Elsa to Fleming's house in Pasadena, where Einstein enjoyed the wonderful gardens and birds. They stayed there for five days before moving into Fleming's suite in the Athenaeum, a short walk away from the Caltech Physics Department. Einstein was greatly pleased by the Athenaeum apartment, although the *Los Angeles Times* reported that Elsa was keen to find a bungalow with garden and kitchen similar to that which they had rented during their previous visit. She searched for one for a while until presumably being overruled by her husband.

The annual New Year's Day Rose Parade wound past the Einsteins, who were stationed in the Flemings' garden. The theme was that of the countries participating in the 1932 Olympics, to be held that summer in Los Angeles. The marching brass bands, some consisting of girls, paid tribute to the Einsteins as they passed. On the following day, Einstein was invited to inspect the German float, a depiction of the Black Forest. He was photographed shaking hands with one of the young girls dressed in folk costume who were perched in front of a representation of a log cabin. He was surprised at the depth of affection for Germany expressed by the local populace, and the hostility towards France, which he attributed to unsophistication and misinformation. In his travel diary, he remarked that the ravages of the Great Depression had resulted in 10% local unemployment and, a few days later, that President Hoover had refused to meet a delegation of the unemployed. Einstein himself, while working behind the scenes in specific areas, kept out of overt political controversy at the request of Millikan, who was absent on Einstein's arrival as he was lecturing at the University of Wisconsin. Indeed, Einstein kept much more out of the public eye than in the previous visit, soaking up the Californian sun on his balcony and playing the violin in the Athenaeum apartment.

Although Einstein had found the ocean voyage very relaxing, he had not been idle. Once he arrived in Pasadena, he looked back in his travel diary to notes he had made on a new idea for his unified theory, which he was keen to report to Mayer, who had been left behind in Berlin. Most of his first week at Caltech was however taken up with discussions on cosmology. Willem de Sitter, Einstein's friend from Leiden and long-term cosmological protagonist, had arrived a few days before Einstein; he left on 13 January. He too was staying in the Athenaeum. On 6 January he gave a lecture at the nearby Mount Wilson base laboratories, which Einstein attended. He described three possible models of the universe, two of which expanded forever and the third of which alternately expanded and contracted. The subsequent discussions between Einstein and de

Einstein. Brian Foster, Oxford University Press. © Brian Foster (2026). DOI: 10.1093/oso/9780198794875.003.0022

Sitter resulted in a joint publication that gave rise to a highly influential model of the universe in which both the cosmological constant and the curvature of space were fixed at zero. This can in a sense be regarded as marking the conclusion of their decade-long discussion on cosmology; its long-term importance was in its simplicity, which allowed it to be very predictive and form a sort of baseline against which more complex 'Big Bang' models of the universe could be evaluated. Neither Einstein nor de Sitter postulated a Big Bang singularity, although such a thing had been proposed by Alexander Friedmann in 1922. Georges Lemaître, who was shortly to arrive at Caltech, had constructed such a Big Bang model in two papers in 1931.

On 8 January Richard Tolman gave a seminar on cosmology at the Caltech physics department. The *Los Angeles Times* reported that:

> Dr Einstein accepted Dr Tolman's invitation to interrupt at any time. After the 'father of relativity' had commented in German for more than ten minutes, Dr E. C. Watson suggested that Dr Tolman interpret Einstein's comments. 'Oh I think everyone here understood Dr Einstein,' retorted Dr Tolman as he glanced apprehensively about the lecture hall. A gale of laughter convulsed the audience.

On Saturday 9 January, Einstein received a reply from his old antagonist Alfredo Rocco to his letter of protest at the requirement of the Italian Fascist government that all university professors take an oath of allegiance. This letter was one of Einstein's last actions relating to his membership of the League of Nations Committee on Intellectual Cooperation, from which he was about to resign in favour of Hugo Krüss. Einstein thought Rocco's answer in justification of the oath to be philistine nonsense and was depressed about the situation in Europe that it indicated. It was therefore a relief later that day to set off to visit their friend Samuel Untermyer in Palm Springs. However, Untermyer was ill and confined to bed so that it seems that he arranged for them to stay instead at the El Mirador Hotel nearby. The Einsteins spent three days there, experiencing both hot sunshine and a sandstorm. That Sunday evening, Einstein asked to join the string trio who were playing for the diners. Much to Einstein's pleasure, he played pieces by Bach, Mozart, Beethoven, and Händel with them.

Einstein seems to have again taken up the interest in the paranormal that his friend Upton Sinclair had aroused during his previous stay in Pasadena. In his travel diary he described at some length his interactions with a young medium, who was making an enormous income from her seances. That Gene Dennis was young and pretty was no doubt an attraction for him. Einstein remarked in public that she had told him things about himself that no-one knew but himself. Armed with Einstein's endorsement, she went on to a successful engagement at a Los Angeles theatre in which she demonstrated her telepathic skills. This endorsement, together with the foreword he had written some years previously to Upton Sinclair's book on the subject, stirred up a storm of controversy in the American press after Einstein's return home, into which Sinclair and others were drawn.

Not as young as Gene Dennis but still very beautiful was the film star Pola Negri,[1] who was staying at the El Mirador while recuperating from a medical procedure. Tony Burke, an English staff member, reported that she and the Einsteins had a long conversation in German. Although he

[1] One of the brightest stars of silent movies, Negri had had a tempestuous affair in the 1920s with Charlie Chaplin. Unlike many, she had made a successful transition from silent films to the 'talkies'. Her early interest in Hitler seems to have grown when she returned to Germany a few years later, where she starred in the movie *Mazurka*, which became one of Hitler's favourites. There were even rumours that she had had an affair with him, causing her to sue and win a libel case against a French magazine. Living in France in 1940, she fled back to the US after the Nazi invasion, where she died in 1987 at the age of 90.

spoke no German, Burke ghosted a short article for Negri on the meeting, which made headlines across the country. *The New York Times* on 14 January reported that Negri asked Einstein whether he had met Hitler; with a twinkle in his eye, Einstein said no but he had seen photographs, and they were sufficient!

After his return to Pasadena on 13 January, Einstein mostly had discussions with colleagues at Caltech, visiting several of their laboratories. At the weekend, he has happy to entertain Ehrenfest's daughter Galinka, and indeed met her on several subsequent occasions during their stay. She had been working with children in the Netherlands and was embarking on a successful career in art and illustrating children's books. On Sunday 17 January they visited a local Mexican market together, whose colourful display delighted Einstein. They also enjoyed a puppet show. That evening the Einsteins paid one of their regular visits to Lili Petschnikoff's house near the Hollywood Bowl. Pieces by Bach, Mozart, and Schumann were played with great enjoyment while Elsa busied herself mending socks that she apparently intended to present to the maids at the Athenaeum.

Millikan returned from his travels the following week and on 22 January, gave a talk on his work on cosmic rays. He and his former student, now research fellow, Carl Anderson, had set up a Wilson cloud chamber in their laboratory to allow them to photograph the tracks of cosmic rays. These were made visible by the water droplets that condensed around the atoms ionised as the cosmic rays passed by. The chamber was filled with saturated water vapour that became supersaturated on expansion by a piston, thereby causing the droplets to condense around any disturbance in the chamber.

Einstein was unimpressed by Millikan's results, some of which he had already heard during Millikan's visit to Berlin a few months earlier. He characterised Millikan's theory that cosmic rays were energetic photons, originating from deep space, that then interact with the Earth's atmosphere as 'weak', although this hypothesis is indeed correct for some cosmic rays. Many of the results were perfectly baffling to physicists, who at that time thought that there were only two types of elementary particle, a positive proton and a negative electron. On his arrival at Caltech, Einstein had heard about Harold Urey's discovery of the deuteron, chemically identical to hydrogen but with twice its mass, via its characteristic spectroscopic lines. It had been assumed that hydrogen was too light to have such isotopes, which were thought to be caused by an extra proton and electron being bound in the nucleus. Robert Oppenheimer, who had a joint appointment between Berkeley and Caltech, gave a lecture at Caltech on 26 January about his theoretical work on the possibility of a 'neutron', a neutral particle that might be bound inside the atomic nucleus. The discovery of the neutron by James Chadwick in Cambridge, first tentatively indicated towards the end of February and confirmed in May, rapidly made it clear that there were no electrons inside the nucleus, just protons and neutrons; the deuteron was such a bound state of one proton and one neutron.[2] A few months after Einstein's visit, Anderson used his cloud chamber to discover the positron, the anti-particle of the electron with positive rather than negative charge. In his lecture, Millikan had spoken of his plans to float balloons into the stratosphere to measure the cosmic-ray flux. The next few decades saw a veritable avalanche of new particles, many originating from the interactions of cosmic rays.

---

[2] Oppenheimer was discussing the 'neutron' first postulated by Wolfgang Pauli in 1930 to explain how the decay of radioactive elements could conserve energy, momentum and spin. After Chadwick's discovery, it was realised that Chadwick's particle could not be Pauli's 'neutron', as it was much too heavy and was essentially an electrically neutral copy of the proton. In the following year, Enrico Fermi coined the term 'neutrino' for the particle first postulated by Pauli.

Einstein evinced interest in none of these remarkable discoveries despite the fact that some of the work was going on under his nose at Caltech. After listening to Oppenheimer's lecture, Einstein noted in his diary that it was obvious that these results represented a dead end. In fact, it was his work on unified theories that was the dead end. Had he not been convinced that quantum mechanics was flawed and had he been more interested in its application inside the nucleus, he would surely have asked himself what force was binding the newly discovered neutron and proton together. An electrically neutral neutron could not be bound by electromagnetic attraction and the gravitational force was much too weak. Simply putting such questions from his mind, Einstein continued to focus on the unification of gravity and electromagnetism. This could be interpreted as representing the triumph of his mathematical and philosophical convictions over the physical intuition that had driven him to his great discoveries. All of physics was the poorer for Einstein's lack of interest in the extraordinary new worlds of nuclear and particle physics that were beginning to open up.

On 19 January Einstein had given a talk at Caltech on his current work on unified theories. His impression, probably optimistic, was that the audience was interested by what he had to say. Even if his younger colleagues were indifferent to his laborious inching towards unification, Einstein's work on this and on the cosmological implications of general relativity was of great interest to the press. The *Los Angeles Times* reported on a lecture he gave on 29 January on cosmology and evidence for the curvature of space-time. He discussed Hubble's observations of the spectra of galaxies being 'red-shifted' to lower wavelengths, evidence for their velocity away from the observer, but which were compatible with zero curvature. Einstein's linking 'Gravitational and Electrical Powers with Mathematical Chains' was headlined in the *Los Angeles Times*. A non-technical talk he gave at the University of California Los Angeles on 15 February, translated by Tolman, attracted a large crowd of students and faculty.

Although Einstein remarked in his diary that he was pleased that this visit was much quieter than his previous one, he could not entirely escape public duties. On 20 January, he had given a speech on disarmament at Whittier College.[3] His remarks in German, translated by his Caltech secretary, returned to his theme of the importance of a world body that could guarantee, by force, if necessary, disarmed nations against attack by an aggressor that refused to accept its arbitration. On 25 January he gave a speech, which Tolman translated, at a major dinner organised in his honour by Millikan. The other guests of honour, historian Charles Beard and former President of Cornell University Jacob Schurman, joined Einstein in attacking the economic distortions caused by the Treaty of Versailles. A visit to the home of Gustav Struve, the German consul, the following evening was not a success; Einstein's sensibilities were offended by Struve's children's ballet to the music of Händel's famous *Largo*; he was only grateful that the composer had not seen it. On 1 February he attended a meeting on disarmament under the auspices of the University of Southern California, chaired by its president, Rufus von Kleinschmid, whose flippant and off-colour remarks offended both Elsa and Einstein. His own talk followed that of a former marine. Einstein thought that the officer had spoken well, but characterised his own remarks as pearls before swine. On 4 February he listened to a speech by Schurmann on the Manchuria situation subsequent to the Japanese invasion a few months previously. Einstein regretted that Schurmann's remarks were clever but lacked any moral content.

[3] Einstein's travel diary states 18 January but in fact the meeting took place on 20 January.

These sometimes-tiresome duties were interspersed with much pleasanter activities; his work, of course, but also sunbathing on his balcony and playing the violin in his apartment. Indeed, according to Marianoff, he went further:

> Einstein played his violin anywhere and everywhere. He played it outside under the eucalyptus trees, in the courtyard, and while he walked up and down the porticoed halls. The strains of it as he strolled would come from the pittosporum trees at the back of the gardens. He was oblivious to where he played.
>
> Some of the faculty as well as the graduate students were also housed in this building. No one ever intruded on these morning peregrinations, or, in fact, ever neared him. One of the students said that at certain times the strains would take on a wild note without regard to harmony – as though he were obeying some inner demand not meant for an outer world'.

On Friday 29 January, Einstein attended a concert devoted to the music of Jewish composers, as a guest of the conductor Artur Rodzinsky. Einstein considered the Los Angeles Symphony's performance of Ernest Bloch's *Israel* Symphony to be excellent. He greatly enjoyed the piece; Bloch was one of very few contemporary composers who gave him pleasure. The concert also contained a piece by Goldmark and Louis Gruenberg's *Enchanted Isle*. The well-known pianist and composer, Mischa Levitzki, played the Piano Concert No. 2 by Saint-Saëns.[4]

Einstein later that evening attended[5] a service at the Temple B'nai Israel synagogue on the invitation of Rabbi Henry Radlin. He had sent a telegram welcoming the Einsteins on behalf of the Pasadena Jewish community when they first arrived in Pasadena a year previously. A member of the Temple congregation recalled that Einstein 'was at a Friday night service and sat on the bema (dais) with his wife and his hair was very wild and sort of high collar . . . He walked down the aisle shaking . . . hands.' Einstein enjoyed the performance of the choir, composed of five women, in particular the final Schalom Alechem, sung by well-known soloist and the Temple's musical director Pearl Hassler. On Sunday 31 January, Einstein attended a service at the Scott Methodist Episcopal Church in Pasadena, a church mostly patronised by African Americans, in memory of the Jewish philanthropist Julius Rosenwald, who had strongly supported education charities for African Americans. Both Rabbi Radlin and Einstein gave addresses, the latter pleading for racial equality and world peace. The message on racial equality he had already forcefully put in an article written for the February issue of the magazine of the National Association for the Advancement of Coloured People (NAACP). Although he enjoyed the singing, Einstein was somewhat bemused by the incongruous performance of a German popular song relating to two lovers.

After a dinner hosted by Professor Schmidt on 2 February, Einstein played a Vivaldi Sonata with Frau Götz, wife of a Caltech physics faculty member and a good pianist. On 9 February, Einstein listened intently to a recital by the violin prodigy Ruggiero Ricci, at that time thirteen years of age, organised by a rich financier who had taken Ricci under his wing. The concert was held privately within the Athenaeum for Einstein's benefit. Probably remembering Yehudi Menuhin, three years older than Ricci, who had been brought up in San Francisco, Einstein remarked in his diary that the area seemed to be swarming with musical prodigies.

---

[4] It was widely rumoured that Saint-Saëns had Jewish heritage, which was given sufficient credence that he was included in what the *Los Angeles Times* labelled an all-Jewish concert. Although his music was banned by the Nazis for this reason, modern scholarship can find no substantiating evidence in his ancestry.

[5] This was either on 29 January or 3 February, with the earlier date being more probable. The chronology in Einstein's travel diary becomes impossible to follow at this stage and other evidence from contemporary newspapers, etc., is inconclusive.

Elsa also pursued some of her own interests during their visit. She had brought with her some of the bronzes and ceramics that Margot had exhibited a year before in Berlin. Shrewdly exploiting the publicity around her husband's visit, Elsa arranged for the pieces to be exhibited at a local Pasadena art gallery and for this to be covered by the local press. The exhibition opened on 2 February and ran for the remainder of the month.

Elsa fell ill with influenza in February, so Einstein was alone at a reception at the Sieroty residence in aid of Zionist causes on the afternoon of 13 February. Einstein drew their attention to a village in Palestine that he and Elsa particularly supported. Einstein would have certainly been relieved when the musical element of the Oneg Shabbat, a traditional celebration of the Sabbath, began. Sol Babitz, a young violinist who had studied with Carl Flesch in Berlin, performed first and impressed Einstein with his talent. Mildred Nehamkin, a coloratura soprano formerly of the Cleveland Grand Opera Company, also sang and gained plaudits from Einstein. The pianist was Sidney Cutler.

While Elsa was confined to her bed, she decided that Einstein needed a rest. She called Lili Petschnikoff and asked her to take him out for a relaxing drive. Lili, as always accompanied by her aged mother and aunt, drove to an inn at Palos Verdes, a beautiful peninsula on the coast close to Los Angeles harbour. She and Einstein sat on the grass admiring the ocean view. They talked about some of the people she had known, including Ibsen and in particular their mutual friend the conductor Bruno Walter. In a conversation a few days earlier, she had given Einstein the full story of her earlier life, her husband's infidelities and her escape from wartime Germany. As they were talking, Lili noticed that a crowd was beginning to gather, alerted by Einstein's easily recognisable hairstyle. Despite fleeing back towards the car, they were intercepted; Einstein was forced to sign several autographs before they could make their escape.

On the evening of 18 February, Einstein attended the Pasadena Civic Auditorium for the first concert at the newly opened venue, given by the amateurs of the Pasadena Civic Orchestra. The programme consisted of Schubert's 'Unfinished' Symphony, several short pieces and concluded with Listz's Piano Concerto No. 1 in E-flat major. The soloist was local resident Mildred Marsh. Not only the playing but the acoustics of the new hall attracted the approval of the *Los Angeles Times* reviewer.

Elsa had recovered by 24 February, when she and Einstein arrived at the famous Arrowhead Springs Hotel in the foothills of the San Bernadino mountains about 90 km from Pasadena. They stayed there until 26 February, attending the National Orange Show. As they left the hot sunshine and entered the huge tent, they were feted with German songs, to which drunken movie stars beat out the time waving their beer glasses, greatly to Einstein's amusement. In a few words of greeting, he praised the California sunshine and its ability to produce such fruit in the middle of winter. Having been presented with a box of prize-winning oranges, the Einsteins returned to their hotel.

Having returned from San Bernadino, they began the busy final weekend of their visit. In the afternoon, Einstein attended an event at the Filmarte Theatre in Hollywood arranged by the German Olympic Committee to raise funds to support German participants at the Los Angeles Olympic Games due to start in a few months. A movie and a lecture on the theme of the musical activities of Frederick the Great of Prussia entertained the audience of mostly businessmen and German expatriates. That evening, Einstein addressed, in German, an enormous meeting of the student bodies of the eleven southern California colleges and universities in the Pasadena Civic Auditorium. The aim was to give the students an overview of international relations, prospects for peace and good government. A fifteen-minute part of the event was broadcast along the west coast on radio. Einstein returned to his favourite theme of the necessity of a world body with teeth that would allow confidence that countries that disarm would be protected from aggression. He advocated that the US, German, English, and French governments should demand an immediate cessation of Japanese aggression in China, under threat of a blanket economic boycott. Robert Millikan and Charles Beard also spoke at the event.

Sometime during the following day, 28 February, Einstein enjoyed playing again with the Zoellner String Quartet members. That evening saw the final major event of Einstein's visit. He and Elsa finally made themselves available to the Los Angeles Jewish community in the banquet hall of the Hebrew Sheltering Home for the Aged. Before attending the banquet, the Einsteins were moved by visiting the old people gathered at the Home's synagogue. The immense hall of the Home was crowded with 600 participants who had paid $2 per plate for their places. The modest price was set to allow as many as possible to attend at a time of great economic deprivation. Greatly moved by the various speeches of welcome, including one from Los Angeles Mayor John Porter, and by the presentation to him of a certificate of donation for tree planting in the Einstein Forest in Palestine, Einstein threw away his prepared remarks and pleaded for better relations and friendship between Jews and Arabs in Palestine as well as for enthusiastic support for the Zionist cause. He was once again moved by the violin playing of Sol Babitz, this time accompanied by Alfred Kaufmann on the piano. The Jewish International Orchestra also provided musical entertainment.

The Einsteins now prepared to return to Germany. Just before they did so, Einstein gave a wide-ranging interview to a *Los Angeles Times* journalist, apparently in English, which touched on his early life as well as his advocacy of world peace. His efforts to learn English appear to have borne fruit, since the interviewer was complementary about his grasp of the language. On 4 March, the Einsteins sailed for Europe via the Panama Canal on board the Hamburg–American Line's *San Francisco*. Once again, Einstein opted to avoid New York and the concomitant publicity and hustle, hoping to enjoy a tranquil voyage home. However, his peace of mind had been troubled by further news of his son Tete. Already on 29 February, he had replied to Mileva's news about Tete's condition, writing that he had realised that something was wrong, since his son had not written for a long time. He expressed his strong opposition to psychoanalysis, which he said had not helped others in his circle and which he characterised as exceptionally dangerous. Once again, he brought up, typically undiplomatically to Mileva, the hereditary nature of Tete's complaint and gave the general impression of a resigned determination to let Tete's fate take its course. He suggested that Tete stay in Caputh for as long as he liked over the summer, considering that his own equable temperament, combined with sailing and the peaceful nature of his summer retreat would have a salutary effect on the young man. Around February/March 1932, Heinrich Zangger wrote three letters to Einstein. Zangger was struggling with influenza that had led to a heart complaint. He urged Einstein, as soon as he returned from his stay in California, to bring Tete, whom he characterised as seeming depressed and tired, to stay with him in Berlin. Instead, in the middle of April Tete arrived in Florence unannounced and stayed with Maja for eight days. He wrote to his father that he needed to be in the country for a while and hoped to visit Caputh soon, where he wanted to discuss psychology. While in Florence he had begun to write verse following an ancient Greek hexameter form. Maja recalled that, during his visit, Tete played only Mozart on the piano with a disturbing metronomic exactitude. He argued endlessly with Maja that Beethoven was a 'false' artist.

Although Tete returned to his studies in early May 1932, Zangger again became concerned that his course was too strenuous. Indeed, in a letter that Mileva wrote to Einstein that month, mostly devoted to their financial troubles now that most of their apartments were without tenants, she mentioned in passing that it was impossible to stir Tete into doing any work. In June she wrote again, begging Einstein to come to Zurich to discuss their financial problems and visit Tete, since he could not come to visit Einstein in his current condition, nor indeed even write letters[6]. On 15 September 1932, Zangger wrote to Einstein telling him he was very worried about Tete. He had invited him and some friends to his house for the evening. These friends of Tete were clearly worried about him, too. For the first part of the evening, Tete sat white and withdrawn, hardly interacting with anyone. It was only after 2 a.m. that he came to life and danced with some of his friends, smiling. Zangger and Michele Besso were in close contact, so that it was probably discussion of this encounter that prompted Besso to write to Einstein on 18 September, pleading with him to interact more with Tete, to let him travel with him to the US and generally to hold him closer. This was something that Besso thought would be good for them both, as he had found in his relationship with his own son, Vero. Zangger thought that Tete was convinced that his relationship with his father had completely changed and that Einstein was no longer interested in him.

Tete's conviction may have resulted from letters that Einstein sent to Zurich in the summer of 1932. In one of these, Einstein asked Tete to sign a notary-certified letter not to contest Einstein's will, of whose contents none of his Zurich-based family were aware. In fact, Einstein left most of his estate to Elsa and his adopted daughters. The letter to Mileva asked for a similar undertaking and pulled no punches; Einstein accused her of not repaying her debts, of continually trying to extract money from him, and of dishonesty. He sent a copy of this letter to Hans Albert. Although Einstein's intention was to ensure that the fact that he had left the Nobel Prize money to Mileva's side of his extended family was considered in any ultimate division of his estate, the result was to unbalance Tete's precarious equilibrium. It also angered Hans Albert. When Tete did not give any clear answer to Einstein's question about the will, he wrote to his son saying that since there seemed to be tension between them, they should not meet at the moment, although a subsequent letter from Tete mollified him and he renewed his invitation to visit. Hans Albert's reply was less accommodating:

> You will have to admit that among those who are close to you, or at least used to be, I am the one who has given you the least trouble, certainly in pecuniary matters. I am beginning to understand that I was really an idiot in trying to earn my keep as quickly as possible. For while all the weak and sick are being pampered and cared for, it does not seem to be enough for you that I take care of myself. Apparently, you do not even consider me worthy of receiving a modest token of remembrance when you die, after you were stolen from me in life. If your paternal feelings toward me have really shrunk to nothing, it would be best briefly to tell me this. In that case, you can be sure that I will never bother you again.

Mileva also replied, in more measured terms, admitting that she did owe Einstein money but was currently unable to repay it owing to her desperate financial situation. She concluded by begging Einstein to try to imagine how deeply she had been affected by Tete's long illness.

Zangger was greatly concerned that Einstein seemed not to care that pressing Tete on the matter of his will was worsening Tete's mental health. He suggested that Fritz Haber, with whom Zangger was in contact with respect to Haber's serious heart condition, might once again mediate, as he had in 1914, between the two branches of the family. Zangger was in contact with Prof Maier at the Burghölzli clinic and encouraged him to write to Einstein setting out forthrightly what he saw as Tete's condition. Maier wrote to Einstein on 6 October with the news that Tete, who had been admitted to Burghölzli, had had a crisis. Einstein, seeming to realise that he had gone too far, wrote to Tete telling him to forget completely about the will. He invited him to

[6] In fact, Tete was writing to Einstein fairly regularly during this period, and indeed saying that he was feeling better, which implies he may have been writing without Mileva's knowledge.

come to Berlin, excusing himself from coming to Zurich because of the important work that he was doing with his assistant Mayer. Although he could not take Tete with him to California this year, he hoped he would be able to come the following year to Princeton. By the time Tete received this invitation to Caputh, it was too late. He was in no condition to travel anywhere. He was released after a couple of weeks but was still in a highly nervous condition at home, often telephoning Zangger, who had to hurry to the apartment to reassure him. By November 1932, Tete had been admitted to the Knonau Klaesi clinic, a few kilometres south of Zurich but soon had to be readmitted to Burghölzli when his condition again worsened. Einstein promised to visit Tete before he left for his annual trip to California at the end of 1932, provided the doctors thought this would be helpful. In fact, he did not make this trip, perhaps because the doctors did not recommend it or possibly because he was convinced that Tete was under the bad influence of both Mileva and Hans Albert and his wife. Hans Albert and his family were now living in Zurich, having moved there in early 1931 when Hans Albert obtained a position as a researcher in the Hydraulics Institute. Einstein had organised this by writing to Eugene Meyer-Peter, Director of the institute and one of Hans Albert's former tutors.

Returning now to the Einsteins' voyage home from California, the *San Francisco* called in at Rotterdam on 3 April 1932 and remained at anchor for two days. Despite Elsa's almost desperate attempts to convince Einstein to take her with him for a flying visit to Leiden, he refused to contemplate it, no doubt wishing to avoid the inevitable social engagements that would entail. Instead, Einstein wrote to Paul Ehrenfest, asking him if he would like to make the trip to Rotterdam. Ehrenfest naturally rushed there on 4 April; he and Einstein spent some time discussing Einstein's 'quantum box' thought experiment and another based on the Compton Effect of light scattering from an electron. In his letter inviting Ehrenfest, Einstein had attempted to dissuade him from moving from Leiden to California, which he was apparently contemplating. The Einsteins then continued their voyage in the *San Francisco*, calling in at Bremen on 6 April, where Einstein told *The New York Times* how much he had enjoyed his stay in California. Disembarking at the final destination of Hamburg, the Einsteins travelled back to Berlin, where they arrived on 9 April.

On his return, Einstein received news that his close friend Moritz Katzenstein had died of cancer on 23 March. Einstein's eulogy at the interment of Katzenstein's ashes was a heartfelt tribute to a friend for whom he had had great affection. He mentioned the almost child-like pleasure that Katzenstein took in the lakes and islands as they sailed by them during the Berlin summers and how they would talk about their hopes and plans for research. He ended his eulogy by saying that he was 'grateful to destiny that he had this good, tireless and highly creative talented man as a friend'.

An entirely happier communication awaited Einstein in Berlin. Emil Abderhalden, the new President of the Kaiserlich Leopoldinisch-Carolinische Deutsche Akademie der Naturforscher, Halle, more colloquially known as the Leopoldina, informed him that the Academy had elected Einstein as a member in March. The oldest continuously existing scientific academy in the world, founded in 1652, the Leopoldina became the German National Academy for Science in 2008, although it had been generally viewed as such for centuries before that. Einstein, who usually hardly noticed such honours, was highly gratified by this long-overdue election. He had been nominated in 1926 at the same time as Lise Meitner, Max von Laue, and others. However, from what was presumably a combination of anti-Semitism, antagonism to relativity from influential members and a dislike of his political stance from the conservative President, he was not then elected.

On Sunday 24 April, David Reichinstein came to visit Einstein in Haberlandstrasse. Reichinstein was a physical chemist who was an old friend of Einstein from his earliest Zurich days. By 1932, his angular personality had managed to make enemies across a broad swathe of German chemists. Einstein however had a soft spot for him and in particular valued what he saw as his highly original approach to his subject. In 1930, he had written a foreword to a book by Reichinstein on interface processes. Reichinstein was keen to elicit Einstein's help in gaining a position, as he was at that time unemployed, although he had a visiting position at the Technische Hochschule in Berlin and was living in Charlottenburg. He asked Einstein to write to a professor at New York's Columbia University. In passing, he mentioned a 'pamphlet' he had written about Einstein, which he expanded on in a follow-up letter that evening. Einstein with his usual hypersensitivity about biographies of himself, replied immediately that he considered it tasteless to write biographies of persons who were still alive. He considered such a production could only be excused because of Reichinstein's current financial position but would still greatly prefer that the work should not be published until after his death. If it must be published, he forbade any publication in Germany.

Reichinstein replied to Einstein's letter in a slightly hysterical way, since he had misinterpreted a remark to imply that Einstein thought that he was producing an attack on him. He reminded Einstein of their more than twenty years of shared experiences and pointed out that Einstein could not have such an objection to biographies since he had written an introduction to Reiser's biography (written under an alias by Einstein's son-in-law Rudi Kayser). He refuted the idea that he was writing purely for money but instead wanted to highlight Einstein's influence both on him personally and on science in general.

The correspondence continued at a frantic pace. Reichinstein tried to outflank Einstein by giving copies of the book to Elsa Einstein and Dima Marianoff, the latter seeming to have been a surreptitious collaborator or at least encourager of Reichinstein. Elsa objected to only one sentence, which Reichinstein promised to change. Marianoff gave a copy to Einstein, who took it with him to Oxford. By 26 May, he had read enough to be infuriated by it. He wrote an angry letter to Reichinstein, informing him that he had reached the limit of his tolerance. He now took a much harder line, threatening him that if he published it any form, anywhere, then their relationship would be terminated. However, as if to balance the stick with the carrot and to compensate for the time that Reichinstein had spent on the book, Einstein promised to redouble his efforts to find him a position worthy of his abilities.

Over the next few months, Einstein was as good as his word. He wrote to various colleagues but found that inside German faculties, there was usually at least one person with whom Reichinstein had already picked a quarrel and who torpedoed the initiative. Einstein approached the doyen of physical chemists, Fritz Haber, but he knew this was hopeless as Reichinstein had a long-running feud with Haber. From September, Einstein began sending Reichinstein regular sums of money to tide him over until he could find a position. The following month, Einstein even wrote to the Polish president Ignacy Moscicki, who was also a chemist.

If Einstein had hoped that his activities as a referee, advocate and source of funding would be enough to prevent Reichinstein from publishing, he was to be sadly disappointed. At some point that year, probably the late summer, Reichinstein had his book printed and bound in soft covers, publishing it himself. It is unclear what distribution it had; copies are rare today. Einstein's final letter was in October, when he conveyed Moscicki's slightly hopeful response to Reichinstein at an address in Kaunas, Lithuania. While there, Reichinstein prevailed on the publisher S. Glatt to bring out a second edition of his book in 1932, again bound in paper covers and also rare today. It can be presumed that by this point Einstein had learned that Reichinstein had broken his word and cut off communication with him. A third edition followed in 1933, printed in Prague, to where Reichinstein had fled after the Nazi takeover, once again self-published by the author. On 13 August 1933, Einstein wrote to a French publisher asking them not to publish what he characterised as Reichinstein's miserable biography. An English translation was published in London by Edward Goldston Ltd., together with a Prague publisher, which appeared in 1934. In a letter at the end of 1935 to a member of staff of the *American Physics Teacher*, to whom Reichinstein had presumably offered the US rights, Einstein continued to fulminate about the worthless book and requested that it not be published. In this at least he was successful, and no US edition appeared. From his final haven of Zurich, Reichinstein however published another book in 1941, in which he discussed the *Religion of the Learned*, which included a section on Einstein. It is unlikely that Einstein would have viewed this volume with any greater favour than he viewed Reichinstein's biography.

Einstein hardly had any time in Berlin after returning from the US before he was once more allowing Elsa to pack his suitcase, complete with obligatory dinner jacket, to return to Oxford. In 1932, he occupied rooms to the right of the main entrance to Christ Church in the main 'Tom' quad. His friend Margaret Deneke called on him with a welcoming bunch of flowers. As they talked, she noted that students were streaming out of a lecture through the Quad and was shocked to see how much they stared at Einstein, although he didn't seem to notice. She invited Einstein to dine on the following evening, 1 May, although their 'regular' chamber music partners including Marie Soldat were not due to arrive until 4 May.

It is not known whether Einstein took the walk with thousands of others down the High Street to Magdalen Bridge to hear the choristers' traditional welcome to May Morning from Magdalen Tower at daybreak. It was apparently a fine morning although it had begun to rain by the time the Deneke sisters arrived in their car to pick Einstein up. Undeterred by the rain, however, Einstein preferred to walk the just over one mile to Gunfield, so Margaret got out of the car and accompanied him. On the way they discussed Albert Schweitzer, whom Deneke had just visited in Africa, where she had volunteered as a nurse at his hospital, and pacifism. On arrival, Margaret showed Einstein the films she had made on her Africa visit. She and another dinner guest, the composer Ernest Walker, played his *West African Duet*, which they often played in conjunction with the film at Deneke's fund-raisers for Schweitzer's work. It contained many African folk tunes, as well as a quotation from Schumann's *Kinderszenen*, 'Strange Lands and People'. Einstein immediately recognised this, calling out that this was black Schumann with a laugh. Einstein seems not to have brought his own violin with him and was decidedly unimpressed with the one that Margaret had borrowed for him. Indeed, he pronounced it so bad that it could only be used to play Mozart if the pianist played very quietly. Margaret played one sonata with him, while Walker played two. At the end of the evening, Einstein pronounced himself satisfied that, although the violin's tone left much to be desired, it was difficult to spoil Mozart completely, which had such a wealth of beauty! Margaret lent him Schweitzer's reminiscences, *Aus meinem Leben und Denken*.

Einstein next travelled to Cambridge, where on 6 May he delivered the Rouse Ball lecture at the Senate House. *The New York Times*, reporting the lecture as the 'House Ball lecture,' gave his title as 'Electricity Theory in the Framework of the General Theory of Relativity'. He gave the current status of his work on unified theories, outlining his current modification of Theodor Kaluza's five-dimensional theory to use only four dimensions but described by a five-component vector. The lecture was attended by many of his Cambridge friends, including Ernest Rutherford and James Jeans. Einstein gave the lecture mostly in English, with occasional lapses into his native tongue. In giving the vote of thanks, Arthur Eddington joked that the complexity of the mathematics was as nothing compared to the complexity of German sentence structure. Indeed, *The Cambridge Review* speculated on how many other lecturers could fill a hall when either his language or his subject matter or possibly both must have been incomprehensible to many of them. It is probably with this trip that Einstein's surely apocryphal relativistic query at the railway station: 'Does Oxford stop at this train?' is associated.

After Einstein's return from Cambridge, he attended another dinner party at the Denekes, this time with Marie Soldat, Margaret Reid on viola rather than Frau Schulz,[7] and Arthur Williams once more playing cello. Before dinner, Einstein insisted on examining a violin that Margaret was considering buying, giving a negative opinion. He recommended getting a very cheap violin and letting his Berlin luthier 'adjust' it by removing wood in strategically chosen positions. After dinner they played Mozart and Haydn quartets. In Brahms' Piano Quintet Op. 54, where they were joined by Margaret Deneke, Einstein enjoyed his scramble through. He complained to his colleagues that he had had too many rests to count and that he had faked so many passages that he felt more like a banker than a musician. Nevertheless, Marie Soldat thought that it had been a more enjoyable evening than the previous year, partially because Einstein himself was more relaxed and that they had got used to the idea of the famous scientist playing music with them. They certainly lost track of time, as Einstein at 11:20 p.m. realised that he would be locked out of Christ Church. The day was saved by a quick telephone call to the porter's lodge and Helena Deneke driving Einstein back to college. He took away with him a gift from Donald Tovey, with whom he had enjoyed playing the previous year but who was out of Oxford at that time. His book *A Companion to the Art of Fugue* carried the dedication 'This copy with Companion as proxy for the Editor in default of capacity to get to Oxford even in a relative way in time to meet Dr Einstein. – May 11th 1932.'

The remainder of Einstein's stay passed quietly. He attended dinners at Christ Church, rapidly using up his store of rather halting jokes in his improving but still broken English. He even at times attended meetings of the college's governing body, although he was always careful to bring a pad of paper with him on which he worked surreptitiously during the meeting. There were also many discussions with academic colleagues, which even at times got in the way of playing music. A scheduled dinner and music-making at the Deneke's on 19 May had to be cancelled when Einstein arrived at Gunfield that afternoon to announce that he had to entertain a group of mathematicians instead.

One meeting that was to be fateful for Einstein's future life occurred during this visit. He had met Abraham Flexner, one of his predecessors as Rhodes Lecturer, during his last stay in Caltech. Flexner was in the throes of founding what was to become the Institute of Advanced Study (IAS) at Princeton. Only loosely associated with the university, it was intended that members would be able to concentrate on research without being bothered with administration or teaching. Having discerned Einstein's possible interest during their conversation in Caltech, Flexner took advantage of a European trip to visit Einstein in Oxford. They went for a long walk together in Christ Church Meadow, at the conclusion of which Flexner offered Einstein a position at the new institution. Einstein, however, was still reluctant to leave Berlin.

By 22 May, Einstein had completed his second stay in Oxford and had arrived in London, where he met the prominent pacifist, Lord Ponsonby. Together they travelled to Geneva, where, although uninvited, they intended to make an appearance at the ongoing disarmament conference being held under the auspices of the League of Nations. Already in session for many weeks, the endless sterile discussions made it clear that the conference was heading for failure. This probability had catalysed Einstein's visit, encouraged by the Joint Peace Council, a pacifist organisation. On 23

[7] Perhaps her argument in 1931 with Soldat had made her reluctant to repeat quartet playing with Einstein.

May, Einstein and Ponsonby made a brief appearance at the Plenary Meeting of the conference in the Peace Palace, causing a ripple of turning heads as they entered. They then proceeded to the press room, where both read statements criticising the proceedings of the conference and suggesting that avoiding war could not be achieved by better regulation. They then answered questions from the reporters. Although some characterised both Ponsonby's and Einstein's performances as ineffective, Einstein at least had the personal magnetism that compelled attention. In an article in the *Pictorial Review*, Konrad Bercovici gave a striking picture of how Einstein's presence electrified the somnolent atmosphere of the conference, inspiring respect from the normally unruly press corps. He went on to report on Einstein's behaviour that night in his hotel room at the Bergues Hotel:

> The hotel clerk told me that Einstein had been playing the violin uninterruptedly until I arrived. 'What did he play?' I asked. The clerk shrugged his shoulders. 'I couldn't make it out. He made his violin do the weirdest things: yell and yowl and screech. I thought at times he was trying to break it. And I heard him curse and fume while he was playing.'

Einstein was back in Caputh by 26 May. However, he was not able to enjoy playing his violin and sailing on the Berlin lakes in a relaxed atmosphere. The political situation was darkening markedly. Just as Einstein returned, Chancellor Heinrich Brüning was dismissed by President Paul von Hindenburg. Since March 1930, he had ruled through Presidential Decree, effectively side-lining the Reichstag, which still had a Social Democratic majority. Von Hindenburg had grown weary of what he saw as Brüning's socialist tendencies in trying to cope with the consequences of the Great Depression. Explicitly requiring a shift to the right in the government, von Hindenburg appointed Franz von Papen as Chancellor. His main qualifications as far as von Hindenburg was concerned were that he had been a Lieutenant-Colonel in the war, was rich, and was an aristocrat. Since he had almost no support in the Reichstag, Papen convinced von Hindenburg to dissolve it, thereby ensuring he could not be ejected until at least after the election, fixed for 31 July. Papen was also the choice of Kurt von Schleicher. General von Schleicher was convinced that Germany could only survive if a strong man were in charge. He ceaselessly manoeuvred to become that dictator, exploiting his influence on the aging president through his friendship with von Hindenburg's son, Oskar, who controlled access to his father. In May 1932, Schleicher became Minister of Defence in von Papen's government.

Antonina Vallentin left a graphic picture of the atmosphere in Berlin. She made what was to be her last visit to Caputh in June 1932.[8] She had been told by General Hans von Seeckt, former head of the German Army and now a nationalist MP, that she should warn all her Jewish friends, in particular Einstein, to leave Germany as their lives were in danger. She found it hard to credit this nightmare as she made her way to the Einstein house through the sleepy village of Caputh on a warm overcast spring day. Elsa was disturbed by the visit on 4 June of Abraham Flexner, who had sat with Einstein on the terrace trying to convince him to accept a position at the IAS in Princeton. Elsa agreed that the position sounded perfect for Einstein and the salary offer extremely generous. With von Seekt's warning ringing in her ears, Vallentin asked whether Einstein had accepted. She was distraught to be told that he had not as he couldn't face a definitive departure from Germany.[9]

Vallentin poured out her concern about their safety and impressed Elsa to such an extent that

---

[8] Although Vallentin dates the visit as May 1932, other particulars, such as the visit of Abraham Flexner, fix it to early June.

[9] Flexner's recollection of his visit differs from Vallentin's account. Flexner, in a letter to Millikan stated that Einstein was enthusiastic about accepting his offer as he shook hands with him on his departure from

she began to look around the house as if already contemplating the pain of having to give it up. When Einstein joined them and Elsa repeated Vallentin's worries, Einstein's calm almost gave her second thoughts. Then Herta, the maid, entered and reinforced Vallentin's warnings by relating a recent conversation with the village butcher, who had asked her with contempt how she could stomach living with the Jews. As they ate their dinner, some of their shock abated. Elsa told Einstein that tomorrow he must go and see Flexner, who would remain in Berlin for a few more days. As Vallentin said goodbye to Elsa, she realised that she would never again experience the calm of this house.

Having been escorted to the bus stop by Einstein, Vallentin changed onto the train at an almost deserted and quiet Potsdam station. As she waited, she once more began to think that surely her warnings to the Einsteins had been overdone. Then a group of young men, loudly calling to each other, erupted onto the platform. They were wearing swastika pins in their lapels. Vallentin decided to move forward to first class, but the coaches were not divided so she was still within earshot of the youths.

> They were still crowded at the windows, producing great bursts of laughter and calling out about pigs and drunkards. The train moved off, the noise drown by the din of their young voices. I was alone, late at night, with about fifteen young Nazis. 'A German woman does not use make-up,' said one holding the *Völkischer Beobachter*. I was wearing lipstick. 'A German woman does not smoke.' I had lit a cigarette. I had an hour of this in front of me. I was not very reassured.
>
> From my place at the back of the car, I watched the young Nazis. For the most part, fair, open faces, smiling. Sons of the good bourgeoisie, many among them well dressed, even on this summer's day, with clean, starched detachable collars. Their delicate hands had never experienced manual labour. Good children, pampered, the pride of their families. The first thing that struck me was the language emanating from those innocently rosy-cheeked mouths. A language that I had never heard before, which I only knew as the one used in the trenches in Remarque's book.[10] Obscene words spoken with a candid smile. Insults pronounced earthily with a hint of tenderness. Suave voices that inflected insolently. Those sons of well-off families spoke with the drawling accent of the working class. They avoided the proper words, the correct pronunciation, grammar. When by chance they used a proper phrase, they immediately corrected themselves by substituting a slang expression
>
> [...]
>
> They smiled complacently, they were a little drunk, sleepy like kids who stay outdoors too long. They wore watches to show they were important people. From the end of the wagon 'Jewish pig' could be heard, followed by the name of the Berlin police chief. The little blond lad in front of me sneered: 'He won't be around for much longer, he will be toast soon.' He was their whipping boy. The conversation became general and turned to the treatments to be inflicted on this Jewish pig, who dared to bully them. The young boys with open faces outbid each other in imagination. The prefect was not the only one for whom they reserved unimaginable torture. His wife was also included. They described female anatomy with crude phrases. They shouted appalling words aloud, employed others at whose meaning I could only guess. They promised each other to walk her naked into flames both from the front and behind. They roared with laughter. It was an unreal atmosphere, composed both of hints of refined sadism and the innocent malice of children. It seemed as if the images they evoked were independent of them as individuals. At the same time, hell seemed to have opened up under their feet to vomit fire, blood and brimstone.

Caputh. However, Flexner had reason to exaggerate Einstein's enthusiasm, as he was defending himself from Millikan's charge of having 'poached' Einstein away from Caltech. Nevertheless, Einstein's letter to Flexner of 8 June implies that he had by then decided to accept the offer in principle.

[10] *All Quiet on the Western Front*

Having been further terrified by a direct approached by a large red-headed Nazi youth who asked insolently for a light for his cigarette, Vallentin was enormously relieved when the train pulled into the central station and she could flee. The following morning, she telephoned to recount her experience to Elsa, following this conversation up with a letter in which she reiterated her arguments that Einstein should accept the position in Princeton. Although both Elsa and even more so Einstein were reluctant to believe that they would have to leave Germany, Einstein had written to Flexner on 8 June clarifying details of the offer. On 12 July, Flexner wrote to Einstein to confirm the arrangements. One of the most important points for Einstein was the offer of a position for Mayer, independent of Einstein, at the IAS. Indeed, Flexner agreed to begin supporting Mayer even before Einstein took up his position, from October 1932. It was agreed that Einstein would take up his IAS appointment in October 1933 and would be expected to be resident in Princeton from October until mid-April. On 30 July, Einstein wrote to Flexner with his final acceptance. By August, it was already being reported in the press that he would be spending half the year in Princeton.

The date of Einstein's letter of acceptance to Flexner may not have been accidental. On 31 July, Germans voted for a new Reichstag. July saw the height of electioneering, which featured widespread street battles between Nazi Brownshirts and their socialist and communist opponents. Hitler exploited the despair engendered by the Great Depression to convince the electorate that he had the answer to their problems. Isolated in bucolic Caputh, most of this passed Einstein by, but the rise in popularity of the Nazis and the danger they posed to Jews was unmistakable. On 30 June, there were riots in the University of Berlin that led to a pitched battle in the streets outside between Nazi students on one side and socialist and communist students on the other. The Nazis demanded that the university should cease to teach Jewish students. Ominously, they added that for the time being they would not insist on action against Jewish professors, of whom Einstein was the most prominent.

On 26 July, Hans Albert and his wife Frieda, together with Einstein's first grandson, Bernhard Caesar, then not quite two years old, arrived in Caputh for a visit. The three generations of Einsteins were photographed smiling on the steps of the Caputh terrace. Away from the camera lens, however, all was not well. Elsa had invited their guests over Einstein's objection as he had reverted to his dislike of Freida. He accused her in a letter to Tete of producing a poisonous atmosphere towards Elsa, despite all Elsa's efforts to make their stay a success. He told Tete that he would forbid any future invitation to Caputh.

Einstein made two visits to Belgium during the summer of 1932. The first was to Brussels on 3 July to attend a meeting of the organising committee of the Solvay Conference planned for 1933. Here he met Niels Bohr, Paul Langevin, Peter Debye, and Abraham Joffe, among others. They were received by the King and Queen at Laeken, where the Queen took a group photograph. Einstein stayed an extra day in order to play music with his royal friend (see Figure 22.1). At the end of July, Queen Elisabeth wrote to Einstein, enclosing copies of the group photograph together with others taken the following day. She thanked him for his explanation of some points of physics during their long walk together in the Laeken gardens and hoped that they would be able to play music again at his next visit, although she was currently unable to play, having injured her thumb. She closed by wishing him a glorious summer with lots of sailing and his body bathed in air and sunshine.

Queen Elisabeth's wish was mostly unfulfilled; the letter did not reach Einstein until mid-September, since he left Caputh for Leiden early in August, once the results of the 31 July election were announced. The Nazi party gained 123 seats, obtaining a total of 230, making them by far the largest party in the Reichstag. Nevertheless, von Hindenburg, who had a visceral dislike of

Figure 22.1 Einstein with Queen Elisabeth of the Belgians in the palace grounds at Laeken in the early 1930s.

Hitler, whom he often referred to as 'that Bohemian corporal', ignored the result and persisted in supporting von Papen through presidential decree. Einstein, and in particular Elsa, appreciated the danger in the triumph of the Nazi party at the polls. There could be no doubt that the thugs of the SA, Hitler's Brownshirts, would be further emboldened. Neither was there a shortage of credible threats explicitly aimed at Einstein. In June, a Nazi deputy to the Prussian Diet had told the chamber that 'When we clean house, the exodus of the Children of Israel will be a child's game in comparison [. . .] a people that possesses a Kant will not permit an Einstein to be tacked onto it.'

Einstein's hastily arranged departure in August incorporated his second journey to Belgium that summer. However, he first went to Leiden, his traditional 'safe haven' from political disturbance in Germany. There he picked up the scientific papers that he had diverted to Ehrenfest's address to avoid the newly introduced German tax on imported printed matter. He discovered Ehrenfest in a very disturbed state of mind. The deterioration in Ehrenfest's mental condition can be tracked graphically from his research notebooks. Originally these were exceptionally neat and logically arranged, with topics numbered and separated. However, by the 1930s his handwriting had become almost illegible and, with a brief exception in 1932, the notebooks' contents had become unrelated to physics and mostly concerned with household finance. Although there is no explicit record of their conversation, that it must have been worrying is substantiated by a letter that Ehrenfest drafted on 14 August, after Einstein's departure. This was addressed to his closest friends, including Bohr, James Franck, Herzglot, Joffe, Pieter Kohnstam, and Tolman. In it Ehrenfest opined that his life was a complete waste, he could no longer keep up with theoretical physics and he must resign his chair by the autumn of 1933 at the latest. He was desperately worried about the burden that the care of his son Wassik, born in 1918 with Down syndrome, would place on his children after his death. The draft, which was never sent, clearly presages what actually happened: that he would first kill his son and then himself. Einstein must have been concerned at his friend's state of mind since he returned to Leiden at the end of August. Here he had meetings with various university officials in an unsuccessful attempt to lighten Ehrenfest's teaching and administrative load.

Einstein remained in Leiden until 12 August, when he took the train to Spa. He was surprised to be asked for a Belgian visa as he crossed the Belgian border at Maastricht and was only allowed to pass after the intervention of a friendly Belgian consul. He stayed in the Hotel Jamar in Spa. He was subsequently joined by Elsa and remained until the end of August. It had become clear in the interim that the Nazis were unlikely to gain power in the near future. Einstein made the acquaintance of Emil Vandervelde, former Foreign Minister of Belgium, who was also staying at the Hotel Jamar. Einstein reported how charmed he was by Vandervelde in his response to Queen Elisabeth's July letter. She and the royal family were at Haslihorn, their Swiss holiday home near Lake Lucerne, where the King indulged in his love of mountaineering.

Einstein's involvement with disarmament and pacifist activities continued throughout the summer of 1932. On 30 July, with the encouragement of the League Committee on Intellectual Cooperation, with which he still had residual dealings, Einstein wrote a public letter to Sigmund Freud questioning whether there was any way of delivering mankind from the menace of war. Freud wrote a long response in September, to which Einstein replied at the end of 1932. Given the identity of the correspondents, the letters received some attention at the time but thereafter were rapidly forgotten as they were overtaken by events. At the end of August, an International Conference against Imperialist Wars took place in Amsterdam. He was urged by many of the luminaries of the pacifist movement to attend, including his friend Romain Rolland. However, Einstein had been informed that it was likely to be dominated by communist apologists for Soviet Russia and, convinced that pacifism had to be politically neutral, he declined to attend. He did however send a message to be read out to the conference.

Matters related to violin playing also impinged on Einstein's summer vacation. On 23 August, he answered a request for a testimonial for Leo Portnoff, who was a violin pedagogue born in Russia. Portnoff had studied with Joachim and subsequently taught and conducted in Berlin before emigrating to Miami, Florida. Einstein wrote:[11]

> The struggle for complete Bow-Controll [sic] and tone-beauty is one making demands upon two of the five senses supplied to us by nature.

[11] The idiosyncratic English translation of Einstein's original German testimonial was provided by Portnoff for Einstein to sign. Someone, probably Elsa, has added some corrections to the typescript in long-hand, one of which is also incorrect.

> Pursuing to this goal, Professor Leo Portnoffs [sic] invention 'EQUI-TONE-SCOPE' (originally called bowpressure control [sic]) suspends the ears and employs the eyes instead.
>
> In matters of concentrating work in tone study and economy of bow are the eyes naturally a stricter watchman than the ears.
>
> Also the other diveces [sic] for developing of the left hand and the wrist of the right hand are excellent.
>
> As a class teaching method – very useful.
>
> It is a new revolutionar [sic] way and should be noticed from [sic] the violine [sic] playing world.

By 4 September Einstein was back in Caputh, ready to welcome a variety of musical guests. Ruggiero Ricci, whom Einstein had met a few months previously in California, was undertaking a European tour and visited Caputh on 13 September. The Caputh Visitors Book contains an inscription by Ricci: 'In remembrance of a few wonderful moments', underneath which Einstein added a note implying that Ricci had played the *Fuga* from Bach's Solo Violin Sonata in C Major, BWV1005. Ricci gave a recital in the Philharmonie on 22 September, during which he played Bach once more, together with Henri Vieuxtemps's A minor Concerto, Beethoven's Romance in F major, Tchaikovsky's *Sérénade mélancolique*, and various short virtuoso pieces. Despite several encores, the reception by the audience, which included Einstein, was cordial rather than enthusiastic; the critical consensus was that Ricci, at age eleven, was not yet quite the mature artist. On 24 September, Einstein's friend Erich Kleiber, Musical Director of the Berlin State Opera, and his American wife visited Caputh. He had also attended Ricci's recital. A musical collaborator of Einstein, violinist Margarete Herrmann, left a quotation by Goethe on 29 September. It can be assumed that the two took the opportunity to play some violin duets together.

A welcome guest at Caputh on a weekend in late September was Konrad Wachsmann, the architect of the house. He had just won the prestigious Prix de Rome, which entailed an extended period of study in the Eternal City, so his visit to Caputh was both a celebration and a farewell. There was a definite tinge of autumn in the air as he and Einstein took an extended walk in the forest surrounding the house, followed by the traditional German Kaffee und Kuchen with one of Elsa's delicious Swabian cakes with vanilla sauce. Ilse and Rudi Kayser arrived in time for coffee. As a parting gift, Einstein gave Wachsmann a guide to Rome, in which he had written 'Dear Konrad, the most beautiful and profound experience that a person can have is a sense of the mysterious', which was a quotation from an address that he had just written at the urging of the League of Human Rights. He must have recorded it for distribution on a disc shortly before Wachsmann's visit. He also presented Wachsmann with the manuscript of the address. It is worth quoting it in full here as a snapshot of Einstein's beliefs on the cusp of the Nazi takeover, which would modify at least the one relating to pacifism:

> It is a particular blessing to belong to those who are both permitted and able to devote their energies to contemplating and investigating objective, non-transitory phenomena. How happy and thankful I am that I have benefited from this blessing, which to a large extent frees one from a dependence on the personal and the behaviour of one's fellow human beings. However, this independence must not blind us to the recognition of the duties that continuously bind us to humanity, past, present, and future.
>
> Our situation on this earth seems strange. Each of us makes a brief stay, involuntary and uninvited, without knowing either reason or purpose. In daily life we feel that we are there only for the sake of others, those we love and numerous others to whom destiny binds us.
>
> I am often oppressed by the extent to which my life depends on the work of my fellows, and I know how much I owe you.
>
> I do not believe in free will. Schopenhauer's words: 'We can do what we want, but cannot will what we want', accompanies me in all situations in life and reconciles me to the actions of others, even if they

> are rather painful to me. This realisation of the non-existence of free will stops me from taking myself and my fellow human beings too seriously as actors and judges and from losing my sense of humour.
>
> I have never aspired to comfort and luxury, and even to a large extent have contempt for them. My passion for social justice has often brought me into conflict with people, as has my dislike of any attachment and dependency that I did not consider absolutely necessary.
>
> I always respect the individual and have an insurmountable aversion to violence and to the herd mentality. For all these reasons, I am a passionate pacifist and anti-militarist, rejecting all nationalism, even if it only manifests itself as patriotism.
>
> Privileges arising from position and property have always seemed to me unjust and obnoxious, as has an exaggerated cult of personality. I am committed to the ideal of democracy, even though the disadvantages of a democratic form of government are well known to me. Social equity and the economic protection of the individual always seem important goals of society to me.
>
> Although in everyday life I am a typical loner, the consciousness of belonging to an invisible community striving for truth, beauty and justice has avoided any feeling of loneliness.
>
> The most beautiful and profound experience that a person can have is a sense of the mysterious. It underlies religion as well as all deep work in art and science. Anyone who has not experienced this seems to me, if not as good as dead, then at least blind. Religion consists of the feeling that, behind the world of experience, something inaccessible to our consciousness is hidden, whose beauty and sublimity reaches us only indirectly as a faint reflected glow. It is in this sense that I am religious. For me, it is sufficient to intuit these secrets with astonishment while humbly attempting to reconstruct a pale reflection of the sublime structure of reality.

It would have also been around this time that Wachsmann had a conversation with Einstein about his using music as a catalyst or restorative relating to his scientific work. Elsa had told Wachsmann her stories on this subject. Wachsmann had never heard Einstein play either the piano or the violin; although he was a keen musician, he did not aspire to Einstein's knowledge and standards. Their tastes were also rather different. He was aware that Einstein, while accepting for example Schoenberg's inner compulsion to write in the twelve-tone idiom, had no understanding of it, nor of the rhythmic drive of Paul Hindemith. Einstein did however express interest is some relatively modern music of the romantics, including Max Bruch and the songs of Gustav Mahler. When Wachsmann asked about the role of music in his work, Einstein replied that, at first, he improvised and, if that didn't do the trick, he sought solace in Mozart. If, however, the improvisation cleared his thoughts, then he required the clear structures of Bach to clarify them further.

On Sunday 16 October, Einstein gave a talk at the Beethovensaal in Berlin organised by the Union of German Students to benefit their scholarship fund. His title was 'The Universe'. The event was sold out and hundreds were turned away at the door. During the talk, Einstein announced that the universe was ten billion years old, rather than the then-accepted figure of three billion years. This subsequently caused scepticism from some of his colleagues at Caltech; the currently accepted figure is 13.8 billion years. Although *The New York Times* reported that he did not employ any mathematical formulae and spoke simply and lucidly, they opined that Einstein's explanations were delightfully unintelligible to many of his listeners. Some press reports of his lecture also included confirmation of his spending five months of the year in Princeton and his offer to resign from his Berlin position. However, this offer was refused by the Prussian Ministry, who, for a few months longer, still valued Einstein's services. Instead, they offered to use his salary during his stay in Princeton to establish an 'Einstein fund' that would be used for scientific purposes. Einstein gave another talk, for the benefit of the Jewish Student Aid Society, in the Prinzregententstrasse Synagogue on the evening of 19 November, this one entitled 'The Problem of Space in Modern Astronomy'.

On 10 November, Einstein was one of the principals at a reception for Eva, Viscountess Erleigh, who had returned from a trip to Palestine earlier in 1932. As the daughter-in-law of the Marquess of Reading and daughter of Alfred Mond, Lord Melchett, she was a member of two of the most distinguished British Jewish families. Her views on Palestine and Zionism were therefore listened to with respect, both by the Jewish attendees at the reception and by members of the German government and foreign policy establishment at subsequent events.

The revealing diarist Harry Graf Kessler left a glimpse of Einstein in the theatre at this time. Wachsmann reported that although Einstein might regularly go to three or four concerts a week, a trip to the theatre was much rarer and normally made under the encouragement of Elsa, who rarely accompanied him to concerts. Kessler's diary entry for 15 November 1932 describes the Einsteins at a festive performance of *Gabriel Schillings Flucht* by Gerhart Hauptmann at the National Theatre. The invitations for the event were issued by the German government in celebration of Hauptmann's seventieth birthday, which fell on that day. The author was the guest of honour along with ministers, ambassadors and other worthies. The Einsteins and Heinrich Mann were in the stalls, directly in front of Kessler, former head of the army von Seeckt and Hugo Simon, banker and art collector. At the end of the first act Simon leaned over to ask Einstein what he thought of the play, to which Einstein replied 'Well, so what?' The play, written in 1907, was by then considered extremely old-fashioned.

Einstein's strong engagement with pacifism in 1932 continued into the late autumn. He had conceived the idea of gathering a group of dependable pacifist intellectuals, artists and scholars to continue the pressure on the political establishment that the transparent failure of the Geneva disarmament conference made even more necessary. He had discussed this with Langevin, whom he hoped could be persuaded to lead the group. However, this initiative seems to have come to nothing by 1933, despite another discussion between him and Langevin in Antwerp as he and Elsa were leaving for California in December.

The final entry in the Caputh guestbook was by von Laue. Since he had also written the first entry but none in between, despite his frequent visits, it seems that either Einstein, von Laue, or both, considered this likely to be the last entry in the book. It reads (a translation with poetic licence):

> Einstein is seen to be a great man,
> The whole wide world knows what he can,
> His greatest feat yet has escaped it
> He has transposed me (!) into a poet!

It must have been sometime in December that Gustav Bucky, who was now the doctor of Ilse and Margot Einstein, visited Caputh. The account of Bucky's son, Thomas, aged thirteen, illustrates a very different aspect of Einstein. Rather than the proselyte of pacifism, Thomas Bucky described the playful Einstein who as a young father in Bern had made toys for his children. To overcome the young Bucky's shyness, Einstein produced a yo-yo, at that time the 'must-have' toy for schoolchildren in Berlin. Unfortunately, his attempts to demonstrate it failed miserably, whereas Thomas was able to do a variety of tricks. The young Bucky subsequently scoured the toy shops of Berlin to find the best yo-yo to send to Einstein as an early Christmas present.

In his previous visits to the US, Einstein had left visa issues to be sorted out by the shipping agents. His German diplomatic passport ensured that all formalities could be bypassed. In 1932, however, various problems arose. The Weimar Republic was in its death throes; for example, at the celebration for Gerhard Hauptmann's seventieth birthday described above, a medal had to be presented to him twice since there were two governments of Prussia, one ingoing and one outgoing, that refused to have any dealings with each other. No government help for Einstein's trip meant that he had to deal with visa formalities himself, travelling on his Swiss passport to the US for the first time since 1921. This might have been unproblematic, had not the Women Patriot Corporation (WPC), an organisation formed after the First World War to campaign against left-wing and feminist organisations, decided that Einstein was a dangerous radical pacifist, exceeding even Stalin himself in affiliation with anarcho-communist international groups. In a long document that charged Einstein with every conceivable radical transgression including an inability to speak English, they demanded that the State Department refuse his visa application. This submission subsequently became the first entry in the voluminous FBI file that J. Edgar Hoover was to compile over the following decades on Einstein and his activities.

Einstein had heard of the WPC's initiative, which he had likened in an interview to the cackling of the Capitoline geese that saved the Roman Republic in 390 BC. He went on with biting sarcasm:

> Never before has any attempt of mine at an approach to the beautiful sex met with such an energetic rebuff, even should perchance such have ever been the case, then certainly, not by so many all at once. But aren't they perfectly right, these watchful citizenesses? Why should one admit to one's presence one who devours hardboiled capitalists with the same appetite and relish as once upon a time the ogre Minotaurus in Crete devoured luscious Greek maidens – a person who in addition is so vulgar as to oppose every war, except the inevitable one with his own wife?

Einstein was surprised to receive a request to attend the US Consulate General on 5 December. He indicated that he would prefer to deal with the visa by post, but the officials were insistent that he must attend in person. Taking Elsa with him, he arrived at the office and was taken through the various questions on the visa form. Having never seen this before, he became increasingly annoyed at what he considered the impertinence of the visa officer at enquiring about his political affiliations. After 45 minutes he became so irate that he left the office, he and Elsa threatening that unless they received their visa by 12.00 p.m. the following day, they would call off their trip to Caltech. Elsa fulminated that she would unpack their six trunks (which were apparently full of reading material for Einstein on the voyage) and never visit the US again. They returned to Caputh and awaited developments.

Fortunately for all concerned, the Einsteins' visa was issued before the noon deadline on 6 December. The Consul-General, who had been away during the week of Einstein's interview, came in for a torrent of criticism in the American press over the visa saga, much to his own consternation. Once the State Department had forwarded the WPC complaint, he was legally bound to consider it before issuing the visa; the questions Einstein was asked were all completely routine. It was this aspect that nettled many of the correspondents, who were outraged that such probing into visa applicants' political beliefs made the US look ridiculous in the eyes of the world.

An illustration of the importance of violin playing in Einstein's life took place on the day the visa was issued. Despite the uncertainty around his departure, Einstein nevertheless found time to write a letter of recommendation for Erich Kielow, who was a luthier established in nearby Potsdam. Kielow's wife, without her husband's knowledge, brought one of his violins to Caputh. Elsa, seeing her outside carrying a violin case, admitted the unknown woman, who begged Einstein to play Kielow's instrument. He obliged and provided Frau Kielow with a letter pronouncing it one of the most beautiful violins that he had ever played.

In September, the new Reichstag had met for the first time and had attempted to pass a motion of no confidence in von Papen's government. He had managed to get a message to von Hindenburg just in time to pre-empt this, asking the President to dissolve the Reichstag once again. In the ensuing November elections, the Nazis lost votes and 34 seats. For a while it seemed that the tide had turned on the Nazis' seemingly inexorable march to power. However, Defence Minister von Schleicher's ambition drove him on remorselessly; in December 1932, he ousted von Papen to become Chancellor himself, convincing von Hindenburg that he could split the Nazi party and forge a parliamentary majority. Now consumed with hatred for his former friend, it was von Papen's machinations to destroy von Schleicher that set the scene for Hitler's assumption of power in January 1933.

Early on 10 December, the Einsteins were picked up from Caputh and driven to the Lehrter Station by Janos Plesch in his opulent limousine. He, together with Ilse and Margot and their husbands waved them off on the 8:40 a.m. train. Philipp Frank wrote that, as they locked up the house before entering Plesch's car, Einstein remarked to Elsa that she should take a good look at the place, as she would never see it again. Elsa didn't believe him; as Antonina Vallentin wrote, Elsa didn't seem to suspect that the tickets for the voyage were single, not return.

Chapter 23

Final visit to Caltech; Hitler comes to power; Exile around Europe (1933)

The New York Times ran a piece on 11 December 1932, just after the Einsteins had embarked on the *Oakland* in Bremerhaven. The article covered the controversy over their visa application but also took the opportunity to question Einstein about his plans for his stay in Pasadena. He commented that he could now remove the cosmological constant from his equations of general relativity since the universe did indeed appear to be expanding. He also remarked that there is always a poetic element in scientific thinking and that the creation of science and appreciation of music require similar processes of thought.

On their peaceful voyage, Einstein not only worked on his unified theory equations but read several of the volumes loaded in one of Elsa's thirty pieces of luggage. These included his son-in-law Rudi Kayser's biography of Spinoza, published a few months before, which, in a touching letter to Tete written on 27 December, Einstein characterised as a glowing and spirited book on a great and splendid man. He also exercised his growing command of the English language by reading a book by Lewis Brown on the development of religion in cultures. Modern history on his reading list included a pre-publication copy of a book by the pacifist Ludwig Bauer, *Der Welt am Sturz*, which he characterised as masterly. They arrived at the head of the Panama Canal on 30 December. After seeing in the momentous year 1933 on board, the Einsteins arrived in Los Angeles at lunchtime on 9 January. Richard Tolman once again came aboard to escort the Einsteins, while Robert Millikan, Arthur Fleming, the German Consul, and others waited on the dock to greet them before they were driven to Fleming's luxurious apartment in the Caltech Athenaeum. The questions from the press about his scientific plans elicited little response from Einstein, who deferred a question on the origin of cosmic rays to Millikan. Millikan's long controversy with Compton as to whether these were photons or charged particles was settled a few weeks later in favour of Compton's hypothesis that they were predominantly charged particles. Einstein meanwhile found himself answering questions on the Great Depression, opining, somewhat bizarrely, that mechanisation should be reduced to correlate with the availability of labour. It may be wondered if he recalled his sarcastic inscription on his own photograph: 'To Charlie Chaplin, the great economist'.

Einstein spent the day after their arrival mostly relaxing in the Athenaeum. However, he did meet Abbé Georges Lemaître, who was also a visiting scientist at Caltech, before dinner. They were reported to be deep in conversation while walking in the grounds of the Athenaeum. They had first met at the 1927 Solvay Conference. On the following morning, Einstein attended a lecture given by Lemaître on cosmic rays, in which he explained his hypothesis of a 'primeval atom', in which all the matter of the universe was concentrated into a single particle. This idea can be thought of as the precursor of the Big Bang theory of the universe. Lemaître thought, rather similarly to Millikan, that cosmic rays were remnants of that original cosmic explosion. Einstein did not much care for Lemaître's theory that the universe began with a singularity. For him it savoured too much of a divine creation. In subsequent discussions, he convinced Lemaître to modify his original isotropic

Einstein. Brian Foster, Oxford University Press. © Brian Foster (2026). DOI: 10.1093/oso/9780198794875.003.0023

homogenous universe to introduce anisotropy. Although Einstein hoped that this would obviate the necessity for an initial singularity, Lemaître quickly showed that this was not the case.

On 17 January, Lemaître gave another seminar, devoted to the cosmological constant. Whereas he was convinced that this had a real significance in cosmology, Einstein, who had been in part relieved when Hubble's observations established that the universe was not static, disagreed and was happy to remove it from his equations. The two had not resolved their dispute when Lemaître left Caltech in February to return to Brussels for the start of the teaching semester.

As soon as she had arrived in Pasadena, Elsa had contacted her old friend Barbara Siebert to arrange a shopping expedition on 10 January. She also asked Lili Petschnikoff to organise a musical evening for Einstein on 11 January, requesting Lili to pick them up from the Athenaeum directly after the end of Lemaître's lecture on cosmic rays. She said that they could stay until 9:30 p.m., allowing plenty of time for music, and that a simple supper would sustain them. This was the first of several musical evenings at Petschnikoff's house during their stay in Pasadena; another took place on Sunday 29 January, as recorded in Einstein's travel diary, which he began only on 28 January and ended on 16 February. This partial record of his stay is even more laconic than his previous diaries.

On his arrival in Los Angeles, Einstein had issued a statement that implied that, in addition to his usual scientific collaborations, he wished to contribute to promoting international good will. This was a response to the funding of his visit by the Oberlander Trust of the Carl Schurz Foundation, whose statutory purpose was 'to bring about a better understanding of German-speaking peoples by the American people, and vice-versa.' Despite this motivation, he made rather fewer public appearances than in previous years. There was indeed a clear tendency for the local press interest to decrease during this final visit, with news about his doings gradually migrating from the front page to inside pages. This may be partially related to the knowledge that his future American stays would be on the east rather than the west coast.

Einstein's first public appearance was arranged for 22 January, when he was invited to speak by the Southern California Student Conference Against War. His speech was scheduled to be given in English. However, it was cancelled on 21 January. A press release labelled as from 'the Athenaeum' pleaded the avalanche of speaking requests that he had received and that a nationwide radio broadcast of remarks scheduled on 24 January would allow everyone to hear his views. In fact, this was somewhat disingenuous as his remarks at the subsequent meeting had a rather different thrust. It seems that it was the hand of the conservative Millikan, who was in any case irritated at having lost his prize physicist to Princeton, that had pulled Einstein back from what promised to be yet another controversial declaration on pacifism. Millikan's communications to the Oberlander Trust confirm this interpretation; an editorial in the *Los Angeles Times* shows that Millikan's views were widely shared by the power brokers in the city. Einstein did eventually publish at least part of what he had intended to say in 1950 in his book *Out of My Later Years*. While recognising the geographical reasons for American exceptionalism, Einstein had intended to call on American youth to ensure that their government resisted isolationism and fully engaged with efforts to build a functioning international order.

Two days later, on 24 January, Einstein was one of three keynote speakers at a symposium on 'America and the World Situation,' organised by the Southern California College Student Body Presidents Association. He delivered his speech, which was broadcast nationally on the radio, in English, the first time he had given a public address in that language. He expanded on his earlier remarks that the current crisis was caused not by war debt but by the too rapid mechanisation of industry. Another cause impeding the understanding of constructive ideas, he asserted, was the black dress suit that creates an atmosphere of euphonious oratory. Since he was wearing such a suit, it can be assumed that this remark was tongue in cheek.

Another aspect of Einstein's broader political activity was his concept of a high-powered committee of influential intellectuals to act as a form of 'world conscience' in highlighting unacceptable behaviour in international affairs. He had discussed this with his friend Paul Langevin. However, it seems that Einstein, having had no response, had already discounted the idea of Langevin chairing the group. In an interview with the Managing Editor of the Jewish Telegraph Agency, reported on 22 January, Einstein talked about finding twenty-five individuals, selected by himself, who would be intellectuals of the highest order and powerful personalities of international renown known for their liberal views. Later in the summer he admitted the difficulty of finding such people willing to participate, again citing Langevin's apparent lack of interest as a critical factor. It was probably this idea that he discussed with the distinguished astronomer, George Ellery Hale, when Einstein visited him on the afternoon of 29 January.

On 30 January, Einstein went to the Biltmore Hotel for a lunch for more than 250 people in aid of Zionist causes. It was addressed by US Federal Circuit Court Judge Mack and was apparently a brilliant success. He later accompanied his radical friend, the author Upton Sinclair, to a lecture by a young journalist on Russia.

In Germany, on the same day but as yet unknown to Einstein, Hitler was appointed Chancellor of the German Reich by President von Hindenburg. The manoeuvrings of von Schleicher and von Papen had finally led to an impasse that caused von Schleicher to resign as Chancellor on 28 January. With great reluctance, von Hindenburg felt he had no option but to call on Hitler. Von Papen convinced von Hindenburg by getting Hitler to agree that there would be only two other Nazi cabinet ministers in addition to himself. Von Papen was to be Vice-Chancellor. He boasted that within two months, they would have pushed Hitler so far into a corner that he'd squeak. It is notable that Einstein makes no mention of Hitler's appointment in subsequent entries in his diary. This may be in part explained by the tendency of newspapers such as *The New York Times* on 31 January to emphasise the constraints they believed, in common with von Papen, had been placed on Hitler's power. These were soon shown to be illusory. In fact, within two months, rather than being pushed into a corner, Hitler had completely dismantled the Weimar Republic.

On 31 January, Einstein visited Charles St John, the Mount Wilson astronomer, who was confined to bed with a heart complaint. In the afternoon he gave his second lecture on his latest idea for a unified theory, known as 'semi-vectors'. This caused some significant head scratching in the press reports, as apparently it also did in Einstein's audience. In the evening he attended a concert in the Los Angeles Philharmonic auditorium by his friend Fritz Kreisler, who played Bach, Tartini, and Mozart's G major Concerto, K216, as well as a host of encores. It was the Tartini 'Devil's Trill' Sonata however that Einstein characterised as being wonderful. Several movie stars, led by Charlie Chaplin, arrived late and stood respectfully in the aisle until a break in the programme. One of them, child star Karol Kay, was photographed with Kreisler and the Einsteins in the Green Room after the concert (see Figure 23.1).

Over the weekend of 3–5 February, Einstein and Elsa were driven to Palm Springs, where they visited their friend Samuel Untermyer, staying once again at the El Mirador Hotel. Also enjoying the sun and the desert air were Alfred Hertz and his wife. Hertz was the chief conductor of the Hollywood Bowl and former Musical Director of the San Francisco Symphony Orchestra, while his wife Lilly Dorn was a professional soprano from Austria. The *Los Angeles Times* reported that a local gardener had interested Einstein with his own explanation for the Great Depression, which revolved around his observation of the growth of date palms and that feminine qualities had caused

Figure 23.1 The Einsteins with Fritz Kreisler after one of his concerts. From left to right: Karol Kay, Fritz Kreisler, Elsa and Albert Einstein.

the depression, which would end as masculine qualities began to dominate. On Sunday, the Einsteins were driven to the nearby San Jacinto Mountains and a lake, probably Lake Hemet, via the newly opened Highway 74. The Hertzs drove the Einsteins back to Palo Alto on Monday morning.

Back in Caltech, Einstein that afternoon heard a lecture by future double Nobel Laureate Linus Pauling on benzene. By applying quantum-mechanical ideas, Pauling was able to correct previous approaches, including that of August Kekulé, and give a predictive solution for many compounds containing so-called benzene rings. Einstein also met with Robert Oppenheimer, himself tacking problems using sophisticated quantum-mechanical techniques. Einstein recorded no reaction to these meetings in his diary, other than characterising Oppenheimer as brilliant. He did however consider some work that Tolman had explained to him about electrodynamics at the end of his busy day as weak.

The following day, 7 February, was also busy. In the afternoon, Einstein had further discussions with Tolman, followed by attending a lecture from Langer on experimental research on cosmic rays. In the evening, he and Elsa were invited to dinner at Charlie Chaplin's residence. Einstein in his diary simply recorded that he played a Mozart quartet there. Chaplin, in his autobiography, goes into somewhat more detail:

> [Einstein] warned me that he was bringing three musicians. 'We are going to play for you after dinner.' That evening Einstein was one of a Mozart quartet. Although his bowing was not too assured and his

technique a little stiff, nevertheless he played rapturously, closing his eyes and swaying. The three musicians, who did not show too much enthusiasm for the Professor's participation, discreetly suggested giving him a rest and playing something on their own. He acquiesced and sat with the rest of us and listened. But after they had played several pieces, he turned and whispered to me 'When do I play again?' When the musicians left, Mrs Einstein, slightly indignant, assured her husband: 'You played better than all of them!'

[...]

A few nights later the Einsteins came again for dinner and I invited Mary Pickford, Douglas Fairbanks, Marion Davies, W. R. Hearst and one or two others. Marion Davies sat next to Einstein and Mrs Einstein sat on my right next to Hearst. Before dinner everything seemed to be going pretty well; Hearst was amiable and Einstein polite. But as dinner progressed I could feel a slow freeze-up until neither one of them exchanged a word. I did my best to enliven conversation, but nothing would make them talk. The dining-room became charged with an ominous silence and I saw Hearst looking mournfully into his dessert plate and the Professor smiling, calmly engrossed in thought. Marion in her flippant way had been making quips and asides to everyone at the table but Einstein. Suddenly she turned to the Professor and said elfishly: 'Hallo!' then twiddled her middle fingers over his head saying 'Why don't you get your hair cut?' Einstein smiled and I thought it time to disperse for coffee in the drawing-room.

During the rest of that week, Einstein had various social engagements, including another with Upton Sinclair. They were mostly related to the economic situation. On Friday 10 February he attended a colloquium by Ralph Fowler, Lord Rutherford's son-in-law, who reported on recent results from Rutherford's laboratory. On the morning of Sunday 12 February, Einstein noted in his diary that he listened to a radio broadcast of a matinee performance in New York of Bruno Walter conducting the New York Philharmonic. The concert commemorated the fiftieth anniversary of the death of Richard Wagner. The programme was all Wagner, including a *Faust* overture and excerpts from the operas. Friedrich Schorr was the baritone soloist. Other than Einstein's friendship with the conductor, it is difficult to reconcile his making time to listen to this radio broadcast with his often-expressed detestation of Wagner's music. That evening he attended the inaugural dinner of the League of Labor Palestine, at which he was 'raffled off', presumably referring to a signed photograph, for $70.

During the following week, Einstein attended a seminar on what he referred to as Arthur Eddington's 'two-electron theory'. This was presumably part of Eddington's own ambitious project to unify all of physics, to prove no more successful than Einstein's. Eddington considered Paul Dirac's work on relativistic quantum theory to be erroneous and had developed a highly mathematically complex alternative that found little acceptance among the physics community. It was the abstruse mathematics that Einstein commented on as being its only interesting feature.

On the following weekend, the Einsteins travelled to Hope Ranch, the estate of Ben Mayer, in Santa Barbara. Mayer was a trustee of Caltech and President of the Union Bank. He had also played host during an earlier visit by Queen Elisabeth of the Belgians, who had planted a tree. Einstein sent her a letter containing a poem on the subject, to which she replied in March with another poem. While his pivoted around the enclosed twig from her tree, her reply showed her concern at the political developments in Germany that had by then become cataclysmic. It was during this stay at Hope Ranch that the iconic photograph of Einstein happily riding a bicycle was taken.

For once, the headline concerning physics in the *Los Angeles Times* of 21 February was not related to Einstein, although he was mentioned in the report. Carl Anderson had discovered a positively charged electron, produced by cosmic-ray particles and observed a few months previously from photographs of a cloud chamber. It had now been confirmed by Patrick (later Lord) Blackett and his colleague Beppo Occhialini in Rutherford's group in Cambridge. Anderson told the reporter that he had received a telegram that morning suggesting the name 'positron' for the new particle, which Anderson considered likely to be adopted. The report also contained the information that the discovery caused the normally placid Einstein to become excited. It would be good to think so, but the only indication that Einstein paid any particular attention to the discovery was when he was questioned about it at a press conference on 10 March, just before his departure for Chicago. Then he identified it as being of fundamental importance for understanding the constitution of matter.

On Saturday 25 February, Einstein was scheduled to speak at a banquet at the Biltmore Hotel, held to establish an organisation for the development of the arts, literature, and science. He was spotted by reporters in Palm Springs at Untermyer's house on 26 February, apparently dressed in Elsa's hat in an attempt to evade detection. The Einsteins were driven back to Caltech on Monday 27 February, where Einstein and Auguste Piccard had a discussion in advance of Piccard's lecture on 1 March at the Pasadena Civic Auditorium, when he described his two previous balloon ascents to the stratosphere. That evening, Einstein attended a banquet in honour of Lord Marley, a British Labour politician who chaired the Parliamentary Committee for the Aid of Jews in Europe. Also on 27 February, Einstein wrote to his friend Grete Lebach.

To understand Einstein's letter to Lebach, it is necessary to outline the developments in Germany since Hitler's appointment as Chancellor. He had quickly called new elections, due to be held on 5 March. The campaign was violent and now that the Nazis had their hands on the levers of power, strongly biased in their favour. The police were instructed by Prussian Interior Minister Hermann Goering to turn a blind eye to the murders of left-wingers and Jews by SA thugs. Hitler warned of an impending Communist revolution. Communist Party meetings were banned, and Jewish organisations were raided. One week before the elections, one day after Einstein's letter to Lebach, the Reichstag was burned down, ostensibly by a deranged Dutch left-winger, Marius van der Lubbe, who was probably set up by the Nazis. Using this as an excuse, Hitler convinced von Hindenburg to sign an emergency decree suspending sections of the constitution protecting personal liberties and allowing the federal government to usurp the powers of the states. Although the March election did not deliver the Nazi majority that Hitler had sought, it did give him a majority if the Centre Party could be duped into supporting him. It and other right-of-centre parties, in the absence of the outlawed Communists, provided the two-thirds majority necessary to pass the Enabling Law that established Hitler's de facto dictatorship on 23 March 1933.

Einstein's letter to Grete Lebach is striking. He began it by reporting that he did not believe that it was safe for him to set foot again in Germany and that he had cancelled a seminar that he was scheduled to give at the Prussian Academy. The rest of the letter is concerned with how to allow them to meet outside Germany as soon as possible. All of the arrangements he mentions were rapidly changed by events. He thought that Elsa would travel to Berlin alone and that he would visit Tete, who Einstein wrote was doing somewhat better but who was still being treated in the Burgholzli clinic in Zurich. He talked of going to Lake Lugano, Holland, or his eventual destination, Belgium. Wherever it was would be influenced by how easy it would be for her to reach. In any case, it would be isolated, and he would be accompanied only by his assistant, Mayer. He finished the letter with hopes of seeing Lebach very soon and regrets that their favourite tryst, his sailing boat, had been left in Caputh.

On 3 March, in response to reports appearing in the press that his American friends feared for his safety if he returned to Germany, Helen Dukas released a Delphic statement saying that Einstein was not immediately returning to Germany. On the same day, he was reported as having sat patiently for several hours for the sculptor and fellow Swabian Frederick Schweigardt to create a bust to be displayed at University of California Los Angeles. Einstein and Elsa seem to have returned to Palm Springs that weekend; Auguste Piccard and his twin brother Jean were also guests of the hospitable Untermyer.

With the situation in Germany continuing to deteriorate, Einstein decided to make public what he had divulged already to Grete Lebach. On 10 March, he called a press conference at Caltech, issuing a statement in which he characterised the current Germany as a state in which political freedom, tolerance, and equality before the law were no longer respected. He said that he would not be returning there but looked forward to a time when Germany would have recovered from its current sickness. Just after he left the press conference, a little before 6 p.m., the Long Beach earthquake struck southeast of Caltech. Although more than 100 people were killed and the earthquake was measured at 6.4 on the Richter scale, Einstein was reported to have been deep in conversation in the Caltech grounds with a colleague,[1] neither of whom noticed the earthquake.

Einstein and Elsa left Pasadena for the last time on the afternoon of 11 March. Seen off by a large party, they caught the train to Chicago. Many, particularly Millikan, whose anger at Flexner's coup lasted for years, realised that Einstein's Princeton appointment meant that it was unlikely that he would spend much time at Caltech in future. During the long journey, Einstein seems to have decided that they would cancel their reservation on the SS *Deutschland* and instead book passage to Antwerp on the SS *Belgenland*, on which they had previously sailed in 1930. The Einsteins arrived in Chicago at 9:30 a.m. on Einstein's fifty-fourth birthday, 14 March. A lunch in his honour at the Standard Club was presided over by Arthur Compton. Einstein spoke of three major problems for the world: light, economic distribution and peace. Perhaps the nature of the photon was of more concern to him and Compton than the rest of the audience; he agreed to leave the distribution of goods to economists. For the maintenance of peace, he suggested that the only solution was that individual citizens had to find the will to insist on peaceful settlement of all issues. He also pleaded for support for the Hebrew University in Jerusalem and the Jewish Telegraph Agency. The former good cause reflected the fact that he had allowed himself to be convinced by Chaim Weizmann in November 1932 to rejoin the Board of Governors and to once more fundraise for the university. That evening the Einsteins departed for New York, arriving at Grand Central Station at 1 p.m.

Einstein had specified that there should be no formal reception but nevertheless a large crowd had gathered, forcing him to stop the car and pose for photographs before proceeding to the Waldorf-Astoria. In the afternoon they visited several friends, including Felix Warburg, who was too ill to attend that evening's banquet, at which Einstein was presented with one of only seven extant copies of the first edition of John Napier's *Rabdologiæ* of 1617. This contains the first reference to the decimal point. Einstein also received the Community Church's Gold Medal for Services to Humanity, the only other recipient being Gandhi.

Einstein's address followed a message from his old sparring partner Judah Magnes, the Chancellor of the Hebrew University, that the university faced bankruptcy. Einstein appealed for funds for the university as well as the Jewish Telegraph Agency, whose fifteenth anniversary the banquet was celebrating. He attacked nationalism and mentioned the persecution of individuals in Germany. He was nevertheless hopeful that the current situation would only be temporary.

The two extra days that Einstein's change of ship had granted were rapidly filled with engagements. They were invited to tea at the residence of German Consul-General Paul Schwarz. Schwarz, who was of Jewish heritage, had met the Einsteins' train at Albany and escorted them

[1] Coincidently, the colleague in this anecdote was Beno Gutenberg, one of the most distinguished seismologists of the twentieth century.

to New York city. It is believed that, over tea, Schwarz counselled Einstein against returning to Germany. Schwarz was summarily dismissed one month later by Hitler's government both for his Jewish blood and in particular for his reception of the Einsteins. On Thursday 16 March, Einstein attended an event at the Waldorf-Astoria organised by pacifist groups. In response to a question as to whether Hitlerism constituted a danger to world peace, Einstein responded:

> This danger is not imminent but where the danger comes in is that Hitlerism is contagious. This form of political thought and action has unfortunately become fashionable, for there are too many ignorant human beings in the world. I think, however, that the moral intervention of the intelligent portion of mankind against the excesses of Hitlerism would be highly desirable. In this connection American public opinion and the American press could be of much help [. . .] Any direct action would be harmful. I do think, however, that the German people would not remain insensible to expressions of astonishment from abroad at the treatment accorded to many German pacifists, Socialists, Communists and liberals in general, as well as to representatives of the best and highest forms of German culture. It would be a great mistake to engage in anti-German agitation as such.

He went on to welcome the famous Oxford Union debate held in the previous month, in which the motion 'This House will under no circumstances fight for its King and country' was carried by a convincing majority. He returned to his thesis that 2% of the population following suit would make wars impossible.

The final question was 'What do you think of pacifists who are pacifists in times of peace but not in times of war?' He replied 'I am sorry to say that 90 per cent of pacifists belong to this category'. Before many months had passed and with only the threat of war, his own position would have shifted to join this majority.

The Einsteins began their final full day in New York with a ceremony redolent with new hope for the future. The wife of the editor of the Jewish Telegraph Agency, Jacob Landau, had given birth a few days before. Einstein and Elsa acted as godparents for the baby at his naming ceremony. Einstein inscribed a photograph to the young Albert Landau. After the ceremony, they drove to Princeton, where they had lunch with a future colleague of Einstein, Oswald Veblen, met the Dean of Mathematics, inspected Einstein's future office and then spent a short time looking at possible housing for their stay in the autumn. The Einsteins boarded the *Belgenland* the following day, 18 March, at 12:30 p.m. A group of female pacifists met them and Einstein agreed to receive them on board. Some of his responses to questions make it clear that he had not yet changed his position on pacifism. His responses included that the only way to accomplish disarmament was for one nation to disarm completely; it would then not be attacked. When this was questioned with the example of the Japanese assault on China, which shortly before had led to Japan leaving the League of Nations, Einstein replied 'Nothing has happened to China. Nothing will happen to China. They are still in Manchuria.' He went on to dismiss any possibility of conflict between Japan and the US.

The trans-Atlantic crossing was not the usual relaxing and restorative time for Einstein. The uncertainty of their future weighed heavily on him. Even on the day they left, *The New York Times* had printed that their final destination after Belgium would be Zurich. If Einstein had informed them of that, he soon began to have second thoughts. The plan for Elsa to travel alone to Berlin had also to be given up, perhaps in response to reports that reached him during the voyage. These described the looting of Einstein's Caputh home on 20 March by Nazi SA men.[2] They were ostensibly searching for weapons secreted there by Communists. This was likely to have been precipitated

[2] There is some controversy about this raid. It was reported on 21 March in *The New York Times* as a 'special cable' from Berlin and by the Jewish Telegraphic Agency the following day. There is no mention of it in

by the reports of Einstein's comments in the US on the evils of 'Hitlerism', which had drawn vituperative attacks on Einstein in several German newspapers. He issued a statement about the incident in which he wrote: 'The raid on the home of my wife and myself in Caputh by an armed crowd is but one example of the arbitrary acts of violence now taking place throughout Germany. These acts are the result of the government's overnight transfer of police powers to a raw and rabid mob of the Nazi militia',

In a letter to his sister Maja on 21 March, Einstein remarked that he and Elsa were now birds of passage without a home. He continued that the political carnival in Germany would last for a considerable time, which would make his return impossible. It seems likely that it was the report of the Caputh raid that decided Einstein that he could not return to Germany and must cut his ties with it completely. The final straw may have been the passage by the Reichstag on 23 March of the Enabling Act that established the Hitler dictatorship. The *Belgenland* docked at Le Havre on 26 March. Einstein was brought up to date by a delegation of the International League against Anti-Semitism who came on board. He provided them with a signed statement in which he said:

> Brutal acts of violence and oppression against persons of liberal opinions and Jews, acts which have taken place and still are taking place in Germany, have aroused the conscience of all countries remaining faithful to the ideals of humanity and political liberties. The International League Against Anti-Semitism has acquired great merit by effecting the union of all peoples not contaminated with the virus. We may hope the reaction will be sufficient to preserve Europe from regression to the barbarism of epochs long past. May all friends of our civilization, so gravely threatened, concentrate their efforts to eliminate this psychological malady from the world.

As they approached Antwerp, Elsa wrote to Maja Einstein. The letter shows just how frightened she had become by Einstein's outspoken statements against the Nazis such as the one quoted above:

> Dear Maja! I don't know what gave us the idea that Tete was staying with you! We realise now that he can't be. We couldn't tell from your letter. We had a report from Zangger. I find what he says about Tete's condition very depressing. It's a sad outlook for the future! We are living through such sadness. Maja – I just got a letter from the children. I can't stop crying. It was delivered by messenger. Letters are being opened there. Maja – what these children are having to suffer because Albert gave an incredibly crass interview in New York. Against my will, I was on my knees in front of him. In vain! Maja, life is difficult and ghastly. Don't write anything about politics to the children, nothing about Albert's interview. My God, all of our friends have either run away or are in prison. The newspapers are censored, you can't find out anything. Maja, what times are these! I have been so miserable and sick in the past few days that I can hardly carry on. We will be landing in Antwerp in 10 minutes. I just wish we were already sitting in a quiet corner somewhere, I am also so frightened about the landing!!!!! Oh God!

Einstein added a postscript to Elsa's letter:

---

contemporary German newspapers, but they were completely dominated by the following day's opening of the new Bundestag in a great ceremony held in the Garnisonkirche in Potsdam. The Einsteins' maid, Herta, interviewed by Friedrich Herneck, remembered nothing about any raid; Herneck was convinced that it had not happened. Herta, however, was in Haberlandstrasse when the raid reportedly took place and her recollection of what happened at Haberlandstrasse in March/April is demonstrably incorrect. Einstein was certainly convinced at the time that the raid had occurred and by 11 April Elsa confirmed in a letter to Antonina Vallentin that although the Caputh house was undamaged, it had indeed been searched for weapons.

> Dear Maja, I made a mistake in thinking that Tete was with you. It was probably because you wrote so much about him, or possibly it was an unconscious wish. He is not doing so badly but he is depressed and seems to lose the thread of conversations. All the best! Now we are going to look for a hide-out for the summer.

The remark about Tete losing the thread of conversations indicates that Einstein had received a forwarded letter that Tete sent to Berlin on 26 December 1932. In this Tete had remarked that he had just been playing with great pleasure some sonatas by Haydn and that when they had played Haydn together, he had thought his music too simple. He then abruptly changed the subject to say that he hadn't written for so long because he had been ill and hadn't written to anyone but was now feeling somewhat better in Burghölzli.

As the *Belgenland* docked in Antwerp on 28 March, Einstein decided that he needed to take urgent personal action. On ship's stationary, he wrote a letter to the German Consul-General in Brussels, asking what steps he needed to take to renounce his German citizenship and requesting that the reply to be sent to his uncle Caesar Koch's address, who was by then living in Brussels. On another piece of ship's stationary, he wrote to the Secretary of the Prussian Academy of Sciences, resigning his professorship and his membership of the Academy. In the letter, he acknowledged his gratitude for the support that it had given him over nineteen years but said that, in the current circumstances, the unavoidable ties to the Prussian government that his position entailed were intolerable.

The Einsteins were met by the Mayor of Antwerp, who boarded the *Belgenland* to greet them, and a large crowd that included a deputation of professors. Among these was Théophile de Donder, a colleague and friend of Einstein, professor of mathematical physics at the Free University of Brussels, and Arthur de Groodt, who was professor of histology and embryology at the University of Ghent. De Groodt, who had previously played host to Einstein's old friend Tagore, had telegraphed offering the hospitality of his home to the Einsteins. This was the rather grand Cantecroy Castle in Mortsel, about three kilometres to the southeast of Antwerp. De Groodt had renovated the castle and made it a centre of arts and music. A large crowd cheered enthusiastically when the Einsteins disembarked. After a brief session in which he answered questions from the press, he and Elsa were driven to Cantecroy, a much less exhausting and much more comfortable destination than trying to reach Uncle Caesar in Brussels, or indeed Ehrenfest's house in Leiden.

It did not take long for Juliette Adant, de Groodt's wife, to convince the Einsteins that they should remain in Belgium. Einstein knew that he could rely on help at the highest level from Queen Elisabeth, although she and the King were in Palestine at that time. The de Groodts spent much time in the little resort town of Le Coq sur Mer (De Haan in Flemish) and were able at very short notice to facilitate the renting of the Villa Savoyard. This was a rather imposing house close to the beach. On 1 April, the Einsteins moved in. By a stroke of good fortune for Einstein, the occupant of the other half of the house was the concertmaster Michel Robert, who quickly agreed to accompany Einstein on the piano and to be pianist in a trio together with Einstein and a cellist, Jeanne Doetsch, the wife of the owner of the Strand Hotel. Music was no doubt an essential soothing influence in those troubled and confusing times.[3]

[3] Einstein seems to have had a violin with him on this stay in the US and therefore also in Belgium. In his travel diary he waxes lyrical about the tonal properties of a violin by his local Berlin luthier, Julius Levin, that he was playing at a musical event after a dinner on 15 February. This must be the violin that he mentioned, in a reference for Levin written in July, as having been carried with him daily over a period of several years.

That this period was troubled, particularly for Einstein's family, can be seen from many sources. Einstein was in the sights of the German government, as is confirmed by the following quotation from Josef Goebbels, addressing a Nazi Party meeting on 31 March:

> 'When American and English Jews attack the German Government, we cannot hinder the German people from attacking German Jews. We did not plan to open this question immediately. We had more important things to do. We endured the Jews, and what was their gratitude? German Jews can thank the wandering Jews Einstein and Feuchtwanger.'[4] At this point there were cries of 'Hang them!'

The servile acceptance by the mainstream German newspapers of this typical Nazi trick, implying that the Jews' troubles were all of their own making, is exemplified by the quote from *Deutsche Allgemeine Zeitung* reported in *The New York Times*: 'Feuchtwanger's idiotic articles and Einstein's political feeblemindedness, dispensed abroad, have produced a very different effect for their coreligionists than those heroes imagined.'

Dima Marianoff in Berlin was already feeling very nervous, as Russian citizens were all suspected to be communists and were being picked up off the streets. On the morning of 10 March, he had a premonition that led to him taking his passport to have it franked with permission to leave Germany. On his way back to Haberlandstrasse, he was confronted by a crowd of young Nazis but bluffed his way through them by pretending to be Italian. On telephoning home, he was warned by Margot that the Nazis had already called twice looking for him. He made his way to Charlottenburg station and escaped to Paris. He was joined at the start of April by Margot.[5]

The Einsteins' maid Herta described the visit of two policemen to the Haberlandstrasse apartment, probably in early April. They were looking for Marianoff and searched his bedroom. Herta recalled that, at the end of May, uniformed SA men arrived while Ilse and Helen Dukas were packing up the household. The house was thoroughly searched while the three women were confined to the library. Fortunately, Margot, on the advice of Marianoff, had already shipped most of Einstein's papers to the French Embassy, from whence they could be forwarded unmolested in the diplomatic bag to Paris and thence to Marianoff. However, carpets, pictures, and silverware were stolen from the apartment. By 1 June the three women had cleared the apartment and sent the remaining contents to storage, probably in Paris, before they were eventually shipped to Princeton.

Margot left Paris and arrived at Le Coq to be reunited with her mother on 13 April. Ilse left Berlin alone; by 5 April she was in Scheveningen, close to The Hague in the Netherlands. Elsa rushed to join her and from there wrote to her friend Antonina Vallentin on 11 April. Einstein's Berlin bank accounts had been frozen on 3 April, leaving Elsa's sister, who they were subsidising, without support. That Elsa's nerves were shattered is demonstrated in the letter:

> . . . I am tired and at the end of my tether. But I have to go on! I have suffered so much because of the children. My husband took no account of them, stayed true to his beliefs and spoke out loudly and clearly. This has almost shattered my children. They were petrified there. Now they are safe. Nothing

---

[4] The author Lion Feuchtwanger, most famous for the novel *Jud Süß*, had also, like Einstein, been in the US when the Nazis had taken power. As a long-standing critic, he too had not returned to Germany and his house was also ransacked by the Nazis.

[5] It is impossible to reconcile the dates given by Marianoff and Herta, two unreliable witnesses. The latter was insistent that there were no raids on the Haberlandstrasse apartment until April, when she says Marianoff departed for Paris. Marianoff gives his departure date as 10 March and since he had an exit visa stamp in his passport, perhaps for once he can be relied on. It would make sense that he was targeted early in the Nazi regime as his activities on behalf of the Soviet Union were widely suspected by the authorities. It may well be that after Marianoff had fled, there was a period of quiet until mid-April.

> happened to them, but Ilse's nerves are shattered and she is in a pitiful state. I see things looking black for her in future. The tragedy of my husband's story is that all the German Jews are making him responsible for all the terrible things they are going through. They believe that his activities have influenced the repression and, in their bigotry, accepted the propaganda and turned themselves away from him and hate him. So we get more hate mail from Jews than from Nazis . . . Ilse is with us [and] wants to go back to Berlin as she must break up our big establishment and also Caputh. She shivers however in case they find out that she has been with the 'great traitors'. Her nerves have gone and I could cry every time I look at her. Please God she won't go back there. But she can't be stopped as Rudi is taking care of his 84-year-old father and she will not be separated from her husband. This conflict has destroyed her self-composure.
>
> [. . .]
>
> My husband has accepted a Professorship in Paris. He has also taken one in Madrid. And in Brussels. And in Oxford and Leiden. All without duties. Whenever he has time, he will go there. All these universities want to get him and, since he feels he has ties to them all, he accepted them all. He has had the Oxford and Leiden jobs for years. He has agreed to spend a month every year in Madrid. He will certainly 'temporarily' lecture in Paris at some point . . .

That Elsa's remarks above about the mindset of German Jews were well informed is illustrated by the German Jewish Press. Issue after issue of the *Jüdische Rundschau* disassociated itself desperately from attacks on Germany and the German government from foreigners and the foreign press. There was little if any reporting of the attacks on Jews and Jewish property in Germany, except indirectly in appeals for respect for legal rights. Einstein was ignored until the issue of 31 March, when reporting on his comments in the US was limited to those remarks praising the contribution of German culture to the world and against any blanket condemnation of the German people, ignoring his statement reported in *The New York Times* on 11 March in which he had identified the psychological illness of the German regime. Even the boycott of Jewish shops and businesses called on 1 April, justified in part by allegations of Jewish propaganda against Germany from overseas, elicited a muted response. The front-page story remarked that this could mark a new beginning for Jews in Germany, who could gravitate back together since both religious and assimilated or baptised Jews were being tarred with the same brush. On 15 August the *Jüdisch-liberale Zeitung* characterised Einstein's interventions in Jewish affairs as disastrous.

If the German-Jewish Press were not sympathetic to Einstein, it is hardly surprising that German scientific academies were even less so. Einstein's letter of resignation from the Prussian Academy had been received with satisfaction by the academy authorities, who were about to instigate proceedings to have him expelled. The situation began to get out of hand when the Academy, or rather one its secretaries, Ernest Heymann, issued a statement on 1 April saying that the Academy had taken note of the newspaper reports of Einstein's involvement in atrocity mongering in France and the US. Before it could act against him, he had resigned. The Academy considered Einstein's agitation abroad a serious contravention of the long-standing tradition that the Academy's strong ties to the Prussian state required its members to avoid political controversy. The statement concluded that for these reasons the Academy did not regret Einstein's resignation.

Einstein was incensed by this statement, as indeed were many of his friends and colleagues in the Academy, although only Max von Laue had the courage to protest against it at the next Academy meeting. Einstein issued another statement denying that he had ever participated in atrocity mongering and had only commented on the official policy to destroy Germany's Jews by economic methods. He reiterated that he stood by every word that he had uttered and called on the Academy, having taken part in his defamation to the German public, to bring his denial to the attention of both its members and the public. Unsurprisingly, two further letters from the Academy and a response from Einstein did not lead to a meeting of minds. Einstein concluded his final letter by remarking that the Academy's letters to him had shown him how right he had been to resign.

On 8 April, the Bavarian Academy took up the cudgels. It sent Einstein a letter asking him in the light of his resignation from the Prussian Academy, and the exchanges between them, what his intentions were with respect to the Bavarian Academy. Einstein replied that although the reasons for the breach between him and the Prussian Academy did not necessarily entail his leaving their academy, too, he wished to resign. His reason was that in his view all German academies had failed in their duty to cherish scientific life in the country. Einstein's membership of the Leopoldina was also terminated in 1933, although he neither explicitly resigned nor was informed about his expulsion. Against his name in the register for 1933 is simply written 'cancelled'. He wrote to von Laue after his resignation from the Prussian Academy asking that he should ensure that Einstein's name was removed from all German academies of which he was a member. It may be that von Laue telephoned the Leopoldina with this information, or that its secretary, viewing the controversy with respect to the Prussian Academy, simply quietly removed Einstein's name without further formality.

A letter from Max Planck, sent via Ehrenfest, had argued that resignation from the Prussian Academy was the only honourable way out to save his friends from having to defend the indefensible, as Planck saw Einstein's views. He characterised them as being separated by a deep gulf from his own but reiterated their long-standing personal friendship. Always a loyal and conservative Prussian, at this early stage in the Nazi dictatorship, Planck was determined not to rock the boat and protect the Academy's position with the new government. The promulgation of a law in April that no Jew could hold any official position, including university professorships, caused an enormous exodus from Germany, including leading figures such as Max Born and Lise Meitner. Planck was greatly distressed by this, particularly by the dismissal of Fritz Haber. In May, Planck went to see Hitler, using his position as President of the Kaiser Wilhelm Society to formally congratulate him on his appointment as Chancellor. In 1947, Planck described this meeting. When Planck broached the topic of the Jewish exodus, Hitler maintained that he had nothing against Jews except they were all communists. When Planck tried to interject that not all Jews were identical, Hitler interrupted to contradict him, insisting that all Jews were the same and he had to treat them all identically. When Planck insisted that excluding the Jews would have a drastic effect on German science, Hitler flew into a rage, slapping his knee and talking more and more rapidly until Planck had no option but to wait for a pause in the diatribe and then take his leave.

If German academia had abandoned Einstein, the rest of the world rushed both to his defence and succour. As indicated in Elsa's letter to Vallentin quoted above, universities all over the world fell over each other to offer Einstein positions now that he had definitively left Germany. Some of these were pushed by scientific friends and collaborators. Langevin was very influential in convincing the French government to establish a special chair for Einstein. On 8 April 1933, French Minister of National Education Anatole de Monzie wrote to Einstein on behalf of the French government, enquiring whether Einstein would accept a chair of mathematical physics at the Collège de France, which would be specially created for him. The Spanish Ambassador in London, Ramón Pérez de Ayala, conveyed the offer of a chair at the Universidad Central in Madrid through their mutual friend, a former academic there, Abraham Yahuda.[6] Einstein's initial letter of acceptance

[6] Abraham Shalom Yahuda was deeply involved in the efforts to get Einstein to Madrid. Yahuda also shared Einstein's feud with Judah Magnes. In 1920, Yahuda had accepted an appointment in the nascent

on 12 April was qualified by the fact that he would only be able to spend a few weeks per year in Madrid and that he would be unable to come until the spring of 1934. He also enquired about the possibility of positions for other excellent scholars from Germany. Concerned about the situation of the many Jewish refugees fleeing Germany, Einstein was discussing the possibility of setting up a scheme to accommodate some of them, possibly in England. He asked his friend Solovine to keep an eye open for any German Jewish scholars passing through Paris. In explaining his embarrassment of chair offers, he said 'I now have more Professorships than ideas in my brain. The devil always shits on the biggest pile!'

One scholar who was much on Einstein's mind at this time was his valued assistant, 'Mayerchen', who by early April had joined Einstein in Le Coq. Apparently contained within his offer from Madrid was that of a full Professorship for Mayer. Directly after receiving it, Einstein wrote to Flexner to try to leverage this offer to improve Flexner's to Mayer, which was only an Associateship at the Princeton Institute for Advanced Study. He told Flexner that he felt duty bound to propose Mayer to Madrid and that doing so would deprive him of the most valuable collaborator that he have ever had.

This letter crossed with one from Flexner, also dated 13 April, in which he complained about newspaper reports that Einstein was accepting a number of professorships scattered across Europe. Ominously foreshadowing the direction in which their relationship would develop, Flexner fustily reminded Einstein that his Institute was supposed to be a haven for scholars to study in peace, not a way station for itinerant academics. In a postscript he noted with relief that one of the mooted positions, at the Free University of Brussels, was an Honorary Professorship, to which he had no objection. He was unaware that, honorary or not, Einstein had already committed to giving a course of lectures in Brussels on spinors, his latest idea for a unified theory, in April and May. Einstein claimed that these further obligations were no different to his half-year commitment to his former Berlin position and did not affect his agreement with Princeton. While these extra obligations would certainly not contribute to his peace and quiet, he felt it his duty to accept them. Flexner seems to have accepted Einstein's explanation but partially called his bluff with respect to Mayer, not granting the full professorship but agreeing that he should have tenure. This seems to have mollified Einstein; his next reference to Mayer was in July, when he enquired about the $100 honorarium that Flexner had promised to Mayer until they arrived in Princeton, admitting that he had not had any further negotiations with the Spanish authorities on Mayer's putative professorship.

Einstein did his best to settle down to work with Mayer, enjoying walks on the beach in Le Coq and the relative quiet of an out-of-season resort. His musical activities with his neighbours were also, as usual, an important part of his life. However, as he complained to Solovine, it was raining people and letters. Elsa complained that she spent half her time dealing with refugees who turned up at their door expecting help; she had to listen to their various sad tales of hardship and despair. Although Einstein's Berlin accounts had been frozen, he had for many years been (illegally) depositing money in the Netherlands via Ehrenfest. They were therefore not destitute, but certainly unable to help all the callers. It was also very difficult to tell who might be genuine and who Nazi agents. There were persistent rumours that the German government had placed a price upon Einstein's head. Queen Elisabeth, who had been informed of Einstein's arrival in Belgium, sent a postcard from Palestine on 10 April. It was probably her and the King's influence on the

Hebrew University to teach Biblical studies and Arabic language and literature. He was pushed out of his appointment a few months later because of his views on the shared culture of Arabs and Jews.

government that resulted, to Einstein's irritation but also amusement, in the despatch of burly detectives to Le Coq to ensure his safety.

The King and Queen returned from their visit to Palestine on 25 April. The Queen lost no time in inviting Einstein to the palace to play chamber music. They played together in Laeken on 3 and 10 May. On the latter date, Mozart Quartets were tackled with cellist Alexandre Barjansky, and a violist. Barjansky was impressed by Einstein's musicality; he wrote to Margot Einstein that 'Einstein's playing of Mozart was unique. Without being a virtuoso and perhaps because of that, he reproduced the depth and tragedy of Mozart's genius so naturally with his violin.'

Einstein's lectures in Brussels, where he interacted closely with de Donder and Georges Lemaître, were scheduled to end on 27 May. Einstein was also committed to spending several weeks in Oxford that spring. On 1 May he wrote to Lindemann enquiring whether he could come in June, since he had agreed to accept an honorary degree and lecture at Glasgow on 20 June. Lindemann replied that term would have ended on 16 June, at which point students and fellows would begin to disperse. Einstein therefore rearranged his Brussels lectures so that he could leave for Oxford on 21 May. In his letter, Einstein told Lindemann that he had it on good authority that the Nazis were rapidly gathering war material, particularly aeroplanes, and that if they were given one or two years, the world could expect something nice from Germany. He also offered to give up one third of his Oxford salary to support some of the young refugee scientists who Lindemann was gathering in Oxford. Future Nobel Laureate Hans Bethe, dismissed from the University of Tübingen because of his Jewish heritage, was a name that Lindemann was considering, as were the brothers Fritz and Heinz London. They had been recommended by Arnold Sommerfeld, whom Lindemann had recently visited in Munich; although Einstein admitted not being very familiar with Bethe's work, Bethe did arrive in the UK later that year, initially to Manchester University.

Max Born, suspended from Göttingen since January, was another refugee looking for a position. In a letter to Born at the end of May, Einstein explained the organisation among British universities that Lindemann had set up. He thought it likely that all the refugees with reputations could be accommodated; what concerned him were the young ones, who could not. He concluded his letter by musing upon the German character, which he said had surprised even him, never an admirer, by its brutality and cowardice. He wished Born a pleasant stay in the mountains. Born and his family had taken a cottage in the skiing village of Selva in the Dolomites while he worked out his future. It was here in June that he received an offer from Rutherford and Blackett to go to Cambridge as Stokes lecturer. Shortly afterwards, Lindemann knocked on the door of their cottage with an offer from Oxford, which Born had to refuse. Born spent two years in Cambridge and was eventually appointed to a chair in Edinburgh, where he spent the rest of his career.

Einstein was also concerned about the position of von Laue. Although not a Jew, he had been outspoken in defence not only of Einstein but also other distinguished Jewish physicists who had been dismissed from their positions and the Prussian Academy.

Einstein had been in close contact with Ehrenfest since his return to Europe. Indeed, on 30 March Ehrenfest had written to Einstein, urging him and Elsa to live with him in Leiden. Einstein presumably found isolation in Le Coq more attractive than being surrounded by his academic colleagues in Leiden, not to mention Ehrenfest's chaotic housekeeping. Ehrenfest certainly travelled to visit Einstein on several occasion. On 10 May he reported on a recent trip to Berlin, where he had talked to both von Laue and Haber. In his view, Haber was under surveillance. Einstein knew how hard Haber would take his rejection by a government that he had served all his life. Einstein compared it to the difficulty in letting go of a cherished theory disproved by the facts, whereas for him it was completely different, having never believed in Germany.

To aid Jewish academics fleeing from Germany and Jewish students who were being barred from universities, Einstein considered the idea of setting up some sort of 'Refugee University' in England that could be a safe haven, but it became clear that the logistical difficulties were too great. In these circumstances, the Hebrew University was an obvious alternative. In his address at the New York dinner in March, he had said:

> The significance of the University in Jerusalem for the Jewish people will be heightened by the fact that the Jews in Eastern Europe are being barred from the sciences and the practice of scientific professions [. . .]. Many talented Jews are lost to culture because the way to learning is barred to them. It will be one of the foremost aims of the University in Jerusalem to alleviate this misery.

Weizmann had been ousted as leader of the Zionist movement at its Congress in 1931 and therefore had had more time to insert himself into the Hebrew University's affairs. Indeed, he had ambitions himself to become its academic head. However, a plan that he had submitted to the university to spend six months per year there researching in chemistry had been, to his great irritation, ignored. On 6 April, he had telegraphed an invitation to Einstein to spend some of his time at the Hebrew University. However, this fell on stony ground; the front page of the *Palestine Post* of 19 April carried the story that Einstein had refused to join the Hebrew University because of his long-standing conflict with the university authorities.

On 27 April, Einstein wrote to the London office of the university, resigning forthwith from the supervisory board. He based this decision on information he had received from reliable third parties that, despite Weizmann's reassurances, the situation had not improved. Principal among these third parties, it transpires, was Abraham Yahuda. Weizmann reacted on 3 May with a long letter in which he chastised Einstein for both his public criticisms and his making them before he had informed Weizmann of his intentions. He was clearly deeply hurt by Einstein's criticism of him personally, which he characterised as undeserved, untimely, and unjustifiable.

Einstein replied to Weizmann on 7 May, insisting that, with people such as Magnes in charge, and he listed several others he considered also both incompetent and malevolent, the university would never amount to anything. He compared the university to a house with an infestation that required complete sanitation to eliminate the vermin. He also accused Weizmann of having broken his word to sort out this intolerable situation several years previously. Weizmann replied to this letter with great sadness, reproaching Einstein for not having confronted him with the charge on breaking his word already in 1928. He gave up his intention of coming to see Einstein in Le Coq because Einstein insisted that he had already made up his mind on the subject. In letters later that month to Jacques Errera, a Jewish professor at the Free University of Brussels, Weizmann characterised Einstein as a prima donna beginning to lose her voice and thought that he ought to be really ashamed of his actions.

On 9 May, Einstein wrote to Lindemann, saying that he needed to postpone his arrival in Oxford until 26 May, as he needed first to travel to Zurich, where his son was very ill. Before he disembarked from the *Belgenland* in March, Einstein had written a touching letter to Tete, part of a concentrated spurt of letter writing that had included his resignation from the Prussian Academy. He wrote that Zangger had informed him of Tete's condition and that he was pleased that he was sufficiently better to be able to play Mozart again. Having not seen him for a long time, he hoped to see him again soon. Despite this wish, it may be doubted that he would have been able to make the trip, given the demands on his time, not to mention the danger of him travelling near to Germany. Indeed, he had also been reluctant since he had lost faith in Prof Maier, the doctor in charge of Tete's case at Burghölzli, whom he considered too optimistic. That this reluctance had suddenly changed is indicated by his letter to Lindemann.

Figure 23.2 Einstein with his son Eduard (Tete) on 22 May 1933. At this, their last meeting, they played violin and piano sonatas.

The tone of urgency in Einstein's letter to Lindemann had been catalysed by Mileva. On 24 April, she had written to Einstein, having discovered his address from Maier. She urged Einstein and Elsa to accept the use of her apartment in Zurich for as long as they wished, offering to confine herself to her room and if Einstein desired, avoid meeting him completely. Tete was still in hospital at that point, so that, as she pointed out to Einstein, she was in any case absent from the apartment for several hours a day while visiting him. On 8 May, she wrote again, imploring Einstein to visit Tete, ending her letter with 'You give your help to half of the world – why not your own son?' By mid-May of 1933, Tete's condition had improved sufficiently for him to be discharged from Burghölzli and to return home, accompanied by a nurse. However, he continued to oscillate into and out of the clinic, a pattern that was to continue until Mileva's death.

Einstein left Belgium early on 21 May and met Tete on the afternoon of 22 May in Burghölzli. Given that Tete had been discharged some days before, the venue seems puzzling. It may be explained by that fact that, a few weeks before, Einstein had written to Zangger that he was determined to avoid both Mileva and Hans Albert. He was no longer on speaking terms with either because of their reaction to his will the previous year. This is highlighted by Hans Albert's angry letter referred to in Chapter 22, which elicited an equally brutal response from Einstein. In this he suggested that they should cease corresponding as there was a complete breakdown of trust between them. On the other hand, Mileva's letters of 24 April and 8 May do seem to have moved Einstein greatly; in his reply to the latter he offered to stay in her apartment if he could be of any help. Nevertheless, from circumstantial evidence it seems most likely that he stayed in a hotel. It may be that the meeting in Burghölzli was convenient because Tete was undergoing some outpatient treatment; or he may have been readmitted if his discharge had not been a success. Certainly by the following month he was once more an inpatient, pleading to be allowed to return home.

A poignant photograph exists of Einstein and Tete at their meeting (see Figure 23.2). Einstein is staring reflectively into the distance, holding his violin under his arm, bow dangling from his right hand while Tete studies the piano score, brows wrinkled in concentration. That evening Einstein consulted with Zangger at his home and on the following evening met a group of old friends from school and the Polytechnikum. At some point in the visit, he relented and met Hans Albert. It is unclear if he was introduced to his new grandson, Klaus Martin Einstein, who had been born in March. Hans Albert and Einstein seem to have resolved their differences over the inheritance and Einstein's will. On 24 May, Einstein left for Oxford. He was never to see Zangger, Besso, Tete, or Mileva again.

Chapter 24

Final stays in UK; Farewell to Europe and arrival in Princeton (1933)

On 26 May 1933, Einstein crossed the channel via Ostend and arrived that evening in Oxford. He had asked Lindemann to meet him at Victoria Station in London in his Rolls-Royce, although he omitted to tell him on what day he intended to arrive! Given that the customs officials seemed to be expecting Einstein's arrival, Lindemann probably arranged for them to telephone him once his distinguished visitor had arrived at Dover. Einstein immediately felt at home back in Christ Church. Since it was raining heavily, he was grateful for the fire burning in his study grate. Margaret Deneke described his rooms, which Einstein characterised as amusing:

> This year he had been moved to the first floor of the first staircase on the right after entering Tom Gate. It was smaller than the room he had had last year; there was a curious built-in seat to which one ascended by steps to the window. The Professor showed us his fine view of the Cathedral with a touch of possessory pride.

Having unpacked and settled in, Einstein's thoughts returned to the sad scene he had just left behind in Zurich. At least he hoped he had made his peace with Hans Albert; on 30 June, he wrote to him apologising for his previous harsh letters:

> I realise that I have de facto done you an injustice in incorrectly and to your detriment judging your certainly awful behaviour towards me. Your main motive was your undeserved feeling of neglect and a desire to somehow get revenge for that. But this is completely erroneous on your part. My life is filled with more stress and responsibilities than that of the average person and therefore I can't be expected to fulfil the same claims on me personally as such a normal individual. That the women who live with me are in a different situation isn't relevant. And even if it were, it would be petty to attach any importance to it. We are after all men and have to stand on our own two feet and thus find our way.

On 1 June, Einstein donned his evening dress and made an appearance at the Oxford Union Society. The Union had only a few months previously hosted its famous debate on the motion 'This House will under no circumstances fight for its King and country,' which was carried by a large majority. Had he been there when the debate took place, Einstein would no doubt have supported it enthusiastically. Now that Hitler had come to power, it is unclear what his attitude would have been had the debate been held on this day, after he posed for the traditional photograph with the committee, headed by the President, Keith Steel-Maitland. Directly behind Einstein stood Michael Foot, President in the first term of that academic year. Foot subsequently became a Cabinet Minister and Leader of the Labour Party.

Einstein had called at Gunfield, the Denekes' house, as soon as he had arrived in Oxford to offer his condolences. Their mother, whom Einstein had met and highly valued on a previous visit, had died; the funeral had taken place the day before his arrival. He regretted not being able to attend and encouraged them to keep up the tradition of musical events at the house that their mother

Einstein. Brian Foster, Oxford University Press. © Brian Foster (2026). DOI: 10.1093/oso/9780198794875.003.0024

had started. They had cancelled Marie Soldat's visit that year but nevertheless were catalysed by Einstein's presence to organise a musical remembrance of their mother on 8 June:

> Jelly d'Aranyi led two Mozart quintets and Ernest Walker played the two birthday pieces he had composed for mother. Our family memorial music was quite simple; we had the same chatty supper, we cheerfully set about arranging the chairs and desks and tuning up. All the players had known her well and loved her. The G minor Mozart Quintet was her first love – Einstein buried his face in his hands during the piano pieces and then turned to me, [and] said 'Really beautiful'.

Margaret Deneke also recalled that she had accompanied Einstein to a recital in the Town Hall by the celebrated African-American singer Roland Hayes, who had just given two highly acclaimed recitals at the Wigmore Hall in London:

> Einstein [. . .] was delighted with the lovely voice and artistic mastery – he pointed out to me however that all lyrics that dealt with a passionate love element failed to be convincing in his rendering. No African could experience that intensity; the feeling was just not in his make-up and even so great an artist as Roland Hayes was unable to dramatise it.

This unfortunate generalisation is in contrast to Einstein's subsequent indefatigable support of African-American artists in the US.

On 12 June, Einstein accompanied Margaret's sister Helena to a quartet concert in the Sheldonian Theatre. He thought it a good performance and he enjoyed it immensely but remarked that English artists could not make Haydn smile.

Amidst all this musical activity, Einstein did find time for physics. On 2 June, he gave the vote of thanks after the Junior Scientific Club's annual Boyle Lecture at the University Museum by his old friend Lord Rutherford. Rutherford described the new particle accelerator devised the previous year by John Cockcroft and Ernest Walton and work that he himself had done with Mark Oliphant. They had used higher-current proton beams with much lower energy, but which were still capable of splitting the lithium nucleus. He also referred, although without naming them, to experiments with the neutron that James Chadwick had discovered in 1932. He mentioned the discovery of the positron by Carl Anderson and confirmed by Patrick Blackett. Given this astonishing string of discoveries with which to regale his admiring audience, it is unsurprising that Rutherford's performance, larger than life at any time, somewhat cowed Einstein. In his still halting English, he nevertheless moved the vote of thanks for the lecturer. An undergraduate who attended reported that:

> As he delivered his speech, it seemed to me that he was more than a little doubtful about the way in which he would be received in a British university. However, the moment he sat down he was greeted by a thunderous outburst of applause from us all. Never in all my life shall I forget the wonderous change which took place in Einstein's face at that moment. The light came back into his eyes, and his whole face seemed transfigured with joy and delight when it came home to him in this way that, no matter how badly he had been treated by the Nazis, both he himself and his undoubted genius were at any rate greatly appreciated in Oxford.

On the evening of 10 June, Einstein gave the Herbert Spencer Lecture at Rhodes House. This was the first time he had delivered a lecture in English at Oxford, although he had written it in German and thanked his colleagues at Christ Church for translating it. It was also a contrast with his previous Oxford lectures in being non-technical. There were no equations, and the content was clearly and directly stated. Its significance is as the apotheosis of Einstein the mathematician. His thrust was that the Greeks, in particular Euclid, developed the logical scientific method but did not realise the connection between their abstract concepts and the world of experience. Kepler and Galileo, and in particular Newton, had taken this logical deductive system and assumed that experiment and experience would uniquely define theories of the world. General relativity, however, exploded this, as, starting from completely different theoretical premises, it explained all of Newtonian physics and went further. Einstein concluded that:

> Our experience up to date justifies us in feeling sure that in Nature is actualized the ideal of mathematical simplicity. It is my conviction that pure mathematical construction enables us to discover the concepts and the laws connecting them which give us the key to the understanding of the phenomena of Nature.

Although he went on to discuss the problem of how atomic 'discontinuities' could be treated in a continuous field theory and the quantum mechanical approach to this, which was in his view incorrect, the quoted sentences are the crux of his argument.

This was not the first time, even at Oxford, that Einstein had enunciated such an approach. In May 1931, in his final Rhodes Lecture 'Latest Developments of the Theory of Relativity,' he was reported to have said that a physical basis for his unified structure was lacking, and one could only be guided by considerations of mathematical simplicity and logical form. Abraham Pais's view on the arguments stated in the Spencer lecture was: 'I cannot believe that this was the same Einstein who had warned Felix Klein in 1917 against overrating the value of formal points of view "which fail almost always as heuristic aids"'. Intellectually, it wasn't the same Einstein. He had by now ceased to employ the powerful physical insight behind all of his great triumphs, including, at least partially, general relativity. The irony is that, while he extolled the ideal of mathematical simplicity, in practice his work with Mayer on unification used some highly complex and abstruse mathematical methods.

Einstein's letter to Elsa on 11 June shows that, once again, he had succumbed to the charms of Oxford. The letter is worth quoting at length as it touches on several of the matters that were concerning him at that time:

> Dear Elsa, My time here is coming to an end. It was a beautiful stay; I am even starting to get used to wearing my dinner jacket, just as at an earlier time was the case with the toothbrush.[1] Having said that, I never wore any socks the whole time, even for the most splendid occasions, a lack of civilisation that I concealed by wearing high boots. Ayala came to visit me yesterday, a really sweet man; I am looking forward to spending a few days with him. He made no mention whatever about the Spanish chair. I think that we will be able to let this uncomfortable subject slip away soundlessly into oblivion. I have strenuously forbidden any formal events in London. This embarrassed Ayala, who is clearly afraid of Yahuda, who wants to make capital from my presence. In his desperation, he brought you into the discussion by offering to provide you with an English document to help regain our German funds; however, I wasn't falling for that one. I am definitely going to Paris and think that you should come with me this time. I have already written to Langevin that they should release me from any teaching duties, because I can achieve much more doing other things. My plan for next year is to go *only* to Paris and the Belgian coast. Hopefully it will work with the boat.[2] Transporting it by water is not as easy as you think, mostly because of the bridges. Nernst has let me know that I should go nowhere near the German border, as the anger against me is enormous. Lindemann has just got back from there. Through him, I have been able to make good use of the document that you absolutely didn't want me to take with me. Here, things and people from the continent are observed with cool condescension. The English are very reserved in conversation, almost as much as me. I get on very well with them and am

[1] This is a joking reference to their early relationship, when Elsa had tried to convince him to use one.

[2] This is a reference to his yacht *Tümmler*, still tied up in Caputh.

starting to be familiar with their way of thinking. Everyone is afraid of the competition from 'educated' Jews. We suffer more from our strengths than our weaknesses. Weizmann is trying by every possible trick to change my mind. You would be appalled if you knew how I have 'shown him the cold shoulder.' Nothing decent can come out of the Jerusalem University in the foreseeable future. I am not going to do anything more for this cause, and so soon will be in 'splendid isolation' from both Jews and Gentiles, which will make the rest of our lives much easier. I will and must get on with Flexner, on the basis of a frosty reserve.

Einstein's final lecture at Oxford was held on 13 June at Lady Margaret Hall. The Philip Maurice Deneke lecture had been founded in memory of the father of the Deneke sisters. In a moment of weakness after one of their musical evenings the previous year, Einstein had accepted a proposal from Marga Deneke to deliver the lecture. Einstein entitled it 'Einiges zum Atomistik' ('A few things about Atoms'), which gives quite a lot of information. Firstly, the lecture would be in German, in contrast to the Spencer lecture two days previously. Secondly, it would be about atomic theory. Finally, its vagueness implies that Einstein had not given much advanced thought to what he was going to say. In fact, about a month before, he had sketched an outline, into which the vestigial report in *The Times* fits nicely.

Einstein began with a historical summary of the development of the concept of the atom, from the Greeks to the present day. He cited examples – from chemistry the law of multiple proportions, and mineralogy the law of rational indices of crystal faces – as early evidence for the atomic theory that was purely geometrical and independent of any atomic kinematics. This was reinforced by the ideas that gas properties could be described by molecular kinematics. In turn this led to estimates of molecular size. In many of these developments, in particular Brownian motion, Einstein himself had played a leading role. He gave the example of the Wilson cloud chamber as the most direct route in which atoms could be visualised.[3] He ended with what he called the quantum catastrophe, that the modern way of describing atomic physics, initiated by Bohr, was deeply flawed. There was an unresolved conflict between particles and waves that was currently circumvented by talking about atomic phenomena in terms of particles but computing them using waves. With a typically musical flourish, thinking of the allowed intervals between the energies of quantum states, he referred to the musicality of the quantum rules. *The Times* report quoted him as concluding that the deeper one searched, the more there is to know; he believed that this would always be the case as long as human life exists.

Marga Deneke wrote that: 'The Deneke lecture was packed and many of our best friends failed to get seats. Sir Charles Sherrington[4] took the Chair. Whilst Dr Einstein was speaking and using the blackboard I thought I understood his arguments. When someone at the end begged me to explain points I could reproduce nothing. It had been the Professor's magnetism that held my attention.' The following morning, she cycled to Christ Church and delivered the fee, £25, in cash. 'Einstein said "What a large packet of lovely money, and it was a rubbishing lecture." He asked me to take the money to Barclay's Bank and dispatch it to a friend who was in distress.'

Einstein's exit from the Prussian Academy meant that his normal method of publication via the Academy's *Proceedings* was no longer available. Ehrenfest offered to steer his publications through the Royal Netherlands Academy of Sciences, of which both were members. On 14 June, Einstein wrote to Ehrenfest, thanking him for looking after the publication of a paper by him and Walther

[3] In fact, such tracks trace the path of charged particles, such as muons or electrons, not neutral atoms.

[4] Distinguished physiologist and President of the Royal Society from 1920–1925.

Mayer on semi-vectors and the Dirac equation. Einstein returned to the situation in the Hebrew University, writing:

> My biggest regret is that Jerusalem University is completely messed up. Bad elements are at the helm, good ones have been weeded out. Weizmann is very clever but too ready for any dirty compromise when he could really sort it out himself. Without a complete and far-reaching clear out, every penny that goes there is a penny wasted; one can't recommend going there to any reputable person. Isn't this a nuisance? If there is any gleam of radical improvement, I will let you know. I am fighting for this with a brutality that would astonish you.

Einstein was indeed being brutal, at least to Weizmann. In a letter at the beginning of June, he had told him there was no point in meeting – Weizmann lived in London – until the situation with the university was clearer and that unless it was solved, their generation would have to give up on the idea of a Jewish university and he would wash his hands of the whole thing. A few days later he returned to the fray, telling Weizmann that unless Judah Magnes and his cohorts were removed, they would be taking money from donors under false pretences. He saw no point in their meeting or in his taking part in committee discussions on the subject unless this was done. Weizmann continued to try to bring Einstein around to a more positive view. On 8 June, he wrote regretting that Einstein could not spare time to meet him and offering to come to Oxford at any convenient occasion. In this letter he broached a scheme he was already carrying out, with the financial support of Israel Sieff, to establish a rival department of chemistry at Rehovot. This small settlement about 25 kilometres south of Tel Aviv already hosted a Hebrew University agricultural outstation. He suggested that Einstein do something similar for physics, thereby outshining the university departments and eventually subsuming them. In fact, physics hardly existed in the Hebrew University and chemistry was small and weak. By this means, Weizmann hoped to outflank Magnes, whom he considered hopelessly biased against science. Eventually, this initiative grew into the prestigious Weizmann Institute, one of the world's leading scientific research centres. Einstein, however, refused to contemplate getting involved with setting up an institute, something for which his experience as a Kaiser Wilhelm Director had taught him he was totally unsuited.

Einstein spent the period 15–23 June in London, with brief excursions back to Oxford and to Glasgow. He stayed at first in the Spanish Embassy in Belgrave Square with his friend Ambassador Ramón Pérez de Ayala. Despite his assertions to Elsa recorded above, he accompanied Pérez de Ayala to the English Association's annual dinner at the Criterion on 16 June. The toast from the guests was proposed by his old friend Sir Arthur Eddington. Perhaps with Einstein's increasing abilities in English in mind, Eddington remarked on how, much as he loved it, the English language was inadequate to express one's thoughts. Also on 16 June, Einstein met Rufus Isaacs, Marquess of Reading, probably the most distinguished British Jew, who had been Viceroy of India and briefly Foreign Secretary. Einstein liked him very much.

On 22 June, his final evening in London, Einstein stayed with his friend Abraham Yahuda, who lived just north of Regent's Park. Weizmann believed, incorrectly, that Yahuda had poisoned Einstein's mind against the Hebrew University. Yahuda had apparently insisted on meeting Fritz Haber when he was in London and had launched into a violent tirade, insisting that both Magnes and Weizmann should be thrown out. On 18 June, Yahuda probably accompanied Einstein to a meeting held at the home of Sir Philip Hartog. He was a distinguished university administrator who was to be Weizmann's chosen instrument to square the circle of the future of the Hebrew University in the light of the extraordinary feud between Einstein and Magnes. Hartog was already deeply involved with the affairs of the university, being a member of a 'Structure Committee' set up the previous year to try to put the university on a sounder organisational basis. This committee, of which Weizmann was joint chair with Hartog, also consisted of Sir Herbert Samuel and Norman Bentwich, both of whom Einstein knew from his visit to Jerusalem in 1921. It had already done a considerable amount of work. As Weizmann had just left London for the US, Hartog took the lead at the meeting with Einstein.

At the meeting at Hartog's house, which was also attended by Selig Brodetsky, Professor of Applied Mathematics at Leeds University, Hartog listened to Einstein's complaints about the Hebrew University. Einstein said that he would be prepared to re-engage with the university if an independent inquiry were set up to look at the shortcomings of its administration. He stated that if the conclusions of the enquiry satisfied him, he would be willing to accept the directorship of mathematics and physics. Hartog, with Einstein's agreement, volunteered to head such a committee. Hartog cabled a report on the exchanges to Weizmann on 4 July, who, in consultation with Magnes, approved the formation of a 'Survey Committee' under executive action on 25 July.

Einstein had been incandescent with rage at reports that Weizmann had stated in New York that the Hebrew University was Einstein's place and he could then cease to be a wanderer throughout the universities of the world. In a speech at the American Zionist Conference on 5 July, Weizmann, having learned from Hartog's telegram about the 18 June meeting with Einstein, was reported as saying:

> It gives me great pleasure as president of the Board of Governors of the Hebrew University in Jerusalem to announce that Professor Albert Einstein has agreed to give the university his active cooperation in the administration and direction of the Physics Institute there. In accordance with the view of the plans already adopted by the board of governors and in accordance with the view of Professor Einstein, there will be certain reforms introduced in the university which will strengthen the structure. Proposals to that effect will be made to the next meeting of the Board of Governors which will be held in Zurich beginning in August.'

The Jewish Telegraph Agency's interpretation of this can be seen from their headline: 'Dr. Einstein to Head Physics Faculty at Hebrew University, Says Weizmann'. With justification, Weizmann insisted that he had been misinterpreted. However, a variety of confusing reports from the Jewish Telegraph Agency throughout July continued to insist that Einstein would be going to Jerusalem.

Slowly the heat was taken out of this situation. Haber was in close contact with Weizmann and was being encouraged by him to take up a position at his new institute in Rehovoth. Weizmann encouraged Haber to intervene with Einstein to try to moderate his invective. However, in August Haber received a long diatribe from Einstein which reiterated all his complaints about Weizmann and Magnes, characterising the latter as a romantic, incapable and deadbeat American rabbi. Einstein advised Haber to stay well away from Jerusalem unless and until Magnes and his friends had been ejected. He also asked Haber to tell Weizmann that he would only restore his lost trust in him when his deeds matched his words. Weizmann finally managed to meet Einstein in London in October, just before he left for Princeton. He had the impression that Einstein had become calmer and reported that Einstein had also had a friendly discussion with Hartog. The Survey Committee of the Hebrew University was appointed with Hartog as chair, the other members being Dr Redcliffe Salaman and Professor Louis Ginsberg. They made their preparations to depart for Jerusalem in the autumn.

Einstein arrived by train in Glasgow during the evening of 19 June, apparently surprising the authorities who were expecting him on the following day. On the next evening, 20 June, Einstein gave a lecture entitled 'The Evolution of the General Theory of Relativity'[5] to 1,500 people packed into the university's Bate Hall. Although he gave the lecture in English, it seems unlikely that many of the non-academic members of the audience understood much. After briefly noting that it was reasonable for him to talk about the subject since writing about the work of another was better left to professional historians, whereas he was in a better position to understand his own mental processes, he launched into the following:

> After the special theory of relativity had shown the equivalence of formulating the laws of nature of all so-called inertial systems (1905) the question of whether a more general equivalence of the coordinate systems existed was an obvious one [. . .] . The simplest [method to include gravity in the framework of the special theory of relativity] was, of course, to keep the Laplace scalar potential of gravity and to extend the Poisson equation by adding, in such a way as to comply with the special theory of relativity, a term differentiated with respect to the time.'

The incipient closing of Scottish eyelids can be imagined. He went on to discuss the necessity of forming invariant differentials of the second order in the Riemann metric. At least he spoke for only twenty minutes and, interestingly, he did not emphasise the primacy of mathematical beauty as he had done only a few days earlier in his Spencer Lecture in Oxford. If his argument may have been obscure, his concluding words would have been inspiring to the whole audience:

> Once we have recognised the validity of this mode of thought, our final results appear almost self-evident: any intelligent undergraduate can understand them without much trouble. But the years of searching in the dark for a truth that one feels, but cannot express; the intense desire and the alternations of confidence and misgiving, until one breaks through to clarity and understanding, are only known to him who has himself experienced it.

On the following day, Einstein returned to Bate Hall for a more formal occasion. He was photographed in animated conversation, walking beside Eduard Herriot, three times French premier, in the procession of honorary graduands. Both, somewhat incongruously in Einstein's case, were made Doctors of Law. Einstein returned to London, from whence he sailed back to Le Coq on 23 June.

Einstein complained of being unwell on his return, with symptoms relating to his heart and liver. This caused him to postpone a projected visit to Paris to discuss particulars about his appointment to the Collège de France. He wrote to Langevin on 26 June asking him to convey his thanks to the government, since Einstein himself could not write, having not received an official offer. He intimated to Langevin that he feared the worst in Europe since France and England could not agree on a joint policy and the English did not want to see the true situation.

The news that the Nazis had seized his boat on 11 July would have been a blow, although he would not have recognised it from the descriptions in the press. The *Badischer Beobachter* on 15 July described it as a large motorboat.

[5] The booklet published by the Glasgow University printers is identical to the text published in the *Glasgow Herald*, which must therefore have been supplied with a copy of Einstein's remarks in advance.

By the second week in July, Einstein had recovered from his indisposition sufficiently once again to pick up his violin case and travel to Brussels to play music with the Queen. For once, music may have taken second place to political matters. Catherine Barjansky had written to Einstein on behalf of the Queen to ask if he could come soon as King Albert wished to talk to him urgently. Einstein immediately contacted Barjansky to arrange to visit on 13 July. The King's dilemma related to two imprisoned conscientious objectors to military service, whose supporters were pressing Einstein to intervene on their behalf. Immediately after returning to Le Coq the following day, Einstein wrote to King Albert, saying that he had been unable to get the matter of the prisoners out of his mind. In a clear pointer to how the Nazi takeover was resulting in an evolution of his previous pacifist positions, he told the King that, given the threat from Germany, it was clear that the Belgian armed forces were a vital and currently essential protection and in no sense a threat to others. He was against criminalising the individuals refusing service and suggested instead that they be offered alternative service in dangerous and unpleasant tasks, such as in hospitals or lunatic asylums. In no circumstances would he, as someone enjoying Belgian hospitality, take any official position contrary to the Belgian government on the matter.

The conversation with King Albert catalysed important developments in Einstein's pacifist and anti-war views that he had held for so long. Shortly after his letter to the King, he wrote to Alfred Nahon, the French pacifist who had urged him to intervene on behalf of the Belgian prisoners. In this letter, he clarified his new position, that it would be irresponsible to encourage refusal of military duties in countries such as France and particularly Belgium in the circumstances brought about by the current regime in Germany, which was clearly re-arming and preparing for war. However, he qualified this by stating that he still believed in refusal of military service should the international situation change for the better. His old collaborators Lord Ponsonby and Romain Rolland did not consider this to be a credible intellectual position, but Einstein struggled bravely to find a coherent new set of principles that could deal with the advent of the Nazi regime. He seemed to believe that the Nazis would be rapidly undone by their own stupidity. Although many would agree that this was indeed an important contributory factor in their eventual demise, it took much longer than Einstein had hoped. As reported in Chapter 25, by the spring of 1934 he could write to the violinist Yehudi Menuhin with a cogent exposition of his new position with regard to pacifism.

It is easy to be astonished by the startling naivety of idealists such as Ponsonby and Rolland, who considered Einstein to be an apostate. With the benefit of hindsight, it can be seen that Einstein had the clear-sightedness to appreciate the true nature of the Nazi regime and the courage to disappoint his long-standing friends and collaborators by changing his mind about a crucial issue. As he noted in a letter written in September: 'To prevent the great evil, it is necessary that the lesser evil – the hated military – be accepted for the time being. Should German armed might prevail, life will not be worth living anywhere in Europe.'

The flood of German refugees washing up in Le Coq showed no sign of abating, as Einstein informed Dima Marianoff in a letter on 16 July 1933. A Jewish doctor, Katz, brother of one of Elsa's oldest friends, had arrived. Einstein wondered if Marianoff could find a place for him in the refugee organisation in Paris. Einstein expressed himself very impressed by Marianoff's hard work for the refugee cause in France and thought that Katz's organisational prowess could be a significant help.

Einstein felt powerless to help the refugees, except by using his influence in individual cases, as he told Gustav Bucky, who had written to him from his new home, New York. Bucky had expressed sympathy at the attacks on Einstein in the German press. It was probably Einstein's growing concern about the plight of particularly the young refugee scientists and his worries about the English averting their eyes from the danger from Germany that induced him to make another ferry crossing to the UK. A somewhat mysterious intermediary[6] had put Einstein back in touch with Major Oliver Locker-Lampson, MP. Locker-Lampson, a bizarre character who had previously flirted with fascism and admired Hitler, had written to Einstein shortly after his return to Europe, offering him shelter in one of his English homes. In June, Locker-Lampson wrote again, once more offering refuge and, catalysed by Yahuda, to arrange British citizenship for Einstein, who politely declined both offers.

Nevertheless, by 19 July, Einstein had decided to accept Locker-Lampson's hospitality and return to England. He made a brief visit to meet, in particular, Winston Churchill. This decision came at some cost, since at his last meeting with Queen Elisabeth, Einstein had agreed to return to Brussels for more music-making on 22 July. In his apology for having to cancel, he regretted that it was always music that had to give way to less pleasant duties and hoped that she would be able to find a replacement who would not have to rely so much on good luck when serving the Muses.

Einstein caught the ferry to Dover on 21 July. He stayed with Locker-Lampson at his country cottage in Esher, Surrey. He was driven twenty-five miles to have lunch with Winston Churchill at his country house, Chartwell, on the following day. Lindemann, already Churchill's *éminence grise* in scientific matters, was also at the meeting. In a letter to Elsa later that day, Einstein characterised Churchill as being both important and clever and that he and his colleagues were well informed and determined to act soon. He was impressed that, mostly thanks to Lindemann's indefatigable efforts, twenty Jewish refugee researchers had been found places in UK. He was also impressed that Locker-Lampson seemed to be genuine and, unusually for a politician, not out to make political capital from helping the cause of German Jews.

Einstein stayed in England for several more days, making contact via Locker-Lampson with other influential politicians. They had tea with the former Leader of the Conservative Party and Foreign Secretary Sir Austen Chamberlain, and paid a visit to former Liberal Prime Minister David Lloyd George. It was at Churt, Lloyd George's country estate, that Einstein signed the visitor's book. Against 'Address', after a pause, he wrote 'None' – an event that so struck Locker-Lampson that he used it in his address to the House of Commons the following day.

On 26 July, with Einstein conspicuous in the Visitors' Gallery in his white summer suit, Locker-Lampson introduced to the House of Commons a Bill under the 'Ten-Minute Rule' entitled 'Nationality of Jews'. Such bills had no chance of becoming law except in the highly unlikely event of the government timetabling their further progress. Locker-Lampson used the opportunity to highlight the plight of German Jews, using Einstein as a prominent example. Locker-Lampson began by saying:

> I am not personally a Jew. I do not happen to possess one drop, so far as I know, of Jewish blood in my veins, but I hope that I do not require to be a Jew to hate tyranny anywhere in the world. I hope I only require English birth and breeding to loathe the oppression of a minority anywhere. It, [sic] is un-English, it is caddish to bully a minority and it is the duty of this House to consider the circumstances of people who are no longer citizens of a State which we recognise. I am not anti-German, and many of us in this House are almost pro-German. But I was one of the few people on this side of the House who, after the War, pleaded for fairplay for Germany. I felt that the Great German people had been misled by

[6] Locker-Lampson was a friend both of King Albert and Abraham Yahuda, either of whom might have urged Einstein to leave Belgium for the safer England for a while.

> their leaders. I thought it was a terrible humiliation that the French should have sent black troops and quartered them in the house of Goethe . . .[7]
>
> [. . .]
>
> The most eminent men in the world admit that he is the most eminent. But there was something beyond mere eminence in the case of Professor Einstein. He was beyond any achievements in the realm of science. He stood out as the supreme example of the selfless intellectual. And to-day Einstein is without a home. He had to write his name in a visitors' book in England, and when he came to write his address, he put 'Without any.' The Huns have stolen his savings. The road-hog and racketeer of Europe have [sic] plundered his place. They have even taken away his violin.[8]
>
> [. . .]
>
> My Bill is designed to promote and extend citizenship in Palestine for Jews deprived of citizenship elsewhere.

The bill was put to the vote and passed but, as expected, made no further progress.

Einstein might well have hoped for some quiet time to work with Mayer on his return to Le Coq at the end of July. However, Abraham Flexner continued to pepper both him and Elsa, whom Flexner seemed to believe had some influence on Einstein, with letters. Although Flexner was on vacation at his remote Canadian holiday home near Magnetawan, Ontario, he reacted strongly against the newspaper reports of Weizmann's remarks about Einstein joining the Hebrew University faculty. He also pressed Einstein to arrive at Princeton by 1 October to take part in a faculty meeting the following day. Einstein wrote on 29 July that Weizmann was simply lying about the Hebrew University position. He also said that he couldn't arrive at Princeton until mid-October as he had various commitments in London. Furthermore, Mayer, whose position Einstein was pleased to note was now confirmed, couldn't arrive until then either, since they couldn't be separated for so long. Einstein pointed out that, in any case all he could say to a Faculty meeting, were he present, was that he and Mayer would be working on semi-vectors and the theory of electricity. In the same letter, he reassured Flexner that he would not be taking UK citizenship, a story that Einstein asserted had been made up by a reporter. In fact, his new friend Locker-Lampson had initiated the story, which appeared in both British and French newspapers. By the end of August, Flexner surrendered and telegraphed that an arrival by mid-October would be acceptable .

On 2 August, Einstein took part in a reception at the Coeur Volant restaurant in Le Coq. Local celebrity James Ensor, who had recently been ennobled by King Albert, took part. Ensor, who lived in Ostend, was a painter of great distinction, although by this time he spent most of his time on musical activities. French Minister of Education Anatole de Monzie also attended the meal. He was to present Ensor with the Légion d'Honneur at a reception in the Kursaal that evening. He presumably took advantage of Einstein's presence to discuss the Paris chair that Einstein had in principle accepted. Others in quite a large party included author Marcel Abraham. Ensor and Einstein did not have much to say to each other, although at the end of the meal Ensor gave a highly florid and laudatory address. It concluded with 'There is only one Einstein who reigns in heaven', which is unlikely to have been appreciated by its subject. However, Einstein apparently enjoyed his meal as newspaper reports characterised him as a gourmet with a particular relish for lobster!

---

[7] The mental disconnect evident between the first part of this paragraph and its end is notable.

[8] One at least was not taken and in fact there is no record of them having stolen any of those stored in Haberlandstrasse.

The Belgian Royal Family were in residence at the Royal Pavilion in Ostend from late July until 10 August. Although Le Coq and the Pavilion were only a brisk walk apart, there is no official record of the Queen and Einstein having met in this period. There is a reference to a concert in which Einstein played the violin in the Queen's presence in the imposing Kursaal on 10 August. However, there is no mention of any such concert in either local or national newspapers or in any Einstein correspondence. Indeed, another concert took place in the Kursaal that evening, professional musicians playing as part of a Beethoven festival. Perhaps Einstein and the Queen made a last-minute arrangement to snatch a couple of hours of making music together before the Queen left for their family holiday in the Swiss Alps later that day. The Kursaal lies between Le Coq and the Pavilion and would certainly have had a good piano. This could be the origin of the reported 'concert'. It would be pleasant to think that this did happen, since Queen Elisabeth and Einstein would never have the opportunity to play together, or even meet, again.

Einstein wrote to his friend, the violinist Adolf Busch, in August to congratulate him on his stance with respect to the Nazi regime. Already on 20 April, Busch, who was a Gentile, had withdrawn from the celebrations to mark the Brahms centenary in his native city, Hamburg. His reason was that Rudolf Serkin, his pianist partner and later son-in-law, had been refused permission to play because he was a Jew. Busch had been a staunch opponent of the Nazis since they first became prominent, withdrawing from Germany to Basel as early as 1927. Einstein congratulated him as being among the wheat rather than the chaff of German intellectuals and artists.

Another musician who joined the flood of Jewish exiles from Germany was Einstein's old friend, the luthier, former doctor, and author Julius Levin. He went first to Paris and then later moved to Brussels. It was he who had removed wood from the belly of two of Einstein's violins with the spectacular results he had noted in Palo Alto. Einstein informed Levin that Queen Elisabeth was very impressed by the violin.[9] At the end of July, he wrote Levin a testimonial on the grounds that his skill was such that he would be economically self-supporting in any country. Throughout August and September, Einstein put him contact with various friends in Brussels and Paris, in particular Professor Jacques Errera, with the aim of setting Levin up in a workshop where he could pass his secrets on to the next generation. Levin, who was over 70 when he fled Germany, made several unsuccessful attempts to establish himself in Brussels. His health gradually broke down and he died there in 1935.

On 15 August Antonina Vallentin arrived in Le Coq. Her presence would have been a balm to Elsa, who was becoming increasingly concerned about her and her husband's safety. Sometime in late August, Einstein broke his silence to his sister Maja. His short letter indicates his uncertainty about his future. He mused about what Princeton would be like, and what he would find when he returned to Europe the following year, if he did return. At least his 'Mayerchen' would be accompanying him, but he regretted leaving his yacht and his girlfriends behind. He was livid that Hitler

[9] Just before departing for Princeton in October 1933, Einstein had his Levin violin evaluated at the famous Hill Brothers establishment in New Bond Street. The report was unfavourable: 'I [and] my partners are unanimous in the conclusion that there is very little to be said in its favour [. . .] a really experienced professional violinist would reject the instrument after a very short trial [. . .] The construction of the instrument is faulty, being ill-conceived both as regards shape and work [. . .] In conclusion let me beg you to dismiss from your mind the idea that this violin is an improvement . . . we can only assume that the enthusiast in whom you are interested has never lived with really fine fiddles, otherwise, it would be difficult to understand how he could have been so misled! I am, Yours faithfully, Alfred Hill.' It can be assumed that Einstein did not forward this evaluation to Dr Levin.

had seized his yacht. He concluded by musing that there were only two fixed points. One was the stars; the other, significantly, was not physics, but mathematics.

On 23 August, Einstein wrote to Locker-Lampson to thank him for his hospitality during his recent visit to England. He had been able to enjoy some relaxing and quiet days in Le Coq. However, his equanimity continued to be disturbed:

> Out of all the terrible things taking place in the world today, what disturbs me the most is that there is as yet no decisive condemnation of German rearmament. I cannot free myself of the thought that every day of temporising without any action will eventually cost the lives of thousands of innocent people.

It was not long before Einstein's quiet days in Le Coq were ended. On 30 August, the Jewish philosopher and writer Theodore Lessing was assassinated by Nazis in Marienbad, Czechoslovakia. As a long-standing critic of both von Hindenburg and Hitler, he had fled there when the Nazis had taken power. He was shot through the window of his study while writing. There was no doubt who was responsible for the atrocity and since Einstein was an even more prominent critic of the Nazis, the vague threats against him that had catalysed his police protection became suddenly much more concrete. This was exacerbated by his support of the *Brown Book of the Hitler Terror*, which was published on 1 September. It caused a sensation by detailing the Nazi's responsibility for the burning of the Reichstag and many of the outrages being perpetrated in Germany. Einstein was not directly involved in the preparation of the book and was mentioned in it mostly in the context of the Nazi's effect on German science. However, he had agreed to chair the international group supervising its publication. Furthermore, it reprinted his statement made on arrival at Le Havre, in which he characterised the Nazis as a psychological disease that all friends of civilisation should strive to eliminate from the world.

Early in September, both Einstein and Elsa wrote to their friends and namesakes the Alfred Einsteins, who had fled Germany for Brussels. At that stage, Einstein was determined to remain in Le Coq, where he hoped to greet his distant cousins, whom he urged to remake their lives in the US since Europe had little future. Elsa also pressed them to come to Le Coq quickly, as she was convinced that Einstein would have to flee to England. In a letter to her friend Vallentin, Elsa admitted that she was so afraid that she did not sleep or even get into bed at night, but lay on top of the sheets, jumping at every sound from outside. She had heard (probably via an article in *The New York Times*), that the Nazis had offered 20,000 marks for Einstein's head. By 7 September, Einstein was sleeping in a neighbouring house that offered better opportunities for defence against assassins. He did not go outside without his police protection.

At some time in early September, Grete Lebach, whom Einstein had been so desperate to meet a few months previously, paid a visit to Le Coq. Elsa's suspicion of their relationship, which had been extreme in Berlin, had not lessened. On 10 September Einstein felt it necessary to write to her to reassure her that the relationship between him and 'Frau L.' remained, as always, within the appropriate limits.

Another female visitor, with whom appropriate limits seem not to have been observed, was an old friend from music-making in Einstein's Zurich days, Antonia Stern. She made her way to Le Coq, probably between mid-June and September. The evidence that she and Einstein had a brief affair is circumstantial but, given Einstein's track record, quite persuasive. A letter from him in June is relatively cool and uses the normal *Sie* form, although it seems to be reacting to a rather intimate note from Stern. The next letter, written four months later, on the day before Einstein embarked for New York, uses *Du* and referred to the 'women's war' that Antonia had ignited in Le Coq, about which Dima Marianoff had given him an update. He proposed they draw a veil over it, which he remarked would hide a harmless, if rather roguish, grin on his part when they met again. He also proposed that she should gain some weight, around her ribs, which makes the world a little more comfy. They continued to correspond infrequently until a gap after 1939, although the use of *Du* had ceased by 1936.

At least partially motivated by the strain his presence was putting on the Belgian police, as well as perhaps the female members of his family, Einstein decided to leave Le Coq and rely on his new friend Locker-Lampson. With only one day's notice to Locker-Lampson, Einstein left Le Coq on 9 September, boarding the Ostend ferry for Dover accompanied by his violin and a reporter working for the *Daily Express* who had interviewed him the previous day. Being about to be reunited with both her daughters, Elsa chose to remain in Le Coq. From Dover, Einstein travelled first to Locker-Lampson's house in Westminster. To keep his whereabouts unknown, Locker-Lampson prevailed upon his former cook to host Einstein that evening. On the following day, he was taken to an isolated shooting lodge that Locker-Lampson owned on Roughton Heath on the Norfolk coast. There were numerous stories in the UK press that indicated he was near Cromer. Such reports were hardly ideal for someone trying to secrete himself from a threat of assassination. Locker-Lampson provided a bodyguard of sorts, consisting of himself carrying a rifle and a gamekeeper with a 12-bore shotgun; supplementary protection was provided by two attractive female secretaries.

Conditions at the little group of three cabins in the fens were basic but Einstein's wants were few. One of the cabins contained a piano. He settled down to work and soon appreciated the peace and quiet. He remarked to his friend Yahuda that he had not had so much blessed peace and quiet for decades. This was clear from his dating his letter as 'God knows when'. He went to the village shop occasionally and received a very few guests. Einstein recorded how grateful he was to Locker-Lampson for turning away the hordes of bishops, ministers and other potential visitors. One who got through this barrier was Walter Adams, secretary of the newly formed Academic Assistance Council. This organisation had been founded in April by William (later Lord) Beveridge, with the aim of helping academics to escape persecution.[10] Lord Rutherford had agreed to be its President. Einstein agreed to talk at the first meeting of the Council, to be held in London in October. As soon as he had agreed, Locker-Lampson swung into action and took over the organisation. His first move was to book the enormous and prestigious Royal Albert Hall.

Ilse and Rudi Kayser had joined Elsa in Le Coq from Berlin shortly after Einstein's departure. He tried to dissuade Rudi from visiting him in England and suggested that they could communicate perfectly well by letter. It seems that Kayser had the idea of writing another book about Einstein, who, predictably, was not enthusiastic. Sometime before 27 September, Kayser came to see Einstein anyway, staying in a nearby hotel. He advised Einstein to remove any direct reference to the Nazis from the remarks he was preparing for the Albert Hall meeting.

Paul Ehrenfest wrote to Einstein for the last time around the start of September. It was a brief letter in which Ehrenfest advised Einstein on the appropriate contribution for a presentation to de Sitter on the twenty-fifth anniversary of his appointment to his chair in Leiden. He apologised that at that moment he couldn't write a proper letter. He was in fact prostrate by depression. His marriage had disintegrated, partially because of his wife's long absences in Russia and partly because of an affair with a woman based in Amsterdam, which he had promised his wife to give up as a condition of her not asking for a divorce. He proved unable to fulfil this condition and himself began divorce proceedings in July. On 24 September, he wrote to his first graduate student, Jan

[10] The Council still exists, now known as the Council for Assisting Refugee Academics (CARA) and is active, for example, in helping academics fleeing from the war in Ukraine in and after 2022.

Burgers, asking him to take care of his son Pawlik, who was just starting to study physics. The following day he visited another former student in Amsterdam before going to the clinic where his fifteen-year-old son, Wassik, who had Down syndrome, was institutionalised. His care at the clinic was very expensive and Ehrenfest was desperate not to saddle his children with 'having to work themselves to the bone just to keep their idiot brother alive.' He drew a pistol and shot Wassik in the head and then killed himself.

It is unclear when Einstein heard of the tragic end of his friend. One of Ehrenfest's children would presumably have written to Le Coq. He knew about it by 1 October, when he wrote to Heinrich Zangger that 'Alsberg[11] has sadly killed himself; as has my dear old friend Ehrenfest in Leiden.'

Dima Marianoff described a visit to Einstein towards the end of his stay in Cromer. Marianoff made the trip in order to try to support himself in Paris via some journalism. He asked Einstein to give him a popular explanation of his theories, resulting in an article that was published in November in *L'Intransigeant*. It is worthwhile quoting from this as representing Einstein's beliefs at this time, which Marianoff stated that he recorded verbatim:

> My essential problem, he said, and the aim of my research is to reconcile quantum theory with the principle of relativity. This is extremely complicated, because the methods of quantum theory, which have been really fruitful (Schredinger (sic)), in their present state do not satisfy relativity. Rather, they employ the principle of action at a distance, in direct contradiction to the principle of relativity, which explains so clearly the phenomena of electromagnetism.
>
> The current theory of relativity is based on the notion of the field, which was introduced into physics by Faraday and Maxwell. All my efforts have been directed towards explaining the phenomena of the quantum and atomic structure of matter using field theory. My work in recent years has been mathematical, related to this goal, but which also has implications for the theory of knowledge.

The last recorded visitor to Einstein's rural idyll was the famous sculptor, Jacob Epstein. Having arranged with Locker-Lampson to have a week of sittings with Einstein, Epstein arrived at the camp on 27 September. He was excited by the potential of Einstein's appearance, noting his wild hair floating in the wind and his resemblance to the ageing Rembrandt. They used the cabin containing the piano for their sittings, which gave Epstein so little room that he asked for the door to be removed. The strapping secretaries obliged. Once he had arranged with Einstein to not veil his features with smoke from his pipe, Epstein made good progress. In his autobiography he noted:

> At the end of the sittings he would sit down at the piano and play, and once he took a violin and went outside and scraped away. He looked altogether like a wandering gypsy, but the sea air was damp, the violin execrable, and he gave up. The Nazis had taken his own good violin when they confiscated his property in Germany.
>
> Einstein watched my work with a kind of naive wonder and seemed to sense that I was doing something good of him. The sittings unfortunately had to come to a close, as Einstein was to go up to London to make a speech at the Albert Hall and then leave for America.

It can be seen that Epstein shared the Hills's estimation of the quality of Einstein's violin. He also considered the Einstein bust one of his best works; it was acquired for the nation in 1934 and exhibited at the Royal Academy Summer Exhibition that year and in the Tate Gallery. A copy was donated to the University of Cambridge Fitzwilliam Museum by Lady Jeans, wife of Sir James Jeans, the distinguished astronomer.

[11] Max Alsberg was a famous lawyer in the Weimar Republic who, as a Jew, had been debarred from practising in Germany by the Nazis. He fled to Switzerland, where he committed suicide on 12 September 1933.

During Epstein's stay, Einstein took an afternoon off to pay a visit to Sir Samuel Hoare, an influential Conservative politician who was at that time Secretary of State for India. Hoare was sympathetic to providing refuge for Jews fleeing Germany and would subsequently as Home Secretary play an important role in the evacuation of Jewish children to the UK from Germany in the famous Kindertransport. No doubt they discussed the situation of German Jewish scientists over tea. Einstein does not seem to have been accompanied by Locker-Lampson, presumably as he and Hoare were at daggers drawn over India policy.

Einstein travelled back to London with Locker-Lampson and Epstein on 1 October. Locker-Lampson busied himself with arrangements for the public meeting in the Albert Hall to be held on 3 October. He was surprised to receive a telegram from Abraham Flexner forbidding Einstein's participation in the meeting. He decided not to tell Einstein about this remarkable communication, which Einstein would certainly have ignored. He only heard about it a few days later just before he sailed for the US.

The *Daily Mail*, under the Nazi sympathiser Lord Rothermere, had been agitating against both this event and Einstein personally, whom it labelled as a pacifist and a communist. Einstein, perhaps encouraged by the very right-wing Locker-Lampson, issued a press release in mid-September in which he insisted that he was not and never had been a Communist:

> In my opinion any Power must be the enemy of mankind which enslaves the individual by terror and force, whether it arises under a Fascist or a Communist flag. All that is valuable in human society depends upon the opportunity for development accorded to the individual.

The *Daily Mail* professed to sympathise with the Jews in Germany but disingenuously argued that pointing out the Nazi atrocities would be unlikely to improve their situation. The Nazis were incandescent about the *Brown Book*; Einstein's position on the committee that had approved its publication was enough for them to blame him for all its censures. The tension was increased when a threat to Einstein's life was reported to the police. As a result, a large contingent of uniformed officers surrounded the Albert Hall on 3 October and a considerable number of students were recruited to act as stewards within.

In fact, the meeting passed off without trouble. An astonishing attendance of 10,000 was recorded by *The New York Times*; the tiers of balconies around the dome must have been packed with standing attendees. Rutherford opened proceedings by naming the four organisations that had come together to sponsor the meeting and said that its purpose was to collect a fund to be used for the relief of students, university teachers, and professionals. He outlined the seriousness of the problem in terms of numbers and said that the main contribution should be financial and the provision of refuge in universities for scholars faced with destitution.

Einstein was loudly cheered as he rose. He spoke in a quiet, accented, but perfectly understandable, English. After paying tribute to the British people for their support, he disarmed the criticisms of those such as Lord Rothermere by saying 'It cannot be my task today to act as the judge of the conduct of the nation which for many years has considered me as her own.' He talked about the blessings of freedom:

> We are concerned not merely with the technical problem of securing and maintaining peace but also with the important tasks of education and enlightenment. If we want to resist the powers which threaten to oppress intellectual and individual freedom we must keep clearly before us what is at stake and what we owe to that freedom which our ancestors won for us after hard struggles. Without such freedom there would have been no Shakespeare, no Goethe, Newton, Faraday, Pasteur or Lister.

After extolling the virtues of the solitary life that he had just led in Suffolk, he recommended such an existence to others, suggesting that some of the fleeing young scholars could be given occupations such as lighthouse keepers. This suggestion was to provide much raw material to newspaper cartoonists. Einstein's peroration was:

> Only through perils and upheavals can nations be brought to further developments. May the present upheavals lead to a better world. Above and beyond this valuation of our time, we have this further duty to care for what is eternal and highest among our possessions – that. which gives life its import and which we wish to hand on to our children – purer and richer than we received it from our forebears. Toward these purposes you have effectually contributed with your blessed services.

In fact, one version of Einstein's speech had contained much more pointed criticism of German rearmament, which he was prevailed upon by Rudi Kayser and others to excise. The speech as given was published verbatim by many newspapers, including both *The New York Times* and *The Times*. Parts were reused in a pamphlet published shortly afterwards entitled *Europe's Danger – Europe's Hope* by the Friends of Europe organisation. This was dedicated to warning of the danger to peace from Germany. In this pamphlet, Einstein was much more explicit in criticising the German government than he had been at the Albert Hall. He used it to return to his conviction of the importance of collective security, concluding:

> I know very many people who are willing to approve in principle opinions such as have been expressed here, but who will fall back the next moment into the ruts of ordinary political action. But he who is seriously convinced that the future growth of our civilisation is bound up with the preservation and development of the human personality, must be prepared to make sacrifices for his convictions. The sacrifice I am thinking of is the partial abandonment of State sovereignty by the separate States, and the yielding of the principle of egoism in favour of international security. Here lies the hope of Europe and the Western world.

Einstein spent the remaining three days of his stay in England in London.[12] On one evening he visited his friend Yahuda's wife, although Yahuda was away at the time. It was likely during this period that he met Weizmann, which resulted in a partial repair of their fractured relationship. He also arranged to meet Alfred Einstein. Another old friend came to London in an unsuccessful attempt to see Einstein; Lindemann, who had been unable to attend the Albert Hall meeting, was thwarted by Locker-Lampson, who seemed to have become very proprietorial over Einstein. Lindemann probably wanted to be reassured that the newspaper reports of Einstein's accepting numerous professorships would not preclude his arrangements with Oxford.[13] Einstein did meet Rabbi Perlzweig, chair of the World Union of Jewish Students to discuss the plight of displaced Jewish students from Germany. On 7 October Einstein took the boat train to Southampton, apparently accompanied by Yahuda, where he joined the Red Star liner *Westernland* for the trip to New York. Elsa, Dukas, and Mayer had already boarded at Antwerp.

[12] Marianoff reported that he and Einstein travelled back to Ostend after the Albert Hall speech. His report of an enormous storm they passed through is graphic, but such a trip is ruled out by other known facts and dates. His account of the Albert Hall event is also atmospheric but riddled with inaccuracies; for example, he states that Ramsay McDonald was on the platform. McDonald was at that time Prime Minister and was certainly not there.

[13] Yahuda had indeed written to Lindemann in August, apparently feeling himself authorised to speak for Einstein, clarifying that Einstein had only accepted positions in Madrid and the Sorbonne and those at a salary reduced to reflect his brief projected stays there.

Once again, Einstein's trip across the Atlantic, which would turn out to be his final crossing, was disturbed by concerns and trepidation. Einstein was uncertain about whether he would find Princeton congenial. He also heard from Elsa that Ilse was planning to return to Berlin. Einstein wrote a very concerned letter in which he pointed out that as Einstein's daughter she was in severe danger, especially after the furore over the *Brown Book*. He urged her not to cross the German frontier. On 14 October, he wrote to the representatives of the organisation that had run the Albert Hall event to congratulate them on its success and suggest the names of several deserving recipients of their help. They were preoccupied with trying to extract the proceeds, which apparently exceeded £8000, from Locker-Lampson's eccentric grip.

Einstein was also concerned about Mayer's future. Despite the concession from Flexner that Mayer's position at Princeton would be permanent, he wrote to Ambassador Pérez de Ayala submitting Mayer's name for a Professorship that was associated with Einstein's position in Madrid. He suggested that were Mayer appointed, he could arrive in Madrid in advance and prepare for Einstein's arrival in spring 1934.

The *Westernland* sighted land on the morning of 17 October after a passage with only three days of rough weather. Two trustees of the Institute of Advanced Study came on board via a tug at Battery Park. Einstein was pictured walking down the gangplank carrying his violin, with Elsa, Mayer, and Dukas following him. By once again disembarking early, they avoided the press scrum and reception party awaiting them in New York. That the reception party was led by the Einstein's old friend Samuel Untermyer as well as Mayor O'Brien cut no ice with Flexner, who was determined to quarantine Einstein from any press contact. On 16 October, he had cabled to Einstein onboard the *Westernland* that Untermyer was involved in a political quarrel involving the forthcoming Mayoral election and that Einstein's dignity and that of Institute was endangered. Already on 26 September he had telegraphed to Le Coq that it was essential that Einstein make no statement of any kind to the press when arriving in the US, as he was concerned that American Nazis might target him. The party was driven quietly to Princeton, where they took rooms in the Peacock Inn. Immediately on arrival, Einstein wrote to his cousin in Madrid, Kuno Kocherthaler, who had long been in charge of various funds that Einstein had salted away outside Germany. Einstein referred to the investment of 30,000 marks, equivalent to roughly $10,000, a very considerable sum in those days. Margot, meanwhile, had remained alone in Le Coq. In the next few months she began to study with a sculptor in nearby Bruges. Having written his letter, Einstein seems to have strolled down to the local shop where he bought a newspaper. His life in Princeton had begun.

Chapter 25

Life and music in Princeton and New York (1933)

Hardly had the Einsteins checked in at the Peacock Inn in Princeton than Elsa began to write to her friends. On the day after their arrival, 17 October 1933, she wrote a letter to the 'other' Einsteins, the musicologist Alfred and his wife Hertha. They had been reunited in Le Coq a few months before, and the two men had also met in London just before the Albert Hall event. They had clearly discussed their options in a future in which both were forced to leave Germany. Elsa remarked on the very high opinion that her husband had of his namesake and how they could be sure that he would do anything he could to help them.[1] Elsa's first impressions of Princeton were that of a university town reminiscent of England. Despite their being happily ensconced in the cosy Peacock Inn, all was not well with the party. Elsa reported that 'Prof. Chum' (i.e. Mayer) was indescribably unhappy and had had enough of working with Einstein. He was happy to accept a drop in salary and a higher workload in order to become a professor independent of Einstein, who, as mentioned in Chapter 24, was attempting to translate his academic entanglement in Spain into a position there for Mayer. Elsa concluded by asking Hertha's help in dissuading Ilse and Rudi Kayser from returning to Germany in November.

One of the Einsteins' first visitors was Otto Nathan, a German Jewish economist who held an associate professorship in Princeton University. As one of the very few German Jews on the faculty,[2] Nathan had felt it appropriate for him to introduce himself to the Einsteins and welcome them to Princeton. Nathan was to become one of Einstein's closest associates and indeed became one of his executors. His fellow executor, Helen Dukas, thought that in the early days in Princeton they would have been lost without Otto Nathan.

At the end of October, Elsa shared her opinion of Princeton with her friend Antonina Vallentin. She compared it directly with Oxford and Westminster, including the many bells that rang over the town. Princeton appeared entirely a park, with impressive trees that now showed their full autumn colours. By then, the Einsteins had moved out of the hotel into a house, 2 Library Place. Elsa thought their new home wonderful, a century-old patrician's house surrounded by a large garden. She worried that everything was so pleasant that it couldn't possibly last. Indeed, she was concerned by the fact that there was no word from either of her daughters, Margot still in Le Coq or the Kaysers, who were now living, temporarily, in Amsterdam.

Meanwhile, Einstein was also gaining his first impressions of Princeton. His primary concern was not the scenery or the size of their garden, but to what extent he could work uninterrupted.

---

[1] Alfred and Hertha moved to Italy, where they lived until they finally arrived in New York in January 1939.

[2] There was an unofficial *numerus clausus*, or limit, on the number of Jewish students at around 3%, the percentage of Jewish inhabitants of the US. It was applied by surreptitious means such as rejection of candidates with obviously Jewish names. The termination of Nathan's professorship at Princeton in 1935 was widely ascribed to anti-Semitism.

Einstein. Brian Foster, Oxford University Press. © Brian Foster (2026). DOI: 10.1093/oso/9780198794875.003.0025

Some sort of attempt to satisfy the press seems to have been made by the organisation of a photo call at 11:30 a.m. on 18 October. After the assembled reporters had waited for an hour, they were told by Professor Eisenhart, the university's Dean of the Graduate School, that Einstein refused to appear for the cameras. He must have relented afterwards however, as the existence of a photograph of a tousle-haired Einstein dated 18 October testifies. Apparently only a few photographers were admitted later that afternoon. This partial cooperation seems to have had a positive effect, as further press coverage was rather restrained, although a steady stream of pressmen and uninvited locals continued to knock on the Einsteins' door. Abraham Flexner, the Director of the IAS, was determined to maintain Einstein's privacy, advising Elsa that she should not even bother to answer the invitations that showered in on them. Some however could not be ignored. The new president of the university, Professor Dodds, invited them to dinner, leading to a scene witnessed by Otto Nathan in which Einstein refused to dress despite Elsa's entreaties. Eventually he gave way, although, as he had in Oxford, effecting a private and inconspicuous rebellion by wearing no socks. This event may have been that reported as taking place on 1 November, a dinner in Einstein's honour at which he and the other professors of the IAS, Oswald Veblen, James Alexander, and John von Neumann, met the Bamberger siblings. James Bamberger and Caroline Bamberger-Fuld had donated the $5M that funded the institute.

As the weeks passed, Einstein's initial apprehension about the US in general and Princeton in particular was assuaged. On 20 November, Einstein wrote to Queen Elisabeth. Before describing his impressions of Princeton, he excused himself in an unusual moment of self-contemplation:

> I would have written to you long ago, had you not been the Queen. Although I really don't know why this should be an obstacle. Answering this is however a matter for psychologists, for we others prefer to look outwards rather than into ourselves; because in the latter case we see only a dark hole that tells us absolutely nothing.

He commented that he had been careful to avoid publicity since arriving, not because he was afraid of his own safety, but because he saw no opportunity to do any good. He expressed his gratitude that the Queen had received and helped his friend Levin, the luthier, comparing her kindly behaviour to that found only of queens in fairy tales. He continued: 'Princeton is a wonderful corner of the world, at the same time an exquisitely amusing little village full of Philistines, puny demigods on stilts. I have, however, achieved a pleasant seclusion by various offences against polite society.'

A few days after his letter to the Queen, Einstein wrote to Tete for the first time since his arrival in Princeton. He characterised the town as quiet and beautiful, a little university with know-it-all professors, football players, and continually bellowing students. However, it was better than he had anticipated. He also remarked on the good reports he had had from Frau Sterne about Tete's accompanying her on the piano and that he often thought about his and Tete's musical interactions. He hoped that they would be able to play together again next spring, as he had to give a lecture in Geneva and would come to Zurich.

Einstein's violin playing quickly became another of the talking points among the locals. Halloween fell just a couple of weeks after their arrival. When neighbourhood children gathered outside their door, Einstein appeared with his violin which he proceeded to play for them. Quite whether his audience considered this trick or treat is not recorded. A similar scene occurred around Christmas. Snow was falling when carollers arriving at the Einstein porch. They were greeted by the professor, who asked if he could borrow the violin they were using and accompany them. Of course, they agreed.

One of Einstein's first concerns on arriving in Princeton was to establish with whom he could play music. Already in early November, Emil Hilb, who had entered the Einsteins' life three years previously on their first visit to Caltech, had gathered some distinguished musicians to play string quartets with Einstein. They met in Library Place on 9 November to play quartets by Haydn, Mozart, and Beethoven. The famous virtuoso violinist Toscha Seidel played first violin, Einstein second, Bernard Ocko viola, and Ossip Giskin cello (see Figure 25.1). Seidel had been a pupil of Leopold Auer in St Petersburg at the same time as Jascha Heifetz and Mischa Elman. He played the 'Da Vinci' Stradivarius of 1714. Hilb knew Seidel[3] from Hollywood, where both had worked on movie soundtracks, Hilb as conductor and Seidel as soloist.

Seidel and Einstein got on well at a personal level. Seidel, perhaps understandably keen to ingratiate himself with someone whose fame greatly exceeded his own, went out of his way to intimate a deep interest in relativity. Einstein apparently spent some time drawing diagrams of Lorentz-contracting rods to help his understanding. Seidel invited the Einsteins to spend that most American of festivals, Thanksgiving, at their New York home. Elsa seems to have been less

Figure 25.1 A string quartet at Library Place, Princeton on 9 November 1933. From left to right the musicians seated are: Ossip Giskin, Toscha Seidel, Einstein, and Bernard Ocko. Standing behind Einstein is Elsa Einstein. To her left is Emil Hilb, and to her right is Helen Dukas.

[3] At this period, Seidel lived mainly in New York, although shortly afterwards he moved to California to concentrate on Hollywood work. His 'Da Vinci' Strad sold in 2022 for more than $15m.

thrilled by the virtuoso's company than her husband: 'Herr Seidel was kind enough to invite us to Thanksgiving (30.11) and to stay overnight. It is however too much of a blessing. Herr Seidel is a wonderful violinist but three days of Herr Seidel is a bit too much of a good thing.'

As he acclimatised to his new home, Einstein's thoughts were still often with his colleagues who had either fled Germany or were seeking to leave and had no financial support. On 21 November, he wrote to Lindemann in Oxford asking whether he could renounce his Oxford stipend and if so if it could be used to support German refugee physicists. At this point, he was still intending to return to Oxford in the spring of 1934. On 17 December he replied, thanking Lindemann for organising this. Einstein supported not only fellow physicists. He wrote to Lindemann again a few weeks later in January 1934 to suggest that a place might be found in Oxford for a historian, Professor Mayer from Berlin, who was an expert in the history of socialism. He tried to circumvent Lindemann's political prejudices by assuring him that Mayer was not himself a socialist.

A reminder of the situation from which so many scholars were fleeing can be seen from a lecture given by Nobel Laureate Johannes Stark, President of the German National Physical Laboratory, that November. Stark did not doubt that many Jews had done useful work, provided they approached it with a 'Germanic' mindset, and was reported to have maintained:

> It is the German spirit to strive towards objective truth. Almost all great scientists are German; they are the real creators of science, particularly physics. In contrast, the Jewish spirit is oriented not towards facts but ego. It inclines towards the creation of theories and dogmatism. While the German is satisfied with facts, the Jew marshals propaganda for his teachings. [. . .] But particularly in the last three decades, the Jewish spirit has impressed its dogmatic stamp on physics and mathematics. The exponent is Einstein. The National Socialist government wishes for truth in science. The state's general measures are intended merely to push back the Jewish influence in science. The days when only those who were among the leading Jews would have a chance of becoming a professor are gone.

Einstein's relationship with Flexner came to dominate his early months in Princeton. As Director of the IAS, Flexner was determined to establish it as one of the premier centres of research in the US. His strategy for this seems to have had two components. One, which he managed with outstanding success, was to attract scholars of the very highest level to work there. Veblen was a distinguished mathematician, whose contributions to differential geometry touched on Einstein's interests. Alexander was Veblen's protégé and an accomplished mathematician. In December 1933, Einstein's esteemed colleague Hermann Weyl took up a professorship at the IAS. Although Einstein and Weyl often disagreed, each admired the other's abilities. While Weyl was not Jewish, his wife was. In October, he resigned his chair at Göttingen, formerly that of David Hilbert, in disgust at Nazi policies and finally, after significant wavering, accepted an appointment at the IAS. John von Neumann was appointed when it was thought unlikely that Weyl would come to Princeton. Although not having yet risen to the stratospheric heights he would eventually attain, von Neumann was one of the most promising mathematicians in the world.

The other strand in Flexner's strategy for the IAS was one that is far less easy to understand today. It seemed to focus on keeping the IAS out of the newspaper headlines. As the foremost attractor of headlines in the scientific world, Einstein's every activity became of crucial importance to Flexner. He was concerned at the threat he perceived to Einstein's safety from American Nazis. His advice both to Elsa and to Einstein was essentially to become academic hermits. While this may well have had some attractions for Einstein, it certainly had none for Elsa, who was a social animal who missed her cosy coffee meetings with Berlin friends and relatives. Both Einsteins, however, became alarmed when the extent to which Flexner intended to intrude himself into their lives became apparent. Flexner's attitude was made clear in an internal IAS note: 'I am beginning to weary a little of this daily necessity of "sitting down" upon Einstein and his wife. They do not know America. They are the merest children, and they are extremely difficult to advise and control.'

The first indication of problems after their arrival in Princeton was a letter to Elsa from Flexner on 14 November complaining that Einstein had given an interview to the *Newark Ledger.* Flexner argued that he received as many interview requests as Einstein and had never given a single one nor ever allowed his photograph to be taken. In fact, Einstein had had what he thought was a private meeting with a few young men in the 'Albert Einstein club' in Newark. Unfortunately, the club had given out some details to the press and this had resulted in a full-blown article in the local newspaper. They apologised profusely when Einstein complained about the publicity.

The second half of Flexner's letter concerned the benefit concert that Einstein was preparing to give in New York in aid of scientific refugees from Germany. Flexner wrote:

> My fear is that the proposed benefit concert at the Waldorf Astoria will not be kept quiet and I base this fear that New Yorkers who have no interest whatever in Professor Einstein's scientific reputation have taken the pains to tell me about the concert . . . My belief is that, if Professor Einstein plays at the Waldorf Astoria before a limited audience, all New York will know it in a week, and he will be absolutely besieged with similar requests . . . There is only one course to pursue and that is the course which I have pursued without a single exception since the Institute began, namely, to say 'No' . . .

The mixture of incredulity at the naivety and astonishment at the effrontery of this missive can be gauged from Elsa's reply, written the same day. She told Flexner that her husband knew exactly what he was doing and would not allow anyone, whether it was the institute or its donors, to interfere. She insisted that the concert was necessary and said that, during his entire time as a member of the Prussian Academy of Sciences, none of the Presidents had dared to suggest rules governing his way of life. Flexner's reply to this was also immediate. He suggested that neither of the Einsteins understood America and he summoned the President of Princeton to support his point of view. He considered that Einstein's appearing in the press would only increase anti-Semitic feeling in America and 'The questions involved are the dignity of your husband and the Institute according to the highest American standards.' At this point, Einstein himself joined battle. After complaining that Flexner had not as yet confirmed the pension arrangements for Professor Mayer, he told Flexner that his interference in his private affairs was intolerable, that no self-respecting man could let such a thing pass, and that he had no intention of so doing.

Flexner now beat a hasty retreat, promising from that moment on to cease any intervention with those seeking to intrude on Einstein's time. He could not refrain, however, from two parting shots. The first was to reiterate that all members of the IAS had a duty to the institution and to Princeton University, which meant that some of their freedoms must be curtailed. The second was that, in order to avoid an increase of anti-Semitism, American Jews should be seen but not heard. Einstein probably shook his head in resignation when he read this, recalling for example the tragic experiences of Fritz Haber, who had espoused the same philosophy.

The flurry of angry letters now paused. However, despite his renunciation of further involvement, Flexner continued to interfere behind the scenes with respect to the proposed benefit concert, as will be discussed in Chapter 26. Furthermore, he had already intruded in the Einsteins' private lives in a way that put everything else into the shade.

Rabbi Stephen Wise, well connected politically and an old acquaintance of Einstein's from his first US visit, was concerned that the Roosevelt administration had so far had little engagement with the Jewish question in Germany. He wrote to his old friend Judge Mack to suggest that an invitation to The White House for the Einsteins would be a small way to raise the profile of the

Nazi threat. Mack warned that Flexner would oppose the proposal but Wise persevered, passing the idea to one of Roosevelt's advisers. An invitation to visit the White House was dispatched to the Einsteins, using the IAS address.

It is clear that, at this point, Flexner was intercepting Einstein's mail. On 3 November, he took it upon himself to reply to the President without informing Einstein. Citing his usual concerns of protecting Einstein's ability to work and not inciting the New York Nazis, Flexner declined the invitation. Part of his reply does beg the question as to whether he was completely rational: 'In addition, if the newspapers had access to him or if he accepted a single engagement or invitation that could possibly become public, it would be practically impossible for him to remain in the post which he has accepted in this Institute or in America at all.' Surely part of the reason for Flexner's attitude was his visceral dislike for the president and his policies, but his obsession about publicity is illustrated by his refusal of the donation of a copy of Jacob Epstein's bust of Einstein unless it could be kept secret and exhibited without any danger of publicity.

Einstein would have been entirely unaware of the President's invitation had not Eleanor Roosevelt expressed disappointment to Henry Morgenthau Sr, who knew both Einsteins and found it strange that they would refuse such an invitation. He wrote to Elsa Einstein, who informed her husband. On 16 November, he wrote to Eleanor Roosevelt explaining that they had received no such invitation. He assumed that the invitation was now cancelled but expressed his great desire to meet President Roosevelt and to avoid the bad impression that his apparent refusal would have created. Eleanor Roosevelt replied on 4 December, no doubt to Einstein's private address. The invitation was reissued and was gratefully accepted. They were invited to spend the night of 24 January 1934 at The White House.

If Einstein had been furious with Flexner before, he was now incandescent. On 30 November, Einstein wrote to Flexner and sarcastically declined an invitation to have dinner with the donors to welcome the Weyls to the IAS, remarking that 'After the way you behaved towards us last week, you will hardly be surprised.' Shortly afterwards, he began to draw up a memorandum of complaint about Flexner's behaviour that he intended to be sent to the Trustees of the Institute. He listed a whole range of unacceptable behaviour by Flexner, including the telegram to Commander Locker-Lampson forbidding him to lecture at the Albert Hall, the *Newark Ledger* article altercation, a late evening visit to Einstein's house, presumably directly after The White House invitation arrived, advising him not to accept such an invitation were it to be proffered, abusive phone calls alleged to have been made by Flexner to the agency arranging the ticket sales for the New York benefit concert, etc. Einstein concluded the draft by asking for assurances that Flexner's interference would cease; if this was refused, he requested a discussion of how he could withdraw from the IAS with the minimum embarrassment to both parties.

Einstein sent the above draft on 9 December. It is unclear whether it was addressed to the Board of Trustees or to Flexner personally. In either case, Flexner would have seen it first. It seems to have finally catalysed a realisation that he was behaving unreasonably; or he may have been simply frightened that unless he gave way, he would lose his prize researcher. Whatever the reason, he and Einstein had a two-hour conversation on 11 December that cleared the air. Flexner reported to the institute's paymasters, the Bamberger siblings, that Einstein was now thoroughly happy and satisfied. Flexner clearly saw their meeting through rose-coloured spectacles. In the next month, January 1934, Einstein received a question from Flexner as to what he should put in the *IAS Bulletin* about his activities for the coming year. Einstein responded by asking what a woman would say in response to a request that she should give a programmatic declaration for the children she expected to produce in the next five years? He suggested that Flexner write in the bulletin that Einstein would carry out basic research in the quantum theory of fields. Despite desultory sniping of this nature, Flexner and Einstein pursued an uneasy truce for the next few years, with hostilities only recommencing in earnest towards the end of the decade.

At the same time as he was dealing with one academic opponent, Judah Magnes once more intruded on Einstein's attention. The Hartog committee arrived in Jerusalem in November 1933. Hartog said that he had never worked so hard in his life as during the succeeding two months. Recovering his strength by the consumption of a large number of oranges, he interviewed many of the staff of the Hebrew University and travelled to various outstations to see some of the research being conducted.

The Hartog report was published in April 1934, although Einstein received his copy early in 1934. Its recommendations included improvements in the situation of the junior staff and a clear division between the administrative and academic sides of the university. No sooner had the report appeared than Magnes began to cultivate the American contingent on the board of governors, always his strongest supporters, in order to, as Chaim Weizmann commented, bury it. A special meeting of the American governors was called in New York on 12 May. Weizmann, who was forbidden to travel by his doctors because of exhaustion, pleaded with Felix Warburg that the American governors should not take up any firm position until the meeting of the full board in the summer. Weizmann was therefore startled to learn that the meeting had decided to reject the report, succumbing to pressure from Warburg and Magnes. Weizmann reported that:

> Magnes himself proposed the expression of confidence in himself, which was unanimously adopted. All our people – Ratnoff, Kalisky, Wechsler etc. – broke down and submitted to Warburg and Magnes. I received a cable from Warburg that their decision was unanimous and that they had all signed the Minutes. He suggests that the European members should also sign, and in this way a meeting of the Board of Governors will not be required this year. In order to satisfy Einstein, Warburg first of all invited him for lunch and then a compromise was reached according to which an academic committee consisting of Bergmann, Fekete, Epstein, and Adler[4] would be elected, to assist Magnes, giving Magnes the right of veto. This commission is just a smoke-screen; they aren't independent people and Magnes' veto could always annul their decision, even if they allowed themselves to take any.
>
> [...]
>
> Einstein attacked me at the beginning for not throwing Magnes out, but Warburg only had to pat him for everything to change. He's changing his mind for the fourth time during the past year and a half.

Einstein also had to explain his *volte face*, shamefacedly, to his friend, Magnes's implacable opponent, Abraham Yahuda.

Magnes illustrated once more what a dangerous man he was to cross by taking the opportunity of the American governors' meeting to attack Weizmann's proposition for his new Sieff Institute in Rehovot. The governors censured Weizmann for supporting a private institution rather than the adjacent agricultural outstation of the university. Weizmann had been careful, as he had intimated to Einstein, to ensure that the institute was independent of Magnes's control. Magnes apparently accused Weizmann of stealing the university's money to fund his institute.

In his comprehensive survey of Einstein's involvement with the affairs of the Hebrew University, Ze'ev Rosenkranz observes:

> Ironically, Einstein, for whom the Survey Commission had been established, now proposed a compromise under the influence of the Americans which would weaken the Hartog Report's demands. He suggested that the responsibility for academic affairs should be transferred to a committee of four senior faculty members with a veto right for Magnes. This was proposed because of the fear of Weizmann

[4] All Professors in the Hebrew University.

> becoming sole dictator of the university. The proposal met with enraged reactions from the European Governors and Einstein finally withdrew it. The obvious switch in Einstein and Weizmann's views is also ironic: In 1933, Weizmann advocated retaining Magnes as Chancellor under pressure from the American Governors. At that point, Einstein demanded Magnes' removal. A year later, they had completely reversed their positions.
>
> The solution to the internal crisis of the university was finally achieved in two stages: At the end of 1934, Magnes lost his 'dictatorial' powers and his authority was transferred to the new Executive Committee. A year later, the post of Chancellor was abolished and Magnes was appointed to the mainly representative post of President. Hugo Bergmann was appointed as the first Rector, i.e., its first Academic Head. As Einstein's original demands had been met, he agreed to return to formal involvement in the university's affairs.

With the appointment of Magnes to the largely honorific position of President, Einstein finally had won. The price he paid, however, was a heavy one in terms of the perception of his judgement and reliability. Friends of long standing, such as Weizmann, considered that he had not behaved as a gentleman and acted as if deranged, while Fritz Haber considered him to have been unbalanced.

With all of the above strife to distract him, it is amazing that Einstein found any time for his research. However, he had the ability to exclude the inessentials and return to his work despite distractions. Pais recounts a story told to him by Helen Dukas that illustrates this capability, albeit somewhat later in his American years:

> During a speech by a high official at a major reception for Einstein, the honored guest took out his pen and started scribbling equations on the back of his program, oblivious to everything. The speech ended with a great flourish. Everybody stood up, clapping hands and turning to Einstein. Helen whispered to him that he had to get up, which he did. Unaware of the fact that the ovation was for him, he clapped his hands, too, until Helen hurriedly told him that he was the one for whom the audience was cheering.

The last paper that Einstein published with Walter Mayer was received at the *Annals of Mathematics* office on 8 November, so represents work begun in Europe but concluded in Princeton. The title, 'Representation of Semi-vectors as Normal Vectors Subject to a Special Form of Differentiation,' indicates that it continued Einstein and Mayer's programme of exploring semi-vectors (essentially spinors) as a mechanism for application to a variety of problems, including general relativity and the solution of the Dirac equation, on which they had published a few months earlier. There is no physics involved, simply mathematical manipulation, at times invoking that other old favourite of Einstein's, distant parallelism.

As noted earlier by Elsa Einstein, Mayer was by now desperate to escape from Einstein's tutelage. It is unclear why this occurred at this particular juncture, but it may be that Mayer had been unhappy for some time. Now he saw the first opportunity for escape thanks to Einstein's efforts with Flexner, which meant that Mayer had a permanent position at the IAS. He could follow his own independent research directions, and he lost no time in so doing, leaving Einstein to do his own calculations on unified theories. This would certainly have caused a major upset to Einstein's work as he had come to depend on Mayer not only as a calculator, as he called him, but also a sounding board for his ideas. He also seems to have become closer to him than any other of his collaborators since Grossmann. The combination of Mayer's defection and the multitude of other activities that Einstein undertook meant that he published no further research paper until 1935, when he collaborated with Nathan Rosen and Boris Podolsky on one of the most influential of all his papers.

On 18 December 1933, the Einsteins attended a banquet to celebrate the centenary of Alfred Nobel's birth. It took place at the Hotel Roosevelt in the presence of three other Nobel Laureates: former Secretary of State Frank Kellogg (Peace, 1929), Sinclair Lewis (Literature, 1930), and

Irving Langmuir (Chemistry, 1932). The main purpose was to celebrate all twelve living American Nobel Laureates and to promote the cause of international peace. The latter was briefly jeopardised by Lewis, who departed the meal in anger at some flash photography from the press, returning only after a twenty-minute sojourn in his hotel room. Having composed himself, in his speech he extolled the recent repeal of Prohibition, of which he seems to have taken considerable advantage that evening, and to excoriate the conservative attitude of American professors of literature. The other nine absent Laureates sent telegrams; these and one from President Roosevelt were read out, the President's by Henry Morgenthau Sr. Kellogg in his address opined that war was not imminent and could be avoided by the development of conciliation and arbitration procedures. The greatest threat to peace was the arms race.

In his speech, Einstein attempted to solve the paradox of how the inventor of dynamite would fund prizes to further science, literature and peace. He found the solution in Nobel's remarkable ability to combine creativity with the will-power and organisation to found a great industry. Nevertheless, Nobel was at heart a creative person:

> Nobel's soul might have been depressed because his most important creative achievement benefited those powers which, as a human being, he considered hostile and destructive. Thus his testament appears as a heroic attempt to place his life work at the feet of the good and life-giving gods, and in this way to dissolve a painful discord in his soul – a deed of the most noble self-liberation.

Einstein had been an inveterate concert attender while in Berlin. The musical life of Princeton was naturally nowhere near as rich, but he began to explore it. As would be expected in a prestigious university town, there were a large number of amateur music groups of all types, some of a high standard, several of which Einstein would subsequently join. A rather special concert was arranged for him on 14 December, when he arrived at the First Presbyterian Church to listen to the Westminster choir, a choral group from the Westminster Choir College based in Princeton, who were about to depart on a European tour that included Russia. Einstein was particularly moved by a choral arrangement of the Largo from Dvorak's 'New World' Symphony, which he recalled having himself played in an orchestra in Berlin. They also sang a Bach Chorale, Singet dem Herrn, BWV 225. Einstein was reported to have shed a few tears of pleasure at the performance and thanked the choir profusely before walking home alone.

If the music scene at Princeton was necessarily provincial, the town is nevertheless handily situated approximately halfway between arguably the two most important music centres in the US at that time: New York and Philadelphia. New York was principally served by the prestigious New York Philharmonic for orchestral music and the Metropolitan Opera. Principal conductor of the New York Philharmonic from 1928–1936 was the towering figure of Arturo Toscanini, a long-time friend of Einstein. His stand against dictators was unequivocal; his Beethoven cycle in the 1933–1934 season represented a defiant signal to the Nazis that Beethoven's music was universal and not to be harnessed for any one country's propaganda. Artur Bodanzky directed the Met for German opera, while Tullio Serafin covered Italian repertoire. The Philadelphia Orchestra at the time of Einstein's arrival at Princeton was experiencing the final death throes of the leadership of Leopold Stokowski, who had been musical director since 1912. In 1930 he had decimated the orchestra's personnel and from around 1934 had engaged in a war of attrition with the orchestra's board. He repeatedly threatened resignation and eventually departed in January 1936.

The pattern of the orchestral subscription concerts was similar to what Einstein had experienced when in Los Angeles: most major concerts were given two performances, one in the evening followed the next day by a matinee. However, concert attendances and income had suffered significantly during the Great Depression. Many orchestras were forced to lay off players, reducing both wages and recording activities. The New York Philharmonic was no exception, reporting a loss of $150,000 from a turn-over of $686,000 by the middle of the 1933–1934 season. Two rounds of salary reductions were enacted, the pension fund was raided, sponsorship from a toothpaste company sought, and a national appeal begun. This was supported by New York's new Mayor, Fiorello La Guardia, and Eleanor Roosevelt among others. Those donating in the name of Toscanini received a personally autographed thank-you note from him; his own salary (of an eye-watering $110,000 in 1931–1932) was also under threat. By April 1934, more than half a million dollars had been collected. However, most of this was absorbed by previous debts, so that serious plans for a merger of the Phil and the Met were drawn up. The plan was eventually vetoed by Toscanini and both institutions struggled on.

A number of the world's greatest musicians and composers arrived in New York directly after Einstein. Fritz Kreisler landed on 27 October 1933, sailing on the same ship as Sergei Rachmaninov. Einstein's old acquaintance from a former transatlantic voyage, author and humourist Hendrik von Loon also arrived that day on another ship. Kreisler played first in New Haven, Connecticut before a Carnegie Hall recital with pianist Carl Lamson on 4 November. Although there is no indication that Einstein attended, it is unlikely he would miss this chance to listen to his old friend. The auditorium was not only sold out, 300 people sat on the stage to listen to a programme that included the Bach Chaconne, Grieg's C minor Sonata, and Chausson's Poème. While the reviewer swooned over Kreisler's musicality and tone, he also noted a few lapses in intonation and hoped that in future recitals, a few more modern pieces might be included. Kreisler gave a repeat of this recital in Brooklyn on 27 November. Rachmaninoff's sold-out Carnegie Hall recital on 9 December was also lauded by the critics, although his transcription of the Bach E major violin partita was characterised as ragged and with un-Bachian harmonies. The remainder of his recital, which included several of his own preludes and his Variations on a Theme of Corelli, as well as pieces by Debussy, Borodin, and Schubert, was much more successful.

Another old friend of Einstein's, the conductor Bruno Walter, arrived in New York. He was engaged to conduct the New York Philharmonic in a series of concerts from October until early December, having left Germany in March 1933 after the cancellation of all his concerts by the incoming Nazi government. At some point towards the end of his stay in New York, he and his wife met the Einsteins. Elsa reported to their friend the violinist Lili Petschnikoff:[5]

> What a splendid man [Walter] is. Albert was again delighted with him. The two played together and had so much fun. I liked the woman better than I used to. The bad times have also changed her and she seemed to me milder and more docile than before. [Einstein and Walter] played wonderfully together. I heard him say to the pianist Friedberg, who was also present in the next room, that [Einstein] was to him a musical miracle. He was quite delighted, I noticed that clearly, with Albert's playing. He couldn't believe it possible that a person who has not studied properly is able to give so much. It's a pity that the two see each other so rarely, otherwise they would become close friends, they are so similar.

A few days after this initial wave of musicians, Arnold Schoenberg and his family arrived in New York. Unlike Kreisler, who was beginning a North American tour and would eventually return to his palatial Berlin home, or Rachmaninoff, who had lived in New York since 1918, Schoenberg was starting a new life in the US. Like Einstein, he had been reluctant to leave Europe but, also like Einstein, he resigned from the Prussian Academy, in his case of the Arts, before he could

[5] Elsa had heard from mutual acquaintances that Petschnikoff had expressed sympathy with the Nazis. Elsa told Petschnikoff that she was sure that this could not be correct, given her close relationship with Einstein, and assured her that their friendship was unimpaired by such rumours.

be ejected. He fled to France in May 1933, where he waited to see how things would develop in Germany. The termination of his contract with the Prussian Academy in September precipitated his acceptance of a position, as he thought, based in Boston. While in France, he made a gesture of defiance to the Nazi regime. In a ceremony in Paris conducted by Rabbi Louis Germaine-Levy on 24 July 1933, Schoenberg returned to the faith of his ancestors, having converted to Christianity in 1898. The declaration was witnessed by Dimitri Marianoff and the artist Marc Chagall, but what the relationship was between the composer and Einstein's son-in-law is a mystery. Perhaps they gravitated towards each other as two newly arrived German Jewish artist exiles. They were not close acquaintances, since, in a letter to his pupil Alban Berg, Schoenberg wrote:

> It wasn't until 1 October that my going to America became the kind of certainty I could believe in myself. Everything that appeared in the newspapers both before and since was founded on fantasy, just as are the purported ceremonies and the presence of 'tout Paris' at my so-called return to the Jewish faith. (Tout Paris was, besides the rabbi and myself: my wife and a Dr. Marianoff, with whom all these dreadful tales probably originated).

Clearly Marianoff, always on the lookout for a way to make money, had sold the story to the Paris press.

Schoenberg arrived in New York from Le Havre on a temporary visitor's visa. His friend Artur Bodanzky, director of the Met, was on the same liner. Perhaps more infamous than famous in Europe, he was at least known there, whereas when he arrived in the US, there was no welcoming committee. However, Fritz Kreisler was waiting for the Schoenberg family at customs. Schoenberg had obtained a position teaching at a 'Conservatoire' in Boston, which turned out to have been founded in the previous year and had only a few pupils. He was required to teach both in Boston and New York; the commuting in the atrocious winter of the east coast was rapidly too much for the composer's asthma and weak heart, so that by January he was too ill to conduct his *Pelleas und Melisande* with the Boston Symphony Orchestra. He was nevertheless keen to meet Einstein, with whom he had carried on a sporadic and slightly bizarre correspondence over the last decade. They finally met in the spring of 1934.

On 8 January 1934, the Einsteins were driven to Mischa Elman's apartment in New York. Einstein and Elman played, no doubt the Bach Double Concerto, accompanied by the famous pianist and composer Leopold Godowsky. Having long wanted to meet Einstein, Godowsky had visited him in Princeton shortly after his arrival. He had been concerned by his ignorance of relativity, but they discussed only economics. Godowsky was enchanted by the scientist and became enthusiastic to help organise Einstein's musical events. Elsa described the scene in Elman's apartment: 'My husband played wonderfully with Mischa Elman. He is a great artist, but I think he was also amazed by my husband's playing. However, the most moved was Godowsky, who is the most sensitive of musicians.'

At the same time as Einstein's musical rendezvous with Elman, another virtuoso was arriving in New York. Yehudi Menuhin, not quite eighteen, began a tour of the US in January 1934. He too had made his statement against Nazism by refusing to play in Germany. He informed reporters that he was vehemently opposed to Hitler. He was scheduled to perform the Beethoven concerto with the New York Philharmonic under the baton of Toscanini. Rehearsals took place in Toscanini's suite at the Hotel Astor with the maestro at the piano. After repeated interruptions from the telephone, which Toscanini asked his wife to intercept, they had reached the beautiful second movement when the telephone rang for a third time. On the third ring, the maestro rose from the piano stool, walked to the telephone installation and ripped it out of the wall before returning to the piano and continuing the rehearsal.

Menuhin's concerts were on 18–19 January. *The New York Times* review compared this performance with his debut with the concerto in New York in 1927, when he was eleven years old. He had clearly gained in the profundity of his interpretation and the reviewer enthused that two musicians of genius, one mature, the other young, had interpreted Beethoven with wonderful mutual understanding. Einstein and Elsa had planned to go to Menuhin's concert, but they were prevented by a delay in the arrival of a visitor. Fortunately, Menuhin's performance of the same piece was broadcast only a few months later on the radio. Einstein responded on 11 March 1934 to a letter from Menuhin to Elsa, which had expressed indignation at Elsa's suggestion that he did not play in support of worthy causes. Einstein wrote: 'The day before yesterday I heard you "on the Radio" playing Beethoven and Bruch – it was a wonderful experience for me, playing that no other living violinist can match.'

Menuhin's letter to Elsa has also asked about Einstein's stance on pacifism, which Menuhin himself espoused. After the compliment reproduced above, Einstein had also tried to clarify his new stance in response to the threat of Nazism and other European dictators:

> I am still the same red-hot pacifist as before. However, I have realised that the remaining European democracies cannot disarm until convincing international guarantees against aggression have been achieved and as long as the German regime constitutes a threat to its neighbours. I am unhappy to find myself in this situation, but nothing is more dangerous than turning a blind eye to a common and threatening reality. It seems to me that too many honourable pacifists seem to believe that disarmament or even calling for disarmament is enough; therefore one must always reiterate that true security can only be achieved by the establishment of an international authority of sufficient strength. This way is not particularly congenial to me, as I detest any sort of force. However, I am convinced that no other effective route exists. When such a security situation is established through the creation of an international 'police force' and has been in being for a sufficient time, then security will be eventually established by familiarity and the police troops can fade away more and more. They can never be wholly dispensed with.
>
> As long as there was no direct threat in Europe from the systematic psychological preparation of war by a dictator who suppressed every pacifist tendency, a simple refusal to serve could have been successful; I myself proposed this until the changed circumstances made this seem for the time being hopeless. Today, however, going in this direction could only result in the liberal states being crushed by the dictatorships.

Menuhin was clearly very excited to receive this letter from his hero Einstein and replied on 14 March while travelling between concerts:

> Dear and beloved friend Prof, Einstein!
> On receiving your letter I experienced a great thrill, that it would be difficult for me to describe! It was like a dream coming true the idea of counting Albert Einstein among my friends! And that in spite of my unworthiness on the contrary because of your modesty!
>
> I am looking forward infinitely to the time when we will meet together quietly and I will be able to benefit by your discourses. I understand your reasoning perfectly, namely that one cannot sit like an ostrich and believe that by shutting one's eyes and ears one thereby also shuts off the peril. However, it seems terrible to start the new run for armaments again and let the big munitions iron coal manufacturers gain power and prestige. If at least France England and the rest were saints whom one could trust, but they are not! However as you said, one must choose and as terrible as it is one must face facts and act accordingly!
>
> Would it be possible for you to attend my concert this Sunday in NY? We would love *nothing* better and Mother will have two seats in her box for you and Mrs Einstein.
>
> Could you give me your picture? I would cherish nothing more and it will be with me and serve as an inspiration wherever I will be.

My parents and sisters join me in sending you our deep love and hoping we will see you and your wife on Sunday.

Please thank Mrs Einstein for me and tell her that we realize our misinterpretation of her letter completely.

Your loving friend,
Yehudi

Unfortunately the Einsteins had guests that weekend and so could not attend Menuhin's concert at Carnegie Hall. It was sold out and remarkable for the New York premier of what was advertised as a newly discovered Mozart violin concerto, the 'Adelaide' in D major. Einstein's namesake, the Mozart scholar Alfred, was sceptical from the outset and later judged it to be a fraud, composed by Marius Casadesus, violinist brother of the pianist Robert. Undaunted by such doubts, Menuhin gave it its first New York performance, accompanied by pianist Walter Bohle. Menuhin also played J. S. Bach's D minor partita, Paganini's first concerto in D major, Op. 6, and two Spanish dances by Pablo de Sarasate.

Einstein's frequent presence in New York that January was predominantly for musical reasons. He not only played with Elman and Godowsky, but also he was in rehearsal. While Menuhin was by now a seasoned performer despite his youth, Einstein, approaching his fifty-fifth birthday, was preparing for his debut as a performing musician on the New York concert stage.

Chapter 26

The New York Einstein concert, and others; Death of Ilse Einstein (1933–1934)

The first written mention of Einstein playing in a charity concert was in a letter from Elsa Einstein to Emil Hilb on 3 November 1933. Hilb's former employment as musical director for Paramount in Hollywood seems to have been terminated, presumably a symptom of the contraction enforced by the Great Depression. In the 1932–1933 season, he directed the Denver Symphony Orchestra but by the autumn of 1933 had arrived in New York, where he took a room in the Windsor Hotel. Elsa had informed him of Einstein's acceptance of the Princeton position in January 1933, which she ascribed entirely to the availability there of employment for Mayer. Apparently without a job himself, Hilb may have decided to hitch his wagon to the great Einstein caravan and to try to exploit his connections therein to raise his profile in New York. Whatever his motives, he and Elsa were the driving force for the Einstein charity concert.

In the 3 November letter, Elsa wrote:

> My husband is completely comfortable with playing in a string quartet. However, I haven't yet been able to convince him to appear as soloist with an orchestra. His modesty forbids it. I hope however that when the musicians play with him on Thursday 9th they will give him more confidence. I am pleased that you have organised such good musicians to come. It will give him real pleasure.

The musicians in question were Toscha Seidel, Leon Barzin, and Ossip Giskin in the string quartet described in Chapter 25. Hilb was also in attendance and no doubt they discussed possibilities for the concert, which was first intended to take place in December at the Waldorf-Astoria Hotel.

Hilb persisted with suggesting that an orchestra be involved in the concert. His next idea was to programme an orchestral piece to avoid Einstein's reluctance to play with an orchestral accompaniment. He suggested Mozart's Eine Kleine Nachtmusik. Elsa however was quick to reject this, claiming that it would take up too large a fraction of the programme. She also betrayed her understanding of Hilb's desire to raise his profile by playing a part on stage in an event guaranteed to have press coverage. With typical tactlessness, she pointed out that had a conductor such as Bruno Walter or Stokowski been engaged, then there would have been no objection.

Hilb took charge of all the arrangements, except for finances, over which Elsa kept tight control. He soon discovered that the Einsteins were demanding associates. An unsigned 'postscript', probably written by Elsa, arrived on 7 November full of criticism for Hilb's proposals for printing, publicity, etc. This was as nothing however compared with his interaction with Abraham Flexner. In the middle of his confrontation with the Einsteins about their public activities, Flexner called the Knabe company, which Hilb had employed to print and distribute the tickets for the concert. Hilb happened to be visiting the company when Flexner rang, probably on 24 November. Overhearing Flexner's apparent haranguing of Neuer, the Director, Hilb took over the call himself. From the letter that Hilb wrote to Flexner a week later, it can be inferred that Flexner had abused the Einsteins to both Hilb and Neuer, neither of whom he knew, taxing them with publicity seeking. Hilb defended Knabe for displaying letters from Einstein, expressing his gratitude for

Einstein. Brian Foster, Oxford University Press. © Brian Foster (2026). DOI: 10.1093/oso/9780198794875.003.0026

the piano that Hilb had arranged should be sent to their Princeton house, in the company office window.

Flexner did not react well to Hilb's letter, claiming that he had never in his life received such an impertinent communication, which he also characterised as replete with misstatements and misunderstandings. He denied having threatened to fire Einstein and objected in particular to the IAS logo on Einstein's letter being posted in Knabe's window. He repeated his usual position that Einstein did not understand America and objected to what he considered Elsa's misrepresentation when she had assured him that the concert would be small and private. He accused Hilb and his associates of giving Einstein bad advice. This fiery exchange however represented the pinnacle of Flexner's overt opposition to the concert, which he realised he could do little about and which was in any case soon subsumed in The White House invitation confrontation detailed in Chapter 25.

At the start of December, it became clear that the original date of 15 December would have to be revised as there was insufficient time to ensure that all the influential, and rich, people necessary could be recruited to attend. Hilb had considerable difficulty in dissuading Elsa from the idea that the concert would somehow organise itself. He was willing to take it on but insisted on having a free hand and Elsa's full confidence. *The New York Times* of 7 December carried a notice of the postponement until mid-January. The official reason given later was that Seidel's concert schedule allowed insufficient time for rehearsals. Hilb meanwhile was already having other problems with the artists. The pianist Harriet Cohen, who had often played for the Einsteins in Berlin, was apparently intent on maximising the publicity for the event and may have been informing the press of the arrangements. Hilb wrote to her in an attempt to reduce her excitement.

Hilb was arranging rehearsals throughout December, mostly in Princeton. At least two took place, one on 15 December. They were rehearsing a chamber-music programme, since Einstein was still reluctant to play with an orchestra, despite Hilb's cajoling. Hilb also acted as an intermediary between Einstein and the various Jewish organisations that were involved in the concert. He asked Einstein to provide a message for the Zionist Society in New York for the *Chanukah* festivities, which were due to begin on 12 December. He also encouraged the Einsteins to attend an event to mark the close of *Chanukah*. The annual Maccabee Festival in Madison Square Garden took place on 23 December. Einstein, who attended as a guest of honour, characterised it as a demonstration of Jewish solidarity. He was amused to be impersonated on stage during a tableau of Jews arriving in Palestine and presented his autograph to the impersonator. Governor of New York Herbert Lehman, who was Jewish, gave a speech in which he reminded listeners that the Macabee Festival commemorated the victory of Jews over the Roman suppression of their faith and that German Jews were currently undergoing a similar trial. He called on all American Jews to support their German brethren. Einstein's benefit concert was aimed at this audience.

On the same day as the above festival, the Yeshiva College of New York announced that it was inviting Rudi Kayser, Einstein's son-in-law, to join the faculty as Professor of German Language and Literature. Einstein had been exerting himself greatly to find a position for Kayser, who at this stage, however, was very reluctant to leave Europe, and did not accept the offer.

On 14 December Hilb told *The New York Times* that the concert would now take place in the grand ballroom of the Waldorf-Astoria on 15 January 1934. Hilb had asked Henry Morgenthau[1] to provide a message to be circulated to likely donors:

[1] Although Hilb is ambiguous as to which Morgenthau is meant here, it is very likely that it is Henry Morgenthau Sr, rather than his son. Morgenthau Jr had just taken office as Secretary of the Treasury on 1

> Professor Albert Einstein desiring to help his needy friends in Germany will participate in a concert to be given on Monday evening, January 15th, at 8.30 p.m. in the grand ball-room of the Waldorf-Astoria. It is to be a Chamber Recital in which a number of prominent artists are taking part.
>
> The cause is so good and the man so prominent and well-beloved, that I urge every one to help him make this concert a great financial success.
>
> Yours sincerely (signed) Henry Morgenthau.

Hilb considered that in order to ensure a full house, this invitation would also need to be given to the press.

By 20 December, however, arrangements had changed again. The ballroom of the Waldorf-Astoria was no longer to be the venue, which was now switched to the house of Adolph Lewisohn, an extremely rich New York Jewish philanthropist who had been born in Hamburg. This seems to have been the decision of Elsa, who had a very low opinion of Hilb's business acumen, considering him to be as unworldly as all artists and that money flowed through his hands like water. She not only thought this, but told Hilb explicitly in a letter in which she also warned him that she was extremely parsimonious when other people's money was involved. Hilb managed to cancel the Waldorf-Astoria booking without penalty, the only cost being the provision of a signed Einstein photograph for the hotel director. Hilb recommended that Henry Morgenthau should be left undisturbed in charge of the publicity and ticket sales, as he did not like interference and would be pleased to do it on his own. Morgenthau's influence, presumably through his son, the Secretary of the Treasury, was also required to reverse a decision of the Treasury Department that the concert could not be held free of federal taxes.

Hilb once again brought up the possibility of Einstein playing the Bach Double Concerto with a small orchestra, which he assured Elsa that he could arrange without charge. He also proposed to find a company to produce a ten-minute film associated with the event, for which he thought they could charge a fee in excess of $4000. Paramount refused to pay more than $2000 but Hilb was able to find a private donor willing to pay the full amount. On 12 January Elsa wrote to Hilb that 'My husband is in agreement with the filming as long as there is a contract in our hands which secures both him but more importantly the $4000'; she further stipulated that the film company must be first class: 'You can imagine how unwilling my husband is to do this, but if we can really get $4K for such a good cause then he feels he has to do it'; she returned to the subject of their visit to Mischa Elman's and continued: 'Tonight we should have gone to Jascha Heifetz[2] – he rang again yesterday – but you can have too much of a good thing.'

Uncertainty about the exact programme continued until the last minute. Elsa was concerned that perhaps Harriet Cohen would have too much to do if she accompanied the Bach Double Concerto, the Trio, and her own solo pieces. Hilb had foreseen this prevarication and arranged with the printer to produce the programme with a day's notice. He travelled to Princeton for a brief conference with Elsa on 13 January. It was brief because the Einsteins had luncheon guests that day and subsequently were entertaining various university professors to dinner. Clearly their social calendar was expanding, despite Einstein's reluctance. Even at this meeting, Hilb's hints

---

January and, submerged by an avalanche of measures trying to alleviate the Great Depression, is unlikely to have had time to organise Einstein's ticket sales.

[2] This reference relates to a concert to benefit destitute German professionals that Heifetz had organised on 12 January at his home in New York. Heifetz played Paganini, Schubert, and Mozart, the baritone Lawrence Tibbet sang a selection including Brahms and Schubert lieder and the pianist José Iturbi accompanied both. The event raised $15,000.

about replacing Harriet Cohen with a small orchestra for the Bach Double Concerto were not acted on.

The Einsteins seem to have been in a state of frequent oscillation between Princeton and New York in the days prior to the concert. They seem to have greatly preferred to sleep in their own beds at night rather than stay over in New York and to have had a car and driver at their disposal. On 16 January Einstein was photographed sitting at a type-setting machine for the *Jewish Daily Bulletin*, which until then had only been printed in a condensed version and only distributed by mail subscription. An expanded edition sold on the street was now to be inaugurated. Einstein issued a statement emphasising the importance of such newspapers at a time when considerable numbers of the Jewish people were in danger. Afterwards he and Elsa attended a luncheon at the Commodore Hotel, where he was the guest of honour. The rehearsal for the concert was scheduled directly after lunch, at 2:45 p.m. at Steinway Hall, about ten minutes' walk from the Commodore.

On 17 January the Einsteins lunched with the Morgenthaus, after which there was a rehearsal at the Lewisohn house from 4–6 p.m. Clearly nervous, Einstein was reported to have had another practice session thirty minutes before the guests began to arrive at 8 p.m. The ballroom was crowded by 264 donating guests who had paid $25 each. They included many of Einstein's old friends, such as the Lehmans, the Warburgs, and Rabbi Wise as well as representatives of New York plutocracy such as Mrs Astor. The composer George Gershwin was also there. The guests began to arrive at 8 p.m.; the concert began at 8:45 p.m.

The New York Times reported at some length on the event, including a statement by a member of the organising committee that Einstein didn't object to criticism of relativity but drew the line at comment on his ability as a violinist. Appropriately, however, it was the *Jewish Daily Bulletin* that carried a more detailed and informal article:

> I think that it can be said without fear of successful contradiction that Albert Einstein has a more masterly stroke on the violin than on the linotype machine.
>
> The two hundred, and more of us, who sat in our best clothes – perhaps I should be speaking only for myself – in the ballroom of Mr. Adolph Lewissohn's [sic] residence last Wednesday evening and heard the world's greatest geographer of the stellar spaces tickle the strings of his fiddle as directed by Bach, Beethoven and Mozart were delighted not only in the performance but in the infectiously naive delight of the performer.
>
> It wasn't only Albert Einstein who played. There were others, quite a number of others and some of them virtuosi of the first water. There was Toscha Seidel, who counts his admirers by the thousands and who was content enough to play second fiddle for a night, and second fiddle in both senses. There was Harriet Cohen, pianist, who had performed the night before in Carnegie Hall and then there was Leon Barzin, orchestra conductor, who worked at the humble viola, while Ossip Gisken was at the 'cello. In addition, there was, for the first number, the Bach Concerto No. 3 for two violins, a string orchestra of ten pieces which Emil Hilb conducted.
>
> But all eyes were on Einstein and ears seemed to be attuned only to his solo passages, and even when his instrument was heard with one or more others, the ear tended instinctively to extract from the musical design only the passages struck by him and to savor them for themselves – as notes of music having a particular freightage of meaning because brought into sound by him. I think that it may truthfully be said that while the average ear heard in full the Bach Concerto No. 3 for two violins and the Allegretto from the Trio for piano, violin and 'cello by Beethoven [Op. 70 No. 2] and the Mozart String Quartet in G Major [K387], eye reinforced ear when the great mathematician struck bow on strings. And eye caught his smiles of pleasure at certain passages, in the Bach and Mozart pieces especially, and ear suspected chuckles of delight even when there were none, for there were times when he looked as if might be, or should be, chuckling. Those who sat in the first row made perhaps a more accurate count of chuckles.

The other performers sensed the emotional value of the concert and sought by no device to take glance away from the concentrated gaze at that haloed, tousled head. Once or twice, I believe, Emil Hilb, in conducting the first number, caught himself giving the cue to Einstein as well as to members of the string orchestra playing with the quartet, but remembered that Einstein needed no cue, that even if he came in a note late – and there was no player more vigilant to the demands of the score, vigilant without being keyed up – there would not be the slightest detraction from the beauty and the value of the performance. At the end of the first movement of the Bach Concerto, however, Einstein relaxed a bit to smile his delight to friends who were smiling at him, and to express his pleasure at the music, but seeing that the other performers were ready and that Mr. Hilb was all but raising his arms to start the second movement, he pulled himself into an attentive posture, ready to do his share.

So notable a virtuoso of the violin as Toscha Seidel might have been forgiven had he made a more flourishing descent upon the strings than he did or squeezed from a run of notes all its exhibitionist opportunities, but he bore in mind not only that he was playing quartet, and not solo, but that he was playing quartet with Albert Einstein as first violin. Concert virtuoso do not always play well in the harness of a quartet, but Mr. Seidel had no difficulty in curbing himself and I believe – from the expression that sometimes I observed flit across his face – that he was deriving a pleasure in the occasion, beyond the joy in the gift of the music itself. Needless to say, there was delighted applause and when Einstein trooped in, at the beginning, with violin tucked under his arm, there was an immediate response, on the part of both auditors and collaborating performers; all rose, while Einstein bowed and smiled his thanks and recognition of the ovation.

Curiously enough, the only solo performance of the evening was given by Miss Cohen who played two of Bach's choral preludes and, as an encore, two additional Bach pieces. Perhaps nothing better illustrates the friendly informality of the occasion than Miss Cohen's saying, just before she played the encores, that she would play them provided she could remember them, and then bowing her head in a kind of concentrated inner gaze. Her memory, needless to say, served admirably.

Now, without reflection on Miss Cohen's performance as soloist, there was the hope that Einstein would play a solo, and he had promised newspapermen he would do a Schubert Sonatina. No solo was announced on the printed program and when the last notes of the Mozart quartet had been heard, the audience rose, applauding and preparing to go. Einstein was standing with his fiddle under his arm. Mr. Morgenthau begged the standers to be seated, as Prof. Einstein was going to play a solo. I overheard the phrase Schubert Sonatina and was ready for a treat, but the audience was for the most part pressing forward to shake Einstein by the hand and by the time any showed an inclination to sit down and hear some more music, the violin had been taken from Einstein, and there was no Schubert Sonatina that night. He had practiced for three hours before the concert and had played in the first number, standing.

Professor Einstein's scientific achievements belong to the world, but his violin playing is his private affair, his private relaxation. If he chooses to play the violin for the benefit of friends in distress in Germany, his playing remains a private affair which need not concern the music critics who pass upon the performances of professionals. Furthermore. I have no particular standing as critic, and do not choose to make any further answer to the question 'But how did he play?' than to say this: that he played competently and casually and correctly, without pretentiousness, without a single flourish. He played with pleasure, with a pleasure no less distinct than that which his playing gave. He played as a delighted amateur. By his playing, his mere appearance, he fixed himself more deeply in the affection of those who were acquainted with him and created a fresh set of admirers. He radiated simplicity and grandeur, and the radiations would have been no less powerful had he appeared in sweater and old pants.

It can be seen from this report that Hilb had indeed organised a small string orchestra to replace Harriet Cohen's accompaniment. This seems to have been a last-minute decision that Elsa believed had not been agreed with Einstein, as can be seen from her letter to Hilb two days later:

> Many thanks for your letter. Frau Bucky didn't call you yesterday because my husband needed some rest and we wanted to give you our feelings as completely as possible the following day. The concert was a real success and I am really grateful to Herr Morgenthau for all his efforts. The programme was also very good and well chosen. Cohen played divinely; I don't mind her practising intensely as long as she wants – maybe she isn't particularly naturally talented. Anyway what she finally produces is wonderful and gives pleasure. I really do sympathise with her because of all her physical problems. The orchestra was a mistake only because they played much too loudly. I was really outraged because both soloists were enormously affected by it. Do you know what several people said to me? 'We would so much have liked to have heard the soloists!' The Double Concerto with a modest piano accompaniment is really a solo piece. But the incredibly loud accompaniment drowned everything else. I could have wept, sitting there. My husband was really taken by surprise and considers it really criminal that, at the last moment of the last day you should come up with this. If you had wanted an orchestra then they really had to be there for the last rehearsal. Then they could have really ensured that the solo parts could have been respected. I am no expert in music, but I have enough sensitivity to know how much was lost through the loud accompaniment. We often talked at the start about whether to have an orchestra or not. Once Frau Cohen was playing, my husband decided to do it with her and there was no further question of an orchestra. Frau Cohen had taken on the accompaniment and you had no right at the very last moment to do anything else. That was my husband's decision; he was very displeased even if he didn't show it to you at the time.

Hilb, no doubt somewhat sheepish that he had finally succeeded in manoeuvring himself onto the stage after all, vigorously defended himself from this charge. He maintained that he had told her about the orchestra and that she had agreed provided it was free. He also claimed that he had informed Einstein that an orchestra would be provided during the Steinway Hall rehearsal the day before the concert and that he had not objected. Hilb contended that Elsa in the front row was particularly badly placed and that others in the audience had been able to hear the soloists perfectly. He implied mischievously that perhaps she got her impression during the *tutti*, when Bach intended soloists and orchestra to play the same notes. Elsa nevertheless stuck to her guns, choosing in a subsequent letter to interpret the *Jewish Daily Bulletin* report reproduced above as evidence that because the reporter regretted there was no Einstein solo piece, Einstein could not be heard in the Bach.

Elsa's letter continued:

> Now to the Movies. My husband doesn't want to do it, despite the $4000 being so nice and enticing. It is entirely possible that the gentleman is offering to do it just to give my husband pleasure and do him a good turn. However, the thought of being shown in every movie theatre in the US is just unbearable for my husband. Herr Laemmle offered my husband $13,500 3 years ago to do a twelve-minute film recording. We didn't accept. Frau Petschnikoff was at that time the intermediary and she did all she could. Of course sometimes one can do nothing about it. If one is disembarking from a ship then one cannot avoid being captured. That is however something completely different. Here one is talking about a dedicated film, approved by my husband [. . .]. Now you will have had just about enough of such events and will be happy to get back to your own work. However you have done a great job and many people will be grateful to you. That is at least something.

Apparently the Einsteins did not return directly home to Princeton after the concert but stayed with their friends the Buckys in the city, where Einstein could recover from the strain from his first and it turned out also his last appearance on the New York concert stage. A frequent correspondence between Elsa and Hilb continued for some time after the concert but increasingly Hilb reverted from impresario to his former role of purveyor of German culinary delicacies. Elsa particularly enjoyed the gherkins, while Hilb promised a delivery of speciality sausages. Irrespective of arguments over the musical qualities of the event, financially it had been a success, with several thousand dollars available for distribution to Einstein's needy German colleagues.

Elsa's next letter to Hilb described their visit to The White House. This took place from 24–25 January. No doubt as a sop to Flexner's concerns, the publicity was kept to a minimum. There were only two exceptionally brief notices in *The New York Times*, one heralding the visit and the other mentioning that the Einsteins had viewed the President's collection of maritime prints and that they had discussed their mutual love of sailing. Elsa was jubilant that they managed to avoid any press photographs of them being taken with the Presidential couple. She remarked that they had had three sessions with the President, taking tea with him when they arrived, a dinner in the Residence which the Spanish Ambassador also attended, and finally a meeting in the West Wing. Both Einsteins used White House headed notepaper to write to various friends. Einstein sent the following poem[3] to Queen Elisabeth of the Belgians:

> In the capital, home of glory,
> Where is forged the nation's story,
> A statesman who can turn the tide,
> Labours happily and with pride.
>
> Last night we talked of friends unseen
> And so thought fondly of the Queen,
> What must be said about that meeting
> Is the reason for this greeting!

Sadly, this was the last message Einstein was to address to Elisabeth as Queen of the Belgians. On 17 February 1934, King Albert was climbing, alone, on a precipitous rock formation that towers above the River Meuse near the village of Marche-les-Dames. He slipped and fell to his death. A few days later, Einstein wrote to the now Queen Mother expressing his sympathy:

> Seldom in my life have I been so shaken as at the news of this heavy blow that destroys your harmonious existence and which leaves an unfillable gap in the handful of devoted fighters trying to halt the fall of Europe into darkness . . . I know how it feels when objects of love are lost forever. But I also know that strong characters such as yours, in service to objective goals, and especially through the consolations of art, find a purpose in life that can, to some extent, grant a respite from the rude interventions of blind fate.
>
> In heartfelt sympathy, I take your hand,
> Your Albert Einstein.

Shortly before sending this letter, Einstein had written to a Jewish violinist who had contacted him asking for help to find a job. He had been born in Germany and arrived in the US in 1923. His employment had been predominantly among the German-American community. The rise of Hitler had resulted in these offers of employment drying up; the violinist recounted examples of Nazi sympathy among the German-American community. Einstein promised to write and sign a letter of thanks to anyone who employed the violinist for more than a month. On 19 February at Philadelphia's Municipal Convention Hall, Einstein, together with 12,000 others, attended a pageant on the course of Jewish history.

---

[3] Translated with poetic licence.

Einstein's perennial desire to play with high-class musicians is illustrated by his invitation to the Hilger sisters, Maria, Elsa,[4] and Greta, who lived in nearby Freehold, to spend the day at Library Place at the end of February. They had been born in German-speaking Bohemia, where the string-playing Hilgers studied under the renowned pedagogue Otakar Ševčík. The sisters formed a piano trio that Einstein had heard a few months earlier playing the Beethoven 'Triple' concerto in a radio broadcast. They had lunch at Library Place and then joined Einstein for two music sessions, separated by tea.

Another musical event that Einstein did not find so enjoyable took place on the evening of 6 March. Arnold Schoenberg gave a lecture at Princeton University titled 'My Method of Composing with Twelve Tones Which Are Only Related with One Another'. The lecture took place in the Frick Chemical Building before a large audience, which included Einstein. As well as expounding his reasons for writing as he did, the composer was anxious to defend himself against a number of charges, in particular that he was a musical 'engineer'. He concluded the lecture by saying that if he wrote bad music, it wasn't because he was an engineer, but because of lack of talent.

Schoenberg would have been delighted finally to make Einstein's acquaintance. They had at least their Jewishness and support for Jewish causes in common but there is no doubt that Einstein would have agreed with almost nothing in Schoenberg's lecture except the final sentiment. In later years, Schoenberg was often referred to in American newspapers as 'the Einstein of Music', but the original would not have cared for the comparison. Schoenberg had told Emil Hilb that Einstein had been at his lecture, causing Hilb to write to Elsa asking what Einsteinian simplicity had made of Schoenbergian complexity. She replied that he had said: 'if Mozart could hear this he wouldn't only turn in his grave, he would stick his tongue out [. . .]. My husband believes that Schoenberg is utterly convinced that everything he says is correct and true. So, he isn't a swindler, just mad.'

The dedication of the Einstein Institute of Mathematics and Physics took place at the Hebrew University of Jerusalem on 6 March. The *Jewish Daily Bulletin* took the opportunity to editorialise on the event, hoping that it would cause the institution to rededicate itself to the Einstein's ideals and spirit, thereby hoping that his sympathy and interest in it would be rekindled.

Early in March, escaping at the end of the Oxford term, Erwin Schrödinger arrived in Princeton. He stayed at the Graduate College, a mere ten-minute walk away from the Einsteins' house. He had accepted a financially generous invitation from Princeton University to stay for a month and give some lectures. In fact, the university was considering him for the vacant chair of theoretical physics, which was indeed offered to him at the conclusion of his stay. He and Einstein spent considerable time working together; Flexner recalled finding them engrossed at the blackboard. While there is no record of what they discussed, given their shared scepticism about quantum mechanics, it can safely be assumed that this dominated their interactions.

Schrödinger certainly asked Einstein about his impressions of Princeton, which were very positive. The position that Lindemann had obtained for Schrödinger as a lecturer at Magdalen College was not a permanent one; in any case, he did not care for Oxford and neither did Oxford care for him. Schrödinger's ménage à trois scandalised the Oxford dons. As drily remarked by John Gribben in his biography of Schrödinger, it was in this period considered a bit odd in Oxford

[4] Elsa Hilger's Pietro Guarnerius cello had been stolen from her locked car in New York City on 2 January. The cello was found and returned to Elsa two years later. She went on to become the first woman section leader in a major American orchestra, the Philadelphia; she was a legend among cellists, dying at the age of 101 in 2005.

colleges to have one wife, let alone two. Schrödinger's attitude to Oxford academics is distilled in remarks he made to Max Born: 'He [Schrödinger] disliked college life and high-table dinners. He regarded a society without women as detestable and barbaric. Once he said: "These colleges are academies of homosexuality. What a queer type of men they produce!"' Schrödinger was deeply concerned about his financial situation, particularly ensuring a reasonable pension for his (legal) wife. The Princeton chair offered an excellent salary, but the associated widow's pension was wholly inadequate. After spinning out the negotiation as long as he could, Schrödinger finally rejected the university offer in June.

Another consideration for Schrödinger was that Einstein encouraged him to believe that he might get a position at the IAS. Directly after refusing the Princeton University offer, Schrödinger wrote to Flexner exploring the possibility of an appointment at the IAS. Flexner replied in July in a guardedly positive manner, remarking on the financial constraints of the time and implying it would be better to reopen negotiations with Princeton University. Schrödinger coming to the IAS continued to be a possibility for the next year, with Einstein urging Schrödinger's merits on Flexner in the summer of 1935. Flexner continued to hide behind the difficulty of trumping the university offer. In the end, Schrödinger, unwisely, took up a position in Graz in 1935 from which he had to flee precipitately after the Austro-German Anschluss three years later. Thus, Flexner's reaction, had he been confronted in Princeton with Schrödinger's colourful domestic arrangements, must be left to the imagination.

A factor that would have reduced Flexner's flexibility to hire Schrödinger occurred at the January meeting of the IAS Trustees. They agreed to provide $10,000 for a year's salary for Paul Dirac, who had previously visited Princeton before Einstein's arrival, and, unsurprisingly, found its similarity to Cambridge congenial. He decided to spend a sabbatical in 1934–1935 at Princeton. Einstein does not seem to have been involved in this decision. Dirac arrived at the end of September 1934. He gave a masterly (at least in content if probably not in delivery) lecture series on relativistic field theory transcribed by Boris Podolsky and Nathan Rosen. However, the visit was more productive on a personal rather than a scientific level. On his first full day, Dirac wandered into the town centre for lunch in one of the restaurants frequented by the faculty. Here he found his colleague Eugene Wigner and Wigner's sister Margit at lunch. In 1937 she became Dirac's wife. In addition to courting Manci, as she was known, in his own inimitable style, he also was deeply involved in the unsuccessful campaign to get his Cambridge colleague and close friend Piotr Kapitsa released from the Soviet Union. Kapitsa, a brilliant Russian experimental physicist of whom Lord Rutherford was deeply fond, had been working in Cambridge since 1921. He had had been detained in the Soviet Union on Stalin's orders after his regular visit to his mother in the summer of 1934. Despite Dirac's office being only one away from Einstein's, more or less their only work together during Dirac's visit was on this campaign, which Einstein strongly supported.

On 7 March 1934, the Einsteins stayed overnight at the Gubernatorial Mansion after dinner with Governor Lehman. More than fifty influential New Yorkers were also invited. On 13 March, a slight acquaintance who had first met Einstein at a reception in Caltech during one of his stays there, arrived at Library Place bearing a book as a present for Einstein, whose birthday fell on the following day. Leon Watters was a Jewish chemist who had become interested in medical matters, particularly food and drugs. He developed a method for sterilising the catgut that became universally used in medical sutures. The company he set up to market this made him a very wealthy man. Both Einstein and Elsa became friendly with Watters, who prevailed upon Einstein to visit the Hebrew Technical Institute, a vocational school in New York set up to support Jewish immigrants, and of which Watters was a strong supporter and Vice President. The Einsteins then visited Watters luxurious Fifth Avenue apartment and admired its wonderful view over Central Park. This subsequently became a favourite haunt of Einstein. Watters put his chauffeur-driven limousine at the Einsteins' disposal whenever they cared to call for it.

For Einstein's fifty-fifth birthday on 14 March, Elsa baked a cake that apparently sported the full complement of candles, which she intended to light for their solitary celebration. She seems to have spent the day in completing Einstein's tax return.

On 19 March Einstein attended a dinner to celebrate the sixtieth birthday of his old friend Rabbi Stephen Wise at the Hotel Astor in New York City. He gave a speech in which he praised Wise for his leadership of the Jewish people, tolerance, and deep understanding.

Throughout the first quarter of 1934, the plan was still for Einstein to return to Europe for six months. Lindemann in Oxford was expecting him, despite Einstein's renunciation of his stipend in favour of German Jewish refugees. On 27 February, a concert was held in Carnegie Hall in honour of the Einsteins 'on the eve of their departure for Europe'. The concert was organised by the American Jewish Physicians Committee and raised $10,000 for the Medical Unit of the Hebrew University in Jerusalem. Einstein was in the audience this time, rather than on stage, but Toscha Seidel was once again a star performer, playing Brahms' D minor Violin Sonata, Op. 108, the Beethoven Romance Op. 40, a movement from Édouard Lalo's *Symphonie Espagnol*, and Kreisler's *Sicilienne* in the style of Francoeur. Ossip Gabrilowitsch played with Seidel in the Brahms and also accompanied the soprano Hulda Laschanska in Schubert and Schumann songs as well as two of his own composition. He also played Chopin and Glazunov solo piano works. The concert ended with a performance of Richard Strauss's *Morgen!*, Op. 27, No. 4, in which both Gabrilowitsch and Seidel accompanied Laschanska.

Another event that was also a farewell for the Einsteins' expected European journey but principally a welcome to residency in the state of New Jersey took place in Newark on 25 March. This was hosted by Governor Moore; the concert was attended by 6,500 guests. Distinguished opera stars Giovanni Martinelli and Ernestine Schumann-Heink sang and Gabrilowitsch played the piano. Einstein addressed the crowd in German, which was translated by Rabbi Silberfeld. He strongly supported President Roosevelt's 'New Deal' legislation then beginning to pull the nation out of the Great Depression. He also thanked supporters for their contributions to the charitable aims of the event, to support the resettlement of German Jews in Palestine. Surrounded by a strong security presence, the Einsteins walked to a subsequent dinner, at which Einstein smilingly declined to give another speech even though he had prepared one. Others filled the gap, and all enjoyed the subsequent brief musical entertainment by soprano Viola Philo. Einstein chased after the presumably attractive performer in order to kiss her on both cheeks in thanks.

As late as 1 April, *The New York Times* was reporting that the Einsteins would be leaving for Europe on 3 April, a date they had set back in February. Another charity concert was arranged to say farewell in Carnegie Hall, originally set for 27 March but then moved to Sunday 1 April. However, at some point towards the end of March, Einstein asked Schrödinger to tell Lindemann that he would not be coming to Europe after all. There were many reasons for his decision. Although affairs in Germany had quietened down somewhat since the previous year, there was certainly some residual danger from the Nazis that would require security precautions that Einstein would find irksome. He knew that once he arrived in Europe, the various countries from which he had accepted chairs would expect him to make an appearance. He had intended to go first to Paris, thereby satisfying the Collège de France, but then Oxford was still expecting him, as was Madrid, and he still had ties with Leiden. At the age of fifty-five, he decided he simply could not face chasing around Europe, evading Nazi threats, and being lionised by faculties across the continent. By now, he had satisfied himself that Princeton was the place for him; he enjoyed its relaxed and peaceful atmosphere, the hospitality and friendliness of the inhabitants, and the ability to concentrate on his work without any lectures or students. Given the state of Europe, he could expect the great figures of physics to come to him, as had already happened with Schrödinger and was about to happen with Dirac. Einstein knew that a leading experimentalist in nuclear physics, Enrico Fermi, had been approached by Flexner[5] to be a visiting professor in 1935–1936. In any case, he was completely isolated in his work on unified theories and increasingly interacted only with the assistants who replaced Mayer. As he remarked on 22 March in reply to Max Born's query as to whether he would come to Europe in the summer: 'If at all possible, I am going to fritter the summer away somewhere in America. Why should an old fellow like me not enjoy relative peace and quiet for once?' As the years went by, his reluctance to leave the East Coast increased to an immovable inertia.

Despite the sudden lack of any necessity for farewells, events continued, not least because they had important fund-raising functions. The first day of April was busy. The Einsteins went to the East Village for a seder dinner for the second day of Passover given by the National Labor Committee for the Jewish Workers in Palestine. Nine hundred stalwarts of the Jewish Community, including Rabbi Wise, attended, with an overflow of hundreds outside in the street listening to proceedings through loudspeakers. The event raised $15,000 for a fund to settle Jewish refugees in Palestine.

After the seder, the Einsteins travelled downtown to Carnegie Hall for the concert, which began at 8:30 p.m. It was entitled 'A Tribute of Music to Science'. The organising committee was chaired by the pianist and composer Einstein had met at Mischa Elman's, Leopold Godowsky. However, Elsa was certainly also 'assisting' Godowsky with the organisation through the inevitable involvement of her musical factotum, who appeared on the front page of the souvenir programme as 'Program supervised by Emil Hilb'. Elsa had asked Mischa Elman to play but was disgusted by what she considered his mercenary attitude. She had also written to Yehudi Menuhin, but he had a prior engagement in Montreal just before the originally planned date. She wondered whether, if all else failed, they could get Seidel yet again. All else does appear to have failed, as in the end no violin virtuoso played at the event. Instead there were a host of top-rank singers led by Emmanuel List of the Metropolitan Opera as well as a plethora of pianists and both the Kroll String Sextet and the Musical Art Quartet. Elsa had been particularly keen to sign up the latter.

For something supervised by Hilb with Elsa as *eminence grise*, the programme was surprisingly eclectic and hardly what Einstein himself would have chosen. Perhaps the most unusual item of all was Godowsky's *Contrapuntal Paraphrase on Weber's 'Aufforderung zum Tanz'* for no less than three pianos. Among the items least likely to win Einstein's approval were two songs by Wagner, and songs by Debussy, Meyerbeer, Wolf, and Delibes. An oasis of classicism in terms of Emmanuel List's rendition of 'In diesen heil'gen Hallen' from Mozart's *Die Zauberflöte* and two Schubert songs served as contrast to the ultra-Romanticism of Schoenberg's *Verklärte Nacht* that followed, ending the programme. Schoenberg himself, making his only visit ever to Carnegie Hall, was in the audience.[6]

After the intermission, Einstein was called to the stage to receive a cheque for $6,000 that was to be devoted to the support of Jewish refugee children arriving in Palestine, as well as a testimonial

[5] Fermi was attracted, certainly by the generous salary and enquired whether he would be able to access the experimental facilities of the university physics department, but in the end was unable to leave Italy.

[6] Even in Carnegie Hall, rarely can so many simultaneous chords have been heard as in Godowsky's piece. Schoenberg may have appreciated the harmonic progressions, which seemed to owe more to his pupil Webern than to Weber.

gift to himself signed by one hundred prominent musicians thanking him for his efforts for music. He spoke briefly:

> It gives me great joy to thank the prominent artists, who tonight have given their efforts toward a noble work of philanthropy. Art is one of man's greatest blessings; self-sacrificing charity is by far the most important and the most beautiful thing we can accomplish. The artists we have heard here have given us both.
>
> I must thank them personally also, for my name has been linked to this evening's entertainment. A sense of humility would have prevented me from permitting such an association had not the cause involved been the alleviation of terrible needs – the cause of rescuing German-Jewish children from a hostile environment, to save them from spiritual and material misery, and to enable them to grow up in happy and healthy surroundings.[7]

The evening ended with a speech by Godowsky, who probably repeated the sentiments he had delivered to a *Jewish Daily Bulletin* reporter during rehearsals for his three-piano piece: 'His long devotion to the art of music has been a noble inspiration to musicians everywhere. We of the musical world are proud to claim Professor Einstein as one of our most revered leaders.' Einstein apparently greeted well-wishers in the green room and signed their programmes until 1 a.m. Among them was Schoenberg, who was photographed, together with Godowsky, arm in arm with Einstein (see Figure 26.1).

The New York Times took advantage of Einstein's recent prominence in the city's musical life to publish a feature interview with him on the subject by Alfons Goldschmidt. Goldschmidt was an economist with Marxist leanings who was an expert on South America and had been the economics editor of the *Berliner Tageblatt*. Einstein had made his acquaintance in Berlin and knew him well enough to write the foreword to his book *Whither Israel* which was privately published in 1934. Goldschmidt had fled the Nazis when they came to power and was now supporting himself in the US via journalism. The article covers Einstein's views on a wide range of musical topics:

> One might almost say that Einstein's musical sense is a variation of his mathematical sense. The same rhythmic force that rules mathematics and physics drives him when he plays the violin, so that his playing is more than relaxation only. His idea is that the constructive logic of mathematics is identical with the rhythm and march of music to a clear conclusion.
>
> Although Harold Bauer, the pianist, is reported to have called Einstein the best musician outside of professional ranks, Einstein does not regard himself as a virtuoso. He merely insists upon the right to play because, as he says, 'I feel the creative process in the composer.'
>
> [...]
>
> Great musical events remain permanently with Einstein. He has never forgotten, for instance, a rendition of Beethoven's Tenth sonata and Bach's Chaconne by the violin maestro Josef Joachim. Einstein himself plays the Chaconne, as he plays all his violin music, with a beautiful tone, with forceful expression, with a deep feeling for purity and a high technical perfection. In a discussion of music he told me: 'As a boy I was enthusiastic over Schumann, but later I found that he did not possess the last word in musical originality. Then, at the age of 15 the pure musicality of Bach and Mozart was revealed to

[7] It is ironic that, beneath *The New York Times* story reporting Einstein's remarks, an advertisement with an enormous headline 'GERMAN • The Seas' Highest Ideals Hold Sway' appeared, extolling the virtues of crossing the Atlantic with the Hamburg-American line. Another bizarre occurrence was a letter from *The New York Times* music critic Olin Downes. This was one of those years in which Passover and Easter coincided and the concert took place on Easter Sunday. Downes, in regretting not being able to be present, told Einstein that on Easter Day 'Light and truth and love itself rose triumphant from the tomb.' Only in a postscript does he admit that Einstein's religion may not see Easter in quite the same way!

Figure 26.1 Einstein with on his left Arnold Schoenberg and on his right Leopold Godowsky, after a concert at Carnegie Hall, New York on 1 April 1934.

me. Many of those who claim a love of Mozart are not sincere. They do not understand the melodious many-sidedness, the delicacy of detail, of this composer, or the natural gayety and grace that flow from diversity to make the entirety.' Thus the enthusiasm for a pure, crystallic music does not signify that Einstein dislikes richness and variability of melodies. On the contrary, he has a mind attuned to melody, and the variability of Mozart's music is only the melodious way to the highest simplicity in music. Mozart is for Einstein the culmination of musical demands, a hundred melodious brooks conducting to the great stream of simplicity.

To my question as to how he regarded Beethoven, Einstein replied: 'Beethoven overwhelms with his violent strength, so to speak, with unhewn blocks. I always need a little running start. in order to attain to Beethoven.' This is especially true of Beethoven's symphonies; Einstein has attained a satisfying friendship for the composer's chamber music.

And Wagner? 'The start I need to get to Wagner is much longer still. In order to get through to the undoubted beauties of Wagnerian music I have to dissociate myself first from Wagner's own personality – though I admit the rare beauty of his *Tristan und Isolde*.

[...]

He did not favor the sentimental [violin] concerto of Mendelssohn, and he expressed astonishment that musical people received it again and again with apparent enthusiasm.

As for Tchaikovsky, Einstein said that he relished the Russian composer in so far as his works were born of the simple folklore spirit, but rejected his music as soon as it became 'too sweet,' touching the realm of 'Kitsch.'

> In music, briefly, Einstein wants the same clarity as in his study. He likes to live and work in cool rooms and believes that the fetid, cloying atmosphere of overheated places is dangerous for men engaged in mental work.

Einstein contributed to the new music magazine *Tempo* in its April 1934 edition, in which he gave some unusually personal insights:

> It seems to me that in our time we do not sufficiently appreciate the significance of an active participation in music as a means of development and of finding true happiness. A constant activity in music will contribute much toward the building up of a well-rounded character and the enriching of the soul, and offers an escape in hours of disillusionment and despondency. Because of the possibility afforded by it to explore and relive depths of emotion, music offers compensation for what one may miss in personal relationships. I am personally most grateful to Bach and Mozart as well as to some of the old Italian masters.

Einstein gave an address at the New Jersey state legislature on 10 April, probably having accepted an invitation from Governor Moore extended at the March reception described above. After visiting the Governor's office, Einstein was escorted to the Capitol building. He once more spoke in German, again translated by Rabbi Silberfeld. He described his joy in being able to work in a free society such as the US and stressed the importance of the cooperation of government and non-scientists in the support of scientific research. He considered that this was nowhere as strongly practised as in the US. On 19 April he made another political statement at the 1934 Princeton Conference on The Cause and Cure of War, advocating that the US should join the League of Nations. In his view, now that the 'aggressor' nations Germany and Japan had left the League, American involvement could make it a real force for peace.

Having decided to stay in the US, Einstein became even more deeply involved in trying to help exiled German Jews. He tried to get his cousin Alfred Einstein assistance from the emergency committee for the aid of displaced German scholars, but despite being a leading musicologist, this failed, since Alfred had never been an academic. Einstein also tried somehow to distil the various European chairs that he was now intending to renounce into resources for others. His stipend from Oxford he had already successfully converted with Lindemann's help. The Spanish position was a prime candidate, as he thought that he had in fact already indicated to the Spanish Ambassador at their White House dinner that he would not be able to take up the offer. However, this message did not get through to the Spanish government, who telegrammed to enquire when they could expect him.

Einstein recruited his friend Abraham Yahuda, who had been instrumental in arranging the initial Spanish offer, to see if it could be transferred to a worthy candidate. His first suggestion in May was Max von Laue, who had been outspoken in Einstein's defence inside Germany and who, although Aryan, could therefore expect the enmity of the Nazi authorities. Von Laue however showed no interest in such a possibility. In June, Einstein suggested Max Born, whom Einstein knew to be looking for a permanent job as, like Schrödinger, he was filling a temporary lectureship position, although in Cambridge rather than Oxford. However, Born preferred to stay in UK and eventually was appointed to a chair in Edinburgh. Einstein's final suggestion was Leopold Infeld, who at that time was working on a form of unified field theory with Born in Cambridge, where he had a one-year Rockefeller Fellowship. Born must have suggested Infeld as a candidate for the Spanish position. Infeld even got as far as being interviewed by Pérez de Ayala, the Spanish Ambassador in London. However, the wheels for such a strangely alchemical appointment, converting Einstein into another physicist, unsurprisingly turned exceedingly slowly and eventually ground to a halt. This turned out to be fortunate for Infeld as shortly afterwards the Spanish Civil War broke out.

Einstein was only contracted and indeed paid by Princeton for the months of October–May, the IAS's academic 'year'. He also had special permission to leave in mid-April if he wished. As he mentioned to Born, once he had taken the decision not to return to Europe, he began to look around for a pleasant place on the East Coast in which to spend the hot and sultry summer months. His friends the Buckys suggested Watch Hill on the coast on the Rhode Island–Connecticut border. They rented a sizeable house, Sunnymere, also called The Studio, for $500 for the period 1 June–1 October. The house was perched above the sea but only a couple of minutes' walk from a beautiful beach in one direction; in the other direction, most importantly from Einstein's viewpoint, was an excellent marina with sailing boats for hire. Also in the party were Helen Dukas and Leon Watters. The latter preferred to stay in the adjacent and very grand Ocean House Hotel.

Before the Einsteins could retire to this peaceful paradise, however, Elsa received a communication from Paris. Margot Einstein had joined her husband Dima Marianoff there during the winter. Her sister Ilse and Rudi Kayser had been living in the Netherlands but Ilse's health, always poor, began to deteriorate markedly. Her prognosis was not improved by her conviction that her problems were psychosomatic so that she would only undergo psychoanalysis rather than consult a physician. In an attempt to alleviate her symptoms, Ilse and Rudi travelled first to Cap d'Ail near Nice but as she continued to fail, they went to Paris so that she could be nursed by her sister. When Elsa learned from Margot that Ilse was now dangerously ill, she was stricken with concern and began to make plans to sail with the first ship to France. Elsa told her friend Antonina Vallentin that she had forbidden Einstein from accompanying her as the situation was too dangerous there for him. She sailed on the luxurious French Line ship, SS *Paris* on 19 May, arriving in Le Havre on 25 May.

Elsa asked Watters, whose limousine had brought them from Princeton, to look after Einstein for a while in her absence. He took Einstein back to his Fifth Avenue apartment and, as instructed by Elsa, ensured that Einstein lay down for a rest in the afternoon. As he lay on the settee, Watters played Liszt's *Lorelei* on his player-piano. After the prescribed period of relaxation, Watters asked his friend whether it had been restful. Einstein replied that the sofa had been but not the music, which was too sugary.

On Elsa's arrival in Paris she was appalled by Ilse's condition. She was emaciated and in constant pain. Margot, who was exceptionally close to her sister, looked little better, worn out with anxiety and her efforts at nursing Ilse. Elsa quickly looked for other opinions, including flying a doctor from Berlin. At the end of June, Elsa contacted Heinrich Zangger to see if he could organise a place for Ilse in a sanatorium near Chamonix. However, before anything else could be done, Ilse died in Paris on 11 July in a manner that Elsa characterised as the most awful that a vengeful fate could inflict. The cause of death was still mysterious; Einstein was told that she had died from intestinal tuberculosis but he remarked to Zangger that the only thing that seemed certain to him was that she was dead.

Both Elsa and Margot were prostrate with grief. It was only with difficulty that Elsa was talked out of the idea of taking Ilse's ashes back to Princeton. The associated bureaucracy would have taken weeks and, in any case, Rudi Kayser wished for her urn to be laid to rest close to their home in Aerdenhout, near Haarlem in the Netherlands. On 19 August they journeyed to what Elsa characterised as a beautiful paradise of a spot near Haarlem, where Ilse's ashes were interred. The fact that Elsa stayed in Europe for so long after Ilse's death was at least partly because Rudi Kayser was inconsolable. Elsa and Margot lodged in a small boarding house in the seaside town of Zandvoort, next to Aerdenhout, until 24 August. After the news of Ilse's death reached him, Einstein wrote to both Elsa and Margot, pleading with Margot to return with her mother to the US. Bucky at Einstein's request facilitated the formalities and arrangements for Margot to enter the US on a tourist visa. After a heart-rending farewell to Rudi, Margot and her mother embarked at Antwerp on the Red Star SS *Pennland*.

On 24 May, just after Elsa had left for Paris, Einstein attended a concert in a private house organised by Sidney Matz, a local manufacturer of laxatives, for the benefit of the committee for refugees in Geneva. Einstein had been collected and taken to the house, where a small group, mostly of artists and intellectuals, including Eddie Cantor had gathered; Einstein was driven back afterwards. He was very pleased by how little effort the event cost him and by the goodly sum raised in cheques sent to him by the attendees. Elsa recounted to Hilb that Einstein had decided to add a codicil to his will stipulating that, after his death, his legs should be taken to banquets so that they could continue to raise money for charitable causes.

Einstein and Helen Dukas arrived at Watch Hill early in June, somewhat in advance of the Buckys. The marina's boat-rental yards were still closed, so Einstein was not yet completely absorbed in sailing, as he became later in their stay. He was therefore very grateful to receive a violin from Oswald Schillbach, a New York luthier known to Gustav Bucky, at the ridiculously small price of $35. In his letter of thanks written in July, Einstein praised the violin's tone and contrasted it with that of his own.[8] His attitude to his old friend Levin's abilities seems to have been changed by the damning appraisal of his violin by the Hill brothers. He now complained to Schillbach that his current instrument had been altered by an amateur, an old and ill man of 72, formerly a doctor and author and now a refugee in Belgium. His alterations, thinning the violin's table, resulted in a weak and inelegant tone, so he was very grateful for Schillbach's violin, which he characterised as having a true and powerful sound. Perhaps if he had owned this instrument at the January concert, Elsa would have had less trouble hearing him above Hilb's orchestra!

Once he had acquired a small sailing boat, Einstein settled into a delightful routine of sailing, eating, and sleeping, with some violin playing on most evenings. He made the acquaintance of Leonora McKim and her husband William. As Leonora Jackson, she had been a protégé of Joseph Joachim and a stalwart of the concert platform around the world at the turn of the century. She played for both the German Imperial family and Queen Victoria. Her career continued to great acclaim until her short-lived first marriage to Michael McLaughlin in 1907, after which she left the concert platform. She and her second husband, an enthusiastic organist, were vacationing in Watch Hill. Leonora was still an excellent violinist, playing the 'Goetz' Stradivari of 1695. Einstein and she played together at least one evening a week. Einstein inscribed a photograph to her in remembrance of their Bach and Händel evenings.

The *Los Angeles Times*, still interested in Einstein's doings despite his desertion of Caltech for Princeton, reported that Einstein could be found in his cottage playing his beloved violin in his shirt sleeves. Einstein told Elsa that he relished the absence of mathematicians and intellectual show-offs. He remarked that homesickness for Caputh was avoided by the fact that there were just as many midges in Watch Hill. The weather was cool, with a refreshing sea breeze and quite

[8] Just after Einstein's arrival in Princeton, a luthier named Oscar Steger presented him with a violin with an inscription on the label that read: 'Made for the Worlds [sic] greatest scientist Profesior [sic] Albert Einstein by Oscar H. Steger, Feb. 1933/ Harrisburg PA'. Since his letter to Schillbach clearly refers to the instrument altered by Levin, the Steger violin must have been inferior. Despite this, it sold at auction in 2018 for $516K. Anon. News in Brief. *The Strad*, 2018, 129: 13.

often foggy. The high humidity opened up the seams of Einstein's new violin, so that Bucky had to take it back to New York for Schillbach to repair. Meanwhile, Einstein sailed every day, acquiring a reputation among the locals as a rather incompetent sailor. On several occasions he ran aground and had to be assisted by local boys to refloat his vessel. Once he and Leon Watters were stuck on a sandbar and, despite several attempts by locals to help, were forced to await the assistance of the rising tide. They didn't return to Watch Hill until 8.00 p.m., when Einstein had to be restrained from dashing off in his wet sailing gear to play his violin at a musical appointment with Leonora McKim. He remarked to Watters that he played the violin with his hands, not his breeches. Watters nevertheless prevailed on him to change into a white linen suit before he departed with his violin. As his vacation wore on, Einstein decided that life in the US was so delightful that he could hardly imagine ever returning to Europe.

However, concerns from the outside world disturbed Einstein's idyll in Watch Hill. At the end of June, he remarked on the ability of Hitler and colleagues to fill the newspapers: Hitler's elimination of Ernst Roehm and other dissident elements in the leadership of the Nazi private Brownshirt army, the SA, was headlined. Einstein misinterpreted these actions, certainly catalysed by the German Army, as evidence that Hitler was becoming their puppet, whereas as gradually became clear, the reverse was true. As well as disturbances from the newspapers, various visitors interrupted Einstein's relaxation from time to time. He took many of them sailing. One such was Felix Warburg, fresh from his coup in favour of Magnes with the American Trustees of the Hebrew University. Apparently considering Einstein's small sailing boat insufficiently grand, he returned at a later date with an enormous steam yacht and took Einstein for a cruise lasting several days. Watters flitted from New York to his luxury hotel room, once bringing with him a cine projector which no doubt entertained the Buckys' two sons. When in Watch Hill, he always strolled to the porch of Einstein's house at 9.00am, where they took coffee together. Hermann Bernstein, editor of the *Jewish Daily Bulletin* and a former US Ambassador to Albania was another visitor. He wanted to engage Einstein as an intermediary between Jews and Pope Pius XI in a possible papal intervention in favour of German Jews. Einstein however refused to become involved in what he characterised as quixotic diplomatic somersaults.

In addition to his concern for his adopted daughter, that for his son Tete again impinged on Einstein's peace of mind. On 21 April, Tete, accompanied by a carer, had spent just under three weeks with his aunt Maja in Florence. She had been shocked by the deterioration in his physical condition and his large gain in weight. He seemed mentally less disturbed than on his previous visit. However, he was clearly suffering and low in spirits, to the extent that he, although he played Maja's piano, he did not seem to take any pleasure in it.

Mileva had written to Einstein enclosing an article on the treatment of mental diseases with insulin. Einstein's reply was guardedly positive although he counselled waiting until more experience had been gained. At some point during the summer, Tete wrote to his father explaining that he had discovered the nature of time and requesting Einstein's expert opinion. His father replied that Tete was speaking of experiences with psychological time whereas his theories related to that in the physical universe. Throughout 1934 and into 1935, Tete wrote a number of extremely long, rambling letters on various aspects of psychology and psychiatry that must have disturbed his father. In August, the faithful Zangger wrote requesting a decision on Tete's treatment; he also warned Einstein that urgent action needed to be taken to avoid the foreclosure of mortgages on Mileva's various Zurich properties, which would probably result in their being seized by the banks.

Einstein was saddened to learn of the death of Marie Curie on 4 July 1934. Her long exposure to radioactivity had finally resulted in a pernicious anaemia that proved fatal. Although her work had never been of much relevance to his, they had served for many years together on the League of Nations Intellectual Cooperation Committee. He had valued her intellect and supported her when she had problems with the press over her extra-marital relationship with Paul Langevin. He praised her in the American press:

> One of the most remarkable scientific personalities of our time. Her ingenuity and her extraordinary energy enabled her to solve some of the most important problems which led to the discovery and to the scientific understanding of the radioactive phenomenon. Not only in the range of her profession, but also in nonscientific matters, she was an unusually independent character, standing up whole-heartedly for justice and for progress in politics and in social matters.

However, it was the death of Ilse that really overshadowed Einstein's summer. He went to meet Elsa and Margot when they landed in New York on 3 September. Fellow passengers on the *Pennland* included their Princeton neighbours the Blackwoods and their son Andrew Jr. The Reverend Andrew Blackwood was a theologian at the Princeton Theological College. Elsa had brought with her several trunks containing Einstein's personal papers that had been rescued from Berlin via Dima Marianoff. Elsa was concerned that as a non-citizen, she would not be allowed to bring the trunks into the US. She therefore requested Blackwood to pretend that they were study materials that he had brought back from Europe. Blackwood agreed to this, and Einstein's papers and books were landed and despatched to Princeton, where Einstein eventually picked them up from the Blackwoods' house in the autumn.

The three Einsteins returned to Watch Hill. Both Elsa and Margot were worn out from their experience. Margot was initially even harder hit than her mother. However, by January 1935 Elsa could write to the Alfred Einsteins that Margot had come back to life and was working again, producing wonderful things, catalysed perhaps by her terrible experiences. It was different for Elsa. It would be no exaggeration to say that she never really recovered from Ilse's death. Only a couple of weeks after arriving in Watch Hill, she fell seriously ill, with what was diagnosed as severe neuritis and sciatica. She was initially unable to move and only slowly regained her ability to stand up. Even by late October, her left leg remained useless. On 3 October, the party motored down to New York, where they stayed for two days with the Buckys before they returned to Princeton on 5 October in time for the start of the IAS semester. On 8 October Einstein was back in New York to receive an honorary doctorate from Yeshiva College. In his acceptance speech, he praised Jewish institutions such as Yeshiva for combating materialism and sustaining the spiritual concepts that kept the Jewish community vital. He then returned to his study in Princeton and his beloved physics. The coming semester was to see his last major contribution to that subject, and indeed one of his most important in a career replete with astonishing achievements. It would not be in the area of arid preoccupation with unified field theories but in that other major thread of his life, quantum mechanics.

Chapter 27

Einstein–Podolsky–Rosen and gravitational waves; Death of Elsa Einstein (1934–1936)

The defection of Mayer, Einstein's 'calculator', on their arrival in Princeton, had been a major blow. Einstein had come to rely on Mayer's expertise and determination is tackling the obscure branches of mathematics into which his work on unified theories had increasingly led. Since he had pushed Flexner considerably to obtain a permanent position for Mayer and given their strained relations in the first few months after his arrival, Einstein did not feel that he could approach Flexner with a request for another assistant until some considerable time had elapsed. Flexner made it clear that there was nothing to stop him attracting into his orbit some of the bright young men (they were all men until considerably later in the IAS's history) who were either studying or had temporary positions at the Institute. Two such were Boris Podolsky and Nathan Rosen.

Podolsky had already worked with Einstein on quantum mechanics in 1931 in Caltech. Having returned in 1931 to his native Russia for two years, he had also worked with Dirac on an early paper on quantum electrodynamics. He returned to the US in 1933 with a fellowship to work at IAS. Rosen had obtained a doctorate from the Massachusetts Institute of Technology (MIT). As a student, he had already written seminal papers on aspects of quantum mechanics, including the description of the hydrogen molecule in terms of products of the two-electron wave functions. His solution, although better than some previous attempts, was not particularly close to the experimental values. It did however give him experience in problems solvable in terms of products of wave functions that were not necessarily separable into two individual wave functions, a precursor of the ideas that he was to develop with Einstein and Podolsky. He arrived at Princeton University in 1932. Rosen's Master's thesis had been on that old Einsteinian favourite, distant parallelism. No doubt in order to introduce himself[1] to the great man, he asked Einstein's opinion of it. Early in 1934, Rosen discussed applying for a one-year fellowship at IAS. Supported by Einstein, his application succeeded.

However, it was quantum mechanics, not general relativity, that was the subject of Einstein and Rosen's first collaboration. This was much more in the forefront of the thinking of the average young physicist at the time than was general relativity. Rosen and Podolosky probably first worked together on the transcription of Dirac's lectures on quantum mechanics mentioned in the previous chapter. Einstein in turn had once again been catalysed into thinking about quantum mechanics by the visit of Schrödinger. The stage for a fruitful collaboration was therefore set by the winter of 1934–1935.

Before turning his attention back to quantum mechanics however, Einstein had returned to his roots. He had agreed to give the prestigious Gibbs Lecture of the American Mathematical Society, which was to be delivered in Pittsburgh at a joint meeting with the American Physical Society

[1] Another entrée to Einstein's world was provided by Rosen's wife Hanna, who was a fine pianist. She joined the ranks of Einstein accompanists.

Einstein. Brian Foster, Oxford University Press. © Brian Foster (2026). DOI: 10.1093/oso/9780198794875.003.0027

and the American Association for the Advancement of Science. This took place from 28 December 1934 to 1 January 1935. He decided to entitle his talk 'An Elementary Proof of the Theorem Concerning the Equivalence of Mass and Energy'. By considering the collisions of two particles and assuming only the Lorentz transformation and the principles of conservation of energy and momentum, he was able to show that E was indeed equivalent to mc^2.

Getting to Pittsburgh involved changing trains at least once, an operation fraught with uncertainty for an Einstein now unused to travel. Normally, Elsa would have accompanied him, but she developed a high fever, forcing her to request that Leon Watters replace her. Einstein and Watters arrived in Pittsburgh the day before the conference began. Einstein stayed with Edgar Kaufmann, a rich department-store owner. Watters arranged a news conference for the following morning, 28 December. Einstein was asked about his views on determinism in quantum physics, where he displayed his usual conviction that determinism would eventually be rescued from quantum-mechanical probabilities. Interestingly, he was also asked about the possibility of using the equivalence of energy and mass to extract energy from the atom. Just like his eminent experimental colleague, Lord Rutherford, Einstein was sceptical: 'I am not a prophet, but I feel absolutely (or at least nearly) sure that it will not be possible to convert matter into energy for practical purposes. You must employ a lot of energy to get any energy out of the molecule, and the rest is lost.' This made the newspaper headlines: 'Atom energy hope is spiked by Einstein,' and 'Man can't harness atom as source of energy [Einstein] says.'

Einstein's lecture itself took place that afternoon in a venue than could hold 400–450. Naturally all tickets were allocated; it was reported that 3000 people waited outside and that tickets were going on the black market at $50 each. Einstein spoke in English, occasionally asking the audience for a German word to be translated, for just less than an hour. Equally predictable were the newspaper headlines: 'Einstein Lectures! – Crowd of 450 Gapes – Then Grows Sleepy'. Einstein attended a dinner in his honour on the following evening before he and Watters, who had also developed a fever and was looked after by Einstein, returned to Princeton on 30 December in time to greet the New Year of 1935 at home.

The exact genesis of the Einstein–Podolsky–Rosen (EPR) paper is unclear. Surprisingly, none of the authors left any description of it. Asher Peres, a student of Rosen's many years later, reported that Rosen had talked to Einstein during the traditional 3 p.m. IAS tea gathering about the problem of wave functions of separated quantum objects. Einstein was immediately interested. Podolsky joined the conversation and the collaboration began. As Einstein confirmed in a letter to Schrödinger (see below), Podolsky was assigned the task of writing the paper. This seems slightly strange, since Podolsky's native language was Russian and he had only moved to the US at the age of 17, whereas Rosen was an American Jew born in Brooklyn. Even the title of the paper, 'Can Quantum-Mechanical Description of Physical Reality Be Considered Complete,' with its missing article, betrays the author's origin. Podolsky wrote the paper and despatched it to the *Physical Review* on 24 March.

The EPR paper is under four pages long. It begins by proposing that the validity of a theory should be assessed by testing its predictions against experiment and by whether every 'element of physical reality' has a counterpart within the theory; this second requirement is called 'completeness'. The agreement of quantum mechanics with experiment is not contested; its completeness is. The authors recognise that the term 'element of physical reality' requires definition, and in a key statement they define it: 'If, without in any way disturbing a system, we can predict with certainty (i.e., with probability equal to unity) the value of a physical quantity, then there exists an element of physical reality corresponding to this physical quantity.'

The authors illustrate what they mean by considering a particle with a definite value of its momentum. Since, for a non-interacting particle, momentum conservation implies that this value can be predicted with certainty, its momentum is an 'element of physical reality'. The standard quantum-mechanical description of this is a quantum-mechanical 'state' that has a single definite value of momentum. Thus there is a one-to-one correspondence between the element of physical reality and the theoretical description. Furthermore, quantum mechanics predicts that if the position of such a particle is measured, then any result is as likely as any other. Therefore, in this case, the particle's position should not be an 'element of physical reality'. In quantum mechanics, momentum (P) and position (Q) are examples of 'non-commuting' physical quantities and cannot have simultaneous[2] definite values: a definite value of the momentum leaves the position completely indeterminate, and vice versa. Thus, according to quantum mechanics, momentum and position cannot both be, simultaneously, elements of physical reality of any system.

However, in the second part of the paper the authors propose a counter-example consisting of two systems that interact for a given time and then cease to interact. The standard quantum-mechanical description of this situation is that a measurement of some physical quantity of one of the systems (system 1), made when the two systems no longer interact, allows the value of that quantity of the other system (system 2) to be predicted with certainty. Hence, the authors argue, this quantity of system 2 is an 'element of physical reality'. However, if the experimenter chooses instead to measure a different physical quantity of system 1, then, by the same argument, this different property of system 2 would be an element of physical reality. The experimenter's choice of what physical quantity to measure on system 1, made when the systems no longer interact, can in no way affect system 2. Therefore, both these physical quantities of system 2 must be elements of physical reality. However, if the two quantities measured on system 1 do not commute, for example, momentum (P) and position (Q), the laws of quantum mechanics forbid this, which precludes quantum mechanics from being a complete theory.

The authors recognise that the inferred values of P and Q for the second system rely on measurements on the first that cannot be made simultaneously, and so write:

> One could object to this conclusion on the grounds that our criterion of reality is not sufficiently restrictive. Indeed, one would not arrive at our conclusion if one insisted that two or more physical quantities can be regarded as simultaneous elements of reality *only when they can be simultaneously measured or predicted*. On this point of view, since either one or the other, but not both simultaneously, of the quantities P and Q can be predicted, they are not simultaneously real. This makes the reality of P and Q depend upon the process of measurement carried out on the first system, which does not disturb the second system in any way. No reasonable definition of reality could be expected to permit this [authors' italics].

On 4 May 1935, *The New York Times* carried an article headlined 'Einstein Attacks Quantum Theory'. It contains a summary of the EPR paper, several direct quotes, and a commentary from Boris Podolsky. The 7 May issue contained a 'Statement by Einstein' saying 'Any information upon which the article 'Einstein Attacks Quantum Theory' in your issue of May 6 [sic] is based was given to you without my authority. It is my invariable practice to discuss scientific matters only in the appropriate forum and I deprecate advanced publication . . . in the secular press.' Einstein's annoyance with Podolosky is palpable and similar to his falling out with Hans Reichenbach for a similar transgression in 1929. It is not clear whether he ever forgave Podolsky, who left the IAS

---

[2] The meaning of 'simultaneous' here is not related to temporal simultaneity, which Einstein, of all people, knew was frame dependent. What is meant is 'by the same measurement'.

a few months later for a faculty position at the University of Cincinnati. He and Einstein never collaborated again; indeed, Asher Peres thought that they never spoke again.

Einstein had another reason to be irritated with Podolsky. For whatever reason, he had not seen[3] the text of the paper before submission and he was disappointed by it. On 19 June, he wrote to Schrödinger that 'After many discussions, for reasons of language, Podolsky wrote the paper. It did not come out as well as I had intended; the essential point is, one might say, buried beneath the erudition.' It seems clear that Einstein's comments relate to the formulae and specific example discussed in the second part of the EPR paper. His concern about completeness and/or non-locality could be adequately illustrated by the much simpler example of the collapse of the wave function of an electron striking a detector, which he had discussed at the 1927 Solvay Conference.

After the publication of EPR, Einstein immediately began to try to clarify what his position really was. In the 19 June letter to Schrödinger, he used an example of a ball that could be placed in one of two boxes. He introduced what he called a 'separation principle' by which the contents of the second box are independent of anything that happens to the first box. While he admitted that this was an imperfect analogy to the quantum-mechanical situation, he maintained that since the ball is 'in reality' in one box or the other, the quantum-mechanical description that, until a measurement is made, it has equal probability to be in either, is an incomplete description of reality. He went on to use a two-particle system, as in the EPR paper, to amplify his point. Here he demonstrated that the same physical situation was described by two different wave functions and therefore again concluded that the quantum-mechanical description was incomplete.

Schrödinger's answer arrived on 13 July. This catalysed a further lengthy letter from Einstein, who began his reply on 8 August by saying that Schrödinger was the only person with whom he could discuss these things. In the August letter, Einstein tried to extend his 'ball in a box' example to a system that is demonstrably quantum mechanical:

> The system is a substance in a chemically unstable equilibrium, such as a pile of gunpowder, that can spontaneously combust due to internal properties, with an average lifetime of the system of around a year. This could be quite readily described quantum mechanically. At the start, the psi [wave]-function characterises a sufficiently well-defined macroscopic system. Your equation however ensures that this is no longer the case after a year. The psi-function then describes much more a sort of mixture of a not-exploded and an exploded system. By no kind of interpretation can this psi-function be made to furnish an adequate description of what has actually happened; in reality there is no intermediate state between exploding and not exploding.

Einstein proposed that sense could be made of this if the psi (wave)function represented an ensemble of possibilities. In his reply on 19 August, Schrödinger refuted this. While he may have rejected Einstein's interpretation, his 'gunpowder' analogy clearly hit home; the idea of amplifying a quantum-mechanical process to give distinct macroscopic effects is the key concept underlying the famous 'Schrödinger cat'. This was discussed in Schrödinger's quantum-mechanical review, which appeared in three instalments between November and December 1935. The 'quantum entanglement' of wave functions was elucidated in the December article. This triad of concepts, EPR, Schrödinger's (perhaps it should be Schrödinger–Einstein's) Cat[4] and entanglement, sets the scene for an intellectual explosion even more powerful than Einstein's gunpowder.

---

[3] It seems impossible to believe that Podolsky submitted it to the journal without sending copies to his co-authors for comment. Presumably neither got around to looking at it before Podolsky's deadline to submit it.

[4] A discussion of the genesis of the Schrödinger–(Einstein) cat is given by Fine.

The EPR paper had an immediate and strong impact on the 'high priest' of quantum mechanics, Niels Bohr. Bohr's collaborator Rosenfeld recalled that it came as a bolt from the blue and that its effect on Bohr was remarkable. He and Bohr worked night and day to formulate a response, which was announced by Bohr's letter to *Nature* on 29 June. In typically obscure language, he called foul because the EPR definition of quantities that partook of 'reality' did not include the measurement apparatus. He submitted a paper to *Physical Review* with the same title as EPR on 13 July. His paper built on the idea that the measurement of a property, such as position, for one of the two EPR particles already implies setting up initial conditions that would preclude the measurement of, for example, momentum. In other words, although he agreed that there was no mechanical disturbance of the second particle by a measurement of the first, there was an 'influence', through the initial conditions that affects the outcome of the second measurement, whatever the spatial distances involved. In 1949, Bohr, re-reading what he had written, admitted that he was deeply aware of the obscurity that must have made it very difficult to understand his argumentation. Schrödinger, in a letter to Edward Teller, compared arguments similar to that of Bohr to the standpoint of a savage who believes that he can harm his enemy by sticking a needle into his image.

Nevertheless, Bohr's rejoinder was good enough for the overwhelming majority of practising physicists, who were unconcerned with whether quantum mechanics was complete but only with whether it worked. It did, triumphantly; at least for particle velocities far from the speed of light. Nevertheless, EPR worried the top echelons of quantum mechanists for a time. Schrödinger in his July letter to Einstein told him:

> I am now having fun and taking your note to its source to provoke the most diverse, clever people: London, Teller, Born, Pauli, Szilard, Weyl. The best response so far is from Pauli who at least admits that the use of the word 'state' for the psi-function is quite disreputable. What I have so far seen by way of published reactions is less witty. [. . .] It is as if one person said, 'It is bitter cold in Chicago'; and another answered, 'That is a fallacy, it is very hot in Florida.'

In recollecting the period many years later, Bohr believed that he had eventually convinced Podolsky to recant his heresy. Rosen however remained recalcitrant. Bohr also described the reaction of some of his eminent colleagues at the time of the EPR publication:

> Dirac came here after [EPR appeared]. He said that now – Dirac had been very, very curious – he has said it several times, he said: Now we have to start all over again, because Einstein has proved that it does not work [. . .] Yes at that time [Dirac thought that somehow the whole scheme of quantum mechanics had to be given up] but we convinced him that it was not so. Rosenfeld and I, we talked about a week here at Carlsberg. [. . .] it was very interesting. [Dirac] is a complete logical genius. But that is the kind of errors he makes.

After the initial exchanges, the EPR paper was essentially forgotten. Abraham Pais writing in the early 1980s summarised EPR by opining that the only part that would survive would be the final sentence of the penultimate paragraph quoted above: 'No reasonable definition of reality could be expected to permit this.' This, Pais believed, poignantly summarised Einstein's views of quantum mechanics. Indeed, between its publication and about 1975, EPR was cited in other publications in single figures per year.

However, Pais's obituary of EPR has proven premature: first, the EPR two-particle example in terms of momentum and position was recast in terms of spin components of two particles that fly apart from a common source. This made investigation by experiment much more feasible. Secondly, and crucially, John Bell showed that some of the quantum-mechanical predictions of the spin correlations in such a two-particle system differ quantitatively from any theory that respected the EPR idea that measurement on one particle should not affect the state of the second particle, which can be arbitrarily far from the first. Experimental tests of 'Bell's inequality' began in the early 1970s and have continued with increasing refinement. The results confirm the predictions of quantum mechanics and hence the idea of quantum entanglement. The 2022 Nobel prize was awarded for the experimental work in this area. Furthermore, quantum entanglement has become the basis of a quantum-devices and quantum-computation industry that is already a multi-billion-dollar worldwide enterprise and is set for further exponential growth. It is therefore unsurprising that, as of 2023, the EPR paper had been cited more than 11,000 times; by the end of 2024, this had risen to 25,000.

Einstein's views on the central issues of the EPR paper evolved somewhat with the years. Despite what is often assumed, the final paragraph of the EPR paper: 'While we have thus shown that the wave function does not provide a complete description of the physical reality, we left open the question of whether or not such a description exists. We believe, however, that such a theory is possible' does not imply that he, or even necessarily Podolsky, believed in some form of 'hidden-variables theory' of quantum mechanics. Einstein's view in later life was that it made no sense to make progress towards a 'final' theory by tinkering with quantum mechanics. Rather, a completely new approach was necessary. His final viewpoint can be best summed up from an article he wrote in 1948: 'The paradox forces us to relinquish one of the following two assertions:

(1) the description by means of the psi [wave]-function is complete

(2) the real states of spatially separate objects are independent of each other.'

Although many aspects of the quantum mechanics of the EPR problem remain deeply mysterious, it would be fair to summarise the conventional wisdom as relinquishing the second of Einstein's assertions. 'Spooky action at a distance', to use Einstein's famous phrase, really seems to be how the universe works.

Einstein's domestic situation was just as entangled as quantum mechanics during the genesis of the EPR paper. As discussed in Chapter 26, Elsa had returned from Watch Hill with very restricted movement in her left leg due to sciatica. Her physical condition gradually improved, and she reported it as back to normal by January 1935. However, mentally she never recovered from the death of her beloved daughter. When Emil Hilb wrote to her in January 1935 asking whether he had offended her, she excused her long silence by telling him that:

> I lead a completely retired life. In the more than three months in which I have been back in Princeton, I have only left the house three times. I much prefer to be left alone in this isolation; I am really no longer good company. It is completely different to last year, when I was often in New York and didn't avoid people.

Einstein however was as busy with musical engagements as ever: Hilb sent his usual German culinary delicacies in January via a taxi and was rewarded with an invitation from Elsa to visit Library Place, which however had to be scheduled around Einstein's engagements to play string quartets.

Margot's husband was still missing from the Einstein family. Dima Marianoff had been living in Paris since fleeing Berlin in March 1933. He busied himself by working for various Jewish refugee charities as well as selling articles to newspapers. He had also founded a school at Born in the south of France in which German Jewish exiles in the professions such as law and architecture could retrain for agricultural pursuits, with a view to emigrating to Palestine. By 1935, the initial wave of Jewish refugees from Germany had died down, leaving Marianoff without much to do and his school in some financial difficulties. Both Einstein and Elsa had written to him regularly and sympathetically since they arrived in Princeton. He decided on 1 March to apply for permission to enter the US, which cannot have been easy as by now he no longer had a proper passport, only the so-called Nansen passport. These had been issued from 1922 onwards, following the decision of the Soviet government to remove citizenship from all Russians living abroad. They had to be renewed annually. Einstein, as he did for so many people, wrote to the immigration authorities on Marianoff's behalf and also intervened when his application stalled. Marianoff sailed from Cherbourg on what turned out to be the penultimate voyage of the RMS *Olympic*, sister ship to the *Titanic*, on 6 March, 1935. On the evening of 12 March, he was met at the New York docks by Elsa and Margot and taken to Princeton[5].

It was not long before Marianoff found himself in some difficulties. One of his aims in coming to the US was to find financial support for his farm school in France. Einstein was as usual ready to help a good Jewish cause, even to the extent of attending a fund-raising banquet organised by a committee interested in Marianoff's school. Einstein had spoken only a few weeks before at a Passover banquet together with his friend Rabbi Stephen Wise, while in February he had been guest of honour at a banquet in Philadelphia to benefit Bikur Cholim Hospital and Home for Incurables in Jerusalem. He was photographed at this event somewhat bemusedly holding a large silver trophy presented to him by Judge Charles Klein. Elsa was sufficiently recovered also to make the trip to Philadelphia. However, on 7 April, banquet fatigue had set in, and Einstein had declined to speak at a dinner in honour of the founder of the Deborah Tuberculosis Sanitorium, apparently because he was too sleepy. Marianoff's dinner was scheduled for 15 May. However, Einstein learned that he had been awarded the Franklin Medal by the Franklin Institute of Philadelphia, whose awards ceremony on 15 May took precedence. A further unfortunate repercussion came from the head of the National Farm School, Herbert Allman, who thought that Marianoff was trying to usurp his school's position. He wrote to Einstein inviting him and Marianoff to visit; Marianoff accepted and came alone towards the end of May, just before his departure to Bermuda with the rest of the Einstein family (see below). No doubt he clarified that his particular interest was in developing schools particularly for the Jewish community, both in France and the US.

As it transpired, Marianoff was probably fortunate that the award of the Franklin Medal took precedence. Einstein's speech fatigue had set in once more. Despite agreeing beforehand to give an address at the presentation ceremony, Einstein refused to speak, explaining that, despite hoping for inspiration, none had arrived and therefore he had nothing to say. As *The New York Times* remarked, the audience was obviously disappointed. His muse deserting him was perhaps no surprise as two nights previously he had been guest of honour at a banquet at the Waldorf-Astoria, where he had delivered an address in German in aid of a fund for the children of German exiles, which raised $6,000.

---

[5] Marianoff's description of his entry to the US in his book conflates, with typical inaccuracy, two separate voyages, one in March and the other in September 1935. Having returned to France sometime in late July, presumably on business related to his farm school, he had almost been deported on arrival back at New York because his passport did not have the requisite six months' future validity. He was saved by the intervention of an acquaintance he had made on board the new luxurious 'Blue Riband' holder SS *Normandie*. This was the distinguished financier and statesman, Bernard Baruch, introduced to Marianoff by Einstein's old friend Jacob Landau, founder of the Jewish Telegraphic Agency, who was also aboard the *Normandie*.

Elsa Einstein kept up her correspondence with Antonina Vallentin, even though in May 1935 she admitted that she was very tired and hardly ever wrote letters now, having rather to dictate them, usually to Helen Dukas. A few months previously, she had dissuaded Vallentin from approaching Einstein with her idea to write a biography of him. Elsa was adamant that Einstein would never agree, even though she was sure he would admire her recent biography of the poet Heinrich Heine. She explained:

> You know him well enough, he doesn't like himself and he will never deign to pose for you for his biography. He would tell you that he had some very deep, wonderful thoughts and wrote them down. A few pages are immortal. But otherwise everything is rubbish. He himself is a person like any other, with many faults and it would not be worthwhile for him to tell you about himself. I'm certain this is the answer you would get if you came here for this purpose.

It seems that Vallentin took her advice, although that did not stop her eventually writing a biography without Einstein's collaboration.

By May 1935, Elsa was at least reassured that Margot had recovered from the loss of her sister, which had catalysed a burst of creativity with her sculpture. She seemed to be strong and resilient, although Elsa made no mention of her reunion with her husband. Elsa told both Vallentin and her other correspondents, the Alfred Einsteins, that she would never return to Europe. With Margot in Princeton, there was no longer anything for her there except sad memories. Her attitude reinforced Einstein's own hardening determination that they would never return.

Meanwhile, Elsa reported to Vallentin that Einstein had laid a great egg, by which she meant produced another great advance in physics, although she admitted that no one would either realise or believe it. She was presumably referring to a paper that Einstein wrote with Nathan Rosen. They had now returned to their beloved general relativity, whose dwindling band of practitioners had been further depleted with the death of Einstein's colleague and friend from Leiden, Willem de Sitter, in November 1934. A correspondence with another old friend, Ludwig Silberstein, catalysed the Einstein–Rosen paper entitled 'The Particle Problem in the General Theory of Relativity'. It tried to construct a theory of massive charged particles by excluding singularities from Maxwell's equations and the Einstein equations. Although the theory had attractive elements, the abstract admitted that the many-particle problem, which would decide the value of the theory, had not yet been treated. It turned out to be yet another failure but, not unusually for Einstein's failures, contained an idea that has subsequently proven very influential. The two space-time planes that Einstein and Rosen required for their particle theory were linked by what came to be called 'Einstein–Rosen bridges'. This construct had first been postulated by Ludwig Flamm in 1916 in one of the earliest papers investigating general relativity, contemporary with the famous Schwarzschild solution. This had been forgotten however by Einstein and Rosen, who rediscovered what subsequently came to be known as 'wormholes'. The possibility thereby inherent of at least information transfer faster than the speed of light has been probably more influential among science-fiction writers than scientists.

Georges Lemaître visited Princeton at the end of January 1935; he stayed until June. As Schrödinger had, he resided in the Graduate College, quite close to Einstein's house in Library Close. Many years later, Lemaître recalled the debate on the cosmological constant and the controversy with Ludwig Silberstein, who claimed to have proved general relativity to be incorrect:

> Finally, in 1935, I visited Einstein again, for the last time, at Princeton and he told me, 'Your mobility is remarkable.' I was too relativistic to not draw some conclusions on that about Einstein himself, but I refrained from any comment. In the same period, the occasion presented itself to me to organize a meeting between a few professors of the Institute, to which Einstein came to present the research he

was conducting at that time, which he had not yet spoken about. The theory was rather bizarre and was received with considerable coldness. I had the impression of senility being rumoured.

In reality, it was one of the most interesting events. Einstein had embarked on a dead-end road following a communication received by letter from his friend Silberstein, a well-known scholar of relativity. He wrote him that he had found a solution to the equations on gravitation, where two point singularities harmonized themselves with each other. According to the approach of Einstein, such a situation radically excluded the most natural way of conducting his research and led him onto imaginative roads.

A short time later, he let Silberstein publish his solution and it was then that one could recognize the claw of the old lion in the masterly way in which he dismantled the subtle paradox of Silberstein and demonstrated a line of immobile singularities, which played the role of a rod linking the two masses, preventing them in a completely natural way from precipitating towards each other.

One day discussing the Dirac equation, which is the origin of semiconductors and spinors, he told me, 'The equation of Dirac – it is an authentic miracle.' Naturally I resumed the discussion concerning the cosmological constant and for a moment I had the impression of being able to stump him: 'Still', he told me, 'if you manage to demonstrate that the cosmological constant is not zero, it would be important.'

[...]

No more than others, on subjects much more dear to his heart (I refer to Born, Pauli, Heitler, Bohr), was I able to convince him, or, I must admit, to grasp his thinking in a more precise fashion.

Einstein's satisfaction with his life in the US was such that he decided to become a citizen. Having refused a resolution in Congress in March 1934 that a special law be passed in his favour, and a similar suggestion by President Roosevelt, he and the family chose to follow standard procedure. This meant that they needed to leave the country and re-enter with the appropriate visa declaring that they wished to take up citizenship. The Princeton term having ended on 3 May, the Einsteins booked passage to Bermuda. Einstein, Elsa, Margot, Dima Marianoff, and Helen Dukas sailed with the *Queen of Bermuda* from New York on 25 May. Also on board was Governor Herbert Lehman of New York and his family. The Governor of Bermuda met the *Queen of Bermuda* when it docked on 27 May. He was able to give Einstein some tips on hotels at which his party might stay. Einstein, however, when he inspected them, pronounced them much too luxurious and carried on to the far side of the town. Here he found a cottage with rooms to let where he installed the party.

Einstein spent most of his stay in Bermuda on boats, including one owned by a German chef at whose restaurant they had eaten and who invited him for a day's sailing and dinner. The Einstein party, without Margot, who stayed on the island, left on 1 June, docking in New York on 3 June, when all obtained the forms required to eventually become US citizens.

Einstein and Elsa travelled directly from New York to the summer house that Elsa had rented from 1 June, the White House in Old Lyme, Connecticut. Elsa wrote to their friend Otto Nathan that:

Bermuda was unsurpassably beautiful. The island is a dreamland. Here in the White House, all is peace and quiet, an unearthly tranquillity. Only on very remote tracks in the mountains does one sometimes experience such a deep silence. Particularly at night it is almost unearthly. We are still completely alone except for the caretaker. However, in the next few days a couple of houses in the neighbourhood will be occupied and then probably it will be a bit less abandoned and lonely, but also not so peaceful. We live wonderfully, so delightfully that Albert is ashamed and until now we have only eaten at the servants table in the kitchen. In such a mansion as this, you don't eat, you dine! The dining room looks as if Albert should go around in white waistcoat and patent-leather shoes and resembles Count Metternich's. We have never been there, but that is how it seems to us.

[American aviator Charles] Lindbergh rented this house two years ago for $10,000 for three months. We got it for $750, which is still a lot for a summer house but in comparison to the beauty of the place is nothing. The grounds are 20 acres, there is a tennis court, a swimming pool and all sorts of summer delights that we didn't ask for but have the use of anyway. In this house there would certainly have always been a butler and a maid for the lady. All this brings Frl. Dukas into ecstasy. There is an enormous garage next to the house with space for eight cars. Our cat has displaced the noble mastiffs and the saddle horses.

Come and stay with us and see our noble surroundings for yourself. Probably this is the only chance as next year our housing costs will have gone up and we won't be able to afford the luxury of such a big place.

Margot has stayed in Bermuda; she couldn't bear to part from the beauty of the island. I hope Princeton isn't too hot and too lonely. If you miss us too much, then don't hesitate, jump on a train and come straight here. Although a telegram before you arrive would be nice.

Except for an interruption in early June, when Einstein travelled to Harvard to receive an honorary degree, he spent an idyllic summer music-making and sailing on the Connecticut River, which ran just below the grand White House estate. He had the usual run-ins with tide and weather, for example, getting stuck on a sandbar and having to be towed off by a passing power boat. Two neighbours, Joseph Copp, an amateur violinist and his wife, a professional pianist, were frequent musical partners. Copp picked Einstein up on most evenings and drove him to their home. Mrs Copp recalled that Einstein '[would come in] to the music room and put his violin case on the sofa and open the case in a great rush to get to his violin. He was so naïve and charming and utterly guileless and he didn't put on airs.'

Many visitors were received, some more welcome than others. Those welcomed included several family members. Hans Meyer, who was a journalist, was Einstein's nephew, son of Elsa's sister Paula. He had tremendous difficulties in getting permission to enter the US, which Elsa believed was due to the anti-Semitism and Nazi sympathies of the Consul-General. Once again these were solved by Einstein intervening personally, shooting off letters from Old Lyme before his sailing trips. Meyer sailed from Cherbourg on 25 September with the SS *Majestic*, arriving at New York one week later.

The Einsteins' son-in-law, Rudi Kayser, sailed from Rotterdam on 14 September on the SS *Statendam*. He had still not fully recovered from Ilse's death on his arrival at Old Lyme. Einstein had exerted himself not only to get him into the US but also to find him an academic position, as indeed he had done once before, only for Rudi and Ilse to refuse to leave Europe. By early 1936, Rudi had recovered his emotional balance but had not yet found a position. Elsa bewailed how difficult it was for a German Jew to find a position in an American university, which she attributed to widespread anti-Semitism. Einstein's influence had diminished potency within arts faculties but nevertheless Kayser was able to obtain a position first with the New School in New York and then with Hunter College. Elsa was pleased that, in addition to a new life in the US, he also found a new wife. In 1936 he married Eva Urgiss, who had also fled from the Nazis and arrived in New York a month after Kayser.

The welfare of musical partners from the Berlin days were on the Einsteins' minds that summer. On 25 August Elsa wrote to the virtuoso violinist Boris Schwarz, who had played with Einstein in Berlin. She had become very fond of the young Schwarz and enquired whether he would like to escape the Nazis and come to the United States. If so, she and Einstein offered to do all they could to help. Schwarz replied enthusiastically in the affirmative and the slow process of obtaining an entrance visa began.

Another Berlin musical friend who visited Old Lyme was the flamboyant cellist and playboy Francesco von Mendelssohn. He arrived in New York on 10 September 1935, on the SS *Majestic*, apparently with ten cellos, which must have made an interesting scene at the customs warehouse. His entry papers described his occupation as a stage director. Also travelling with him was his sister Eleanora Jeszenszky, who described herself as an actress, and the composer Kurt Weil and his wife. They had all travelled to New York to work with Max Reinhardt, who arrived on 3 October. Francesco was characterised as Reinhardt's 'right-hand man' in the stage production of Franz Werfel's *The Road of Promise* at the Manhattan Opera House,[6] scheduled to begin in mid-December. Reinhardt attended the opening night of his movie of *A Midsummer Night's Dream*, starring some of Hollywood's biggest names, on 9 October. There was hope among the organising committee that Reinhardt would agree to curate a festival in conjunction with the upcoming World's Fair to be held in New York, which eventually took place in 1939. The wooing process included a gala dinner at the Waldorf-Astoria on 8 October to welcome Reinhardt and try to drum up money for the production of *The Road of Promise*; both Mayor Fiorello La Guardia and Einstein gave speeches. On his way back to Princeton from his idyll at Old Lyme, it can be imagined with what feelings Einstein greeted the resumption of the gala-dinner circuit, particularly as he had already given an address in honour of Reinhardt in June, when he had been greeted in the city after leaving Hollywood en route to a summer stay in Italy. However, Einstein did have an ulterior motive, other than supporting his old friend Reinhardt, a fellow Jewish exile: Dima Marianoff was also working with Reinhardt on the staging of *The Road of Promise*.

Marianoff had arrived back from France just before Reinhardt, whom he had presumably known in Berlin. Elsa worried about his future in a letter to Antonina Vallentin:

> Dima is working for Reinhardt on the new piece by Werfel. This blessing will however not last very long. Once the play is finished, what then? He doesn't understand a word of English. I am fond of him and would like to smooth his way in life here. I know that you and countless others know awful things about him. For the most part, I believe these horrible things. Nevertheless: I am fond of him and want to look out for him as much as I can.

Elsa was as good as her word and supported Marianoff financially. It is interesting to speculate whether the 'horrible things' that Elsa mentions were related to the possibility of Marianoff being a Russian spy. The fact that this was written to Antonina Vallentin, whose former employment in the German Ministry for Foreign Affairs would have given her access to confidential information, certainly adds to the likelihood.

In late July, Otto Nathan called in for a day to Old Lyme on his way to board ship for a trip to Europe. His position at Princeton had not been renewed, which Einstein considered was clearly due to the anti-Semitism rife there. Both Elsa and Einstein were relieved that he had found a teaching position in New York so that they would be relatively close to each other. He had already become their most trusted adviser on financial and legal matters, a position that would deepen further with the years. Their other close friends, the Buckys, did not join them until August but Einstein and Bucky exchanged letters throughout July. They discussed ideas for various inventions, in particular taking out patents for an airplane horizon indicator and an electrostatic microphone, which they discussed with Wall Street patent lawyers. Towards the end of July, David Sarnoff, the President of the Radio Corporation of America (RCA) and his wife visited Old Lyme to discuss the microphone idea. Sarnoff was very keen to establish a mutually profitable relationship between RCA and Einstein. Another visitor in September was Felix Frankfurter, Professor of Law

---

[6] Reinhardt's colleagues Weisgal and Bel Geddes had first tried to obtain permission to stage the production in the Kingsbridge Armoury in Manhattan. Einstein was roped in to call his friend Governor Lehman to try to get permission. Unsurprisingly, Einstein was unable to convince the Governor of New York to sign over sole use of the armoury for an unspecified duration for a theatrical presentation, however worthy.

at Harvard and a trusted adviser to President Roosevelt, who in 1939 appointed him as only the third Jewish Associate Justice of the Supreme Court.

Not so welcome visitors were members of the press. Artist and journalist S. J. Woolf arrived in July and convinced Einstein to sit for a portrait and interview, which Einstein insisted however was not for publication. Luigi Pirandello, who had won the Nobel Prize for Literature the previous year, arrived in New York on 20 July aboard SS *Conte di Savoia*, two weeks after his luggage and his secretary arrived on the *Normandie*. He denied the story that Mussolini had forbidden him to travel on a non-Italian ship. Pirandello, who was an acquaintance of Marianoff, gave an interview to the press on his arrival in New York. He defended Mussolini's bellicose tone towards Ethiopia, which Italy invaded two months later. Indeed, the playwright stated that, as *The New York Times* headlined, a modern state has the right to use force to civilise a barbaric people. He also opined that the Italian people were absolutely united in following the Duce. Such sentiments would have elicited no sympathy from Einstein, whom he visited at Old Lyme towards the end of July.

That July Einstein busied himself not only with his inventions with Bucky but also catching up on his correspondence and reading. This included a fine sketch of the life of his friend Walther Rathenau by Alfred Kempner, who wrote under the penname Alfred Kerr. He encouraged Kerr to find a good publisher and offered to write a foreword if that would help. He also enjoyed reading a book on philosophy, probably Phosophie ein unlösbares Problem, by Ludwig Freund. In the Old Lyme library, he came across Stuart Chase's *A New Deal*, which came to be identified with President Roosevelt's economic programme and may have been the inspiration for its name. Einstein was moved to write to the author expressing his admiration for its clarity and simplicity. Some weeks later, having with great effort deciphered Einstein's signature, Chase replied effusively and enclosed another of his books. For lighter reading, Einstein enjoyed writings by Mark Twain.

Einstein received a letter from Frieda Busch, wife of his friend the violinist Adolf Busch. She thought she detected a certain homesickness in his previous letter and admitted that both she and Adolf also felt this but that it was impossible to return to Germany. They were living in Basel, where Adolf spent most of his time composing. He had just completed a requiem for four boys' voices, choir, and orchestra. She hoped that Einstein would at least return to Europe where she promised that he and Adolf would once more play violin duets together as in the old days.

On 5 July, Einstein wrote to his old flame Grete Lebach. The letter is worth quoting because it leaves little doubt of the depth and importance of their relationship and also tends to substantiate Frieda Busch's impression of Einstein's homesickness:

> My dear little Lebach, We are sitting here in a delightful isolated spot near the coast. The adversity of circumstances and the delusion of my fellow human beings mean that I become more and more of a saint; age also helps and ensures that I don't fall out of character.
>
> [...]
>
> Our beautiful Kaputh [sic] times, those carefree times with unforgettable adventures stick in my memory as the golden age. Do you still remember the night landing near the Caputh inn, threatened by the calm and the enemy ship, which we only narrowly escaped? And then your adventure after we landed? The divine and human world order would be quite unbearable without the romance and tragicomedy that well up from sources that cannot be staunched. And our walks in Le Coq under the watchful but discreet eyes of those badly bred strange guardian angels, who devised for us a completely different protection than was officially intended? And the very curious goings-on of the exiles in Le Coq with the associated gossip and the little intrigues of the beautiful sex? Our lovely farewell, the momentous nature of which we, thank God, could not see? I am writing this with the beloved brown pen,

whose predecessor left me under such lovely circumstances. This pen with its three little gold rings has remained my true companion. With it, I have spent countless hours in quiet labour and I look after it jealously. It will not be taken away from me until my eyes close for the last time. With it, I wrote a piece of work a couple of months ago that, after interminable arduous toil, hopefully represents a first step towards a deeper understanding.

I follow with fearful anxiety the unsteady ship of your endangered Austria. Hopefully it will weather the storm so that you and your good and dear husband have a safe place in life after so many years of disappointment. With heartfelt kisses from your AE.'

Although Austria, as Einstein feared, was not to prove a safe haven, the two were able to sail together once again, as described in Chapter 28.

At the end of August, Einstein was reminded by Helen Dukas that he must write to his sister Maja. In his letter, he lamented both the current state of physics, in which each distrusts what the other thinks is a promising advance, and the state of the world, which looked as if it is in a permanent decline. He thought often in the past of 'Papa' (Jost) Winteler and the visionary correctness of his political opinions, but never with the force and clarity that the present situation evoked. Finally he expressed pleasure that Hans Albert had visited Maja and Paul in Florence. He called him good and true-hearted but could not resist a side-swipe at his marriage, which he compared to Margot's cat. This had just had kittens, whose father he characterised as less blue-blooded. Maja's domestic arrangements were also giving Einstein cause for concern. The house near Florence required substantial repairs that were beyond their means so that they were contemplating returning to Switzerland. Einstein warned them sternly against such a step and threatened never to help them again were they to take it. He offered financial assistance, even though he was severely worried both by Mileva's financial difficulties with her Zurich houses and by the requirement to lodge financial bonds in favour of the individuals such as Boris Schwarz whom he was supporting to emigrate to the US. These were required by the federal government should the immigrants fail to become self-supporting.

The reference to increased housing costs in Elsa's letter to Otto Nathan quoted above was to the Einsteins' purchase of a house at 112 Mercer Street in Princeton. The previous year, Elsa had thought about moving to another rented house but had eventually decided against it; the problems with the Library Place house in the meantime catalysed a move. Two blocks closer to the IAS than Library Place, the Mercer Street house was an unpretentious three-bay wooden construction built in 1838, set back from the road and with a long half-acre rear garden. In a letter to her friend the wife of Alfred Einstein, Elsa described the house:

The house we bought is about 120 years old [sic], so from an American perspective, ancient. Completely in contrast to Caputh, which seemed so American-hypermodern. It is in pure Colonial style, situated in the middle of the prettiest part of Princeton, with a wonderfully beautiful garden to the rear that has a proper meadow at the end, which I am just converting into a little grove, mostly of birches. I got some of my confiscated possessions out, after great complications and even greater expense, but they are there and look as if they have always been in and belonged to the old colonial house. [. . .] The house is so beautiful, not too large but big enough that we have enough space.

Elsa and Einstein were to remain there for the rest of their lives.

Elsa completed the purchase of 112 Mercer Street, which was bought in her name, on 24 July 1935. She commuted back and forth from Old Lyme as she supervised the installation of their belongings and the alterations required to the house, which included the construction of a study for Einstein on the second floor, with a large picture-window overlooking the garden. Watters also donated an electric refrigerator, which Elsa found particularly thoughtful. Although she was far from satisfied at the quality of the work carried out in her absence, she decided at the end of September that the house was ready for occupation. Einstein left Old Lyme on 27 September to spend a week with the Lehmans at their home in Port Chester. Elsa remained to wind up the household by 1 October. She returned to Princeton and their new home once she heard that the massive 'Biedermeier' furniture from Haberlandstrasse, as well as the furniture from Ilse's last apartment, had arrived at the docks.

As Elsa was supervising the moving vans outside 112 Mercer Street, sometime between 21–29 October, she became conscious of an impairment of her vision, which was later diagnosed as a macular oedema, a swelling behind the retina caused by blood haemorrhage. This was merely a symptom of a much more serious disease of the kidneys and heart, which forced Elsa to stay in bed for almost five weeks. She was devastated not to be able to enjoy the delightful garden that she hoped would be the centre of her future life. She also regretted not being able to arrange the house to her liking, instead leaving it to Margot, who she thought was dreading the prospect of having to take care of another invalid after the trauma of Ilse's death.

Not long after giving the after-dinner speech to welcome Max Reinhardt reported above, Einstein was once more on duty. On 22 October he shared the platform with former New York Governor and Presidential candidate Al Smith to raise money for German exiles, this time predominantly for non-Jews and political exiles. Einstein was reported as saying that:

> . . . to leave these victims to their misery would be a heavy blow to all those who believe in human solidarity, and would encourage those who believe only in force and oppression and act accordingly. That German Fascism is directed most against my Jewish brothers is a well-known fact. The reason given is to purify the Aryan race in Germany. As a matter of fact, no such Aryan race exists and the myth of the same has been invented solely to motivate the persecution and robbery of the Jews.

In November, similar remarks were made by economist Franz Oppenheimer at a dinner in his honour given by the New York Zionist Club at the Park Central Hotel, at which Einstein also spoke.

Max von Laue became the second friend of Einstein to visit Princeton in 1935. He sailed from Bremen to New York in mid-September and arrived at the IAS as a visiting scholar at the end of the month. On his arrival in New York he had stayed for a couple of days with Richard Courant, the eminent mathematician from Göttingen who had fled the Nazis in 1933. Courant drove von Laue to Princeton on 31 September. He would have stayed with the Einsteins, but they were in New York and after they returned Elsa wrote to von Laue that they were in the middle of moving but as soon as they had resolved the chaos, she would contact him. Sadly, she fell ill before she could do so. Von Laue therefore stayed on in the Nassau Club. In October, he lectured on thermodynamic equilibria and X-ray diffraction.

The third Einstein friend was Wolfgang Pauli, who left Zurich for Princeton with his new wife, Franca, on 20 September 1935. Pauli at this time was digesting the importance of the discovery of the positron and gave a series of lectures on the subject while at the IAS. Pauli bought a new car in order to import it back to Zurich, using it to show his wife the sights of the US, motoring across the country to the west coast. Once back in Zurich in early June, Pauli wrote to Werner Heisenberg that his experiences in Princeton convinced him that it was necessary to pursue a particular idea mathematically rather than by discussion. Such a predilection may be a reflection of Einstein's influence.

Einstein had begun over the summer in Old Lyme to push the candidacy of Carl von Ossietzky for the Nobel Peace Prize. Von Ossietzky, a pacifist, had been imprisoned by the Weimar authorities before Hitler came to power and was thrown into a concentration camp directly thereafter.

Einstein was hesitant about how best to achieve success but eventually on 27 October he wrote to the Nobel Committee espousing von Ossietzky's cause. Although no Prize was given in 1935, von Ossietzky was awarded the 1936 prize. This award caused the Nazis to pass a law forbidding German citizens from accepting any Nobel Prize. Von Ossietzky died in 1938 following severe beatings by the SS.

Elsa's condition having not markedly improved in December, she was admitted to the Montefiori Hospital in the Bronx for observation. The Director of the hospital, Dr Bluestone, tried a new treatment that seems to have, at least temporarily, worked wonders. By 10 January 1936, she was back at home in Mercer Street, attended by a nurse, from where she wrote to Bluestone in appreciation, telling him that he had given her back her life. To her friend Vallentin she confided that she had been afraid that she was going to die but the thought of leaving Margot alone had given her strength. She was moved by Einstein's solicitude and how he seemed to depend on her, which she would never have guessed.

On 15 January 1936, Einstein lodged his paperwork for United States citizenship at the District Court in Trenton, NJ. He made three pledges:

> ... to renounce forever all allegiance and fidelity to any foreign prince, potentate, state, or sovereignty, and particularly, by name, to the prince, potentate, state or sovereignty of which I may be at the time of admission a citizen or subject; I am not an anarchist; I am not a polygamist or believer in the practice of polygamy ....

It is remarkable that Einstein certainly did not comply with the first undertaking; he remained a Swiss citizen until his death. Of the second, a sizeable fraction of his future fellow citizens would harbour substantial doubts. As for the third, while he may have followed the letter of the law, he certainly transgressed in conviction and habit. Other remarkable things about the declaration can be found in the data that Einstein supplied. Demonstrating his notorious tendency to forget anniversaries, he not only wrote down the birthdates of his wife and two sons incorrectly but also his marriage date, which was wrong by two years and two months. He would be eligible to become a citizen in 1938, although he did not in fact complete the final steps until 1940. It may be that he waited until his stepdaughter Margot was also eligible. Together with Helen Dukas, all three were sworn in at Trenton on 1 October 1940.

On 1 February 1936, the great violinist Bronislaw Huberman wrote to Einstein. Huberman had been a typical nineteenth-century child prodigy. Born into an impoverished Jewish family in Częstochowa, Poland, his father was an unsuccessful lawyer. A keen amateur musician, his father soon realised Bronislaw's gifts. After studies at the Warsaw Conservatoire starting at age six, Huberman moved first to Paris and then to Berlin accompanied by his father, who by then had given up his legal job to exploit his son's earning potential. Bronislaw was taught by Joseph Joachim, who was astonished by his playing. However, he stopped taking lessons after the age of twelve, pushed by his father, of whose violent temper he lived in fear, into a full concert career. Johannes Brahms heard him play his violin concerto in Vienna in 1895 and was so moved by the boy's performance that he signed a musical quotation from the concerto, adding the words 'As a friendly memento from your happy and grateful listener'. A tour of the US in 1896–1897 brought Huberman the colossal fee of 100,000 guilders (more than $1M in today's currency). By 1898, burned out, he withdrew from public appearances for four years, after which he resumed his phenomenally successful concert career. His and Einstein's paths had crossed in the 1920s, when Huberman had become a strong supporter of a pan-European state. An outspoken opponent of the Nazis from well before Hitler's assumption of power, he refused to play in Germany and moved his teaching activities to Vienna in 1934.

Huberman's letter to Einstein concerned the founding of a first-class symphony orchestra in Palestine. He had toured Palestine early in 1934, playing with an orchestra of local musicians whose brass and woodwind sections in particular left something to be desired. They were starting to welcome refugee German players into their ranks. This gave Huberman the idea of systematically trying to recruit German Jewish musicians who were in the process of being thrown out of their positions. He began to devote some of his concert earnings to the cause and to tour first Germany and then other countries which he considered to be vulnerable to German aggression, in particular his native Poland. He carried out hundreds of auditions, often conducted blind since he knew many of the Polish musicians and needed to avoid his selection being swayed by emotion. By the beginning of 1936, he had the core of the orchestra in place but still lacked funds. Part he raised by a whirlwind tour of forty-two concerts through the US between February and April. He now appealed for Einstein's support, which he considered a matter of great importance for both Palestine and international Jewry. He offered to come to Princeton to discuss it with Einstein sometime after 7 February, when he was giving a piano trio concert[7] in the New York Town Hall with Artur Schnabel and Emmanuel Feuermann.

Shortly after 7 February, Huberman visited Einstein. They combined business with pleasure by playing together as well as discussing the Palestine Symphony Orchestra. Einstein was only too pleased to agree to chair the American Association of Friends of the Palestine Orchestra. He suggested that they should arrange a benefit concert and volunteered to bring his violin. Perhaps believing that his own concerts in aid of the orchestra were saturating that particular market, Huberman instead suggested a dinner for wealthy possible patrons. Einstein sent out printed invitations in his and Elsa's names for the event 'to meet Mr Bronislaw Huberman' to be held on 30 March in the Hotel Gotham on Fifth Avenue. The dinner was a great success, giving Huberman the funds required to launch the initial concerts. By then it had become known that the friend of Einstein and Huberman, the great conductor Arturo Toscanini, had agreed to conduct the first concerts in Palestine. Toscanini was also a convinced anti-fascist campaigner, who had refused to appear in Germany after the Nazi takeover. In late December, the maestro arrived in Palestine to conduct rehearsals in the concert hall in Tel Aviv, whose renovation had been completed just in time. On 27 December 1936, the final rehearsal was made open, given the insatiable demand for concert tickets. That evening, thousands waited outside, some climbing onto the roof of the concert hall in order to be able to hear snatches of the music. The programme, which consisted of Rossini's *Scala di Sieta* overture, Brahms' Second Symphony, Schubert's 'Unfinished' Symphony, Mendelssohn's Nocturne from *A Midsummer Night's Dream*, and Weber's *Oberon* overture, received an ovation of thirty minutes. The orchestra went from strength to strength. After the foundation of the state of Israel in 1948, it became the Israel Philharmonic, one of the world's great orchestras. It is estimated that, including player's families, Huberman and his vision saved more than 1,000 Jews from the Holocaust.

On 11 February, Einstein was one of the principal speakers at the opening of the new premises of the New York Museum of Science and Industry in the Rockefeller Center. A sophisticated show was initiated by Sir William Bragg, Director of the Royal Institution in London, who sat in Faraday's chair in his laboratory in the Royal Institution's premises. Linked by radio, Bragg lit a candle

[7] The concert was not a great success according to *The New York Times* review published the following day, particularly the Brahms Op. 8. This was the first time the musicians had performed together, and it showed in their differing ideas on interpretation. Matters improved in the Beethoven 'Ghost' trio and the final piece, the Schubert B-flat major trio, was thought by the critic to be both stylish and pleasing.

which, shining on a photoelectric cell, produced a signal that was transmitted across the Atlantic via radio and initiated the illumination of banks of high-brightness mercury-vapour lamps framing the museum's entrance. Einstein gave an address reminiscent of that he had given at the Berlin Radio Exhibition in 1930, when his world had been so very different. He encouraged listeners to understand the basic principles of operation of radio and the telephone by for example visiting the museum. Other speakers included Mayor La Guardia, the aviator Amelia Earhart, and Einstein's old friend Robert Millikan, who connected via radio from California.

On 20 March, Einstein wrote to Queen Elisabeth of the Belgians. She had experienced another tragedy in August 1935, when her daughter-in-law Queen Astrid had been killed in a road accident. Their mutual friend Catherine Barjansky had written to Einstein, telling him how distraught the Queen was at this further bereavement coming so close to the loss of her husband King Albert. Einstein's letter tried to give solace by describing the spring sunshine at Princeton and its symbolism in growth and renewal. He reminded her of the unchanging verities and beauty of their mutual love of music. 'Mozart', he mused, 'has remained as beautiful and tender as he always was and always will be.'

In April 1936, Einstein wrote to his son Tete, whose letters continued to reflect his mental disturbance, often long and rambling and concentrating on bizarre topics, his most recent one on nervousness. Einstein's letter is worth quoting in full as it touches on many aspects of his life at that time:

> Dear Tetel, Once more a good long time has passed since I wrote to you last. In the meantime, I was pleased to get a very complimentary report on [Hans] Albert's work. He's a good solid chap. I am sitting in my study with a blanket over my knees but still freezing because they never heat properly in deference to the Spring! However I like it here a lot, because one can just live for oneself. I am only plagued by letters. Other than that, I work with the same young man on problems so difficult that one is astonished at one's own courage. When I run out of steam, I look out of my big window, from which I can see a lawn and trees, a substantial piece of the sky and an enormous tower in the distance, a university building that looks just like that of an English university. People here only really have respect for England. When one says that one is Swiss, they find it strange because they think that only cheese and chocolate come from there.
>
> A few days ago I got the news of the death of my dear old friend Prof. Stern. He was a man good and honest like few others. I don't think he understood much about people, which is why he quickly turned to the study of all humanity. Therefore he succeeded, for the latter deceives one only impersonally.
>
> I often read the *Basler Nationalzeitung*, sometimes also that of Zurich. Fraulein Dukas gets them sent by her brother.[8] However, I rarely read a book. Science eats one up completely, particularly after the flexibility of youth has departed. In the end you are constrained to follow along familiar trains of thought, but the compensation is that you are more independent of those around you. I hear scarcely anything anymore directly from Germany. Everything that was familiar to me there is either dead or scattered across the world. However, Professor Pauli from Zurich is here, an extremely clever young physicist, to whom I may seem like an old fossil. He has the position in ETH that I had 23 years ago, when you saw the light in this best of all possible worlds.
>
> Freud will shortly be eighty. I have now indeed learnt enough to realise that he was correct with his main thesis. I have been told of a few simple cases from quite reliable sources, the interpretation of which seems unquestionable. In this area you saw more clearly than I very early on. This is probably connected with the fact that my private life has been so thoroughly and for so long practically out of my head (not only suppressed but also forgotten) that I have no living material to hand.

[8] Helen Dukas' brother was a lawyer in Zurich who looked after Mileva's tangled financial affairs.

Einstein subsequently wrote to Freud to congratulate him on his birthday, expressing similar sentiments about the correctness of his theories as he had written to Tete.

The mention of Hans Albert's work above stems from a letter that he had sent his father the previous year, in which he broached the possibility of moving to the US. Einstein replied that it was very difficult to find positions at the moment but asked him to send him his work so that he could show it to possibly helpful persons. It was presumably feedback from them that had pleased Einstein. In the same letter to Hans Albert, Einstein remarked that he hoped his thesis work would soon be published. 'I would be happy to pay for that, but not Tetel's insane nonsense. I am almost sure that this stuff with the Vienna doctor is a total swindle.' The Vienna doctor to whom Einstein referred was Manfred Sakel, to whom Mileva had taken Tete in 1934 for insulin-shock treatment, which was repeated in 1936 by Max Müller in Münsingen. The treatment made no noticeable improvement in Tete's condition, nor did a repeat in Burghölzli in 1942. Electroshock therapy, administered in 1944, was no more successful.

Although Einstein's attention in physics had moved from general relativity towards unified theories, he maintained an interest, which was aroused by an unusual visitor to Princeton in 1936, Rudi Mandl. Mandl was a Czech Jew, who had escaped captivity in Siberia as an Austro-Hungarian prisoner of war, graduated as an electrical engineer and eventually made his way to the US. At the time he met Einstein, he was working as a restaurant dishwasher in Washington DC. His interest in science was undimmed by his prosaic occupation and he managed to interest someone in the Washington offices of *Science News Letter* in his enthusiasms. Among several more outlandish ideas, Mandl had realised that general relativity predicted that the bending of light around a massive object could mimic the way that a magnifying glass bends light. Intrigued but not competent to evaluate Mandl's idea, the staff member offered to pay Mandl's fare to Princeton and arranged a meeting with Einstein on 17 April.

In fact, Einstein had had the idea of gravitational lenses as early as 1912, even before his formulation of general relativity, but had discounted it at the time since he, and several subsequent authors who had also picked it up, such as Arthur Eddington, had thought it too small an effect to be observable. Einstein had subsequently forgotten all about this idea so was intrigued when Mandl brought it to his attention. However, his interest quickly subsided and had not Mandl continued to pepper him with letters on details of the calculation, would certainly have let it drop. Mandl persisted however, in the interim moving to New York. Since their correspondence dried up, Mandl wrote to his original sponsors, enquiring whether Einstein had in fact published their idea. The journal asked Einstein, who, thus stimulated, wrote a brief note which was published in *Science* on 4 December 1936.

Acknowledging Mandl's role in the publication at the outset, Einstein showed that a perfect gravitational lens focuses to a ring centred on the point joining the star to the observer, unlike a perfect optical lens, which focuses a parallel monochromatic beam of light to a point. He concluded that, although the effect had several intriguing features, it would be very difficult to observe for stars.

Einstein's paper immediately aroused the interest of Fritz Zwicky, who had known him when Einstein was a visitor at Caltech. Zwicky realised that the lensing of massive objects such as galaxies would be much more likely to be visible than the effect of a single star. He published two papers on the subject. However, it was only in 1988 that the first such ring was finally observed. Gravitational lensing observations have now become commonplace and are among the most frequently used tools in astrophysics.

The last piece of research that Einstein carried out with Rosen before the latter's departure for Ukraine was also in the area of general relativity, in particular the existence of gravitational waves. This arose from Einstein's desire to find shortcomings of his own theory that might point the way to a unified theory. In the original exposition of general relativity, Einstein had shown that gravitational waves could exist and in a subsequent paper shown that dipole radiation (e.g. in which two spheres oscillate backwards and forwards along a line connecting their centres) could not; the quadrupole term, equivalent to a pair of oscillating dipoles, is the lowest non-zero order of gravitational radiation. In the spring of 1936, Einstein and Rosen convinced themselves that in fact gravitational waves could not exist after all. They wrote a paper and sent it off to the *Physical Review* at the end of May. Convinced that he had made a major discovery, Einstein wrote from his summer retreat to both Myron Mathisson and Max von Laue, telling them of this result.

On 23 July, Einstein received a letter from the Editor of *Physical Review*, including a referee report with a long list of detailed criticisms. Einstein was outraged and drafted a letter 'We (Herr Rosen and I) sent you the manuscript for publication and did not authorise you to show it to other experts before publication. I see no need to reply to the (no doubt erroneous) comments of your anonymous expert. Because of this incident, I prefer to publish the paper elsewhere' (author's underlining). Einstein never submitted another publication to *Physical Review*.

Einstein's indignation was understandable in that for his entire career after 1905, his papers had never been refereed. Once Planck was satisfied with his bona fides and he had had a paper excepted by *Annalen der Physik*, future papers were waved through. After his move to Berlin, he used the *Proceedings of the Prussian Academy*, where, as a member, his contributions were published without question. This is a major reason why his papers are strewn with minor calculational and typographical errors. Even in US journals, the practice of anonymous peer review, now universal, was in its infancy. It is however a pity that Einstein, in high dudgeon, ignored the referee's comments, since they were correct.

It was around now that Leopold Infeld, despairing of obtaining an academic position in Poland because of rampant anti-Semitism, arrived in Princeton to take Rosen's place as one of Einstein's co-workers.[9] Infeld had spent a year from 1933–1934 in Cambridge, working with Max Born, before returning to Lviv. At his first meeting with Einstein, they were joined by Einstein's old sparring partner in the tensor manipulation that had led to general relativity, Tullio Levi-Civita. He had arrived in Princeton that summer for a six-month stay. The three discussed Einstein and Rosen's proof that gravitational waves did not exist. This astonished Infeld, who shared the common view that the close analogy between gravitation and electromagnetism meant that both would be mediated by waves. Einstein gave him a copy of his and Rosen's paper, which Infeld tried to understand by approaching the result in his own way, which turned out to be simpler than Einstein's. Both were pleased when, a few days later, Infeld showed Einstein his simplified proof. At afternoon tea, however, Infeld mentioned his result to Howard Robertson, professor of theoretical physics at Princeton University. Robertson, mirroring Infeld's own initial reaction, did not believe it and after an afternoon's work on the blackboard, showed Infeld his error. Infeld immediately went to tell Einstein. He showed no surprise, telling Infeld that he himself had found a different mistake in the reasoning. Unfortunately, he had arranged to lecture on the subject that afternoon in the IAS. Having disproved his own proof, he was left with no alternative but to outline it and then show why it was wrong, concluding with the words 'If you ask me whether there are gravitational waves or not, I must answer I do not know. But it is a highly interesting problem.'

Not long afterwards, Einstein convinced himself that gravitational waves could exist, with help from Robertson, communicated to him via Infeld. In fact, the reason that Robertson was both so interested and familiar with gravitational waves has only recently come to light through the

[9] Although not assistant, as that position was still nominally filled by the apostate Mayer.

research of Daniel Kennefick. The anonymous *Physical Review* referee whose report Einstein had spurned was none other than Robertson himself. Einstein changed his and Rosen's paper, which eventually appeared under the title 'On Gravitational Waves' in the rather obscure *Journal of the Franklin Institute*. The journal no doubt sprang to Einstein's mind after the *Physical Review* debacle as he had recently published his Franklin Medal lecture in this journal under the title 'Physics and Reality'. Rosen was somewhat disgruntled when he discovered the paper's eventual destination, which Einstein had omitted to mention to him. He was even more disgruntled when he managed to read it as he thought it dodged their original question by instead discussing a different type of gravitational wave.

Even though Einstein had reiterated the reality of gravitational waves, that did not mean that he believed in those whose discovery by the Ligo Collaboration led to the 2017 Nobel Prize for Rainer Weiss, Barry Barish, and Kip Thorne. Rather, Einstein thought that they did not necessarily carry away energy. At least, however, after several excursions into the wilderness, Einstein had returned to the right path, whose final destination was not to be revealed until long after his death.

Although Elsa's health had improved since her discharge from hospital, she remained an invalid, which severely constrained their plans for the summer of 1936. Her doctors recommended altitude, which given Einstein's passion for sailing pointed to a lake somewhere in the mountain ranges inland from the east coast. The choice fell on Saranac Lake in the Adirondacks in upstate New York. Accompanied by Margot and Helen Dukas, they rented Distin's Cottage from 1 July. It was set back from the road, surrounded by trees and very close to Lake Flower. A large solarium-type room was ideal for Elsa to sit quietly. Sadly, she suffered from chronic insomnia and rarely felt well enough even to go outside, being nursed assiduously by Margot.

As well as her physical discomfort, Elsa was also troubled by an unwanted lodger in Mercer Street. This young woman had lived in Pasadena and was an acquaintance of the Einsteins' violinist friend, Lili Petschnikoff. Elsa had invited this mutual friend for a brief stay, but she had remained and now had been in residence for almost a year. Elsa was constrained in trying to get rid of her since Einstein had become very fond of her. She asked Petschnikoff's advice on her background and history. The advice must have been good as by September she was able to report that the unwanted guest had finally left the house.

One bright moment in what was for Elsa a sad summer was the arrival of Boris Schwarz in July. On 22 July, Einstein wrote to welcome him to the US. He had been busy trying to find a position for Schwarz with one of the major east-coast orchestras, asking Otto Nathan to help. Elsa also wrote, first welcoming Schwarz and then in August thanking him for the beautiful flowers that he had sent her. She was too ill even to sign her letters, which Helen Dukas signed. Einstein's efforts to find Schwarz a position were initially unsuccessful, but he managed to support himself with temporary work.

Einstein meanwhile spent most days sailing on the small Lake Flower. As usual, tales of his misadventures on the water entered into the local folklore; apparently, he had to be rescued already on his first day. On 7 August, newspapers reported his opinion, no doubt offered tongue in cheek, that the violin played by an elderly guide at the lake was a genuine Stradivarius. He demonstrated by giving an impromptu concert with it to a small crowd that had gathered as he disembarked from his afternoon's sailing.

Another event that summer that had lightened the gloom of Elsa's illness had been the remarriage of Leon Watters. She had written twice to congratulate him and his new wife at the end of June. On 10 September, as her stay in Saranac came to an end, she wrote again to Watters, informing him that she had passed a very sad summer and was confined to her bed. After congratulating him once more, she concluded: 'I think you will be a most caring and loving husband. How I would like to send Albert to learn from you, but, oh God, it wouldn't do any good.'

As in the previous year, Einstein left their summer house early in order to spend some days with Judge Lehmann at Port Chester, New York. He participated in fund-raising for the New York Guild for the Jewish Blind, before returning to Princeton on 17 September. On his return he found, buried in piles of unopened mail, the announcement of the death of his old friend and collaborator on general relativity, Marcel Grossmann. Grossmann had suffered for many years with multiple sclerosis, diagnosed by their mutual friend Heinrich Zangger, and had taken early retirement from ETH in 1927. Einstein immediately wrote to Grossmann's widow, telling her of his gratitude to Grossmann for his friendship when they were students, Grossmann perfectly organised and gregarious, Einstein untidy and unworldly. In particular, he was grateful for Grossmann's and his father's help in arranging his employment at the Swiss Patent Office. Had this not happened, Einstein was convinced that he would have never developed intellectually. Finally, in an unusual display of emotion, he told her: 'But one thing is really nice: we were friends and stayed that way throughout our lives.'

Elsa rallied temporarily after their return to Princeton. One of the last people to visit her was Boris Schwarz, their reunion put off again and again by her health. Finally on 6 November, Elsa dictated an invitation to him and on 12 November, Schwarz arrived at Mercer Street with his violin. Their meeting was an emotional one, with Elsa shedding tears. Schwarz played for her accompanied by another of Einstein's young collaborators, Valentin Bargmann. Bargmann was an excellent pianist who coincidently had been a pupil of Schwarz's father in Berlin. Presumably Schwarz and Einstein also played together, probably Einstein's beloved Bach Double Concerto that they had played together so often in Berlin.

Subsequently, Elsa's condition rapidly worsened. On 3 December, she dictated her last will and testament to Einstein, in which she left the Mercer Street house to Margot, along with her other possessions and the contents of her personal bank account. Already on 18 December, Einstein knew that Elsa was dying. Nevertheless, he worked hard on problems with Infeld every afternoon in his study as his wife lay in the bedroom beneath, nursed by Margot. Her groans were audible in Einstein's study but did not penetrate his concentration. Einstein did spend substantial time at her bedside, reading to her, but was able to turn his mind back to physics as soon as he closed her bedroom door behind him. Her immune system weakened by her long illness, she recovered from one lung infection only to contract another. Einstein wrote to his sister Maja that Elsa had struggled hard but saw death coming with clarity and at the end, welcomed it. On 20 December 1936, Elsa Einstein died.

Chapter 28

Refugees arrive, including Maja Einstein (1936–1939)

Einstein's outward reaction to Elsa's death was described differently by various witnesses, ranging from a relative indifference and injunction to bury her[1] quickly, to tears of grief. To his sister Maja he wrote on 5 January 1937 that 'Now she is released but the nightmare of her suffering still weighs on Margot and me just as before and at night we still have a painful expectation of hearing her groans and whimpers.' On the pevious day, he wrote to Hans Albert: 'About 14 days ago, my wife died after a long and painful illness. Everything comes at the same time to make my life difficult. But as long as I can work, I may and will not complain, because this is the only thing that gives my life a true meaning.'

From the written evidence, the initial feelings with which Einstein and Elsa had begun their affair in Berlin before the war had never been comparable to those that either Marie Winteler or Betty Neumann had roused. Their relations soon cooled to be more like a partnership: Einstein appreciated Elsa taking all domestic concerns onto her shoulders, freeing him to concentrate on his work; Elsa really loved him and, although hurt by his many infidelities and frequent apparent indifference, knew that she occupied a special place in his life and enjoyed her status as the 'Frau Professor', basking in social circles in Einstein's reflected glory.

It was probably in that social field that most changes occurred to Einstein's life as a widower. Elsa's housekeeping and domestic duties fell almost seamlessly into the hands of the two other women in Mercer Street: his step-daughter Margot and secretary Helen Dukas. What was lost was Elsa's fondness for society and meeting people. Once his initial sadness had been dissipated by a rapid return and immediate absorption in his work, Einstein developed a routine centred on his research that could only with great difficulty be interrupted. The incessant round of receptions and dinners that he had tolerated in aid of good causes began to fall off; no longer could Elsa nudge him towards acceptance of invitations that his initial reaction was to refuse. His health, which at that time was in fact quite good, began to be used with increasing frequency as an excuse to refuse any invitations that involved travel much beyond Princeton or at the furthest New York or Philadelphia. Indeed, by the end of the decade he used his health as a reason to excuse himself from almost all dinners outside of Princeton. More slowly, his attendance at attractions such as the theatre began to fall off, as did concert attendance outside Princeton.

Even the frequency of Einstein's music-making began to decrease, since Elsa had often taken the initiative to invite prominent musicians to play with her husband. This is evinced in a letter Einstein wrote to Queen Elisabeth on 11 August 1937:

[1] Marianoff thought she had been buried. In fact she was cremated; the rapidity was to avoid any publicity or official fuss.

Einstein. Brian Foster, Oxford University Press. © Brian Foster (2026). DOI: 10.1093/oso/9780198794875.003.0028

> Life in a small circle brings plenty of excitement, whether it is kittens or beautiful scientific ideas, which I still seek after with the same delight as when I was young. And sometimes it really is possible to lift the veil that shrouds the eternal truth just a little. For that I enjoy with gratitude every opportunity for solitude that a life among people with a completely foreign past and view on life offers. Hopefully you are more attentive to your violin than I am to mine. When we squeak by ourselves it is rarely uplifting and playing with others is rare. So I seek refuge in the richer piano, despite the fact that I have never found the energy to play it from sheet music. However, fortunately no judge exists when there is no plaintiff!

Einstein's involvement in other aspects of music-making continued. Back in May 1936, he had received a letter from England from Alex Cohen, a cellist who was an enthusiast for the music of Ernest Bloch. Cohen and a group of friends were proposing to found an Ernest Bloch Society. He told Einstein that, in his opinion, he and Bloch were the two greatest living Jews but that Bloch had a much lower profile and his music was rarely played. He hoped that the proposed society could put that right. Citing support from musical luminaries such as Jean Sibelius and Havelock Ellis, he hoped that Einstein would agree to be Honorary President, a position that he emphasised had no duties but would attract other supporters. Although Einstein and Bloch had never met, Einstein had both heard and played quite a lot of his music. For example, his sheet music collection contained a copy of the famous *Baal Shem* suite for violin and piano, which he had acquired while in Berlin. Furthermore, they had mutual friends in Alexandre and Catherine Barjansky, with whom Einstein had often played and socialised when visiting Queen Elisabeth. Bloch had been so impressed by Alexandre's playing that he composed the *Schelomo* rhapsody for cello and orchestra in 1916 and dedicated it to him. This inspiration came not only from Barjansk's playing but also from a sculpture of Bloch as King Solomon – King Schelomo in Hebrew – created by Catherine Barjansky. Einstein had liked what he had heard of Bloch's music, approving of its often distinctively Jewish character. On 5 June he wrote to Cohen agreeing to become President of the Bloch Society.

The announcement of the formation of the Bloch Society was not made until November 1937. The society was based in London, so most of Einstein's colleagues who were nominated as Vice Presidents were British. They included Sir Thomas Beecham, Havelock Ellis, Dame Ethel Smyth, Ralph Vaughan Williams, John Barbirolli, Sir Henry Wood, Sir Donald Tovey, Sir Arnold Bax, and Arthur Bliss. Serge Koussevitzky and Romain Rolland represented the non-British musical world. An inaugural concert was announced at the Aeolian Hall on 10 December, to be followed by two more concerts. The artists included Max Rostal, Solomon, the Griller String Quartet, Franz Osborn, Louis Kentner, Nikolai Graudan, Hansi Freudberg, and Sela Trau, all of whom were Jewish. During the concerts, Rostal, the violinist, made his first public appearance as a viola player. Bloch, at that time living again in Switzerland, did not attend. In May 1938, the Society issued an annual report, which documented an impressive number of performances of Bloch's works both in the US and in Europe. His opera *Macbeth* was being revived and revised to be performed in Italy, with hopes that performances could also be arrange in the US. Bloch's Violin Concerto received its first performance in Cleveland on 15 December 1938.

The fact that Margot Einstein was available to take over many of Elsa's domestic duties was a result of the breakdown of her marriage to Dima Marianoff. In the months immediately prior to her death, Elsa had been saddened to note the unmistakable signs of marital discord. Margot and Marianoff had been separated for many months when he arrived from Paris in March 1935. Elsa mentioned in a letter that Margot and Dima's relationship so often brought concerns and that it would be easier if he were not so dependent on her. The latter was one aspect of the surprising trouble that Marianoff, who already spoke three languages, had with learning English. Elsa wrote to the Alfred Einstein family, who were in Italy in April 1936, that they had lots of trouble with Dima and that something wasn't right, although she found him personally to be a very sympathetic person. Indeed, Elsa and to a lesser extent Einstein, were fond of Marianoff and during 1935 and 1936, provided him with significant monetary support. Although Marianoff had found an apartment in New York where he and Margot lived together for some time, Margot's habitual weekend stays in Mercer Street expanded to full-time attendance on her invalid mother. In fact, this was probably a convenient excuse for Margot, as she seems to have no longer wished to live with her husband. He in his turn admitted that he could not support a wife. His position in the Broadway production of *The Road of Promise*, which eventually was renamed *The Eternal Road*, was probably terminated as the production ran into delays. Rehearsals began in December 1935 but had to be called off and then resumed several times[2]. The first night on 7 January 1937, was however a triumph.

By August 1936 it seems clear that Margot and Dima's marriage was effectively over. In that month, Einstein wrote to the lawyer Leon Ziguelnik about a substantial debt of $450 that Marianoff had run up. He offered to pay part of it but informed Ziguelnik that he should no longer contact him or Margot as they had passed all matters relating to Marianoff to a firm of attorneys. On 18 August Einstein wrote to Marianoff that he had read his letter to Margot and what he had to say in reply should not be construed as an accusation. The couple were divorced on 6 December 1937. Marianoff departed to the west coast, hoping given his experience with film to obtain employment in Hollywood. He was living in Los Angeles in June 1942 when he reaped the benefit of his trip to Bermuda with the Einstein family and became a US citizen. By then Marianoff had disappeared from Einstein's sight until two years later, when the publication of his biography of his former father-in-law caused him to reappear with a bang.

Despite the gradual decrease in Einstein's 'corporate' philanthropy, his willingness to help individual acquaintances could always be relied upon. The violinist Boris Schwarz was established enough by the spring of 1937 to be able to bring both his parents over to join him. Father and son gave a concert together in Philadelphia which Einstein attended, driven by Schwarz, who called in on his way from New York. It was with some difficulty that Helen Dukas convinced Einstein to change from his pullover and baggy trousers into something more presentable. The Schwarz duo also played in a tribute to Einstein at the Jerusalem House in New York on 8 April 1940. Boris Schwarz went on to have a successful career, first as an orchestral musician, including being concertmaster of the Indianapolis Symphony Orchestra and then as a first violinist in the NBC Symphony Orchestra in New York. While in New York, he took degrees in musicology, becoming a prolific and respected author as well as conductor and head of the music department of Queen's College of the City College of New York. He was a regular visitor to Princeton until Einstein's death.

[2] As the delays began, Max Reinhardt the director oscillated back and forth between New York and Hollywood. To accommodate the production, the interior of the Manhattan Opera House had to be ripped out and refurbished. The whole enterprise cost the colossal sum of around half a million dollars; through his support of the concept, Einstein played an important role in attracting the necessary funding. In this febrile atmosphere, other members of the production team became stressed; for example, Francesco von Mendelssohn fell out with Kurt Weill. Von Mendelssohn eventually became an alcoholic, in and out of institutions, asylums, and prison cells for his outrageous behaviour. Although probably syphilitic and possibly lobotomised – certainly his character changed completely to a bored indifference to life – he survived until 1972.

Bruno Eisner was another Berlin musical acquaintance whom Einstein helped at this time. Eisner had had an extensive career as a soloist based in Berlin and had appeared with Einstein in several concerts. Along with other Jewish professors in the Berlin Hochschule für Musik, he was relieved of his duties in April 1933. Gradually his ability to perform became more and more restricted. He played for a while with the Jewish Cultural Organisation in Berlin, which facilitated performances by Jewish musicians. However, they were not allowed to perform music by the 'Aryan' composers Bach, Beethoven, Brahms, Mozart, and Schubert. Eisner decided to escape to the US and asked Einstein to provide him with the necessary affidavit. He was able to leave Germany in September 1936, arriving, as many other refugees did, via Havana in Cuba. His wife followed him more than a year later. Eisner remembered that:

> When I came as a visitor to NY in 1936 [Einstein] took a room for me with Dr Talmey and his family and paid the rent for a month. Dr Talmey was an optician, a small man with a pointed beard and a mousey face. Had he been a bit taller he would have been a dead ringer for Don Quixote. His passion was to talk about art. He had played a historic role in the life of Einstein. He was a poor student in Munich at the time that the Einsteins also lived there. In gratitude for the fact that he often went there for free meals, he taught Einstein algebra. Einstein was grateful to him for the rest of his life. I often met Einstein at the Talmeys and even after the Dr died he often visited Mrs Talmey and often took her to concerts.

Einstein gave Eisner advice on how he should go about establishing himself in the US. He advised him to contact their mutual friend Leopold Godowski for help. However, he warned him about the pervasive anti-Semitism in academia in the US and the chaotic way in which academic jobs were filled. He suggested that he join a labour union, as Boris Schwarz had already done. Eisner found work initially with the New York Young Men's Hebrew Association, going on to positions with the New York College of Music and universities in Philadelphia, Indiana, and Colorado.

Since his arrival in Princeton from Poland in 1936, Leopold Infeld had been working with Einstein on a very general problem, whether the equations of general relativity were sufficient to determine the motion of massive bodies. They made good progress in this highly non-trivial investigation, collaborating also with Banesh Hoffman. Most theories, such as Maxwells' equations for the electromagnetic field, do not fix the motion of the source particles. For electromagnetism, these are determined by a combination of Maxwell's equations, Newton's laws, plus the Lorentz force law, which is an additional element that has to be inserted 'by hand'. For general relativity however, the situation is different. The fact that the curvature of space-time is constrained by the Einstein equations means that additional massive bodies distort space-time. Such distortions can only be matched smoothly, a requirement of all field theories, in severely restricted ways. These restrictions force the bodies to move in a particular manner which turns out to be a slightly modified version of that predicted by Newton's laws. Even more remarkably, and at the heart of Einstein's conviction that electromagnetism and gravity could be unified, is the fact that if Maxwell's equations were fused with general relativity, similar constraints produced the Lorentz force in a completely natural way.

In February 1937, Infeld was immersed in calculations with Einstein. He was distracted however by his uncertain future. His fellowship would expire in the summer. Einstein reassured him that he would get a one-year extension but at the faculty meeting in March various financial constraints, together with a conviction among his colleagues that his work with Einstein would be fruitless, meant that Infeld was turned down. He and Einstein discussed various options, including a hint that Einstein could do another benefit concert on his behalf. None was acceptable to

Infeld however until, racking his brains in his room, it occurred to him that he and Einstein could write a popular book together. Any book with Einstein as an author was assured an enormous readership and Infeld was sure that he could support himself until publication with an advance from the publisher. After discussing it with colleagues, he summoned up his courage to broach the idea with Einstein. However, overcome by nervousness, he could hardly stammer out the words, until Einstein, amazed, insisted. Finally he said:

> 'The greatest men of science wrote popular books, books still regarded as classics. Faraday's popular lectures, Maxwell's *Matter and Motion*, the popular writings of Helmholtz and Boltzmann still make exciting reading.' Einstein looked silently at first, stroking his moustache with his finger. Then he said quietly: 'This is not at all a stupid idea. Not stupid at all.' Then he got up, stretched out his hand to me and said: 'We shall do it.'

This was the genesis of *The Evolution of Physics*, which Einstein and Infeld completed in the summer of 1937, after Infeld had made two visits to Einstein's sailing retreat on Long Island. It was published in January 1938.

The *Evolution of Physics* is perhaps the apotheosis of a journey Einstein began with Max Talmey in his childhood in Munich, when Talmey recommended the popular science books of Aaron Bernstein. From this experience, Einstein had always understood the importance of explaining science in a way simple enough for young people and other non-experts to understand. Throughout his career, he had devoted significant time and effort to such endeavours; his lectures to working people in Berlin, attended by non-scientists such as Berthold Brecht, are an obvious example. Another is the enormous number of articles in the popular press that he either wrote himself or that resulted from interviews with him. He was a very active participant with Infeld in producing the thrust and plan of the treatment, if not the actual language. The book, while being in the tradition of Bernstein, is more similar to books of contemporary luminaries such as Sir James Jeans. In particular, his best-selling *Stars in Their Courses*, published in 1931, would have been familiar to both Einstein and Infeld.

In their book, the authors eschew all equations but make liberal use of diagrams. They wrote that thought and ideas, not formulae, initiated every physical theory. The leitmotiv of the book, mostly unspoken, is the belief in an objective physical reality that can be explored via theoretical constructs. Starting from a discussion of mechanics and dynamics according to the erroneous ideas of Aristotle, the book describes optics. The mechanical reductionism characteristic of the physics of Einstein's youth began to be weakened by the introduction of the concept of the electromagnetic field. This concept led Einstein and Infeld to the introduction of special relativity, the demise of the aether, and general relativity. Only after all this had been discussed did the authors turn to the quantum. They illustrated the concept by appealing to that bedrock of American life – money – which is quantised in units of one cent. After discussing particle–wave duality, they move into areas in which Einstein would have felt less comfortable. Probability waves for particles are described and the impossibility of predicting the precise course of a particular electron that contributes to the production of a diffraction pattern. The Einsteinian conclusion is a typical question: 'Will the further development be along the line chosen in quantum physics, or is it more likely that new revolutionary ideas will be introduced into physics? Will the road of advance again make a sharp turn, as it has so often done in the past?'

The book was of course a best seller, although, somewhat surprisingly, the first print run was of only 6000 copies. However, it was reprinted many times; already by 1940 it was in its seventh printing. In 1938, Cambridge University Press brought out a UK edition and Flammarion a French edition, translated by Einstein's old friend Maurice Solovine. Reviews were uniformly positive, with *Nature* considering it a book of real distinction. *The New York Times* was highly complementary about the content but not so enthusiastic about the prose:

> Though not the simplest mathematical equation appears anywhere to frighten away the man who has forgotten everything but his multiplication table we miss the turn of the phrase – the poetic analogies that elevate the writings of Jeans and Eddington to the rank of literature.
>
> [...]
>
> In this book we have no poetry, very little philosophy, but a remarkably successful attempt to show why physicists have been forced by sheer logical necessity to give us a world in which there are no forces in the old sense and in which electrons leap about so erratically that they have to be handled statistically.

Neither author was a native English speaker and Infeld admitted in his autobiography that his original draft was riddled with errors. Three native English speakers were kind enough to help him correct it. It is perhaps unsurprising that such a compositional committee was unable to reach the poetical heights expected by a science journalist such as Waldemar Kaempffert, the reviewer quoted above. The success of the book made Infeld's career. He was offered a temporary appointment at the University of Toronto that was quickly made permanent, although he returned to his native Poland after the Second World War.

The rebuff over Infeld's renewal clearly irked Einstein and in April of 1937 he wrote to Abraham Flexner asking to be allowed to appoint another assistant, since Mayer was no longer collaborating with him. Flexner immediately reassured him that he was perfectly free to do so and to name anyone he liked, either inside or outside the IAS. Indeed, Rosen, although he already had his own funding as a Fellow, had been considered to be Einstein's assistant. This was clearly news to Einstein, given his efforts to find a solution to Infeld's funding problem. Now that he had solved that with their joint book, he could look elsewhere and appointed Peter Bergmann, who had already been working with him since the previous year as a Fellow. They were constructing a variation of the Kaluza–Klein approach, to which he had returned yet again in 1937. Bergmann retained his position as Einstein's assistant until 1940, when he was succeeded by another Fellow already working with Einstein, Valentin Bargmann. His tenure lasted until 1945. Ernst Straus, John Kemeny, Robert Kraichnan, and Bruria Kaufman, the first woman, were the researchers who successively filled this position until Einstein's death. After his paper with Bergmann in 1937, he published on average less than one paper a year, usually in the area of general relativity and unified theories. Most, but not all, were co-authored with his assistants. None are of especial note. As he remarked to Solovine at the end of a letter about the French translation of *Evolution of Physics* in April 1938: 'Nowadays I am greatly valued as a labelled museum-piece and curiosity but that is par for the course. I'm still working hard, supported by some daring young colleagues. I can still think, but the ability to work has decreased. Anyway, being dead isn't so bad either.'

Other concerns with respect to Flexner and the future of the IAS began to affect Einstein towards the end of 1937. As in the case of the Hebrew University, Einstein was not above the grubby business of academic politics when the matter involved was important enough. On 3 November, he wrote to a member of the IAS Board whom he knew well, Samuel Leidesdorf. His concern was the influence of Princeton University over the IAS. He had heard that a young man had recently been refused entry to the university because he was an African-American. Einstein was concerned that, knowing the anti-Semitic views of many faculty, such racism could also be extended to discriminate against other minorities such as Jews. He was also concerned about Flexner's successor. The head of Swarthmore College, Frank Aydelotte, who was also a trustee of the IAS, had long been considered a natural successor to the seventy-five-year-old Flexner. However, Einstein had been informed that the President of Princeton had stated that his appointment would not find favour with the university. Recently, Flexner had backtracked from supporting Aydelotte. In a subsequent letter to Leidesdorf, Einstein said that the faculty strongly endorsed Aydelotte's candidacy and ruled out several others whose names had been mentioned. Their main concern however was to have elected representatives from the faculty added to the Board of Trustees. This had been brought up two years earlier but rejected by Flexner. In fact, Flexner was merely reflecting the strong opposition of the IAS's donors, particularly James Bamberger, who thought that faculty members were far too unworldly to have a share in running the institute. Nevertheless, Flexner unilaterally appointed two professors to the board, no doubt assuring the donors that he had selected 'sensible' members. They, however, had not been elected by the faculty.

Two years passed without any progress. On 15 March 1939, the faculty, led by Einstein among others, wrote to Flexner to insist that they should be consulted about both faculty appointments and that of the director. Flexner as usual evaded any response, so members of the Board of Trustees were contacted directly. These included Aydelotte, who was asked to suggest to the Board that the selection process for Director be made open. This put Aydelotte in an awkward position as he had already been designated as Flexner's successor by the Board but was unable to mention this because of confidentiality. Relations between the faculty and Flexner went from bad to worse; eventually in August 1939, Flexner resigned, effective from October. In his advice to his successor, he remarked about the faculty: '. . . you are dealing with intriguers. I freely confess I was a baby in their hands.' A more unlikely baby it would be difficult to imagine. That October, the IAS moved out of the university's Fine Hall into its own building and Aydelotte's much more flexible leadership completely changed the atmosphere for the better.

During his time in Princeton, Einstein became deeply engaged with what became the civil rights movement. In the 1930s, Princeton accepted no African-American students; the first to be admitted since the nineteenth century were only accepted under the officer training programme in 1942. The town itself was strictly segregated, with an African-American quarter around Witherspoon and Jackson Streets. Segregation stretched across all aspects of life. However, McCarter Hall in Princeton was privately owned and therefore was not constrained by the official segregation. It became the preferred venue for concerts involving African-American artists. It was here in 1935 that Einstein first heard the great Paul Robeson sing. Coincidently, Robeson had been born in Princeton; he considered it the northernmost town in the South. He and Einstein became friends and subsequently collaborators in many of Robeson's initiatives to better the lot of African Americans, perhaps most notably his American Crusade Against Lynching in 1946, discussed in Chapter 30.

It was also in McCarter Hall, on 16 April 1937, that Einstein attended a sold-out recital by the extraordinary black contralto, Marian Anderson, who a year before had become the first African American to sing at The White House. Accompanied by Finnish pianist Kosti Verhanen, her programme included works by Händel, Schubert, Ravel, Strauss, and Sibelius. She concluded with a selection of spirituals. The recital elicited rave reviews, including 'superlative' and 'complete artistic mastery of a magnificent voice . . . Seldom is a voice like this combined with such a perfect intellectual and emotional understanding of the music. Anderson's vocal technique is unsurpassed in variety of colour; contrast of range and marvellous effects in dynamics' from the *Daily Princetonian*. However, such admiration did not extend to providing accommodation in the racially segregated central Princeton hotels. Hearing of this, Einstein, who had first met the singer backstage after one of her Carnegie Hall recitals, was outraged. He offered Anderson hospitality in Mercer Street. This began a friendship that lasted until Einstein's death. Whenever she performed nearby, Anderson would stay at Mercer Street.

Just before his summer vacation in 1937, Einstein wrote to his old friend Michele Besso. He told him with some apparent pleasure that his life was now that of a confirmed bachelor, enjoying investigating his problems with excellent young physicists. He remarked on how curious it was that, throughout his long career, he had worked exclusively with Jews. After bringing Besso up to date on his latest ideas on unified theories, he confirmed that he was thinking of Besso as his earliest collaborator: 'I often think about our old days in the Patent Office with pleasure and can't get it into my head that I left it almost 30 years ago. At any rate, that is 10^9 seconds, which makes one wonder that one couldn't have achieved more in that time.'

In 1937 Einstein decamped from Princeton somewhat later than usual, in mid-June, expecting to escape the enervating summer heat. His chosen resort this time was Huntington Bay, on the north coast of Long Island, where he rented a house with grounds of four acres. The large water frontage gave access to the sheltered waters of Northport Bay as well as Huntington Bay itself. Unfortunately, that summer was particularly hot and sticky, even on the coast, so that both Einstein and Margot felt exhausted until the autumn arrived and they were back in Princeton. As usual, Helen Dukas accompanied them and Einstein's correspondence, much to his irritation, was periodically forwarded to her. The Bucky family was also there. One unusual visitor was Einstein's old flame, Grete Lebach. The reason for her trip was ostensibly to recover from some business setbacks in Vienna, where she and her husband Willi now resided. These were apparently resolved before she returned. In fact, she may well already have been diagnosed with breast cancer before she travelled to Le Havre, choosing the French port to avoid Nazi Germany. She embarked on the SS *Champlain* on 4 June. She arrived in New York on 11 June and stayed in Huntington Bay for a considerable time. She was photographed sailing with Einstein on his yacht *Tinev*[3] on the Bay (see Figure 28.1). Einstein pressed her to emigrate to the US, no doubt emphasising the danger to Jews of the growing Nazi sympathies in Austria. However, she returned to Vienna. Einstein sent her a touching note after her departure: 'My dear little Lebach, meanwhile the sun, seagulls, the waves and *Tinev* but especially me send their best wishes. Today Dr Bucky sat in your place in *Tinev*.'

Lebach's decision to return to Vienna turned out to be fatal. The Anschluss, when the Nazis overran Austria the day before a plebiscite vote on union was scheduled, took place on 12 March 1938. The union was confirmed by a non-secret vote a month later, characterised by shameless intimidation, that recorded 99.7% in favour. Almost immediately, Austrian Jews began to be ejected from official positions and have their property confiscated; many were arrested, including Willy Lebach. Already on 16 March SS men arrived at the offices of the Bachwitz AG, a large publishing company owned by Grete's mother. They forced the entire family to give up their shares. Grete's brother-in-law committed suicide. A friend of hers, Alice Kornitzer, who happened to be a US citizen and therefore able to travel freely, escaped to the Hague, from whence she wrote to Einstein alerting him to Grete's danger and begging him to help. Einstein, by now a veteran of rescuing Jews from Nazi tyranny, swung into action, promising to take care of Grete financially and to take her to live with him in Mercer Street. This offer was subsequently also extended to Willy Lebach. The State Department forwarded affidavits in the diplomatic pouch and Einstein even enlisted aid

[3] Einstein at some point acquired his own small sailing boat for use initially on vacations and eventually on the lakes around Princeton. It was very simple, around five metres in length with a single mast and no engine. The name, *Tinev*, is German slang, originally derived from Yiddish, which roughly translates as 'worthless'. However, *The New York Times* report implies that for this vacation he was still using a borrowed or rented yacht, to which he attached this generic name.

Figure 28.1 Grete Lebach and Einstein sailing in Huntington Bay, summer 1937.

from the Secretary of State, Cordell Hull. Einstein had got as far as getting quotes for their travel tickets. However, before everything could be finalised, Grete died of cancer.[4]

Another visitor to Huntington Bay in the summer of 1937 was the British physical chemist *manqué*, C. P. Snow. Snow, a Fellow of Christ's College, Cambridge, had given up research after a calculational error was found in one of his papers in *Nature*. With the independence provided by his Fellowship, which did not require research activity, he had increased the involvement in literature and journalism that he had begun while still researching. In the summer of 1937, he travelled to the US and stayed for a while in Princeton, visiting his friend Infeld, whom he had met during Infeld's year in Cambridge working with Max Born. Infeld arranged for Snow to meet Einstein

[4] This tragic story had at least a partially happy ending. Willy Lebach, having had his entire wealth confiscated, was allowed to leave Austria in October 1938. His step-son, Grete's son from an earlier marriage, Theo, spent several months in Dachau but was released. Both, with significant help from Einstein, finally arrived in the US; on their ships' immigration manifest, each gave their US address as Einstein's home in Mercer Street; Willy arrived in November 1938, Theo in May 1940. Both received significant help from Einstein in finding jobs and supporting themselves in their new homeland.

and drove him to Huntington Bay. It was a sweltering day; Snow drank soda water continuously as he, Einstein, and Infeld chatted:

> I was a friend of G. H. Hardy,[5] Infeld began. Einstein smiled with pleasure. Yes, a fine man. Then, quite sharply, he asked me: was Hardy still a pacifist? I replied, as near as made no matter. 'I do not understand', he said sombrely, 'how such a fine man can be so unrealistic.' Then he wanted to know if I also was a pacifist. Far from it, I explained. I was by that time certain that war was inevitable. I was not so much apprehensive about war, as about the chance that we might lose it. Einstein nodded. About politics that afternoon, he and I and Infeld were united. About politics in the widest sense, I don't think there has been a world figure in my time who has been wiser than Einstein. He wasn't much interested in political techniques, and brushed them off too lightly: but his major insights into the world situation, and his major prophecies, have proved more truthful than those of anyone else.

In January of 1937, after having broken the news of Elsa's death to Hans Albert, Einstein had answered his earlier query about the possibility of emigrating to the US. He had told his son that he was keeping his eyes open for a position for him but that it is was very difficult unless he was actually in the country to follow up possibilities. He had recommended that Hans Albert should send his doctoral dissertation, which had recently been published, to a professor Einstein had known in Caltech, Theodor von Karman, who had many contacts in Hans Albert's field. Einstein had advised however that he should not bring his wife with him to the US. This might have been to allow Hans Albert more flexibility to travel around but was more likely to be a continued reflection of Einstein's antipathy to his daughter-in-law.

Einstein also wrote to Tete in May in which he mentioned Hans Albert's plans:

> I live now as a lonely old man, gradually preparing my farewell. It has its advantages – there is a right and a possibility for quiet dispassionate observation on the vanities of life, that once seemed so important.
>
> Albert wants to come here in the autumn, to try to build a nest here. He's right, but it isn't easy. Here everything is a frantic and selfish race, a proper dance around the golden calf, a wild and horrible dance.

Hans Albert boarded the TSS *Veendam* at Rotterdam and arrived at New York on 12 October 1937. He was met by his father in a chauffeur-driven limousine, either that of his friend Leon Watters or Gustav Bucky's. Hans Albert stayed in Princeton for some weeks. He and his father got on well and enjoyed playing a great deal of music together. Hans Albert then spent six weeks driving across the United States, including visits to Caltech and Berkeley and to the Department of Agriculture in Washington. Einstein encouraged him to keep a travel diary, as Einstein himself had on some of his travels, but had little hope. As he told Tete, Hans Albert suffered from both a writing and a speaking block!

On arriving in Washington, Hans Albert was interviewed by a senior member of the newly established Soil Conservation Service (SCS), who was very impressed and recommended him to his superiors. Feeling positive about his experiences, Hans Albert returned to Zurich in late January 1938, leaving his father to advance his claims to a position.

Before long, a letter from the SCS arrived in Princeton expressing interest in engaging Hans Albert. Einstein enlisted von Karman to travel to Washington to negotiate with Dobson, the head of the SCS, to whom Einstein also wrote. This culminated in Hans Albert being offered a position with the SCS based in South Carolina. He accepted and arrived with his family on 13 June 1938. Einstein had paid essentially all the expenses related to Hans Albert's earlier visit and the relocation

[5] The eminent mathematician, Sadleirian Professor of Mathematics and Fellow of Trinity College, Cambridge.

of the family from Switzerland. He was greatly relieved that his son and grandsons had joined the many others whose escape to the US had been made possible through his name and resources. Another relative for whom Einstein had vouched and supported financially arrived soon after. Hans Moos was Einstein's second cousin.

The plans of Hans Albert and family to leave Zurich possibly catalysed the resumption of correspondence between Einstein and Heinrich Zangger. They had not communicated for four years, a gap that can be attributed to a bizarre misunderstanding by Einstein of something that Zangger had written to him just after the Nazis came to power. Einstein interpreted what was certainly an innocent remark by Zangger as somehow anti-Semitic. Zangger's initial letter on 3 February concerned the plight of young children of a mutual acquaintance but Einstein could do nothing to help. Zangger wrote again to recommend that Einstein talk with Prof Meyer from Baltimore who was an expert on schizophrenia and could possibly help with Tete's treatment. Tete was about to undergo yet another fruitless course of insulin. Mileva's finances had again worsened. Zangger, who himself was in poor health, having had two bouts of lung infection and heart trouble, had had to intervene to prevent her furniture from being sold off by creditors. The depressed state of the housing market in Switzerland had meant that the rents on the properties Mileva had bought no longer covered the repayments on the mortgages that she had taken out. The only solution was to sell several at the current depressed prices. Einstein's somewhat exasperated response was that he had already provided Mileva with substantial funds and left matters in the hands of his representative, Dr Dukas, Helen Dukas's brother. He suggested that Mileva and Tete could move to the family home in the Balkans, where living would be cheaper. This was hardly practicable given Tete's need for a resident male carer and his treatment in the Burghölzli clinic. Zangger also gave Einstein news of Besso, whose lecturer position at ETH had been terminated since he was not attracting sufficient students.

Einstein and Margot took the train to New York on 8 February 1938 to attend a concert at the home of Frederick Muschenheim. He lived in the Hotel Astor, of which corporation he was president. The recital was given by their old friends Adolf Busch and Rudi Serkin. They played pieces by Brahms, Mozart, and Schubert to an invited audience that also included Jascha Heifetz, Arturo Toscanini, and Francesco von Mendelssohn. Heifetz thought the recital wonderful, a sentiment echoed by Einstein:

> Wonderful. That is Germany, that is the true, the best Germany. How lucky we are to meet it wherever we hear such music, and what a testimony it is, if such were needed, against all the imbecilities of 'Race'. Can anyone imagine a more perfect unity, a purer fusion than that created by those two – the fair and the dark, the Aryan and the young Jew?

Busch himself was depressed by the current state of the world:

> Sometimes I feel like giving up my violin. It seems to me at such moments as if it were a sin to make music, 'only' music, in a world one ought to teach in a way it would better understand. Sometimes I think we oughtn't to leave politics to the politicians. I'd be glad to let Chamberlain play my violin if he would let me tell the truth to the English.

To general laughter, the consensus seemed to be that the cure might be worse than the disease.

On 6 April 1938, Einstein attended a tribute by Jewish actors at the Public Theatre on Second Avenue in New York to raise funds for the Council for Jewish Organisations for the National Palestine Appeal. On 17 April, he fulfilled an increasingly rare speaking engagement at a seder and dinner organised by the National Labour Committee for Palestine for 3000 guests at the Hotel Astor. He shared the platform with a leading member of the UK Labour party, Herbert Morrison, who was also Jewish. Einstein spoke in German with simultaneous translation. *The New York Times* reported that:

> While urging continued effort in the upbuilding of Palestine as a Jewish homeland, he expressed himself as opposed to the British proposal for a division of Palestine between Jews and Arabs and the creation of a separate Jewish State. He feared that the setting up of a political Jewish State in Palestine might lead to the development of a 'narrow nationalism within our own ranks, against which we have already had to fight strongly even without a Jewish State.'
>
> [. . .]
>
> 'We are no longer the Jews of the Maccabee period.' he continued. 'A return to a nation in the political sense of the word would be equivalent to turning away from the spiritualization of our community which we owe to the genius of our prophets.'

In May, Einstein received a letter from Betty Neumann, his former secretary, with whom he had had no contact since the end of their torrid affair in 1923. She had just lost her job as an X-ray technician in a hospital in Graz and, like so many others, asked Einstein's help to emigrate to the US. She managed to find work at the Jewish Hospital in Leipzig in the summer until the paperwork arrived. She received it just before Christmas 1938. In the following March, before embarking in the SS *Gerolstein* in Antwerp, she wrote to Einstein to mark his sixtieth birthday and in gratitude for everything he had done for her. She arrived[6] in New York on 11 May, where she stayed with a cousin in Hoboken until she found an apartment on W 94th Street in the city. She visited Einstein in Princeton on several occasions and Einstein assisted her to find a suitable job by providing a reference. After some work without salary, she eventually obtained a position in the autumn of 1939 in St Francis Hospital which she retained for several years. She and Einstein met occasionally thereafter, for example, in 1940 at the Buckys home in New York and probably in New York in 1944. However, there is no indication that they resumed any intimacy, as evinced by Einstein using the formal Sie mode of address in his letters.

On 6 June 1938 Einstein received an honorary degree at the Swarthmore College Commencement ceremony. Swarthmore is a few miles west of Philadelphia; Einstein stayed overnight at the residence of the President, Aydelotte, shortly to become head of the IAS. In the evening before the ceremony, in what was likely to have been a significant factor in his accepting this invitation, Einstein played Haydn, Beethoven, and Mozart string quartets with members of the Swarthmore faculty. In his speech, Einstein told the graduating class about what he considered the barbarity occurring abroad and how it was related to moral degeneracy. He criticised the isolationist tendency of the US as looking on passively while the innocent were killed in Europe.

In the week after his Swarthmore address, Einstein left Princeton with Margot and Helen Dukas for his vacation home, renting a cottage on Nassau Point belonging to his friend Dr Moore. Hans Albert and his family arrived in New York on 13 June, having embarked on the SS *American Trader* in London on 3 June. Einstein arranged for a car to meet them and bring them to Long Island. Hans Albert began his new position on 1 July and had to go through registration procedures in advance of that date in Washington, so he and his family stayed only for two weeks. Einstein was for the first time able to see both his grandchildren in an environment where they could romp. He thought the elder, Bernhard, rather introverted but was enchanted by the high spirits of the

[6] Curiously, on the same ship being carried to safety were Arthur and Grete Mayer, brother and sister-in-law of Einstein's former assistant, Walther 'Mayerchen'. It seems likely that they too were beneficiaries of Einstein's help and support.

younger, Klaus, who he thought resembled a grandchild of Sancho Panza. He could not resist sniping at his daughter-in-law, who he said was egotistical and had a suspicious manner. He accused her of only having stayed for a couple of days but leaving an enormous mess behind her. After Hans Albert and his family had departed, the Einsteins were joined by the Bucky family. As usual, Einstein spent as much time as possible in his yacht, although he was more communicative than usual. His sister Maja must have been astonished to get two letters in the space of a single month. This may have been related to news that she had conveyed to him of two old flames.

In his first reply to Maja, he wrote:

> I received the greetings of Anneli and the Marangoni[7] with pleasure, a light-beam from vanished youth. I really can't imagine that they too can have become old and fat like me. The water, the sailing ship and mathematics have however remained young, as has Mozart [. . .] I am very pleased that you have got to know Alfred Einstein and his family. They really are outstanding people.[8]

He mentioned that he had recently sent some money to the children of Mathias Winteler, the son of his old teacher Jost. Mathias had died in 1934; his children in Canada were apparently in some danger due to illness. Einstein asked Maja whether any of the European family were in contact with them. Maja presumably replied by sending him Marie Winteler's address. In September he wrote once more, asking for Anneli's address and saying that he had heard back from Marie.

Einstein wrote[9] to Marie in August; she replied directly on receipt of his letter on 18 August. She wrote her long letter in a cemetery overlooking Zurich and in sight of her brother Mathias's grave. She explained that Mathias had been ill for many years but his girlfriend had stuck by him to the end. He had been her favourite brother because of his fine character. Mathias's partner had presumably left Switzerland with their children after his death and the family had been wondering for a long time what had become of them. However, Marie was in no condition to help them, as she could not even afford to have her piano moved from Bern to Zurich.

In her second letter in February 1939, Marie replied to Einstein's offer to pay for the transport of her piano with the happy news that she had managed after all to be reunited with the instrument through the kindness of someone from Aarau whom she had met recently. She was overjoyed by this and was now playing Beethoven, Schubert, Mozart, and Mendelssohn. She regretted not having any Bach to hand but had Chopin nocturnes waiting for her attention. She wrote that she could almost play herself to death, but also had to make a living.

Arnold Schoenberg wrote to Einstein on 20 August, pleading the case of Dr Oscar Adler, a friend of his schooldays who had given him his first lessons in music and played chamber music with him. He characterised him as first and foremost a musician, among the leading violinists in Vienna, even though he made his living as a doctor. Having begun with characteristics sure to appeal to Einstein, Schoenberg then made the mistake of remarking on Adler's last book, *An Introduction to Astrology as the Science of the Occult*, and of Adler's fascination with occultism. After praising another of Adler's books on musical theory, Schoenberg listed several ways in which Adler could be helped to emigrate to the US, which he hoped that Einstein would be more inclined to support than he had been in the case of Adolf Loos. Sadly, he had even less success in enlisting Einstein's support for Adler. Schoenberg was not the first person to bring Adler's case to Einstein's attention; in fact, he was the third. Because of these previous correspondents, Einstein had already

---

[7] Anneli (Anna Meyer-Schmid) and Einstein's probable first girlfriend, Ernestina Marangoni.

[8] The Alfred Einsteins had moved near to Maja in Florence.

[9] Unfortunately there are no copies of Einstein's letters to Marie.

read Adler's book and was thus able to reply promptly on 28 August. He agreed that the book was well written but admitted that he could in no way support a subject such as astrology, which he considered absurd. Indeed, he considered the fact that Adler's book had considerable literary style made it more dangerous, as more likely to convince the unwary that astrology had validity. He advised Schoenberg to try to find Adler a position as a musician and offered to write in support of any such position that could be found, provided it did not imply that he was supporting even indirectly the subject of astrology. There is no record of Schoenberg replying and he seems to have given up trying to bring Adler to the US. Adler did escape Vienna for the UK, where he remained for the rest of his life. Obviously discouraged at his lack of persuasive powers, this was Schoenberg's final attempt to enlist Einstein's support and the two do not appear to have corresponded again.

Around 14 September 1938, Einstein, Margot, and Dukas left Peconic and returned to Princeton. Margot was due to have a serious operation the following month but according to Einstein was very brave about it. He had written a message for a time capsule that was to be buried on the site for the forthcoming World's Fair. It was deeply gloomy, bewailing the capitalist economic model and its attendant fear, insecurity, and apparently endemic wars, despite the scientific advances that allowed radio communication and powered flight. He ended by trusting that posterity would read these statements with a feeling of proud and justified superiority. *The New York Times* headlined this dystopian view of the current world situation as Einstein was hopeful for a better world!

The global situation did indeed seem dire. Fighting in the Spanish Civil War, with the involvement of Italy and Germany on Franco's side and Russia on that of the Republicans, intensified in 1937. On 26 April 1937, Guernica was bombed. In July 1937, the second Sino-Japanese war began; by August the Japanese had occupied Beijing. Italy left the League of Nations in December 1937. Hitler's insatiable appetite for territorial aggrandisement began with the Anschluss with Austria in March 1938. In September 1938, the Sudeten crisis was 'resolved' by the Munich agreement and Germany occupied the German-speaking areas on Czechoslovakia's western border.

The great cellist Pablo Casals was a passionate supporter of the cause of the Republican government in Spain. In March 1937, Einstein received a letter from the North American Committee to Aid Spanish Democracy, inviting him to lend his name to a proposal to invite Casals to the US to give a series of concerts in aid of the Republican cause. Einstein was only too happy to add his name to the list of supporters. Unfortunately, the initiative was unsuccessful.

Another cellist who entered Einstein's life in November 1938 was George Barati. He and his Hungarian colleagues in a string quartet had been invited to Princeton a few months before as instructors in the music department, where he befriended and became a pupil of the composer Roger Sessions. Barati met Einstein at the marriage of John von Neumann, also a Hungarian, on 17 November. The Hungarian quartet were invited to play at the wedding reception. While playing, Barati became distracted by a figure in the front row of the audience wearing a formal dinner jacket and black tie but no socks. It was of course Einstein. A few weeks later, he and Barati became acquainted in the back of Barbara Rahm's car when she picked both up for an evening of string quartet playing at the Rahm's home. These evenings became regular events until Barati volunteered for the army in 1943.

Louis Rahm was a professor of mechanical engineer at Princeton University, which is probably how Barbara and Einstein became acquainted. She had had a brilliant start as a soloist until her marriage in 1929. Her husband disapproved of her having her own career so she became a wife and mother of two sons at least until their divorce in 1947, when she returned to music. The level of her accomplishment can be judged by the recommendation of her teacher, the most famous of all violin pedagogues, Leopold Auer, who in 1920 said that, at the completion of her training, she was destined to become one of America's leading violinists.

In their string quartets, Barbara Rahm clearly took the first violinist's chair, while Einstein took the second. Barati recalled further instances of Einstein's sartorial peculiarities; in winter he believed in the virtue of many layers of clothing, discarding sweater after sweater as they warmed to their musical work in the Rahms' well-heated home. He remarked on Einstein's preference for Haydn, Mozart, Schubert, and early Beethoven, rejecting a suggestion to tackle the Beethoven Rasoumovsky string quartets as being too difficult. The seriousness with which Einstein took his musical activities is related by Barati: when two physicists requested an urgent consultation with him during one of their quartet sessions Einstein said: 'Please don't bother me. I told you when I play music I don't want to be disturbed. Leave me alone.' Barbara Rahm recorded a variation on the ubiquitous 'counting' story that does have a ring of truth about it:

> I remember one evening when we had difficulty with a Beethoven adagio movement marked in 12/8 rhythm and usually counted in 4. I said that we should take it very slowly and count 12. Professor Einstein looked abashed and said, quite innocently, 'Oh I don't think I can count 12.' There were guests that night and I think it afforded them some quiet amusement.

Kurt Weill had not made the breakthrough on Broadway he had expected with his score for *The Road of Promise*. This finally happened in 1938 with his next project, *Knickerbocker Holiday*, a political anti-New Deal allegory which had a long run at the Ethel Barrymore Theatre and was eventually made into a movie, although much of Weill's score was cut from the latter. Weill invited Einstein to the opening night, which was for the benefit of German refugees. Despite this and his deep interest in the political message, Einstein wrote to Weill on 19 November, regretting being unable to attend due to pressure of work.

An aspect of the darkening international situation that hit close to home was an action of the Italian government. On 1 September 1938, in a move that signalled the growing subservience of Mussolini to Hitler, all Jewish immigrants who had arrived in Italy since 1919 were to be forced to leave the country. On the following day, all Jews were excluded from teaching at or attending any recognised Italian educational establishment, with an exception only that those already taking a course were allowed to complete it. In his letters to his sister Maja over the summer of 1938, Einstein had already advised her to think about leaving Italy and coming to the US. On 14 December, he wrote to Maja again, saying that it was now time for her to leave. He told her how to get a visa and instructed her that once she had it, she should tell him and he would make all the travel arrangements. Maja's husband, Paul Winteler, as an 'Aryan', was not in danger in Italy. His health was also fragile, so they decided that he would not accompany her on a visit that they both expected would be of only a few months duration. In his letter, Einstein suggested that Paul could learn the lifestyle of a confirmed bachelor, or he might consider taking in his sister Marie, with whom Einstein told Maja he had exchanged several letters.

In his final letter before Maja set out for Princeton, Einstein remarked on the shocking death of his grandson Klaus. Einstein had found him a sweet child, much more balanced a personality than his introverted elder brother Bernhard. Hans Albert had written a brief note on 6 January 1939, giving his father the news, as he had been too devastated to telephone. He told his father that Klaus had had a cold over New Year and that it had developed into diphtheria, snatching him as rapidly as it had Zangger's young daughter Trudy, who had died in 1918. The funeral was to take place on the following day. Einstein immediately wrote to console his son on his bereavement.

Before she left for the US, Maja travelled to Switzerland to see old friends and family. She went to Luzern, Filzbach, Aarau, and Zurich. In the latter she visited her former sister-in-law, Mileva, and was pleased to find that the hatred that Mileva had formerly held for all Einstein's family seemed to have subsided, at least in her case. At the end of February, Maja travelled to Paris and from there to Le Havre, where she took ship aboard the SS *Paris*. She arrived in New York on 8 March and was met and embraced by her brother. They had not seen each other for more than six years, when the world had looked very different from the one they now inhabited.

Two months earlier, Maja's former neighbours, distant cousins and fellow exiles from Mussolini's race laws, Alfred Einstein and his family, had disembarked from the *Conte di Savoia*. Within four months, Alfred had been appointed visiting professor of music at Smith College in Northampton, Massachusetts.

Two days after Maja's arrival, on 10 March, Einstein's old friends Adolf Busch and Rudi Serkin gave a concert in Princeton's McCarter Theatre. They would have played a run-through of their New York concert of the following evening at the Town Hall: Mozart's D Major Sonata, K306, Busoni's Sonata in E minor, and Beethoven's C minor Sonata, Op. 30, No. 2. It is not known whether Maja had recovered enough from her journey to attend but certainly Einstein did. So also did the novelist Thomas Mann and his wife. They had fled the Nazis in 1933, first to France and then Switzerland before arriving in New York in February 1938. In September 1938, they had moved to Princeton, where Mann became an honorary professor, a post he occupied for two and a half years. Mann and Einstein were thrown together as fellow exiles. They had both received honorary degrees from Harvard at the same ceremony in 1935 and had collaborated in nominating Carl von Ossietzky for the Nobel Peace Prize. However, they were not close. Mann was not Jewish, although his wife was. After the concert, Busch, Serkin, and Einstein were invited back to the Manns' house.

Einstein's musical tastes have often been mentioned, in particular his love of Bach and Mozart, as well as Baroque composers such as Vivaldi and Händel. Sometime in 1939, he answered a questionnaire on his musical tastes which provides some further information. Although he called Händel's music perfect, he qualified that by saying that he found it in some ways flat. He loved the melodic invention of Schubert but found a lack of structure in some of his larger-scale works. Schumann he found very original on a small scale but lacking formal structure; Mendelssohn he considered very able but with a lightness that could lead to banality. He found some of Brahms's songs and chamber music really important, but most not convincing. Wagner's invention was noteworthy but his lack of structure decadent. His musical characteristics he found so offensive that he could only hear his compositions with disgust. In Einstein's view, Richard Strauss was learned but without any spiritual conviction, concerned only with outward show. Modern music left him cold; Debussy he found colourful but formless. The logic of modern music escaped him. Although he loved Godowsky on a personal level, he preferred to remain silent about his compositions.

Einstein's light-hearted view of his own playing at this time is illustrated by one of his favourite types of communication, a humorous poem. This one was written in April 1939 to his old musical factotum, Emil Hilb, and loosely translates as:

> If you really like to fiddle,
> It surely isn't right to diddle
> Others by your own performance
> When that is laughably indecorous.
>
> The dilettante though is entitled,
> However badly or benighted
> He plays, provided he ensured
> That open windows are secured.

Maja had arrived just in time for Einstein's sixtieth birthday celebrations on 14 March 1939. To Einstein's great satisfaction, these was much lower key than had been his fiftieth birthday in Berlin; there were certainly no public events in Hitler's Greater Germany. Einstein did however receive many postal and telephonic greetings, including from the heads of the Jewish communities in the US. Another message came from Max von Laue, risking the wrath of the Gestapo. He wrote early in January because a trustworthy person had presented himself who could be relied on to ensure the letter's delivery. After recalling the circumstances of Einstein's fiftieth birthday, he was sure that he would now be at peace with himself and secure in the knowledge that his work would stand as long as there were still civilised people on earth. He concluded by hailing Albertus Magnus, referencing the thirteenth-century polymath Bishop of Regensburg but also of course playing on the literal translation of the Latin, 'Albert the Great'.

The New York Times editorial was typical of the press reaction to Einstein's birthday:

> Congratulations will pour in upon Einstein today, his sixtieth birthday. They ought to pour in on the country too. Thanks to the barbarity and the bigotry of the present German Government, the most distinguished of living mathematical physicists, one of the great figures in the history of intellectual achievement, is a resident of the United States and a citizen-to-be. If the dazzling gift of Einstein were not darkened by a tragedy which has brought despair to fifteen hundred university professors, American science would be justified in cabling a warm 'thank you' to the Fuehrer who has out done the Middle Ages in fanatic cruelty and so far blighted German research that not for a generation after his evil influence has passed can it expect to recover something of its former glory.

Einstein was not at home during the day of his birthday, presumably to avoid reporters, but apparently a small dinner party was held in Mercer Street that evening.

Another tribute to Einstein's birthday was made in a radio broadcast by J. R. Oppenheimer, in which he made some interesting and prescient remarks. To understand Oppenheimer's words, it is necessary to go back to the autumn of 1938, when Otto Hahn and Fritz Strassman in Berlin had discovered the fission of uranium into two smaller fragments. They reported details to Hahn's former colleague Liese Meitner, who had fled Germany for Sweden. Meitner and her nephew Otto Frisch, who was working with Niels Bohr, deduced that the process must have been accompanied by an unprecedentedly large release of energy. Bohr had been invited the previous year to spend the first months of 1939 at the IAS. He sailed from Gothenburg on 7 January with his son Erik and colleague Leon Rosenfeld. They arrived in New York on 16 January and were met by Fermi among others. Rosenfeld went straight to Princeton but Bohr accompanied Fermi to stay with their mutual friend, Alfred Lee Loomis. Loomis was a multi-millionaire lawyer and investment banker. After making his fortune during the great stock market crash, he used it to indulge his fascination with first physics and later biology. He funded a laboratory in his compound at Tuxedo Park in suburban New York that did world-class work, resulting in him eventually being elected to the National Academy of Sciences. He was to play a major role in uranium research.

On the day after his arrival, Rosenfeld gave a seminar on Frisch and Meitner's work. Bohr was irritated by this when he arrived in Princeton after his stay with Loomis and Fermi because he was keen to ensure Frisch and Meitner's priority and their work had not yet been published. Bohr went to see Einstein, taking with him John Wheeler, a former student, who by then was on the faculty of Princeton University. Wheeler had attended Rosenfeld's seminar. Bohr told Einstein the news about fission, which quickly reached Fermi, probably via Isidor Rabi, who had attended Rosenfeld's lecture.

Once he heard of the Frisch–Meitner work, Fermi began experiments to try to detect the released energy from uranium fission. He discussed the issue with Bohr on 26 January, when they both attended a conference on theoretical physics organised by George Gamow and Edward Teller at George Washington University in Washington, DC. Bohr told the conference about Hahn and Strassman's results, implying that uranium had been split. This 'fission' produced barium and 200 MeV of energy, by far the largest energy release hitherto observed in nuclear processes. Fermi also reported on his work at Columbia. Bohr and Fermi discussed the possibility that neutrons were emitted in the process and whether this might precipitate further fissions, causing a chain reaction.

There had been some amateurish attempts to keep details from the press by expelling them from the discussion room when the physicists began to realise the possible import of the discovery. That notwithstanding, the news made it into *The New York Times* of 31 January, where the headline implied that vast energy had been freed from the uranium atom. Once these reports appeared in the west coast newspapers, Oppenheimer, then on the faculty of the University of California Berkeley, called Gamow, trying to find out what had been discussed. Apparently in a subsequent discussion with Luis Alvarez, Oppenheimer proved that fission was impossible, only for Alvarez to take him into his laboratory and demonstrate the experimental proof. After a few minutes head scratching, Oppenheimer was able to renormalise his opinion and work out that neutrons must be accompanying the fission.

Things began to move very fast. On 7 February, Bohr submitted a paper to *Physical Review*, following up on a previous letter to *Nature*, in which he pointed out that the rare U^{235} isotope was much more likely to undergo fission than the common U^{238}. This paper appeared next to several experimental investigations of fission; such was the excitement that some members of one group had dashed away from the Washington conference before Fermi had finished his talk to start experimentation. In the April issue of *Physical Review*, Fermi and others reported experimental results on uranium fission. In July, Anderson, Fermi, and Szilard published a measurement of the number of neutrons produced, which built on the result published in *Nature* in April by Joliot-Curie and his collaborators;[10] Fermi and the others concluded that their result of 1.5 neutrons per fission was enough in principle to sustain a chain reaction, although probably not using water, as they had done, to slow down, or 'moderate', the neutrons; very slow, 'thermal' neutrons were much more likely to cause uranium fission. By June, Bohr and Wheeler had written a paper on a quantum-mechanical theory of the fission process by treating the nucleus as a liquid drop of nuclear material, which was published in September.

Oppenheimer's celebration of Einstein's birthday was broadcast on a local Oakland, CA station on 14 March. In it he said:

> . . . it is for its return, and not its abstruseness, that our children, when they learn of Einstein's work in college or in school, will prize it. It will be part of their thought, as for us to-day the work of Newton and of Pasteur; they will be able to trace its history in the development of countless new and powerful technological methods. In fact it was some of the early work of Einstein on the theory of relativity that first pointed the way to vast and hitherto untapped sources of energy. We know now that most of the sun's heat comes from these sources. And it is one of the most spectacular projects of contemporary atomic physics to make this energy available terrestrially, and thus to solve as far as human wants are concerned the problem of mechanical and electrical power.

[10] Joliot-Curie et al.'s number of 3.5±0.7 neutrons per fission was slightly high, although they admitted that their uncertainty was an under-estimate; Fermi's estimate of 1.5 was too low, although quoted without an uncertainty and was valid for the particular properties of their experimental set-up. What was crucial for the possibility of a chain reaction was that both results were greater than 1. The accepted number is 2.5.

The atomic age was approaching fast, in which, as will be chronicled in Chapter 29, certainly Oppenheimer and to some extent Einstein were to play seminal roles.

Two other events of note marked Einstein's birthday. The first was the publication of a new Einstein biography, apparently and surprisingly written with the subject's collaboration. Gordon Garbedian, a science correspondent of *The New York Times*, published *Albert Einstein – Maker of Universes* on Einstein's birthday. In the foreword, the author expressed his gratitude to Einstein, who he said had given generously of his time and energy to answer innumerable questions and help in the completion of the book. Given Einstein's great and often expressed distaste for biographies of himself, this statement must be taken with some scepticism. There is no correspondence between Einstein and Garbedian in the Einstein Archive, nor any mention of his name. Einstein's strong objection to the idea of Antonina Vallentin, or anyone else, producing a biography of him had been conveyed by Elsa as recently as 1933, as mentioned in Chapter 27. His violent reaction to Dima Marianoff's book, which appeared in 1944, suggests that nothing had changed in the intervening period. Neither is Garbedian's book any more accurate than Marianoff's would prove to be. Einstein's collaboration, for example, didn't extend to correcting a fictional visit to Lord Haldane's country seat at Cloan, which in any case Garbedian placed in Dorset rather than Scotland. Clearly Einstein did not read the book in advance; reviewers pointed out various flights of fancy that it contained, including Einstein's thoughts as he entered the Zurich Polytechnikum as a student in the previous century. Garbedian also exhibited a bizarre fixation on the facial hair of various persons mentioned. The book contains long passages attacking Hitler and the Nazi regime. The most likely explanation of its genesis is that Garbedian obtained an interview with Einstein on the basis that he would produce a piece of work with a strong anti-Nazi component, which blossomed with considerable imagination on the part of the author into the published volume.

Contrary however to the above suggestion is a report in *The New York Times* that Einstein had released a statement through his biographer, H. Gordon Garbedian. In the statement, Einstein expressed his gratitude for the ideal conditions in the US and his happiness at being able to become a US citizen in a year's time. Why Einstein should choose to release such a statement through Garbedian is difficult to fathom.

The second birthday event was rather less surprising. On 13 March, Einstein announced, yet again, a breakthrough in his search for a unified theory; or rather, this is how the press interpreted his answer to written questions submitted to him. In fact, his statement was that, although his previous attempts had not been successful, one year ago he had had a new idea and with two colleagues was developing the results to a point when they could be compared with experimental facts. *The New York Times* placed this 'news' on its front page, interpreted as Einstein being about to turn the key to understanding the universe after twenty years of labour. Einstein's careful phrasing was more accurate; this work with Bergmann on Kaluza-like theories turned out to be yet another dead end.

The spring of 1939, despite the relative low-key birthday celebrations, was exceptionally busy for the increasingly reclusive Einstein. On 21 March, he and Maja attended a benefit concert in aid of the Palestine Symphony Orchestra at Carnegie Hall. The music was an eclectic mix, including several Jewish songs as well as opera excepts. Joseph Lhévinne played four Chopin preludes, a valse, and the Scherzo in C sharp minor. The highlight for Einstein would have been the performance of Paganini's D Major Concerto by his acquaintance from Berlin days, now grown into a young man, Ruggiero Ricci. Louis Persinger, Menuhin's first teacher, accompanied Ricci on the piano. The evening raised $11,000 for the orchestra. On the same evening, Einstein gave an impassioned plea on the radio for support for the United Jewish Appeal for Refugees and Overseas Needs, of which he was the honorary chairman. He concluded the appeal with the words:

> I urge my listeners to support the United Jewish Appeal with all the energies at their command, that we may rescue our persecuted brethren from their peril and calamitous distress and lead them to a better future. Thus you will have an active share in averting the danger that now threatens all mankind, the danger of a reversion to the barbarism of ages long past.

Einstein and Maja called in at Newark airport on their way back to Princeton from the Carnegie Hall event. They were driven by Peter Bucky, son of his close friend Gustav Bucky, and were shown around the airport's facilities. Einstein was particularly interested in a new altimeter and stunned the technical expert into silence when the supposed unworldly professor clearly knew much more about the basic operating principles than he did. The expert was not to know that Einstein and Peter Bucky's father had spent considerable effort in 1934–1935 on the very question of altimeter design.

Another concert in aid of the Palestine Orchestra was given at a private house by Huberman's close friend from childhood, the great pianist Arthur Rubinstein. It is not clear when this concert took place but it is likely to have been around 1938–1939. Einstein attended and would no doubt have been surprised and pleased to hear Rubinstein play an arrangement of the famous Chaconne from Bach's second partita for solo violin. This event was probably the origin of a photograph of the two that Rubinstein proudly displayed on his Steinway grand in the study of his west-coast mansion. He liked it because he thought that it made Einstein look like a broken-down musician and him like an old philosopher.

The World's Fair in New York had been absorbing planners for several years in advance of its opening on 30 April 1939. There had been an intention that science education should feature prominently in the exhibits but, unsurprisingly, it was very much marginalised by commercial interests, to the great irritation of the scientific consultants. Science, when it appeared, was mostly relegated to entertainment. Einstein had agreed to be honorary chairman of the science advisory committee. Despite initially refusing, he was finally prevailed upon to take part in the opening ceremonies by delivering a lecture on the cosmic rays that were to be harnessed to trigger a light show. Unfortunately, the event degenerated into farce when firstly the sound system failed so that Einstein's speech was inaudible[11] and then by the light show blowing a fuse. Further elements of farce were provided by an event at the 'Congress of Beauties' in which a macaw named Einstein had been trained to remove the bra of one of the participants to the beat of tom-toms from the neighbouring Seminole model village.

Undaunted, Einstein returned to the fair on 28 May as the main speaker at the opening of the Palestine Pavilion. Between 50,000 and 100,000 people crowded into the square in front of the pavilion to hear Einstein's speech. Such a large crowd was also partly a protest against the newly announced policy of the British government, departing from the Balfour Declaration by limiting Jewish immigration into Palestine. Although several of the speakers, including Mayor Le Guardia of New York, attacked the British government's new policy, Einstein spoke only of the struggles of the Palestinian Jewish people, whose inner security and traditions were expressed in the peace and nobility of the pavilion.

Einstein left Princeton, first to visit the World's Fair yet again, with old friend Judge Lehman, on 11 June, moving on for his by-now traditional sailing vacation on Long Island. However, he seems no sooner to have arrived in Peconic than he left again for yet another visit to the World's Fair, for a speaking engagement with his friend Rabbi Wise at a lunch in one of the pavilions. He even

[11] The only parts that could be heard can be accessed on a video available at https://www.pond5.com/stock-footage/item/124143011-1939-albert-einstein-gives-speech-new-york-worlds-fair.

appeared *in absentia*, being one of ten influential people, including Thomas Mann, who appeared on film to suggest a resolution to the current international crisis. In addition to the distractions of the fair, this vacation was to be different to his previous ones, with repercussions that would change the world forever.

Chapter 29

War and the atomic bomb (1939–1945)

Einstein's colleague Leo Szilard had arrived in the US on 2 January 1938 and spent some months travelling around the country working with a variety of researchers. He said he did nothing but loaf until after the Munich crisis, in which British Prime Minister Neville Chamberlain abandoned Czechoslovakia to Hitler's tender mercies. He then resigned his position in Oxford and committed himself to the US. In November, he moved into a hotel opposite Columbia University in New York. He gained permission to use a laboratory in Columbia which he furnished by raising money from his many contacts with rich industrialists. He began to work with Walter Zinn, Herb Anderson, and Enrico Fermi. Fermi, whose wife was Jewish, had reacted to the Italian anti-Semitic laws in the same way as Einstein's sister Maja. The award of the Nobel Prize in December 1938 for his work on nuclear reactions mediated by neutrons gave him and his family an excuse to leave Italy to deliver his lecture in Stockholm. They travelled from Stockholm to Southampton and thence to New York, where they arrived on 2 January 1939. Here Fermi took up a visiting professorship and started a new life in the US.

As soon as Szilard heard about the discovery of uranium fission, he was certain that a weapon was feasible and became obsessed with the idea that Hitler's Germany might make one first. He attempted in April 1939 to get the principals to agree not to publish results in scientific journals but instead to circulate them privately within the community. Szilard contacted the research groups in the US, the UK, and France. Bohr agreed to halt publication of his paper with Wheeler; Patrick Blackett, John Cockcroft, and Paul Dirac in the UK agreed with Szilard's proposition; Frederick Joliot-Curie in France did not. As a result, many papers appeared in journals in the ensuing months, as described in Chapter 28. Szilard also did his best to accelerate research into the possibilities of fission for military purposes but was thwarted by, for example, Bohr and Wheeler's initial conviction that a chain reaction was impossible. Nevertheless, he was able to convince affluent businessman Lewis Strauss, who later became a central figure in the nuclear story, to arrange the loan of 230 kg of uranium oxide for his and Fermi's use.

Despite this success, Szilard's fellow Hungarian friend and colleague, Eugene Wigner, a professor at Princeton University, believed that it was essential to get US government involvement if further progress was to be made. Overcoming Szilard's initial reluctance, he convinced Szilard to consult the head of department at Columbia University, who was also Dean of Science, George Pegram. Pegram knew the Assistant Secretary to the Navy, Charles Edison (son of Thomas Alva Edison) and telephoned him, explaining the situation. Edison catalysed contact with Rear-Admiral Stanford Hooper, Director of the Technical Division in the office of the Chief of Naval Operations, to whom Pegram wrote to set up a meeting with Fermi. Pegram characterised Fermi by saying that no man was more competent in the field of nuclear physics. The meeting took place on 17 March in the Navy Department in Washington. The attendees included Ross Gunn, technical adviser to the Naval Research Laboratory (NRL) and representatives from the Naval and Army Ordinance and engineering. Fermi talked about the potential of nuclear power both as a weapon and as a source of energy. He was disappointed with the outcome, even though he was told that the Navy wished to

Einstein. Brian Foster, Oxford University Press. © Brian Foster (2026). DOI: 10.1093/oso/9780198794875.003.0029

keep in contact with his work at Columbia. It was in fact the possibility of nuclear propulsion that had attracted the interest of Gunn, who extracted $1,500 from naval engineering to get research under way at the NRL.

The apparent failure of a direct approach to the armed forces forced Szilard and Wigner into a rethink. This they carried out in consultation with Edward Teller, another Hungarian exile who had been brought to the US by George Gamow following Gamow's defection from the Soviet Union in 1934. If US government support was not forthcoming, they thought they might inhibit any German plan to exploit fission by restricting German access to uranium deposits. Those in the Belgian Congo were of exceptional purity. Szilard was aware that Einstein was a friend of Queen Elisabeth, so they sought to bring him into their councils. On 17 July, they tracked him down to Dr Moore's cottage in Peconic, although when they arrived in Wigner's car, no one seemed to know its location. Eventually a boy told them where Einstein lived, and they arrived at the porch to find Einstein in his typical sailing garb. He invited them in, where they explained to him the latest developments in the fission research that was being carried out mostly in Columbia but also in Paris by Joliot-Curie. Having talked to Bohr in January and having been queried about the subject during his birthday celebrations, Einstein was certainly not totally ignorant of the basic developments, but neither was he interested in the details. He listened carefully however and agreed that, were an uncontrolled chain reaction to be possible, it was vital that the US have the knowledge before the Germans.

When the matter of a letter was broached, Einstein, no doubt reluctant to involve the Queen in politics, preferred to write to the Belgian Ambassador via a covering letter to the State Department. Einstein dictated a letter to Wigner in German, which he and Szilard took back to the mainland to translate and type up. However, Szilard's friend Stolper, whom he had taken into his confidence, urged him to meet Alexander Sachs, to whom Stolper had explained the situation. Sachs, Jewish and originally from Russia, was an economist and banker, a vice-president of Lehman Brothers. Sachs had had an important role in Roosevelt's first administration and was a close friend of the President. When Sachs heard of the Einstein letter, he convinced Szilard that it would be much better to go straight to the President and that he, Sachs, could guarantee to get it to Roosevelt. Szilard, whose instinct was also to communicate directly with the most important people, immediately agreed. Wigner was away on a trip to California, so Szilard turned to Teller for advice. Teller agreed that the President was the best recipient, so Szilard redrafted the latter and mailed it to Einstein, who then asked Szilard to return to Peconic so that they could discuss this altered draft.

On 2 August, Szilard, who couldn't drive, was chauffeured to Long Island by Teller, in the continued absence of Wigner. The letter as finally agreed with Einstein informed the President that recent research made it highly likely that a chain reaction in uranium could be brought about, giving rise to new elements and a source of power and that there was a remote possibility that this could lead to a devastating weapon. By mentioning that the German government was now restricting the export of uranium from Czechoslovakia, the letter raised the spectre of Hitler getting to any possible bomb first. Indeed, Joliot-Curie et al.'s *Nature* paper in April, as discussed in Chapter 28, had been read with interest by several German scientists. They passed on to the German government their assessment that uranium could be used to develop a weapon. A secret conference was held on 29 April, at which it was decided to initiate a research programme and to stop the export of uranium from the Czechoslovakian mines and divert it to Berlin.

Einstein doubted whether Sachs was the best intermediary for this letter to Roosevelt. Sachs himself proposed a couple of other sensible names but, perhaps by now having sized up Szilard, also suggested the aviator Charles Lindbergh. Szilard immediately jumped at a celebrity name.

Unfortunately, Lindberg, while certainly famous, was also becoming notorious for right-wing views and sympathy for the Nazis. Furthermore, Roosevelt and he mutually despised each other. Once this bizarre idea had been dismissed, the choice, as he had no doubt hoped, returned to Sachs. Szilard and Sachs drafted a somewhat longer alternative letter and despatched them both to Einstein, asking him to select which he preferred. Einstein selected the longer one, signed them both and returned them to Szilard by post. At Sachs's request, Szilard wrote a detailed technical memorandum to accompany the letter, which was completed by mid-August. He gave both to Sachs and returned to his uranium research to await the President's reaction.

Einstein meanwhile had put the letter from his mind and was out sailing at every possible opportunity. Maja and the rest of the occupants of the Mercer Street household were also in Dr Moore's cottage, Maja often accompanying Einstein on his boat. Certainly the Buckys were also in Peconic, as well as, for some time, Hans Albert and his family. Maja had been to stay with them in South Carolina after the tragic loss of their son. When they arrived in Peconic, Frieda was pregnant with their third child, a son who died a month after his premature birth on 14 October.

Although mostly preoccupied with sailing during the daytime, Einstein was always on the lookout for opportunities to play music in the evenings. David Rothman, who ran the local store, was an active amateur musician who played the violin. Margot Einstein went to Rothman's shop to buy a sculptor's chisel. Rothman apparently recognised her from newspaper pictures and insisted she take the chisel free of charge, in honour of her and her stepfather's arrival in town. On the following day, Einstein arrived in the shop. Rothman offered to turn down Mozart's Symphony No. 40 that was playing on the record player, but Einstein insisted on hearing it to its conclusion. The music may have limited Rothman's ability to understand Einstein's heavily accented English when he asked if he had sandals in stock. Rather than showing him sandals, Rothman took him into the garden to show him his sundial! Having established that they both loved music and could share a laugh, a friendship was forged. The two spent the evening together and, Rothman having mentioned a string quartet that regularly gathered in his house, Einstein asked to play with them.

Einstein came to Rothman's house regularly twice a week for their music sessions. Rothman quickly realised that he wasn't up to Einstein's standard of playing and instead organised some of his musical acquaintances into a string quartet that met regularly throughout Einstein's stay. David Rothman may not have been a great violinist, but he was a driving force behind the Peconic and Southwold music scene. He ran a local orchestra and his wife was an accomplished amateur pianist.

It was through Mrs Rothman that two exiled English musicians impinged briefly on Einstein's musical life. Dr Mayer, a psychiatrist, was her doctor and analyst. He ran the Long Island Home, a sanatorium for patients with mental disorders. The composer Benjamin Britten and his new partner the singer Peter Pears had arrived in Canada in April 1939. Shortly afterwards they moved to New York. Elizabeth Mayer, who had been born and brought up in Germany, was a great supporter of artists of all descriptions. She met Britten and Pears and offered them accommodation in the Mayer home, Stanton Cottage, in Amityville, Long Island. Britten and Pears's friend the poet Wystan Auden, who had arrived in the US some months before them, was also a frequent visitor to Amityville. Christopher Isherwood, who had emigrated with Auden but, unlike the others, remained in the US for the rest of his life, also subsequently became a friend of the Mayers.

Britten and Pears arrived in Amityville at the start of September. It must have been shortly thereafter that they were introduced to the Rothmans by the Mayers and arrived in Southwold for a musical visit, since by mid-September, Einstein was back in Princeton. Rothman recalled evenings in which Britten and Pears shared in the music-making. Britten was a good violist and may have played in the quartet. Britten and Pears apparently began the evening with some Schubert songs. On another evening, Britten and Elizabeth Mayer played the piano, Einstein the violin and Pears sang. Rothman asked Einstein his opinion of the young Englishmen, to which Einstein replied: 'You know, they are very talented, and they will go very far.' Pears's and Britten's abiding memory was of Einstein's uncertain intonation.[1]

While Einstein was sailing and playing music on Long Island, his letter to Roosevelt languished in Alexander Sachs's possession. He was waiting for an opportunity to obtain an interview in the White House. He was determined that he would read the letter to Roosevelt in person, rather than hand yet another piece of paperwork to a Chief Executive who was deluged in it. Unfortunately, the President had more urgent matters on his mind than possible future super weapons. Having blind-sided Britain and France by concluding the Molotov–Ribbentrop pact on 23 August, Hitler sent German troops into Poland on 1 September. On 17 September, the Russians invaded from the east. The Second World War had begun.

Despite the international situation, or indeed because of it, Szilard and Wigner grew increasingly impatient at the silence from Sachs. In the last week of September, they visited him in New York and were dismayed to find that he had been unable to schedule an appointment with Roosevelt. Indeed, he had decided that it was better to wait until the President was less burdened. Szilard wrote to Einstein that he and Wigner had decided to give Sachs ten days more before they looked for an alternative intermediary. The rapid German blitzkreig crushed Polish resistance so that by 6 October, the conquest was complete and Poland was divided between Germany and Russia. In the subsequent period of uneasy quiet, Sachs saw his chance and on 11 October he arrived at The White House. After an interview with Roosevelt's chief military aide, General Edwin 'Pa' Watson, who read the Einstein letter and glanced through Szilard's memorandum, he was admitted to the Oval Office.

Over a glass of fine Napoleon brandy, presumably placing the interview in the early evening, Sachs began. Knowing Roosevelt well, he first arrested the President's attention with a story about the Emperor of the French refusing to listen to a technological innovation that could have conquered England in the Napoleonic Wars. He then read to him his own precis of the Einstein letter and Szilard memorandum, in which he emphasised first the possible peaceful uses of fission before discussing the possibility of a bomb. Sachs concluded by quoting from a report by Francis Aston, Nobel Laureate in 1922, who wrote:

> Personally I think there is no doubt that sub-atomic energy is available all around us, and that one day man will release and control its almost infinite power. We cannot prevent him from doing so and can only hope that he will not use it exclusively in blowing up his next door neighbour.

---

[1] Rothman's memories are somewhat contradictory. The author Dan Rattiner wrote that Rothman recalled that Britten spent the summer staying with them and had almost no contact with Einstein. However, Britten was certainly in Amityville, not Southwold, for the remainder of the summer of 1939 but subsequently was a regular visitor to the Rothmans until he and Pears returned to England in the spring of 1943. Rothman also arranged for Britten to conduct the local orchestra he ran and consoled him when the audience at their concert was tiny. Britten apparently was so depressed that he considered giving up composition and working in Rothman's store; a course from which, thankfully for the world of music, Rothman dissuaded him. Rothman's recollection that Britten spent most of his visits upstairs building model aeroplanes with his son Bobby, then 14, was probably accurate. Britten's fondness for young boys was notorious and he was certainly enamoured with Bobby, to whom he dedicated a song and with whom he corresponded for several years after his return to UK.

Roosevelt immediately got the message, saying 'Alex, what you are after is to see that the Nazis don't blow us up.' He called in Watson with instructions to take action.

As a good bureaucrat, Watson, in discussion with Sachs, set up a committee. The chair was to be the lead government physicist, head of the National Bureau of Standards, Lyman Briggs, with military representation at field-officer level. Sachs saw Briggs the next day, and agreed with him that the new committee would meet with Szilard and associates. Reporting back to Roosevelt directly after meeting with Briggs, Sachs pronounced himself as satisfied that action was being taken. If he was satisfied, so was Roosevelt.

The first meeting of the Uranium Committee was held at the Bureau of Standards on 21 October. The attendance was Briggs and an assistant, Sachs, Szilard, Teller, Wigner, the two military representatives, Lieutenant Colonel Adamson and Commander Hoover, and Richard Roberts, standing in for Merle Tuve of the Carnegie Institution in Washington. Tuve had been working on nuclear physics for many years using a Van de Graaff generator to accelerate subatomic particles. The committee meeting did not go well. The military contingent, who were ordnance experts, were openly hostile. Roberts was also sceptical of the possibilities of a chain reaction in natural uranium. He touted the possibility of separating out the U^{235} isotope, which Bohr believed was the fissionable nucleus, via the high-speed centrifuges that Jesse Beams had developed at the University of Virginia.

Exasperated with all this technical jargon, Hoover launched into a long diatribe on how it was morale that won wars, not weapons, only to be interrupted by Wigner. Innocently the physicist remarked that this was a very interesting idea. Since it was morale that won wars but weapons that cost all the money, perhaps it was time for the army budget to be reappraised? Hoover quietly asked how much the physicists needed. Teller requested $6000 for the first year, which he said was required to purchase the pure graphite that he and Fermi wished to use to moderate the neutrons from the uranium fission. Szilard, no doubt unwilling to break the unanimity of the physicists, said nothing but wrote to Briggs a few days later that in fact just the graphite would require at least $33,000. Teller later remarked that his colleagues never forgave him for starting off the nuclear programme with such a pittance. It was indeed the $6000 figure that was mentioned in Briggs' report of the meeting to the president, which was submitted on 1 November, with the title 'Possible Use of Uranium for Submarine Power and High Destructive Bombs'. However, Szilard's lobbying seems to have had a good effect in that the report recommended that the government should also procure the four tonnes of pure graphite and, if subsequently justified, fifty tonnes of uranium oxide. Roosevelt read the report and instructed Pa Watson to keep it on file, where, much to the chagrin of Szilard, Fermi, and collaborators, it remained until the following year.

Sachs had by now got the bit between his teeth. His letters and memos make it clear that he had been starstruck by Einstein and never lost an opportunity to portray himself and Einstein as leading a lonely charge to convince the US government of the importance of uranium research. In fact, most communication between Sachs and Einstein went through Szilard, who wrote often to Einstein in this period to keep him in touch with developments. Teller's $6000 request had finally been paid to Szilard and Fermi in February, 1940, after prompting from Sachs to Pa Watson. Watson sent a copy of Briggs's memo to the President, written on 1 November, which Sachs had apparently not seen. Sachs was dismayed by the drafting of the report, which he thought much too academic in tone, and promised to write a memorandum on progress in research over the last few months. He went to see Einstein in Princeton to discuss the troubling lack of progress with him.

Szilard had meanwhile also become frustrated with the slow progress. He was especially concerned since word had reached him through Peter Debye that German efforts on uranium had intensified. Debye, who had won the Nobel Prize for Chemistry in 1936, had fled Germany on 16 January 1940. He travelled to Genoa to embark on the SS *Conte di Savoia* on 23 January, arriving in New York on 1 February, from whence he travelled to Cornell University to take up a visiting lectureship. As Director of the Kaiser Wilhelm Institute for Physics and President of the German Physical Society, Debye was in a position to know what was going on with the German work on uranium.[2]

Szilard prevailed on Einstein on 7 March to write a letter to Sachs, for forwarding to the President. Einstein pointed out the increased German activity and also questioned whether or not papers giving details of uranium research should be published in scientific journals. Sachs duly forwarded Einstein's letter with a covering letter to the president via Pa Watson. Sachs requested a meeting with the President to bring him up to date. Watson however called Adamson and Hoover into his office and was advised that nothing should be done until Fermi and Szilard came up with some results from their $6000 grant. Nevertheless, Watson forwarded Sachs and Einstein's letters to the President, who wrote to Sachs on 8 April. He told Sachs that he had asked Watson to convene another meeting of the Uranium Committee that month, to which he and Einstein would be invited. A briefing for this committee, presumably drawn up by Sachs and the physicists, requested $100,000 be authorised to hire some young physicists, to purchase the raw materials and to further research into separating the U^{235} from the bulk uranium.

The Uranium Committee met again on 27 April, in Briggs' office. Einstein had declined the opportunity to attend, pleading a bad cold and shyness to Sachs. He had however written a letter urging that the work of Szilard and Fermi be accelerated and that it be set up as efficiently as possible. He suggested, presumably on prompting from Sachs, that some sort of not-for-profit company could oversee the research in a flexible way. Sachs pressed this idea on Watson. However, he made the mistake of tacking on to his requests relating to uranium a much broader memorandum on the organisation of science in relation to military requirements. It seems likely that this was sufficient to confirm Watson in a belief that Sachs was somewhat of a crank from whom the president should be shielded as much as possible. The Administration had already been influenced on the organisation of science and technology by a group of researchers and industrialists much more highly powered than Sachs, led by the formidable, indeed legendary, Vannevar Bush.

Bush was already involved in uranium research as Director of the Carnegie Foundation, which funded the Atomic Physics Observatory (APO) inside the famous Department of Terrestrial Magnetism. Using the APO's Van de Graaff accelerator, Roberts, Hafstad, Meyer, and Wang had confirmed uranium fission and the emission of delayed neutrons early in 1939. Bush, although an engineer rather than a physicist, understood clearly the potential importance of uranium research.

[2] Debye's position vis-à-vis the Hitler regime has been a cause for controversy. There have been allegations that he was a Nazi and might indeed have been a spy. Certainly he had made anti-Semitic statements in the past, as reported in Chapter 10. He took care not to resign as Director of the Kaiser Wilhelm Institute, taking leave of absence and thereby allowing his daughter and sister-in-law to continue living in his Berlin official house. Einstein forwarded a letter that he had received in June 1940 to the President of Cornell University. This accused Debye of being a Nazi sympathiser. Einstein passed the letter on, which he considered his duty, while writing that he had the utmost respect for Debye as a physicist. However, in an interview with an FBI agent, Einstein gave a negative assessment of Debye's character, saying that he was not to be trusted and was a careerist. Debye providing information on the German uranium project would tend to exonerate him from these charges, as does a detailed study by van Ginkel.

He also understood even more clearly how Washington worked. He gained the ear of Roosevelt's right-hand man, Harry Hopkins, and convinced him that his own ideas for organising scientific research were more sensible than those of Hopkins. Bush was ushered in by Hopkins to see the president on 12 June. He had walked his plan for the National Defense Research Council (NDRC) around the military and scientific centres of power beforehand and gained agreement, as Hopkins informed the president. In ten minutes, Bush was leaving the Oval Office with his plan initialled with the crucial 'OK – FDR' – the first of many such endorsements that Bush was to obtain over the next five years.

The establishment of the NDRC with Bush at its head brought a sudden end to Sachs's and Einstein's involvement with uranium matters. Bush moved with typical energy to reshape the Uranium Committee, fortified by the access he had gained to the president's special fund and the promise of as much money as he needed. Although he retained Briggs as chair, he removed the military members, although Ross Gunn, from the ONR, was for a while retained. A number of physicists were recruited. However, none of the foreign scientists were included for security reasons. All the members – Harold Urey, George Pegram, Merle Tuve, Jesse Beams, Gunn, and Gregory Breit – were American citizens; indeed only Breit was born outside the US, in Russia, but had lived in the US since 1923. The key players – Fermi, Szilard, and Wigner – were not US citizens; nor was Einstein. Although the first three were eventually accepted into the fold, Einstein, partly by choice but also partly because of his left-wing views,[3] was not. Indeed, Robert Oppenheimer was also almost excluded from participation in the uranium research because of his radical politics. However, Ernest Lawrence insisted on Oppenheimer's participation. Lawrence was using cyclotrons on the west coast to produce plutonium, which subsequently morphed into an effort to separate uranium isotopes via electromagnetic deflection.

The Uranium Committee became the S-1 committee in order not to broadcast its functions so obviously. Information sent from the UK, clearly demonstrating the Americans were far behind, convinced Bush to order a review of activity by the National Academy of Sciences. Its favourable report fed into the creation of the Office of Scientific Research and Development (OSRD), with Bush as director. The OSRD subsumed NDRC, now under James Conant, while atomic bomb research and S-1 reported to Bush through Conant. Bush had a blank cheque from the Budget Bureau. Policy matters on the bomb project were restricted to the President, Vice-President Wallace, Bush, Conant, Secretary of War Henry Stimson, and the Chief of Staff of the Army General George C. Marshall.

Meanwhile, in July 1941 the Maud Committee in the UK had concluded that a bomb was possible after Otto Frisch and Rudolf Peierls had estimated the amount of U^{235} required to be a few kilogrammes. This conclusion was brought informally to Washington by Oliphant in late August. He rapidly convinced the OSRD hierarchy that top priority should be given to the project. On 6 December 1941, Bush decided that a reorganisation was necessary to accelerate progress, particularly in techniques to separate out and concentrate U^{235}. On the following

[3] The army, presumably requested by NDRC, asked the FBI in July 1940 if Einstein or Sachs were security risks. The Director, J. Edgar Hoover, replied with a copy of the unsubstantiated charges brought by the Women's Patriot Corporation to try to stop his visa being issued in 1933, together with various other of his activities, particularly in favour of the Republican government of Spain. Einstein's FBI file contains a cornucopia of often bizarre allegations and gossip gathered over the succeeding decade.

day, Japanese planes bombed Pearl Harbor and the US found itself a belligerent in the Second World War.

For the remainder of the war, Einstein was isolated from the activities directed towards the development of the atomic bomb. He was no longer even able, in the unlikely event that he had so wished, to follow the development of the science in journals; Bush, immediately he had taken over responsibility in June 1940, had forbidden the publication of research papers relating to sensitive fission research. Einstein did participate in the war effort however. By December 1941, a consensus had developed that fast neutron capture was the most promising route to a bomb. This would require either the separation and concentration of U^{235} from the dominant U^{238} in uranium ore, or the production of sufficient quantities of plutonium by neutron irradiation of uranium ore. The latter had to wait for the operation of the first nuclear reactor, achieved in Chicago by Fermi and colleagues in December 1942. Rapidly thereafter other reactors were built, first at Oak Ridge in Tennessee; subsequently an enormous plant was constructed in Hanford, Washington State. Efforts to separate out U^{235} however had been going on since the 1939 discovery of fission and Bohr and Wheeler's theory that U^{235} was the critical component. Bush reached out to Einstein via Aydelotte, Director of the IAS, asking Einstein to work out some aspects of the gaseous diffusion method that was one option to concentrate U^{235}. Einstein happily did so, without detailed knowledge of the reason for the request but probably knowing enough to put two and two together. Although pressed by several physicists to enlist further help from Einstein by revealing more about the work on the bomb, Bush refused on security grounds, no doubt having had sight of the FBI report referred to above.

Although the army and its Manhattan Project had no further use for Einstein, the navy did. A young chemist, Stephen Brunauer, had volunteered for the Navy after Pearl Harbor; on the strength, according to his recollections, of recognising that TNT and dynamite were explosives, he became head of the High Explosives and Propellent section of the Bureau of Ordnance in November 1942. In May 1943, Brunauer, having established that Einstein was not working for the Army, decided to see whether he could enlist him in the Navy's group of civilian consultants. When Brunauer visited him, Einstein was delighted to accept. He was placed on a contract that ran until June 1946, specifying that his naval work would be carried out for the nominal fee of $25 per day. Einstein worked on the problems despatched to him in Princeton by the Navy Department in Washington. After 1944, however, his contact with Brunauer seems to have ceased.

The question of how the communication between Einstein and Washington was carried out deserves a passing mention. It illustrates once again the tendency of those who knew Einstein to magnify their degree of contact and intimacy. The distinguished physicist George Gamow was at this time also a consultant to Brunauer's section. He had earlier explained the alpha-particle decay of radioactive nuclei and subsequently developed Georges Lemaître's work into what became the Big Bang theory of cosmology. In his autobiography, he recalled travelling to Princeton every two weeks with a portfolio of projects that Einstein inspected and on which he gave his verbal opinion. Brunauer however explicitly contradicted every aspect of that narrative. Probably Gamow did take some material to Einstein on one or two occasions and lapses of memory many years later and wishful thinking provided the rest.

Also in the summer of 1943, Einstein received a letter from Bush inviting him to accept a consultancy in Division 8, the Explosives Division, of the OSRD. Einstein replied that he was unsure if such an appointment was compatible with his work for the Navy. This letter seems to have confused the bureaucracy, and no reply appears to have been sent. Indeed, Bush either forgot having received this letter or indeed never received it, since, after the war, he denied ever having been told of Einstein's work with the Navy.

Other than these contacts, Einstein spent the war like most US citizens, an accolade that he, Margot, and Helen Dukas acquired on 1 October 1940. His grandson Bernhard left an account of Einstein's daily life at around this time:

> At Princeton, my grandfather Albert had a daily routine, which essentially never changed. In the early morning, he would dress in the same clothes but without socks. After breakfast, his housekeeper and secretary, Helen Dukas, would do the dishes while he would play the violin. He played the violin beautifully with a natural grasp of the composer's meaning. When he became too weak to play the violin, he would play the piano with improvisations, making it up as he went along.
>
> [. . .]
>
> After playing the violin, Grandfather Albert and Dukas would work together on his extensive volume of mail for about one hour. He would dictate while she typed. I sat in on several of these sessions; it impressed me that he answered all of his letters, even the 'crackpot' ones. Some of them he would think about and answer later. He was very meticulous about his responses; everything he wrote had a natural clarity, needing corrections only on rare occasions. Then Grandfather would walk to his office at the Institute for Advanced Studies in Princeton University. He never drove a car. He would then come back between noon and 1 p.m. and perhaps bring someone home with him for lunch. I remember meeting his assistants, Professors Bergmann and Bargmann, during these luncheons.
>
> [. . .]
>
> After lunch, Grandfather would read *The New York Times*. He would give a running commentary on politics and world events with succinct and appropriate comments. He then took a nap. After his nap, Grandfather would again go to the Institute. Upon his return home, we would have a light supper. He would take a mixture of medicines during the meal. In the evening, there would be more music, or he would retire to his room.

Music and physics continued to offer solace and escape. On 7 May 1940, Einstein attended a recital by his old friends Adolf Busch and Rudi Serkin hosted by his neighbours Katya and Thomas Mann in their imposing home near Library Place. On the same day, the 'Narvik' debate on the British and French defeat against the Germans in Norway began in the British House of Commons. Busch would have just been tuning up as Leo Amery sat down after delivering his famous admonition to Prime Minister Chamberlain. Pointing at Chamberlain, Amery quoted from Oliver Cromwell's speech to the Long Parliament: 'You have sat too long here for any good you have been doing. Depart, I say, and let us have done with you. In the name of God, go.' On 10 May, Winston Churchill became Prime Minister.

During the Battle of France, Einstein made several headlines on his own account. On 20 May, in a startling *volte face* compared to many of his previous pronouncements, he was the driving force behind a telegram to President Roosevelt which contradicted the recent call for the US for peace by the American Association of Scientific Workers. The telegram, signed by sixteen other Princeton faculty, urged the President to assist the democracies against fascist aggression.

Also on 20 May, Einstein was credited with the discovery of an aria by Mozart which received its first performance in Princeton at a concert given by the Westminster Choir College. The discoverer was not of course Albert but his friend Alfred Einstein, who was now teaching in Smith College in Massachusetts. However, he appeared approximately monthly in Princeton to teach at the music faculty. Both Einsteins must have been amused by this confusion, which had been going on ever since they lived near to each other in Berlin. For Alfred, there would have been a positive side, since his unearthing of the aria '*Io, ti laschio*' K. Anh. 245, would have been unlikely to make the news pages of *The New York Times* other than by this case of mistaken identity. It would be nice to think that they attended the concert together.

On 15 May Einstein had addressed the American Scientific Congress, a prestigious event held in Washington DC, which delegates from all over the Americas attended. It was as much a diplomatic occasion as a scientific one. Secretary of State Hull opened the meeting and Under-Secretary of State Sumner Welles headed the US delegation. Einstein's address was headlined as his first since 1934 and that it was eagerly awaited in Washington. Only afterwards, and in smaller type in *The New York Times* headline, was the Secretary of State mentioned.

In fact, Einstein did not report on his studies, or only in the most general sense. His address, 'Considerations Concerning the Fundaments of Theoretical Physics', was a thoughtful and masterfully concise historical survey of those fundamentals and how they had changed with time. He first discussed the foundation due to Newton, conserved masses interacting gravitationally by instantaneous action-at-a-distance. Describing the difficulties this scheme had with light and electromagnetism, he lauded in particular the achievement of Faraday in postulating electromagnetic fields and that of Maxwell in achieving a unified understanding of electromagnetism. He then described the twin but independent revolutions of relativity and quantum theory. While each seemed sovereign in its own areas of applicability, joining them together to form a new foundation for physics seemed to Einstein daunting. He was compelled to admit that: 'Thus it is probably out of the question that any future knowledge can compel physics again to relinquish our present statistical theoretical foundation in favor of a deterministic one.' He concluded that:

> It is agreed on all hands that the only principle which could serve as the basis of quantum theory would be one that constituted a translation of the field theory into the scheme of quantum statistics. Whether this will actually come about in a satisfactory manner, nobody can venture to say. Some physicists, among them myself, cannot believe that we must abandon, actually and forever, the idea of direct representation of physical reality in space and time; or that we must accept the view that events in nature are analogous to a game of chance. It is open to every man to choose the direction of his striving; and also every man may draw comfort from Lessing's fine saying, that the search for truth is more precious than its possession.

Einstein was to live to see the 'translation of the field theory into the scheme of quantum statistics' via the formulation of quantum field theory by Julian Schwinger, Richard Feynman, Freeman Dyson, and Shin'ichirō Tomonaga. However, its continued reliance on 'God throwing dice', not to mention the many infinities in calculations that were swept under the carpet by the 'renormalisation' procedure, did nothing to dispel Einstein's conviction that this was not the new foundation for physics that he sought.

On 22 September 1940, Einstein sent a message to the inauguration of the 'Wall of Fame' at the World's Fair site in New York. The wall contained a list of immigrants, Native Americans, and African Americans who had made distinguished contributions to American life. Einstein's message pointed out the debt that the country owed to African Americans for the discrimination with which they had been, and still were, burdened. He particularly mentioned African-American music, whose songs and choirs he characterised as 'the most beautiful cultural contribution that America has as yet gifted to the world. However, we don't have to thank those whose names stand on this 'Wall of Fame' for this wonderful gift, but the children of the people, who bloom anonymously like the lilies of the field.'

At the end of January 1941, Einstein appeared in a public concert in aid of the American Friends Service Committee for British refugee children. By this time he had become friendly with the distinguished pianist couple Robert and Gabriele (Gaby) Casadesus. They had been on tour in the US when France fell in 1940, fortunately having brought their two sons with them. They remained, renting a house in Princeton near to that of Einstein. Always quick to hear of pianists in his vicinity,

Einstein soon introduced himself and began playing Mozart sonatas with Gaby. With such a highly accomplished pianist, Einstein felt confident in once more appearing on stage in public. Since the concert was inside a programme that also contained a children's play, Einstein assumed a juvenile audience. He therefore chose simple, easily understood pieces, including two manifestly aimed at children, *Old Indian Song* and *Russian Dance*, by Frida Bucky, wife of his old friend Gustav. He was therefore discomfited to find the Present Day Club in Princeton packed with adults and reporters.

The story in *Time* on 3 February 1941 was rather light-hearted, indeed sarcastic:

> Corona-haired Scientist Albert Einstein looks more like a concert violinist than most concert violinists do. To many a ruthless young mathematician, fiddling is the best thing Oldster Einstein does nowadays. In any event, he fiddles well enough to be heard in public. One sleety afternoon last week he made one of his rare semi-public concert appearances, in his adopted town, Princeton, N. J. Two things prompted gentle Dr. Einstein to brave reporters and photographers: he thought most of his audience would be children, whom he likes; the occasion was a benefit for the American Friends Service Committee's work for British refugee children.
>
> Dr. Einstein played a simple program sitting down: two movements of a Mozart sonata, an Indian song and a 'Russian Dance' by Frida S. Bucky, a Bach minuet for an encore. His able accompanist – pretty Mme. Gaby Casadesus, wife of Concert Pianist Robert Casadesus (unpronounceable, rhymes roughly with 'has a canoe') – rippled discreetly at the piano. Dr. Einstein proved that he could play a slow melody with feeling, turn a trill with elegance, jigsaw on occasion. The audience applauded warmly. Fiddler Einstein smiled his broad and gentle smile, glanced at his watch in fourth-dimensional worriment, played his encore, peered at the watch again, retired. The refugee children stood to get about $200.

In addition to the pieces by Frida Bucky, the Mozart sonata was that in E minor, K304 and the encore a Bach minuetto. Gaby Casadesus remembered that:

> [Einstein] went through playing very well and he was not disturbed [by the noise of the newspaper reporters cameras] . . . Don't forget to say he was a true musician. Naturally he did not have the time to practice, so maybe he would not be able to do things that were difficult in technique, but, anyway, he played very well.

In May 1943, the distinguished Hungarian pianist Andor Foldes and his wife Lili arrived at Mercer Street. They had met Einstein at a concert in Princeton some weeks earlier when he had come to the green room and asked Foldes if he might find the time to play some sonatas with him. Foldes immediately agreed and subsequently he and his wife returned to Princeton from their home in New York. Foldes and Einstein played Vivaldi, some Italian Baroque sonatas whose composers were unknown to Foldes, Biber (of whom Foldes shame-facedly admitted he had also never heard), and finally Mozart. In the pause that followed, Foldes asked if they shouldn't now try a Beethoven sonata. He was astonished by Einstein's answer: 'I don't care for Beethoven much. He tugs at my heartstrings, he exposes too much of himself in his music.' Taken aback, Foldes went to the piano and played a short Beethoven sonata. When he stood up at the end with a questioning look, Einstein smiled and said 'You played it very beautifully.' However, when Foldes pressed him to play Beethoven, he gently but firmly declined. Foldes cherished the memory of his afternoon of making music with Einstein for the rest of his life.

Another exiled central European musician was introduced to Einstein by mutual friends, Robert and Gabby Casadesus. Bohuslav Martinu was born in Polička in Bohemia in 1890, son of a shoemaker who doubled as a fire watcher. The latter occupation gave the Martinu family the right to live in the church tower of the town. The sound of the bells in his tower home filled his childhood imagination. Often ill and very shy, he began to compose at the same time as showing talent as a violin prodigy, to the extent that the local community subscribed the money to send him to further his musical education at the Prague Conservatory. Martinu continued his dreamy isolation, neglecting his studies and violin practice but reading widely and becoming fascinated with musical analysis, particularly new music. His feats of memory were prodigious, echoing Mozart in the ability to hear a piece at a concert and subsequently reconstruct the score. These remarkable talents were insufficient to stave off expulsion from the Conservatory in 1910 for 'incorrigible negligence'. Returning to his home village, he qualified as a teacher and continued to compose. The end of the First World War gave him the opportunity to compose a piece celebrating the foundation of the Czechoslovak state, which brought him to national attention. He joined the Czech National Orchestra as a violinist and received tuition in composition from Dvorak's son-in-law, Josef Suk. In 1923, he moved to Paris, receiving sporadic composition lessons and writing a number of ballets. He married Charlotte Quennehen in 1931; they remained in Paris until 1940, when they fled the advancing German army, initially to Limoges and eventually Marseilles. Although not Jewish, Martinu had managed to offend the Germans by ignoring an order to return to Czechoslovakia and particularly by his collaboration with the Czech National Council, the effective government in exile based in London. In January 1941, with financial support from Paul Sacher, an admirer of Martinu who was conductor of the Basel Chamber Orchestra, the Martinus fled into neutral Portugal, from whence two months later they embarked for New York.

Initially based in St Hubert's Hotel in central New York, and in receipt of a small grant from the Czech government in exile, Martinu began to interact with the east coast musical scene. In the summer of 1941, he visited Tanglewood, where a performance of his piano quartet took place. There he met an old friend, Serge Koussevitzsky, conductor of the Boston Symphony Orchestra. In November, Koussevitzsky conducted the first performance of Martinu's *Concerto Grosso*. The reviews and reception were very positive, giving the composer confidence that his music was understood in his new home. In the summer of 1942, Martinu returned to Tanglewood, this time as one of the staff, sharing compositional teaching with Aaron Copland. It was here that he met Robert Casadesus and Gregor Piatigorsky. In 1943, he received a commission from Mischa Elman[4] for a violin concerto. It is likely that one of the three remarked on Einstein's violin playing to the composer.

Such a remark would have fallen on fertile ground, for Einstein's name, but, more surprisingly, also his work, were familiar to Martinu. He had developed a great interest in physics and had read widely. He considered the concept of an 'inner harmony' in the universe, of which Einstein often spoke, as vitally important in his search for musical ideas. Particularly from 1943 onwards, his notebooks are full of references not only to Einstein's work but also to that of Max Planck. In November of 1943, Martinu began work on 'Five Madrigal Stanzas', which was explicitly tailored towards its dedicatee, Einstein. The music itself is mostly tuneful, peaceful with slightly unsettling harmonic progressions. The violin part is rather simple, without high positions, double stops, or rapid passages, clearly written with an amateur in mind. However, it is far from trivial to play and requires a sustained tone and great musicality. The piano part, intended for Robert Casadesus, is more complex with the occasional virtuoso section. Presumably introduced

[4] Martinu surprised Elman with his taciturnity when they met to discuss the commission. After getting only monosyllabic responses to his general conversation, interspersed with long silences, Elman decided to let his playing speak for him. After a thirty-minute private recital, Elman put down his bow and waited for a response. The composer merely sat impassively until he eventually rose and left the nonplussed violinist. Nevertheless, the concerto eventually arrived and Elman considered that Martinu had captured his violinistic personality perfectly.

by Casadesus, Martinu visited Einstein in Princeton in December 1943, when he presented Einstein with a photostat of the manuscript of the *Madrigal Stanzas*, as yet unpublished, inscribed 'Avec toute mon admiration. B. Martinu, November 1943.' In return, he received a copy of *The Evolution of Physics* by Einstein and Infeld, inscribed to him by Einstein. Apparently Einstein played the pieces with Robert Casadesus, although there is no record of him having done so in public.

Einstein and Martinu remained on friendly terms until well after the war. Martinu became a part-time professor at Princeton on Koussevitzsky's recommendation in 1948. He travelled to Princeton every Thursday to lecture and work in the well-stocked music library. He gave Einstein a copy of the now-published Madrigal Stanzas, inscribed 'Albert Einstein with my deepest admiration – Princeton March 1950 – B. Martinu.' He continued his deep interest in physics, copiously annotating his copy of *The Evolution of Physics*. In 1946, he wrote to a friend that he was continuing his research into modern physics but the deeper he read, the more he got lost and the less he understood. Nevertheless, his curiosity about physics continued undamped. Given Einstein's lack of interest in modern music, it can be safely assumed that, when they met, physics was more likely to have been the subject of conversation than music.

The outbreak of the war had severely reduced Einstein's options to help his fellow Jews escape the cauldron of Europe. This was exacerbated by an unofficial but highly effective campaign within the State Department to reduce the influx, particularly of Jewish refugees, by bureaucratic impediments ostensibly designed to bar radicals and left-wing elements. These obstacles also hindered other figures from Einstein's past who wished to escape the European nightmare. Marie Winteler, although she was a Gentile and in relatively safe Switzerland, was nevertheless in significant distress. In December 1939, she wrote to Einstein again, hoping he had received her postcard with thanks for the money he had sent her a month previously but understanding that with the outbreak of war, marine communications would be disrupted. She had promised to write once her move to a cheaper area of Zurich was completed. Unfortunately, her new landlady was an alcoholic and not of high moral standing, which was disturbing her peace. This was now the only thing she sought from life after so many trials and tribulations. Only her beloved music kept her going. She had no idea what the new year would bring, least of all how she could escape from her perilous financial situation. Without help, she thought that a workhouse might be her fate. In concluding she sent Einstein best wishes for the new year and continued health, hoping that his scientific work would continue to bear fruit.

Marie's next letters were written in June and September 1940. In June, she informed him that her debts of a few hundred francs were very pressing and reminding him of his previous offers to help her. She even reminded him of her mother's kindness to both Einstein and Maja as a further appeal to his good offices. In September, she began by sadly noting that she had received no reply to her previous letter and begging for an answer as to whether he could help her and her son emigrate to America. She reminded him of his promises of help and that she and her son were at the end of their resources. They were only kept going by her son working at a labourer's job that was likely to wear him out. If Einstein couldn't help them get to America, she begged him to send them some money to tide them over the present emergency. She finished by telling him that she had not eaten a midday meal for one and a half years. Her next letter was not until 1944 and seems to have been precipitated by the death of a mutual friend, 'Max W'. She writes:

> Dear Albert, who knows what kind of spectre drives me to write to you. I have written several times over the years and also sent you a photograph at your request. However, there was never an answer from you. Whether the war was to blame, or you, or me, I don't know, maybe even those around you. Now I want to wait and see whether this letter reaches you.

She concluded by hoping that Einstein continued in good health and that this terrible war would soon be over. There are no further letters extant. As mentioned previously, Marie never made it to America. She spent her final years, still loyal to Einstein's memory, in a sanatorium for the mentally unstable in Meiringen near Bern, where she died in September 1957.

Einstein's summer vacations away from steamy Princeton continued during the war years, although the delicate state of Margot's health meant that a mountain climate, rather than Long Island, was preferred. Beginning in the summer of 1940, the Einstein household decamped to Saranac Lake, in the Adirondack range in the north of New York State. Its altitude of about 1500 feet ensured escape from the enervating heat and humidity of the east coast. They stayed in one of the six houses in the Knollwood Club complex, which had been founded in 1900 as a refuge for well-heeled New York Jewish families to escape from the growing anti-Semitism in some holiday resorts. In additional to these substantial houses, there were communal buildings; in particular, the boathouse housed Einstein's *Tinev*, which he had asked his friend David Rothman to transport from Peconic Bay to Knollwood. It arrived safely, allowing Einstein to continue to indulge his passion for sailing, in which he was often accompanied by his sister Maja. Rothman continued to send him a pair of sandals every summer, even though Einstein remonstrated that the previous pair were still perfectly serviceable. He also told Rothman how much he missed Peconic and their music-making, but the requirements of his womenfolk had to take priority.

The Einstein family stayed at Knollwood during every summer until 1946, except that of 1943. In that year, Einstein's new consultancy with the Navy in Stephen Brunauer's department described above was sufficiently pressing to induce him to stay in Princeton over the summer months. In accepting the consultancy in May, 1943, he had asked whether it would interfere with his usefulness if he spent the summer at Lake Saranac. The density of communications between them that summer, invitations to meetings as well as Brunauer giving Einstein access to library facilities of the National Defense Committee in Washington, all point to Einstein's realising that a 1943 trip to Saranac was not an option. Nevertheless, he resisted all Brunauer's blandishments to travel to Washington, citing his health and the fear of being molested by what he called snobbish people. In conjunction with his friend John von Neumann, who was also about to become immersed in the Manhattan project from which Einstein remained excluded, Einstein suggested mechanisms to improve the destructive efficiency of torpedoes. These are detailed in several letters to Brunauer.

The usual guests visited Saranac during the Einsteins' vacations. The Bucky family were regulars. Other guests included Einstein's old ally in his fights over the Hebrew University, Abraham Yahuda. He and his wife Ethel visited both in the summer of 1940, shortly after he had arrived in the US, and in August 1941. During his visit in 1940, he and Einstein discussed the theological writings of Sir Isaac Newton, many of whose original manuscripts Yahuda had acquired in London in the previous decade. Einstein was fascinated by Newton's drafting and redrafting of these thoughts on the meaning of biblical passages, which he considered gave an insight into the development of his theories of physics. In the following years, Einstein was able to assist Yahuda in finding various unpaid academic positions in the US and often wrote covering letters supporting the publication of Yahuda's views in prestigious newspapers. On some matters, such as the position of Amin al-Husseini, the Grand Mufti of Jerusalem, who made mischief for the Allies during the war by promulgating Nazi propaganda, Yahuda was uniquely well informed.

A photograph taken at Knollwood in August 1941 shows a smiling woman in a polka dot dress standing next to Einstein and the Yahudas. This is Margarita Konenkova, whose husband Sergei had, in 1935, sculpted a bust of Einstein. Margarita, through the sculpture connection, became friendly with Margot Einstein. Sometime later she and Einstein began a passionate relationship, which deepened during the war years. On Margarita's side, there was a reason for this. She was highly likely to have been a Russian spy who had been tasked with getting as close to the physicists involved in the atomic bomb work as possible. There is no evidence that she extracted any useful information from Einstein, who in any case knew no details about the Manhattan Project. She did however arrange for him to meet a Russian intelligence agent attached to the New York Consulate. That there was great affection at least on Einstein's side is reflected in the letters that he wrote to her after she and her husband returned to the Soviet Union after the war. A typical sign off was 'Warm greetings and kisses from your A.E.' None of her letters to Einstein have survived.

Hans Albert and his family were also regular guests at Knollwood, often staying for many weeks. Hans Albert enjoyed sailing with his father, Frieda swam; Evelyn, whom Hans Albert and Frieda had adopted after losing two sons in quick succession in 1941, was just over one year old during this visit. She captivated her adopted grandfather with her antics in the water. Einstein's grandson Bernhard recalled staying in Knollwood:

> I visited my grandfather for two successive summers in the early 1940s in Saranac Lake, New York, at a resort called Knolwood (sic). It was owned by six wealthy New York families. To me it was a paradise with many recreational facilities. At Knolwood my grandfather and I frequently went sailing together. He usually said very little during sailing. But on one afternoon when there was practically no wind, he became talkative. For the three hours on the lake, he talked continuously about soap bubbles and their physics and mathematics. I did not understand much of what he said; nevertheless, I was very fascinated and the three hours went by very quickly. Except for more sailing, Grandfather would continue his work routine as if he had not left Princeton.
>
> During one of my stays at Knolwood around 1941, Wolfgang Pauli came to visit. Pauli was a distinguished physics professor from ETH in Zurich, Switzerland. Grandfather invited Pauli to Princeton University as a visiting professor. After breakfast my grandfather and Pauli would discuss for a long time. In the afternoon, my great aunt Maja and Pauli would play chess on the veranda for hours while they would devour large amounts of candy. Today this memory always makes me smile. Perhaps Pauli's spin theory of atomic particles was conceived during this visit in the middle of candy intoxication.

In fact, Pauli spent several vacations in the Lake Saranac area, beginning with this one in 1941, his first full year at Princeton. He had not been specifically invited to Princeton by Einstein, as Bernhard intimates, but rather had contacted Frank Aydelotte in order to extricate himself from a dangerous position in Switzerland. Once France had fallen in June 1940, Switzerland was completely surrounded by the Axis powers and there were widespread fears that it might be invaded by the Germans.[5] Born an Austrian, after the Anschluss with Germany, Pauli's passport had to be converted to a German one. Despite characterising himself as in German law being three-quarters Jewish, he at least escaped having the dreaded 'J' designation on his passport that would have identified him as a Jew. He urgently tried to become naturalised in Switzerland. Although he satisfied the twelve-year Swiss residence requirement and enjoyed strong

[5] Indeed, already in May 1940, Mileva described the panic in Zurich, which was filled with military personnel and anti-aircraft guns; tens of thousands of the inhabitants had fled the city. In November, she pleaded with Einstein, now he was an American citizen, to arrange for Tete and her to be brought to America. Einstein, knowing this would be very difficult to arrange since US immigration laws forbade immigration from the mentally ill, had written to Tete and Mileva from Saranac in the summer of 1941 to reassure them that he thought it unlikely that the Germans would invade. Perhaps to distract Tete from such worries, he wrote to him at length about literature, for example, praising Nietzsche for his style but deprecating his content. By 1944, Einstein could write to them with relief that Switzerland was no longer surrounded by the fascists. This time his literary recommendations were Tolstoy's *War and Peace* and Aeschylus's *Prometheus Unbound.*

support from the ETH Zurich authorities, the Swiss police considered him 'insufficiently assimilated' in Switzerland and denied his application. Their characterisation reeks of thinly veiled anti-Semitism. Denied Swiss citizenship, Pauli found a refuge from possible German invasion in the IAS in Princeton. Originally given leave of absence from ETH for one semester, winter 1940–1941, he was granted an extension until the end of 1941. Once the US entered the war in December 1941, Pauli was unable to return to Zurich. His German passport, once expired, could not be renewed in the US. He could not leave the US in any case as he was in principle an enemy alien. As his absence from his teaching duties at ETH lengthened, attempts by the ETH authorities to have his employment terminated intensified, leading to increasingly acerbic correspondence and threats by Pauli of legal action. Having become a US citizen in January, 1946, Pauli eventually returned to Switzerland three months later, his path no doubt smoothed by the award of the Nobel Prize in November 1945. He eventually became a Swiss citizen in 1949.

Initially, Pauli worked with Einstein's assistant, Valentin Bargmann, with whom he had previously collaborated in Zurich. Pauli and Einstein naturally interacted frequently, particularly since, as the war progressed, more and more of the faculty vanished to carry out war work at various distant locations, in particular Los Alamos. Pauli and his wife Franca were frequent visitors to Mercer Street, where she befriended Maja, with whom she read Dante, and Margot. In fact, Einstein and Pauli saw more of each other socially than scientifically, as their research interests rarely overlapped. Pauli's scornful remarks on Einstein's previous attempts at unified theories have been noted in previous chapters. However, they did share an interest in general relativity and a collaboration was catalysed by Einstein's renewed enthusiasm for unification via extra dimensions as pioneered by Kaluza and Klein.

At the end of 1942, Einstein and Pauli attempted to show that, in the absence of matter and energy, Einstein's field equations led uniquely to flat, Euclidean space-time. That this was one solution was obvious but to show that it was the only possible one was much more difficult. In their paper, which was submitted in January 1943, they adopt a curiously modern viewpoint. By considering Kaluza's five-dimensional theory, they noted that in such a theory, fields corresponding to non-singular solutions were not point-like but linearly extended in a four-dimensional space. In some sense, Einstein and Pauli were using ideas of modern string theory twenty years before it began to be systematically constructed. In another similarly modern technique, they choose to solve their problem in n dimensions rather than four or the five of Kaluza's theory because the solution is simpler. They concluded that the Einstein equations in the absence of mass-energy do not have a time-independent solution for any massive charged particle. This, although interesting, is not quite the proof that they were aiming at, which was eventually accomplished three years later by André Lichnerowicz after several exchanges of letters with Pauli.

The philosopher Bertrand Russell arrived in Princeton briefly towards the end of 1943 after a complicated itinerary around other American seats of learning. From two of these, City College of New York and the Barnes Foundation, he was dismissed, in 1940 and 1942, respectively. The former decided that his views on sexual morals as expressed in his book *Marriage and Morals* made him unfit to teach the young. Given Einstein's views on this matter, it is unsurprising that he took Russell's side. Einstein joined in a chorus of protest from academics at Russell's treatment, publishing a statement which trumpeted that great spirits were always opposed by mediocrities. The mediocrities won. Russell spent a few months in Princeton, during which time he attended a meeting at Einstein's house which also included some of the physics faculty. They argued points related to physics and philosophy. Only rarely did politics intrude, but Russell recalled suggesting that the victors would be generous to a defeated Germany. Einstein was outraged at the suggestion.

As it became clear that the Allies would soon win the war, the question of post-war arrangements became of increasing importance. Einstein naturally was very concerned about what would become of the Jews and their putative homeland in Palestine. On 15 February 1944, Professor Philip Hitti testified to the House of Representatives Foreign Affairs Committee on a resolution for the US to aid the establishment of a Jewish 'commonwealth' in Palestine. He presented a damning indictment of such a move, which he characterised as the initiative of political Zionists from New York. Einstein and his friend and neighbour Erich von Kahler, a well-known author and essayist, attempted to counter Hitti in an article published by the *Princeton Herald* on 14 April. The discussion of the historical arguments about the priority of the various settlers in Palestine was of mind-numbing obscurity. They also countered Hitti's contention that the superior development of the land so obvious in the Jewish areas of Palestine was solely the result of the influx of capital from rich Jews in the US. The argument went back and forth in the pages of the *Princeton Herald*. Einstein and Kahler noted that Jews were powerless to prevent the annihilation of millions of their people or to stop doors to safe refuges from the Nazis being closed. This attests to the fact that the existence of the Holocaust was by then known in the US. A condensed version of Einstein and Kahler's arguments was published later that year in a pamphlet entitled *A Test Case for Humanity* by the Jewish Agency for Palestine in Jerusalem.

Few Jewish families with members in Europe escaped the Second World War without tragedy. Einstein's was no exception. His cousin Robert Einstein, son of Jakob, the business partner of Einstein's father Hermann, had trained as an electrical engineer. He lived in Rome where he prospered, met his wife and subsequently moved to Florence. They had two daughters and adopted two nieces. In 1943, the Germans occupied the villa, forcing the Einsteins to move into the nearby farm buildings. As the Allies advanced from the south, Robert went into hiding on the advice of local partisans. His wife, imagining that as a Gentile she and the family would be in no danger, remained behind. She was wrong. On 3 August 1944, retreating German soldiers arrived at the Einstein villa at Troghi, about 16 km east of Florence, looking for Robert. In his absence they turned on his family, first interrogating and then shooting his wife and daughters before setting the house alight. The two nieces survived. Robert, who was hiding in nearby woods, saw the smoke and ran back to discover the bodies of his family. He made strenuous attempts to trace the perpetrators of the atrocity. Major Wexler of the Inspectorate-General of the US Fifth Army wrote to Einstein on 17 September at Robert's request, informing him of the murders, which he was investigating. Robert himself wrote to Einstein on 27 November, asking for his help in identifying and punishing the murderers. They were never found; despair eventually took its toll and Robert committed suicide on 13 July 1945. In his final letter, he requested that his body be buried next to his family.

At the end of 1943, another of Einstein's old friends had arrived in Princeton after fleeing from the Nazis. Niels Bohr had had a much closer shave than Wolfgang Pauli. On 29 September 1943, Bohr and his family had been spirited away by the Danish resistance after crawling on their hands and knees across a beach to a fishing boat, which transferred them to a trawler. On 5 October Bohr was flown to the UK. After being briefed on the UK activities towards the atomic bomb in the Tube Alloys project, Bohr and his son Aage took ship for the US from Glasgow, arriving in New York on 6 December. They travelled to Washington DC, where Bohr met Lord Halifax, the British Ambassador. On 22 December, Bohr spent the day with Einstein and Pauli in Princeton. Apparently he found Einstein in the usual crowded teatime gathering of physicists. When Einstein spotted Bohr he called out how pleased he was to see that he had come to sort out the mess the Americans were making of the uranium business, thus demonstrating that the government were right to be worried about Einstein as a security risk. After Christmas, Bohr and his son left for Los Alamos in New Mexico, where the headquarters of the Manhattan project had been located.

Having experienced the magnitude of the Manhattan Project, Bohr returned to Europe with the conviction that its success would completely change the political situation worldwide. He believed that stability required sharing the secrets of the bomb with the other great wartime ally, the Soviet Union. After his return to London on 14 April 1944, he met with Sir John Anderson, the minister tasked by Churchill with responsibility for the British atomic bomb efforts. Bohr conveyed a message from the President via Roosevelt's adviser, Supreme Court Justice Felix Frankfurter. This message asked that Bohr's views be communicated to Churchill. Despite Frederick Lindemann (now Lord Cherwell) warning him that Churchill was too occupied with current war matters to focus on post-war issues, Bohr insisted on meeting the Prime Minister.

Bohr accompanied Cherwell to Downing Street to see Churchill on 16 May. It did not take long for Churchill, preoccupied with the impending Normandy landings, to become exasperated by Bohr's slow and elliptical manner. Most of the meeting was apparently taken up by Churchill arguing with Cherwell. Bohr told R.V. Jones, a physicist deeply involved in intelligence and weapons matters during the war, that the Prime Minister had scolded him and Cherwell like schoolboys and said he couldn't see what was worrying them; the new bomb was just bigger than the old and would make no difference to the principles of war. Such a frightening misconception was largely shared by Roosevelt; neither leader had much if any appreciation for technical or scientific matters.

Although Bohr's meeting with Roosevelt on 26 August 1944 was much less fractious, the end result was equally nugatory. Roosevelt seemed to understand and sympathise with Bohr's point of view. However, at the Quebec conference two weeks later and particularly afterwards in the president's upstate New York family mansion, Hyde Park, Churchill convinced Roosevelt that the Manhattan Project should remain top secret. Indeed they discussed the desirability of ensuring that Bohr did not leak anything to his old Russian friend Peter Kapitza. Churchill's memo to Cherwell after leaving Hyde Park shows the extent of Bohr's failure:

> It seems to me that Bohr ought to be confined or at any rate be made to see that he is very near the edge of mortal crimes. I had not visualised any of this before, though I did not like the man when you showed him to me, with his hair all over his head, at Downing Street.

At least Bohr escaped confinement in the Tower of London. A few months later, Einstein took up the cudgel. Possibly news in November 1944 of the death of Sir Arthur Eddington, Quaker and convinced pacifist, reminded Einstein of his own abhorrence of weapons and war. Otto Stern, winner of the previous year's Nobel Prize in Physics, whom Einstein had known since his Prague days, had been a regular visitor that year. He was a member of the Carnegie Institute in Pittsburgh. While not in the Manhattan Project team, he was an advisor to one of its main centres, the 'Metallurgical Laboratory' in Chicago, then led by Szilard. As such, he moved in circles close enough to be well informed about its progress. Like Bohr, Stern became deeply concerned about the situation with regard to the new weapons that would develop after the war. He communicated his worries to Einstein, who, the following day, wrote to Bohr, echoing Bohr's own concerns as expressed to Churchill and Roosevelt.

Knowing that Einstein's least word would be amplified enormously by the press, Bohr was petrified that Einstein would express his concerns publicly and that Bohr would be blamed for recruiting him to his crusade and supplying him with information. He rushed down to Princeton and spent the day with Einstein. During this conversation, of which there is only a very brief summary by Bohr, it seems that Bohr told Einstein what he knew about the Manhattan Project, at least in general terms. Einstein characterised their conversation as having resulted in a cloud of heavy secrets descending upon him. Bohr prevailed upon Einstein not only to keep quiet himself but also to convince Stern to do likewise. Einstein wrote to Stern after Christmas, saying as little as possible about his information from Bohr while telling him that any public statement at that juncture would be counter-productive.

The year 1945 was a momentous one not only for the world at large but also in Einstein's personal life. In April, he formally retired from the IAS, although this resulted in no noticeable change in his pattern of work.

Meanwhile, many of Einstein's colleagues were working day and night in New Mexico on the Manhattan Project. The urgency to beat the Germans to a possible bomb had receded as the prospect of their success in the shattered economy of the Third Reich diminished. Neither was there a Japanese threat: although Japanese physicists were also well aware of the possibilities of producing a bomb from U^{235}, they had scarcely any supply of uranium ore and did not even get as far as developing a successful method to extract the U^{235} from it. In fact it is now clear, although it certainly was not at the time, that there had never been any real possibility of a German or Japanese bomb. The British, much in advance of the Americans in the early stages of the project, had rapidly realised that they did not have the enormous industrial capacity necessary to isolate enough U^{235} or plutonium to manufacture a device in a timely manner. It was even less likely that blockaded Germany, short of all sorts of raw material, could do so. Nevertheless, there was a substantial German research effort to this end, led by Werner Heisenberg and others, which made considerable progress. However, at a conference on nuclear matters in June 1942, Heisenberg was questioned by Albert Speer, who as armaments minister was in a position to divert resources towards nuclear projects. Heisenberg admitted that it would take at least three or four years before any bomb could be manufactured. Speer was convinced that the war would be decided long before then, so in the autumn of 1942 he ordered the direction of this research to be moved towards producing a reactor that could ultimately be used for propulsion. Even this never actually achieved criticality before the US soldiers arrived at the site at Haigerloch near Hechingen, ironically enough the birthplace of Elsa Einstein. The German stocks of uranium ore, in principle enough to produce many functioning U^{235} warheads had the capacity to extract that isotope existed, were released by Speer in the summer of 1943 to be used to replace tungsten in solid-core ammunition.

Szilard was the next person to take fright at the implications of a successful atomic bomb. Having done more than anyone to set the atomic project in motion, he worried that it was rapidly running out of control. He discussed the matter with Einstein in March 1945. This must have been an interesting meeting since Szilard was very limited in what he could reveal, whereas Einstein had by now had similarly oblique discussions with enough colleagues, in particular Bohr and Stern, that he would have been aware of the general situation. Einstein was sufficiently disturbed by Szilard to draft another letter to President Roosevelt. Dated 25 March 1945, its purpose was to ask Roosevelt to read Szilard's memorandum, which of course Einstein was not officially permitted to read himself! His letter reflects this complexity by admitting that he didn't know the substance of Szilard's concerns. He had to imply what they were about by spending the bulk of this letter referring to his previous 1939 letter to Roosevelt about uranium. Einstein described Szilard's worries as relating to insufficient contact between the scientific experts and the cabinet-level policy makers. With that he signed off by commending Szilard's memorandum, which he enclosed. Quite what the President would have made of it will never be known; he died suddenly in Warm Springs, Georgia on 12 April.

The new president, Harry S. Truman, knew less about the Manhattan Project when he took the oath of office than even Einstein did; only two weeks later was he fully briefed by Secretary of War Stimson. Einstein's letter and Szilard's memorandum were shown to Truman, who asked his old friend from their Senate days, James Byrnes, to deal with it. Szilard, who had arrived at The White House to see Truman, got no further than the Presidential Appointments Secretary.

Byrnes, who had been favourite to be nominated as Vice President by Roosevelt until Truman was selected, was about to be appointed Secretary of State. He was singularly unqualified to deal with atomic issues, despite, or perhaps because of, eventually and uniquely, serving in all branches of the US government: as Representative, Senator, Justice of the Supreme Court, Secretary of State, and finally as Governor of his home state, South Carolina. Three weeks after the unconditional surrender of Germany and the end of the war in Europe, on 28 May, Szilard and his colleague Urey met Byrnes at his home in Spartanburg, South Carolina. They handed over the memorandum, which Byrnes read.

Szilard's memorandum began by explaining that the work on uranium had now reached fruition, that an atomic bomb could be detonated imminently, and that the War Department was certainly considering the use of the bomb against Japan. It went on to advise against the precipitate use of the bomb since any major country, such as Russia, would be able to make its own bombs from natural uranium with an expenditure of $500 million, much less than the US had so far invested, so that the US would lose its current advantage. This would precipitate an arms race that would be uniquely dangerous for the US population, concentrated as it was in big cities vulnerable to the new weapon. Szilard spelled out the damage the bomb would do in graphic terms. However, he diluted his memorandum's impact by going on to list a complicated series of rather ill-defined questions relating to cooperation with Russia and setting in place a system of controls for uranium issues, followed by recommendations for the composition of a cabinet committee to liaise with the scientists.

Szilard's encounter with Byrnes was not a meeting of minds. When Szilard expressed his concern that if the bomb were used, Russia would soon obtain it, Byrnes replied that General Groves, the director of the Manhattan Project, had told him that there was no uranium in Russia. Szilard pointed out, as had Einstein's 1939 letter to Roosevelt, that Czechoslovakia had rich uranium ore deposits to which Russia would have access. Szilard rapidly became convinced that Byrnes was only interested in the bomb as a diplomatic weapon with which to pressure the USSR. On leaving Spartanburg, he noted that he had rarely been as depressed. Byrnes was just as caustic, rejecting the idea of scientists advising the Cabinet and complaining that Szilard's demeanour and wish to participate in policy-making had made a bad impression.

Szilard, as dogged as ever, did not give up. In July, he organised a petition against the use of the bomb on Japan that was signed by all the leading physicists (but none of the chemists) at the Metallurgical Laboratory. Against his better judgement, he sent it to the president through the 'usual channels'. Groves sat on it until it was too late; it never reached Truman, although it certainly would have had no effect, as the decision to recommend use of the bombs had already been taken by the Interim Committee organised by Secretary of War Stimson to advise him on such matters. At dawn on 16 July at the Alamogordo testing site in the New Mexican desert, a plutonium bomb was exploded.[6] The success of the test was reported to the President in positively lyrical terms. Oppenheimer's quotation from the *Bhagavad Gita* is too familiar to require repetition. Another witness, General Thomas Farrell, was equally apocalyptic: '[It] warned of Doomsday and made us feel that we puny things were blasphemous to dare tamper with the forces heretofore reserved to the Almighty.'

Before dawn on 6 August, the *Enola Gay* bomber dropped the U^{235}-bomb, 'Little Boy,' on Hiroshima. Three days later, the plutonium-based 'Fat Man' was dropped on Nagasaki. It would

[6] Groves hesitated to authorise the test as he was concerned about how to explain to Congress the loss of what he estimated to be $1 billion worth of plutonium if it failed.

be no exaggeration to say that the rest of Einstein's life outside physics was dominated by the aftermath of Hiroshima and Nagasaki. As time went on, not only did he revert to his pre-Nazi pacifism but he began to minimise in his own mind his involvement in any war work, in particular the development of the atomic bomb. The question of the importance of his involvement is now clearer than during his lifetime, when, for example, *Time* published its famous front cover depicting Einstein, the mushroom cloud and $E = mc^2$. The thoughts of Oppenheimer, the centrality of whose contribution to the project was never in doubt, are illuminating:

> Einstein is often blamed or praised or credited with these miserable bombs. It is not in my opinion true. The special theory of relativity might not have been beautiful without Einstein; but it would have been a tool for physicists, and by 1932 the experimental evidence for the inter-convertibility of matter and energy which he had predicted was overwhelming. The feasibility of doing anything with this in such a massive way was not clear until seven years later, and then almost by accident. This was not what Einstein really was after. His part was that of creating an intellectual revolution, and discovering more than any scientist of our time how profound were the errors made by men before then. He did write a letter to Roosevelt about atomic energy. I think this was in part his agony at the evil of the Nazis, in part not wanting to harm any one in any way; but I ought to report that that letter had very little effect, and that Einstein himself is really not answerable for all that came later. I believe he so understood it himself.

Bush concurred with Oppenheimer's assessment:

> An advisory committee was set up to coordinate efforts and exchange information, under the chairmanship of Lyman Briggs, then Director of the National Bureau of Standards. It included prominent physicists and representatives of the Army, Navy and Bureau of Standards. It resulted from numerous conferences of those who were then thinking of possible military uses. Also there was a letter written to President Roosevelt and signed by Einstein urging support. How much this had to do with the formation of the committee is not important; the committee was formed and did good work. The letter may have stirred the President's interest; I just do not know. He never mentioned it to me, and I feel he did not really grasp what might be involved until much later. When he did grasp it, he certainly supported the effort vigorously.

Similar remarks were made by General Groves, Isidor Rabi, and Eugene Wigner.

Einstein, as has been documented above and despite his later denials, was deeply involved in both the initiation of US work towards the bomb and attempting to influence its direction. However, his influence was ultimately peripheral. The real turning point in the US came when Oliphant brought the news of the estimate of Frisch and Peierls of the mass required for criticality and the Maud report to the attention of the US authorities, in particular Bush. Einstein's letter to Roosevelt in 1939 began the process but, had it never been written, the effect on the development on the bomb would have been marginal.

Einstein was at Knollwood on Lake Saranac when he was told the news of Hiroshima. His dismayed comment was 'Oy vey', which requires no translation. It sets the tone for his endeavours for peace during the rest of his life.

Chapter 30

Warrior for peace; The Brandeis University affair; Death of Mileva Einstein (1945–1949)

Directly after the dropping of the bomb on Nagasaki, the US government published an official account of the development of these weapons. *Atomic Energy for Military Purposes* became known as the Smyth Report after its author, Professor Henry Smyth, chair of the Department of Physics at Princeton University. This made the first public reference to Einstein's letter to President Roosevelt in 1939. Although this reference is fleeting, its implication that Einstein had initiated the atomic bomb programme was immediately seized upon. Within a week of the publication, *The New York Times* published a summary of the report that mentioned the Einstein letter as catalysing the first government committee.[1] An editorial published a few days later recognised the world-changing significance of the dawn of the atomic age, urging similar organisational efforts as the Manhattan Project to solve many other pressing problems. It also recognised the likelihood that the bomb's 'secrets' would soon become known in other countries. Einstein agreed with this analysis and urged the US government to pre-empt this by sharing knowledge with a world government to be given jurisdiction over military matters.

The thesis that there was no nuclear 'secret' was widely discussed. The view that the research on the bomb, although not the actual engineering techniques, should be shared with, in particular, the Soviet Union, was also presented to the US government. However, neither resonated. This was despite advocacy for sharing information from none other than retiring Secretary for War Stimson. General Groves, at that point in effective charge of what passed for US nuclear policy, was convinced that it would take the Soviet Union decades to develop its own bomb. Congress moved swiftly to regulate; the May–Johnson Bill would have given the military full control of nuclear matters and imposed a stifling level of secrecy. There was a strong reaction to this by scientists associated with the Manhattan Project, which led to the foundation of the Federation of Atomic Scientists. Testimony by Leo Szilard played a decisive role in defeating the bill. In August, 1946, Congress passed the McMahon Act which, as well as establishing the much more civilian-oriented Atomic Energy Authority (also known as the Atomic Energy Commission (AEC)), ended collaboration with America's partner in producing atomic weapons – the UK.

It may be that the Hiroshima and Nagasaki bombs had stunned Einstein somewhat, as he eschewed all comment until more than a month afterwards, when he made the remark referred to above. Being isolated at Saranac Lake at least gave him opportunity for contemplation and some

[1] It is interesting that this article begins with a number of errors, in particular that only plutonium was used in the two bombs. This is clearly contradicted in the Smyth Report, which describes the development of both a uranium and a plutonium bomb. The Smyth Report itself, for obvious security reasons, was extremely oblique in many of its references, nor was it without errors, for example, it stated that President Truman was aware of the Manhattan Project when a US Senator, whereas in fact he was not told even as Vice President; it also drastically downplayed the prevalence of radioactive fallout.

Einstein. Brian Foster, Oxford University Press. © Brian Foster (2026). DOI: 10.1093/oso/9780198794875.003.0030

relief from the press. He listened to the radio, hearing the broadcast of Raymond Swing, whose views on the establishment of a world government in the atomic age coincided with his own. Einstein's letter to him at the end of August resulted in an interview which appeared in *The Atlantic Monthly* in November. In it, Einstein went out of his way to minimise his own part in the development of the bomb. He counselled against entrusting control over atomic weapons to the newly establish United Nations. Its constitution, with a veto for each of the 'Great Powers', could not effectively deliver such a role. A world government was the only solution in Einstein's viewpoint. While he was logically correct, there was never the slightest chance of such a thing happening. A poll in August 1945 showed that only 15% of the US population supported giving the control of atomic weapons to an international organisation.

While Einstein was proposing a world government, Oppenheimer was working assiduously to influence the US government towards adopting measures that would avoid the arms race that he foresaw. At the end of 1945, the US and Soviet governments had agreed to the creation of a United Nations Atomic Energy Commission (UNAEC), which was by unspecified means to take control of the situation. To agree on a policy to implement this, Secretary of State James Byrnes appointed an Atomic Energy Safeguards Committee, chaired by his Under-Secretary of State Dean Acheson, in January 1946. Other members of the committee included Vannevar Bush and James Conant, General Groves, and Stimson's Assistant Secretary of War John McCloy. Scientific advice was provided by a board of consultants, chaired by David Lilienthal but dominated by Oppenheimer. The latter wrote most of what became known as the Acheson–Lilienthal Report; it was Oppenheimer's suggestion to control the raw material – uranium and thorium – at source, rather than try to police activity inside states such as the Soviet Union, which he considered both unfeasible and unacceptable to the Soviets. Licences would be issued to countries wishing to develop nuclear power for peaceful purposes. Another recommendation was the destruction of the US's atomic arsenal and the transference of the knowledge of how to build the bomb to the Soviet Union.

President Truman was certainly unconvinced by these recommendations. In order to defend American interests, he appointed someone he considered a hard-nosed businessman, Bernhard Baruch, to negotiate within the UNAEC to ensure that the US would only give up its bombs if it was clear that the Soviets would be unable to produce their own. Baruch modified the Acheson–Lilienthal recommendations in a crucial way, not only insisting on a comprehensive programme of inspections of nuclear facilities but also stripping away the Security Council veto on punishment for violation of the agreement. Both of these changes were unacceptable to the Soviet Union, which abstained when the matter was put to a vote, killing the plan on 30 December 1946.

Einstein despaired as 1946 saw the inauguration of Truman's 'get tough' policy with the Soviet Union, marked by milestones such as Churchill's famous 'Iron Curtain' speech in Fulton, Missouri in March. The administration became convinced of a Soviet masterplan for world domination, catalysed further by the revelation that it had obtained atomic secrets from spies in Los Alamos, and the despatch of Soviet troops into Manchuria and Iran. Einstein's concern was such that he gave an interview to Michael Amrine of *The New York Times*, resulting in an article published on 23 June under Einstein's name entitled *Only Then Shall We Find Courage*. It broadly advocated sharing knowledge of the atomic bomb and a world government. On 1 July, the US conducted an atomic-bomb test on Bikini Atoll, while the process of rearming western Germany, as a buffer against the Soviet Union, also began to gather speed, much to Einstein's dismay. This was the background to his decision, against his lifelong aversion to committee work, to agree to become chairman of the Emergency Committee of the Atomic Scientists.

The Emergency Committee, which was mostly inspired by that tireless campaigner for arms control, Leo Szilard, had Harold Urey as Einstein's deputy and included Robert Bacher, Hans Bethe,

Thorfin Hogness, and Victor Weisskopf among the other trustees.[2] It first began to gather funds, issuing an appeal for $200,000 under Einstein's signature in May. This received wide coverage in the press and later in the newsreels after Einstein agreed to be filmed. In order to facilitate Einstein's attendance at trustee meetings, which he attended surprisingly assiduously, an office was established in Princeton; several meetings took place in Einstein's home. The seriousness with which he took the situation can be gauged by the fact that he also attended a meeting of trustees, including dinner, in New York. In August, the Emergency Committee gained a legal basis by incorporation in the state of New Jersey. This was a springboard to a further campaign in the autumn, culminating in a conference in Princeton from 17 November that included a fund-raising lunch at the Institute of Advanced Study (IAS). Einstein's remarks at the subsequent symposium emphasised the urgency of the work they were doing. It may be that it was around this time that a man stopped Einstein in the street to ask whether the world would be totally destroyed in the next war by nuclear bombs? Einstein reply was that that would be too bad as then we could no longer listen to Mozart. A similar encounter resulted in a question as to what weapons would be used in a third world war. Einstein's famously replied that he didn't know, but he knew what they would use in the fourth – stones.

In January 1945, Einstein had nominated Wolfgang Pauli for the Nobel Prize, which he had duly won. Indeed, Einstein was the prime mover of the offer in June to Pauli of a permanent position in the IAS, although Pauli eventually returned to Zurich, as noted in the previous chapter. On 19 December 1945, a dinner was held in Princeton to honour Pauli. Einstein's speech at the celebratory dinner made a deep impression on Pauli, who, on Einstein's death, recalled it in a letter to Max Born:

> Dear Born, Einstein's death has also hit me personally. Someone who was so friendly, so fatherly, is no more. I will never forget the talk that he gave about, and for, me at Princeton in 1945, after I had won the Nobel Prize. It was as if he were a king who was abdicating and was installing an 'adopted son' as his successor.

By the end of 1945, the ghost town that Princeton had become had begun to fill up again as colleagues and friends returned home. One such was John von Neumann, who had been a major figure in the Manhattan Project and one of the few physicists to be trusted by its leader, General Leslie Groves. Von Neumann began to use this wartime experience to set up an electronic stored-program computer at the IAS.

New arrivals were Dorothy and Saxe Commins, who spent summers in Princeton from 1946 to 1953, when they moved there permanently. Dorothy was a concert pianist; her husband was editor-in-chief of the Random House publishing company. Just after they had first arrived in Princeton, Dorothy met Einstein in the street. Continuing the habits of a lifetime, having discovered her identity, he immediately proposed that they play music together that evening. Saxe went to Mercer Street to pick Einstein up:

> ... he was standing at his door holding his violin. He handed Saxe some pages of music – Bach and Corelli and a small collection of Dutch folk melodies arranged for violin and piano by Roentgen, the brother (sic) of the discoverer of X-rays. We played the Bach Concerto in D minor,[3] then turned to

---

[2] Bacher and Hogness had played leading roles in the Manhattan Project.

[3] There is no concerto for violin and orchestra by Bach in D minor. Either the A minor concerto is meant, or more probably the Double Concerto in D minor, one of Einstein's favourite pieces, with Dorothy playing both the continuo and one of the two violin parts.

> Corelli. Once he stopped and said in German, 'Could we play that section again, I've made so many mistakes?' In truth, he loved the piece and wanted to play it over and over.

The above encounter was the first of many. Einstein enjoyed the company of both Commins and often asked Saxe to help him compose a communication in English. On one occasion, Dorothy went for tea to Mercer Street:

> Professor Einstein talked to me about music. I said, 'The scene outside at this almost twilight hour seems to evoke the same dream quality I feel in the music of Debussy's *Pelleas and Melisande*.' 'That is interesting,' said Einstein. 'But alas! Debussy is an enigma to me. You know my great love is Mozart.'

Another musical friend, Paul Robeson, had met Einstein occasionally during the war. Afterwards, they were thrown together by the new impetus to the struggle against racism given by the return of the many African-American servicemen. Having fought against the racism inherent in Nazi ideology, they were not about to submit to it when they returned to their homes. This attempt to reassert their rights as US citizens produced an upsurge in violence and lynching. A particularly egregious example was the virtual destruction of an African-American community by five hundred Tennessee state troopers in February 1946. Einstein joined the National Committee for Justice in Columbia, Tennessee, set up by Eleanor Roosevelt. A few weeks later, Einstein made a conspicuous exception to his rule of declining all honorary degrees by accepting one from Lincoln University. Founded in 1854, Lincoln was the first institution to offer higher education to African-Americans. After receiving his honorary doctorate, Einstein gave a speech to the assembled university in which he said that his visit was in a worthwhile cause and that they were not to blame for the separation of races but white people were. He added that he did not intend to be quiet about this. He was not, but reports of his words were muted or absent. In its story on the Lincoln College event, the only quote of the *New York Times* was that Einstein had said he believed there was a great future for the Negro and had asked the students to work long and hard and be patient. The remaining 80% of the article highlighted criticisms of Lincoln University's objectives and how it was run.

The end of summer 1946 saw the particularly gruesome lynching of an African-American former serviceman. During this disturbance, the counsel for the Columbia, Tennessee victims, future Justice of the Supreme Court Thurgood Marshall, only narrowly escaped lynching himself. Robeson approached Einstein to chair a new group, the American Crusade Against Lynching. Einstein agreed and had consented to attend the rally in Washington DC on 23 September. In the end, he pleaded ill health but sent a supportive message addressed to President Truman. Robeson and a delegation were admitted to the Oval Office after the rally. When Robeson pressed the president for action against lynching, he received an icy response, along the lines that this was a political matter and must wait for a propitious time. The involvement of Einstein at least ensured wider news coverage than would otherwise have been the case. Einstein was deeply engaged in the fight against racism and continued to work with Robeson over the next few years, giving speeches in which he excoriated the southern attitude to race as blatantly in violation of the founding principle of the country. He supported many civil-rights organisations, including the Civil Rights Congress, which was considered a communist front by J. Edgar Hoover's FBI, which devoted considerable resources to monitoring the Civil Rights Congress, Einstein, and even more bizarrely, Helen Dukas. The Ku Klux Klan, in contrast, was not put under FBI surveillance until 1964.

Einstein's plea of ill health for not attending Robeson's rally was only partially true. The annual vacation in Saranac Lake had been cancelled in 1946 because his sister Maja had had what was probably a minor stroke in May. In the second half of September, Einstein travelled alone to Deep

Creek Lake in western Maryland. Apparently, he had been invited there by John Steiding, a chance acquaintance. He stayed in the lakeside villa Mar-Jo Lodge, which belonged to a physician, Dr Frank Wilson. It is unclear what the connection between Einstein, Steiding, and Wilson was; it seems likely that Einstein was feeling unwell, mentioned this to Steiding, who then suggested the trip to Maryland in conjunction with an opportunity to be examined by Wilson. No doubt the promise of sailing on the lake was decisive; Einstein was spotted on several occasions sailing either with the head of the local yacht club, Frank Muma, or alone in Muma's small sailing boat. It is possible that Wilson's examination gave Einstein the first indication of the problem that was to lead to surgery in 1948 but this did not detract from his enjoyment of what he characterised as the sunny days in Wilson's fairytale castle. Despite the best efforts of Wilson and Steidling, Einstein's presence in the area soon became general knowledge, leading the rabbi of the local synagogue to invite Einstein to attend. His response was classic Einstein: 'Despite being something like a Jewish saint I have been absent from a synagogue so long that I am afraid God would not recognize me and if He would it would be worse.'

Einstein's son Hans Albert and his family had spent the summer of 1946 back in Europe, mostly in Switzerland where they had visited both his mother and his brother. Einstein supported the trip by sending his son $500. Tete was particularly taken with Evelyn, Hans Albert's adopted daughter, who was then five years old. His letter to Einstein recording this is in 1947 is deeply affecting:

> We now hear so many amazing things about you in the newspapers. I would also like to visit you sometime if it would suit you – despite someone's joke that 'I have never been to America; on the sea journey there, one is too much in the hands of God!' I have never learnt so much personal about you previously as I have from your big political activities, but I enjoy thinking of the old times when we read Schopenhauer together. I have to be in a clinic at the moment, but maybe I can come afterwards. Please don't worry about the clinic, I'm not so completely befuddled yet.
>
> Friendly greetings from
> Your Teddy,
> Zurich-Switzerland
> Currently in Burghölzi.

Just after Einstein's return from his vacation in Maryland, he was delighted to welcome his old friend Maurice Solovine to Princeton. Einstein insisted that he stay in Mercer Street, if for no other reason than it was impossible to find any other accommodation in the town after term had begun in October. They had not seen each other since the early thirties although they had corresponded, mostly about the French translation of Einstein's various publications. Solovine was one of only two of his old friends with whom he kept up a (fairly) regular correspondence, the other being Michele Besso. Einstein's fondness for Solovine is clear from an unusually personal remark in a letter of 28 March 1949, replying to Solovine's congratulations on his seventieth birthday: 'The best, which endures, are a couple of upright friends whose hearts are in the right place and who understand each other, as we do.' Their mutual acquaintance Paul Langevin died in Paris on 19 December. Solovine remarked on this sad loss in a letter after his return to France.

Einstein was hit even harder by the death of Max Planck in October 1947. Although he had little time for those of his old colleagues who had remained in Germany and compromised with the Nazis, nor for the learned academies with which he had formerly been associated, he did not hide his continuing respect and admiration for his senior colleague and mentor Planck. His tribute written for the US Academy is eloquent:

> ... even in these times of ours when political passion and brute force hang like swords over the anguished and fearful heads of men, the standard of our ideal search for truth is being held aloft undimmed.

> This ideal, a bond forever uniting scientists of all times and in all places, was embodied with rare completeness in Max Planck.
>
> Even the Greeks had already conceived the atomistic nature of matter and the concept was raised to a high degree of probability by the scientists of the nineteenth century. But it was Planck's law of radiation that yielded the first exact determination – independent of other assumptions – of the absolute magnitudes of atoms. More than that, he showed convincingly that in addition to the atomistic structure of matter, there is a kind of atomistic structure to energy, governed by the universal constant h, which was introduced by Planck.
>
> This discovery became the basis of all twentieth-century research in physics and has almost entirely conditioned its development ever since. Without this discovery it would not have been possible to establish a workable theory of molecules and atoms and the energy processes that govern their transformations. Moreover, it has shattered the whole framework of classical mechanics and electrodynamics and set science a fresh task: that of finding a new conceptual basis for all physics. Despite remarkable partial gains, the problem is still far from a satisfactory solution.
>
> In paying homage to this man the American National Academy of Sciences expresses its hope that free research, for the sake of pure knowledge, may remain unhampered and unimpaired.

In January of 1946, Ruggles Smith, the son of the founder of Middlesex University in Waltham, Massachusetts, approached Dr Isaac Goldstein, a prominent New York rabbi with a proposition. Middlesex University was a medical school, the only one in the United States that did not impose a quota on Jews. It had run into financial difficulties and Smith hoped that Goldstein, who was known to be interested in the idea of establishing a university open to all religions in the US, would be willing to form a new group of trustees to take over the institution. Goldstein lost no time in visiting Waltham and, impressed with what he saw, began recruiting prominent Jews to support the project. He turned to Einstein, visiting him twice in January and obtaining his enthusiastic endorsement. He told Goldstein that he would do anything in his power to help and that such an institution would always be near to his heart. On 5 February, Goldstein visited Einstein again and obtained his agreement to the formation of the Albert Einstein Foundation for Higher Learning, with Goldstein as president. This was to be the instrument to create the new institution. On 7 February, Goldstein was also elected president of the Board of Trustees of Middlesex University.

In the succeeding months, Goldstein energetically raised funds for the new university, arranging for potential donors to visit Einstein in Princeton in May and June. Einstein's promise to do all in his power to help the venture did not however extend to attending the first fund-raising dinner held in the Waldorf-Astoria on 20 June, although he did send a message of support. The dinner raised $350,000. By this point, tensions had already begun to arise between Einstein and Goldstein. In March, Einstein had written to Goldstein recommending Otto Nathan as an excellent person to help the project. One month later, however, Einstein wrote again to complain that he was not being kept informed of progress and that Nathan had not been able to meet Goldstein. This catalysed Goldstein to bring Nathan into his counsels, Einstein having nominated him as his voice in any business matters. Einstein's concern was to avoid at all costs a repetition of his experiences with the Hebrew University of Jerusalem. He was therefore particularly exercised about the appointment of the academic head of the putative university. The trustees agreed that any proposals should be first discussed with Einstein for his approval. As the summer progressed however, further tensions arose between Goldstein and Einstein. These were no doubt exacerbated by Einstein's old friend Rabbi Stephen Wise, who knew Goldstein well as a fellow New York rabbi and did not have a high opinion of him. In June, Wise wrote to Einstein suggesting that he should only let himself be associated with the university if Nathan were at the side of Goldstein and could give him the benefit of his own wisdom as well as Einstein's in university affairs. Wise offered to ensure this happened.

It may have been these concerns that led Einstein to refuse Goldstein's offer to name the university after him, an offer surely to be expected since the fund-raising was proceeding under Einstein's name. That his modesty was not so overwhelming as to resist any such academic memorial is shown by his subsequent willingness to allow Yeshiva University to give his name to their medical school. Goldstein may have breathed a private sigh of relief, since there was a significant body of opinion that wished to name the university after a native-born American. The name of the distinguished Supreme Court justice, the late Louis Brandeis, was finally selected.

Einstein wrote to Goldstein on 1 July, insisting that a board of distinguished scholars should be set up to advise on academic appointments and that he should be consulted through Nathan. Goldstein was quick to agree. However, the campaign director of the Albert Einstein Foundation, Boris Young, had offended Einstein by the proposal to invite Justice Jackson to speak at another fund-raising dinner in October. Einstein, who clearly had a much less ecumenical view of their operation than either Young or Goldstein, emphasised that this was a Jewish cause and Justice Felix Frankfurter would be much more appropriate. He eventually relented and wrote to invite both Justices. However, on 2 September, he wrote to Frankfurter to withdraw his invitation, telling him that he had decided to withdraw his support from the whole project, since Goldstein had operated behind his back and committed other irregularities.

Einstein's reaction was triggered by two things. Firstly, he had heard (incorrect) rumours that Abram Sachar, a relatively undistinguished historian but experienced administrator and politician, was being considered for the post of university president. Secondly, on 29 August, Young had written to tell Helen Dukas that Cardinal Spellman, Archbishop of Boston, had agreed to deliver the Invocation at the October fund-raising dinner. Young and Goldstein were delighted; Einstein was incensed. His objection was not so much religious as political. Spellman had just returned from a visit to Spain, where he had praised the Franco regime to the heights. On 2 September Einstein wrote to Goldstein highlighting these two issues, which he considered two new breaches of confidence; he withdrew his support from the project. He further remarked that he could not allow his name to be used for fund raising by any organisation in which Goldstein played an important part .

Goldstein was naturally dismayed by Einstein's letter and wrote an abject letter of apology to him, in which, in Wise's words to Nathan, he offered to immolate himself. Goldstein told Einstein of his intention to resign as president of the foundation and as chair of the board of trustees of the new university. In his letter, he gave a comprehensive refutation of Einstein's charges of going behind his back to approach Sachar as well as laying out all the ways in which he had bent over backwards to consult Einstein on all matters academic. He not unreasonably pointed out that he had not thought it necessary to consult Einstein about speakers at a dinner that Einstein had no intention of attending. Goldstein having resigned his positions, Einstein relented and reaffirmed his support to the new university. Nathan subsequently had discussions with another member of the trustees, a prominent Boston lawyer, George Alpert, whom he found to be an agreeable man. This however was not the view of Rabbi Wise, who wrote to Nathan that he would not be able to work with Alpert for long, as he would clash with Einstein. In this, Wise proved to be a better judge than Nathan.

Nathan now became a member of the board of the Albert Einstein Foundation and was deeply involved with plans for the new university. Between December 1946 and January 1947, he went on a fact-finding tour of US institutions and then went on to the UK, where he met with his friend Harold Laski, a distinguished left-wing educationalist and prominent member of the Labour Party, which was then in government. Nathan decided that Laski would make an excellent head of Brandeis University. When he returned to the US, he convinced Einstein to support this and furthermore that he should ask the trustees for the authority to make the appointment. The story now becomes murky, involving the complex academic politics that Einstein professed to hate but in which he seemed to become continually entangled. Although the board met and apparently gave Einstein the requested authority, it was later declared to have been inquorate. Einstein, thinking himself authorised to do so, wrote to Laski in April 1947, asking him whether he would accept the position. Laski quickly replied that the centre of his activities was in London and that he could not accept the offer.

Over the next few months, Alpert became increasingly estranged from Nathan, who was playing a leading role in the continuing search for a president of Brandeis and was also manoeuvring for an appointment for himself at the new institution. Alpert decided to use the Laski episode to lever both Nathan and Einstein out of Brandeis' affairs. He was successful, leaving Einstein with a rancour that continued to burn against Alpert, Sachar (who had been the cause of the initial break and who ironically was appointed[4] the first President of Brandeis), and the institution itself. Einstein indignantly refused the offer of an honorary degree and spurned all of Sachar's attempts at reconciliation. He complained bitterly when, in May 1953, he discovered that Alpert had been made chair of fundraising for the Albert Einstein College of Medicine campaign. At least he seems to have forgiven Rabbi Goldstein for his imagined transgressions when they met by chance at a reception for the Friends of Hebrew University in 1951.

In January, 1947, Erwin Schrödinger hit the headlines of *The New York Times* with a claim more usually associated with Einstein. Schrödinger was characterised as a scientist in Dublin who claimed to have achieved a unified field theory. Einstein initially refused comment but before long was pouring scorn on Schrödinger's claims. His idea was basically a variation on Einstein's work with his assistant Ernst Straus on non-symmetric metric tensors, also influenced by a paper by Arthur Eddington. Einstein had been bouncing his ideas in this area off Schrödinger since the previous year, so it is perhaps unsurprising that he was irritated by Schrödinger producing a paper that certainly owed much to Einstein's own work. In a rather terse letter, Einstein told his old friend that their two approaches were essentially identical, only differing in the addition of a cosmological element in which Einstein saw no merit. Pressed by newspapers to respond, Einstein issued a statement in which he said:

> Schrödinger's latest effort . . . can therefore be judged only on the basis of its mathematical-formal qualities, but not from the point of view of 'truth' (i.e. agreement with the facts of experiment). Even from this point of view I can see no special advantages over the theoretical possibilities known before, rather the opposite.
>
> As an incidental remark, I want to stress the following. It seems undesirable to me to present such preliminary attempts to the public in any form. It is even worse when the impression is created that one is dealing with definite discoveries concerning physical reality. Such communiques given in sensational terms give the lay public misleading ideas about the character of research. The reader gets the impression that every five minutes there is a revolution in science somewhat like the coup d'etat in some of the smaller more unstable republics. In reality one has in theoretical science a process of development to which the best brains of successive generations add by untiring labour, and so slowly lead to a deeper conception of the laws of nature. Honest reporting should do justice to this character of scientific work.

[4] The appointment was a conspicuous success. Sachar remained President for twenty years and established Brandeis as a world-class institution.

The final paragraph seems particularly bizarre from a source of precisely such behaviour in the past and who would be guilty of it again in the near future, although the fault and exaggeration were generally, and understandably, more with the reporters than the researcher. On 27 December 1949, the front page of *The New York Times* carried the headline 'New Einstein Theory Gives a Master Key to the Universe – Scientist, after Thirty Years' Work, Evolves Concept that Promises to Bridge Gap between the Star and the Atom'. The concept in question was essentially a development of the work that had caused the Schrödinger–Einstein controversy. Both Einstein's and Schrödinger's papers were soon forgotten as arid mathematical exercises. Schrödinger was reported as having remarked that he believed that he was right, but that he would look an awful fool if he was wrong. Although neither he nor Einstein was 'wrong;' neither had any prospect of making predictions allowing confrontation with experimental data, a weakness that left their work more in the realm of mathematics and philosophy than physics. However, they continued to discuss it sporadically until 1951.

On 5 June 1947, former Army Chief of Staff and now Secretary of State George Marshall gave a speech in which he put forward what came to be known as the Marshall Plan to deliver aid to the devastated and bankrupt states of Europe. Einstein was not able to accede to a request to join a committee in its support since, although he applauded the delivery of aid to Europe, he could not support the Marshall Plan as then constituted. He thought that it ought to be delivered through the United Nations in order to avoid the widespread impression that the Marshall Plan was an American political act directed against the Russian bloc.

On 17 June, Einstein attended the bicentennial celebrations of Princeton University, at which President Truman and the only other living former president, Herbert Hoover, were given honorary degrees. In his address, the President called for a program of universal training, which appeared to echo Einstein's own calls to educate the public about atomic issues and warfare. However, they were not discussing the same sort of training: Truman's proposal for military training was anathema to Einstein. Given that Princeton gave the astonishing number of 111 honorary degrees across the bicentenary celebrations of 1946–1947, it is unlikely that President Truman shared the stage with Albert's distant cousin and old friend Alfred Einstein. Since Princeton apparently did not award a doctorate in music, the eminent musicologist, who taught at Princeton once a month, was given a D. Litt. degree, probably at an earlier ceremony in 1947. Albert Einstein was however at both Truman's and his namesake's ceremonies and congratulated the latter on the honour.

On 15 July Einstein and the Federation of Atomic Scientists issued a statement to mark two years since the test explosion of the atomic bomb, in which they reiterated support for negotiations through the United Nations to make the world safe for atomic energy. Two days later, Einstein took part in a radio broadcast making the case for a world government. In the autumn of 1947, Einstein supported a suggestion by Szilard for an international meeting of scientists that raised no resonance with Secretary of State Marshall but seemed to be supported by Andrei Gromyko, the Soviet Ambassador to the United Nations. However, Gromyko's eventual reply was that Soviet scientists had decided they did not wish to participate. It seems unlikely that Stalin consulted them. That November, Einstein published another article in *The Atlantic Monthly* entitled 'Atomic War or Peace', in which he bewailed the fact that there had been no progress in the two years since his previous article. Although he blamed the US for its disingenuous proposals on international control of atomic weapons, he also criticised the Russians for their attitude, expressed by Mr Gromyko, that international sovereignty was unacceptable. Russian scientists authored a response entitled 'Dr Einstein's Mistaken Notions', in which they lambasted the capitalist system of which they alleged Einstein was a well-meaning dupe. Einstein replied in an article published in February 1948, although it was clear that there was no prospect of agreement.

Antonina Vallentin, an intimate of Elsa Einstein in the Berlin days, appeared in Princeton in the spring of 1948. She arrived at Mercer Street to find Helen Dukas and Margot Einstein little changed but was shocked by Einstein's appearance. The elasticity had gone from his step and his face was that of an old man (see Figure 30.1). His laughter was forced, no longer as in Berlin days welling up from a deep personal amusement at the foibles of his fellow human beings. He confessed to being very frightened by the current state of world affairs. As she left the house, catching sight of the invalid Maja walking uncertainly to her bedroom supported by a nurse, she had a premonition that she would not see Einstein again. It is doubtful that she mentioned her idea of writing his biography, which she had aired to Elsa Einstein thirteen years previously, who had said that her husband would never collaborate on such a project. In fact, by now Einstein had more or less given up opposing the literary outpourings of his friends. His successor at Prague, Philipp Frank, was now in the US teaching at Harvard. He had published his scholarly biography in 1947, which Einstein must have considered an enormous improvement over the notoriously inaccurate pot-boiling volume of Dima Marianoff and Palma Wayne that had appeared in 1944. Two more biographies, aimed at young persons, written by Catherine Owens Peare and Elma Levinger, were published in 1948. He would a few years later give his help and partial collaboration to Carl Seelig's biography. Vallentin's book appeared, initially in French, in 1954, immediately following her biography of that other stalwart of arms control and world government, H. G. Wells.

Mileva Einstein's life in Zurich continued to be one of pain and difficulty. In 1947, she broke her leg in a fall on the ice on her way to visit Tete in the Burghölzli clinic and she was in and out of hospital for several months. In many ways it was the least of her troubles. Her financial affairs

Figure 30.1 Albert Einstein in 1947.

deteriorated considerably as the houses she had bought with Einstein's Nobel Prize money had continued to be unprofitable and were gradually sold off. Finally, the only remaining one, in which her apartment was situated, was sold off in a transaction complicated by continual reference back to Einstein and Nathan in America. They were trying to wind up the company they had founded in which to vest the house's ownership. Eventually the house was sold to a Herr Siegmann in late 1947, who agreed that Mileva should have life tenancy of her apartment, which was notarised by the Zurich authorities. However, barely a month later, Siegmann sold the house to a publisher, Frieda Ehrler, who lived nearby. Early in 1948, she attempted to terminate the tenancy agreement, leaving Mileva anxious that she might be turned into the street with no prospect of finding alternative accommodation.

By far Mileva's biggest worry however was Tete. He was still oscillating between residence in Burghölzli and her apartment. Her concern for his future led her to hold onto the money that Siegmann had paid for her house, even though it was owed to Einstein. He made several unsuccessful attempts to convince Mileva to send it to the US, where he intended to use it in a trust fund to provide for Tete's future care. However, by January 1948, he reluctantly conceded that it would be better to leave it to hand in Zurich, insisting only on the full particulars of the sale so that he could ensure that the appropriate taxes should be paid. Although Mileva forwarded the transaction details, she clearly now began to distrust Einstein's motives, in particular the influence of the women who shared his household. She wrote to the Director of Burghölzli to give her permission for Tete to be placed in the hands of a legal guardian, Dr Meili, early in January 1948. She gave permission on the understanding that the guardian would allow her to visit Tete whenever she wished. She then added the following extraordinary paragraph:

> Perhaps it would be possible that [the guardianship office] could help us somewhat in getting my poor son's father to provide him with a little more money. He doesn't show the slightest interest in the poor, greatly suffering boy. At the beginning of the war his sister went to visit him, leaving her husband and home behind. She told me at the time that her brother had a salary of $400 a week. He has three to four women with him, this sister, a stepdaughter, a secretary and I don't know who else, all of whom have a bad reputation and 'fleece' this otherwise well-respected physicist. I am only interested in this because it breaks my heart that there is so little money for my poor son that we can't manage. However, we depend on it. Because my homeland, a part of Yugoslavia, was so to speak destroyed in the war and there is hardly anything to get from there. I am writing to you about this unpleasant matter. It should remain just between us, but perhaps you too could help in this sad affair so that our son's sad existence can be better secured. I would be very grateful.

While the malice Mileva exhibits in this extract towards Elsa's daughter Margot and Helen Dukas, whom she associated with Elsa, is perhaps not surprising, that towards Maja is. The two had been on good corresponding terms for many years and, as recently as 1940, Mileva had written a friendly postscript to her in a letter to Einstein. Her accusation that Einstein showed no interest in Tete was also grossly unfair. The letter quoted above illustrates the extent to which her concern for Tete's future after her death was warping her judgement.

Einstein explained the dire state of Mileva's finances in a letter to Hans Albert in January 1948, largely to warn him that unless he and Nathan could bring order from the chaos, the provision for the care of Tete would have to come from Einstein's US savings and would adversely affect Hans Albert's inheritance. The depressing nature of the letter was somewhat alleviated by a postscript in which Einstein reported that his former assistant Valentin Bargmann had recently visited Zurich and rescued a violin, presumably a spare that Einstein had kept there to make music on his trips to visit his children in his Berlin days. He offered to send it to Hans Albert when he could find someone going to the west coast. It was intended for Einstein's grandson Bernhard, who had taken up playing the violin.

In the spring of 1948, Tete was well enough to stay again with Mileva. On 17 March he wrote to his father with belated birthday greetings and apologies for being too poor to be able to send a present. He informed his father that he was reading Alfred Einstein's new book *Mozart – His Character: His Work* and finding it very interesting. However, by May Tete's condition had sharply deteriorated. He had an episode of uncontrolled rage that led to the apartment being in a state of chaos when Lisbeth Hurwitz visited. Mileva had a breakdown which led to a stroke and paralysis of the left side. She had had several minor strokes in the preceding months. On 27 May she was taken to the Zurich Cantonal Hospital. The ambulance staff found 87,000 Swiss francs on her person, the proceeds of the house sale. It was now Mileva's turn to be submitted to the guardianship bureau so that her affairs, including the 87,000 francs, could be looked after. On 2 June she assented to this in the last document to bear her signature, in which she agreed that Dr Meili should also be her legal guardian and that all her property should be used for Tete's benefit. Her words were transcribed for her by a member of the hospital staff but her voice can be clearly heard in her statement that Tete would be completely alone when she was no longer there.

Shortly after Mileva had been taken to hospital, Tete was well enough to write a quite coherent letter to his father. He informed him of Mileva's illness, although he thought that she merely had high blood pressure. He described her whereabouts and that she was in an insalubrious ward and asked his father if he could arrange for her to be moved somewhere better. He also relayed Mileva's plea for Einstein to come to Switzerland.

Perhaps Einstein did intervene. At any rate, Mileva was moved to the Eos private clinic, although whether this was an improvement seems doubtful from subsequent events. On 13 June, Lisbeth Hurwitz and her mother visited Mileva, who complained about Einstein, the hospital, the staff, and seemed confused. Her buzzer had been disconnected since the staff said she used it continually without real need. Tete was often at her bedside. Meanwhile, a friend of Mileva's, who had met Hans Albert when he had visited Zurich two years previously, had written to him to inform him of his mother's stroke. She asked him to come to Zurich as soon as possible as his mother was dangerously ill; she reported that Mileva talked often about how devotedly Hans Albert had cared for her when he was a young boy. On 3 June, Hans Albert began a correspondence with his father about what he should do, as he feared his mother had not long to live and he had neither time to take a ship nor money enough to fly. His letter crossed with one from Einstein informing him of Mileva's illness and her and Tete's complicated financial situation. Einstein answered Hans Albert's letter the following day, advising him not to travel as there was nothing he could do for his mother. If, however, he did decide to fly, Einstein's savings were at his disposal but they could be better used to help those left behind after his mother's death. Hans Albert replied that he agreed with his father and that in any case he had no passport and getting one would take at least three weeks. Hans Albert and his father had several more exchanges after Hans Albert received letters with further indirect pleas from Mileva for him to come to Zurich; Einstein consistently both promised to pay for a trip and dissuaded him from making it. In the end, Hans Albert remained in Berkeley.

Lisbeth Hurwitz continued to visit Mileva during her illness. On 25 July she found Mileva to be sinking fast, passing in and out of consciousness. Tete visited her every day; she held his hand and pressed it. On the day before her death, she was completely conscious and seemed herself. On 4 August 1948, before Tete could arrive, Mileva Einstein-Marić ended her long struggle and lonely road, alone.

Einstein received a telegram from the guardian, Dr Meili, on the day of Mileva's death. He immediately wrote to Tete to try to console him, although some of his sentiments, such as partial freedom came from work, in which we could feel ourselves part of the human race, seemed more applicable to himself than to Tete. He also wrote to Hans Albert that same day, reiterating that he had done the right thing, difficult as it was, by not travelling to Zurich. His final words were striking and not particularly comforting:

> Tetel is the most to be pitied, because he exists in this pitiful situation alone, without any helping hand. If only I had known, he would never have been born. Dear Adu, keep your chin up and don't let yourself get depressed. You have always been such a strong character. With best wishes, Your Dad.

The complexities of Mileva's financial affairs were immense, a confusion exacerbated by Frieda Einstein's flying out to Zurich to look after Hans Albert's rights. Not a little friction between father and son was caused by these complications, which were eventually resolved two years later by Hans Albert agreeing to an unequal partition of Mileva's estate so that Tete's care could be assured. In fact, Hans Albert was Tete's heir; on Tete's death, he received back the greater part of what he had generously initially given up.

Once he had internalised the loss of Mileva, Tete seems to have turned back somewhat towards his father and their earlier life. In the next few months, he wrote two letters that seem relatively well balanced, although tinged with melancholy. In the first, he reminisced about trips he and his father had made together in the mountains, while admitting that his memory of those times was not now very good. Nevertheless, he intended to write down memories of Einstein, mentioning his inventive work on compasses with Anschütz-Kaempfe and the refrigerator with Szilard. He remarked that Mozart wrote two operas in which statues came to life – *Don Giovanni* and *Ascanio in Alba*. His father answered some questions on Mozart in his reply, now lost, stimulating Tete to produce a long list of fragmentary discussions about Mozart's operas, remarking that Alfred Einstein knew all about these. He closed by remarking that he was still in Burgholzli, where Dr Maier had told him that he had what he called functional difficulties.

Dr Albert Schweitzer, physician, humanitarian, theologist, expert on Bach and the organ, devoted his life to his hospitals in Gabon. He and Einstein met in Berlin in the early 1930s and may have played music together. Certainly, Schweitzer remembered their encounters fondly. However, their contact was brief; it was not until April 1948 that they began to exchange letters, after a projected trip to the IAS in Princeton had to be cancelled. In lieu of their reunion, Schweitzer described the day-to-day work of his hospital, which was treating many patients with the leprosy that was rife in the area. He remarked on his nightly practice on a piano fitted with ersatz organ pedals to improve his skills as an organist. Einstein did not answer until September but gave a handsome tribute to Schweitzer's work, remarking that if there were more people like him in the world, it would not have got itself into such a dangerous international situation.

The world outlook had indeed continued to darken in 1948. In January, Gandhi was assassinated; Einstein penned a heartfelt tribute. In February, Stalin moved to consolidate communist control of the Czech government. Initially popular after the cessation of hostilities, the communists faced defeat in elections scheduled for May. The Czech Prime Minister, Klement Gottwald, was encouraged by Stalin to mount a coup, which he achieved by packing the police with communists, catalysing mass demonstrations, and taking over Prague. President Edvard Beneš acquiesced in the exclusion of non-communist elements from the government, which in May passed a new constitution through the parliament, essentially making the country a one-party state. These events only confirmed the US administration in the conviction that Stalin was bent on world domination and led directly to the Marshall Plan, the foundation of the North Atlantic Treaty Organization (NATO), which Einstein abhorred, and the increasing militarisation of the US, including a programme of universal military training and the reinstitution of the draft for young men, which Einstein opposed. A further result was the erection of a West German state, rumours of which catalysed Stalin in June to impose a blockade of the western part of Berlin controlled by France, the UK, and the US. Tensions mounted, with discussions within the US administration advocating a preventative nuclear war against the Soviet Union. The situation de-escalated when the US and UK proved able to supply the city by air; Stalin called off the blockade in May 1949.

The end of the British mandate in Palestine on 15 May 1948 also saw the proclamation of the state of Israel in Tel Aviv. Although a committed Zionist for many years, Einstein was also a rather unorthodox one; many of his views on, for example, partnership with the Arab inhabitants of Palestine were anathema to mainstream Zionist opinion. His views on the necessity or desirability of a Jewish state also fluctuated. He was however resolutely against the terror groups, such as Irgun, led by Menachem Begin, that had grown up towards the end of British rule. Of the later regime and of the British government in general, he was unsparingly and consistently critical; indeed, a few months later he described the British Foreign Secretary Ernest Bevin as the Jewish people's worst enemy. Once Israel was established, Einstein put any doubts behind him and supported it strongly. On 18 May, he wrote to his old friend Chaim Weizmann, congratulating him on having been nominated as the first president of the new state.

In this period of growing militarism and anti-communist hysteria, Einstein continued his dogged campaign to convince the American people that co-operation not confrontation, and education rather than militarism, were the only sensible policies in a nuclear age. He was increasingly a voice crying in the wilderness in favour of world government. He set out his ideas in a contribution to a handbook for college debaters. In response to a letter from the secretary of the War Resisters' League, he set out how his thinking had changed since his pre-war advocacy of individual refusal of military service. He believed that such individual action was only practicable if it could be freely chosen in all countries without violent repression and it was only likely to be successful under the influence of a powerful quasi-religious movement. Only a world government could ensure the conditions necessary for the former requirement. On 11 April, the Emergency Committee of the Atomic Scientists issued a statement, signed by all trustees including Einstein, endorsing world government. On 17 June Einstein spoke by telephone to a meeting at Carnegie Hall attended by two thousand people, called to counter the growing threat of war. This was followed by an initiative of French and Polish 'intellectuals' to call a World Congress of Intellectuals in Wrocław, Poland.

Einstein wrote an address to the Wrocław meeting, which he gave to Nathan, who attended the meeting in person. The address had been carefully drafted and contained calls for a supra-national authority. The organisers of the meeting however, who were under heavy communist influence, insisted on changing the wording of Einstein's address, in particular to excise any mention of world government. Nathan indignantly refused to make any such changes. As a result, not only was Einstein's address not conveyed to the congress but also the organisers substituted an earlier letter that he had written to them which was more to their taste, implying that this was his message. When Einstein discovered this, he angrily concluded that there was no future in expecting collaboration from colleagues in the Eastern bloc who were not free to express their own opinions. He ensured that his real text was published in *The New York Times*.

During 1948, internal tensions inside the Emergency Committee of the Atomic Scientists became increasingly evident. The formation of a Foundation for World Government in July, backed by a $1 million donation by an elderly widow, Anita Blaine, and headed by Stringfellow Barr, exposed some of these tensions. Barr solicited the support of the committee, which issued a statement endorsing the foundation. Einstein was appointed a member, presumably chair, of a committee to work with the foundation, which appears to have offered to effectively merge with it. However, the foundation's flirtation with the presidential bid of Roosevelt's former Vice

President, Henry Wallace, which was widely considered, incorrectly, to be a communist front, complicated the issues. In fact, the foundation never had a formal association with Wallace's bid, although Wallace was endorsed by Einstein personally, who was completely disillusioned with the aggressive policy of the Democratic nominee, President Truman. The complications with Wallace led the Emergency Committee to disassociate itself from the foundation, to Einstein's disappointment.

This organisational debacle, which weakened the sizeable number of disparate organisations advocating various forms of world government, was the death knell for the Emergency Committee. By the end of 1948, the difficulty of both convening the committee and, if quorate, getting agreement, led to it becoming increasingly ineffective. Part of the problem was that Einstein was not a particularly good chairman; his own views were sufficiently strongly held that he often found it impossible to craft a compromise with which he could live among the widely differing opinions of his colleagues. On 10 November, a subset of the committee, including Szilard but not Einstein, met in Chicago to recommend that it should become inactive[5] as of 1 January 1949. Einstein accepted this and wrote to the *Bulletin of the Atomic Scientists*[6] admitting that, although the 'emergency' had if anything intensified, a longer-term strategy, involving strong support for the *Bulletin*, was necessary in future. In June 1951, Einstein agreed to the formal dissolution of the committee, which was confirmed at a meeting of the committee at his home on 8 September. While there is no doubt that that the Emergency Committee had done much good and influential work and left a strong legacy, Einstein was sufficiently disappointed with these developments that his future interventions on matters of world peace and government were made as a private individual. In August, he wrote to his old friend Conrad Habicht to commiserate on the death of his brother Paul, co-worker almost forty years previously on their Maschinchen. A few weeks later, his former mathematical assistant, Walther Mayer, died. The cap to this *annus horribilis* for Einstein was the breakdown in his own health, which had been worsening for several years.

Einstein's digestive problems dated back to his student years. As discussed above, he was feeling very unwell in the summer of 1946. In January 1947, his diet was blamed for his increasing emaciation. He told Erwin Schrödinger that once he had rectified the deficiencies in his diet, he put on fifteen pounds within four weeks and began to look human again, rather than like a ghost. This was however only a temporary respite. By autumn 1948, Einstein was once more in intense discomfort with abdominal pains. An old medical friend from the Berlin days, Rudolf Ehrmann, in conjunction with another, Gustav Bucky, diagnosed an intestinal cyst and recommended surgery. Einstein was admitted to the Jewish Hospital in Brooklyn, where he was operated on by Rudolf Nissen on 31 December. Although several intestinal growths were found and removed, an aneurysm or swelling the size of a grapefruit was discovered in his aorta artery above the stomach. Nissen took what steps he could to treat this but surgical techniques of the day and the age of the patient militated against further surgery. Although Einstein was told about this discovery and that it would be likely to eventually prove fatal, he did not remark on it in letters to, for example, Hans Albert at the time. Indeed, when warned by his doctors that he should live carefully lest the aneurysm burst, he seemed unconcerned by the prospect of his own mortality.

---

[5] In fact, the committee continued to hold meetings occasionally, for example, in March 1950 to consider the formation of a 'Citizen's Committee' to discuss the basis for a rapprochement with Russia.

[6] The *Bulletin* and the Emergency Committee were in principle independent; the *Bulletin* was founded first, in late 1945 by Eugene Rabinowitch and Hyman Goldsmith. The two organisations cooperated closely and evolved towards each other; eventually the *Bulletin* absorbed the assets of the Emergency Committee.

Einstein initially recovered from the operation quite quickly, as noted by daily bulletins on his heath published in *The New York Times* until 5 January 1949. By 7 January he could write to intimates from hospital. On 13 January he was discharged and returned to Mercer Street and by 16 January he was dealing with some routine correspondence from home. Nevertheless, he felt weak and was convinced by his friends, principally Gustav Bucky, to convalesce in Lido Key, Sarasota, on the west coast of Florida. Bucky recommended the resort from personal experience. He and Helen Dukas accompanied Einstein, leaving Princeton around 10 February and returning on 27 February. Einstein seems to have truly relaxed and enjoyed the sun, warmth, and sea. According to the local newspaper, he spent most of his time strolling on the beach and visiting the Jungle Gardens. He also went to a local drugstore to have a surgical support fitted to help with the painful hernias that his operation had caused.

Perhaps the wide range of physicians consulted during Einstein's illness included a fellow Schwabian exile, Dr Else Toller from New York. She had been born in Memmingen, some fifty kilometres south of Ulm, Einstein's birthplace, in 1892. She graduated in medicine from the University of Munich in 1918. Presumably, like so many others, she had fled the Nazi regime and found a new home in the US. Two days before Einstein's seventieth birthday on 14 March 1949, she addressed some verses to him whose intimate nature and tone suggest that she and Einstein had a romantic relationship. Both her verses and those written by Einstein in reply use the intimate *du* form that he so rarely employed. Sadly, this exchange of verses is the only clue to their relationship.

A female former admirer, Antonia Stern, who had had a brief affair with Einstein while he was staying in Le Coq in 1933, re-established contact in February 1949 after a decade of silence. She congratulated Einstein on his recovery from surgery, commiserated on the death of Elsa, reminded him of their meeting in Le Coq and his many visits to her parents' house in Zurich in his youth before filling in her own story. She had left Switzerland for Paris, where she had been interned during the war and only fortuitously escaped being transported to a concentration camp. She wanted Einstein's help in publishing a book she had written on Hans Beimler. Beimler, a communist who had lost his life in the Spanish Civil War, was the other passion of Stern's life, after Einstein. She believed that he had actually been killed by his communist comrades under Stalin's orders rather than by Franco's fascists. She hoped that Einstein could recommend a left-wing – but not Stalinist – publisher. Einstein was supportive of her book and recommended that she talk to his friend Solovine, who also lived in Paris. When Stern visited him, however, he had little hope that such a book would find a publisher in France.[7] Stern also sent Einstein some of her poems, together with an extract from her memoires as an internee in Paris. Einstein was discouraging about publication of this, noting that her experiences, shocking though they were, were similar to those of countless others who had suffered during the war and that people were more likely to want to forget such things than read about them. Another letter in May mentioned Jean-Paul Sartre, whose opinion of her book she had sought; she expressed surprise that Einstein seemed to have a low opinion of him.

Shortly after his return from Florida, Einstein celebrated his seventieth birthday. The IAS and the university did him proud, with 300 colleagues attending the official celebrations in Princeton. After some mild sarcasm at what it saw as Einstein's political naivety, *The New York Times* ended its

[7] Stern eventually succeeded in getting some of her biography of Beimler published in Germany by Mosaic in 1957, as part of a book *Das Gewissen entscheidet* edited by Annelore Leber in collaboration with future West German Chancellor Willy Brandt.

editorial encomium with the sentiment that the world congratulated the scientist, philosopher, and kindly, impulsive humanitarian. He remained at home on his birthday, still recovering from his operations, receiving Robert Oppenheimer, the Director of IAS, Oppenheimer's predecessor Frank Adyelotte, and President of the American Chemical Society Linus Pauling. Niels Bohr, Arthur Compton, and the mathematician Jacques Hadamard, a leading figure in the peace movement, prepared speeches for radio broadcast under UNESCO sponsorship on Einstein's contributions. Hadamard, who had first met Einstein during his visit to Paris in 1922, concluded his remarks by saying:

> To [the pleasure of intimate conversations] was added that of a common love of music. It was a delight to have that great scientist reveal himself in my home as a first-class violinist with a rare musical gift and to see the pleasure which he took in taking part in our musical reunions.

On 19 March, the birthday symposium took place in the Frick Chemical Laboratory Auditorium in Princeton. It was entitled 'The Theory of Relativity in Contemporary Science'. The speakers were Oppenheimer, on 'Relativity in the Atomic Domain', Isidor Rabi on 'The implications of Relativity for Modern Experimental Physics', Eugene Wigner on 'Invariance in Physical Theory', 'Bob' Robertson on 'The Present State of Relativistic Cosmology', Gerald Clemence on 'Relativity Effects in Planetary Motion', and Herman Weyl on 'The Theory of Relativity as Stimulus to Mathematical Research'. The more senior participants repaired to the Princeton Inn for lunch. Einstein, who had entered quietly to prolonged applause, remained silent until after lunch but contributed animatedly during Robertson's talk. The Princeton authorities went out of their way to keep the press, other than local Princeton reporters, away from the event, although *The New York Times* correspondent was invited on condition he not attempt to interview Einstein.

Rabi's talk included the assertion that:

> . . . relativity is as vigorous today as it was 44 years ago, when, like Athena, it sprang from [Einstein's] jovian brow. [He] created the theory of relativity and in the same year recreated the quantum theory which Max Planck had suggested five years earlier. Since then these two theories have become inseparably intertwined, although not yet welded into a unity. Without these children of Einstein's imagination, one natural born, the other adopted, there would be no modern experimental physics. If we gave relativity back to Einstein, he would have to take along with it a major portion of the most interesting results of experimental physics and practically all the promise of the future.

Although the talks were not published, foiling the wish of Senator Claude Pepper to write them into the Congressional Record, the symposium did give rise to a volume that Georges Lemaître considered to be an important one for the history of science. *Albert Einstein: Philosopher-Scientist*, edited by P. A. Schilpp, contains twenty-five essays on Einstein's science and philosophy by some of the greatest names in both areas, topped and tailed by his own contributions. The first is particularly important, his own recollections of his life in 'Autobiographical Notes'. The last is a somewhat acerbic response to what he saw as criticisms contained in his colleagues' contributions.

On 19 April, Einstein's old friend Rabbi Stephen Wise died. Almost his last act had been to visit Brandeis University, where he pronounced himself well satisfied with the progress that had been made, expressing the wish to President Sachar that Einstein could be reconciled to the institution. This gave Sachar an opening to write to Einstein. Ostensibly, his letter was to solicit Einstein's opinion of his former collaborator, Nathan Rosen, as a faculty member, but the majority was taken up with a progress report and an attempt to try to reestablish relations. Einstein's reply was cold, both on the subject of Brandeis and indeed on Rosen. He ignored the peace overtures completely and remarked that Rosen, while a well-informed theoretical physicist, was not the top quality man Sachar was seeking.[8]

Einstein passed the summer of 1949 quietly at home in Mercer Street. Maja's increasingly fragile health meant that he spent a great deal of time reading to her by her bedside. His most regular companion in walking to the IAS was Kurt Gödel. Like so many, Gödel had arrived in the US as a refugee from the Nazis. Although not Jewish, he had fallen foul of the Viennese authorities and was in danger of conscription into the German army when he fled in 1939. He and Einstein became firm friends, sharing many interests, including philosophy and mathematics. Indeed, Gödel became interested in general relativity and, reputedly as a present for Einstein's seventieth birthday, produced an exact solution of Einstein's equations for a universe in which matter is assumed to consist of a fluid of moving dust particles interacting gravitationally. When the cosmological constant is tuned to a specific value, solutions have bizarre properties, such as allowing the possibility of time travel. Einstein is thought to have wondered whether this solution implied that general relativity must be incorrect! Around this time, he remarked to the economist Oskar Morgenstern that he only went to the IAS to have the privilege of walking home with Gödel.

Gödel's famous incompleteness theorem is a cornerstone of theoretical computer science. Another Princeton academic friend of Einstein, John von Neumann, was at the heart of the nascent computing revolution. On 12 September 1949, *The New York Times* announced that a little-known corporation, International Business Machines (IBM), had produced a robot brain that could calculate the orbits of the planets using Newtonian plus Einsteinian physics. The 'brain' in question could multiply two 14-digit numbers in a fiftieth of a second and had a memory storage capacity of 500,000 digits. Von Neumann explained the applications of such machines at a three-day IBM seminar attended by 75 scientists at Endicott, New York on 17 November.

The atomic age, now complemented by the computer age, took another decisive turn on 23 September, when President Truman announced that Russia had exploded its own atomic bomb. Although the physicists had warned the politicians that this was inevitable, it came as a shock to many in government, who had thought they would be sole possessors of the atomic 'secret' for at least a decade. After the discovery of the espionage activities of Klaus Fuchs in 1950, many imagined that betrayal of US secrets must have allowed the Soviets to catch up, ratcheting up anti-communist hysteria. The conditions in which Senator Joseph McCarthy would rise to prominence were in place.

Indian Prime Minister Jawaharlal Nehru, together with his daughter, Indira, visited the US that autumn. On 5 November they arrived in Princeton. Before being entertained by the Princeton academic establishment, the two, joined by Nehru's sister, who was ambassador to the US, spent one and a half hours with Einstein in Mercer Street. Einstein had tried to sway Nehru to support the foundation of the state of Israel in 1947, writing to him praising the de jure, if certainly not de facto, abolition of untouchable status in India and drawing parallels with the persecution of the Jews. Nehru remained resolutely deaf to Einstein's pleas then. The topic of their conversation two years later is not recorded but given that India recognised Israel the following year, it seems likely that this formed part of their discussion.

---

[8] He was to give a much more positive assessment at the end of 1951, when Rosen was under consideration to be the founder of a new physics department at Technion in Israel. He was appointed in 1953 and spent the rest of his career there.

On 28 November Einstein broadcast a radio address to the United Jewish Appeal in Atlantic City. He once again blamed the 'divide and rule' policy of the British for the fact that the Jewish and Palestinian communities could not live peacefully side by side in an undivided Palestine. As 1949 ended, Einstein looked forward to a new decade, which he would have realised was likely to be his last. He undoubtedly saw the future with considerable gloom and despondency given an international situation in which the arms race, which he had spent so much time and energy warning against, was beginning to get into full swing, with all its concomitant dangers for the survival of humanity.

Chapter 31

The end of violin playing; Death of Maja Einstein; Einstein the political activist (1950–1954)

The new decade began with further darkening of both the international and internal US situations. On 31 January, 1950, President Truman announced that he had decided to authorise the development of the hydrogen bomb. The brainchild of Edmund Teller and Stanislaw Ulam, it greatly increased the explosive power of a fission bomb by exploiting the extreme conditions found after the explosion to induce fusion. The fusion occurred in deuterium locked into lithium deuteride in a vessel separated from the plutonium of the fission bomb. Such a device was equivalent to megatons of TNT rather than the kilotons typical for fission bombs. The hydrogen bomb was successfully tested in February 1952. The Soviets tested a similar device, developed largely through the work of Andrei Sakharov, in November 1955.

Senator Joseph McCarthy began his meteoric rise to national prominence when he gave a speech on 9 February 1950, in which he claimed to have a list of communists working for the State Department. He saw the opportunity to have his name permanently in the headlines by leading a crusade to root out communists and 'fellow-travellers' from all aspects of American life. When he became chairman of the Senate Permanent Committee on Investigations in 1953, he initiated a blizzard of probes and accusations.

Einstein was depressed by these developments. On 1 February Thomas Mann,[1] who had left Princeton in 1941 for Pacific Palisades near Los Angeles, joined with Einstein and Linus Pauling to make a stand against anti-communist hysteria. They protested against a sentence of imprisonment for contempt of court imposed on the attorneys for a group of leaders of the American Communist Party. Stories began to appear in the New York newspapers accusing Einstein of being a communist spy.

On 15 February Einstein appeared via a film recording in a television program moderated by Eleanor Roosevelt, in which David Lilienthal and Robert Oppenheimer also took part. He warned that the development of the hydrogen bomb brought the extinction of life on earth into humanity's grasp. He warned that an arms race was futile and that the only way forward was mutual trust and cooperation, together with some form of restricted world government.

Later in February, Einstein found time to congratulate Clarence Adler, a concert pianist who included Aaron Copland among his students, for setting up Music Careers Inc., an organisation to help young musicians make careers for themselves. He agreed to write a statement of support that Adler could use in his publicity and began to refer the many musicians who approached him for help to Adler's organisation.

[1] Mann and his family, in fear of persecution as McCarthyism reached a climax, in a striking reversal of his flight from Hitler, moved back to Europe in 1952 and settled in Switzerland.

Einstein. Brian Foster, Oxford University Press. © Brian Foster (2026). DOI: 10.1093/oso/9780198794875.003.0031

As another birthday came and went on 14 March, Einstein wrote to his cousin Paul Moos thanking him for his birthday greetings. In closing, he reassured Moos that he was enjoying his final years:

> I have retained my good sense of humour and don't take either myself or others too seriously. No matter how foolish or cruel people's actions may be. One has to get out of the habit of measuring it against an ideal. Their excuse lies in the non-existence of 'free-will.' All are reflex movements whose relative complexity conceals their true nature from us.
>
> Appreciate your days of serenity and don't let the appearance of external circumstances fool you. Enjoy the fact that few care about you and believe me – it has its good sides. Better an intelligent member of the audience than the actor in the spotlight.

The loving nature of Einstein and his sister Maja has been highlighted on numerous occasions. Their relationship had become ever closer as they aged and, particularly in Maja's case, sickened. It was evident to outside observers even during the war years. In the evenings during their vacations at Saranac Lake, Einstein would often read to Maja. Einstein's friend Alice von Kahler recalled that:

> They loved each other very much. Even before she became ill, he read aloud to her every night. On vacation in Saranac Lake, NY, I sat in on these evening sessions. Once he was reading Herodotus, and there was a not-so-interesting passage, and I said, 'Maybe we should skip that part.' Einstein was absolutely appalled. 'How can you say such a thing! We might miss something important! We are not going to go through this without reading every line.'

Maja had never given up hope of returning to Florence, for which she often expressed her profound homesickness. When the war ended, she began to plan her return to be united with her beloved husband Paul, who had been waiting patiently for her throughout the war years. Tragically, her health deteriorated too rapidly for these plans to remain other than dreams. She had advanced arteriosclerosis, which first struck a devastating blow when she was unable to play the piano and share music-making with her brother. This she had treasured as, for her, the most beautiful thing in the world. Her sight began to fail, and she rapidly became unsteady on her legs and required assistance to walk.

It was Maja's sickness that had ended their joint holidays in Saranac Lake, the last of which was in 1945. By 1946, Einstein wrote to Winteler that she had to be considered an invalid but that it was important not to tell her that, in order to maintain her spirits. Einstein was reduced to writing dual letters to Paul: one to be shown to Maja and the other solely for Paul's consumption. In the first one, plans for Maja's return to Italy were discussed; in the second, Einstein told his brother-in-law that Maja was much too sick to travel, although he did not remove all hope that she might recover enough to make the trip, which might even be by aeroplane.

Sadly, no recovery took place and Maja slowly faded. She became confined to her bed, although on rare occasions she could be carried down to the dining room. During the day in the warmer months she spent considerable time on the first floor balcony, which overlooked the garden and gave her the contact with the natural world that she had always craved. Her greatest comfort though was Einstein's reading to her. He read what he considered the greatest works of literature. In April 1947, he read Xenophon's *Cyropaedia*, which he characterised as truly delightful. Also that year, in a more light-hearted vein, he chose one of his favourite works, *Tristram Shandy*. In 1948, he read the argument of Ptolemy against Aristarchus's contention that the earth moved around the sun. He also regaled Maja with some Aristotle. However, among her favourites were the works of Einstein's old acquaintance George Bernard Shaw, to whom he wrote in September 1950:

> Dear Mr Shaw, I am enjoying so much reading again your dramatic works that I feel the strong wish to thank you for the beautiful hours you are giving me. I am thanking you also in the name of my invalid sister, to whom I am reading every evening for an hour.

Maja also loved to listen to readings from Bertrand Russell, whom Einstein admitted was also a favourite of his. He described Russell as having an extraordinary style and to have remained a rascal even to a great age.

In his March 1950 letter to his cousin Paul Moos referred to above, Einstein remarked that he had not shown the letter to his sister[2] as she could no longer stand or walk, her coordination had been lost, and her mental powers were failing. The fact that in the 1950 US census, Helen Dukas is recorded as living in the nearby 108 Mercer Street rather than 112, Einstein's house, as recorded in the 1940 census, is presumably related to Maja's illness. Possibly she had given up her bedroom in the Einstein household in favour of a nurse.

Einstein continued to read to Maja even after her memory had deteriorated to the extent that she could no longer remember what she had listened to the previous day. Einstein had written to Maurice Solovine in July 1950 to say that Maja was sinking, although there was no particular condition to which this could be ascribed. By March of 1951, she could hardly speak recognisably, although other indications were that her some aspects of her cognition were unimpaired.

The end came after Maja had a fall in which her arm was broken. The lack of movement required to permit setting the bone when she was in hospital allowed pneumonia to set in. She developed a high fever and lost consciousness, leading to death on 25 June. Einstein wrote to her husband Paul Winteler with news of her passing, reassuring him that she had always longed to return to him and their home Samos in Florence, and that his behaviour throughout their long separation had been exemplary. His regular and sympathetic letters were her greatest joy. He told Paul that he had continued to read to Maja every night until the very end. To Hans Albert, Einstein described the funeral, which only he, Margot, Helen Dukas, and another friend, who drove them to the crematorium, attended. He decided that a similar arrangement was what he wished for his own funeral. A few months after Maja's death, he admitted to Michele Besso that he missed her enormously.

During Maja's illness and particularly after her death, another woman came to play an increasingly important part in Einstein's life. Johanna (also known as Hanne) Fantova wrote that, in the evening and often late into the night, Einstein enjoyed reading to Maja and her, mainly from Sigmund Freud, Arthur Schopenhauer, and Bertrand Russell. He interspersed the reading with his own comments. Fantova had a very early connection with Einstein. Her mother-in-law, Berta Fanta, had hosted a salon in Prague that Einstein had often attended when he was a professor there in 1911–1912. Fantova's future husband Otto had written the script for a movie on general relativity that brought him into contact with Einstein again in the early 1920s. In 1925 he married Johanna; in 1929 the couple visited Einstein in Berlin. Otto suggested that Johanna should catalogue Einstein's books, which were scattered throughout the apartment. It seems that Einstein found the young bride attractive, since she was invited to sail at Caputh on several occasions and in 1933 she and Otto visited the Einsteins at Le Coq for an extended period. After the occupation of the Sudetenland in 1938, she and her husband left Prague and fled to the UK, losing all their possessions. Otto, who had become a respected handwriting expert, seems to have been detained by the British authorities; he died suddenly in Oxfordshire in 1940. Johanna had left him there

---

[2] Most of the letter was addressed to Maja and, assuming Einstein would be too busy, Moos had asked her to reply with news of their extended family.

and emigrated to the US in May 1939. Taking Einstein's advice, she qualified as a librarian at the University of North Carolina at Chapel Hill. In 1944, she was appointed to a position in the library at Princeton, eventually becoming curator of maps there.

The return of Margarita Konenkova to Russia at the end of the war left an emotional gap in Einstein's life which Johanna was able to fill. A flavour of their relationship can be gained from an index card (presumably a reference to her cataloguing activities) that Einstein sent, inscribed (freely translated): 'I could have sent a letter/ But I thought this card was better/ My blessings to both Mozart and you/ To our wager's terms I remain true./ A kiss, and sleep well! Your Ele.'[3] By the time of his operation at the end of 1948, Fantova had clearly become very important to Einstein. She was the first person to whom he wrote from his hospital bed, and they exchanged several letters while he was recuperating in Florida. Once he returned to Princeton, they resumed what became a daily routine for the rest of his life. Weather permitting, they sailed on Lake Carnegie every day. On most evenings, they talked to each other on the telephone. Fantova was also often a witness to Einstein's meetings and telephone calls with prominent people, many of which she recorded in a journal.

In March 1950, Erwin Schrödinger broke his long silence with a letter, responding to a breaking of the ice by Einstein, who had sent him a reprint of one of his papers on unified theories. Einstein's reply was friendly, noting that Schrödinger still seemed attracted to the idea of formulating a unified theory using an asymmetrical metric tensor, which had caused friction between them three years previously. He pointed out that in the interim he had pondered long and hard on this idea and that he had discovered at least two fundamental problems, which he went on to explain. Schrödinger, no doubt encouraged by the friendly tone, hastened to answer. Before long, they were back to their favourite subject of quantum mechanics. In December, Einstein praised Schrödinger and Max von Laue as the only contemporary physicists who understood that any sensible conception of reality implied that quantum mechanics was incomplete. Schrödinger and Einstein continued a quite intense correspondence both on quantum mechanics and general relativity until Einstein's death. Max von Laue and Einstein also exchanged letters at this time concerning gravitation. Finally there was a flurry of letters on the same subject exchanged between Einstein and his old friend and erstwhile collaborator, Besso.

Einstein had indeed been thinking long and hard about unified theories involving asymmetric metric tensors. He exchanged many letters on the subject with Ernst Straus, his former assistant, now on the west coast. In March he published an article entitled 'On the Generalized Theory of Gravitation' in *Scientific American* that gives a deep insight into his thought processes. After a masterly summary of the history of western physics culminating in special relativity, he explained that the equivalence of gravitational and inertial mass requires an extension from the linear Lorentz group relevant to special relativity to non-linear transformations. Mathematically such groups are the continuous transformations of coordinates due to Carl Friedrich Gauss and Bernhard Riemann, which lead to general relativity. Particles of matter such as the electron, which are sources of the electromagnetic field, are singularities in general relativity and therefore cannot be explained within it. A further extension is required to include electromagnetism: the removal of the requirement that the metric tensor should be symmetric, which Einstein convinced himself was the only natural extension to general relativity. Unfortunately the complexity of the equations resulting from such an extension is daunting. Einstein concluded his article by admitting that it was unclear

[3] Fantova called Einstein by the pet name 'Elefant'. Einstein often decorated his notes to her with little drawings of an elephant.

whether the system of equations resulting from this theory was sufficiently powerful to explain known phenomena and that clarifying this would require enormous effort and probably new mathematical techniques.

Einstein's starting point was diametrically opposed to that of the overwhelming majority of physicists of the time and subsequently. They were happy to explain physics at the molecular level and below by ignoring gravity, the strength of which is negligible at such scales; Einstein was determined to explain all of physics by using only general relativity as a paradigm and ignoring quantum mechanics. Both positions could be rationally defended; neither has so far led to the unified theory of physics that Einstein sought. However, there is no doubt that the majority's strategy has been more productive.

On 14 March 1950, Einstein wrote to Hans Albert with the news that he had decided to transfer $5,000 to him in order to fund Bernhard, Einstein's grandson, while he was following the steps of father and grandfather in studying in the ETH in Zurich. He had been there since the late summer of 1949, having called in to Princeton to see his grandfather en route. Hans Albert was asked to send funds to Bernhard at the rate of $80 per month. Bernhard wrote to his father about the condition of Tete. When he had seen Tete on his arrival in Switzerland, he had been shocked by his condition. In May, he reported that Tete had been well enough to leave the clinic and stay with Father Freimüller, a parish priest in a rural suburb of Zurich who had experience and training to deal with troubled people. Although at first, he had done little more than play ferociously on the piano, gradually he began to communicate normally, particularly with Freimüller's young sons. He was even able to audit some university courses in history and music history. Bernhard reported that he thought it wouldn't be long before Tete could be self-sufficient. Hans Albert, in conveying this assessment to Einstein, was unconvinced. Sadly, he was proven to be correct.

On 29 July, Hans Albert and his family drove from Berkeley to New York to meet Bernhard's ship. He had travelled back from Zurich via Paris, where he spent a month. The reunited family travelled south to visit Einstein in Princeton; the following day, all wrote to Tete with birthday greetings. At the end of the year, Einstein wrote again to Tete, telling him how glad he was to hear that he was attending university courses. He wrote that he himself was getting along pretty well, given that he was almost seventy-two, but he noticed that his stamina was diminishing. He was amazed at how long it took for a complex machine such as a person to run down. He offered to send some English books, if Tete was able to read them. This seems to have been Einstein's last letter to Tete. In 1954, he admitted to Carl Seelig, who had taken an interest in Tete, that he couldn't really analyse why he had stopped writing. He explained that he was afraid of awakening painful thoughts in Tete's mind; more likely, subconsciously, he feared the painful memories it awoke in him.

In April 1950, acting as Chairman of the American Committee for the Hebrew University, the Weizmann Institute of Science, and Technion, Einstein sent out invitations to a dinner to be held in the Nassau Tavern in Princeton on 10 May. The dinner was to celebrate the twenty-fifth Anniversary of the Hebrew University. Exceptionally, Einstein not only attended the dinner but gave a speech afterwards. The aim was to raise $5.5M for the institutions.

The retirement of Alfred Einstein from Smith College was marked by a three-day music festival in April 1950. Madrigals and early music formed the core but also new pieces from composers such as Roger Sessions were featured. Alfred was due to take up a summer teaching position in Berkeley. However, Einstein was saddened to learn that Alfred had been taken ill in June[4] in Oklahoma City,

[4] Einstein's letters are dated at the end of May but given that Alfred's retirement was in June and his daughter noted that his illness had occurred in June, it seems likely that these dates are incorrect.

during an overnight stay while driving to the west coast. Alfred had a simultaneous stroke and heart attack and was rushed to hospital. Einstein wrote both to him and separately to his family to commiserate and to advise Alfred to take things easy and reduce his workload, although he admitted how difficult this was. Although Alfred recovered partially and moved with his family to El Cerrito near San Francisco, he had another stroke the following year and died in February 1952. Einstein wrote to the family to express his deep sadness.

The outbreak of the Korean War on 25 June 1950 ratcheted up international tension to a new level. As the war ebbed and flowed, first with the involvement of a United Nations' force marshalled by the US and then by the involvement of China, Einstein worried that Armageddon was approaching. Having met Prime Minister Nehru of India and knowing Nehru's determination to uphold India's non-aligned status encouraged Einstein to address several letters to him in the succeeding months, urging him to use his good offices to help restore peace. These were to no avail; the war rumbled on until the armistice of July 1953.

Einstein had been president of the Ernest Bloch Society since 1937. His duties were far from onerous. However, in November 1950 he received a letter concerning Bloch from an old friend, Ernest Zeisler, a mathematician and theoretical physicist from Chicago. Einstein had first met Zeisler on his visit to Chicago in 1921, when Zeisler was a graduate student in mathematics working on the mathematical aspects of general relativity. Zeisler's book with Horace Levison, *The Law of Gravitation in Relativity*, published in 1929, was sent to Einstein on publication by their mutual friend, Francis Neilson, at whose luncheon party they had met. Zeisler was also a violinist; writing in 1936, he offered Einstein a guest room and 'peace and quiet and a string quartet' if he should visit Chicago. He had been very active in raising funds for the Palestine Orchestra. Zeisler was also an admirer of Bloch and wished to champion his music. He wrote to Einstein requesting a testimonial. On 15 November, Einstein replied with 'My knowledge of modern music is very restricted. But in one respect I feel certain: True art is characterised by an irresistible urge in the creative artist. I can feel this urge in Ernest Bloch's work as in few later musicians.'

The Christmas and New Year period of 1950–1951 saw Einstein report a milestone in his personal musical world. Until a couple of years before, in addition to playing the violin with various friends and neighbours, Einstein had also played in university orchestras, mostly specialising in early music. Sometime in 1948, he had abandoned playing the violin; he reported this capitulation to age and infirmity to an old friend from his student days in Zurich, Susanne Markwalder. However, he said that he continued to improvise, in a bumbling way, on the piano. Early in the new year, he also wrote to his former chamber music companion, Queen Elisabeth of the Belgians. She had sent him greetings after a decade of war-related silence. To her, he reported that for some time he had been unable to stand the sound of his own playing but that his scientific work remained and would stay with him until the end. To both correspondents, he expressed his inability to understand the American way of thinking, adding to Queen Elisabeth that he bewailed his powerlessness to stop the American people repeating the errors of the Germans and aligning themselves with evil.

Einstein subsequently told Johanna Fantova that playing the violin put too much strain on his body, whereas the piano was much more suitable for improvising and playing alone, so he played it every day. As will be discussed in Chapter 32, it seems he did make an exception to this self-denying ordinance shortly before the end. Fantova remembered him as indulging in his love of Mozart and Bach on the piano, by which she presumably meant that he improvised on themes by those composers.

Just before Einstein gave up playing the violin, the fire to play music burned as brightly as in his youth. This is clear from a story that he told to *The New York Times* on his seventy-fifth birthday.

He met a concert pianist and her brother on a Princeton street. The brother was a violinist with the Colonne Symphony Orchestra on a visit from Paris. Einstein inquired whether he knew the Bach and Vivaldi Double Concertos. On receiving an affirmative answer, Einstein said that he would come to the pianist's house that night. He arrived with his violin and his music and the three played together all night.

On New Year's Day 1951, Einstein wrote to Maurice Solovine about the translation into French of his recent book *Out of My Later Years*. In discussing the article in the book about the relationship between science and ethics, Einstein disagreed with Solovine's viewpoint on the argument that Einstein had advanced but admitted that this was merely his opinion. However, he added 'If you don't agree with me, then I ask in all humility: whose nonsense should be in this book, mine or yours?'

On 16 January 1951, Einstein replied to a letter from the President of the German Federal Republic, Theodore Heuss, inviting him to resume his membership in the re-established eminent German order 'Pour le mérite'. Einstein replied disdainfully that, after the mass murder of the Jewish people by the Germans, no self-respecting Jew could associate themselves with any official German organisation. On the same day he thanked a New Yorker, Bernhard Schnitzer, for a portrait he had sent. He appreciated it but remarked that there can really be no worse person than the subject to evaluate a likeness, since he never saw it himself as he didn't use a mirror when shaving. In a similarly light-hearted vein, he thanked the International College of Surgeons for the beautiful certificate they had sent him, promising that, should he ever kill anyone, it wouldn't be by using their techniques.

Einstein was re-energised in his opposition to the current thrust of American foreign policy by reading a letter sent to *The New York Times* by the author Pitirim Sorokin, a proponent of world government and the head of the Sociology Department at Harvard University. Sorokin, a Russian exile and fervent opponent of communism, denounced President Truman's policy as certain to lead to a disastrous war in which civilisation would be destroyed. On 17 January, Einstein wrote to him requesting his permission to reproduce his letter as an advertisement endorsed by eminent personalities that Einstein and his friend Erich von Kahler would recruit. However, this was subsequently strongly opposed by Otto Nathan, who told Einstein that Sorokin was not taken seriously by fellow scholars. Einstein therefore proposed to Kahler to publish similar sentiments in letters to newspapers of record throughout the US. He promised to reconcile Sorokin to this new plan. It seems that even this amended plan went no further, presumably on Nathan's advice. The tone of Einstein's letters in this period expresses a deep pessimism for the prospects of humanity. When his cousin Paul Moos expressed amazement at the doings of President Truman and his associates, Einstein wrote that he would be happy if the Devil took Truman and his backers, but Satan wouldn't because he valued his representatives on earth too much. He concluded another letter with the desolate sentiment that there was no help for mankind.

On 28 February, Einstein received a postcard from Albert Schweitzer showing patients at his hospital receiving their food ration. Schweitzer remarked on their shared worries at the state of humanity. He thanked Einstein for his kind words on his seventy-fifth birthday and apologised that in his three-week stay in the US in 1949, he had not found the time to visit him. Einstein was even more laudatory in his greetings composed in July 1954 and published in a volume for Schweitzer's eightieth birthday. He wrote:

> I have hardly ever become personally acquainted with someone who combines in one individual such a quantity of goodness and thirst for beauty as in Albert Schweitzer [. . .] Everywhere he avoids rigid tradition. He fights against it whenever anything promising for the individual could result.

> One feels it clearly when he exposes in his classic Bach work the dross and mannerisms by which the musical establishment has obscured the creations of the beloved master and interfered with its simplicity.

His final letter to Schweitzer was written on 6 December 1954, conveying a donation for his hospital from the pupils in a Princeton school to commemorate their teacher who had been killed in an accident. Einstein commented that reasonable people everywhere were able to discern clearly the route out of the confusion of the time and that Schweitzer's quiet example had a profound effect.

In March 1951, Einstein's cousin Luigi Ansbacher sent him a postcard of Isola della Scala. Einstein recalled travelling there with his father more than fifty years before to inspect their electricity substation. Ansbacher had mentioned Guiseppe Verdi, prompting Einstein to say:

> The works of the greatest masters awake in us the feeling that everything must be as it is, and that nothing can be changed without detracting from the piece. It's as if the artist didn't create the work but rather discovered it completely fashioned in another world. Of the musicians of the post-classical period, Verdi is the only one in whose works I find the urgency, self-awareness and lack of artifice that distinguishes a true genius.

On 23 March, Einstein wrote to the head of Yeshiva University to congratulate him on the plans to open a medical college, which he considered to be of great importance to American Jewry. This letter led to invitations to be involved in its genesis. On 15 July 1952, he became Honorary Chairman of the fund-raising campaign. On the day following his seventy-fourth birthday, 15 March 1953, he took part in a ceremony in Princeton that marked the official naming of the Albert Einstein College of Medicine, which opened in September 1955.

Nicholas Harsanyi had left Hungary in 1938 and arrived in Princeton to teach at the Westminster Choir College, whose concerts Einstein often attended. He became friendly with Einstein as a result of his acquaintance with the wife of John von Neumann. Harsanyi was a violinist by training and since Einstein liked nothing better than to play the Bach 'Double' concerto, he was enlisted in the chamber music meetings in Einstein's house. Usually on Wednesdays, these carried on for some time even after Einstein had ceased to play the violin. Harsanyi described Einstein as having an 'accurate but not sensuous' tone and that he was well versed in music up to the nineteenth century. However, Einstein would never have listened to Bartok, he opined.

Although there was a great deal of amateur music-making in Princeton, including the university orchestra and various early-music and chamber groups, it did not have its own professional orchestra. In 1951, Harsanyi not only founded the Princeton Symphony Orchestra and became its first musical director, but also persuaded Einstein to become a vice president. The Orchestra gave its first performance in the McCarter Theatre on 26 April. The programme began appropriately with Brahms's *Academic Festival Overture* and included a Haydn symphony, Chopin's Piano Concerto No. 2, and the Barber *Adagio*.

Just before this concert, Einstein wrote another of the innumerable letters that he sent to aid fellow scholars in difficulty. This one is particularly noteworthy as it concerns David Bohm, one of the most profound thinkers on quantum mechanics of the twentieth century. Indeed, Einstein, who worked with Bohm in Princeton, is said to have remarked that if anyone could come up with a radical new quantum theory, it would be Bohm. He had been a student of Oppenheimer and had the possibly unique distinction of not being allowed to write his own PhD thesis. Bohm had been refused security clearance at Los Alamos because of his brief membership of the communist party as well as his much more enduring Marxist convictions. His work in 1943 on proton–deuteron scattering was directly relevant to the work of the Manhattan Project and was classified. He could neither write nor defend his thesis and was awarded his doctorate after written confirmation from Oppenheimer to the University of California Berkeley that Bohm had completed his studies successfully.

Einstein's letter of recommendation for Bohm to fellow Nobel Laureate Lord Blackett, then at the University of Manchester, was less than frank about Bohm's background. He wrote that Bohm had not been in any way politically active but had refused to answer questions concerning colleagues, which had led to an official indictment. In 1949, Bohm had refused to testify to the House Committee on Un-American Activities and in December of that year was arrested. At his trial in May 1950, Bohm was acquitted of all charges. However, this did not save him from losing his job. The following month, Princeton, under pressure from rich donors, did not renew his contract. Convinced that he was being tailed by the FBI, Bohm fled to Florida. He returned to Princeton and, after failing to get a position anywhere in the US, or indeed at Manchester (despite Einstein's letter), he asked Einstein to write a letter of recommendation to the University in São Paulo, where Bohm had some contacts. Just before he left for Brazil in October 1951, he met Oppenheimer in the street, who expressed surprise that he had not yet left. The interest that Oppenheimer showed in his departure date was caused by his concern that he would be the next to be investigated for 'un-American activities', an investigation in which his relationship to Bohm was to play a part. Bohm left Brazil for the UK in 1957, first to the University of Bristol and then for the rest of his career to Birkbeck College, London.

Bohm's book *Quantum Theory* was completed in 1951 while he was still living in Princeton. It presented an orthodox view of quantum mechanics in a particularly clear way. It received positive reviews, including from Einstein. After several discussions with him, however, Bohm began to have similar doubts to Einstein about the validity of the standard 'Copenhagen' interpretation as championed by Bohr. Eventually this would result in Bohm postulating a 'hidden variables' variant of quantum mechanics that would become immensely influential in the debate on its nature that continues to this day.

Arnold Schoenberg had made attempts to befriend Einstein over several decades, first in Europe and subsequently when both were refugees in the US. Despite their common interest in Zionism, he was unsuccessful, perhaps because their differences in musical appreciation were unbridgeable. According to Allison Portnow:

> Late in his life, Schoenberg assessed his compositional career and the notoriety it brought him by pondering a time when everybody made believe he understood Einstein's theories and Schoenberg's music. In fact, Schoenberg was not the first to build a parallel between himself and the famous physicist. Of the many composers who lived and worked in America during the time period of Einstein's fame, Schoenberg was the one whose music was thought to resemble most the tenets of relativity theory. His strict twelve-tone technique was considered by many at the time (and even today) to be the musical embodiment of Einstein's ideas. In giving each of the twelve tones equal importance (or equal lack thereof, in that none of them adhered to a tonal hierarchy), Schoenberg's dodecaphonic music was thought to have made the relationships between the tones a relative one. For the most part, this view of twelve-tone music borrowed the popular notion of relativity as having made (sic) or proven that everything was relative.

Schoenberg had been ailing for a considerable time; by mid-1951, he had a premonition that his end was approaching. He had a superstitious dread of Friday the 13th, so on Friday 13 July he spent the day in bed. Waking at 11.45 p.m. he seemed relieved almost to have survived the day but two minutes later, murmuring 'Harmony, harmony . . . ', he died.[5]

Einstein received a telegram with New Year's greetings for 1952 from Queen Elisabeth of the Belgians. With unusual promptness, he replied on 3 January, with a bad conscience for not having answered her previous letter. He talked of the hardships and trauma of the twenty years since they had last seen each other. The worst was the disappointment at humanity's current behaviour. The young didn't seem to care, since unlike his and the Queen's experiences in their youth, they had never known a period of rationality and security. Even in the period between the wars, there was hope of a brighter future. He saw the only hope in the founding of a world government. He concluded with the consolation that nature continued in its everlasting beauty and that people still enjoyed life and forgot the human dilemma, just like innocent beasts. He wished the Queen the same pleasures, wondering if music could be enjoyed in a similarly carefree manner. Yes, he concluded, provided one were not a professional!

On 2 January Einstein replied to a question from Jean-Jacques Fehr, a correspondent in Columbia, who had asked his opinion about the 'steady-state' theory of cosmology that had been postulated by Fred Hoyle, Thomas Gold, and Hermann Bondi in a series of papers in 1948. Einstein thought that the idea that atoms spontaneously appeared out the void was insufficiently motivated to be taken seriously and advised Fehr to devote his energies to more pressing problems such as quantum theory or the development of general relativity.

The death in 1952 of Robert Koch – Einstein's cousin, schoolmate in Aargau, and sometime rival for the affections of Marie Winteler – hit Einstein hard. He received the news from Robert's sister Alice. It not only revived memories of their shared youth, but also of Italy, where Alice was now living: 'Italy is really the most luminous memory of my life. I spent the longest time there when I was 16, in the attractive little city of Pavia. Maja was also very happy there. Now I see this all in the light of the [illegible] sun. There is nothing painful there.'

There was news from Italy when on 26 January, Einstein received a letter from his brother-in-law Paul Winteler. He had left the hospitality of his other brother-in-law, Besso, in Geneva and had returned to his and Maja's beloved house, Samos, near Florence. However, on 26 June he came back to Geneva, suffering from influenza. He refused any treatment and on 15 July, he slipped peacefully away. Besso's son Vero wrote to give Einstein the sad news. Paul had been anxious that on his death, the sale of Samos should repay the money that Einstein had advanced to Paul and Maja. Vero also reported that Paul had mentioned a biography of Einstein that Maja had written and that he would search for it among Paul's papers. Einstein, in his reply to Vero, said that he had always intended the money as a gift but had never clarified this to Paul and Maja to prevent them from slipping into further debt to others. Einstein knew nothing of the biography. Fortunately for posterity, Vero must have found it and shipped it to Einstein with the rest of Maja's belongings since it is preserved in the Einstein archives and provided the foundation for much of the earliest chapters in this book.

More memories of Italy were stirred when Einstein received two letters from his old flame from his Pavia days, Ernestina Marangoni. In his reply on 1 October, he recalled with great pleasure the golden time they spent together in Casteggio, adding:

[5] This is such a good anecdote, almost rivalling Mahler's last words 'Mozart – Mozart,' that it is included here. It must be admitted that McDonald gives no source and other books, for example, Stuckenschmidt's, quote Schoenberg's wife Gertrud's report on her husband's death, which contains no such words – simply 'his throat rattled twice', which seems rather more plausible.

> I myself, on the other hand, tended more and more towards loneliness, a trait that tends to increase with age. It's strange to be so widely known and yet lonely. But the fact is that the kind of popularity that has arisen in my case pushes those affected into a defensive position that leads to isolation.

Recollections of the time just after Einstein's Italian idyll were recalled by a letter from the keeper of the Canton of Aargau art collection telling him that a museum was to be built to house it. Einstein replied on 8 January 1952 that, as a former pupil, he was particularly pleased by this news. He continued to believe that the Aarau school was an excellent example. His youthful experience showed very clearly that the decentralisation of the educational system and the freedom of teachers to decide the curriculum and how it was taught was much more effective than regimentation. Since people are not machines, their intelligence withers away unless given the opportunity to develop their own ideas and the freedom to make their own choices.

Yet another stimulus to reminiscence occurred at this time. Carl Seelig and Einstein had met in Berlin just after the First World War and exchanged a few letters subsequently. In February 1952, Seelig wrote to Einstein suggesting that he write a definitive account of Einstein's time in Switzerland. Surprisingly, Einstein agreed. In the next few months, Seelig peppered Einstein with questions and received his permission to talk to his friends, such as Besso, about his earlier life. Seelig became not only an important Einstein biographer but also a conduit between Einstein and his son Tete, whom Seelig had befriended. He visited him frequently following his readmission to the Burghölzli clinic after his period of living in the community ended. Indeed, in the spring of 1952, Seelig offered to take over the guardianship of Tete from Dr Meili, although Einstein, with gratitude, demurred. Seelig nevertheless kept Einstein informed of Tete's condition.

By the end of 1952, Seelig had a draft of his book ready for Einstein's inspection. Although he had made a vow never to read material about himself, in a moment of weakness he had asked Helen Dukas to read him a few sentences and had immediately become so interested that she had continued for hours. Einstein was charmed by the book and the many details of his life contained therein that he had forgotten; he congratulated Seelig on tracking down many of Einstein's contemporaries with whom he had lost touch.

Linus Pauling wrote to Einstein early in May telling him that in January of 1952 he had been denied permission to attend a conference in the UK. Pauling had been an active member of the Emergency Committee of the Atomic Scientists and this activism, together with the suspicion, without apparently any proof, that he was either a communist or a communist sympathiser, led to his being banned from travel by the State Department, who considered that allowing him to travel was not 'in the interests of the United States'. Einstein replied to Pauling's letter, together with a subsequent one requesting his intervention with either the President or Secretary of State, on 21 May. He applauded Pauling's fight for the freedom to travel outside the US. Einstein also wrote to the Secretary of State requesting the granting of a passport, without any apparent effect. The affair caused Sir Robert Robinson, former President of the Royal Society, to write to *The Times* of London expressing his disgust at the attitude of the American authorities. It took the public intervention of Senator Wayne Morse in June to finally convince the State Department to issue a temporary passport limited to Pauling's travel to the UK and France that summer. His request for a full passport was denied again in July 1954. It was only in October 1954, after he was awarded the first of his Nobel Prizes, in Chemistry, that Pauling was finally issued with a full passport to allow him to travel to Stockholm for the Nobel ceremony.

A distant echo of Einstein's musical life in Berlin came from a letter he wrote in May 1952. The splendour of the von Mendelssohn's house in Grunewald had been destroyed by Nazi appropriations. The cellist Francesco von Mendelssohn, his health broken down by alcoholism but not yet institutionalised or lobotomised, enlisted Einstein's help to recover some of the lost treasures. His sister, the brilliant Eleonora, was already dead. In a letter that reads like an index of European art, Einstein wrote to request the restoration of pictures by Rembrandt, Monet, Degas, Manet, Rubens, and Corot that had once adorned the walls of the rooms in which Einstein had played piano trios with Francesco. Despite Einstein's support, Francesco's requests were denied and the pictures remained in the various museums in which they had washed up after the Nazis were swept away.

An even more distant echo came with a letter from Jakob Ehrat, Einstein's fellow student at the Polytechnikum in Zurich. Seelig's contacts with Einstein's former friends and associates had included Ehrat, who was stimulated to write to Einstein and give him news of his family and activities in the preceding almost forty years since they had last met. Einstein replied in similar vein, telling him that he was doing pretty well, having victoriously survived both the Nazi era and two wives.

One by one, Einstein's family, friends, and acquaintances were dying. In November 1952, it was the turn of one of the oldest and most consequential, Chaim Weizmann, President of the state of Israel. On 9 November, Einstein expressed his condolences to Weizmann's widow in a telegram in which he characterised Weizmann as a shining example to future generations. On 14 November, the Jerusalem newspaper *Maariv* began to canvas the idea of Einstein as the new President of Israel. Einstein was contacted by *The New York Times* but refused to comment, given that he had heard nothing official. To Helen Dukas, however, he made it clear that there was not the slightest chance of his accepting such an offer. The idea quickly gained traction to the extent that Israeli Prime Minister David Ben Gurion felt obliged to request his ambassador to the US, Abba Eban, to enquire whether Einstein might accept. Dukas, trying to short-circuit the discussion before it gained a momentum of its own, had tracked Eban down to a New York reception and connected him with Einstein by telephone. Acknowledging Einstein's refusal, the ambassador nevertheless said that he had to make an official offer of the position by letter, which was delivered by Eban's deputy, David Goiten, on 18 November.

Eban's letter contained the following text:

> I am anxious for you to feel that the Prime Minister's question embodies the deepest respect which the Jewish people can repose in any of its sons. To this element of personal regard, we add the sentiment that Israel is a small state in its physical dimensions, but can rise to the level of greatness in the measure that it exemplifies the most elevated spiritual and intellectual traditions which the Jewish people has established through its best minds and hearts both in antiquity and in modern times. Our first President, as you know, taught us to see our destiny in these great perspectives as you yourself have often exhorted us to do!

Einstein letter declining the honour was given into Goiten's hands. He wrote:

> I am deeply moved by the offer of our state of Israel, while at the same time sad and ashamed that I am unable to accept it. Since all my life I have been studying the objective, I have neither the natural ability nor the necessary experience to deal with people and to carry out official functions. Therefore I am really not suitable for this great task, even were my advanced age not increasingly limiting my strength.
>
> This situation saddens me even more since my relation to the Jewish people has become my strongest human attachment ever since our precarious position among the nations became fully clear to me.

Another change of president, that of the United States, was decided that November. Einstein had written to the Democratic candidate, Adlai Stevenson, in October, offering him his support because of his integrity, judgement, and independence. The defeated candidate visited Einstein in Princeton when he received an honorary degree in June 1954. Only then did Einstein admit, according to Johanna Fantova, that he only voted for Stevenson because he trusted Eisenhower even less. In fact, he made it clear in a letter written just after the election to Paul Arthur Schilpp that it was not so much Eisenhower to whom he objected but to those with whom he kept company. In particular, he pointed the finger at John Foster Dulles, who became Eisenhower's Secretary of State, as one of the main orchestrators of the Korean affair, as he called it. He went on to opine that the notion of an atomic strike was dangerous in as much as it encouraged megalomania among many hotheads.

Einstein continued to attend musical events, albeit with decreasing frequency. He went to another concert by Marian Anderson in 1952, which is described by Carl Seelig:

> The only woman who could boast that her hand had been kissed by Albert Einstein is [. . .] Maria (sic) Anderson. This was after her performance of Schubert's *Death and the Maiden*. Statuesque and dignified as a Giotto figure she stood on stage and sang this wonderful song with such feeling that the otherwise somewhat facetious researcher was deeply moved. Also on this artist's programme were children's songs by Frida Sarsen, who had been married since the summer of 1911 to the radiologist Dr Gustav Bucky, originally from Leipzig. These were among Einstein's closest friends. When at the end of 1952 Dr Bucky, who was one of New York's most prominent doctors, had to fight a court battle having been charged by a company with illegal use of one of their photographic inventions,[6] Einstein appeared as a witness in court for him.

A few months later, Bucky won his case.

The concert career of Einstein's friend Paul Robeson had been greatly damaged by the implacable hostility of J. Edgar Hoover and his FBI. This had been evident as early as 1943, when Hoover had added Robeson to the list of those to be put in concentration camps in the event of war between the US and the Soviet Union. He was subsequently barred from almost every US concert hall; his previous annual income of $100,000 dropped to a mere $6,000. Nevertheless, he continued publicly to espouse left-wing views and friendship towards the Soviet Union. In September 1949, Robeson's picnic-concert at Peekskill, New York was attacked by right-wing radicals while state police stood idly by. It was common to see car bumper stickers declaring sentiments such as that communism is treason, behind communism stands – the Jew. In July 1950, Robeson's passport was confiscated, and he was denied motor insurance cover. Undeterred by this threatening atmosphere, Einstein invited Robeson to visit him in October 1952, when they spent an entire afternoon together.

Einstein increased his reputation as a radical left-winger by his open pleas for mercy for Julius and Ethel Rosenberg, who had been convicted of espionage on behalf of the Soviet Union in March 1951. As appeals made their way through the US courts, doubts began to emerge about both the convictions and the severity of the sentence that had been imposed. Einstein wrote to the trial judge, Irving Kaufmann, in December 1952, requesting the commutation of the death sentence. Given that Kaufmann, in his summing up before passing sentence, had blamed the accused for not only accelerating the Russian acquisition of the atomic bomb but also catalysing the Korean War, Einstein cannot have been sanguine about a positive reaction. Indeed, far from showing any sympathy for Einstein's appeal, Kaufmann sent the confidential letter to Hoover to add to the FBI's copious Einstein file. In January 1953, Einstein wrote to the outgoing President Truman, asking for mercy for the Rosenbergs. Both Truman and his successor Eisenhower remained deaf to such appeals, including from Pope Pius XII. Einstein came under pressure to issue similar public appeals

[6] In fact it was Bucky who had sued C. J. Sebo and Coreco Research Corp for infringement of his patent on the Coreco–Bucky medical camera.

for clemency in the infamous 'Doctors' plot' trial then in progress in Moscow. Although he condemned them and similar show trials, he considered that a public statement from him would never reach the relevant Russian authorities. The Rosenbergs[7] were executed in the electric chair on 19 June 1953.

Further controversy followed, related to the case of William Frauenglass, who was a teacher in the New York public school system. Like many other educators, artists and intellectuals, Frauenglass had been called as a witness to a congressional panel and, without prior notice, interrogated about a course he had attended some years earlier. His hearing took place on 24 April 1953. In his acceptance speech on receiving an award on 4 May, Einstein had criticised the effect of the congressional committee inquisitions in suppressing non-conformist opinion. Encouraged by this, Frauenglass had written to him; Einstein replied on 16 May. His letter, slightly amended, was subsequently published in an article on the front page of *The New York Times*. In it, Einstein recommended non-cooperation as practised by Gandhi, even if refusal to testify resulted in jail or economic ruin. The latter proved prophetic for Frauenglass, who had refused to answer questions on whether he was affiliated with any organisations, pleading the Fifth Amendment to the US Constitution, which excuses citizens from answering questions that may incriminate them. Constitutional protection and Einstein's support were unavailing in protecting Frauenglass's livelihood, as he was dismissed from his post on 18 June after twenty-three years of service, along with five other teachers similarly under suspicion of being communists. A further five were suspended. A glance at the surnames of those involved gives a strong impression that were was an admixture of anti-Semitism along with the political intolerance.

The repercussions of the Einstein letter went far beyond the careers of Brooklyn teachers. It catalysed a torrent of hostile editorials from newspapers, including *The New York Times*. Bertrand Russell was moved by this to write a letter to *The New York Times*, which was published in the 26 June edition. The letter delighted Einstein both for its sentiment and its wit:

> In your issue of June 13th . . . You seem to maintain that one should always obey the law, however bad. I cannot think you have realised the implications of your position.
>
> Do you condemn the Christian martyrs who refused to sacrifice to the emperor? Do you condemn John Brown? Nay, more, I am compelled to suppose that you condemn George Washington, and hold that your country ought to return to allegiance to Her Gracious Majesty Queen Elizabeth II. As a loyal Briton, I of course applaud this view, but I fear it may not win much support in your country.

Frauenglass and his family visited Einstein in Mercer Street on 30 June, where their thirteen-year-old son Richard took several photographs. They made notes on what they could remember of the conversation as they drove back to Brooklyn. Richard had told Einstein that his favourite subject was mathematics, but his mother interjected that he was not good at French irregular verbs. Einstein sympathised, remarking that he had not been a good student: and that he could never remember foreign languages, especially Greek, and having to memorise verbs in a particular sequence. Richard's view on the physicist was that he was warm, gentle, had a good sense of humour, liked children, and talked slowly but had a lot to say. Afterwards, Einstein busied himself with writing recommendations for Frauenglass to various potential employers. By February 1954, Frauenglass had turned his hand to psychological counselling and was making enough money to support his family. He wrote to Einstein that he was always in their thoughts and hearts, as well

[7] It subsequently came to light that at least Julius Rosenberg was guilty of being at the centre of an extensive Soviet spy ring. The doctors, mostly Jewish, on trial in the Soviet Union were released after the death of Stalin on 5 March 1953.

as in those of countless others. He could so easily have remained silent but had spoken up for the freedom of the people of America. By doing so, he had had imparted something of himself and his courage.

Several observers remarked on the fact that taking a public stand against McCarthyism seemed to re-energise Einstein from his previous resigned fatalism. Several similar cases to that of Frauenglass absorbed his attention at this time. Einstein and Frauenglass continued to correspond until just before Einstein's death. In his last letter in March 1955, no doubt influenced by the Senate censure vote in December 1954 that marked the beginning of the end of McCarthy's influence, Einstein expressed the hope that slowly the American population was developing a resistance to this sort of demagoguery.

Shortly after receiving Queen Elisabeth's usual New Year's greeting for 1953, Einstein replied, recalling that it was now twenty years since they had last made music together and five years since he had given up playing the violin. He noted that there was now so much wonderful music available from the radio that one's own scratchings were unbearable. He hoped however that her experience was different. He remarked on the strange phenomenon that old age divorced one from the immediacy of the here and now, that one no longer felt intimately involved but was much more of a spectator.

Einstein's seventy-fourth birthday arrived with the usual large number of greetings. To one such he replied that he saw that his cursed birthday had also disturbed his peace. His response to a gift, which sounds like some sort of illuminated screen, perhaps a television set, was:

> In receiving it I compared myself to our venerated Moses. There was the tablet of the Law and the light radiating from it. There is more analogy: Moses' laws were not obeyed and my laws are not believed, both for the reason that they do not correspond very well with human nature.

On 30 March 1953, Einstein wrote a brief appreciation of Pablo Casals for a volume edited by J. M. Corredor:

> The evaluation of Pablo Casals as a great artist does not require my support as there is unanimity among the experts. What really amazes me about him is not only his principled opposition to those oppressing his people but also against the opportunists who are always willing to do a deal with the devil. He has made it clear that the world is more threatened by those who tolerate or abet evil than by the evil-doers themselves.

On 3 April, Einstein addressed his old comrades Maurice Solovine and Conrad Habicht in their guise as members of their Bernese debating forum:

> To the immortal Olympia Academy: In your short active existence you took childlike pleasure in feasting on everything that was clear and clever. Your members created you to make fun of your great, old and pompous sisters. After many years of careful observation, I learned to appreciate fully how much more you did the right things compared to them.
>
> At least all three of us members have proven ourselves to be enduring. Even if we are a bit decrepit, something of your cheerful and invigorating spirit still sheds light on our lonely path of life; because you didn't grow up while we have gone to seed.
>
> You have our loyalty and devotion to our last learned breath –
>
> Your, now only corresponding, Member.

Einstein's former classmate in Aarau, Emil Ott, wrote to him to tell him he had read Seelig's biography. Einstein replied, praising both Seelig personally and his book. Despite his earlier misgivings, Einstein had been captivated by the many memories that the book had evoked. Ott also sent him a brochure he had written in which he had denied the existence of a personal God. Einstein concurred but was astonished that Ott seemed to have come to this conclusion recently, whereas in his experience it normally happened, after considerable internal struggle, to young people. Similarly, in a letter a month later to his old friend Hans Mühsam's cousin Paul, Einstein quoted Voltaire, which he intimated should be spoken in a whisper: 'Just between you [God] and me, you really don't exist.'

Willy Lebach, husband of Einstein's old flame, Grete, wrote to Helen Dukas in May. Although the letter is lost, its contents can be reconstructed from Einstein's reply. He had opened the letter since Dukas had departed to England for several weeks' vacation. Lebach was by now eighty and living alone. He was concerned about his future as his health failed. Einstein counselled him to consider entering a care home, which he said he would certainly have done himself had he not been fortunate enough to be cared for by much younger people. He urged Lebach to contact him if he could help in any way. Lebach outlived Einstein, dying in October 1962 in New York.

Einstein had got to know the remarkable polymath, Manfred Clynes, when he arrived in Princeton in 1952 as a Fulbright Scholar in the psychology of music. Born in Vienna but subsequently living in Melbourne, he was not only a scientist, inventor, and pioneering computer programmer, but also a virtuoso pianist. He had studied at the Juilliard School and performed the Beethoven First Piano Concerto at Tanglewood with Serge Koussevitzky conducting. He spent many evenings at Mercer Street playing in particular Mozart and Schubert, to Einstein's intense pleasure. He presumably also played Bach's *Goldberg Variations*, which was a speciality of his and which he played on a European tour in the second half of 1953. He was one of the first pianists to play in London's newly erected Royal Festival Hall. In May 1953, Einstein wrote him a letter of recommendation in conjunction with this tour, in which he praised him for the freshness of his playing and the fact that his technique was always at the service of the music.

Earlier in 1953, Antonina Vallentin had sent Einstein the manuscript of her biography of him, *The Drama of Albert Einstein*, which was published in the following year. In November 1953, she wrote to him again to try to elicit a response. Einstein replied that he had made it a strict rule since he first came into the public eye never to read material about him personally, thereby avoiding either being puffed up by flattery or depressed by criticism.[8] Nevertheless, he wished her book as much success as if it had had a different, more kosher subject. He was much less positive in response to a letter from Francis Clark, a correspondent from Bournemouth in UK. Clark had written to say that he had found Vallentin's book very interesting, particularly the light it shed on Einstein's philosophy. Einstein in his response ignored Clark's wish to send him a copy of his book on why civilisation was a mistake and why an atomic holocaust might be advantageous. Instead, he responded that Vallentin hardly knew him and the contents of the book were mostly fantasy, sentiments that seem neither true nor fair.

On 30 October 1953, Einstein and Fantova attended a Princeton Symphony Orchestra concert. Einstein was once more very impressed by Harsanyi, the conductor. On 3 November they went to a concert of early music, from the twelfth to the sixteenth century, directed by Safford Cape. Einstein commented that they played as if they had just climbed out of the grave, but it was still very enjoyable. He opined that it was not healthy to hear too much music, or it ceased to become a special occasion, and the musicians become stale because they play so much. Furthermore, there were far too many music competitions.

---

[8] He omitted to mention that he had recently made an exception to this rule in the case of Carl Seelig's book.

On 7 November, Immanuel Velikovsky visited Einstein again. He had moved to Princeton in 1952 and became a regular visitor, regaling his host with his theories of how electromagnetic forces had caused collisions among the inner planets, events reflected in biblical accounts. Einstein showed great restraint in listening to these crackpot ideas, which enjoyed popularity among the general public for some years. In May 1954, he sent Velikovsky a detailed critique of what eventually became a chapter, 'Poles Displaced', in his book *Earth in Upheaval*. In addition to many detailed comments, he concluded 'Catastrophe, yes, Venus, no', by which he indicated disagreement with Velikovsky's central thesis that various biblical catastrophes were caused by a collision of Venus and Earth. He nevertheless thought that Velikovsky had done a useful service in cataloguing various past events that scientists, who had no sense of history, normally ignored.

On 2 December Einstein told Fantova that Max Born had returned to Germany after his retirement in Edinburgh because his pension was so small that he couldn't afford to live in the UK. On 21 December Einstein made a rare trip to New York to attend the wedding of Thomas Bucky, son of his close friend Gustav. According to the groom, he spent some of the reception using the hotel stationary to jot down various equations .

A much more significant event than a visit from Velikovsky also occurred on 7 November 1953. William Borden sent a letter to J. Edgar Hoover. Its contents precipitated what was in some sense the culmination of Einstein's fight against the forces of McCarthyism. Borden wrote that 'More probably than not, J. Robert Oppenheimer is an agent of the Soviet Union.'

Chapter 32

Oppenheimer, Russell–Einstein manifesto; Last things (1954–1955)

It is not difficult to see why J. Robert Oppenheimer would come to the attention of those who were convinced of the existence of a global communist conspiracy to undermine the US. He had never made a secret of his left-wing convictions. He had moved in communist circles; both his wife and his brother had been members of the American Communist Party and it is possible that he was surreptitiously a member himself. He also had a talent for making important enemies. There was tension between him and Edward Teller during the Manhattan Project when Teller essentially withdrew to work on the hydrogen bomb design, which Oppenheimer opposed. This tension became more marked after the war, when Oppenheimer continued to speak against the development of a hydrogen bomb in the various influential advisory committees of which he was a member. Of the others whom he antagonised, the most important was Lewis Strauss, first a member and from July 1953, Chair, of the Atomic Energy Commission (AEC). As both a religious Jew and a right-wing Republican, Strauss was poles apart from Oppenheimer and a variety of frictional interactions over several years increased Strauss' antagonism. As soon as he became chair of the AEC, Strauss began to agitate with President Eisenhower to remove Oppenheimer from involvement with classified nuclear activities. In December 1953, Eisenhower ordered that Oppenheimer be isolated from government activity and on 21 December Strauss called Oppenheimer into his office to tell him that his security clearance had been suspended.

On Christmas Eve, Oppenheimer received a letter detailing the charges from General Kenneth Nichols, head of administration at the AEC, who loathed Oppenheimer and had been present at his interview with Strauss. There were two main groups of charges: the first was related to Oppenheimer's communist links, the second that he had materially slowed the development of the hydrogen bomb, even after its approval by President Truman. He was given thirty days to respond to the charges. Also on Christmas Eve, two FBI agents arrived at Oppenheimer's home in Princeton to seize his classified papers.

After Christmas, Oppenheimer instructed a senior counsel, Robert Garrison, to defend him. Garrison advised Oppenheimer to apply for an extension to the thirty-day deadline, which Garrison achieved by travelling to Washington on 18 January 1954. The news of what most scientists saw as Oppenheimer's persecution spread rapidly among the scientific community. That doyen of scientists in government, Vannevar Bush, long since shorn of influence even in the Truman administration, let alone that of Eisenhower, angrily confronted Strauss. Bush told him that what he was doing was a great injustice, which would certainly redound on Strauss himself. Equally angry, Strauss rejoined that he would not be blackmailed by such suggestions.

Einstein's reaction to the news about Oppenheimer was supportive but also incredulous. He thought Oppenheimer should simply go to Washington and resign from all his positions requiring security clearance, which is certainly what Einstein would have done. He realised however that

Einstein. Brian Foster, Oxford University Press. © Brian Foster (2026). DOI: 10.1093/oso/9780198794875.003.0032

Oppenheimer loved to feel that he was at the centre of power. In a letter to Carl Seelig in the summer of 1954, Einstein remarked that, although he had the highest regard for both Oppenheimer's intellect and the breadth of his knowledge, he had never been particularly close to him, probably because their scientific views were diametrically opposed. He considered Oppenheimer's reaction to the calumnies thrown at him to be far too obsequious; there were limits to what one should put up with in the name of patriotism.

Oppenheimer however did not agree with Einstein. He determined to clear his name; as was his right, he requested a review by a panel of the Personnel Security Board of the AEC. The review, chaired by Gordon Gray, began on 12 April and concluded on 6 May. The day after proceedings began, *The New York Times* caused a sensation by publishing both the letter from Nichols summarising the charges, and Oppenheimer's response. These had been supplied to the paper by Garrison. On the weekend before, Abraham Pais, now a faculty member of the IAS, had been telephoned by the Associated Press informing him of the breaking Oppenheimer story and asking him to elicit a statement from Einstein. Realising that the alternative was a press scrum outside Mercer Street, Pais went to visit Einstein. After Pais was admitted, Einstein appeared at the top of the stairs in a dressing gown. Despite initially saying with a laugh that Oppenheimer should just resign, Einstein agreed to a brief supportive statement, which he then read over the telephone to the Director of the Associated Press. However, the anticipated press scrum still took place the following lunchtime: walking back as usual from the IAS, Einstein was only saved by Helen Dukas charging down the drive to rescue him from the reporters.

It was no surprise that the board concluded on 23 May by a 2–1 vote that Oppenheimer's security clearance should be withdrawn. Strauss surreptitiously steered the course of the hearing from afar; the testimony of Edward Teller that he was uncomfortable with Oppenheimer being in possession of sensitive information, and of General Leslie Groves that, given the current rules, he would not give Oppenheimer a security clearance, did not help. Garrison appealed to the full AEC but, given that this was chaired by Strauss, the outcome was predictable. Strauss tried to minimise any doubt by an intensive programme of lobbying, including buying his fellow commissioners expensive lunches and hinting at lucrative contracts. Indeed, in one case he essentially bribed a commissioner by transferring his own lucrative legal business to him after the vote. On 28 June, the AEC vote of 4–1 confirming the verdict of the Gray panel was announced to the press. The one vote against was from the only scientist on the board, Henry Smyth, who, much to the irritation of Strauss, wrote a dissenting report in which he maintained that no charges had been proven. The verdict ended Oppenheimer's role in US government activities. Many years later, the verdict would be quashed due to a host of irregularities in the proceedings, caused substantially by Strauss's interference and Gray's supine chairmanship in failing to rein in the prosecutor's flouting of legal protocols.

Oppenheimer did remain as Director of the IAS. The overwhelming majority of scientists, the one notable exception being Teller, deprecated the AEC verdict. Forty leading scientists, including Einstein, expressed opposition to the inquisitorial nature of the current climate in the US and supporting Oppenheimer personally in the May edition of *Bulletin of the Atomic Scientists*. Strauss was an influential member of the IAS Board of Trustees. He was absent, perhaps diplomatically, when the chair, Robert Maas, declared on 13 April that the board had full confidence in Oppenheimer. Smelling problems ahead, however, Einstein decided on pre-emptive action. He wrote to his old friend Herbert Lehman, ex-Governor of New York and now junior United States Senator from New York, who was also an IAS trustee. Writing on behalf of his colleagues, Einstein explicitly warned Lehman that they were worried that the well-known antagonism between Strauss and Oppenheimer together with Strauss's influence among the trustees might lead to an attempt to force Oppenheimer out of his directorship. In order to prevent a surprise being sprung on the board, Einstein conveyed his and his colleague's feelings about Oppenheimer's performance. Einstein considered Oppenheimer to be by far the most capable director the IAS had ever had. His breadth of knowledge was a particular advantage. If he were to be dismissed, it would cause indignation among the scientific community and grave damage to the reputation of the IAS because of the enormous respect that Oppenheimer engendered among his colleagues. He concluded by expressing confidence that Lehman would never be party to any grubby political compromise in such a matter. The challenge never came; a political operator of Strauss's experience would have counted noses on the board in private conversations before making any move against Oppenheimer. At least in part because of Einstein's initiative, the votes to remove Oppenheimer were not there. Realising the situation and perhaps trying to make a peace gesture, it was Strauss who in October proposed the renewal of Oppenheimer's contract, which passed the Board unanimously.

At the New Year of 1954, Queen Elisabeth had sent her usual greetings telegram. Einstein replied quickly. His letter was reflective and nostalgic, remarking on how it was their joint destiny to be well known but isolated, and on the dying down of passions with old age. He concluded that there is little that one could accomplish through one's own efforts and that be began to appreciate the role of the understanding but also comforting observer.

No doubt still influenced by the Christmas season, Einstein replied on 5 January 1954 to a laudatory letter:

> I feel somewhat unhappy reading your kind letter of December 17th. Through no fault of mine (I hope) I have become in the minds of some people a legendary figure, somewhat like Santa Claus, who has accomplished more or less impossible things and is able to perform more of them, the sober truth is quite different . . . The only thing I feel sure of is that the actions and even the judgements of men are more determined by their wishes and instincts than by rational thought – a fact to which we have to bow in resignation.

On 12 January, Einstein wrote to the Nobel Committee to second Max von Laue's nomination of Walther Bothe. In his citation, Einstein wrote that Bothe's major contribution was the experimental verification that energy and momentum were conserved in the interactions of individual particles, rather than only on average over many interactions. Although Einstein never let his distaste for quantum mechanics stand in the way of his recognition of individual merit, Bothe's results, limiting the realm in which the statistical nature of quantum mechanics held sway, would have been congenial to him. His support was no doubt instrumental in Bothe being awarded the 1954 Nobel Prize for Physics.

Another Nobel Laureate in Princeton from January to April 1954 was Wolfgang Pauli. He and Einstein renewed their old arguments about both quantum mechanics and field theory. In several long letters to Max Born, Pauli described some of their conversations about a manuscript that Born had sent to Einstein. Pauli reported that Einstein had told him that he did not believe that determinism was fundamental, but rather a criterion for evaluating the correctness of a theory. More important was whether a theory was 'realistic', by which he meant his philosophical prejudice that a state could be described completely independently of how it was observed. Pauli spelled out in great detail Einstein's views on how the solutions of microscopic particles described in quantum mechanics should be interpreted for the macroscopic objects that they constituted. Pauli of course disagreed with these views, even more fundamentally with Einstein's insistence that there was an objective reality independent of the process of measurement. That Pauli still relished his discussions with Einstein on a wide variety of issues in physics is apparent from these letters. This visit was the last time that Pauli and Einstein met.[1]

Einstein's old chamber music partner, renowned pianist Harriet Cohen, sent him a copy of a book, probably the recent reprint of *Music's Handmaid*, that she had written on the history and interpretation of music. She herself often arranged music; a copy of her arrangement for piano of J. S. Bach's cantata *Mein Gott, wie lang, ach lang*, BWV 155, published in 1931 by Oxford University Press, is in Einstein's music library with the dedication: 'To Albert Einstein with love from Harriet Cohen.' In his letter of thanks to Cohen in February, he wrote that he found what she said about the right way of playing Mozart especially convincing and that many famous musicians could learn from it.

Einstein's seventy-fifth birthday arrived, as he wrote to his well-wisher Queen Elisabeth, without any particular effort on his part. Despite being a legend in his own lifetime, he wrote with tongue in cheek, he could still cherish the genuine greetings. He told the Queen that he had become an *enfant terrible* in his adopted country because of his inability to keep his mouth shut but that it behoved the old to speak up since they, unlike the young, had nothing more to lose. He lamented that swords still had no inclination to beat themselves into ploughshares but this time the Europeans hesitated to allow themselves to be dragged into oblivion. He hoped that this would continue. In contrast to previous communications from German organisations, he thanked the Berlin Academy of Sciences for their birthday greetings, which he characterised as a letter that originated from a benevolent and independent colleague who has taken an active part in the developments of the last few decades. He similarly thanked the Mayor of Ulm for his greetings and a presentation book. Franz Rupp, long-time accompanist of Marian Anderson, sent his recording of Schubert's *Trout Quintet*, which Einstein was able to enjoy via the new radio-gramophone presented to him by his colleagues in the IAS.[2] Eugene Anderman sent him a recording of Vivaldi pieces, which Einstein looked forward to listening to after he had overcome the horrors of his birthday. Apparently, Einstein used his new gramophone to play repeatedly a recording of Beethoven's *Missa Solemnis*.

Of old friends, his Aarau schoolfriend Emil Ott congratulated him, as did the daughter of Susanne Markwalder, his landlady when he was a student in Zurich. Leon Watters sent greetings and fruit from his garden in Florida. His former secretary and lover Betty Neumann congratulated him. Max and Hedi Born sent best wishes; both remarked on the important role Einstein had played in their lives and Max said that there was no one for whom he had a more profound admiration. Boris Schwarz arranged to visit Mercer Street with his family; if Einstein could no longer play his beloved Bach 'Double' Concerto with Schwarz, he could still listen to a master play with great enjoyment. The Schwarz family visited on 16 April.

An unexpected gift was a parrot, which arrived through the post! Not surprisingly somewhat disorientated by this journey, the parrot eventually recovered and was named Bibo. Einstein became

---

[1] Pauli died of pancreatic cancer after a very short illness in Zurich on 15 December 1958. Having had a fascination for his entire career about why the value of the fine structure constant, which defines the strength of the electromagnetic interaction, was almost exactly the inverse of the integer 137, he died in hospital room 137.

[2] This replaced the previous record player owned by Einstein, a gift from colleagues for his seventieth birthday. Apparently, Einstein often played along with records by Telefunken in which the solo part was missing. If this is true, he must have had a record player before the one presented in 1949, as he had given up playing the violin by then.

fond of the bird, apparently telling it jokes that it did not appreciate. In return it gave him an infection.

Einstein's letters to his old friends in Israel, the Mühsams, were perennially overdue, but always as a compensation long and chatty. At the end of March, he thanked them for their birthday wishes and said that he had not been ill for a considerable time but the price he had to pay was a diet with no meat, fish, or fat (indeed, he subsisted substantially on vegetarian packet soups that Carl Seelig sent regularly from Zurich). In musing on his current state, he thought that he had become a very religious unbeliever. He noted that he could be as isolated as they were, except for the curse of the post, which always held him in thrall. His birthday had been a catastrophic cascade of letters under which he had almost drowned. Still, he mused, all things pass and only the old gypsy, for the moment, remained.

In a similarly valedictory mood, Einstein replied to a query from his closest friend, Michele Besso, who still worried away at details of general relativity almost forty years after they had sweated together over the perihelion advance of Mercury. After two pages of complex tensor manipulation, Einstein concluded: 'If only everything was as clear. But in my generalised field theory the situation is so complex that it isn't even clear to me whether I should believe in its truth or not. Many are going to break their heads over that after I have finally kicked the bucket.'

On 11 May, Einstein wrote to congratulate his son Hans Albert on his fiftieth birthday. He admitted at the outset that Hans Albert's wife had reminded him of the occasion, but he was glad to avail himself of it to say how much he appreciated having a son who shared his most important characteristic: the ability to ignore the short term and devote his life to a long-term goal. He reviewed Hans Albert's career with satisfaction. Unusually, he reminded Hans Albert of various episodes from his childhood, including his use of the invented word 'Voio-Voio', which originally meant 'curtain' but came to stand for anything that seemed big and impressive but had little substance, such as smoke from the fireplace or a flood of meaningless words.

An example of the lengths to which Einstein would exert himself, in particular for friends, is given by his efforts to find help for Rolf Ehrmann, son of his long-time physician, Rudolf. Rolf had some form of physical handicap that had affected his academic work. When Rolf was seven, a schoolboy in Berlin, he was in the habit of seeking supplements to the sparse food provided by his summer school by walking over to nearby Caputh; Einstein always obliged. In 1949, Einstein had written a recommendation for him, presumably to enter law school. All did not go according to plan; in1952, Einstein wrote to the President of the New York Law School to plead for Rolf's readmittance after he had been expelled for poor performance. This was presumably unsuccessful, as in April 1954, Einstein wrote again, this time to Judge Leibowitz, pleading Rolf's case, which was before the court on charges that included illegally practising law. Einstein's plea may have had a mitigating effect on the sentence, but Rolf was convicted, nonetheless. Einstein then exerted himself further by writing in June 1954 to New York societies for rehabilitating ex-prisoners. He asked them to help find Rolf a position, saying that Rolf had experience in patent law and as a translator. One society's reply emphasised the difficult of placing someone who was not only a convict but also handicapped. However, they must have been successful as, by 1979, Rolf was practising, presumably legally, as a patent attorney in the northern suburbs of Chicago.

On 19 June, Einstein's grandson Bernhard called in at Mercer Street on his way back home from his studies in Zurich. Einstein remarked to Johanna Fantova that he wouldn't set the world on fire but would soon get married and was a nicer man than either his father or grandfather. On 14 July, Einstein received a manuscript from Immanuel Velikovsky, presumably for *Earth in Upheaval*, which was published in 1955. Einstein's view on the manuscript was that it was very learned, but crazy. Presumably he conveyed this view to Velikovsky, who arrived at Mercer Street on 21 July and tried until midnight to convince Einstein of his arguments. A character almost as colourful as Velikovsky arrived a week later. Einstein's old friend from Berlin days, Dr Janos Plesch, had left the UK first for Switzerland and then, on the death of his wife in 1954, came to live in California with his sister. In the interim, he had published a scholarly work on Rembrandt. Einstein enjoyed his old friend's company, writing to Fantova that 'Plesch came yesterday, this time with his sister, who he has swapped for his dead wife. He is still the master of the grandiose lie and indefatigable enterprise despite his battered heart. He is a great trickster but it suits him splendidly.' Plesch visited again in December. Einstein characterised him as being as bubbly as ever and as good-hearted. He thought that he was condemned to roam around the world as his children could not stand to have such a volcano in their vicinity.

In August 1954, Einstein busied himself in sending out invitations to a meal in the Nassau Inn in aid of the Hebrew University. This took place in the context of a two-day conference to discuss the site of the university. After the foundation of the state of Israel, Jerusalem was partitioned between the new state and Jordan. The original Mount Scopus campus of the Hebrew University, towering over the city, had been of major strategic value in the 1948 war. Afterwards, it fell in the Jordanian sector, which meant that the Hebrew University was essentially evicted and had to find temporary accommodation elsewhere. The conference took place from 19–20 September and resulted in pledges of more than $2.5M towards a goal of $10M to cover a building programme at a new site. Einstein gave a rare speech at the conclusion of proceedings, in which he decried the shallow materialistic tendency of the age for the insufficient support and priority given to scholarship. Dr Wise, President of the American Friends of the university, also announced at this meeting that an additional $1M had been raised to establish the Albert Einstein School of Physics at the Hebrew University.

Perhaps as a reaction to his exertions on behalf of his beloved Hebrew University, Einstein became quite ill in the next few months. His doctors diagnosed anaemia, which limited his movements and kept him housebound as he felt unsteady on his feet. In a letter to his old friends in Israel, the Mühsams, he attributed this to a viral infection that was reducing his blood-oxygen levels; it was being treated with cortisone. He complained that the drug was infernal stuff and tended to lead to depression. Nevertheless, he seemed to quite welcome the simplification of his life caused by his illness and limited himself to answering only the most urgent and important communications while carrying on with his physics work. Only by the end of the year did he begin to feel better. He was at least able to enjoy the company of Hans Albert, who visited Mercer Street in October. Keeping up his feuds to the end, Einstein remarked to Hannah Fantova that Hans Albert was very nice when he came without his wife. In December, he wrote to Hans Albert to thank him for a present he had sent. His grandson Bernhard had married Aude, whom he had met in Germany, a few months previously. He had written a 'begging letter' to his grandfather, presumably asking for funds as he established a family home. Einstein was delighted by what he characterised as this demonstration of trust. He thought that Aude was already having a positive effect on Bernhard through the strength of her character. He concluded by commenting on the vagaries of God's own country as he sarcastically called the US, but also with a conviction that it would eventually get back to normal.

Other visitors whom Einstein felt well enough to receive included Werner Heisenberg, who arrived on 30 October. Einstein had never warmed to Heisenberg and certainly held it against him that he had remained in Germany and worked on the German atomic bomb project. He admitted that he was a great physicist but did not consider him a pleasant man. On 3 November Niels Bohr visited. He exhausted Einstein by talking non-stop and being typically elliptical and obscure. He did make it clear however that he considered Einstein's unified theory to be nonsensical. Einstein presumably discussed with Bohr the draft that he worked on in December for Appendix II of the new edition of *The Meaning of Relativity*, which contained his final exposition of his theory. One great physicist who left the scene on 28 November was Enrico Fermi, who died at the age of 53 of cancer, probably caused by his work on the first atomic reactor. Einstein wrote an appreciative obituary for the Italian newspaper *Corriere della Sera*.

When 1 January 1955 brought the now-traditional greetings telegram from Queen Elisabeth, Einstein hastened to reply immediately. He bewailed the shortness of political memory, commenting that yesterday the Nuremburg trials took place, today Germany must be rearmed with the utmost speed. He blamed the behaviour of contemporary Europe, including England, on a new sort of capitalist colonialism. He remarked that as he grew older, he came to appreciate Lichtenberg more and more; he more than anyone had an uncanny sense of the future.

Marian Anderson stayed with Einstein shortly after her historic debut on 7 January at the Metropolitan Opera, the first African-American to perform there. On the invitation of the general manager, Rudolf (subsequently Sir Rudolf) Bing, she sang the part of Ulrica in Giuseppe Verdi's *Un ballo in Maschera* under the baton of Dimitri Mitropoulos. Later that month, Princeton's Friendship Club brought her back for another concert at McCarter Hall, after which she stayed in Mercer Street. She described the occasion in her autobiography:

> In January 1955 I was scheduled to sing in Princeton, and again there was an invitation to feel free to use the Einstein home. Dr. Einstein was not downstairs when we arrived, and I could understand that indeed, I was glad that he was not putting himself out for us. Margot Einstein led me upstairs. I had not been in my room more than five minutes when there was a gentle knock on the door. I opened it, and there stood Dr. Einstein. He had come to say, 'How do you do' and 'Glad to have you back.' This time he did not go to the concert. Afterward, when I left his home, he came out of his room to say good-by. This, though I did not know it, was really good-by.

The year 1955 marked the fiftieth anniversary of Einstein's *annus mirabilis* and thus of both the light quantum and of special relativity. On 16 January, Max von Laue wrote to Einstein to invite him to a celebratory conference in Berlin on 18–19 March. Von Laue would not have been surprised when Einstein replied a few weeks later that he would not be travelling to Europe to participate and that indeed he had decided not to participate in any similar events, which he knew were being planned elsewhere. He concluded his letter by musing that they were much further away from a deep insight into elementary processes than most of their contemporaries (excluding von Laue himself) thought, so that noisy celebrations weren't appropriate. Nevertheless, on 31 January, Einstein received a letter from the Presidents of the West German and East German Physical Societies, Max von Laue and Gustav Herz, respectively, congratulating him on the anniversary.

Carl Seelig wrote to Einstein around the beginning of February, asking for clarification of the genesis of special relativity. Einstein replied on 19 February that indeed the work of Hendrik Lorentz and Henri Poincaré meant that its discovery was ripe but that nonetheless his own work was independent since, although he was aware of Lorentz's 1895 publication, he did not know about his formulation of the Lorentz transformation nor was he at all familiar with Poincaré's work in this area. He said that the realisation that every theory of physics must be Lorentz invariant was particularly important for him at the time, as he had realised that Maxwell's equations took no account of the microscopic structure of light and therefore could not describe all physical situations. He closed this last letter to Seelig by thanking him for the news he had sent on Tete's condition and for the report on a lecture given by Pablo Casals in Zurich that Seelig had attended. Einstein thought that Casals was someone really too good for wretched humanity.

Queen Elisabeth wrote on 19 February. She had heard a ninety-minute broadcast on French radio in celebration of fifty years of relativity, in which a great deal had been said about Einstein as a person, a scholar, and a genius. Louis de Broglie had been one of the speakers. There were even anecdotes about her and Einstein, to which she could have contributed had she been asked. It all made her think fondly of her far-off friend. She lamented the stupidity of the world and in particular the senselessness of rearming Germany. Einstein's reply on 11 March noted with pleasure their agreement on the major issues of the day. Her remarks on the radio broadcast however made him uneasy. He thought himself an involuntary swindler whose protestations only made things worse. In answer to her question on who Georg Lichtenberg was, he had sent her a book about him, an eccentric eighteenth-century professor of physics at Göttingen. His final words to her were typical; a simple explanation of a Lichtenberg story about the relative merits of the sun and the moon in which he painlessly inserted a little science.

Many congratulated Einstein on his seventy-sixth birthday. His old flame Betty Neumann proposed to visit but Einstein refused because of his continued weakness, instead inviting her to write. Von Laue had also written, and his letter contained a variety of biographical questions, as he had agreed to contribute to a volume of the lives of great scientists. Einstein deeply regretted that von Laue had involved himself in such an activity but nevertheless patiently answered various questions about work that particularly influenced him, his activities in 1932–1933 and his role in the development of the atomic bomb. Here he was at pains to emphasise that all he did was to sign a letter provided by Leo Szilard and, had either of them realised that their fears that Hitler would beat the Allies to the bomb were groundless, neither would ever have opened that Pandora's Box .

Perhaps a little hypocritically, Einstein was himself that March engaged in a similar exercise to that for which he had criticised von Laue. With fortunate timing, the last full month of his life coincided with a request from ETH for a contribution to their centenary celebrations. His memory no doubt helped by his friend Seelig's literary endeavours, he wrote a seven-page autobiography, starting with his escape from German militarism at the age of sixteen, crossing the Alps to join his parents in Milan. Having reached the completion of general relativity, he broke off the story with a final paragraph that can serve as an epitaph on his life in physics:

> Forty years have passed since the completion of the theory of gravitation. These have been devoted almost completely to the search for a basis for all of physics by incorporating the theory of the gravitational field within a more general field theory. Many are working towards the same goal. I myself have in the past postulated several promising directions. Indeed, the last decade has led to a theory that I consider natural and promising. However, the fact that I am unconvinced that this theory is physically valuable derives from the insuperable mathematical difficulty inherent in solving the required non-linear field theory. In addition, it seems very doubtful that a field theory can account for the atomic structure of matter and radiation and quantum phenomena. Most physicists would answer 'No' with conviction, as they consider the quantum problem is in principle solved *in another way*. However it may turn out, we are left with Lessing's reassuring words, the pursuit of the truth is more pleasurable than its secure possession.

Einstein expressed similar sentiments in his last letter to his old friend Maurice Solovine, of whom, in 1953, he had written: 'It seems to me that you are not only my faithful translator but also my only really attentive reader.' On 27 February he wrote: 'Anyway, I have found a notable improvement in the generality of the theory of gravitational fields (non-symmetric field theory). But the thus so much simplified equations because of their mathematical difficulty are not yet susceptible to experimental proof.'

On 14 March, the sister Bice and son Vero of his closest friend, Michele Besso, wrote to Einstein. Ten days previously, Michele had had a stroke which partly paralysed him and robbed him of his speech. Although the family had hoped that he would recover, on 13 March his condition had worsened, and it was clear that he was dying. They reported that he was lying in bed reading Seelig's book, on whose cover was a picture of Einstein, which he had lingered over. They told Einstein that Besso had greatly appreciated Einstein's last letter and had charged them with greeting Einstein in his name. Two days later they wrote again to inform Einstein that Michele Besso had faded peacefully away at 3 a.m. on 15 March.

Einstein's reply to Vero and Bice's letter announcing his old friend's demise was consonant with the deep bonds that they had shared. He wrote:

> His end was as harmonious as was his life [. . .]. The ability to live such a harmonious life is rarely combined with such an unusually sharp intelligence as he possessed. However, what I most admired about him as a person was that he succeeded in living for many years not only in peace but in lasting unison with a woman – an endeavour in which I twice failed quite disgracefully. In our student times in Zurich, our friendship was founded on regular musical evenings. He, the older and more experienced, gave me much stimulation. The breadth of his interests seemed to have no limits. The strongest however seemed to be in critical philosophy.
>
> Later the Patent Office brought us together. Our discussions on our shared way home were of unequalled charm – it was as if grubby everyday concerns didn't exist at all. It couldn't be the same later when we communicated by letter. The pen could never keep up with the versatility of his thought, so that his reader could never fill in the missing leaps.
>
> Now he has said goodbye to this strange world a little before me. This doesn't matter. For us believing physicists, the difference between past, present and future is but an illusion, albeit a stubborn one.[3]

On 10 March Einstein listened to Mozart's 'Jupiter' Symphony on his radio. He remarked that it was his best work. Of his operas, Einstein thought *Don Giovanni*, *The Marriage of Figaro*, and *Entführung aus dem Serail* were wonderful. However, the *Zauberflöte* didn't please him as much. Among modern operas, he expressed the view that only Mussorgsky's *Boris Gudonov* was good.

In April 1955, tensions between Israel and the Arab states had mounted dangerously. On 4 April Einstein wrote to the Israeli Consul in New York to propose making an address evaluating the political situation and the attitude of the Western powers to Israel. He thought it could be useful in influencing public opinion. On 11 April, Ambassador Abba Eban and Consul Reuven Dafni visited him at Mercer Street and discussed the content of his address, to be given on Israeli Independence Day. Einstein got as far as beginning to write the address; his hand-written draft breaks off mid-sentence, marking the point when he was incapacitated by his final illness.

In the same month, US–China relations had reached a critical stage, to the extent that the US threatened to launch a nuclear attack on China unless it stopped shelling Nationalist positions on the islands of Quemoy and the Matsu Islands. Deeply concerned, Einstein discussed the situation with Szilard, who drew up a letter for Einstein to forward to Prime Minister Nehru of India. In a letter on 6 April, Einstein asked Nehru at an appropriate point to forward Szilard's letter to the

[3] This sentence is often quoted, probably because of its Delphic and evocative nature. In fact, Einstein had explained precisely what he meant by this in a letter to Besso two years earlier, in which he characterised the 'arrow of time' as being purely a thermodynamic effect, marked in modern parlance by the increase in average entropy. In theories such as special and general relativity and indeed in the quantum mechanics known at that time, all elementary processes were symmetrical to the reversal of the direction of time. Today, we know that in fact the universe could only have formed and evolved to what is seen today because this symmetry, time reversal, is broken in some elementary processes, albeit at a tiny level.

Chinese government. He closed his letter by expressing his appreciation for all Nehru's efforts to promote international understanding.

On 14 February 1955, Bertrand Russell wrote a letter to Einstein that would dominate the short time he had remaining. In it, Russell asked Einstein's advice on getting together a very small number of the most eminent men in science with as wide a range of political views as possible to warn in particular the US and Soviet governments of the disastrous nature of warfare with the H-bomb, which he thought, but they did not, could lead to the extinction of life on Earth. It was Einstein's old friend, Max Born, who had suggested this initiative in a letter to Russell in January 1955, following Russell's broadcast on the dangers of atomic warfare, *Man's Peril*, on British radio in December 1954. A transcript was also published in the BBC's magazine, *The Listener*, on 30 December 1954. Born was stimulated by his recent Nobel Prize award to try to use the prize funds for the second of Nobel's two aims in founding the prize. The first, promoting the progress of the human race, he felt was no longer within his powers, given that his award was for his work on quantum mechanics almost 30 years earlier. The second, the promotion of peace, prompted his letter to Russell.

Russell had also been in discussion with Frédéric Joliot-Curie, who was a communist, and found that they could agree on the general thrust of a manifesto. He did not however agree with Joliot-Curie that the best strategy was to organise a conference of leading scientists; he was concerned that the logistical and organisational problems of a large gathering would dilute its effectiveness. He also worried about more than just atomic bombs; he cautioned of the growing dangers of biological warfare. He suggested commissions in several neutral nations consisting of:

> . . . a nuclear physicist, a bacteriologist, a geneticist, an authority on air warfare, a man with experience of international relations derived from work in UNO, and a Chairman who should not be a specialist but a man of wide culture. I should like their report to be published and presented to all the governments of the world who should be invited to express their opinion on it. I should hope that in this way the impossibility of modern war might come to be generally acknowledged.

Einstein replied immediately, saying he agreed with every word of Russell's letter. He suggested somewhat widening the number of eminent people to be involved to twelve and that it might be possible to include partisans such as Joliot-Curie provided that they were balanced with those of the opposite persuasion. He particularly suggested Bohr, as a citizen of a neutral country, and assumed that Russell would wish him to circulate his letter to those who might be sympathetic, although that was difficult since the US was swept by a political plague, to which researchers were in no way immune. Soviet signatories were essential; Einstein suggested that Leopold Infeld, by then a professor in Warsaw, might be a useful intermediary. He suggested Whitehead and Urey as signatories from the US but emphasised the importance of involving 'neutral' American figures .

The exchange of letters between Russell and Einstein occurred quite rapidly. Russell's reply indicated pleasure that the two were on the same wavelength. He approved Einstein's suggestion to involve Bohr, whom he knew slightly and liked, but admitted he didn't know the 'Whitehead' to whom Einstein had referred in his letter. In fact, Einstein meant Alfred North Whitehead, Russell's co-author of the magisterial *Principia Mathematica*. As Einstein noted in his reply on 4 March, he was grateful to Russell for pointing out to him in this very diplomatic way that Whitehead had died some years previously. Russell enclosed a draft memorandum that he intended to submit to the Indian government, which pointed out India's unique position as a member of the British Commonwealth but also non-aligned and friendly with China and suggested it take an initiative along the lines that had been contained in his original letter to Einstein.

On 2 March, Einstein wrote to Bohr, enclosing a copy of Russell's original letter and inviting Bohr to take a lead in concert with Russell if he approved of the project. He told Bohr that he should contact Russell directly rather than reply to him. Bohr did so on 23 March. However, he was not in sympathy with the initiative, telling Russell:

> It is not clear to me whether a joint declaration from a group of scientists would have the desired effect. The perils are now common knowledge and many of the most competent scientists with special access to information are thoroughly studying the dangers of radioactive effects so generally talked about. My question is therefore not only whether a group like that you have in mind could agree on an approach on sufficiently constructive lines, but also what relevant new information its declaration could contain.

In his letter to Russell of 4 March, Einstein asked to be told of Bohr's response. He also suggested that Albert Schweitzer should be added to the list of signatories as he had enormous moral authority world-wide. Einstein gave Russell carte blanche to organise the initiative with his full support and noted that he was awaiting Russell's orders.

Russell's final letter was written on 5 April. He preferred to limit the signatories to scientists, although he seems to have thought that Einstein had suggested Arnold Toynbee, the British historian, rather than Schweitzer. He enclosed a draft of what became known as the 'Russell-Einstein manifesto'. At the press conference that launched it on 9 July, Russell made some introductory remarks:

> The accompanying statement, which has been signed by some of the most eminent scientific authorities in different parts of the world, deals with the perils of a nuclear war. It makes it clear that neither side can hope for victory in such a war and that there is a very real danger of the extermination of the human race by dust and rain from radioactive clouds . . . The only hope for mankind is the avoidance of war. To call for a way of thinking which shall make such avoidance possible is the purpose of this statement. The first move came as a collaboration between Einstein and myself. . . .

From the long text of the manifesto, which followed closely that of Russell's December BBC broadcast, it is worthwhile to extract a single sentence that has the genuine ring of Einsteinian sentiment:

> We appeal, as human beings to human beings: Remember your humanity, and forget the rest.

Einstein replied to Russell's letter on 11 April, in which he wrote that he was gladly willing to sign his excellent statement. This letter and his signature on the manifesto itself were the last documents he signed.

Helen Dukas kept a diary that traces the course of Einstein's final illness. On 12 April, Einstein complained of a pain in his groin which he connected with his long-standing aneurysm. He forbade Dukas to call a doctor, but she contacted Margot anyway, who was herself hospitalised with sciatica, and asked her to alert Dr Dean. That evening Janos Plesch came to dinner and stayed talking to Einstein until 10 p.m. Plesch left an account of that last conversation, which was wide-ranging.

On 13 April, Einstein reported that the pain was worse when he was lying down. He didn't go to the Institute as usual, as he was expecting the Israeli Consul at lunchtime for a further discussion of his broadcast; Plesch reported this was to be on television and Einstein was very nervous about it. Plesch also returned around lunchtime with his nephew Eugenio, an artist for whom Einstein had agreed to sit for a portrait. They and the Consul left at around 1 p.m. as Einstein was very tired. He ate little and then retired to bed. Dukas cancelled his meeting that afternoon with his assistant Bruria Kaufman, with whom in the previous year he had published his final scientific paper, 'Algebraic Properties of the Field in the Relativistic Theory of the Asymmetric Field' – his last small step in print towards a unified field theory. At 4:30 p.m., Dukas heard Einstein being violently sick in the bathroom; he complained of pains everywhere. She called Dr Dean immediately; he arrived twenty minutes later. He diagnosed bleeding from the aneurysm and called in two other doctors as well as contacting Einstein's old friend Dr Bucky, who arrived from New York at 9:30 p.m., probably driven by his son Thomas. After administering medication, the doctors left at 11 p.m. and Einstein spent a reasonably comfortable night.

On the morning of 14 April the pains returned. Thomas Bucky was despatched to bring a surgeon from New York. Professor Ehrmann, Einstein's long-standing family doctor, arrived and disagreed with Dean's diagnosis. Einstein meanwhile refused any pain-killing injections but after further treatment by Dean, again spent a quite comfortable night. On 15 April, the patient deteriorated further and was taken to hospital in an ambulance. By that evening, having been placed on a glucose drip, he felt better. Nevertheless, Margot called Hans Albert in Berkeley, who arranged to fly out immediately and arrived the following morning. Einstein was very pleased to see him, and they talked for a long time, mostly about scientific matters. As well as Hans Albert, Johanna Fantova was at his bedside, as was Otto Nathan. Although Hans Albert tried to persuade him to have an operation in New York, Einstein indignantly refused, saying: 'Such tastelessness, I will go when I want to!' In any case the surgeon admitted it had little chance of success. Einstein began to feel somewhat better, asked for his spectacles and the papers he was working on and the following day scribbled down some equations as he lay in bed. He talked to those who called in to see him, including Hans Albert. On the evening of Sunday 17 April, Dr Dean checked on him at 11 p.m. and found him sleeping peacefully.

Margot described her last meetings with her step-father:

> Twice I was allowed to see him and speak to him for a few hours. I was wheeled in to him in my chair. I did not recognize him at first – so changed was he by the pain and the lack of blood in his face. But his manner was the same. He was glad I was looking a little better, joked with me and faced his own state with complete superiority; he talked with perfect calm, even with slight humour about the doctors, and was waiting for his end as if for an expected 'natural phenomenon.' As fearless as he was in life, so quietly and modestly did he face his death. He left this world without sentimentality and without regret.

Albert Einstein died quietly in his sleep at 1:35 a.m. on Monday 18 April 1955.

The entire world mourned. *The New York Times* edition of 19 April had the news as the largest item on its front page. One of the most famous and evocative tributes was published in *The Washington Post* that day. It was a cartoon by Herb Block showing a number of heavenly bodies, the central one bearing a large plaque with the words 'ALBERT EINSTEIN LIVED HERE'. One of the tributes closest to his heart would have been the Einstein Memorial concert given by the Princeton Symphony Orchestra on 17 December 1955, in the McCarter Theatre in Princeton, scene of so many concerts he had attended. It was conducted by Einstein's friend, Nicholas Harsanyi. The programme noted:

> Albert Einstein, until his death last April, was also an honorary Vice-President and, like Robert Casadesus, evidenced a profound interest in the mission of the orchestra. In playing the Coronation Concerto of Mozart this evening, Robert Casadesus wishes to honour the memory of Einstein, his friend and associate. We are sure that the members of the Princeton community who are here this evening will gratefully share in this noble tribute to the man who lived so simply among us.

The programme was: Concerto Grosso No 8 in G. Minor, Op. 6 by Corelli; Piano Concerto in D, K537, by W. A. Mozart; Sonatina from Cantata No. 106 *Actus Tragicus* by J. S. Bach; Symphony No.

104 in D (London) by J. Haydn. The programme contained the instruction after the Bach Cantata: 'In tribute to the memory of Albert Einstein, please refrain from applause.'

This chronicle of Einstein's life, science, and music ends with a report of the last time that he is known to have partaken in his favourite pastime, chamber music, which took place long after he had notionally given up his beloved violin. Violinist Robert Mann recorded the story:

> In 1952, the Juilliard String Quartet spent an evening with Dr. Albert Einstein in Princeton, following a Sunday afternoon concert at the university. Aware of Dr. Einstein's lifelong devotion to chamber music both as a listener and participant, we offered to play for him at his home. He graciously accepted. A fervent hope inspired a devious plan that necessitated us bringing along an extra viola and the music of Mozart's two viola quintets. He received us in a genial mood and wearing extremely comfortable attire. As we played both Beethoven and Bartok, he listened a room removed (not wishing for a visual distraction) and, while he did not seem to share the general aversion to twentieth-century music that seemed to infect his fellow physicists, it was evident that Dr. Einstein's heart belonged to the earlier music.
>
> After the Bartok we launched our surprise attack. Producing the Mozart score and the extra viola, we said, 'It would give us great joy to make music with you.' He protested that he had not played for years due to a hand injury. With the insensitive exuberance of youth we persisted, and he, the true scientist recognizing an irresistible force versus immovable object contretemps, gently acceded to the inevitable. Our second violinist handed his violin to Dr. Einstein, and picked up the extra viola himself. Dr. Einstein without hesitation chose the brooding G minor quintet.
>
> Supersensitive to the great man's remark that he had not played in seven years, from the very first sounds we regrouped around Dr. Einstein in complete rapport with his deliberate but purposeful momentum: basking in the warm glow of making Mozart's music with this disarmingly unassuming human being. Slow, slower and slowest crept each successive movement. In direct reaction to the slow tempi, the intensity of our manic happiness grew. Dr. Einstein hardly referred to the notes on the musical score (notes in his mind were never to be forgotten), and, while his out-of-practice hands were fragile, his coordination, sense of pitch, and concentration were awesome. The mood of the Juilliard String Quartet, as we finished the Mozart, was beatific. Time was late; we gathered instruments and reluctantly bid him goodbye.

Chapter 33

Afterword

> Biography, the most interesting perhaps of every species of composition, loses all its interest with me, when the shades and lights of the principal character are not accurately and faithfully detailed.
>
> *Sir Walter Scott*

> One has to know something about history in order not to get lost in meaningless questions.
>
> *Albert Einstein*

The above quotations have been my guiding lights in the preparation of this volume. I have also been encouraged by Einstein's statement that the biographical aspects of scientists had always interested him as much as their ideas, and that he liked to learn the lives of the men who had created the great theories and performed the major experiments, what kind of men they were, how they worked, and how they treated their fellow men. However, I am also painfully aware of another Einstein quotation: 'By the way, it also proves to be the case here that such a biography reveals more about the author than the subject.' It is for the reader to judge to what extent I have adhered to the exhortations of the first three quotations while avoiding falling into the trap of the fourth. It is also of course up to readers to make their own minds up about Einstein; but I feel obliged to put before them my own conclusions about the hero of this book – if indeed hero he was – which have certainly evolved as it was written. In order to structure these final words, I divide them into the three areas on which this book has mostly concentrated: Einstein and music; Einstein the man; and Einstein the scientist. Since this is a drawing together of the book's contents, references are given only to significant pieces of information that are mentioned for the first time.

The surest way for authors to reveal more about themselves than the subject is a failure to put themselves into the environment and attitudes of the time. One of the driving forces behind this book was the thesis that not only was music an important part of Einstein's life but also that it was exceptionally so. The former is obvious, the latter more difficult to establish. Playing music was important to the lives of most of Einstein's contemporaries. In the final quarter of the nineteenth century, as discussed in Chapter 1, there were few other recreations. Scientists seem to have a particular enthusiasm for music. Certainly, many of Einstein's physicist contemporaries were able musicians. Many were pianists; of those close to Einstein, Max Born and Paul Ehrenfest spring immediately to mind. Indeed, to be an accomplished pianist was almost a free pass to Einstein's friendship; often his choice of scientific assistants or collaborators seemed to fall on those who as well as being able scientists, also had facility at the keyboard. It is in this search for musical opportunities that Einstein seems exceptional. Both in his youth and in his old age, if he met someone in the street whom he either knew or suspected to be a pianist or a violinist, then an

Einstein. Brian Foster, Oxford University Press. © Brian Foster (2026). DOI: 10.1093/oso/9780198794875.003.0033

invitation to make music together was almost inevitable. Of his contemporaries, perhaps only Max Planck made music such an important part of his life; although Planck's musical activities were much broader than Einstein's, involving singing, composition, conducting, chamber music, and playing the organ, it is difficult to imagine him breaking off what he was doing to run down the street in pursuit of a promising musical collaborator.

Einstein's taste in music has been extensively chronicled in this book. Mozart was the genius who held him in thrall from his early youth, a bond that only intensified with age. Close behind was J. S. Bach, whose solo sonatas and partitas were a solace when no pianist was available. His youth however admitted a much broader musical spectrum than he would admit to in later life. We have seen how his first extant violin examiner praised his interpretation of Beethoven and how intensively he practised Brahms' G major sonata to give himself the maximum benefit from hearing the great Joachim perform it. His inveterate concert-going as a young man in Zurich and Bern exposed him to a wide range of music. His guiding principle seems to have been structure. Provided a piece of music had a discernible skeleton, he could find merit in it. However, structure wasn't enough: there also had to be beauty, by which he meant melodic beauty. There is structure aplenty in Schoenberg's music; it was his atonality, to Einstein's ear, that made him consider Schoenberg to be mad. He never wavered in his detestation of the assault on the emotions that he considered Wagner to perpetrate. Wagner, Liszt, and the road to the second Viennese school and beyond was not one that he ever trod. The 'impressionism' of composers such as Debussy and Ravel, with their blurring of formal organisation in favour of musical colour, was never to Einstein's taste, despite often-wonderous melodic invention. The remark by Robert Mann in Chapter 32 that Einstein listened to the Juilliard Quartet play Bartok in another room to avoid distraction gives rise to a knowing smile.

Nevertheless, Einstein's attitude to some composers 'in between' the poles of Bach and Berg was both ambivalent and fluctuating. Beethoven he almost seemed to blame for the transition from Mozartian classicism to Schumann, but he was apparently capable of becoming fixated on Beethoven's *Missa Solemnis*. At one time or another, he made approving noises about Modest Mussorgsky and Verdi. His enthusiasm for the music of Ernest Bloch was no doubt driven by the strongly Jewish flavour that imbued much of it, but it had to be swallowed along with substantial dollops of quasi-Wagnerian grandiosity. His friendship with Martinu flourished because of the composer's interest in physics but certainly also brought Einstein into intimate contact with aspects of contemporary music. While Martinu kept a foot also in the camp of the Baroque masters' musical forms, Einstein's performance of the *Madrigal Stanzas* brought him face to face with modern harmonic invention, a considerable distance from that of Bach.

While Einstein's musical taste was considerably broader than he is often given credit for, or indeed than he often himself stated, there is no doubt that he felt most at home in the music of the classical period or earlier. In later life, he played in early music ensembles in Princeton University; his sessions with violin luminaries such as Boris Schwarz concentrated on Baroque composers, in particular Vivaldi and Bach. Similarly, his playing of chamber music such as string quartets would most often be Mozart or Haydn, but he was certainly not averse to tackling Brahms and Schumann in the right company, in which his lapses in technique could be coped with by the rock-solid support of great virtuosi of the day.

The above picture brings us to the final question in this musical synopsis, one that is most difficult to answer: just how good a violinist was Einstein? 'Good' is of course itself a subjective term. If by that we mean technically proficient, then Einstein was the first to admit that his technique was wanting. It could hardly be otherwise given that he had dispensed with the services of a teacher early in his musical career. Self-taught violinists accumulate idiosyncrasies of technique as easily as snow in Antarctica; the violin is not an instrument that can be mastered without expert assistance. Given the painfulness of listening to the apprentice violinist to those with any sort of musical ear, intonation tends to be the first priority of the teacher. Einstein seems to have had enough tuition and to have had a good enough ear himself to be generally speaking an accurate player of the written note. What he seems to have lacked is sufficient attention to his bow-arm. While it is the left hand that defines the note that is played, it is the right arm that determines the noise that is made. Most good amateurs have decent intonation; few have the subtlety of bow arm that creates a sound pleasurable to the listener. Even fewer can achieve such a sound without a teacher to guide them in the arcane complexities of the required muscle control and finger placement.

Although opinions of Einstein's playing vary greatly, ranging from the subtlety of a lumberjack to the voice of an angel, the average seems to settle on an accuracy of intonation coupled with a 'starved' tone, lacking in vibrato. In this particular judgement, it is not helpful to look to the opinions of those great musicians whom one might expect to 'know better'. They find it very difficult to judge those not at their stratospheric level. Some were rather complimentary; but they may have been being polite or wishing to ingratiate themselves so something of Einstein's fame could rub off and boost their careers. Others were judging at their own level, where Einstein could not be expected to compete. Amateurs with whom he played also gave mixed opinions. Victor Weisskopf, a distinguished theoretical physicist and Director-General of CERN from 1961–1965, knew Einstein well after the Second World War through their activities in the Emergency Committee of the Atomic Scientists. Weisskopf opined, rather immodestly, that 'His violin playing was not as good as his physics. He was a real amateur. My piano playing was better than his fiddle.'

It must be emphasised that Weisskopf only knew Einstein at the very end of his time as a violinist, when he himself complained that he could no longer abide the sound he made. Particularly before his days of world fame, the fact that Einstein convinced so many amateurs to play with him so often in so many groups is confirmation that such an activity was pleasurable. This could only have been the case if Einstein's playing was at a good, perhaps even very good, amateur level. He had the courage to play the *Kreutzer* sonata in public; he practised the *Chaconne* from Bach's D minor Partita and performed the Bach 'Double' Concerto in a widely reported New York benefit concert. He also had the foresight to ensure that his playing was never recorded for posterity, so the stature of Einstein the musician, in contrast to Einstein the scientist, will remain uncertain.

Einstein the man confronts us with no less contradictory data than Einstein the musician. Once again, some evidence could support the conclusion of Einstein as a monster; much more, of Einstein as a saint. He was certainly neither. The misconception that Einstein suffered from some sort of personality disorder akin to autism can be conclusively disposed of using the information just cited on music. It has often been said that Einstein used music-making as some sort of alternative to interpersonal relations, a refuge from which he could survey the frightening world of personalities without getting involved. Any such statement could only come from someone who has never played in an amateur chamber music group. Few activities require more social interaction, more interpersonal skills, than rehearsing a Mozart quartet. Each player has their own concept of how the music ought to be played and is rarely afraid of asserting it. Patience at the mistakes of colleagues: 'Can't you count, Albert?'; exasperation at their being unable to play the correct rhythm, are all part of the experience. After the instruments are put away comes the laughter and society of the shared refreshments. Few endeavours wake an appetite as much as an evening of Haydn quartets. We have observed Einstein the young man indulge in sausages, eggs, and beer after musical exertion. Perhaps as he grew older, the food became more sophisticated but no less important. Einstein used his violin, particularly in his youth, as a method of seduction. His violin was a key he could turn to enter social intercourse, not a barrier to keep it out. He was certainly shy, even as a celebrity, and playing music overcame this diffidence. However, Einstein didn't play music because he wanted to escape from anything; he played music because he adored it.

More evidence, if that were required, of Einstein's pleasure in human contact can be found in his relationship with women. We have seen his almost continuous need for female companionship, from his first emotional awakening in Pavia with Ernestina Marangoni, to his final comfortable flirting with Johanna Fantova. That he was a deeply sensual man with a healthy sexual appetite can hardly be doubted. His sexual activity almost certainly started with his tempestuous relationship with Marie Winteler. Although his attraction to Mileva was initially through physics, sex soon became an integral part of it. The almost bewildering procession of Einstein's lovers as this book has progressed testifies to the fact that at any point in his life after leaving school, he was rarely not in a relationship with at least two women simultaneously. Einstein was a serial philanderer who made no secret of his belief that the whole idea of monogamous marriage was unnatural. This can be summed up by a quote from a woman who knew him well in his final years in Princeton, Alice von Kahler:

> He loved women. He was very attractive to them and he was infatuated by the ladies. But he once said to me: 'The whole thing lasts just ten minutes and then it is all over.' Yes, Einstein loved women and he wrote on my most precious photo of him and myself that he regretted I wouldn't sleep with him. I have it in my sleeping room. I would have been interested in him if I wouldn't have had a husband.

Was Einstein's behaviour detailed above extraordinary for his time? Like many aspects of Einstein's behaviour, it was probably in the wings of the statistical distribution of his contemporaries, but far from unique. Many of his distinguished colleagues in physics also had their sexual peccadillos. Paul Langevin's affair with Marie Curie scandalised turn-of-the-century Paris. Oppenheimer was a notorious womaniser who would often be dating several women simultaneously. Paul Ehrenfest had a long-term relationship with a woman that deeply offended his wife. However, all examples pale in comparison to that of Erwin Schrödinger, whose ménage à trois was notorious in an admittedly easily scandalised Oxford.[1] It can perhaps also be mentioned that Hermann Weyl had an affair with Schrödinger's wife, while Weyl's wife was in love with the physicist Paul Scherrer. Considered in such company, Einstein was not an untypical senior physicist.

The idea that, despite his constant amorous adventures, Einstein was never emotionally engaged is widespread. This *canard* has been refuted by the discovery of the correspondence between him and Marie Winteler during their second affair in 1909–1911, discussed in Chapter 8. No one who reads Einstein's letters in this correspondence could doubt that here was a man who loved deeply and was desperate because he could not find a way to live with the woman who he had, probably, always loved. Indeed, it seems likely that it was this affair and the role that Mileva played in thwarting Einstein's desires, possibly by engineering to fall pregnant, that led to the catastrophic deterioration in their marriage and their eventual separation in 1914. The sheer malice of Einstein's letter to Mileva quoted in Chapter 12 setting out the conditions for their continuing to live together can only be the product of a man who felt an anger bordering on hatred. His treatment of Mileva in the ensuing years fluctuated between detestation and a grudging tolerance, eventually

[1] It was Schrödinger's recently unearthed paedophilia, however, that has precipitated an avalanche of moves to relabel buildings and prizes named after him.

subsiding in the final years to something like a distant if exasperated affection. It may well have been the emotional turmoil of the Marie affair, together with the trauma of separating from his children at the Berlin station in 1914, that scarred Einstein so deeply that in self-preservation he learned to bury his emotions and to affect the mien of dispassionate imperturbability that so many remarked on in later life.[2] As for Mileva, she never varied in her obsessive love of Einstein.

Einstein's relationship with his second wife, Elsa, evolved quite rapidly after their early passionate letters to what seems more like a partnership than a love affair. The emotional disruption caused by Mileva's and his sons' departure to Switzerland was probably contributory. Einstein's relations with his sons, who absorbed Mileva's detestation of Elsa, certainly made things difficult. Elsa remarked to her friend Antonina Vallentin about problems with Tete's behaviour: 'This sorrow is eating up Albert. He finds it difficult to cope with it, more difficult than he likes to admit. He has always aimed at being unmoved by any personal matter. He is really, much more than any other man I know. But this event was painful for him.' That Elsa and Einstein's relationship soon ceased to be sexual is hinted at by their separate bedrooms at opposite ends of the house in Haberlandstrasse and by Einstein's seeming indifference when marriage was being discussed as to whether Elsa or her daughter Ilse should be his spouse.

Elsa was certainly a patient and long-suffering partner to Einstein. As she expressed it in a letter to a mutual friend, Hermann Struck:

> One can't analyse him too much or one gets into inconsistencies. A genius like him has such things. You can't think him irreproachable in all respects; nature isn't like that. Where she gives a boundless ability in one area, she takes it away in another, which leads to problems. You have to take him in the round, not try to classify him into a particular category. Otherwise it leads to unpleasantness. God has however included much good in him and I find him wonderful, although life at his side is exhausting and complicated, not just in this way but in every way.

It certainly was. Elsa had to put up with Einstein seducing her friends, flaunting his mistresses at Caputh, taking them sailing with him – an activity to which she was rarely (if ever) invited. He brought a succession of mistresses to Le Coq while Elsa was trying to cope with exile from her beloved Berlin and the danger they were in from Nazi assassins. However, she had compensations. She loved her position in society as Frau Professor Dr Einstein. She loved being fêted by the rich and famous. As was the case for Mileva, Elsa truly loved Einstein, despite his manifold faults.

The one person for whom, throughout his life, Einstein had a constant and deep love was his sister Maja. Although his shortcomings as a correspondent were notorious, it was rare to find a period of several months without some communication between them. He looked after her all his life, often financially, always with advice. As we have seen, in her final years of illness, he was devoted to her care and read for long periods to her every night. In her turn, she was devoted to him, as is tangible from her manuscript detailing the early part of his life. Other family members were also close to him. Ilse Einstein almost became his wife but remained an affectionate step-daughter until her early death. Her sister Margot was particularly an object of Einstein's solicitation; painfully shy,

---

[2] Fritz Haber recorded that Einstein cried bitterly as he waved goodbye to his children in Berlin. Michelmore talked to Hans Albert Einstein on this subject: 'Mileva and the boys were keenly aware of Albert's love for them. It was heart-felt love. "While it was there, it was very strong", remembered the older boy. "He needed to be loved himself. But almost the instant you felt the contact, he would push you away. He would not let himself go. He would turn off his emotion like a tap."'

often ill, very artistic, she thrived behind his protection. It was to her that he left the Mercer Street house as well as the largest legacy from his estate. Einstein's stepsons-in-law partook of his patriarchal affection: Ilse's husband Rudi Kayser was a favourite whose biography was one of very few that Einstein endorsed. Einstein helped him through the tragedy of Ilse's death and in establishing himself in America. The enigmatic Dima Marianoff, probably a Russian spy, was also helped by an Einstein often amused by his escapades, until their relationship broke down over his unauthorised and often inaccurate biography. The extended Einstein family was numerous, and Einstein kept up an irregular correspondence with many of his cousins, half-cousins etc. Some, such as the Kocherthalers, were entrusted with delicate financial matters. He shared in their tragedies and often helped financially, just as he had been helped in his impecunious youth by several of his uncles and aunts.

There is no doubt that Einstein was a lovable man with a magnetic personality. His closest friend from his Berlin days, Moritz Katzenstein, a medic like so many of Einstein's friends, remarked that 'Being with this man [Einstein] and his wife is on every occasion like a tonic to me.' Throughout his life, Einstein attracted friends who seemed willing to go to enormous trouble to help sort out the difficulties into which he was perpetually falling. Many of these friends were made in his youth, although of them, only Michele Besso and Maurice Solovine remained faithful correspondents to the end. He was always grateful to Marcel Grossmann, who had been a pillar of strength through his student days and whose influence procured Einstein's job in the Bern Patent Office. His collaboration in the genesis of general relativity was of the first importance, as was that, more indirectly, of Besso. Grossmann's tragically early death was a great loss to Einstein. Of slightly later vintage, the friendship of Heinrich Zangger was indispensable to both Einstein's career and personal life. Zangger's assistance to Einstein was enormous, ranging from a decisive role in arranging his academic appointments in Zurich to caring for his children medically and by acting *in loco parentis* by taking them into his home during Mileva's many illnesses. He mediated in the decades of civil war between Einstein and Mileva. As Hans Albert once crushingly pointed out in a letter, they knew Zangger much better than they did Einstein and Zangger knew them much better than Einstein did. That Zangger was willing to do all this is another testament to Einstein's attractiveness as a friend, as well as to Zangger's extraordinary kindness.

Others near to Einstein from later years included the larger-than-life medical practitioner Janos Plesch, who fascinated, amused, and irritated Einstein in equal measure. Another doctor, Hans Mühsam and Einstein were close in Berlin until the Betty Neumann affair estranged them; after a lapse of decades they were reconciled. Einstein's letters to the ailing Mühsam and his wife in his final years were the longest and chattiest he wrote to any correspondent outside his family. Yet another medic, Gustav Bucky, was as close as anyone to Einstein in his years in the US. Einstein and the Bucky family shared holidays, Gustav and Einstein had a common interest in inventions, and chauffeuring was provided by Thomas Bucky. No list of those close to Einstein would be complete without the inclusion of his devoted secretary, housekeeper, and factotum Helen Dukas, and his man of business and general advisor Otto Nathan. His executors after his death, they fiercely defended Einstein's reputation against any hint of what they considered scandal.

Among colleagues, Einstein also made many friends. When he delivered the Rhodes Lectures in Oxford, he was characterised as follows: '[Einstein's] sense of humour is scarcely concealed in some of his papers and this combined with his extraordinary modesty and power of lucid exposition make him one of the best conversationalists a physicist could wish to meet.' His closest friend among colleagues other than Grossmann was certainly Paul Ehrenfest, whose emotional insecurity necessitated deep relationships with his friends, of whom Einstein and Niels Bohr were closest.

Sharing a love of both music and physics, Einstein reciprocated Ehrenfest's affection to a level that was unusual for him. They realised that they needed each other. Max von Laue was almost as close as Ehrenfest in terms of friendship and probably closer in terms of his attitude to physics. He shared much of Einstein's scepticism towards the philosophy of quantum mechanics. Von Laue was one of the few German physicists who remained in Germany during the Nazi period whom Einstein continued to respect. The other was Planck, who more than anyone else was responsible for establishing Einstein's stature in the physics community. Of a similar nature was Einstein's relationship with Ehrenfest's predecessor at Leiden, Hendrik Lorentz. They were more like son and father than friends; Einstein respected him as he did few if any others, both in physics and in life. There was a similar relationship between Wolfgang Pauli and Einstein, with Einstein as the father figure; their bickering concealed a deep respect. Max Born remarked that he respected and honoured Einstein more than anyone else he had ever met.

Orbiting somewhere between these inner satellites and a densely populated outer region were people such as Leo Szilard and some of Einstein's assistants such as Ludwig Hopf, Leopold Infeld, and Cornelius Lanczos. Many of these unconsciously amplified their own intimacy with the great man as the years passed. None of these assistants were granted use of the intimate *Du*. One who was so honoured was his fellow-seeker for unified theories, Erwin Schrödinger, with whom Einstein continued a frequent correspondence right up to Einstein's death.

Beyond them were those attracted to Einstein's orbit by his fame, such as Charlie Chaplin; the great musicians with whom he played, Kreisler, Piatigorsky, Rubinstein, etc.; politicians such as Walther Rathenau and Ramsay MacDonald; and literary figures such as H. G. Wells, George Bernard Shaw, Thomas Mann, and Gerhard Hauptmann. Then there were the hundreds of people who owed much to Einstein's help, often indeed their very lives. They were with difficulty deterred from inundating Mercer Street with visits to express their gratitude. This deterrence was important because Einstein treasured the isolation and indeed loneliness of his later years. Infeld felt that 'Through all the stream of events, the impact of people and social life forced upon him, Einstein remains lonely, loving solitude, isolation and conditions which secure undisturbed work.' Plesch quoted Einstein as saying, just before his death:

> I do not socialize because social encounters would distract me from my work and I really only live for that, and it would shorten even further my very limited lifespan. I do not have any close friends here as I had in my youth or later in Berlin with whom I could talk and unburden myself. That may be due to my age. I often have the feeling as if God has forgotten me here.

Einstein was unusual in his attitude to male friendship. It is most clearly seen in his relationship with those who felt closest to him. His stepson-in-law Rudi Kayser wrote that Einstein found a too-obtrusive friendship much more unbearable than hostility. Einstein's other stepson-in-law opined:

> Einstein is always impersonal. He is kind and humorous and companionable just like other human beings, or troubled or sad – but always impersonal. I do not think that he was ever interested solely in the individual. But give him man in the mass, man in a peace movement, man struggling to find a foothold, then he is aroused.

Einstein himself in a rare written moment of introspection while on one of his Asian voyages in the early 1920s felt as if there were a sheet of glass separating him from most emotions. A recipient of Einstein's most passionate feelings, Betty Neumann, nevertheless remembered him saying that it was difficult to love people once you got to know them well. Born had felt close enough to Einstein in his early Berlin days to berate him on his feckless behaviour with regard to

Moszkowski's biography; nevertheless he wrote that, for all his kindness, sociability, and love of humanity, Einstein was totally detached from his environment and the human beings included in it. In 1930 Einstein himself remarked on this:

> My passionate sense of social justice and social responsibility has always contrasted oddly with my pronounced lack of need for direct contact with other human beings and human communities. I am truly a 'lone traveler' and have never belonged to any country, my home, my friends or even my immediate family with my whole heart. In the face of all these ties, I have never lost a sense of distance and a need for solitude – feelings that increase with the years.

In this sense of detachment, Einstein was perhaps in the wings of the statistical distribution of human attitudes but not pathologically outside it. Many men prefer the company of women and are averse to close male relationships. Einstein after all lived almost all his home life in Berlin and Princeton surrounded by women, apart from the transitory presence of the occasional son-in-law. Like many men of genius, he was selfish in protecting the safe space he needed to do his work. This was paramount; when a man, or indeed, but granted more tolerance, a woman, began to make emotional demands that interfered with that, Einstein acted decisively to protect himself. This self-defence mechanism led him at times into a surprising unkindness to friends. The most egregious examples are his behaviour towards Zangger and to a lesser extent Besso when they were taking care of his sons. As was documented in Chapter 14, Einstein's ingratitude and petulance were inexcusable. Even worse was the way in which Einstein essentially cut off contact with Zangger, despite decades of tireless work on Einstein's behalf, because of a careless remark in a letter, which he interpreted as anti-Semitic. It is difficult to avoid the conclusion that, having acquired the services of another 'man of business' in Nathan on arrival in Princeton, Einstein no longer required Zangger and simply dismissed him from his mind. As reported in Chapter 15, it was to Zangger that Einstein had written: 'I have learnt to recognise the mutability of all human relations and learnt how to isolate myself from the heat and the cold, so that an equable temperature is pretty much guaranteed' and to Besso: 'Here, everyone is close to me only to a certain limit, so life goes on almost without friction; this I have learned in life.'

Despite the above quotations, there are many contradictory indications: that beneath Einstein's apparent impassivity lurked a passionate soul. One was his capacity to bear a grudge. Among the first detectable instances of this was his behaviour to Erwin Freundlich, who had pioneered the idea of testing general relativity at a solar eclipse, through which they had established a friendship of many years standing. However, once Einstein became convinced that Freundlich was being dishonest, he broke off their relations, which required at least a decade to re-establish. A similar pattern was repeated with Judah Magnes and the administration of the Hebrew University of Jerusalem, without the initial friendship. Their long-running feud went well beyond an intellectual disagreement to deep disdain and dislike. Towards the end of his life, similar personal quarrels blew up over what became Brandeis University. Einstein rejected all of the many attempts towards reconciliation by the principal protagonists in this dispute. His erstwhile stepson-in-law, Dima Marianoff, of whom he had been fond, was completely cut off after the publication of his Einstein biography. Another example is that of Zangger quoted in the previous paragraph. Finally, what can only be described as the hatred he felt for Mileva in the first years after their separation cannot be ignored.

The history of Einstein's relationships with two other males, his sons, could easily fill a book. To say that they were problematic would be an understatement. Again, they illustrate the often-illusory nature of Einstein's assertion of his own lack of emotional involvement. He was far from unique in his difficulties with his children after a traumatic separation and eventual divorce. Perhaps his uniquely strange characteristic was his almost total neglect of their birthdays, which does evince a complete lack of empathy for the psyche of young boys, especially those for whom their father is almost never physically present. There is little doubt that Einstein felt guilt about his lack of involvement with his sons' upbringing, increasing the tension between them. His relations with Hans Albert once the latter had passed puberty were prickly in the extreme. They grew even worse over Einstein's open and obstinate opposition to Hans Albert's choice of wife. It was only considerably after Hans Albert had moved to the US that they grew closer and felt relatively comfortable with each other. Even then, the long-standing dislike that Einstein felt for Hans Albert's wife was always present.

That Hans Albert was emotionally scarred by his fraught relationship with his father is evident from various remarks that he made in interviews after Einstein's death. The terms of Einstein's will and the preference he showed, at least in Hans Albert's eyes, to his 'second' family prolonged the tension after his death, as did the literary executorship of Nathan and Dukas, which interfered in Hans Albert and his family's right to publish even his own father's letters to him. It is a credit to Hans Albert that he overcame what could fairly be described as a traumatic childhood, as well as the difficulty of being a scientist in the enormous shadow of his father, to become a highly respected and indeed beloved academic engineer. He died at the age of sixty-nine in 1973 and is buried in Woods Hole, MA, where he was serving as a scientist in residence. He had a massive heart attack in Vineyard Sound while anticipating an afternoon's sailing, a passion he shared with his father. He is buried nearby; his tombstone is inscribed with 'A Life Devoted to his Students, Research, Nature, and Music'.

As for Einstein's relationship with Tete, its development, as Tete fought off ill health as a child only to descend into mental illness, has been chronicled in this book in considerable detail. That Einstein cared deeply about him is not in doubt, but it seems that eventually it became too painful for him even to contemplate writing to his son. Einstein's instinct for self-preservation, or selfishness if you will, kicked in; with an inward sigh of relief he delegated contact to the kindness of his biographer, Carl Seelig, who continued to interest himself in Tete's wellbeing until his own death in 1962. A lifelong concern of both Mileva and Albert Einstein had been that Tete's care should be guaranteed after both were dead. Einstein would have been gratified to know that his financial provision had been sufficient, although only to provide his son with a somewhat spartan lifestyle. Tete died in the Burghölzli clinic on 25 October 1965, at the age of fifty-five.

Contradictions abound in Einstein's story; the picture of a man eschewing personal contacts and keeping distance from friends in order to concentrate on his work conflicts with his attitude to the tsunami of letters that were delivered to his address. He complained continually about his servitude to the arrival of the daily mail sack; and it was indeed a sack. Nevertheless, he was assiduous in replying to almost every letter he received, making exceptions only for the (many) abusive ones or those obviously deranged. If there was even some doubt about the latter category he would reply briefly. Every morning for decades, starting as the headlines announcing the eclipse result began to have their effect and lasting to almost the last day of his life, he devoted hours to working through letters with his secretary, for most of that time the indefatigable Dukas. He read many of the (often dauntingly thick) scientific works sent to him, at least until he could ascertain that they were nonsensical or he found a major mistake. Literally hundreds of books were sent to him over these years, often on subjects far removed from physics; again, he was highly likely to read them and send back appreciation, critique, or comment. Correspondents asked his advice on the most bizarre subjects. A good-sized book could be written about Einstein's correspondence. Here there is only room for a single example, by far not the strangest.

A certain Alexander Bogolub wrote to Einstein in June 1950, claiming to have met him during his stay in Bermuda in 1935:

> When I die, I would appreciate it if you would have my body sent to Bermuda, so that I might be interred in the family plot with my deceased wife. I make this request to you because there is no one in this world that I can turn to at this time for this or any other request . . .

Einstein replied: 'I am not a bit interested in my own corpse and therefore you cannot expect that I might be interested in yours. You may be sure, after all, that when the time arrives the people will take care of it in their own interest.'

Another category of letters that arrived begged for Einstein's help in life- or career-threatening situations. In the 1930s, the majority of these were from his fellow Jews trying to escape Nazi persecution. Einstein was tireless in offering his help, frequently financial, to Jews in Europe. He wrote countless letters of support to bureaucrats of many nationalities, to the extent that by the end the value of his letters had begun to be discounted. Often those he helped escape the Nazi nightmare were the continued beneficiaries of his aid in settling in the US and in finding jobs. Certainly hundreds, possibly thousands, of individuals owed their survival to Einstein, a record that might well have seen him nominated for an additional Nobel Prize: for Peace.

The innate goodness of Einstein's character shines out in all these activities and comes nearest to the 'Jewish sainthood' with which he has been retrospectively invested. That he was willing to devote so much of his precious time to philanthropic works but so chary of any emotional involvement with those close to him is an enigma. Perhaps he found that, having devoted his mornings to letter writing and good works, he could instantly switch his intellect to physics, whereas the disturbances caused by emotional involvement with people took longer to subside.

This book has not attempted a comprehensive account of Einstein's views on politics, philosophy, and religion, although all have often been mentioned because of the insight they give into his character. In politics, he maintained a left-wing viewpoint throughout his life. Although he would have resisted being labelled, he was clearly a mainstream socialist and not a communist. Overcoming his initial scepticism, by the end of his life he was a Zionist, albeit one who never lost hope that Jews and Palestinians could peacefully coexist in a common homeland. His pacifism, except during the Nazi period, was a constant conviction. It is a testament to his courage that, once he had decided that the evil represented by the Nazis was such that it had to be resisted by force, he changed his standpoint. Despite being excoriated as an apostate by most of his former pacificist colleagues, he persisted in his stance. Although often characterised as a dreamer, in this he was a realist: to defeat the ultimate evil, some lesser evil was inevitable. Einstein's philosophy was both broad and deep, although not aligned to any particular 'school'. He often chuckled that, to paraphrase, he was the most religious person in the world who did not believe in God. He believed in a force of goodness in the universe, in absolute verities and values; as he himself frequently noted, he believed in 'Spinoza's God'. He never hesitated to speak in support of the values he held dear. In an article in *The New York Times* to celebrate his seventy-fifth birthday he said: 'And I believe that although there are fewer and fewer voices to be heard speaking up for freedom and decency and truth, I must be among them. If my voice has to be the last to speak, then let it be the last.'

To end this survey of Einstein the man, I address two damaging accusations that have been from time to time been thrown at him: that he was a misogynist, and that he was a racist. It is difficult for a biographer considering these subjects to gain an appropriate distance from contemporary mores. The past here is truly another country. That Einstein considered women on average intellectually inferior is established by many quotations. Perhaps the most shocking comes from Moszkowski's biography, in which Einstein is quoted as having said that it was conceivable that Nature may have created a sex without brains. His ex-lover, Betty Neumann, wrote down several similar remarks, including 'Women are able to please by their mere existence, without having to do or to know anything. They live in a perpetual hopeful excitement for something to happen, for somebody to come.'

These beliefs are of course both alien and offensive in the modern world. That Einstein knew well such women as Marie Curie, Hedda Born, and indeed his own first wife, Mileva, makes such attitudes even more difficult to accept. However, it is vital to remember that the casual misogynism quoted in the previous paragraph was the norm for Einstein's contemporaries. Women were only allowed to vote in the US in the second half of Einstein's life; they were not admitted to Princeton as undergraduates until fourteen years *after* his death. Vannevar Bush, born a decade after Einstein, wrote a book detailing sixty years of his work in science and government entitled *Pieces of the Action*, which was published in 1970. The fifteen-page index to its more than 300 pages names precisely seven women, of whom only Lise Meitner and arguably Eleanor Roosevelt are mentioned other than as appendages to their husbands. Women were simply either absent or rendered invisible as leading roles in science, academia, engineering, and industry. Even today, physics is not without its sexism. Despite years of effort by countless people, only roughly half as many young women as men enter UK universities to study physics. Einstein's views were very much in the mainstream at the time. Moreover, he was capable of moving away from the generalisations quoted in the previous paragraph, as shown by the fact that he substantially helped the careers of Emmy Noether and Lise Meitner. His last scientific paper was published in 1954 jointly with his female assistant, Buria Kaufman. He seemed very pleased at their collaboration.

The second accusation that has had some currency is that Einstein was a racist. This is demonstrably untrue. Some remarks he made in his travel diaries about the squalor and degradation in which the native Chinese lived in Shanghai and their resignation in such conditions were simply what any interested observer would note. The depth of poverty in the China of the 1920s was indeed frightening. That he had no prejudice against Asians is indicated by the pleasure he had during his visit to Japan just afterwards and the respect he evinced for Japanese culture and customs, although he never came to terms with Japanese music. His behaviour towards African-Americans in the US is beyond reproach. Princeton in Einstein's day was rife with racial prejudice and was strictly segregated. Einstein often visited the ghetto of Princeton and sat on doorsteps conversing with the inhabitants. He made an exception to his usual refusal of honours such as honorary degrees if they were offered by historically African-American colleges or universities, which he would visit and at which he would regularly lecture. His support for prominent African-American artists such as Marian Anderson and Paul Robeson has been detailed in these pages. Einstein spoke out about the evils of racism on many occasions. This was hardly surprising as he had personal experience of anti-Semitism, about which he was even more outspoken.

I conclude this section with two quotations that characterise Einstein the man. The first was written after Einstein's death by his final collaborator in the quest to avoid nuclear Armageddon, Bertrand Russell:

> Einstein was not only a scientist but a great man. He stood for peace in a world drifting towards war. He remained sane in a mad world, and liberal in a world of fanatics. Of all the public figures I have known, Einstein was the one who commanded the most whole-hearted admiration.

The second quote is from his voluminous correspondence. In February 1950, Einstein received a letter from Rabbi Norman Salit, with whom he and Chaim Weizmann had shared a taxi on their way to a meeting in City College, New York, during Einstein's first visit to the US in 1921. After one of Salit's daughters had died suddenly, he sought Einstein's help for the other, who was inconsolable. He asked for a note giving what comforting explanation Einstein could for this event. Einstein replied with could be considered both a synopsis of his world view and an insight into his own psyche. He does not attempt to assuage the personal grief but writes:

> My dear Mr Salit, A human being is a part of the whole, called by us 'Universe', a part limited in time and space. He experiences himself, his thoughts and feelings as something separated from the rest, a kind of optical delusion of his consciousness. This delusion is a kind of prison for us, restricting us to our personal desires and to affection for a few persons nearest to us. Our task must be to free ourselves from this prison by widening our circle of expansion to embrace all living creatures and the whole nature in its beauty. Nobody is able to achieve this completely but the striving for such achievement is in itself a part of the liberation and a foundation for inner security. Yours very sincerely, Albert Einstein.

In summary, Einstein was a normal man, with some rather unusual personality characteristics and touched with a scientific genius that has rarely (if ever) been reached by humankind. It is this last element that forms the final part of this chapter. As remarked by Sir Roger Penrose in his foreword to the second edition of Abraham Pais's seminal book, *Subtle is the Lord*, Einstein commented that the essential of the being of a man of his type lay in what and how he thought, not what he did or suffered.

We have seen that Einstein's ascent to the highest levels of science was not smooth. He was no prodigy like his idol Mozart, but an average schoolboy with strong and weak subjects. It was predominantly his dislike of authority that placed him apart from his schoolfriends, that made him a disruptive influence in his Munich schoolroom. It drove him to Italy and then Switzerland where in the conducive surroundings of the Winteler household in Aarau he learned to think and to love. It was Mileva and Einstein against the world, even beyond his attainment of security at the Patent Office. The simmering thoughts, beginning to be organised in conversation with Mileva and Besso, surfaced in the great creative leap of the *annus mirabilis* of 1905.

As we have also seen, special relativity was perhaps the least 'original' of the four great papers, in the sense that the ideas it organised had been around for years and other physicists had almost reached Einstein's conclusions. It was two of his great powers as a scientist that made the difference: the courage to imagine a universe different from Newton's and the stubbornness to persist when his ideas seemed to make no impact. The third power, extraordinary imagination, is evident in the other great papers; the motion of pollen grains understood by the behaviour of atoms; the behaviour of light understood by reverting to Newton and 'corpuscles'. Einstein knew light was a wave, but he also knew that sometimes it behaves as a particle. He did not put his head in his hands and think there must be a mistake somewhere, but instead boldly asserted that light must be both – a proposition so extraordinary that Planck, one of the greatest physicists of the time, never accepted it. Finally that year came $E = mc^2$, an equation that would shake the foundations of the modern world and lead to consequences in terms of the release of the power of the atom that he at first could not imagine, and then denied. More than thirty years elapsed before he realised the danger and tried to mobilise the world to control it and harness it for good rather than evil.

Once accepted into the academic world, Einstein continued regularly to produce papers that astounded his colleagues. He soon set himself the task of bringing acceleration into the world of relativity. Realising the intimate connection with gravity, he thrashed around for the right approach without knowing the mathematical tools with which he could have tackled the problem. Years of effort produced the realisation that geometry held the key to how the planets orbited the sun; his old friend Grossmann pointed him to the mathematics he needed to describe a geometry that Euclid had never imagined. There were still mistakes and months of travelling on the wrong path. His other old friend Besso slaved over the orbit of Mercury with him, but the answer was wrong. Then came the final realisation that a particularly attractive option, beautiful in its elegant mathematics, had earlier been wrongly dismissed. A final sprint, run against David Hilbert, the greatest mathematician of his day, proved that it did need a physicist to solve physics problems. General relativity was published, its predictions led to expeditions to the far corners of the earth, and news from a Royal Society meeting in London made *The New York Times* proclaim that the stars were wobbling in their courses, but no one should worry. Einstein became, with his friend Chaplin, the first 'A-listers'.

This was far from the end of Einstein's contributions to physics. Paper followed paper, including returns to his old stomping ground of statistical physics, where his work with Satyenda Nath Bose and the definition of the Einstein A and B coefficients laid the groundwork for that omnipresent scientific workhorse, the laser. General relativity had however produced in its creator's mind a seed of doubt. He convinced himself that it was by following the siren of mathematical beauty and simplicity that he had finally found the Einstein equations. Recent scholarship, however, indicates that it was at least as much his physical intuition that allowed him to see the path forward. As he began to turn his attention more and more towards a unified field theory that incorporated both gravitation and electromagnetism, his approach was an increasingly mathematical one.

The pivot towards mathematics combined with another conviction, that quantum mechanics was not the final word in the understanding of the atom, moved Einstein further and further from the mainstream of physics. His colleague, cosmologist and Catholic priest George Lemaître, wrote that the volume *Albert Einstein: Philosopher-Scientist*, published in 1949 and edited by Paul Schilpp:

> . . . is an important document for the history of science. Maybe it shows that, even in a scientist who maintained prodigious activity until the end, old age still altered a bit the wonderful balance of his great period. Perhaps some faculties age faster than others, perhaps the critical spirit survived in Einstein, even if exacerbated, when the creative genius began to fade? It became difficult for him to follow exactly, to the end, the narrow path passing the same distance from both pitfalls lurking in every type of scientific research: the short-sighted positivism that cannot go beyond experience and the dreamy idealism that loses touch with it.
>
> Maybe the pitfall of Einstein's old age was the tirelessly pursued dream of a perfect theory, which led him to discard anything that did not fit with the aesthetic ideal he had conceived.
>
> The cosmological constant can be compared to those iron rods that escape in all directions of a concrete construction. They are undoubtedly superfluous and inadmissible in a completed construction, but they are indispensable if the construction must later attach to others and become an element of a broader synthesis.

Einstein voiced similar sentiments in an interview with *The New York Times* to celebrate his seventy-firth birthday:

> I believe that Nature is basically simple. I have always been deeply impressed by the order of things and I believe the universe can be explained and reduced to its simplest terms in one formula embracing every manifestation of matter and energy. I know of no way in which the unified field theory can be verified. But I have always had confidence in human reason and its ability to cope with new experience – to seek unexpected harmony and structure.

Einstein's immoveable conviction that quantum mechanics could not be completely correct met the irresistible force of its astonishing success in predicting atomic and sub-atomic phenomena. His response to this in his final years was essentially to ignore the realm in which quantum mechanics had shown itself to reign supreme and to concentrate on a unification of the two areas of physics in which it played no part: the classical electrodynamics of Maxwell's equations and general relativity. He ignored the growing evidence that other forces must exist to explain the subatomic world. This was not necessarily a recipe for failure. Electrodynamics and general relativity do have very similar mathematical properties and can both be expressed beautifully and neatly in the same tensor notation. It might well have been the case that a successful unification of electromagnetism and gravity could have opened the way to unification with the other forces. Unfortunately, it turns out that electromagnetism has much more in common with the weak nuclear force than gravity. Unification there has been successful, producing, in the 'Standard Model' of particle physics, the most accurately tested theory ever devised. Again, it may have been Einstein's conviction that a mathematical approach was most fruitful that led him to ignore the hint of their incompatibility given by the fact that the relative strengths of gravity and electromagnetism differ by 10^{36}.

It is in a sense ironic that Einstein's central problem with quantum physics was his stubborn attachment to strict determinism, an aspect of the Newtonian universe many of whose nostrums he had himself overthrown. He said on several occasions that he accepted the validity of the Heisenberg Uncertainty Principle but that it was a symptom of the incompleteness of quantum mechanics as a theory. The response of the physics community until the advent of John Bell was to shrug their collective shoulders and get on with using quantum mechanics. Developments in the philosophy of quantum mechanics since Einstein's death have been strongly influenced by his point of view and there are still many open questions on the philosophical basis of this strange, wonderful, and extraordinarily successful theory. It is quite likely that Einstein would have disagreed with most, but not all, of the following observation of his old friend Born, who, together with Bohr, was one of the high priests of quantum mechanics:

> I believe that ideas such as absolute certitude, absolute exactness, final truth, etc. are figments of the imagination which should not be admissible in any field of science. On the other hand, any assertion of probability is either right or wrong from the standpoint of the theory on which it is based. This loosening of thinking [Lockerung des Denkens] seems to me to be the greatest blessing which modern science has given to us. For the belief in a single truth and in being the possessor thereof is the root cause of all evil in the world.

Einstein's long-standing isolation in his work on unified theories, combined with the difficulty involved in many of the solutions of the Einstein equations, led to an appreciable fall off in the study of general relativity that began during his lifetime and accelerated after his death. The biography of Einstein published in the *Biographical Memoirs of the Royal Society* summarises the situation:

> The appeal of General Relativity has in recent years been weakened by a growing doubt as to whether continuous differential equations in four-dimensional space-time can possibly provide a solution of some of the problems of quantum theory, such as those relating to the time of decay of a radium atom.

In one of his last letters to his faithful co-worker Besso, Einstein concurred with Whittaker's statement quoted above:

> In this sense, the theory is clearly sufficiently determined by the principles of relativity alone. However, I think it is entirely possible that physics can't be expressed in terms of continuous fields. In this case, nothing will remain of all my castles in the air, including gravitation theory, or indeed any of the rest of physics.

This type of pessimism probably contributed to the fact that, when I was a physics undergraduate in the early 1970s, general relativity was considered an obscure branch of mathematics by most physics departments. The easiest problems had been solved by the pioneers, most of whom were now dead, while the brightest and best young physicists saw the future in quantum mechanics and particle physics. In the UK, researchers such as Fred Hoyle and Hermann Bondi continued to work in general relativity applied to the nascent field of cosmology, but it was the pioneering explorations of black holes and related phenomena by Dennis Sciama and his pupil Stephen Hawking, as well as Roger Penrose, together with others in the US, in particular John Wheeler and Kip Thorne, that catalysed a revival in gravitational studies.

The indirect observation of gravitational waves, the reality of which Einstein had wavered over for decades, by Russell Hulse and Joseph Taylor in studies of pulsars led to the award of the Nobel Prize in 1993. The realisation that gravitational waves could possibly be detected experimentally by among others, Ron Drever and Rainer Weiss, later joined by Barry Barish, gave another boost to general relativistic studies. The discovery of gravitational waves from the merger of two black holes in 2016 marked the birth of a whole new area of astronomy. Weiss, Thorne, and Barish were awarded the 2017 Nobel Prize in Physics in recognition of their seminal roles in this discovery. Thus, general relativity is once more at the core of physics.

Einstein's concern about the demise of general relativity in the letter to Besso quoted above has proven to be wrong. Indeed, the agreement between the observations and the predictions of general relativity in the most extreme conditions of black holes merging is little short of astonishing. Much of the initial work of pioneers such as Einstein, Karl Schwarzschild, Willem de Sitter, and others was in the weak-field approximation. Many theories that give excellent agreement with experiments in such a limit fail when the interactions become stronger. General relativity, however, comes through with flying colours even in the strongest and most violent events in the universe.

Although we still have failed to achieve Einstein's dream of unifying gravity with any of the other forces, the observation above must give today's physicists food for thought. There must surely be some deep truth in Einstein's conviction that geometrical insights play a key role in understanding the fundamental interactions of nature. The assumption of the Standard Model of particle physics that all particles are Euclidean points is another geometrical assumption, one that must be wrong, and which is in some sense responsible for the infinities that have to be 'renormalised' away in standard relativistic quantum theories. If we are finally to succeed where Einstein failed, we will surely need to take more seriously his insights about the key role of the geometry of space and time in the formulation of a unified theory. String theory moves us away from point-like particles and includes its own complex geometry of extra hidden dimensions – a concept that Einstein often returned to after being exposed to the ideas of Theodor Kaluza and Oskar Klein. A highly topical area of current research in string theory is that of the 'holographic principle', which postulates that a gravity theory can be equivalently described in terms of a 'hologram', which is a quantum field theory without gravity in one fewer dimension. This 'anti-de Sitter[3]/conformal field theory correspondence', also known as Maldacena duality, is a concrete realisation of this principle that has emerged from string theory and establishes a specific link between general relativity and a particular quantum-mechanical system. This suggests that Einstein's bugbear – quantum mechanics – and arguably his greatest scientific achievement, general relativity, may be just different sides of the same coin. It is surely the case that an identification of this coin, in other words, an eventual

[3] It is indeed appropriate that the Willem de Sitter mentioned here is Einstein's old friend and colleague from Leiden.

unified theory of physics, will incorporate many of the insights that Einstein pointed us towards a century ago.

Einstein was very careful to ensure that there was no grave or memorial that could become some sort of place of pilgrimage. His funeral was, as he requested, identical to that he had organised for his sister Maja – a minimum of mourners, including Hans Albert, the Bucky family, Nathan, and Dukas; not even Margot attended as she was still hospitalised. As he had promised his eccentric correspondent Bogolub, quoted above, he did not care what became of his body after death, provided it was destroyed. His brain and eyes were removed without permission for 'scientific study'. After the cremation, Otto Nathan scattered the ashes over the Delaware River.

Thus, as he certainly would have wished, it is his science that is Einstein's memorial. His concepts have become ubiquitous in the twenty-first century, embedded either explicitly or implicitly in countless applications: satellite navigation, lasers, photovoltaics, particle accelerators, radiotherapy; all are based on Einsteinian theory. The elegance, breadth, and profundity of his work evokes awe in the physicist of today. His conviction that the basic laws of physics must be economical in concept, elegant in mathematical expression, and unified in description still guides the ongoing search for a fundamental theory of the universe. Physicists would apply to Einstein the epitaph of another man of pre-eminent abilities, Sir Christopher Wren: 'Lector, si monumentum requiris, circumspice' – 'Reader, if you seek a monument, look around you'.

Figure Credits

Figure 3.1 Wikimedia Commons, under Creative Commons CC0 License.
Figure 8.1 Leo Baeck Institute – New York/Berlin.
Figure 9.1 Leo Baeck Institute – New York/Berlin.
Figure 15.1 Wikimedia Commons, under Creative Commons CC0 License.
Figure 17.1 Leo Baeck Institute – New York, Berlin/Pasternak Foundation.
Figure 18.1 Leo Baeck Institute – New York | Berlin.
Figure 18.2 bpk | Bayerische Staatsbibliothek | Archiv Heinrich Hoffmann.
Figure 21.1 Leo Baeck Institute – New York | Berlin.
Figure 21.2 Leo Baeck Institute – New York | Berlin / Emil Hilb.
Figure 21.3 bpk Bildagentur.
Figure 21.4 bpk / Erich Salomon.
Figure 22.1 Leo Baeck Institute – New York | Berlin.
Figure 23.1 Leo Baeck Institute – New York | Berlin.
Figure 23.2 Institute for Advanced Study, Princeton, NJ 08540 USA.
Figure 25.1 Leo Baeck Institute – New York | Berlin / The New York Times.
Figure 26.1 Leo Baeck Institute – New York | Berlin / Emil Hilb.
Figure 28.1 Leo Baeck Institute – New York | Berlin / Lotte Jacobi.
Figure 30.1 bpk / Münchner Stadtmuseum, Sammlung Fotografie / Archiv Landshoff.

Notes and References in the Text

The following notes and references are prefaced by an extract from the book text, allowing the text block to which they refer to be identified. A list of abbreviations for sources frequently cited is given below.

For reasons related to the wealth of Einstein books, no explicit Bibliography is attempted here. However, some remarks on particularly relevant books have been made in the Preface, and an inspection of the following references will give a comprehensive list of those books, journals, newspapers and other sources that have been used in the production of the current volume.

ABP	Reiser, Anton. *Albert Einstein: A Biographical Portrait* (London, 1931).
AEA	—*The Albert Einstein Archives.* http://alberteinstein.info/.
AEPS	*Albert Einstein: Philosopher-Scientist.* Paul Arthur Schilpp, ed. (Evanston, 1947).
AS	Milan Popović, *In Albert's Shadow* (Baltimore, 2003).
BIOG	Seelig, Carl. *Albert Einstein: eine dokumentarische Biographie* (Zurich, 1954).
CHAIM	Weizmann, Chaim. *The Letters and Papers of Chaim Weizmann* (New Brunswick, 1977),
CPE1	—*The Collected Papers of Albert Einstein.* John Stachel, David C. Cassidy, Robert Schulmann, eds., Vol. I (Princeton, 1987).
CPE2	*The Collected Papers of Albert Einstein.* John Stachel, David C. Cassidy, Jürgen Renn, Robert Schulmann, eds., Vol. II (Princeton, 1989).
CPE3	-*The Collected Papers of Albert Einstein.* Martin J. Klein, A. J. Kox, Jürgen Renn, Robert Schulmann, eds., Vol. III (Princeton, 1994).
CPE4	*The Collected Papers of Albert Einstein.* Martin J. Klein, A. J. Kox, Jürgen Renn, Robert Schulmann, eds., Vol. IV (Princeton, 1995).
CPE5	*The Collected Papers of Albert Einstein.* Martin J. Klein, A. J. Kox, Robert Schulmann, eds., Vol. V (Princeton, 1993).
CPE6	*The Collected Papers of Albert Einstein.* Martin J. Klein, A. J. Kox, Robert Schulmann, eds., Vol. VI (Princeton, 1996).
CPE7	*The Collected Papers of Albert Einstein.* Michel Janssen, Robert Schulmann, József Illy, Christoph Lehner, Diana Kormos Buchwald, eds., Vol. VII (Princeton, 2002).
CPE8A	*The Collected Papers of Albert Einstein.* Robert Schulmann, A. J. Kox, Michel Janssen, József Illy, eds., Vol. VIIIA (Princeton, 1998).
CPE8B	*The Collected Papers of Albert Einstein.* Robert Schulmann, A. J. Kox, Michel Janssen, József Illy, eds., Vol. VIIIB (Princeton, 1998).
CPE9	*The Collected Papers of Albert Einstein.* Diana Kormos Buchwald, Robert Schulmann, József Illy, Daniel J. Kennefick, Tilman Sauer, eds., Vol. IX (Princeton, 2004).
CPE10	*The Collected Papers of Albert Einstein.* Diana Kormos Buchwald, Tilman Sauer, Ze'ev Rosenkranz, József Illy, Virginia Iris Holmes, eds., Vol. X (Princeton, 2006).
CPE12	*The Collected Papers of Albert Einstein.* Diana Kormos Buchwald, Ze'ev Rosenkranz, Tilman Sauer, József Illy, Virginia Iris Holmes, eds., Vol. XII (Princeton, 2009).
CPE13	*The Collected Papers of Albert Einstein.* Diana Kormos Buchwald,

	József Illy, Ze'ev Rosenkranz, Tilman Sauer, eds., Vol. XIII (Princeton, 2012).
CPE14	*The Collected Papers of Albert Einstein*. Diana Kormos Buchwald, József Illy, Ze'ev Rosenkranz, Tilman Sauer, Osik Moses, eds., Vol. XIV (Princeton, 2015).
CPE15	*The Collected Papers of Albert Einstein*. Diana Kormos Buchwald, Jozsef Illy, A. J. Kox, Dennis Lehmkuhl, Ze'ev Rosenkranz, Jennifer Nollar James, eds., Vol. XV (Princeton, 2018).
CPE16	*The Collected Papers of Albert Einstein*. Diana Kormos Buchwald, Ze'ev Rosenkranz, Jozsef Illy, Daniel J. Kennefick, A. J. Kox, Dennis Lehmkuhl, Tilman Sauer, Jennifer Nollar James, eds., Vol. XVI (Princeton, 2021).
DEF	Hettler, Nicolaus. *Die Elektrotechnische Firma J. Einstein u. Cie in München: 1876–1894* (Munich, 2003).
EB	Bracco, Christian. Einstein and Besso: From Zürich to Milano. *Rendiconti di Instituo Lombardo Accademia di Scienze & Lettere*, 2014, 148. http://www.ilasl.org/index.php/Scienze/article/view/178.
EIB	Flückiger, Max. *Albert Einstein in Bern* (Bern, 1974).
ES	Rogger, Franziska. *Einsteins Schwester* (Zürich, 2005).
ISAE	*Trbuhović-Gjurić, Desanka. Im Schatten Albert Einsteins: Das Tragische Leben der Mileva Einstein-Marić* (Bern, 1988).
LORD	Pais, Abraham. *Subtle is the Lord: The Science and Life of Albert Einstein* (Oxford, 1983).
LZ	Frank, Philipp. *Einstein: Sein Leben und seine Zeit* (Braunschweig, 1979).
YE	Pyenson, Lewis. *The Young Einstein: The Advent of Relativity* (Bristol, 1985).

Chapter 1

1 *When Albert Einstein's biographer* Seelig, Carl. *BIOG*, 13.

1 *His father's family* Jüdisches Museum Hohenems. Hohenems Genealogie [Online]. https://www.jm-hohenems.at/sammlung/genealogie.

1 *The community thrived* Jewish Bad Buchau. [Online] http://www.judeninbuchau.de/html/jewish_buchau.html.

1 *The location of the house* Chrischerf. Gedenktafel für Albert Einstein am Wohnhaus seiner Eltern in Bad Buchau in Hofgartenstraße 14. Wikimedia Commons [Online]. https://commons.wikimedia.org/wiki/Category:Gedenktafel_Wohnhaus_Familie_Einstein_(Bad_Buchau).

1 *For example, there were 550* Pyenson, Lewis. *YE*, 66.

1 *The two brothers lived together* Winteler-Einstein, Maria. *CPE1*, xlix–l.

1 *Their portraits in oils* AEA [Online]. 23 March 1953 [Cited: 16 July 2024]. Cat. no. 59–981. http://alberteinstein.info/.

2 *Almost a third* Kampe, Norbert. Jews and Antisemites at Universities in Imperial Germany (I): Jewish Students: Social History and Social Conflict. *Leo Baeck Institute Year Book*, 1985, 30: 357–94.

2 *The great scholar* Elon, Amos. *The Pity of It All* (New York: Metropolitan Books, 2002), 33.

2 *At this time* Boyle, Nicholas. *Goethe: The Poet and the Age* (Oxford: Oxford University Press, 1992), 149.

2 *He became Moses* Mercer-Taylor, Peter. *The Cambridge Companion to Mendelssohn* (Cambridge: Cambridge University Press, 2004), 35.

2 *The contemporary attitude* Elon, Amos. *The Pity of It All* (New York, Metropolitan Books, 2002), 63.
2 *They could study law* —. *The Pity of It All* (New York, Metropolitan Books, 2002), 165.
2 *Among the children* Einstein, Albert. *CPE9*, 304.
3 *For example, Martha Novak Clinkscale's* Clinkscale, Martha Novak. *Makers of the Piano*, Vol. 2, *(1820–1860)* (Oxford, Oxford University Press, 1999).
3 *Not only was his head large* Winteler-Einstein, Maria. *CPE1*, lvi.
3 *In fact, a letter* Koch, Jette. *AEA* [Online]. 24 June 1881. Cat. no. 74–844. http://alberteinstein.info/.
3 *Despite his initial disappointment* Winteler-Einstein, Maria. *CPE1*, lvii.
4 *Although active operations* —. *CPE1*, l. Note 9.
4 *On his marriage* Hettler, Nicolaus. *DEF*, 84.
4 *The Müllerstrasse apartment* —. *DEF*, 83.
4 *Surreptitiously observed* Winteler-Einstein, Maria. *CPE1*, lviii.
4 *The illumination of the trenches* Gorman, M. *Electric Illumination in the Franco-Prussian War* (Los Angeles, Sage, 1977), 525–9.
4 *In 1885 Jakob persuaded* Winteler-Einstein, Maria. *CPE1*, li. Notes 11, 12.
5 *Munich was ripe* Hettler, Nicolaus. *DEF*, 26.
5 *. . . while the Residenztheater* Pyenson, Lewis. *YE*, 38.
5 *Pauline loved to play* Reiser, Anton. *ABP*, 32.
5 *Maja reports that* Winteler-Einstein, Maria. *CPE1*, lvii.
5 *'I took violin lessons'*—. *CPE1*, lviii, Note 39.
5 *He loved to sit* —. *CPE1*, lxii.
5 *Curiously, he appears never* Talmey, Max. *The Relativity Theory Simplified and the Formative Period of its Inventor* (New York, Falcon Press, 1932), 165.
5 *The fact that the needle* Einstein, Albert. Autobiographical Notes. In *AEPS*, 8.
6 *In this vessel* Winteler-Einstein, Maria. *CPE1*, liii. See also Albert Einstein, *AEA* [Online]. 11 August 1952. Cat. no. 124–369. http://alberteinstein.info/.
6 *Einstein particularly looked forward* Einstein, Albert. *AEA* [Online]. 28 June 1941. Cat. no. 29–383. http://alberteinstein.info/.
6 *They spoke a broad Schwäbisch* Frank, Philipp. *LZ*, 15.
6 *'Since earliest childhood'* Reiser, Anton. *ABP*, 124.
6 *'. . . where the small inns'* Marianoff, Dimitri, Wayne, Palma. *Einstein: An Intimate Study of a Great Man* (New York, 1944), 35.
6 *'It was our custom'* Rogger, Franziska. *ES*, 14.
6 *The fact that he began* Winteler-Einstein, Maria. *CPE1*, lvii, Note 29.
6 *He characterised his teachers* Moszkowski, Alexander. *Einstein: Einblicke in seine Gedankenwelt* (Hamburg, Hoffman und Campe, 1921), 221.
6 *Even in arithmetic* Winteler-Einstein, Maria. *CPE1*, lix.
7 *His growing interest* Moszkowski, Alexander. *Einstein: Einblicke in seine Gedankenwelt* (Hamburg, Hoffman und Campe, 1921), 220.
7 *It was only at about twelve* Einstein, Albert. Autobiographical Notes. In *AEPS*, 2–4.
7 *It was also rather socially* Pyenson, Lewis. *YE*, 3, 49.
7 *In this establishment* Hettler, Nicolaus. *DEF*, 131–3.
7 *By Alexander* Moszkowski, Alexander. *Einstein: Einblicke in seine Gedankenwelt* (Hamburg, Hoffman und Campe, 1921), 221–3.
7 *'It's an amusing science'* Frank, Philipp. *LZ*, 26.

8 *'At the age of 12'* Einstein, Albert. Autobiographical Notes. In *AEPS*, 8.

8 *This request was* Weinstein, Galina. *Einstein's Pathway to the Special Theory of Relativity* (Cambridge, Cambridge Scholars Publishing, 2015), 32.

8 *This thin volume* Einstein, Albert. Autobiographical Notes. In *AEPS*, 8–10.

8 *In the years between* Hettler, Nicolaus. *DEF*, 110.

8 *'. . . we can be confident'* Pyenson, Lewis. *YE*, 47. Hettler corrects Pyenson's figure of six patents between 1886 and 1893.

8 *An internal walkway* Hettler, Nicolaus. *DEF*, 103; 'Lebenserinnerungen von Aloys Höchtl, geschrieben München 1934,' Appendix, ix–xvii.

8 *Although Talmud gave him* Winteler-Einstein, Maria. *CPE1*, lxi, Note 50.

9 *With great foresight* Weinstein, Galina. Albert Einstein: Rebellious Wunderkind. *ArXiv* [Online]. 21 May 2012 [Cited: 17 August 2016], 10, Note 49. http://arxiv.org/pdf/1205.4509.pdf.

9 *He also suggested that* Gregory, Frederick. The Mysteries and Wonders of Natural Science: Aaron Bernstein's 'Naturwissenschaftliche Volksbücher' and the Adolescent Einstein. In *Einstein: The Formative Years*, Don Howard and John Stachel, eds. (Boston, Birkhäuser, 2000), 32.

9 *'I recommended to him'* Talmey, Max. *The Relativity Theory Simplified and the Formative Period of its Inventor* (New York, Falcon Press, 1932), 164.

9 *'I really only started to learn'* Winteler-Einstein, Maria. CPE1, lviii, Note 39.

9 'Confusingly, Talmud recounts' Talmey, Max. The Relativity Theory Simplified and the Formative Period of its Inventor (New York, Falcon Press, 1932), 165.

9 *'The feeling of profound emotion'* Frank, Philipp. *Einstein: His Life and Times* (London, Jonathan Cape, 1948), 24–5. Curiously, this sentence is omitted from the German version of Frank's biography, since in general the German version is the complete text, and the English version has many omissions.

9 *'Albert does such bad work'* Marianoff, Dimitri, Wayne, Palma. *Einstein: An Intimate Study of a Great Man* (New York, Doubleday, Doran and Co., Inc., 1944), 32.

10 *After political pressure* Hettler, Nicolaus. *DEF*, Section 8.5, 147–57.

10 *The extent of their problem* Pyenson, Lewis. *YE*, 47.

10 *Finally, the loss* Hettler, Nicolaus. *DEF*, Section 10, 184–8.

10 *Maja thought that* Winteler-Einstein, Maria. *CPE1*, lii.

10 *Nevertheless, at this period* Pyenson, Lewis. *YE*, 37.

10 *These negotiations failed* Hettler, Nicolaus. *DEF*, 106.

11 *'The beautiful property'* Winteler-Einstein, Maria. *CPE1*, liii.

11 *Unfortunately for Einstein* Clark, Ronald. *Einstein: The Life and Times* (New York, Avon Books, 1972), 40.

11 *He also studied* Reiser, Anton. *ABP*, 39.

11 *When Albert learned* Fink, Michael, Hieronymus, Bess, Einstein, Alfred. The Autobiography and Early Diary of Alfred Einstein (1880–1952). *The Musical Quarterly*, 1980, 66: 361–77.

12 *Any German male* Winteler-Einstein, Maria. *CPE1*, lxiv.

12 *When asked what his transgression* —. *CPE1*, lxiii, Note 58.

12 *'The talent manifested'* Reiser, Anton. *ABP*, 39.

12 *Another part of the deal* Clark, Ronald. *Einstein: The Life and Times* (New York, Avon Books, 1972), 41.

13 *The impressive double doors* Gregori, Valentina. Einstein, genio scanzonato. *La Provincia Pavese* 28 January 2005. https://ricerca.gelocal.it/laprovinciapavese/archivio/laprovinciapavese/2005/01/28/PT1PO_PT101.html.

13 *Jakob found accommodation* Sanesi, Elena. Einstein e Pavia. *Settanta*, 1972, 3(20): 33–41.
13 *He had access there* Bracco, Christian. *EB*, 5. http://www.ilasl.org/index.php/Scienze/article/view/178.
13 *'You know, my nephew'* Winteler-Einstein, Maria. *CPE1*, lxiv, Note 62.
13 *Bracco points out* Bracco, Christian. EB, 5. http://www.ilasl.org/index.php/Scienze/article/view/178.
14 *In the accompanying letter* Einstein, Albert. Über die Untersuchungen des Aetherzustandes im Magnetischen Felde. *CPE1*, 5–10, Documents 5 and 6.
14 *At the same time* Frank, Philipp. *LZ*, 34–5.
14 *. . . the certificate releasing him* Einstein, Albert. *CPE1*, 20, Document 16.
14 *'Although limited at first'* Winteler-Einstein, Maria. *CPE1*, lxiv–lxv.
14 *'The free life'* —. *CPE1*, lxiv.
14 *Shortly thereafter he moved* Kamp, Johannes-Martin. *Kinderrepubliken: Geschichte, Praxis und Theorie radikaler Selbstregierung in Kinder- und Jugendheimen* (Wiesbaden, Springer-Verlag Fachmedien, 2013), 321, Note 261.
14 *No normal person* Einstein, Albert. AEA [Online]. 26 March 1952 [Cited: 14 July 2024]. Cat. no. 39–16. http://alberteinstein.info/.
15 *Ernestina was clearly struck* Pelizza-Marangoni, Ernestina. Momenti pavesi nella vita di Alberto Einstein. La Provincia Pavese. 14 May 1955.
15 *Einstein had already begun* Reiser, Anton. *ABP*, 42.
15 *Ernestina's family home* Bernini, Fabrizio. *Che bel ricordo . . . Casteggio* (Pavia, Gianni Iuculano, 2002).
15 *He had clearly also become* Einstein, Albert. *AEA* [Online]. 16 August 1946. Cat. no. 57–113. http://alberteinstein.info/.
15 *'Do you remember'* Neustätter, Otto. *AEA* [Online]. 12 March 1929. Cat. no. 30–436. http://alberteinstein.info/.
15 *Luzzatti developed a liking* Winteler-Einstein, Maria. *CPE1*, lxv.
15 *Herzog replied cautiously* Einstein, Albert. *CPE1*, 12–13, Document 7.
16 *'The examination gave me'* Einstein, Albert. Erinnerungen-Souvenirs. *Schweizerische Hochschulzeitung: Sonderheft 100 Jahre ETH*, 1955, 28: 145.
16 *If he graduated* Winteler-Einstein, Maria. *CPE1*, lxv.
16 *Winteler, originally of German* Schulmann, Robert. Tilling the Seedbed of Einstein's Politics: A Pre-1905 Harbinger? In *Einstein and the Changing Worldviews of Physics*, Jürgen Renn and Matthias Schemmel Christoph Lehner, eds. (New York, Birkhäuser, 2012), 62.
16 On 26 October 1895 Einstein, Albert, *CPE1*, 13–15, Documents 8 and 10.

Chapter 2

17 *However, he would have seen* Pyenson, Lewis. *YE*, 10–14.
17 *The town was small* —. *YE*, 9.
17 *However, it was in linguistics* Trubetzkoy, Nikolai. *Principles of Phonology*, 2nd ed. [trans.] Christine Baltaxe (Berkeley, University of California Press, 1971), 322.
18 *Children followed in regular succession* Rogger, Franziska. *ES*, 20.
18 *Einstein slotted in between* Winteler-Einstein, Maria. *CPE1*, lxv.
18 *The free and wide-ranging* Einstein, Hermann. *CPE1*, 17, Document 11.
18 *In addition, there was a link* Einstein, Albert. *CPE1*, 14, Document 9.

18 *'In his inspired teaching'* Pyenson, Lewis. *YE*, 14.

18 *Two other teachers* —. *YE*, 15–16.

19 *'He walked about'* Byland, Hans. Aus Einsteins Jugentagen. Ein Gedenkblatt. *Neue Büdner Zeitung*, 7 February 1928.

19 *He was a keen participant* Einstein, Albert. *CPE1*, 18, Document 11, Note 1.

19 *At this time he was* Moszkowski, Alexander. *Einstein: Einblicke in seine Gedankenwelt* (Hamburg, Hoffmann und Campe, 1921), 225.

19 *he wrote:* Einstein, Albert. Erinnerungen-Souvenirs. *Schweizerische Hochschulzeitung, Sonderheft 100 Jahre ETH*, 1955, 28: 146.

19 *It should be noted* Nelson, John D. Chasing the Light: Einstein's Most Famous Thought Experiment. In *Thought Experiments in Philosophy, Science and the Arts*, Mélanie Frappier, Letitia Meynell, James Robert Brown, eds. (Abingdon, Routledge, 2012), 123–140

20 *'When it was known'* Seelig, Carl. *BIOG*, 17–18.

20 *This was one of Aarau's* Cäcilienverein, Aarau. NK003_Cäcilienverein. *Stadtarchiv Aarau* [Online]. 31 May 2016. [Cited: 24 September 2024.] NK.003-008 *Chronik des Cäcilienvereins*, Band VI (Klebebuch) 1891–1896. https://www.aarau.ch/public/upload/assets/694/NK003_Caecilienverein.pdf?fp=1531983655998.

20 *It was probably at some time* Pelizza-Marangoni, Ernesta. Momenti pavesi nella vita di Alberto Einstein. *La Provincia Pavese*. 14 May 1955.

20 *Jost expressed some surprise* Winteler, Jost. *CPE1*, 18, Document 13.

20 *Ernestina sent Einstein* Pelizza-Marangoni, Ernesta. *AEA* [Online]. 25 December 1895. Cat. no. 81–142. http://alberteinstein.info/.

21 *Not only had Einstein* Einstein, Pauline. *CPE1*, 19, Document 15.

21 *In February he told her* Einstein, Albert. AEA [Online]. 18 February 1896. Cat. no. 95–662. http://alberteinstein.info/.

21 *Einstein was picked out* Ryffel, Johann. *CPE1*, 21, Document 17.

21 *Here they played Schubert* Seelig, Carl. *BIOG*, 17–18.

21 *After the end of the school* Einstein, Albert. *CPE1*, 14, Document 8, Note 5; 17, Document 10, Note 4.

21 *Certainly, Pyenson was misled* Pyenson, Lewis. *YE*, 12.

21 *This ambiguity extends* Einstein, Albert. *CPE1*, 17, Document 10, Note 5.

22 *His first absence* —. *CPE1*, 21, Document 18.

22 *'Einstein's excursions'* Highfield, Roger, Carter, Paul. *The Private Lives of Albert Einstein* (London, Faber and Faber, 1993), 24.

22 *'My mother . . . makes fun'* Fregonese, Lucio. *Gioventù felice in terra pavese* (Milan, Cisalpino, 2005), 97.

22 *In February, the Volta* Dept. of Physics 'A. Volta,' University of Pavia, Italy. Gli Einstein imprenditori pavesi. *Albert Einstein, Ingegniere dell'universo*. [Online] 2005. [Cited: 24 September 2024]. Poster on 'The Einstein Entrepreneurs Pavia' page. http://einstein-pavia.mpiwg-berlin.mpg.de/exhib_SEQ00000225_ZOG00000226.html.

23 *Einstein's lessons* Pyenson, Lewis. *YE*, 10–14.

23 *This involved a further* Hunziker, Herbert. The Physical Tourist: Albert Einstein's Magic Mountain: An Aarau Education. *Physics in Perspective*, 2015, 17: 55; 65–6.

24 *'It is all the same to me, Professor'* Seelig, Carl. *BIOG*, 22.

24 *When a violent thunderstorm* —. *BIOG*, 15–16.

24 *Given his long-standing* —. *BIOG*, 20–21.

24 *Einstein's movements* Einstein, Albert. *CPE1*, 25, Document 20.

24 *There seems to have been* Hettler, Nicolaus. *DEF*, 181.
24 *Maja, on the other hand* Winteler-Einstein, Maria. *CPE1*, liv.
25 *This opposition* —. *AEA* [Online]. 15 February 1924. Cat. no. 73–880, 17. http://alberteinstein.info/.
25 *By the time Einstein returned* Bracco, Christian. *EB*, 6. http://www.ilasl.org/index.php/Scienze/article/view/178.
25 *'A happy man'* Einstein, Albert. *CPE1*, 15–16, Document 22.
25 *'One is struck'* Pyenson, Lewis. *YE*, 19.
26 *On 29 October* Einstein, Albert. AEA [Online]. 29 October 1896. Cat. no. 76–519. http://alberteinstein.info/.

Chapter 3

27 *For the first two years* Seelig, Carl. *BIOG*, 40.
27 *The remaining money* Winteler-Einstein, Maria. *AEA* [Online]. 15 February 1924. Cat. no. 73–880, 17. http://alberteinstein.info/.
27 *Many years later* Einstein, Albert. *CPE13*, 192–3, Document 93.
27 *He attended a number* —. *CPE1*, 46, Document 28.
28 *Unusually for that period* Seelig, Carl. *BIOG*, 29.
28 *'I thank you so much'* Winteler, Marie. *CPE1*, 50–1, Document 29.
29 *As they sat in a corner* Bracco, Christian. *EB*, 6. http://www.ilasl.org/index.php/Scienze/article/view/178.
29 *Einstein subsequently introduced* Overbye, Dennis. *Einstein in Love: A Scientific Romance* (London, Bloomsbury, 2001), 25.
29 *As far as Newton was concerned* Frank, Philipp. *Einstein: His Life and Times* (London, Jonathan Cape, 1948), 49. NB: While the German edition of Frank contains much that is not in the English edition, the quotation from the diary of Newton's student David Gregory referred to here is only in the English edition.
29 *'That Gravity should be innate'* Newton, Isaac. *Four Letters from Sir Isaac Newton to Doctor Bentley*, Vol. Letter III (London, R. and J. Dodsley, 1756), 25–6.
30 *Einstein mentions explicitly* Einstein, Albert. Autobiographical Notes. In *AEPS*, Vol. VII, 62.
30 *This book, recommended to Einstein* Mach, Ernst. *Die Mechanik in Ihre Entwicklung* (Leipzig, F.A. Brockhaus, 1883).
30 *'While it is never safe'* Michelson, Albert. Speech at the Dedication of the Ryerson Physical Laboratory. *Annual Register of the University of Chicago* (Chicago, University of Chicago Press, 1896), 159.
30 *A prediction surpassed Hathaway,* W. S. *The Speeches of the Right Honourable William Pitt in the House of Commons*, Vol. II (London, Longman, 1806), 36
31 *She almost seemed relieved* Einstein, Pauline. *CPE1*, 54, Document 32.
31 *Her memoir* Niggli, Julia. *AEA* [Online]. 1 January 1955. Cat. no. 72–81. http://alberteinstein.info/.
32 *In the first letter* Einstein, Albert. *CPE1*, 55–6, Document 34.
32 *The second letter* —. *CPE1*, 57–8, Document 35.
32 *Einstein seemed initially* Winteler, Marie. *AEA* [Online]. 3 February 1957. Cat. no. 71–183. http://alberteinstein.info/.

33 *She herself requested* Trbuhović-Gjurić, Desanka. *ISAE*, 14–26.
34 *In October of that year* Highfield, Roger, Carter, Paul. *The Private Lives of Albert Einstein* (London, Faber and Faber, 1993), 33–8.
34 *Trbuhović-Gjurić recounts* Trbuhović-Gjurić, Desanka. *ISAE*, 36.
34 *Once, when asked why* Seelig, Carl. *BIOG*, 33.
34 *Again, the sheer pluck* —. *BIOG*, 45.
34 *He used Maxwell* Poincaré, Henri. Sur les rapports de l'analyse pure et de la physique mathématique. In *Verhandlungen des Ersten Internationalen Mathematiker-Kongresses*, F. Rudio, ed. (Zurich, B.G. Teubner, 1898), 81–90.
34 *Many years later he wrote* Einstein, Albert. Autobiographical Notes. In *AEPS*, Vol. VII, 15–17.
34 *The confusion probably* Kolros, Louis. *Albert Einstein en Suisse – Souvenirs.* Basel : Swiss Physical Society, 1956, Helvetica Physica Acta – Supplementum 4. Fünfzig Jahre Relativitätstheorie – Cinquantenaire de la Théorie de la Relativité, Vol. 29. p. 273.
34 *but quotes from* Poincaré, Henri. *Science and Hypothesis.* London : Walter Scott, 1905. p. 90.
35 *Although this letter* Marić, Mileva. *CPE1*, 58, Document 36.
35 *Einstein, despite his delayed reply* Einstein, Albert. *CPE1*, 211–12, Document 39.
36 *'Weber was a typical spokesman'* Seelig, Carl. *BIOG*, 34.
36 *Decades later, Einstein recalled* Einstein, Albert. Erinnerungen-Souvenirs. *Schweizerische Hochschulzeitung: Sonderheft 100 Jahre ETH*, 1955, 28: 146.
36 *However, the ETH regulations* Eidgenossen Polytechnischen Schule, Zurich. Studium am Polytechnikum in Zürich (1896–1900). *ETH Bibliothek* [Online]. [Cited: 25 September 2024.] https://library.ethz.ch/standorte-und-medien/plattformen/einstein-online/studium-am-polytechnikum-in-zuerich-1896-1900.html.
36 *Indeed, several of Grossmann's* Sauer, Tilman. Marcel Grossmann and His Contribution to the General Theory of Relativity. *arXiv.org* [Online]. 22 April 2014, 4, Note 2. https://arXiv:1312.4068v2.
36 *He remarked to Max Born* Seelig, Carl. *BIOG*, 33.
36 *'His studies were carried out'* Winteler-Einstein, Maria. *AEA* [Online]. 15 February 1924. Cat. no. 73–880, 18. http://alberteinstein.info/.
36 *Einstein became a regular* Seelig, Carl. *BIOG*, 40. See also Walter Leich, *AEA* [Online]. 2 March 1950. Cat. no. 60–251. http://alberteinstein.info/.
37 *When Einstein's visits* Trbuhović-Gjurić, Desanka. *ISAE*, 53–4. See also Milan Popović, *AS*, 4.
37 *'I have quite a lot to do'* Winteler-Einstein, Maria. *AEA* [Online]. 15 February 1924. Cat. no. 73–880, 17. http://alberteinstein.info/.
37 *By early 1899* Bracco, Christiano. *EB*, 6, Note 17. http://www.ilasl.org/index.php/Scienze/article/view/178.
37 *The evidence therefore points* Einstein, Albert. *CPE1*, 211–16, Documents 39–41, 45.
37 *Einstein returned to Zurich* Dünki, Robert. Albert Einstein in Zurich: Meldedaten der Einwohnerkontrolle. *Stadt Zürich Präsidialdepartement* [Online]. [Cited: 27 September 2024.] https://www.stadt-zuerich.ch/prd/de/index/stadtarchiv/bilder_u_texte/texte_zu_albert_einstein/albert_einstein_in_zuerich.html.
37 *Einstein did very well* Einstein, Albert. *CPE1*, 214, Document 42.
38 *'In the evenings'* Seelig, Carl. *BIOG*, 41–3.
39 *'I often work with [Mileva]'* Niggli, Julia. *AEA* [Online]. 1 January 1955. Cat. no. 72–81. https://ein-web.adlibhosting.com/aea/Details/archive/110050503.

39 *'I lived on the third floor'* Milentijević, Radmila. *Mileva Marić: A Life with Albert Einstein* (New York, United World Press, 2010), 47–8.
39 *However, he did promise* Einstein, Albert. *CPE1*, 218–19, Document 48.
39 *In his next letter* —. *CPE1*, 221–3, Document 51.
40 *She asked Rosa to send* Schlegel, Katja. Für nur 50 Rappen erfuhr sie Julias Geheimnis – und ihr Interesse an Einstein. *Aargauer Zeitung* [Online]. 14th January 2017. https://www.aargauerzeitung.ch/aargau/aarau/fuer-nur-50-rappen-erfuhr-sie-julias-geheimnis-und-ihr-interesse-an-einstein-130848845.
40 *Anna Schmid (known usually as Anneli)* Einstein, Albert. *CPE1*, 220, Document 49.
40 *Einstein was convinced* —. *CPE1*, 225–27, Document 52.
40 *He returned to this theme* Fizeau, Hippolyte. Sur les hypotheses relatives à l'éther lumineux, et sur une expérience qui parait démontrer que le mouvement des corps change la vitesse avec laquelle la lumiere se propage dans leur intérieur. *Comptes Rendus*, 1851, 33: 349.
40 *Besso moved to Milan* Bracco, Christian. *EB*, 12; http://www.ilasl.org/index.php/Scienze/article/view/178.
40 *It was a combination* Falconer, Isobel. J J Thomson and the Discovery of the Electron. *Physics Education*, 1997, 32: 230.
41 *Although she lodged* Rogger, Franziska. *ES*, 19.
41 *'But if I were to see that girl'* Einstein, Albert. *CPE1*, 233–5, Document 57.
41 *She was at least relieved* Trbuhović-Gjurić, Desanka. *ISAE*, 63.
42 *'Already in his school'* Winteler-Einstein, Maria. *AEA* [Online]. 15 February 1924. Cat. nos. 73–880, 19–20. http://alberteinstein.info/.
42 *He finally became a Swiss citizen* Coradi, Adam. *CPE1*, 239–41, 272, Document 84.
42 *At Zurich, he frequented the Opera* Reiser, Anton. *ABP*, 54.
42 *Although Mileva implies* Einstein, Albert. *CPE1*, 243–4, Document 63 and notes therein.
43 *'. . . before his graduation'* Bernini, Fabrizio. *La Casteggio di Albert Einstein ed Ernestina Pelizza Marangoni* (Pavia, Gianni Iuculano Editore, 2002), 49.
43 *A letter written* Einstein, Albert. *CPE1*, 244–5, Document 64, and notes therein.
43 *Milana Bota, who only two years* Trubuhović-Gjurić, Desanka. *ISAE*, 64.
43 *'After dinner we first took a walk'* Popović, Milan. *AS*, 60.
43 *Both she and Einstein did rather poorly* Seelig, Carl. *BIOG*, 35.
43 *Her mark in the latter* Einstein, Albert. *CPE1*, 247, Document 67.
43 *Einstein and Mileva* —. *CPE1*, 247–9, Documents 67 and 68. The date attributed to Document 68 implies that Einstein left Zurich the day on which Hurwitz wrote the formal letter requesting approval for the marks. The students may have been told informally that day or possible slightly earlier, before the marks were formally confirmed by the President of the Council.
44 *'This necessity (revision)'* —. Autobiographical Notes. In *AEPS*, Vol. VII, 16.

Chapter 4

45 *He continued to be solicitous* Einstein, Albert. *CPE1*, 248–9, Document 68.
46 *His letters from this period* —. *CPE1*, 252–5, Documents 71 and 72.
46 *In a postcard* —. Christie's. *Einstein and Family: Letters and Portraits* [Online]. 11 September 1900. [Cited: 26 September 2024.] Lot 11. https://onlineonly.christies.com/s/einstein-family-letters-portraits/trip-venice-his-father-11/55692.

46 *One such was to investigate* —. *CPE1*, 257–9, Document 74, and notes therein.

46 *He eagerly anticipated* —. *CPE1*, 261–3, Document 76.

46 *He also did some legal work* sull'Oglio, Comune di Cannetto. Amministrazione. *lombardiabeniculturali.* [Online] 16 January 1909. [Cited: 26 September 2024.] Paragraph 1. http://www.lombardiabeniculturali.it/archivi/unita/MIUD0FCFC7/.

46 *The physics studies related* Einstein, Albert. *CPE1*, 57–8, Document 35.

46 *These speculations* —. *CPE1*, 264–7, Document 79.

47 *The lack of any allowance* —. *CPE2*, 8.

47 *'It is a wonderful feeling'* —. *CPE1*, 290–1, Document 100.

47 *Furthermore, Heinrich Weber* Clark, Ronald. *Einstein: The Life and Times* (New York: Avon Books, 1972), 62.

47 *Einstein's disappointment was assuaged* Trbuhović-Gjurić, Desanka. *ISAE*, 69.

47 *He also was paid* Seelig, Carl. *BIOG*, 57.

47 *Einstein apparently did not* Einstein, Albert. *CPE2*, 9–21, Document 1.

48 *Mileva knew that he intended* Popović, Milan. *AS*, 69–72.

48 *'I remember you very clearly'* Lebegott, Edward. *AEA* [Online]. 22 January 1931. Cat. no. 47–379. http://alberteinstein.info/.

48 *It is also surely the case* Frank, Philipp. *LZ*, 42–3.

49 *Even this level of flattery* Einstein, Hermann. *CPE1*, 289, Document 99.

49 *It was not, however* Einstein, Albert. *CPE1*, 277–8, Document 91.

49 *His parents were weighed* —. *CPE1*, 300, Document 107.

49 *Einstein was dragged* —. *CPE1*, 306, Document 112.

49 *Fortunately, Frau Winteler* Rogger, Franziska. *ES*, 23–5.

49 *When Einstein arrived* Bracco, Christian. *EB*, 17–19. http://www.ilasl.org/index.php/Scienze/article/view/178.

50 *Einstein concluded the anecdote* Einstein, Albert. *CPE1*, 281–3, Document 94.

50 *In addition to the usual* —. *CPE1*, 296, Document 104.

50 *He was also thinking deeply* —. *CPE1*, 290–1, Document 100.

50 *His letter to Mileva* —. *CPE1*, 292, Document 101.

50 *After some prevarication* Einstein, Albert, Marić, Mileva. *CPE1*, 293–8, Documents 102, 103, and 105.

51 *She was also* Einstein, Albert. *CPE1*, 301–3, Document 109.

51 *'He really loves children'* Winteler-Einstein, Maria. *AEA* [Online]. 15 February 1924. Cat. no. 73–880, 2. http://alberteinstein.info/.

51 *However, with typical self-absorption* Einstein, Albert. *CPE1*, 304–5, Document 111.

51 *Completing this set* —. *CPE1*, 298–9, Document 106.

51 *Einstein wrote to Winteler* —. *CPE1*, 309–10, Document 115.

52 *It is likely that these* Renn, Jürgen. Einstein's Controversy with Drude and the Origin of Statistical Mechanics: A New Glimpse from the 'Love Letters.' *Archive for History of Exact Sciences*, 1997, 51(4): 315–54.

52 *Einstein was incensed* Einstein, Albert. *CPE1*, 303–9, Documents 110–14.

52 *There are some very evocative images* Barthelts, Marie. *AEA* [Online]. 10 March 1929. Cat. no. 30–384. http://alberteinstein.info/.

53 *Einstein seems to have joined* Flückiger, Max. *Albert Einstein in Bern* (Bern: Verlag Paul Haupt, 1974), 31.

53 *It was presumably at rehearsals* Orchestergesellschaft Winterthur. [Online] https://www.orchestergesellschaft.ch/.

53 *Schnauder was also a composer* Einstein, Albert. *CPE5*, 46, Document 43.

53 *His only contribution* —. *CPE8B*, 989–90.
53 *Sadly, no celebrations* —. Christie's. *Einstein and Family: Letters and Portraits* [Online]. 27 July 1901. [Cited: 26 September 2024.] Lot 12. https://onlineonly.christies.com/s/einstein-family-letters-portraits/einstein-mountains-12/55691.
53 *It is not known if Einstein* —. *CPE1*, 313–15, Document 122.
54 *It is likely that a friend* Seelig, Carl. *BIOG*, 60.
54 *Eventually, Einstein moved out* Einstein, Albert. *CPE1*, 320–7, Documents 126–9.
55 *Einstein told Mileva* Marić, Mileva, Albert Einstein. *CPE1*, 316–28, Documents 125–30.
55 *He wrote that he led a life* Einstein, Albert. *CPE1*, 325–6, Document 128.
55 *In addition to the rehearsals* Cordes, Martin. Albert Einstein in Schaffhausen. Stadtarchiv Schaffhausen [Online]. 27 September 2024, 8–9. http://www.stadtarchiv-schaffhausen.ch/uploads/media/Einstein-Broschuere.pdf.
55 *Although there is no record* Winteler-Einstein, Maria. *AEA* [Online]. 15 February 1924. Cat. no. 73–880, 19. http://alberteinstein.info/.
56 *Einstein was examined* Reiser, Anton. *ABP*, 62–3.
56 *Towards its end* Einstein, Albert. *CPE1*, 331, Document 133.
56 *Einstein picked up the entire* —. *CPE1*, 331, Document 132.

Chapter 5

57 *He plied Mileva* Einstein, Albert. *CPE1*, 332–3, Document 134.
57 *The only other mention* —. *CPE5*, 22–3, Document 13.
57 *In later life, he blamed* Winteler-Einstein, Maria. *AEA* [Online]. 15 February 1924. Cat. no. 73–880, 20. http://alberteinstein.info/.
57 *Einstein had to reassure* Einstein, Albert. *CPE1*, 335–6, Document 137.
58 *'Private lessons in Mathematics'* —. *CPE1*, 334, Document 135.
58 *This had an immediate response* —.*CPE1*, 334, Document 136.
58 *It was the Easter advertisement* Solovine, Maurice. Freundschaft mit Albert Einstein. *Physikalische Blätter*, 1959, 15(3): 98–9. Originally published as the foreword to *Albert Einstein: Briefe an Maurice Solovine* (Dusseldorf, Verlag Johann Fladung, 1956).
58 *'We have in the law'* Pearson, Karl. *The Grammar of Science*, 2nd ed. (London, Adam & Charles Black, 1900), 99.
58 *Paul thereafter decamped* Rogger, Franziska. *ES*, 28.
59 *'We occupied ourselves mostly'* Einstein, Albert. *AEA* [Online]. 6 March 1952 [Cited: 16 July 2024]. Cat. no. 7–205. http://alberteinstein.info/.
59 *'. . . the edifice of electro-dynamics'* Poincaré, Henri. *Science and Hypothesis* (London, Walter Scott, 1905), 244. Other references: Brownian motion, 179; Lorentz electron theory, 242–4; non-Euclidean geometries, 35–50; relativity, 72–88; absolute space and time, 90–1; Fizeau experiment, 170.
60 *'Bern was fascinating'* Solovine, Maurice. Freundschaft mit Albert Einstein. *Physikalische Blätter*, 1959, 15(3): 100–2.
61 *Chavan's wife also became* Flückiger, Max. *EIB*, 136.
61 *The paper was completed* Einstein, Albert. *CPE2*, 22–40, Document 2.
61 *'My main purpose for doing this'* —. Autobiographical Notes. In *AEPS*, 46.
61 *It transpires that much* Gibbs, Josiah Willard. *Elementary Principles in Statistical Mechanics* (New York, Charles Scribner's Sons, 1902).
62 *Einstein had probably drafted* Einstein, Albert. *CPE2*, 41–55, 57–75, Document 3.

62 *'Only Mr and Mrs Einstein'* Talmey, Max. *The Relativity Theory Simplified and the Formative Period of its Inventor* (New York, Falcon Press, 1932), 166–7.
62 *She was glad to escape* Rogger, Franziska. *ES*, 30.
63 *By the end of May* Brenner, Ernst, Friedrich Haller. *CPE1*, 338–40, Documents 140–2.
63 *It seems likely that* Milentijević, Radmila. *Mileva Marić Einstein: Life with Albert Einstein* (New York, United World Press, 2010), 97.
63 *. . . she was certainly nearby* Einstein, Albert. *CPE5*, 5, Document 1.
63 *Einstein therefore moved again* Flückiger, Max. *EIB*, 134, but see plate of Einstein's Einwohnerkontrolle entries between pp. 136–7, which contradicts parts of p. 134.
63 *Sometime in that summer* Einstein, Albert. *CPE5*, 6–7, Document 2.
64 *Hermann was buried* Pais, Abraham. *LORD*, 47 – private communication attributed to Helen Dukas. See also Plesch, J., Plesch, P. H. Some Reminiscences of Albert Einstein. *Notes and Records of the Royal Society of London*, 1995, 49(2): 307.
64 *The marriage date was fixed* Einstein, Albert. *CPE5*, 9–10, Document 4.
64 *Michele Besso had returned* Rogger, Franziska. *ES*, 28–30.
64 *A photograph exists* Flückiger, Max. *EIB*, 35.
64 *Einstein may have taken* Fölsing, Albrecht. *Albert Einstein: Eine Biographie* (Frankfurt am Main, Suhrkamp Verlag, 1993), 137. Fölsing considers this visit to have been in June 1902, although he cites no evidence.
64 *'The evening was spent'* Winteler-Einstein, Maria. *AEA* [Online]. 15 February 1924. Cat. no. 73–880, 20–1. http://alberteinstein.info/.
65 *'I am now a married man'* Einstein, Albert. *CPE5*, 10–12, Document 5.
65 *'I am, if possible'* Einstein-Marić, Mileva. *The Albert Einstein Archives.* [Online] 20 March 1903. Cat. no. 70–720. http://alberteinstein.info/.
65 *That Mileva was still* Winteler-Einstein, Maria. *AEA* [Online]. 15 February 1924. Cat. no. 73–880, 22. http://alberteinstein.info/.
65 *'Mileva was clever and reserved'* Solovine, Maurice. Freundschaft mit Albert Einstein. *Physikalische Blätter*, 1959, 15(3): 103.
65 *This last is interesting* Seelig, Carl. *BIOG*, 72.
65 *Maja blamed this* Winteler-Einstein, Maria. *AEA* [Online]. 15 February 1924. Cat. no. 73–880, 22. http://alberteinstein.info/.
66 *'His job was therefore reduced'* —. *AEA* [Online]. 15 February 1924. Cat. no. 73–880, 21–2. http://alberteinstein.info/.
66 *The secretary of the Society* Flückiger, Max. *EIB*, 71–2.
66 *Einstein gave up yet another* Einstein, Albert. *CPE5*, 17–18, Document 7.
67 *He went on to react* —. *CPE5*, 22–3, Document 13.
67 *A small table stood* Flückiger, Max. *EIB*, 99.
67 *That it was another difficult birth* Popović, Milan. *AS*, Letter from 14th June 1904.
67 *'Albert's salary was only just'* Winteler-Einstein, Maria. *AEA* [Online]. 15 February 1924. Cat. no. 73–880, 22. http://alberteinstein.info/.
67 *Another possibility is* Milentijević, Radmila. *Mileva Marić Einstein. A Life with Albert Einstein* (New York, United World Press, 2010), 104–10.
68 *It is also interesting* Einstein, Albert. *CPE5*, 25–6, Document 17.
68 *Since he reviewed several papers* Stachel, John, Cassidy, David C., Renn, Jürgen, Schulmann, Robert. *CPE2*, 109–11.
68 *It is interesting to note that* Einstein, Albert. *CPE2*, 98–108, Document 5.

68 *The Bach Solo Sonatas* Flückiger, Max. *EIB*, 137–8.
68 *It is typical that it did not* Einstein, Albert. *CPE5*, 27–29, Documents 20–3.
69 *Any disappointment Einstein might* Einstein, Albert. *CPE5*, 29–30, Document 24.
69 *She had saved up enough* Rogger, Franziska. *ES*, 32.

Chapter 6

70 *Almost every day* Einstein, Albert. *CPE9*, 293, Document 207.
70 *He and Einstein were again inseparable* Flückiger, Max. *EIB*, 73.
70 *He unpacked his violin* —. *EIB*, 60.
70 *It is entitled* Einstein, Albert. *CPE2*, 149–69, Document 14.
71 *He had introduced a method* Pais, Abraham. *LORD*, 69–70.
71 *Einstein's closing sentence* Einstein, Albert. *CPE2*, 98–108, Document 5.
72 *These move without* —. *CPE2*, 149–69, Document 14.
73 *However, the cumulative effect* Wheaton, Bruce R. Philipp Lenard and the Photoelectric Effect, 1889–1911. *Historical Studies in the Physical Sciences*, 1978, 9: 299–322.
73 *The inevitable delay* Einstein, Albert. *CPE2*, 170.
73 *Einstein added one line* Seelig, Carl. *Albert Einstein: Leben und Werk eines Genies unserer Zeit* (Zurich, Europa Verlag, 1960), 112.
73 *It was dedicated to his friend* Einstein, Albert. *CPE1*, 281–83, Document 94.
73 *The thesis* —. *CPE2*, 183–205, Document 15.
74 *Unfortunately, Einstein made* —. *CPE2*, 170–82.
74 *Therefore, although relatively unknown* Straumann, Norbert. On Einstein's Doctoral Thesis. *arXiv.org* [Online]. 2 February 2008. ArXiv:physics/0504201v1. https://arxiv.org/pdf/physics/0504201.pdf.
75 *The extremely short time* Einstein, Albert. *CPE5*, 36, Note 2.
75 *By careful experimentation* Brown, Robert. A Brief Account of Microscopical Observations on the Particles Contained in the Pollen of Plants and the General Existence of Active Molecules in Organic and Inorganic Bodies. *Edinburgh New Philosophical Journal*, 1828, 5: 358–71.
75 *As pointed out* Renn, Jürgen. Einstein's Invention of Brownian Motion. *Annalen der Physik*, 2005, 14: 23–37.
75 *Poincaré mentions Brownian motion* Poincaré, Henri. *Science and Hypothesis* (London, Walter Scott, 1905), 179.
76 *'If the motion discussed here'* Einstein, Albert. *CPE2*, 223–36, Document 16.
76 *What they failed* Renn, Jürgen. Einstein's Invention of Brownian Motion. *Annalen der Physik*, 2005, 14: 23–37
77 *Eventually, it was Jean Baptiste Perrin* —. *CPE2*, 215–22.
77 *At some point* Medicus, Heinrich A. The Friendship among Three Singular Men: Einstein and His Swiss Friends Besso and Zangger. *Isis*, 1994, 85(3): 457–9. http://www.jstor.org/stable/235463.
77 *His letters to Mileva* Einstein, Albert. *CPE1*, 281–4, Document 94.
77 *There is little documentary evidence* Flückiger, Max. *EIB*, 76.
77 *An enormous amount* Weinstein, Galina. Einstein Chases a Light Beam. *ArXiv* [Online]. December 2012, 13 and Note 43. https://arxiv.org/pdf/1204.1833.pdf.
78 *Indeed, in his lecture* Einstein, Albert. *CPE13*, 629–41, Document 399.

78 *In an attempt to circumvent* Rynasiewicz, Robert, Renn, Jürgen. The Turning Point for Einstein's Annus mirabilis. *Studies in History and Philosophy of Modern Physics*, 2005, 37(1): 5–35. The authors make a convincing case for the strong interdependence of the three major 1905 papers.

78 *A decade earlier* Lorentz, H. A. *Versuch einer Theorie der elektrischen und optischen Erscheinungen in bewegten Körpern* (Leiden, E.J. Brill, 1895).

78 *He, and independently Irish physicist* Fitzgerald, G. F. The Ether and the Earth's Atmosphere. *Science*, 1889, 13: 390.

78 *The physical source* Brown, Harvey R. Michelson, FitzGerald and Lorentz: The Origins of Relativity Revisited. *Philosophy of Science Archive* [Online]. 10 February 2003 [Cited: 13 September 2017]. http://philsci-archive.pitt.edu/987/1/Michelson.pdf. This paper contains an interesting discussion of the background to the Lorentz–Fitzgerald contraction.

78 *Such an effect had recently* Heaviside, O. The Electro-magnetic Effects of a Moving Charge. *The Electrician*, 1888, 22: 147–8.

78 *In the Japanese record* Abiko, Seiya. Einstein's Kyoto Address: 'How I Created the Theory of Relativity'. *Historical Studies in the Physical and Biological Sciences*, 2000, 31(1): 1–35. See also Einstein, Albert, AEA [Online]. 9 February 1954. Cat. no. 17–199. http://alberteinstein.info/.

79 *Subsequently, Lorentz set up* Lorentz, H.A. Electromagnetic Phenomena in a System Moving with Any Velocity Smaller Than That of Light. *Proceedings of the Royal Netherlands Academy of Arts & Sciences*, 1904, 6: 809–31.

79 *Poincaré had polished* Poincaré, Henri. Sur la dynamique de l'électron. *Comptes rendus de l'Académie des Sciences*, 1905, 140: 1504–8. See also Poincaré, Henri. Sur la dynamique de l'électron. *Rendiconti del Circolo Matematico di Palermo*, 1906, 21: 129–75.

79 *Einstein was unaware* Einstein, Albert. *AEA* [Online]. 19 February 1955 [Cited: 14 August 2024]. Cat. no. 124–425. http://alberteinstein.info/.

79 *Fizeau's experiment* Fizeau, Hippolyte. Sur les hypothèses relatives à l'éther lumineux. *Comptes Rendu de l'Academie des Sciénces*, 1851, 33: 349–55.

80 *Again, as so often* Poincaré, Henri. La Measure du Temps. *Revue de Métaphysique et de Morale*, 1898, 6: 1–13.

80 *Having made a decisive step* Poincaré, Henri. *Science and Hypothesis* (London, Walter Scott, 1905), 90–1.

80 *Later, in* The Foundations of Science —. *The Foundations of Science* [trans.] George Bruce Halstead (Lancaster, The Science Press, 1946), 306–8.

80 *Around 16 May* Flückiger, Max. *EIB*, 134.

80 *This time, Einstein took* Einstein, Albert. *CPE5*, 9–10, Document 4.

81 *'. . . of these the first I can send'* —. *CPE5*, 31–32, Document 28.

81 *The paper 'On the Electrodynamics of Moving Bodies'* —. *CPE2*, 275–310, Document 23.

81 *He began with an example* Weinstein, Galina. Einstein Chases a Light Beam. *ArXiv* [Online]. December 2012, 13 and Note 43. https://arxiv.org/pdf/1204.1833.pdf.

82 *Indeed, only two years later* Einstein, Albert. *CPE2*, 433–88, Document 47.

82 *When this was confirmed* Pais, Abraham. *LORD*, 147 – private communication attributed to Helen Dukas. See also Einstein, Bernhard, 'Recollections of My Grandfather. In *The Fascinating Life and Theory of Albert Einstein*,W. C. Mih, ed. (New York, Kroshka Books, 2000), xii.

83 *Michelmore reported* Michelmore, Peter. *Einstein, Profile of the Man* (New York, Dodd, Mead & Company, 1962), 46.

84 *Huber also conducted* Flückiger, Max. *EIB*, 73–4; 175. Note that the latter contains a reference to Einstein playing chess with Huber. However, since it relies on the recollections of unnamed contemporaries long after the event, it is not clear what credence can be placed on it.

84 *They visited Mileva's maternal* Trbuhović-Gjurić, Desanka. *ISAE*, 69.
84 *This three-page paper* Einstein, Albert. *CPE2*, 311–14, Document 24.
85 *'The idea is amusing'* —. *CPE5*, 32–3, Document 28.
85 *'The celebrations in the little apartment'* Winteler-Einstein, Maria. *AEA* [Online]. 15 February 1924. Cat. no. 73–880, 22. http://alberteinstein.info/.

Chapter 7

86 *On receiving his pay slip* Flückiger, Max. *EIB*, 64–8.
86 *Indeed, he was the first* Pais, Abraham. *LORD*, 150–1, and references therein.
86 *He derived, among other things* Planck, Max. Das Prinzip der Relativität und die Grundgleichungen der Mechanik. *Verhandlungen Deutsche Physikalische Gesellschaft*, 1906, 8(7): 136–41.
86 *After the long period* Winteler-Einstein, Maria. *AEA* [Online]. 15 February 1924. Cat. no. 73–880, 26. http://alberteinstein.info/.
87 *'Having arranged things by letter'* Seelig, Carl. *BIOG*, 92–93.
87 *Thus in the cramped* Winteler-Einstein, Maria. AEA [Online]. 15 February 1924. Cat. no. 73–880, 26. http://alberteinstein.info/.
87 *Wilhelm Röntgen wrote* Einstein, Albert. *CPE5*, 43–4, Document 40.
87 *Arnold Sommerfeld attended* —. *CPE2*, 267.
88 *Mileva also asked* Trbuhović-Gjurić, Desanka. *ISAE*, 101.
88 *Marie Winteler's mother died* Rogger, Franziska. *ES*, 32–6.
88 *Einstein wrote a touching letter* Einstein, Albert. *CPE5*, 44–5, Document 41.
88 *As might be expected* —. AEA [Online]. 25 May 1917. Cat. no. 95–358. http://alberteinstein.info/.
88 *In December 1908* Rogger, Franziska. *ES*, 39–44.
88 *'I am a dignified Federal'* Einstein, Albert. *CPE5*, 46, Document 43.
88 *In fact, Einstein was often* Highfield, Roger, Carter, Paul. *The Private Lives of Albert Einstein* (London, Faber and Faber, 1993), 129.
88 *Clearly, despite Mileva's love* Trbuhović-Gjurić, Desanka. *ISAE*, 21.
88 *'Mileva played the tamburitza'* Popović, Milan. *AS*, 4.
89 *'He also loves music a lot'* —. *AS*, 95.
89 *His lecture on Brownian motion* Einstein, Albert. *CPE2*, 408, Document 43.
89 *Einstein subsequently blamed* —. *CPE5*, 48, Document 46.
89 *This is borne out* Seelig, Carl. *BIOG*, 103–4.
90 *His desultory attempts* Einstein, Albert. *CPE5*, 56, Document 48.
90 *Certainly, on 16 August* —. *CPE5*, 70, Document 54.
90 *In September he wrote* —. *CPE5*, 72, Document 56.
90 *It was completed* —. *CPE5*, 74–80, Documents 58, 60, 61, 63, 64, and 66.
90 *The* Yearbook *review* —. *CPE2*, 432–88, Document 47.
90 *New things included* Kaufmann, W. Über die Konstitution des Elektrons. *Annalen der Physik*, 1906, 19: 487–554.
91 *Indeed, Einstein's self-belief* Bucherer, Alfred. *CPE5*, 133–49, Documents 117, 119, 120, and 128. Einstein's replies, newly discovered, show his excitement at Bucherer's results (see Lots 177–80, https://www.sothebys.com/en/buy/auction/2024/books-manuscripts-and-music-from-medieval-to-modern).
91 *'I was sitting in a chair'* Einstein, Albert. *CPE13*, 638, Document 399.
91 *'As far as we know'* —. *CPE2*, 476, Document 47.

91 *'There then came to me'* —. *CPE7*, 265, Document 31.
92 *Abraham Pais concludes his account* Pais, Abraham. *LORD*, 183.
93 *They played works by Haydn* Seelig, Carl. *BIOG*, 91.
93 *However, Einstein's hopes* Einstein, Albert. *CPE5*, 82, Document 69.
93 *A few weeks later* —. *CPE5*, 85–6, Document 72.
93 *Indeed, in another letter* —. *CPE5*, 86–9, Document 73.
93 *Irrespective of the difficulties* —. *CPE2*, 490–2, Document 48.
93 *Despite the discouraging news* —. *CPE5*, 92, Document 76.
94 *His* venia docendi Flückiger, Max. *EIB*, 119–20.
94 *Even before his* venia Einstein, Albert. *CPE5*, 99, Document 84.
94 *Einstein began his lecture* Physics Department, University of Bern. Informationen zum Studiengang Physik und Astronomie. [Online] 24 August 2025, 7. https://www.philnat.unibe.ch/e17097/e95924/e258233/e261719/Physik_Broschuere_ger.pdf.
94 *Unsurprisingly, he attracted* Flückiger, Max. *EIB*, 122. Note that Seelig mistakenly names the third attendee as Schenker.
94 *It is clear that* Einstein, Albert. *CPE5*, 119–21, Document 101.
94 *On Sundays they went walking* Seelig, Carl. *BIOG*, 85–6.
94 *The tone of this letter* Einstein, Albert. *CPE5*, 114–15, Document 96.
94 *'God knows, Wagner is not'* Seelig, Carl. *BIOG*, 86.
95 *They produced two such papers* Minkowski, Hermann. Die Grundgleichungen für die electromagnetischen Vorgänge in bewegten Körpern. *Nachrichten von der Gesellschaft der Wissenschaften zu Göttingen, Mathematisch-Physikalische Klasse*, 1908, 1908: 53–111.
95 *It was Einstein's dislike* Pais, Abraham. *LORD*, 152.
95 *This discussion rumbled on* Einstein, Albert. *CPE2*, 503–7.
95 *'Gentlemen! The picture of space'* Minkowski, Herman. Raum und Zeit. *Physikalische Zeitschrift*, 1909, 10: 104–11.
95 *Indeed, as modified* Walter, Scott. Hermann Minkowski and the Scandal of Spacetime. [Online] Spring 2008 [Cited: 24 August 2025.] https://shs.hal.science/halshs-00319209/document.
95 *In January 1909* Weinstein, Galina. Max Born, Albert Einstein and Hermann Minkowski's Space-Time Formalism of Special Relativity. *arXiv.org* [Online] 2012 [Cited: 7 June 2018]. https://arxiv.org/pdf/1210.6929.pdf.
96 *Bucherer and Einstein* Einstein, Albert. *CPE5*, 133–9, 147–9, Documents 117, 119, 120, and 128.
96 *'I am always available'* Flückiger, Max. *EIB*, 124.
96 *'Albert Einstein didn't really sympathise'* Winteler-Einstein, Maria. *AEA* [Online]. 15 February 1924. Cat. no. 73–880, 25–6. http://alberteinstein.info/.
96 *For the moment he contented* Einstein, Albert. *CPE2*, 542–53, Document 56.
96 *The talk took place;* —. *CPE5*, 155–6, Document 135.
97 *Kleiner already had Einstein* —. *CPE5*, 95–6, Document 80.
97 *Einstein wrote to Laub* —. *CPE5*, 130–2, Document 113.
97 *Einstein informed Laub* —. *CPE5*, 188–90, Document 160.
97 *He told Ehrat that* —. CPE5, 155–6, Document 135
97 *In April, Einstein wrote* —. *CPE5*, 169, Document 151.
97 *Adler responded to this attack* Frank, Philipp. *LZ*, 129–30.
98 *However, it seems that Kleiner* Adler, Friedrich. AEA [Online]. 28 November 1908. Cat. nos. 73-563–73-565. http://alberteinstein.info/.
98 *Nevertheless, it is a testament* Einstein, Albert. *CPE8A*, 442, Document 331, Note (5).

98 *Furthermore, he wrote to the university* Fölsing, Albrecht. *Albert Einstein: Eine Biographie* (Frankfurt am Main, Suhrkamp Verlag, 1993), 284–9. Fölsing gives a very careful and detailed account of the story of Einstein's appointment, drawing on Adler's letters to his father published in Ardelt's biography.

98 *Therefore, it is reassuring* Pais, Abraham. *LORD*, 185–6.

98 *Kleiner is recorded* Medicus, Heinrich A. The Friendship among Three Singular Men: Einstein and His Swiss Friends Besso and Zangger. *Isis*, 1994, 85: 456–78. http://www.jstor.org/stable/235463.

98 *Einstein wrote to Laub* Einstein, Albert. *CPE5*, 161–2, Document 143.

98 *Only as an afterthought* —. *CPE5*, 206, Document 177.

98 *Eventually Paul Habicht* —. *CPE5*, 51–5.

98 *He was very impressed* Mehra, Jagdish. *The Solvay Conferences on Physics* (Dordrecht, D. Reidel, 2012), xxi.

99 *In fact, the invitation* Dukas, Helen, Hoffmann, Banesh, eds. *Albert Einstein: The Human Side* (Princeton, Princeton University Press, 2013), 5.

99 *At least he did not* Ostwald, Wilhelm. *Wilhelm Ostwald: The Autobiography*, Fritz Scholz Robert Jack, eds., [trans.] Robert Jack (Cham, Springer International, 2017), 483.

99 *'The festivities ended'* Seelig, Carl. *BIOG*, 108–10.

99 *Planck was on vacation* Planck, Max. *CPE5*, 135–6, Document 118.

99 *Ladenburg, who was also Jewish* Mehra, Jagdish. *The Golden Age of Theoretical Physics*, Vol. I (Singapore, World Scientific, 2001), 440.

99 *At some point in August* Einstein, Albert. *CPE5*, 623.

99 *The family also stayed* Seelig, Carl. *BIOG*, 115–16.

99 *In answer to suggestions* Einstein, Albert. *CPE2*, 556–9, Document 58.

100 *Einstein's own talk* —. *CPE2*, 565–83, Document 60.

101 *Planck, as can be deduced* —. *CPE2*, 584–7, Document 61.

101 *Lise Meitner and Born* Hermann, Armin. Einstein auf der Salzburger Naturforscherversammlung 1909. *Physikalische Blätter*, 1969, 25: 433–6. Available at https://onlinelibrary.wiley.com/doi/pdf/10.1002/phbl.19690251001.

101 *In December 1907* Einstein, Albert. *CPE5*, 88–9.

101 *However, Sommerfeld's prejudice* —. *CPE5*, 210–11, Document 179.

101 *In a letter to Edgar Meyer* —. *CPE5*, 206–9, Document 178.

Chapter 8

102 *He was certainly attracted* Reichinstein, David. *Albert Einstein: A Picture of His Life and His Conception of the World* (London, Edward Goldston Ltd., 1934), 43–4.

102 *Although her mother reported* Trbuhović-Gjurić, Desanka. *ISAE*, 104.

102 *'Einstein's soul swiftly took wing'* Reichinstein, David. *Albert Einstein: A Picture of His Life and His Conception of the World* (London, Edward Goldston Ltd., 1934), 27–8.

103 *He replied with a letter* Einstein, Albert. *CPE5*, 181, Document 154.

103 *She ceased to teach* —. *CPE1*, 385.

104 *As an unwelcome reminder* Rogger, Franziska. *ES*, 35, lower right-hand photograph.

104 *There are many quotes* Highfield, Roger, Carter, Paul. *The Private Lives of Albert Einstein* (London, Faber and Faber, 1993), 210.

104 *'I escape my eternal longing'* Einstein, Albert. *CPE15*, 19–20, Document 177a.

105 *Before the discovery* —. *CPE5*, 218–19, Document 188 & footnote [6].
105 *He wrote of their previous* —. *CPE15*, 20–1, Document 198a.
105 *'The bad mood that you'* —. *CPE5*, 237–9, Document 204.
105 *The silence was too much* —. *CPE15*, 21, Document 213a.
105 *. . . and then a letter* —. *AEA* [Online]. 21 July 1910. Cat. no. 95-682 (or 62–812.2); only the envelope survives. https://einsteinpapers.press.princeton.edu/vol15-doc/980
105 *He asked her not to think* —. *CPE15*, 21, Document 218a.
105 *The rumour apparently gained* Overbye, Dennis. *Einstein in Love: A Scientific Romance* (London, Bloomsbury, 2001), 393, note in reference 13.
105 *The meeting to which* Rogger, Franziska. ES, 46.
105 *Readers may judge* Rogger, Franziska. ES, 35, lower left-hand corner photograph.
106 *Paul Albert Müller was born* Klein, Martin J., Kox, A. J., Schulmann, Robert, eds. *CPE5*, 438, Document 377, Note 14.
106 *'I welcome Marie's wedding'* Einstein, Albert. *CPE5*, 380–2, Document 331.
106 *In a later letter to Besso* —. *CPE5*, 435–8, Document 377.
106 *In January 1914* —. *CPE5*, 592, Document 504.
106 *He directed Habicht to reply* —. *CPE5*, 223, Document 192.
106 *Mileva, without consulting Einstein* Einstein, Mileva. Letter to Georg Meyer, SSLB-D-13-01-108. 23 May 1909. National Library of the Swiss Federation, Bern, Switzerland.
106 *When Einstein discovered* Einstein, Albert. *CPE5*, 198–9, Document 166.
108 *This latter conclusion* Meyer-Schmid, Anna. *AEA* [Online]. 23 December 1926. Cat. no. 44–447. http://alberteinstein.info/.
108 *'It's good, by this chance'* Einstein, Albert. *CPE15*, 668, Document 435.
108 *There seems to have been* Winteler-Einstein, Maria. Letter to Anna Meyer-Schmid. 1933. Schweizerische Nationalbibliothek Bern, Schweizerishes Literaturarchiv SLA. SSLB-D-13-01-105.
108 *Erika's husband had been told* Overbye, Dennis. *Einstein in Love: A Scientific Romance* (London, Bloomsbury, 2001), 376–7.
108 *Erika's eldest son remembered* Isoz, Hanspeter. *Wie bitte, Herr Einstein, war das jeztz ganz genau im Paradies?* (Mettmenstetten, privately published, 2016), 115.
109 *His jibe that this was typical* Einstein, Albert. *AEA* [Online]. 27 July 1951. Cat. no. 60–525. http://alberteinstein.info/.
109 *Indeed, her daughter recalled* Isoz, Hanspeter. *Wie bitte, Herr Einstein, was das jetzt ganz genau im Paradies?* (Mettmenstetten, privately published, 2016), 111.
109 *In 1976 she wrote to Helen* Schaerer-Meyer, Erika. *AEA* [Online]. 30 November 1976. Cat. no. 44–451. http://alberteinstein.info/.
109 *It is sad to note* Winteler-Müller, Marie. *AEA* [Online]. 1927. Cat. no. 48–796. http://alberteinstein.info/.
109 *Einstein was touched* Einstein, Albert. *CPE16*, 64, Document 15.
109 *One of the doctors there* Overbye, Dennis. *Einstein in Love: A Scientific Romance* (London, Bloomsbury, 2001), 144.
109 *'Also later on I remained defiant'* Winteler, Marie. *AEA* [Online]. 3 February 1957. Cat. no. 71–183, 6. http://alberteinstein.info/.

Chapter 9

110 *His surviving lecture notes* Einstein, Albert. *CPE3*, 11–129, Document 1.
110 *His notes on the kinetic theory* —. *CPE3*, 179–247, Document 4.

110 *He took the opportunity* —. *CPE3*, 316–400, Document 11.
111 *He grabbed Tanner's arm* Seelig, Carl. *BIOG*, 119–25.
111 *Hopf was also an excellent pianist* —. *BIOG*, 126.
111 *By the summer of 1910* Hoftler, Gisela. 'He Was a Friend of the Greatest Geniuses of His Time – Indeed, He Was One of Them' – Ludwig Hopf (1884–1939). In *Voices from Exile: Essays in Memory of Hamish Ritchie*,Ian Wallace, ed. (Leiden/Boston, Brill, 2015), 113–40.
112 *Tanner had remarked* Reichinstein, David. *Albert Einstein: A Picture of his Life and his Conception of the World* (London, Edward Goldston Ltd., 1934), 26.
112 *He would arrive with his clothes* Trbuhović-Gjurić, Desanka. *ISAE*, 107.
112 *In order to make ends meet* Frank, Philipp. *LZ*, 131.
112 *It seems that at least* Einstein, Albert. *CPE5*, 222–3, Documents 190–2.
112 *His conversations with Einstein* —. *CPE5*, 231–3, Document 199.
112 *The arrival in Zurich* Fölsing, Albrecht. *Albert Einstein: Eine Biographie* (Frankfurt am Main, Suhrkamp Verlag, 1993), 301–2.
113 *In time the two became close* Reiser, Anton. *ABP*, 133–4.
113 *In April 1916* Ebbinghaus, Heinz Dieter. *Ernst Zermelo: An Approach to His Life and Work*, 2nd ed. (Berlin, Springer, 2015), 119–35. Written in co-operation with Volker Peckhaus.
113 *He was accompanied by his wife* Einstein, Albert. *CPE5*, 233–4, Documents 200, 201.
113 *Einstein was clearly fond* —. *CPE5*, 239–40, Document 205.
113 *The third was a request* —. *CPE5*, 237–9, Document 204.
113 *The Einstein's apartment* Trbuhović-Gjurić, Desanka. *ISAE*, 104.
113 *Their musical evenings* Einstein, Albert. *CPE5*, 183, Document 157. See also Scherrer, Lucien. Der mysteriöse Tod eines Helden. *Neue Zürcher Zeitung*, 10 April 2018. https://www.nzz.ch/schweiz/der-mysterioese-tod-eines-helden-ld.1375839.
113 *Einstein would arrive* Seelig, Carl. *BIOG*, 132–3.
114 *They played through a Bach* Trbuhović;-Gjurić, Desanka. *ISAE*, 107.
114 *The latter, who was an engineer* Schulmann, Robert, Kox, A. J., Janssen, Michel, Illy, József, eds. *CPE8A*, 400–2, Document 306.
114 *Hans Albert had been sent* Einstein, Albert. *CPE5*, 248–51, Documents 215–21.
115 *This could not be explained* —. *CPE3*, 248–53, Document 5.
115 *Einstein did not let slip* Eckert, Michael. *Arnold Sommerfeld: Science, Life and Turbulent Times 1868–1951* [trans.] Tom Artin (New York, Springer, 2013), 174.
115 *His performance of Chopin* —. *Arnold Sommerfeld: Science, Life and Turbulent Times 1868–1951* [trans.] Tom Artin (New York, Springer, 2013), 40–1; 62–3.
115 *The papers were submitted* Einstein, Albert, Hopf, Ludwig. Über einen Satz der Wahrscheinlichkeitsrechnung und seine Anwendung in der Strahlungstheorie. *Annalen der Physik*, 1910, 33: 1096–104.
115 *The papers were submitted* — —. Statistische Untersuchung der Bewegung eines Resonators in einem Strahlungsfeld. *Annalen der Physik*, 1910, 338: 1105–15.
115 *Some of this came* Klein, Martin J., Kox, A. J., Renn, Jürgen, Schulmann, Robert, eds. *CPE3*, 268, Document 7, Note 5.
115 *Shortly after Sommerfeld's* Einstein, Albert. *CPE5*, 255–56, Document 226.
115 *However, the Ministry* —. *CPE5*, 253–5, Document 224.
115 *The students particularly* —. *CPE5*, 243–4, Document 210.
116 *The scattering of light* —. *CPE3*, 283–312, Document 9.
116 *Einstein was trying* —. *CPE5*, 263, Documents 233, 234.
116 *Einstein hastened to reply* —. *CPE5*, 259–60, Document 230; 262–3, Document 232.

116 *Einstein must have decided* Klein, Martin J., Kox, A. J., Schulmann, Robert, eds. *CPE5*, 261, Document 177, Note 3.

117 *The pianist in these pieces* Hurwitz, Lisbeth. *AEA* [Online]. 6 March 1954. Cat. no. 38–141. http://alberteinstein.info/.

117 *Einstein described Flesch* Einstein, Albert. *CPE5*, 264–65, Document 236.

117 *The critic of the* Neue Zürcher Zeitung Neue Zürcher Zeitung. Feuilleton. *Neue Zürcher Zeitung*. 10 December 1910, 131(341): p. 1.

117 *The New Year celebrations* Einstein, Albert. *CPE5*, 271–3, Document 245.

117 *The letter is very upbeat* —. *CPE5*, 273–4, Document 246.

117 *After Einstein had fielded* —. *CPE3*, 425–48, Documents 17 and 18.

117 *It is clear from Einstein's letter* —. *CPE5*, 266–7, Document 250; 281–3, Document 254.

118 *In the past, he had tried* Seelig, Carl. *BIOG*, 116.

118 *He first became* Alder, Douglas D. Friedrich Adler: Evolution of a Revolutionary. *German Studies Review*, 1978, 1(3): 263–7.

118 *However, this seems unlikely* Einstein, Albert. Postcard, n.d. ETH Zurich Library [Online] [Cited: 2 March 2019]. The postcard has no date but is of the Hofbrauhaus Munich. The fact that Einstein and Mileva were together in Munich without childcare duties and that they had just stayed with Habicht, point to this as the likely date. https://tinyurl.com/4d4rdb4z.

Chapter 10

119 *When he was about to buy* Einstein, Albert. *CPE5*, 289–90, Document 263.

119 *Lampa, after studying* Institut für Neuzeit- und Zeitgeschichtsforschung. Lampa, Anton. (1868–1938), Physiker. *Österreichisches Biographisches Lexikon* [Online] [Cited: 27 January 2019]. https://www.biographien.ac.at/oebl/oebl_L/Lampa_Anton_1868_1938.xml.

119 *That he succeeded in doing so* Klinert, Andreas. Anton Lampa and Albert Einstein. The Call to the Chair in Physics at the German University at Prague 1909 and 1910. *Schweizerische Gesellschaft ür Geschichte der Medizin und der Naturwissenschaften*, 1975, 32: 285–92.

120 *Nevertheless, according to Frank* Frank, Philipp. *LZ*, 138–39.

120 *His abstraction from everyday* —. *LZ*, 131–2.

120 *He was mistaken* Seelig, Carl. *BIOG*, 144.

120 *They naturally took offence* Frank, Philipp. *LZ*, 140.

120 *He might therefore have* Einstein, Albert. *CPE5*, 295–6, Document 267.

121 *There are several mistakes* Institut für Neuzeit- und Zeitgeschichtsforschung. Lampa, Anton. (1868–1938), Physiker. Österreichisches Biographisches Lexikon [Online] [Cited: 27 January 2019]. https://www.biographien.ac.at/oebl/oebl_L/Lampa_Anton_1868_1938.xml.

121 *On Philipp Frank's first visit* Frank, Philipp. *LZ*, 142–43.

121 *His first public lecture* Illy, József. Albert Einstein in Prague. *Isis*, 1979, 70(1): 82.

121 *'The entire Prague intelligentsia'* Fölsing, Albrecht. *Albert Einstein: Eine Biographie* (Frankfurt am Main, Suhrkamp, 1993), 320; quoting from Kowalewski, Gerhard. *Bestand und Wandel* (Munich, Oldenbourg, 1950), 237.

121 *In addition, Prague had venues* Moskovitz, Marc. *Alexander Zemlinsky: A Lyric Symphony* (Woodbridge, Boydell Press, 2010), 125–38.

121 *As Stone points out* Stone, A. Douglas. *Einstein and the Quantum* (Princeton, Princeton University Press, 2015), 154–6.

122 *In these, he modified* Einstein, Albert. *CPE3*, 459–77, Document 21.

122 *In 1909, while working* Ehrenfest, Paul. Gleichförmige Rotation starrer Körper und Relativitätstheorie. *Physikalische Zeitschrift*, 1909, 10: 918.
122 *After correspondence with Ehrenfest* Einstein, Albert. *CPE3*, 482–4, Document 22.
122 *However, it seems likely* Pais, Abraham. *LORD*, 187–90.
123 *In the final section* Einstein, Albert. *CPE3*, 485–97, Document 23.
123 *A letter arrived* Einstein, Albert, Pollak, Leo. AEA [Online]. 1911. Cat. no. 11–199 – 11–202. http://alberteinstein.info/.
123 *The bending of light* Michell, John. On the Means of Discovering the Distance, Magnitude, &c. of the Fixed Stars, in Consequence of the Diminution of the Velocity of Their Light, in Case Such a Diminution Should Be Found to Take Place in Any of Them, and Such Other Data Should Be Procured from Observations, as Would Be Farther Necessary for That Purpose. By the Rev. John Michell, B.D. F.R.S. In a letter to Henry Cavendish, Esq. F.R.S. and A.S. Philosophical Transactions of the Royal Society, 1784, 74: 35–57. Laplace, Pierre-Simon. Beweis des Satzes, dass die anziehende Kraft bey einem Weltkörper so gross seyn könne, dass das Licht davon nicht ausströmen kann. Allgemeine Geographische Ephemeriden, 1799, 4: 1–6. Longair, Malcolm. Bending space–time: a commentary on Dyson, Eddington and Davidson (1920) 'A determination of the deflection of light by the Sun's gravitational field'. London: The Royal Society, 13 April 2015, Philosophical Transactions of the Royal Society A, Vol. 373: 20140287. Jaki, S.L. Johann Georg von Soldner and the Gravitational Bending of Light, with an English Translation of his Essay on it published in 1801. Springer Nature, 1978, Foundations of Physics, Vol. 8, pp. 927–950.
124 *This work continued until* Mehra, Jagdish. *The Solvay Conferences on Physics* (Dordrecht, Springer, 2012), xx and 1–9.
124 *However, Zangger found his host* Einstein, Albert. *CPE5*, 314, Document 279, Notes 5 and 6.
124 *Haber was about to move* Stern, Fritz. *Einstein's German World* (Princeton, Princeton University Press, 2016), 61.
124 *Hopf became an authority* Holfter, Gisela. 'He Was a Friend of the Greatest Geniuses of His Time – Indeed, He Was One of Them' – Ludwig Hopf (1884–1939). In Voices in Exile, Ian Wallace, ed. (Leiden/Boston, Brill/Rodopi, 2015), 113–40.
124 *The peculiar situation* Frank, Philipp. *LZ*, 141–2.
125 *She signed off touchingly* Einstein, Albert. *CPE5*, 331, Document 290.
125 *He had been in discussion since August* Julius, Willem, Einstein, Albert. AEA [Online] 1911. Cat. no. 75–61 – 75–78. http://alberteinstein.info/.
125 *Einstein replied that* —. *CPE5*, 340–1, Document 297.
125 *'Einstein of course was not'* Eckert, Michael. *Arnold Sommerfeld.* [trans.] Tom Artin (New York, Springer, 2013), 179–80.
126 *The discussion after his talk* Straumann, Norbert. On the First Solvay Congress in 1911. *The European Physical Journal H*, 2011, 36: 379–99.
126 *Einstein did however remark* Einstein, Albert. *CPE5*, 345–50, Documents 303, 305.
126 *Despite the disappointment* Mehra, Jagdish. *The Solvay Conferences on Physics* (Dordrecht, Springer, 2012), xv.
126 *Indeed, he remarked that* Einstein, Albert. *CPE5*, 380–2, Document 331.
126 *Irrespective of their personal relations* Seelig, Carl. *BIOG*, 161–3.
126 *She left for Stockholm* Quinn, Susan. International Year of Chemistry 2011: A Test of Courage: Marie Curie and the 1911 Nobel Prize. *Clinical Chemistry*, 2011, Vol. 57(4): 653–8.
127 *On 23 November* Einstein, Albert. *CPE8A*, 7, Document 312a.
127 *A few days later* —. *CPE5*, 347–58, Document 304–11.

127 *The administrative wheels turned* —. *CPE5*, 414–15, Document 361.

127 *Not surprisingly, he was an object* Medicus, Heinrich A. The Friendship among Three Singular Men: Einstein and His Swiss Friends Besso and Zangger. *Isis*, 1994, 85: 461.

127 *Since he assumed that Lorentz* Einstein, Albert. *CPE5*, 365–8, Documents 317–20.

127 *However, by the time Lorentz wrote* —. *CPE5*, 409–11, Document 359.

127 *Paul Habicht, who had recently* —. *CPE5*, 371–84, Documents 324–6, 328, 330–2.

127 *On his return* —. *AEA* [Online]. 16 December 1911. Cat. no. 74–496. http://alberteinstein.info/.

127 *That Einstein could still be thoughtful* —. *CPE5*, 629, Chronology, ca. 25 December.

127 *They probably returned to Switzerland* Popovic, Milan. *AS*, 108.

128 *He himself was enormously excited* Ehrenfest, Paul, *Diaries of Paul Ehrenfest*, Cat. no. ENB4 11. Diary entries for 23 February–6 March 1912 Rijksmuseum Boerhaave (bibliotheek).

128 *Meitner experienced this when* Kline, Martin J. *Paul Ehrenfest: The Making of a Theoretical Physicist*, Vol. I (Amsterdam, North Holland, 1970), 48–9.

128 *Einstein in later years wrote* —. *Paul Ehrenfest: The Making of a Theoretical Physicist*, Vol. I (Amsterdam, North Holland, 1970), 176–9.

129 *In the first quarter* Einstein, Albert. Lichtgeschwindigkeit und Statik des Gravitationsfeldes. *Annalen der Physik*, 1912, 38: 355–69. Einstein, Albert. Theorie des statischen Gravitationsfeldes. *Annalen der Physik*, 1912, 38: 443–58.

129 *It was never published* Klein, Martin J., Kox, A. J., Renn, Jürgen, Schulmann, Robert, eds. *CPE4*, 3–8.

130 *This sort of mathematics* Einstein, Albert. *CPE4*, 9–108, Document 1.

130 *Debye, who was on* Klein, Martin J., Kox, A. J., Schulmann, Robert, eds. *CPE5*, 447, Document 382, Note 15.

130 *Ehrenfest received this letter* Ehrenfest, Paul, *Diaries of Paul Ehrenfest*, Cat. no. ENB4 13. Diary entry for 3 May Rijksmuseum Boerhaave (bibliotheek).

130 *However, the frequency of the meetings* Einstein, Albert. *CPE3*, 563–97, Appendix A.

131 *He first began to court Paula* Einstein, Albert. AEA [Online] 1912. Cat. no. 72–243. http://alberteinstein.info/. See also Highfield, Roger, Carter, Paul. *The Private Lives of Albert Einstein* (London, Faber and Faber, 1993), 148.

131 *He made a commitment* —. *CPE5*, 456–8, Document 389.

131 *Elsa's reply, which Einstein destroyed* —. *CPE5*, 459, Document 391.

131 *Nevertheless, he was careful* —. *CPE5*, 469–70, Document 399.

131 *Having attended Sommerfeld's lectures* Segre, Emilio. Otto Stern 1888–1969. *Biographical Memoires of the National Academy of Sciences*, 1973, 44: 215–36.

131 *This was Einstein* Fölsing, Albrecht. *Albert Einstein: Eine Biographie*. (Frankfurt am Main, Suhrkamp, 1993), 340. From transcript of interview of Otto Stern by Res Jost, 1961.

132 *This would have an important bearing* Einstein, Albert. *CPE4*, 174–9, Document 7.

132 *For this work, von Laue* —. *CPE5*, 475–6, Documents 402, 403; 482–4, Documents 407, 408.

133 *'the solemn sounds of the organ'* Clark, Ronald. *Einstein: The Life and Times* (New York, Bloomsbury, 1972), 180–1. Clark quotes Marianoff as his source, but this passage does not appear in Marianoff and Wayne's book.

133 *When Frank informed her* Frank, Philipp. *LZ*, 140–42.

133 *He often accompanied Einstein* Illy, Jozsef. Albert Einstein in Prague. *Isis*, 1979, 70(1): 82.

133 *Just before his death* Grüning, Michael. *Ein Haus für Albert Einstein* (Berlin, Verlag der Nation, 1990), 248–9.

133 *The main attraction* Seelig, Carl. *BIOG*, 146.

133 *'I am pleased that our stay here'* Einstein-Marić, Mileva. *AEA* [Online]. 26 December 1911. Cat. no. 7–65. http://alberteinstein.info/.
133 *In February 1912 she wrote* —. *AEA* [Online]. 4 February 1912. Cat. no. 7–258. http://alberteinstein.info/.
134 *'We are well and are all'* Popovic, Milan. *AS*, 106–8.

Chapter 11

135 *Einstein played the latter* Meyer-Gentner, Elisabeth. *'Die Hofburg': Geschichte eines Hauses am Zürichberg*. (Zurich, Theodor Gut Verlag, 1998), 73–6.
135 *Despite this heavy administrative* Kollros, Louis. Albert Einstein en Suisse: Souvenirs. In *Helvetica Physica Acta*, Vol. 29, Suppl 4, Andre Mercier, Michel Kervaire, eds. (Basel, Swiss Physical Society, 1956), 278–9. https://archive.org/details/helvetica-physica-acta_1956_4_supplement/mode/2up.
135 *The good relationship* Einstein, Albert. CPE5, 599, Document 510, Note 3. Gentner-Aichroth, Friedrich. AEA [Online]. 4 December 1913. Cat. no. 81–376. http://alberteinstein.info/.
136 *'After the colloquium'* Seelig, Carl. *BIOG*, 93.
136 *Einstein was able to reassure* Pais, Abraham. *LORD*, 211–13.
136 *He then entered a period* Einstein, Albert. *CPE5*, 505–6, Document 421.
136 *This contrasts with his earlier opinion* Stern, Otto. *AEA* [Online]. 1961. Cat. no. 74–831. Transcript from a tape recording of an interview with Res Jost. http://alberteinstein.info/.
137 *This notebook has been extensively* Norton, John D. A Peek into Einstein's Zurich Notebook [Online]. 21 June 2008 [Cited: 26 August 2025]. https://sites.pitt.edu/~jdnorton/Goodies/Zurich_Notebook/index.html.
137 *As Einstein worked systematically* Einstein, Albert. *CPE4*, 210–69, Document 10.
137 *The paper's title* Einstein, Albert, Grossmann, Marcel. CPE4, 302–43, Document 13.
138 *It is important to remember* Norton, John. General Covariance and the Foundations of General Relativity: Eight Decades of Dispute. *Reports on Progress in Physics*, 1993, 56: 791–858.
138 *During the summer and autumn* Norton, John D. How Einstein Found His Field Equations: 1912–1915. *Historical Studies in the Physical Science*, 1984, 14: 253–316.
138 *Forced then to work* Renn, Jürgen, Sauer, Tilman. Pathways Out of Classical Physics: Einstein's Double Strategy in his Search for the Gravitational Field Equation. In *The Genesis of General Relativity*, Vol. I, Michel Janssen, John D. Norton, Jürgen Renn, Tilman Sauer, John Stachel, eds. (Dordrecht, Springer, 2007), 113–312, in particular 182–4.
139 *Discouraged, Einstein turned* Janssen, Michel, Norton, John D., Renn, Jürgen, Sauer, Tilman, Stachel, John. Introduction to Volumes 1 and 2: The Zurich Notebook and the Genesis of General Relativity. In *The Genesis of General Relativity*, Vol. I, Michel Janssen, John D. Norton, Jürgen Renn, Tilman Sauer, John Stachel, eds. (Dordrecht, Springer, 2007); in particular a very helpful summary is given in the Introduction, 7–20.
139 *The Entwurf paper appeared* Einstein, Albert, Grossmann, Marcel. *CPE4*, 302–43, Document 13.
139 *... however, it seems likely* Sauer, Tilman. Marcel Grossmann and His Contribution to the General Theory of Relativity. In *Proceedings of the 13th Marcel Grossmann Meeting on Recent Developments in Theoretical and Experimental General Relativity, Gravitation, and Relativistic Field Theory, Stockholm University, Sweden, 1–7 July 2012*, Kjell Rosquist, Robert T. Jantzen, Remo Ruffini, eds. (Singapore, World Scientific, 2013), 17, footnote. https://arxiv.org/abs/1312.4068v2.
140 *On 19 October* Trbuhović-Gjurić, Desanka. *ISAE*, 120.
140 *'My big Albert has become'* Popović, Milan. *AS*, 107–8.
140 *Other colleagues and friends* Trbuhović-Gjurić, Desanka. *ISAE*, 120–2.

141 *This paper* Einstein, Albert. *CPE4*, 114–21, Document 2.

141 *A subsequent supplement* —. *CPE4*, 165–70, Document 5.

141 *Einstein added a postscript* Einstein-Marić, Mileva. *AEA* [Online]. 16 March 1913. Cat. no. 80–828. http://alberteinstein.info/.

141 *Helped out linguistically* Einstein, Albert. Déduction thermodynamique de la loi de l'équivalence photochimique. *Journal de physique*, 1913, 3: 277–82.

141 *Their correspondence had also* Einstein, Albert. *CPE5*, 517–18, Document 434; 520, Document 436.

141 *He wrote an effusive thank-you note* —. *CPE5*, 518–19, Document 435 & pp. 520–521, Document 437.

141 *However, he could be stern* Michelmore, Peter. *Einstein, Profile of the Man* (New York, Dodd, Mead & Company, 1962), 46.

141 *Unfortunately, his informant* Einstein, Albert. *CPE5*, 524–25, Document 442.

142 *Somehow Einstein managed* Wohlwend, Hans. *AEA* [Online]. 19 May 1946. Cat. no. 59–30. http://alberteinstein.info/.

142 *He had both lodging* Einstein, Albert. *CPE5*, 592, Document 504.

142 *It is ironic to note* —. *CPE5*, 526–29, Document 445.

142 *The two friends were soon* —. *CPE5*, 531, Document 447.

142 *Ehrenfest also renewed* Kline, Martin J. *Paul Ehrenfest: The Making of a Theoretical Physicist*, Vol. I (Amsterdam, North Holland, 1970), 294–5. See also Ehrenfest's diary entries for this period, Ehrenfest Archive, Cat. no. ENB4 15, Leiden, Rijksmuseum Boerhaave (bibliotheek).

142 *'The liveliest discussions'* Seelig, Carl. *BIOG*, 94.

142 *This work is contained* —. *BIOG*, pp. 161–63.

143 *The following day* Klein, Martin J., Kox, A. J., Schulmann, Robert. *CPE5*, 534, Document 451, Note 2.

143 *Stern's thesis was accepted* Einstein, Albert. *CPE5*, 535–6, Document 452; see also 631, entry for 29 October.

144 *The youngsters were amused* Curie, Eve. *Madam Curie: A Biography* (London, William Heinemann Ltd., 1958), 296.

144 *This letter contains* Einstein, Albert. *CPE5*, 545, Document 465.

144 *He wrote again to Lorentz* —. *CPE5*, 546–50, Document 467; 552–3, Document 470.

144 *Grossmann and Mileva also attended* Trbuhović-Gjurić, Desanka. *ISAE*, 124.

145 *Michel Janssen believes that* Janssen, Michel. What Did Einstein Know and When Did He Know It? A Besso Memo Dated August 1913. In *The Genesis of General Relativity*, Vol. I, Michel Janssen, John D. Norton, Jürgen Renn, Tilman Sauer, John Stachel, eds. (Dordrecht, Springer, 2007), 785–837.

145 *In his letter, Einstein more or less* Einstein, Albert. *CPE8A*, 234–6, Document 178.

145 *Einstein also remarked that* —. *CPE5*, 588–9, Document 499.

145 *The famous* The New York Times *headline* Anon. Lights All Askew in the Heavens. *The New York Times*. 19 November 1919, 17.

145 *They did so and indeed* Einstein, Albert. *CPE5*, 554–5, Document 472.

145 *Directly after Einstein's talk* —. *CPE4*, 474–85, Documents 15 and 16. Document 15 is a very clear short summary of the talk.

145 *Although Besso added* Janssen, Michel. What Did Einstein Know and When Did He Know It? A Besso Memo Dated August 1913. In *The Genesis of General Relativity*, Vol. I, Michel Janssen, John D. Norton, Jürgen Renn, Tilman Sauer, John Stachel, eds. (Dordrecht, Springer, 2007), 785–837.

146 *He does not appear to have* Trbuhović-Gjurić, Desanka . *ISAE*, 124.

146 *In the subsequent discussion* Einstein, Albert. *CPE4*, 486–511, Documents 17 and 18.

146 *Einstein was rapidly* Hofmann, Thomas. Relativitätstheorie 'leicht fasslich'. *Weiner Zeitung* [Online]. 25 October 2015 [Cited: 30 September 2024]. https://www.wienerzeitung.at/nachrichten/reflexionen/vermessungen/782207-Relativitaetstheorieleicht-fasslich.html?em_cnt_page=2.
146 *According to von Hevesy's account* Kragh, Helge. *The Early Reception of Bohr's Atomic Theory (1913–1915): A Preliminary Investigation.* RePoSS: Research Publications on Science Studies 9 (Aarhus, Centre for Science Studies, 2010), 19–20. https://css.au.dk/fileadmin/reposs/reposs-009.pdf.
146 *This was the organ* Frank, Philipp. LZ, 178.
146 *This was not the first* Frank, Philipp LZ, 106–7. Lecher, Ernst. Die Minute in Gefahr, eine Sensation der mathematischen Wissenschaft. Neues Weiner Tagblatt. 22 September 1912, 11–12.
147 *He was back in Zurich* Einstein, Albert. *CPE5*, 557–9, Document 476.
147 *He even implied that* —. *CPE5*, 560–1, Document 478; 572–5, Documents 488, 489.
147 *Thomson in his report* Kragh, Helge. *The Early Reception of Bohr's Atomic Theory (1913–1915): A Preliminary Investigation.* RePoSS: Research Publications on Science Studies 9 (Aarhus, Centre for Science Studies, 2010), 20–1; 32–4. https://css.au.dk/fileadmin/reposs/reposs-009.pdf.
147 *Other than having to toast* Various. *CPE4*, 552–9, Document 22.
147 *Gradually Fokker's musical* Leedy, Douglas. *Adriaan Daniel Fokker. Selected Musical Compositions (1948–1972).* Notes, 1989, 46: 224–226. See also review of book of same title [ed.] Rudolph Rasch; https://www.jstor.org/stable/pdf/940773.pdf.
148 *Before long they were working* Einstein, Albert. *CPE5*, 575–80, Document 490.
148 *After the playing was over* Trbuhović-Gjurić, Desanka. *ISAE*, 125.
148 *There was a generous counter-offer* Ghnem, Robert. *CPE5*, 583, Document 494, Note 1.
148 *Indeed, some months later* Einstein, Albert. *CPE10*, 22–23, Document 16a (Vol. VIII).
148 *'The gentlemen in Berlin'* Kollros, Louis. Albert Einstein en Suisse: Souvenirs. In *Helvetica Physica Acta*, Vol. 29, Suppl 4, Andre Mercier, Michel Kervaire, eds. (Basel, Swiss Physical Society, 1956), 280. https://archive.org/details/helvetica-physica-acta_1956_4_supplement/mode/2up.
148 *On 23 December* Trbuhović-Gjurić, Desanka. *ISAE*, 126.
148 *Directly after Christmas* Einstein, Albert. *CPE5*, 585–7, Document 497.
148 *Einstein's response to this* —. *CPE5*, 593, Document 505 .
148 *He assured Freundlich* —. *CPE5*, 593–5, Document 506.
149 *The result was a very beautiful paper* Einstein, Albert, Fokker, Adriaan. *CPE4*, 588–97, Document 28.
149 *His attitude to these criticisms* Einstein, Albert. *CPE5*, 595–7, Document 507.
149 *Their final joint paper* Einstein, Albert, Grossmann, Marcel. *CPE6*, 6–18, Document 2.
149 *Despite her favourite composer* Trbuhović-Gjurić, Desanka. ISAE, 125.
149 *Einstein completed his teaching* Klein, Martin J., Kox, A. J., Schulmann, Robert. *CPE5*, 636.

Chapter 12

150 *Their furniture was due* Einstein, Albert. *CPE8A*, 11–12, Document 1.
150 *Some indication of the importance* Schrödinger, Erwin. AEA. [Online] 18 March 1930. Cat. no. 22–26. http://alberteinstein.info/
151 *In a letter to Wilhelm Wien* —. *CPE8A*, 31–3, Document 14.
151 *This whirl of intellectual* —. *CPE8A*, 17–18, Document 6.
151 *In the concluding paragraph* —. *CPE6*, 3–5, Document 1.

151 *Einstein once again implied* —. *CPE6*, 19–24, Document 3.
151 *Pauline was operated on* —. *AEA* [Online]. 13 August 1945. Cat. no. 38–456. http://alberteinstein.info/.
151 *In fact, they mostly discussed* Ehrenfest, Paul. *Diaries of Paul Ehrenfest*, Cat. no. ENB4 16 Diary entries for 1–4 June 1914, Ehrenfest Archive, Leiden, Rijksmuseum Boerhaave (bibliotheek).
152 *Einstein began to absent himself* Besso-Winteler, Anna. *CPE8A*, 45, Document 23, Note 3.
152 *Rather than sitting* —. *CPE8B*, 1032–3, Appendix.
152 *As a sort of postscript* Einstein, Albert. *CPE8A*, 44–5, Document 22.
153 *The tone of this sequence* —. *CPE8A*, 45–6, Documents 23 and 24.
153 *Early the following week* Besso-Winteler, Anna. CPE8A, 45, Document 23, Note 3.
153 *Einstein wrote to her* —. *CPE8A*, 47–9, Documents 26 and 27.
153 *Einstein was inconsolable* —. *CPE8A*, 50–1, Document 29.
153 *He remarked to his confidant* —. *CPE8A*, 204–6, Document 152.
154 *'Unbelievable things have now begun'* —. *CPE8A*, 56–7, Document 34.
154 *An expedition from the Lick Observatory* Hentschel, Klaus. Erwin Finlay Freundlich and Testing Einstein's Theory of Relativity. *Archive for History of Exact Sciences*, 1994, 47: 143–201.
155 *Since he had remained a Swiss* Frank, Philipp. *LZ*, 196–203. See also https://web.archive.org/web/20100217085620/http://www.nernst.de/kulturwelt.htm.
155 *Finally, in his September letter* Einstein, Albert. *CPE8A*, 57–9, Document 36.
155 *After this eventually arrived* Schulmann, Robert, Kox, A. J., Janssen, Michel, Illy, József. *CPE8A*, 85, Document 48, Note 3.
156 *He reminded Straneo that* Einstein, Albert. *CPE8A*, 77–8, Document 45.
156 *This was the first of many* —. *CPE8A*, 84–6, Document 48.
156 *In the six or so months* —. *CPE8A*, 95–7, Document 60. See also Documents 62, 64, 66, 67, 69, 71, 74, 75, 77, 78, and 80.
157 *The only mention is a bald statement* Einstein, Albert, de Haas, Wander Johannes. *CPE6*, 150–71, Document 13. See also Documents 14–16, 23, 28.
157 *De Haas, who eventually was appointed* Lorentz, Hendrik A. *CPE8A*, 298, Document 225.
157 *De Haas eventually accepted that* Galison, Peter. Theoretical Predispositions in Experimental Physics: Einstein and the Gyromagnetic Experiments, 1915–1925. *Historical Studies in the Physical Sciences*, 1982, 12: 285–323. https://www.jstor.org/stable/27757498.
157 *Mileva, who had warmed to Clara* Einstein, Albert. *CPE8A*, 128–9, Document 83.
158 *There were painful associations* Highfield, Roger, Carter, Paul. *The Private Lives of Albert Einstein* (London, Faber and Faber, 1993), 122–3.
158 *Most important were the restrictions* Einstein, Albert. *CPE8B*, 858–61, Document 604.
158 *. . . in particular his heart attack* —. *CPE16*, 381–2, Document 237.
158 *He and Zangger did visit* —. *CPE8A*, 166–9, Documents 114–17; see also https://onlineonly.christies.com/s/fine-printed-books-manuscripts-including-americana/his-former-student-42/195364?ldp_breadcrumb=back.
158 *Finally, Einstein received his passport* —. *CPE8A*, 172–5, Documents 120–1.
159 *'I know very well'* Frank, Philipp. *LZ*, 218.
159 *It is not certain why* Janssen, Michel, Renn, Jürgen. Arch and Scaffold: How Einstein Found His Field Equations. *Physics Today*, 2015, 68: 30–6. https://physicstoday.scitation.org/doi/pdf/10.1063/PT.3.2979.
159 *There is no doubt* Janssen, Michel. What Did Einstein Know and When Did He Know It? A Besso Memo Dated August 1913. In *The Genesis of General Relativity*, Vol. I, Michel Janssen, John D. Norton, Jürgen Renn, Tilman Sauer, John Stachel, eds. (Dordrecht, Springer, 2007), 788.

159 *According to Ledermann* Ledermann, Walter. *Encounters of a Mathematician* (London, Walter Ledermann, 2009), 80.
159 *In this, in a tone of some excitement* Einstein, Albert. *CPE8A*, 177–8, Document 123.
160 *A few days later* —. *CPE8A*, 182–4, Document 129.
160 *In his major review* —. *CPE6*, 72–130, Document 9.
160 *Janssen and Renn make* Jansen, Michel, Renn, Jürgen. Untying the Knot: How Einstein Found His Way Back to Field Equations Discarded in the Zurich Notebook. In *The Genesis of General Relativity*, Vol. I, Michel Janssen, John D. Norton, Jürgen Renn, Tilman Sauer, John Stachel, eds. (Dordrecht, Springer, 2007), 839–925.
160 *Hilbert famously remarked that* Reid, Constance. *Hilbert* (Berlin, Springer, 2012), 127.
161 *'The people in Göttingen'* —. *Hilbert* (Berlin, Springer, 2012), 142.
161 *Sommerfeld, an old friend* Einstein, Albert. *CPE8A*, 191–2, Document 136.
161 *The rest of this paper* —. *CPE6*, 214–24, Document 21.
162 *With this assumption* Jansen, Michel, Renn, Jürgen. Untying the Knot: How Einstein Found His Way Back to Field Equations Discarded in the Zurich Notebook. In *The Genesis of General Relativity*, Vol. I, Michel Janssen, John D. Norton, Jürgen Renn, Tilman Sauer, John Stachel, eds. (Dordrecht, Springer, 2007), 889, Note 101.
162 *Einstein referred to such a postulate* Pais, Abraham. *LORD*, 253.
162 *It appeared in print* Einstein, Albert. *CPE6*, 225–9, Document 22.
162 *Hilbert wrote back* Hilbert, David. *CPE8A*, 195–7, Document 140.
162 *Einstein declined however* Einstein, Albert. *CPE8A*, 199, Document 144.
162 *He concluded his letter* —. *CPE8A*, 201–2, Document 148.
162 *When his calculation resulted* —. *CPE8A*, 243–4, Document 182.
162 *Having once failed* —. *CPE6*, 233–43, Document 24.
163 *Einstein sent him a postcard* —. *CPE8A*, 201, Document 147.
163 *On 25 November* —. *CPE6*, 245–9, Document 25.
163 *Einstein, as he remarked to Besso* —. *CPE8A*, 218, Document 162.
163 *Zangger* —. *CPE8A*, 204–6, Document 152.
163 *Sommerfeld* —. *CPE8A*, 206–9, Document 153.
163 *Einstein's euphoria* —. AEA [Online] 28 November 1915. Cat. no. 21–382. http://alberteinstein.info/.
164 *Pais records a second-hand account* Pais, Abraham. *LORD*, 261.
164 *The first recorded letter* Einstein, Albert. *CPE8A*, 222–3, Document 167.
164 *Although the equations* Hilbert, David. Die Grundlagen der Physik. (Erste Mitteilung). *Nachrichten von der Gesellschaft der Wissenschaften zu Göttingen, Mathematisch-Physikalische Klasse*. 1915: 395–408. Note that although dated 1915, the paper was not published until March 1916.
164 *He forwarded Einstein's postcard* Medicus, Heinrich A. The Friendship among Three Singular Men: Einstein and His Swiss Friends Besso and Zangger. *Isis*, 1994, 85: 477.

Chapter 13

165 *He followed this with a letter* Einstein, Albert et al. *CPE8A*, 188–226, Documents 133, 142, 143, 150, 154–6, 158, 159, 162–4, 166, 168, and 170.
165 *Einstein, however, had at various times* Einstein, Albert. *CPE8A*, 234–6, Document 178.
165 *He suggested that this* —. *CPE8A*, 257–8, Document 187.

166 *Therefore, after the war* —. *CPE8A*, 270–1, Document 200.

166 *Instead, he wrote back* —. *CPE8A*, 280–1, Document 211.

166 *After a trip to Lucerne* —. *CPE8A*, 283–4, Document 216.

166 *To serve the musical appetites* Muck, Peter, [ed.]. *Einhundert Jahre Berliner Philharmonisches Orchestra*, Vol. I (Tutzing, Hans Schneider, 1982), 501.

166 *Indeed, Arnold Schoenberg* Schoenberg, Arnold. Style and Idea: Selected Writings of Arnold Schoenberg. In *New Music*, Leonard Stein, ed. (Berkeley, University of California Press, 1984), 137–9.

166 *Daring musically* Ross, Alex. *The Rest is Noise* (London, Farrar Straus & Giroux, 2008), 3–10.

167 *Thus encouraged, he composed* Wilhelm, Kurt. *Richard Strauss: An Intimate Portrait* (New York, Rizzoli, 1989), 77–8.

167 *In August 1914* Ross, Alex. *The Rest is Noise* (London, Farrar Straus & Giroux, 2008), 60.

167 *He had lost his entire life savings* Kennedy, Michael. *Richard Strauss: Man, Musician, Enigma* (Cambridge, Cambridge University Press, 2006), 188–9.

167 *For example, on 18 October* Various. *Berliner Tageblatt*. 1914. Various articles and advertisements between August and November.

167 *Newspaper reports talked about* Dennis, David. 'O Freunde, nicht diese Töne!' First World War Beethoven Reception as Precedent for the Nazi 'Cult of Art.' In *Musik bezieht Stellung: Funktionalisierungen der Musik im Ersten Weltkrieg*, Claudia Glunz, Dietrich Helms, Thomas F. Schneider, Stefan Hanheide, eds. (Göttingen, V&R Unipress GmbH, 2013), 243–62.

168 *In the 'Foundation of the General Theory of Relativity'* Einstein, Albert. *CPE6*, 283–339, Document 30.

168 *'Ehrenfest, aged 35, in Leiden'* Pais, Abraham. *LORD*, 271.

168 *Another enthusiastic convert* Einstein, Albert. *CPE8A*, 262–3, Document 191; see also 266, Document 195 and notes therein.

168 *During those dark days* Born, Max. *Physics in my Generation*, 2nd rev. ed. (New York, Springer Science & Business Media, 1969), 156. See also Seelig, Carl, ed. *Helle Zeit – Dunkle Zeit* (Berlin, Springer, 1956), 35.

168 *He used his time-honoured method* Einstein, Albert. *CPE8A*, 269–70, Document 199.

169 *Schwarzschild remarkably had obtained* Schwarzschild, Erwin. *CPE8A*, 224–5, Document 169.

169 *It was in a letter to him* Einstein, Albert. *CPE8A*, 265–6, Document 194.

169 *He found that as he compressed* Schwarzschild, Erwin. *CPE8A*, 258–60, Document 188.

169 *He ended with the correct prediction* Einstein, Albert. *CPE6*, 358–62, Document 33.

169 *Einstein himself often found* —. *CPE8A*, 255–7, Document 186.

169 *Indeed, he was shortly to produce* —. *CPE6*, 372, Document 35.

170 *Einstein was very excited* —. *CPE8A*, 301–3, Document 227.

170 *Unfortunately, in his lecture* —. *CPE6*, 347–57 Document 32.

170 *In 1918* —. *CPE7*, 11–28, Document 1.

170 *This was the first he had heard* Weinstein, Galina. Einstein's Discovery of Gravitational Waves 1916–1918. *arXiv.org* [Online]. 2 December 2016. https://arxiv.org/abs/1602.04040.

170 *Malnutrition was beginning* Davis, Belinda J. *Home Fires Burning* (Chapel Hill, University of North Carolina Press, 2000), 21, 118.

170 *A minor role* Schrödinger, Erwin. Die Energiekomponenten des Gravitationsfeldes. *Physikalische Zeitschrift*, 1918, 19: 4–7.

171 *Einstein certainly benefited* Einstein, Albert. *CPE8A*, 407–9, Document 309; 561–3, Document 403.

171 *Einstein was alerted to the problem* Besso, Michele. *CPE8A*, 315–16, Document 239.

171 *Einstein did offer to come* Einstein, Albert. *CPE8A*, 311–12, Document 233.

171 *As a medical doctor* —. *CPE8A*, 315–16, Document 237.

171 *Indeed, Lisbeth Hurwitz was of the opinion* Trbuhović-Gjurić, Desanka. *ISAE*, 138–9.

171 *He confided to Zangger* Einstein, Albert. *CPE8A*, 320–1, Documents 241, 242.

171 *She had by now been diagnosed* —. *CPE8A*, 330–1, Document 251.

171 *He had every confidence that* —. *CPE8A*, 337–8, Document 258.

172 *Nevertheless, she was able to return* —. *CPE8A*, 351, Document 271, Note 2.

172 *It isn't clear why* —. *CPE8A*, 409–11, Document 310.

172 *By using the fact that* —. *CPE6*, 363–70, Document 34.

172 *At the end of August* —. *CPE6*, 381–98, Document 38.

172 *Stone considers* Stone, A. Douglas. *Einstein and the Quantum.* (Princeton, Princeton University Press, 2015), 186.

172 *However, Einstein* Rutherford, Ernest, Andrade, Edward. The Spectrum of the Penetrating Gamma Rays from Radium B and Radium C. *The Philosophical Magazine*, 1914, 28: 263–73.

173 *He wrote in the paper* Einstein, Albert. *CPE8A*, 332–3, Document 254. See also 329–31, Documents 250 and 251.

173 *In a letter written after* —. *CPE8A*, 345–7, Documents 268, 269.

173 *As for Ehrenfest* Kline, Martin J. *Paul Ehrenfest: The Making of a Theoretical Physicist*, Vol. I (Amsterdam, North-Holland, 1970), 301.

173 *During his stay in the Netherlands* Einstein, Albert. *CPE8A*, 359–61, Document 273.

173 *Mach had died* —. CPE6, 278–82, Document 29.

174 *To get around this problem* —. *CPE6*, 540–2, Document 43.

174 *This was anathema to Einstein* Schulmann, Robert, Kox, A. J., Janssen, Michel, Illy, József. *CPE8A*, 351–7 and references in Note 1.

174 *A quote from Einstein to de Sitter* Einstein, Albert. *CPE8A*, 475–7, Document 356.

174 *He offered to help* —. *CPE8A*, 394–5, Document 301.

174 *Although he had initially been overjoyed* Adler, Friedrich. *CPE8A*, 402–04, Document 307.

174 *Einstein was moved* Einstein, Albert. *CPE8A*, 409–11, Document 310.

174 *In April, Einstein wrote* —. *CPE8A*, 432, Document 324.

174 *In sending a substantial section* Adler, Friedrich. *CPE8A*, 437–8, Document 329.

175 *His affirmation that he killed* Anon. *Arbeiter-Zeitung*, 18 May 1916, 1–8.

175 *The Emperor initially commuted* Alder, Douglas D. Friedrich Adler: Evolution of a Revolutionary. *German Studies Review*, 1978, 1: 260–84.

175 *He wrote to Besso* Einstein, Albert. *CPE8A*, 441–2, Document 331.

175 *Adler apologised for the long silence* —. *CPE8B*, 828–29, Document 582.

175 *He ended his letter* —. *CPE8A*, 840–7, Document 594.

175 *Typically, Adler ignored Einstein's advice* Adler, Friedrich. *Ernst Machs Ueberwindung des mechanischen Materialismus* (Vienna, Verlag der Wiener Volksbuchhandlung Ignaz Brand & Co., 1918).

175 *. . . that on relativity in 1920* —. *Ortszeit, Systemzeit, Zonenzeit und das ausgezeichnete Bezugssystem der Elektrodynamik* (Vienna, Verlag der Wiener Volksbuchhandlung Ignaz Brand & Co., 1920).

175 *Frank records that* Frank, Philipp. LZ, 288–89.

175 *In May 1917* Einstein, Albert. *CPE10*, 84–5, Document 344a.

176 *'Yesterday I was at the Plancks' house'* Ernst, Sabine. *Lise Meitner an Otto Hahn: Briefe aus den Jahren 1912 bis 1924* (Stuttgart, Wissenschaftliche Verlagsgesellschaft mbH, 1992), 64.

176 *Later that year* Potter, Tully. *Adolf Busch*, Vol. I (London, Toccata Press, 2010), 172–4, 219, 276. See also 'Drei Brahms Abende,' Deutsche Brahms-Gesellschaft, and Konzert-Direktion Hermann Wolff, 'I Brahmsabend', 1 December 1917, in Henschel Collection, Box 9, 1916–17, British Library Special Collections.
176 *In mid-November she had a relapse* Einstein, Albert. *CPE8A*, 372–5, Document 283.
177 *He remarked that he had learned* —. *CPE8A*, 332–3, Document 254.
177 *Hans Albert replied with a sketch* Einstein, Hans Albert. *CPE10*, 35–6, Document 279a (Volume VIII).
177 *Einstein in his reply* Einstein, Albert. *CPE8A*, 367–8, Document 279; 380–1, Document 287.
177 *Einstein recalled that* —. *CPE8A*, 400–2, Document 306.
177 *She had to return to hospital* Trbuhović-Gjurić, Desanka. *ISAE*, 140–1.
177 *She poured out her heart* Rogger, Franziska. *ES*, 53–4.
177 *Einstein blamed Tete's frail health* Einstein, Albert. *CPE8A*, 400–2, Document 306.
177 *An amusing incident* Einstein, Albert. CPE8A, 567–9, Document 406.
178 *Einstein was adamant* —. *CPE8A*, 446, Document 335.
178 *She arrived at the end of August* —. *CPE8A*, 502, Document 371, Note 2.
178 *By February 1918* —. *CPE8A*, 658–9, Document 469, Note 1.
178 *Initially, Einstein was diagnosed* —. *CPE8A*, 400–2, Document 306; 429–31, Document 322.
178 *The diagnosis then shifted* —. *CPE8A*, 567–9, Document 406; 579, Document 417.
179 *'Then I asked him'* Seelig, Carl. *BIOG*, 187–9.
179 *Anna Besso devised an extremely efficacious treatment* Einstein, Albert. *CPE8A*, 497–8, Document 367.
179 *The bohemian Einstein* —. *CPE8A*, 511–12, Document 377.

Chapter 14

180 *'In the theoretical treatment'* Einstein, Albert. *CPE6*, [trans] R. W. Lawson, 247–420, Document 42.
181 *This was not indeed possible* —. *CPE7*, 130–40, Document 18.
181 *'The question of whether'* —. *CPE8A*, 379–80, Document 286.
181 *'[As] is not unusual for him'* Pais, Abraham. *LORD*, 287–8.
181 *He tersely noted in a footnote* Einstein, Albert. Über die formale Beziehung des Riemannschen Krümmungstensors zu den Feldgleichungen der Gravitation. *Mathematische Annalen*, 1927, 97: 99–103.
181 *The influence of the gravitational field* Sauer, Tilman. Einstein's Unified Field Theory Program. In *The Cambridge Companion to Einstein*,Christoph Lehner, Michel Janssen, eds. (New York, Cambridge University Press, 2014), 281–305.
181 *In fact, modern theories* Nordström, Gunnar. Über die Möglichkeit, das elektromagnetische Feld und das Gravitationsfeld zu vereinigen. *Physikalische Zeitschrift*, 1914, 15: 504–6.
181 *Even then, Kaluza in a footnote* Einstein, Albert. *CPE9*, 38–40, Document 26; 46–7, Document 30; 56–7, Document 35; 65–8, Document 40.
182 *He introduced the concept* —. *CPE8A*, 393–94, Document 300 & pp. 425–426, Document 319.
183 *In December, Einstein requested* —. *CPE8A*, 513–14, Document 379 & pp. 570–571, Document 409.
183 *Eventually, the observatory at Potsdam* —. *CPE8A, B*, 563–64, Document 404; pp. 603–604, Document 434 & pp. 613–614, Document 441.

183 *He put his hope in Hans Albert* —. *CPE8B*, 598–99, Document 428 & pp. 614–615, Document 442.
183 *He played the 'Moonlight'* Einstein, Hans Albert. *CPE10*, 156–7, Document 513a (Volume VIII).
184 *'You are completely unable'* —. *CPE10*, 140–1, Document 442a (Volume VIII).
184 *By March he had gained weight* Einstein, Albert. *CPE8B*, 666–8, Document 474.
184 *During his stay in Luzern* —. *CPE10*, 124–5, Document 371b (Volume VIII).
184 *He apologised that* —. *CPE8B*, 714–16, Document 503.
184 *Indeed, he wrote to Hans Albert* —. *CPE8B*, 819–20, Document 576.
184 *This came as a major disappointment* Einstein, Hans Albert, Einstein, Eduard. *CPE10*, 166–8, Documents 557b and c (Volume VIII).
184 *It was not until the late summer* Einstein, Albert. *CPE8B*, 884–5, Document 621.
184 *The only cure was Nature's* —. *CPE8A*, 199, Document 144.
185 *Without Zorka to help her* —. *CPE6*, 347–57, Document 32.
185 *This was the carrot* —. *CPE8B*, 622–4, Document 449.
186 *Exactly two years ago* Einstein, Mileva. *CPE10*, 141–3, Document 461a (Volume VIII).
186 *Things would be easier* —. *CPE8B*, 635–6, Document 457. See also Heinrich A. Medicus. The Friendship among Three Singular Men: Einstein and His Swiss Friends Besso and Zangger. *Isis*, 1994, 85: 469.
186 *He once more attempted* —. *CPE8B*, 666–8, Documents 473, 474.
186 *This enraged Anna* Besso-Winteler, Anna, AEA [Online]. 1918. Cat. no. 83-461. http://alberteinstein.info/.
186 *This altercation clearly made* —. *CPE8B*, 815–17, Document 572.
187 *In this letter, he compared* Zangger, Heinrich. *Seelenwerwandte*, Robert Schulmann, ed. (Zurich, Verlag Neue Zürcher Zeitung, 2012), 294–5, Document 185, letter to Einstein written after 4 March 1918.
187 *The receipt of several valuable* Einstein, Albert. *CPE8B*, 729–31, Documents 514, 515; 754–6, Document 533.
187 *Arriving surreptitiously in Berlin* Goenner, Hubert, Castagnetti, Giuseppe. Albert Einstein as Pacifist and Democrat during World War I. *Science in Context*, 1996, 9: 359, 372.
187 *Overbye states that Ilse was in love* Overbye, Dennis. *Einstein in Love: A Scientific Romance* (London, Bloomsbury, 2001), 342.
187 *For example, back in April* Einstein, Albert. *CPE8A*, 283–4, Document 216.
187 *In May of 1918* —. *CPE8B*, 758, Document 536.
187 *The letter* Lowenthal, Ilse. *CPE8B*, 769–71, Document 545.
188 *In early June, Einstein talked* Einstein, Albert. *CPE8B*, 818–19, Document 575.
188 *Concluding the letter* —. *CPE8B*, 835–7, Document 591.
188 *Some correspondence that Ilse* Bucky, Gustav. *AEA* [Online] May 1918. Cat. nos. 77-984, 77-985. http://alberteinstein.info/.
188 *It is interesting to note* Einstein, Albert. *CPE8B*, 851–4, Documents 598, 599.
188 *To Zangger he remarked* —. *CPE8B*, 855–7, Documents 601, 602.
188 *On 4 November* —. *CPE8B*, 935–6, Document 644.
189 *No doubt having many* Born, Max. *The Born–Einstein Letters*, [trans.] Irene Born (London, Macmillan, 1971), 150–1.
189 *This carried on through October* Einstein, Albert. *CPE8B*, 974–5, Document 676.
189 *Back on duty in Zurich* —. *CPE8B*, 939–42, Documents 648, 649.
189 *By December, Einstein could write* —. *CPE8B*, 958–60, Document 663.
190 *On 15 January* —. *CPE9*, 3–4, Documents 1, 2. For events in Berlin in this period, see Christopher Clark, *Iron Kingdom* (London, Penguin Books, 2007), 622–8.

190 *Nevertheless, the concert* Muck, Peter, ed. *Einhundert Jahre Berliner Philharmonisches Orchestra* (Tutzing, Hans Schneider, 1982), 465. Vol. I - 1882–1922.
190 *'While outside both foreign'* Schünemann, Georg. *Deutsche Allgemeine Zeitung.* 9 April 1919, 2.
190 *They were extremely popular* Seelig, Carl. *BIOG*, 184.
190 *During the musical evenings* Trbuhović-Gjurić, Desanka. *ISAE*, 143–4.
190 *Part of the judgment* Einstein, Albert. *CPE9*, 8–11, Document 6.
191 *It can only be assumed* —. *CPE9*, 83–4, Document 55.
191 *He also saw an eclipse* Stanley, Matthew. *Einstein's War* (London, Viking, 2019). See 252–96 for a full account of the expedition and its analysis. See also Daniel Kennefick, *No Shadow of a Doubt: The 1919 Eclipse That Confirmed Einstein's Theory of Relativity* (Princeton, Princeton University Press, 2019).
191 *Einstein told Zangger* Einstein, Albert. *CPE9*, 91–4, Documents 61, 62, 63.
192 *He turned out to be* Einstein, Mileva, Einstein, Eduard. *CPE10*, 220–2, Document 148a (Volume IX); 226–7, Document 183a (Volume IX).
192 *He also took him sailing* Trbuhović-Gjurić, Desanka. *ISAE*, 142.
192 *Einstein was back in Berlin* Einstein, Albert. *CPE9*, 125–6, Document 84; 128–30, Document 86.
192 *He also commissioned a violin* —. *CPE9*, 170–1, Document 113; 218–9, Document 151; 266–70, Document 189; 272–3, Document 192; 289–90, Document 204.
192 *Since the average of* Kormos Buchwald, Diana, Schulmann, Robert, Illy, József, Kennefick, Daniel J., Sauer, Tilman. *CPE9*, xxxi–xxxvii.
192 *Sponsel makes a compelling case* Sponsel, Alastair. Constructing a 'Revolution in Science': The Campaign to Promote a Favourable Reception for the 1919 Solar Eclipse Experiments. *The British Journal for the History of Science*, 2002, 35: 439–67.
193 *Rather, by concentrating on three* Mie, Gustav. *CPE9*, 97–9, Document 65.
193 *So convinced was Eddington* Earman, John, Glymour, Clark. Relativity and Eclipses: The British Eclipse Expeditions of 1919 and Their Predecessors. *Historical Studies in the Physical Sciences*, 1980, 11: 49–85.
194 *'By an application of the theory'* Einstein, Albert. *CPE7*, 212–15, Document 26.
194 *He was inundated with requests* —. *CPE9*, 273, Document 193.
194 *Offers of very substantial funding* —. *CPE9*, 274–5, Document 194.
194 *Within a couple of weeks* Einstein, Elsa. *AEA* [Online]. 10 December 1919. Cat. no. 9–455. http://alberteinstein.info/.

Chapter 15

195 *Planning permission was not obtained* Hoffmann, Dieter. *Einsteins Berlin: Auf den Spuren eines Genies* (Weinheim, Wiley-VCH Verlag GmbH & Co., 2006), 19–27.
196 *Einstein and Lasker* Einstein, Albert. Foreword. In Hannak, Jacques. *Emanuel Lasker: The Life of a Chess Grand Master* [trans.] Heinrich Fraenkel (New York, Dover, 1991), 7–8.
196 *It was large and spacious* Marianoff, Dimitri, Palma, Wayne. *Einstein: An Intimate Study of a Great Man* (New York, Doubleday, Doran and Co., Inc., 1944), 1, 22.
196 *Although Graf Kessler characterised* Graf Kessler, Harry. *Tagebücher, 1918–1937*, Wolfgang Pfeiffer-Belli, ed. (Frankfurt am Main, Insel Verlag, 1961), 278–80.
196 *'I have learnt to recognise'* Einstein, Albert. *CPE8A*, 407–9, Document 309.
196 *'Here, everyone is close to me'* —. *CPE8B*, 870–1, Document 612.

197 *After great efforts* —. LOT 18: A Room for His Terminally-Ill Mother. *Christie's Sale 16447: Einstein and Family: Letters and Portraits* [Online]. 23 December 1919 [Cited: 26 April 2020]. https://onlineonly.christies.com/s/einstein-family-letters-portraits/room-his-terminally-ill-mother-18/55699.

197 *Looked after by the nurse* Rogger, Franziska. *ES*, 61–3.

197 *One doctor called in* Highfield, Roger, Carter, Paul. *The Private Lives of Albert Einstein* (London, Faber and Faber, 1993), 192.

197 *However, none of them* Marianoff, Dimitri, Palma, Wayne. *Einstein: An Intimate Study of a Great Man* (New York, Doubleday, Doran & Company, 1944), 176.

197 *'My mother died a week ago'* Einstein, Albert. *CPE9*, 451–2, Document 332.

197 *His numbness apparently also gave way* Highfield, Roger, Carter, Paul. *The Private Lives of Albert Einstein* (London, Faber and Faber, 1993), 192.

197 *'We thought that space was straight'* Ehrenfest, Paul. *CPE9*, 413–17, Document 303.

197 *The string-pulling worked* Bezirksamt Berlin-Schöneberg, Ilse Einstein. AEA [Online]. October 1921. Cat. nos. 44–998, 44–999. http://alberteinstein.info/.

198 *Lorentz was able to write* Einstein, Albert, et al. *CPE9*, 246–50, Document 175; 266–70, Document 189; 320–1, Document 229; 355–6, Document 256.

198 *When the inaugural lecture* Kormos Buchwald, Diana, Sauer, Tilman, Rosenkranz, Ze'ev, Illy, József, Holmes, Virginia Iris, *CPE10*, xliii–xlvii.

198 *It was only after significant efforts* Einstein, Albert, Einstein, Elsa. *CPE10*, 247, Document 7; 253–5, Document 10; 265–6, Document 20; 337–8, Document 76.

198 *Ehrenfest's wife Tatiana* Ehrenfest, Paul, Ehrenfest, Tatiana. Begriffliche Grundlagen der statistischen Auffassung in der Mechanik. In *Mechanik*, C. Müller, F. Klein, eds. (Leipzig, Teubner, 1907), 773–860.

199 *'What more could anyone want'* Seelig, Carl. *BIOG*, 224–5.

199 *Despite his previous antipathy* Einstein, Albert. *CPE7*, 426–30, Document 57.

199 *In fact, the conference* —. *CPE9*, 227–8, Document 160.

200 *Einstein formed a great admiration* Einstein, Hans Albert. *CPE10*, 244–5, Document 4; 246–7, Document 6.

200 *'That this insecure and contradictory'* Einstein, Albert. Autobiographical Notes. In *AEPS*, 45, 47.

200 *The background was that* —. *CPE9*, 425–7, Documents 311 and 312.

200 *There had been a disturbance* Anon. *CPE7*, 284–8, Document 33.

200 *That this was not the case* Einstein, Albert. *CPE7*, 344–9, Document 45.

201 *Einstein's opponents continued* Anon. Der Kampf um Einstein. *Vossiche Zeitung*. 24 September 1920, 1–2.

201 *Max Born commented* Einstein, Hans Albert. *CPE10*, 442, Document 161.

201 *Lenard's response in his diary* Schirrmacher, Arne. *Philipp Lenard: Erinnerungen eines Naturforschers* (Berlin, Springer Verlag, 2010), 233–6.

201 *Here Einstein reported that* Einstein, Albert. *CPE10*, 454, Document 170.

201 *After stopping in Konstanz* —. *CPE10*, 464, Document 179.

202 *Moszkowski was significantly older* —. *CPE13*, 362, Document 233.

202 *He was pained* Einstein, Hans Albert. *CPE10*, 465–7, Document 180; 474–5, Document 187.

202 *While it attracted some criticism* von Seimens, Friedrich Carl, Einstein, Albert. *CPE12*, 74–8, Documents 44 and 45.

202 *Her husband was almost as outspoken* Born, Hedwig, Born, Max. *CPE10*, 447–50, Document 166; 459–61, Document 175.

202 *This must have infuriated Hedwig* Born, Hedwig. *AEA* [Online]. 16 August 1946. Cat. no. 65–850. http://alberteinstein.info/.

203 *The Moszkowsi book* Moszkowski, Alexander. *Einstein: Einblicke in seine Gedankenwelt* (Hamburg, Hoffmann und Campe, 1921), 87–8; 135–45; 185–6; 210.

203 *He was thoroughly enjoying* Einstein, Albert. CPE9, 487–8, Document 361.

203 *Its fascination still held* Snow, C. P. *Variety of Men* (London, Macmillan, 1967), 88.

203 *In the article* Einstein, Albert. CPE7, 338–40, Document 43.

203 *'He is confessedly a classicist'* Moszkowski, Alexander. *Einstein the Searcher.* [trans.] Henry L. Brose (New York, E.P. Dutton & Company, 1921), 233–5.

204 *He is known to have played* Goenner, Hubert. *Einstein in Berlin 1914–1933* (Munich, C.H. Beck, 2005), 252.

204 *This was not due to lack* Einstein, Albert. *CPE16*, 35, Document 222a.

204 *In 1920, she arranged* Einstein, Elsa. *CPE10*, 265–6, Document 20.

205 *Their meeting occurred before 1922* Goldschmidt, Heinrich. *AEA* [Online]. 25 April 1922. Cat. no. 43–772. http://alberteinstein.info/. See also Hoffmann, B. *Albert Einstein: Creator and Rebel* (London, Hart-Davis, MacGibbon, 1972), 152–3.

205 *Franz's books carried a bookplate* Mendelssohn-Gesellschaft. Franz von Mendelssohn the Younger: The Chairman: 1865–1935. *The Mendelssohns Society* [Online] [Cited: 10 April 2020]. https://www.mendelssohn-gesellschaft.de/en/mendelssohns/biografien/franz-von-mendelssohn.

206 *Einstein, with his indifference* Prieto, Carlos. *Adventures of a Cello* (Austin, University of Texas Press, 2018), 62–5.

206 *'A pianist asked me'* Janof, Tim. Conversation with Eva Heinitz. *Internet Cello Society* [Online]. 20 December 1997. http://cello.org/Newsletter/Articles/heinitz.htm.

206 *He accused them* Einstein, Albert. *CPE10*, 528–9, Document 232.

206 *She noted that Mileva* Trubuhović-Gjurić, Desanka. *ISAE*, 149.

206 *He had taken up the double bass* Einstein, Hans Albert. *CPE12*, 18–20, Document 232a (Volume X), Document 1.

206 *His father-in-law had transferred* Einstein, Albert. *CPE9*, 419–20, Document 305.

206 *For example, by November* Kormos Buchwald, Diana, Sauer, Tilman, Rosenkranz, Ze'ev, Illy, József, Holmes, Virginia Iris. *CPE12*, 475–9.

207 *The numbers should be multiplied* Ehrenfest, Paul. *CPE13*, 132–3, Document 45.

207 *He suggested that he should* —. *CPE10*, 479–81, Document 191.

207 *When Einstein requested* Schmedeman, Albert. *CPE5*, 523–4, Document 229, Note 3.

207 *Such precautions* Kassel, Heinrich, Einstein, Albert, et al. AEA [Online]. October 1921. Cat. nos. 43–675, 43–676, 43–678, and 43–679. http://alberteinstein.info/.

207 *In 1925, Fanta married Johanna* Gordin, Michael. *Einstein in Bohemia* (Princeton, 2020), 256–9.

208 *'Einstein showed himself'* Anon. *Prager Tagblatt*, 8 January 1921, 3.

208 *'. . .he played quite well'* Fürth, Reinhold. Personal Reminiscences. In *Einstein: The First Hundred Years.* Alan Mackay, James Woudhuysen, Maurice Goldsmith, eds. (Oxford, Pergamon, 1980), 19–21.

208 *Einstein was probably glad* Frank, Philipp. *LZ*, 283–8.

208 *In the end the audience* Einstein, Albert. *CPE12*, 35–6, Document 15, Note 5.

209 *Einstein had a high regard* —. *CPE12*, 46–7, Document 24.

209 *He told Busch that* —. *CPE15*, 28, Document 32a (Vol. XII).

209 *Although Graf Kessler's memory seems* Easton, Laird M. *The Red Count: The Life and Times of Harry Kessler* (Berkeley, University of California Press, 2002), 203–4.

209 *Graf Kessler's diary entry proclaims* Graf Kessler, Harry. *Tagebücher 1918–1937*,Wolfgang Pfeiffer-Belli, ed. (Frankfurt am Main, Insel-Verlag, 1961), 240–1.

209 *Serkin had made* Henahan, Donal. Rudolf Serkin, 88, Concert Pianist, Dies. *The New York Times*, 10 May 1991. D18.
210 *At first Einstein was disappointed* —. *Tagebücher 1918–1937*, Wolfgang Pfeiffer-Belli, ed. (Frankfurt am Main, Insel-Verlag, 1961), 241–5.
210 *However, he reduced his requested* Einstein, Albert. *CPE12*, 89–90, Document 53. See also Weizmann, Chaim, *CHAIM*, 158, Letter 133.
210 *Weizmann refused and ordered* Weizmann, Chaim. *CHAIM*, 174, Letter 145.
210 *In addition to enjoying Weizmann's company* Weizmann, Vera, Tutaev, David. *The Impossible Takes Longer* (London, Hamish Hamilton, 1967), 102.
211 *'Whenever he became weary'* Anon. Prof. Einstein Here, Explains Relativity. *The New York Times*, 3 April 1921, 1, 13.
211 *On 8 April* Cruise, M. J. *CPE12*, 163–4, Document 121.
211 *On 25 April, he and Elsa* Anon. Einstein Idea Puzzles Harding, He Admits as Scientist Calls. *The New York Times*, 26 April 1921, 1.
211 *Brandeis supported* Weizmann, Chaim. CHAIM, 177–83, Letters 154–8.
211 *'. . . received with a tremendous burst'* Einstein, Albert. *CPE12*, 170–1, Document 129, Note 2.
211 *He seems to have had* Zeisler, Sigmund. *CPE12*, 318–9, Document 272.
212 *It is interesting that Einstein* Einstein, Albert. *CPE7*, 496–577, Document 71.
212 *'The following remarks'* Talmey, Max. *The Relativity Theory Simplified and the Formative Period of its Inventor* (New York, Falcon Press, 1932), 177–8.
213 *'No living man plays Mozart'* Brucia, Margaret A. Einstein Comes to Dinner. *The Gotham Centre for New York City History* [Online]. 17 October 2017 [Cited: 3 June 2023]. https://www.gothamcenter.org/blog/einstein-comes-to-dinner.
213 *He was cheered* Anon. Professor Einstein. *The Jewish Chronicle*, 17 June 1921, 26.
213 *Two months before Einstein's visit* Lord Haldane, Richard. *The Reign of Relativity* (London, John Murray, 1921).
213 *Arthur Eddington remarked that Haldane* Sanchez-Ron, Jose M. Einstein's Relativity among British Philosophers. In *Einstein and the Changing Worldviews of Physics*, Jürgen Renn, Matthias Schemmel, Christopher Lehner, eds. (Berlin, Springer Science & Business Media, 2012), 84–5.
214 *After a dinner* Anon. Professor Einstein. *The Jewish Chronicle*, 17 June 1921, 26.
214 *On Sunday, they lunched* Clark, Ronald. *Einstein: The Life and Times* (New York, Avon Books, 1972), 340–1.
214 *He wrote to Ehrenfest* Einstein, Albert. *CPE12*, 194–5, Document 152; 198, Document 155.
215 *He denied any responsibility* —. *CPE12*, 222, Document 180.
215 *This was no doubt a lesson* —. *CPE7*, 620–30, Appendices D, E.
215 *Missner ascribes Einstein's astonishing fame* Missner, Marshall. Why Einstein Became Famous in America. *Social Studies of Science*, 1985, 15: 267–91.
216 *'The war had just ended'* Chandrasekhar, Subrahmanyan. *Eddington: The Most Distinguished Astrophysicist of His Time* (Cambridge, Cambridge University Press, 1983), 28.

Chapter 16

217 *The fact that Hans Albert mentioned* Einstein, Hans Albert. *Einstein, the Man and his Achievement*, G. J. Whitrow, ed. (North Chelmsford, Courier Corporation, 1973), 20–1.
217 *He wrote to his father* Einstein, Hans Albert. *CPE10*, 258–60, Document 15.

217 *'. . . contrary to my habit'* Einstein, Albert. *CPE13*, 295–96, Document 179.

217 *Birthdays were an irrelevance* —. LOT 34: 'Possibly it is better not to think too much.' *Christie's Sale 16447: Einstein and Family: Letters and Portraits.* [Online] 4 October 1924 [Cited: 26 April 2020]. https://onlineonly.christies.com/s/einstein-family-letters-portraits/possibly-it-better-not-think-too-much-34/55715.

217 *'It would be great if you could'* Einstein, Hans Albert, Einstein, Tete. *CPE12*, 154–5, Document 110; 367, Document 311; 402, Document 349.

218 *From this it can be deduced* Einstein, Albert. *CPE12*, 249, Document 206; 258–9, Document 214; 268–9, Document 223; 344–6, Document 296.

218 *Einstein returned to Berlin* Anschütz-Kaempfe, Hermann. *CPE12*, 467–8; 232–3, Document 191.

218 *Any remaining correspondence* Duerbeck, Hilmar. German Astronomy in the Third Reich. In *Organizations and Strategies in Astronomy*, Vol. 7, Andre Heck, ed. (Berlin, Springer Science & Business Media, 2007), 397–8.

218 *He had latterly expressed* Einstein, Albert. *CPE12*, 226–7, Document 221; 386, Document 331; 469; 472.

219 *On Christmas Day 1921* Einstein, Albert, Berliner, Arnold. *CPE12*, 385–6, Document 330; 391–2, Document 337; 393–4, Documents 339, 340.

219 *In return, in 1925* Hentschel, Klaus. Erwin Finlay Freundlich and Testing Einstein's Theory of Relativity. *Archive for the History of Exact Sciences*, 1994, 47: 143–201; see also Kohn, Leo. *AEA* [Online]. 19 March 1925. Cat. no. 11–330. http://alberteinstein.info/.

219 *Einstein arranged to meet Hans Albert* Einstein, Albert, et al. *CPE12*, 302–3, Documents 255–7; 305–6, Document 260.

219 *He wrote to his friend* Rogger, Franziska. *ES*, 64–8.

220 *This was Levi-Civita's doctoral advisor* Goodstein, Judith. *Einstein's Italian Mathematicians* (Providence, American Mathematical Society, 2018), 122–3.

220 *Lisbeth remarked on Einstein's* Trbuhović-Gjurić, Desanka. *ISAE*, 151–2. Note that the second date given, 8 November, is incorrect since Einstein was then in Leiden.

220 *From Leiden he wrote* Einstein, Albert. *CPE12*, 239–40, Document 292.

220 *As he remarked sheepishly* —. *CPE13*, 91, Document 25; 109–10, Document 31; 116, Document 37; 124–30, Document 43; 304–5, Document 190.

221 *He told him that he preferred* —. *CPE12*, 316–17, Document 270.

221 *The then-current conception* Mie, Gustav. Grundlagen einer Theorie der Materie. *Annalen der Physik*, 1912, 37: 511–34; see also *Annalen der Physik*, 1912, 39: 1–40, and *Annalen der Physik*, 1913, 40: 1–65.

221 *Einstein and Grommer's findings* Einstein, Albert, Grommer, Jakob. *CPE13*, 63–74, Document 12.

221 *Since Lenin was somewhat* Eisinger, Josef. *Einstein on the Road* (Amherst, Prometheus Books, 2011), 21.

221 *In March, Einstein moved* Jun Ishiwara, et al. *CPE12*, 287–90, Documents 244–6; 334–5, Document 285; 367–8, Document 312; see also *CPE13*, 213–14, Document 118.

221 *However, on 18 January* Einstein, Albert. *CPE13*, 57, Document 6; 85, Document 19; 91–2, Documents 25 and 26.

222 *In an interesting aside* —. *CPE13*, 148–9, Document 60.

222 *In early March, he wrote to Langevin* —. *CPE13*, 145–6, Document 56; 155–6, Document 63; 161–2, Document 69.

222 *On arrival in Paris* —. *CPE13*, 162–4, Documents 70, 71.

222 *Nordmann characterised his French* Nordmann, Charles. Einstein à Paris. *Revue des Deux Mondes.* 15 April 1922, VII, p. 932.

222 *Others who participated* Einstein, Albert. *CPE13*, 227–52, Document 131.
222 *Hearing of this* Frank, Philipp. *LZ*, 314.
223 *Earlier in his stay* Anon. *Le Petit Parisien*. 10 April 1922, 1; see also *L'Intransigeant*, 10 April 1922, 1.
223 *Einstein left Paris* Nordmann, Charles. Avec Einstein dans les régions dévastées. *L'Illustration*, 1922, 80: 328–31.
223 *They encountered each other* Einstein, Albert. *CPE8B*, 906, Document 6.
223 *He arranged a meeting* Anon. Professor Einstein. *The Jewish Chronicle*, 24 March 1922, 25.
224 *An accomplice followed* Levenson, Thomas. *Einstein in Berlin* (New York, Bantam Books, 2003), 263–70.
224 *Her reply, explaining that* Einstein, Albert. *CPE13*, 374, Document 245; 376, Document 249.
224 *He attended both the funeral* —. *CPE13*, 449–52, Document 317.
224 *However, a few weeks later* —. *CPE13*, 388–9, Documents 262, 263; 392–3, Document 266; 403–4, Documents 278–80; 438–9, Document 314.
224 *It is unclear precisely* —. *CPE9*, 147–8, Document 99.
224 *It was no doubt Katzenstein's* Franke, Kurt. *Jüdische Miniaturen: Moritz Katzenstein* (Berlin, Hentrich & Hentrich, 2005), 10–40.
225 *It appears as if Hans Albert* Einstein, Albert. *CPE14*, 175–6, Document 109.
225 *He had also played* Einstein, Hans Albert. *CPE10*, 258–60, Document 15.
225 *However, Hans Albert was left behind* —. *CPE13*, 135–6, Documents 49, 50; 260–1, Document 142; 295–6, Document 179; 374–6, Documents 246–8.
225 *After exploring various options* —. *CPE13*, 440–6, Document 315.
225 *In the final paragraph* —. *CPE13*, 453–62, Document 318.
225 *Three years later* Stone, A. Douglas. *Einstein and the Quantum* (Princeton, Princeton University Press, 2015), 212.
226 *Hence his preference* Einstein, Albert. *CPE13*, 522–6, Document 370, in particular Note 5.
226 *Busch was as well-known* Lehmann, Stephen, Faber, Marion. *Rudolph Serkin: A Life* (Oxford, Oxford University Press, 2003), 48.
226 *Another visitor in September* Einstein, Albert. *CPE13*, 528, Document 373.
226 *Around 1 October* Einstein, Albert. *CPE13*, 531–54, Document 397. See also Eisinger, Josef. *Einstein on the Road* (New York, Prometheus Books, 2011), 21–49.
227 *Each was very pleased* —. *CPE13*, 593, Document 386; 697–8, Document 421.
227 *The fascinating story* Friedman, Robert Marc. The 100th Anniversary of Einstein's Nobel Prize: Facts and Fiction. *Annalen der Physik*, 2022, 534: 2200305.
228 *'His performance was greeted'* Anon. Dr Einstein's Exquisite Performance. *Kahoku Shimpo*, 3 December 1922. [trans.] Michiko Shidara, with the author's grateful thanks also to Tetsuo Shidara.
230 *'Japanese music, which has developed'* —. *CPE13*, 605–12, Document 391.
230 *Einstein concluded by remarking* —. *CPE13*, 705–13, Document 417 and references therein.
230 *This paper, and the following* —. *CPE14*, 32–5, Document 13.
231 *The Einsteins subsequently visited* Bentwich, Norman. *My Seventy-Seven Years* (London, Routledge and Keegan Paul, 1962), 78. See also Einstein, Albert. *CPE13*, 559, Document 379.
232 *They took the train to Barcelona* Switzerland Embassy, Berlin. AEA [Online]. 9 April 1923. Cat. no. 29–179.13, http://alberteinstein.info/; and Cat. no. 29–179.15, http://alberteinstein.info/.
232 *After lunch with von Sauer* Einstein, Albert. *CPE13*, 554–88, Document 379. See also Glick, Thomas. *Einstein in Spain* (Princeton, Princeton University Press, 1988), 100–49.

Chapter 17

233 *At least the election of Lorentz* Einstein, Albert. *CPE13*, 738, Document 447; *CPE14*, 26–30, Documents 9 and 10; 43–5, Document 19.

233 *Einstein attended a performance* Curie, Marie, et al. *CPE14*, 309–10, Document 192; 391–2, Document 258; 409–12, Documents 273–5; 456–7, Documents 291, 292.

234 *Krüss formally succeeded Einstein* Grundmann, Siegfried. *The Einstein Dossiers: Science and Politics - Einstein's Berlin Period with an Appendix on Einstein's FBI File* (Berlin, Springer, 2006), 175–217. See also Vallentin, Antonina. *Le Drame D'Albert Einstein* (Paris, Pilon, 1954), 147–54; and United Nations Library & Archives Geneva, file R4001-5B-8036-1976.

234 *The Ministry agreed and informed him* Embassy of Switzerland, Berlin. *AEA* [Online]. 26 March 1923. Cat. no. 79–351; Cat. no. 43–15. http://alberteinstein.info/.

234 *Einstein pointed out that* Einstein, Albert. *CPE14*, 15–17, Document 1.

234 *He was told that if* Smith, Adam. *Paper Money* (New York, Dell, 1982), 69.

234 *It was probably during this stay* Einstein, Albert. *CPE14*, 227–8, Document 142, Note 4. For Kamerlingh Onnes's recollection, see *AEA* [Online]. 13 December 1927. Cat. no. 14–394. http://alberteinstein.info/.

235 *'In particular, in my opinion'* —. *CPE14*, 123–36, Documents 75, 76.

235 *His insistence at the time* —. *CPE14*, 329, Document 209; see also *AEA* [Online]. 6 May 1923. Cat. no. 30–27. http://alberteinstein.info/.

235 *In February 1925* —. *CPE14*, 540, Document 345; see also *AEA* [Online]. 2 December 1924. Cat. no. 29–179.17,; and 5 February 1925. Cat. no. 82–351. http://alberteinstein.info/.

236 *He angrily reproached Mileva* —. *CPE14*, 72, Document 37; 75–6, Document 43; 77–8, Document 46; 80, Document 50.

236 *After a flurry of letters* Zürcher, Emil, et al. *CPE14*, 92–3, Document 54; 321–2, Document 202; 323–4, Document 204; 325, Document 206.

236 *'[Hans] Albert is hard at work'* Popovic, Milan. *AS*, 139–40.

236 *Einstein received a letter* Einstein, Albert. *CPE14*, 303–4, Document 186.

237 *Mileva wrote to Einstein begging* Einstein, Hans Albert, et al. *CPE14*, 97, Document 56; 103–4, Document 60; 105–6, Document 62.

237 *Tete was twelve years old* Einstein, Albert. *CPE14*, 111–15, Documents 67–9; 136–7, Document 77; 142, Document 81.

237 *However, he opined* Einstein, Hans Albert, et al. *CPE14*, 905, Appendix 1; pp. 156–7, Document 90; 159–60, Document 92; 161, Document 94; 164–5, Document 97a; 175–6, Document 109; 189–91, Document 118.

237 *Einstein visited Zurich in July* Albert Einstein, et al. *CPE14*, 72–3, Document 38; 97, Document 56; 103–4, Document 60; 110, Document 65; 357–8, Document 233; 466–7, Document 299; see also *AEA* [Online]. 2 June 1923. Cat. no. 30–28. http://alberteinstein.info/.

238 *'I am pleased, dear Albert'* Einstein, Albert. *CPE14*, 408–9, Document 272.

238 *This relatively positive view* Rogger, Franziska. *ES*, 127.

238 *However, Hans Albert declined* Einstein, Albert. *CPE15*, 75–6, Document 25; 78–9, Document 27; 89–90, Document 37; 96–7, Document 45; 116–17, Document 63; 129–30, Document 73; 143–55, Documents 88, 89.

238 *Zangger spent considerable time* Zangger, Heinrich. *Seelenverwandte*, Robert Schulmann, ed. (Zurich, Verlag Neue Zürcher Zeitung, 2012), 424–8, Documents 267–70.

238 *Mileva's sister Zorka* Trubhović-Gjurić, Desanka. ISAE, 192–3.

239 *Hans Albert, to his credit* Einstein, Albert. *CPE15*, 136–7, Document 79; 189–90, Document 105; 195–8, Documents 110, 111; 238–9, Document 135;.

239 *'Albert had such a hell of a time'* Highfield, Roger, Carter, Paul. *The Private Lives of Albert Einstein* (London, Faber and Faber, 1993), 227.

239 *Hans Albert and Frieda married* Einstein-Marić, Mileva, et al. *CPE15*, 325–6, Document 183; 525–6, Document 325; 528, Document 328; 579, Document 366; 704–6, Documents 448, 449; 731, Document 467; 766–7, Documents 484, 485; 769–71, Documents 487, 488.

239 *'Dear Zangger, It is good'* Zangger, Heinrich, Einstein, Albert. *Seelenverwandte*, Robert Schulmann, ed. (Zurich, Verlag Neue Zürcher Zeitung, 2012), 453–5, Documents 289, 290.

239 *'[Hans] Albert used to be'* Popovic, Milan. *AS*, 144.

240 *It is in this context* Levine, Philippa. *Eugenics: A Very Short Introduction* (Oxford, Oxford University Press, 2017).

240 *His father was delighted* Einstein, Albert. *CPE14*, 155, Document 159 .

240 *'How beautifully the lad plays'* Zangger, Heinrich. *CPE15*, 222–4, Document 127.

240 *While he knew his father* Einstein, Eduard, Einstein, Albert. *CPE15*, 292–3, Document 159; 326–7, Document 184; 594–6, Document 379.

240 *He won several prizes* Einstein, Eduard. *CPE14*, 576–77, Document 382; see also *AEA* [Online]. 16 September 1925. Cat. no. 144–466. http://alberteinstein.info/.

240 *Tete's belief* Graeser, Wolfgang. AEA [Online]. 24 December 1925. Cat. no. 43–793. http://alberteinstein.info/.

241 *To Mileva, Einstein wrote* Einstein, Eduard, Einstein, Albert. *CPE15*, 86–7, Document 78; 334–5, Document 190; 454–6, Document 274; 488–9, Document 295; 571–3, Document 359; 632–5, Documents 414, 415; 665–8, Documents 433, 434; 769–71, Documents 488, 489.

241 *When, in 1928, Tete published* Einstein, Albert. *CPE16*, 457–8, Document 299.

241 *Much music was made together* —. *CPE15*, 522–3, Documents 321, 322; 555–6, Documents 345, 346; 798–9, Documents 507, 508.

241 *Tete joined him* —. *CPE16*, 68–9, Document 20; 75, Document 26; 76, Document 27; 81, Document 32.

242 *Such musical offerings* Winteler-Einstein, Maja. *CPE14*, 557–8, Document 364; 572–4, Document 378; 578–9, Document 384; 596–8, Documents 388, 389. See also Rogger, F. *ES*, 69–71.

242 *At least his and Elsa's excursions* Einstein, Albert. *CPE15*, 217–18, Document 124; 228–9, Document 129; 308–9, Document 169.

242 *Despite Einstein's vehement denials* —. *CPE15*, 590–2, Document 376.

242 *Nothing more was said* Salaman, Esther. Memories of Einstein. *Endeavour*, 1979, 52: 18–23.

243 *She left a charming account* Neumann, Betty. When I was Einstein's Secretary. *Muehsam Family Collection 1822–1999*. Leo Baeck Institute Archives, 1132–41 [Online]. 1956. [Cited: 3 August 2020]. https://archive.org/details/muehsamfamily_11_reel11.

243 *Betty returned to Berlin* Mühsam, Hans. *Muehsam Family Collection 1822–1999*. Leo Baeck Institute Archives, 1055–6 [Online]. October 1923 [Cited: 3 August 2020]. https://archive.org/details/muehsamfamily_11_reel11.

243 *He began to meet her* Einstein, Albert. *CPE14*, 165–6, Document 98; 193, Document 121; 207–8, Document 126; 222, Document 135; 227, Document 140; 244, Document 151; 258, Document 161; 264–5, Document 167; 316–17, Documents 195, 196.

243 *Einstein and Mühsam were only* —. *CPE14*, 709–11, Document 457; for their re-establishment of contact, see *AEA* [Online] 15 June 1942. Cat. no. 38–338. http://alberteinstein.info/.

243 *Presumably subconsciously* —. *CPE14*, 321, Document 201; 322–3, Document 203; 372, Document 241; 379–80, Document 247; 386, Document 254; 401, Document 262; 464, Document 294; 463–4, Document 296; 467–8, Document 300; 527, Document 335; 544, Document 350.

244 *However, he resigned* —. *CPE15*, 744, Document 474.

244 *A student, Edwin Sieradz* —. *AEA* [Online] 23 November 1927. Cat. no. 47–758. http://alberteinstein.info/.

244 *Dukas was appointed* Highfield, Roger, Carter, Paul. *The Private Lives of Albert Einstein* (London, Faber and Faber, 1993), 211.

244 *Indeed, he had considerable difficulty* Einstein, Albert. *CPE14*, 109–10, Document 64.

244 *Einstein quietly got her treatment* —. *CPE15*, 640–1, Document 418.

244 *On 15 July 1923* —. *CPE14*, 137–8, Document 78.

244 *Reminded by another music partner* —. *CPE15*, 641, Document 419; *AEA* [Online] 5 December 1926. Cat. no. 39–441. http://alberteinstein.info/.

244 *They all appear to have stayed* Einstein, Albert, et al. *CPE14*, 370, Document 239; 508, Document 326; 512–13, Documents 330, 331; *AEA* [Online] 21 October 1924. Cat. no. 37–573. http://alberteinstein.info/.

244 *The frequency of letters* Zangger, Heinrich. *CPE14*, 714, Document 460.

245 *This had come to a head* Medicus, Heinrich A. The Friendship among Three Singular Men: Einstein and His Swiss Friends Besso and Zangger. *Isis*, 1994, 85: 472–3.

245 *As usual, Zangger's influence* Einstein, Albert, et al. *CPE15*, 44–5, Document 5; 84–5, Document 34; 620–1, Document 401; 664, Document 432; 669–70, Document 436.

245 *'One thing is beautiful'* Sauer, Tilman. Marcel Grossmann and His Contribution to the General Theory of Relativity. In *Proceedings of the 13th Marcel Grossmann Meeting on Recent Developments in Theoretical and Experimental General Relativity, Graviation, and Relativistic Field Theory, Stockholm University, Sweden, 1–7 July 2012,* Kjell Rosquist, Robert T. Jantzen, Remo Ruffini, eds. (Singapore, World Scientific, 2013). https://arxiv.org/abs/1312.4068v2.

245 *The fountain pen that he had used* Einstein, Albert. *CPE15*, 590–2, Document 376; 641–3, Document 420; 700–2, Document 446; 707–8, Document 450.

245 *As he had in 1922* —. *CPE15*, 654 Document 426; 698–9, Document 444; 760–1, Documents 481, 482.

246 *No further correspondence* —. *CPE16*, 68–9, Document 20.

246 *He sold the company* —. *CPE14*, 22–3, Document 18.

246 *Only in the 1980s* Piccard, Auguste. *CPE14*, 618–20, Document 405 for the electron-proton charge experiment. For the reanalysis of Miller's results, see Roberts, T. J. An Explanation of Dayton Miller's Anomalous 'Ether Drift' Result. *arXiv*. https://arxiv.org/abs/physics/0608238.

246 *They were particularly active* Szilard, Leo. *CPE15*, 803, Document 512; 638–40, Document 417.

246 *Quickly, the tour was extended* Einstein, Albert, et al. *CPE14*, 224–5, Document 138; 310–12, Document 193; lxxiv–lxxv.

246 *As Einstein had hoped* —. *CPE14*. For all details of the South American trip, see Einstein's travel diary, 688–708, Document 455; see also lxxiv–lxxxiv; 743–4, Document 474; Eisinger, J. *Einstein on the Road* (New York, Prometheus, 2011), 73–91.

247 *His pleasure at his reception* Einstein, Albert. Epistolario de Albert Einstein a Coriolano Alberini. *Faculty of Psychology, University of Buenos Aires, Argentina* [Online]. 1925 [Cited: 17 August 2020]. https://www.psi.uba.ar/institucional/historia/psicologia/psicologia_reformismo/capitulo_2/epistolario_einstein_alberini.pdf.

248 *Mileva would not have been amused* — *CPE14*, 777–8, Document 489.

248 *It was only because* —. *CPE15*, 31–2, Document 7; 174–5, Document 95.

248 *In December 1924* Porth, D. *AEA* [Online]. 22 December 1924. Cat. no. 44–675. http://alberteinstein.info/.

248 *These musical soirées* Einstein, Albert. Lot 38. *Christies* [Online]. 14 March 1926. https://onlineonly.christies.com/s/einstein-family-letters-portraits/his-favourite-books-don-quixote-bible-38/55719.

248 *One time my mother* Grüning, Michael. *Ein Haus für Albert Einstein* (Berlin, Verlag der Nation, 1990), 489. See also *AEA* [Online] after 6 November 1927. Cat. no. 34–181. http://alberteinstein.info/.

248 *They played chamber music* Campbell, Margaret. *The Great Cellists* (London, Gollancz, 2011).

249 *Born in Prague, Orlick moved* Hoffmann, Dieter. *Einsteins Berlin: Auf den Spuren eines Genies* (Weinheim, Wiley VCH, 2006), 131–2.

250 *'The man who has discovered'* Reiser, Anton. *ABP*, 96–7.

250 *'Certainly, you could call it that'* Graf Kessler, Harry. *Tagebücher, 1918–1937.* Wolfgang Pfeiffer-Belli, ed. (Frankfurt-am-Main, Insel-Verlag, 1961), 521–2.

250 *In thanking Siegfried Ochs* Einstein, Albert, et al. *CPE14*, 568–9, Document 373; see also *AEA* [Online]. Object nos. 44–619, 45–011, 45–012. http://alberteinstein.info/.

250 *'Dear Sir, unfortunately I feel'* Einstein, Albert, Singer, Kurt. *CPE15*, 562, Document 351; see also *AEA* [Online]. 20 July 1926. Cat. no. 44–904. http://alberteinstein.info/.

250 *Since Einstein often spent the night* Herneck, Friedrich. *Einstein Privat* (Berlin, Der Morgen, 1978), 59, 146.

251 *He did not enjoy it* Grüning, Michael. *Ein Haus für Albert Einstein* (Berlin, Verlag der Nation, 1990), 250–1.

251 *In 1931, he attended a lecture* Adler, Meinhard. *Brecht im Spiel der Technischen Zeit* (Berlin, Nolte, 1976), 124. The date of the lecture is given incorrectly as 1930.

251 *In contrast to* Dreigroschenoper Einstein, Albert. *AEA* [Online]. 4 May 1939. Cat. no. 34–27. http://alberteinstein.info/.

251 *Whether he would have attended* 'Sch.'. Wozzeck. *Deutsche Allgemeine Zeitung*, 15 December 1925, 1. See also: *Berliner Tageblatt und Handels-Zeitung*, 15 December 1925, 4; *Vossische Zeitung*, 15 December 1925, 9.

251 *He hoped for a positive* Schoenberg, Arnold. *CPE14*, 622–4 Document 410.

251 *A myth has grown up* Friedrich, Otto. *Before the Deluge* (New York, Harper & Row, 1995), 183.

251 *A similar myth* Levitz, Tamara. Racism at The Rite. In *The Rite of Spring at 100*, Gretchen Grace Horlacher, Severine Neff, Maureen Carr, eds. (Bloomington, Indiana University Press, 2017), 146–78.

251 *After the Berlin premiere* Adorno, Theodor. *Alban Berg: Master of the Smallest Link* (Cambridge, Cambridge University Press, 1997), 10

252 *Einstein wrote to Bloch* Einstein, Albert. *CPE15*, 400–1, Document 237.

252 *At the forefront of a move* Botstein, Leon. Einstein and Music. In *Einstein for the 21st Century,* Gerald Holton, Silvan S. Schweber, Peter L. Galison, eds.(Princeton, Princeton University Press, 2008), 161–75.

252 *Piatigorsky and Einstein became friends* King, Terry. *Gregor Piatigorsky: The Life and Career of the Virtuoso Cellist* (Jefferson, McFarland & Company, 2010), 5–51.

252 *'Scientists' perception of matter'* Piatigorsky, Gregor. *Cellist* (New York, Doubleday, 1965). Chapters 16, 17.

252 *It is probably Harriet* Plesch, Janos. *The Story of a Doctor*, [trans.] Edward Fitzgerald (London, Gollancz, 1947), 207, 355.

253 *'Nature requires a balance'* Anon. Einstein and Mussolini: Fiddlers. *The Literary Digest*, 30 May 1931, 17. Original interview in *The Musical Observer*, New York.

253 *Indeed, it became for a while* Plesch, Janos. *The Story of a Doctor*, [trans.] Edward Fitzgerald (London, Gollancz, 1947), 214.

253 *For the rest of his life* Einstein, Albert. *CPE15*, 617, Document 399. See also *AEA* [Online]. November 1926. Cat. no. 95–824. http://alberteinstein.info/.

253 *The letter to his father* Mehl, Margaret. *Not by Love Alone: The Violin in Japan, 1850–2010* (Copenhagen, Sound Book Press, 2014).

Chapter 18

254 *Einstein grew to respect von Laue* Einstein, Albert. *CPE15*, 816–17, Document 518.

255 *In his subsequent recollections* Stone, A. Douglas. *Einstein and the Quantum* (Princeton, Princeton University Press, 2015), 215–40.

255 *It is this tendency* Einstein, Albert. *CPE14*, 677–8, Document 446.

255 *He wrote to Einstein* —. *CPE14*, 399–400, Document 261.

255 *In particular, the ground-state energies* Einstein, Albert, et al. *CPE14*, 433–42, Document 283.

256 *He asked to see Einstein* Bose, Satyendra Nath. *AEA* [Online]. 8 October 1925. Cat. no. 6–132. http://alberteinstein.info/.

256 *At the end of 1925* Einstein, Albert. *CPE14*, 580–94, Document 385.

257 *Langevin apparently thought that* Frank, Philipp. *Einstein: His Life and Times* (London, Jonathan Cape, 1948), 339–40.

257 *. . . he also wrote in* Einstein, Albert. *CPE14*, 608–10, Document 398; 610–11, Document 399.

257 *Einstein thought this result* —. *CPE14*, 364–7, Document 236.

257 *This 'BKS theory'* Bohr, N., Kramers, H. A., Slater, J. C. Über die Quantentheorie der Strahlung. *Zeitschrift für Physik*, 1924, 24: 69–87.

257 *On 7 June 1924* Bothe, W, Geiger, H. Ein Weg zur experimentellen Nachprüfung der Theorie von Bohr, Kramers und Slater. *Zeitschrift für Physik*, 1924, 26: 44.

258 *The result, published in the spring* Bothe, W, Geiger, H. Über das Wesen des Comptoneffekts; ein experimenteller Beitrag zur Theorie der Strahlung. *Zeitschrift für Physik*, 1925, 32: 639–63.

258 *. . . the probability of the BKS* Bothe, W. Nobel Lecture 1954: Walter Bothe. *Nobel Prizes* [Online]. 1954. https://www.nobelprize.org/prizes/physics/1954/bothe/lecture/.

258 *Heisenberg was a protégé* Heisenberg, W. Quantum Theory and its Interpretation. In *Niels Bohr: His Life and Work as Seen by His Friends and Colleagues*, S. Rozental, ed. (Amsterdam, North-Holland, 1967), 94–5.

258 *He also remarked that* Born, Max. *CPE14*, 21–4 Document 5; see also Stone, A. Douglas. *Einstein and the Quantum* (Princeton, Princeton University Press, 2015), 269.

258 *Pauli's reputation* Peierls, R. E. Wolfgang Pauli. *Biographical Memoirs of the Royal Society*, 1960, 5: 186.

259 *Heisenberg thought the Beethoven* Heisenberg, W. *Werner Heisenberg: Liebe Eltern!*, A.-M. Hirsch-Heisenberg, ed. (Munich, Langen-Müller, 2003), 51.

259 *In it he conjectured* Einstein, Albert. *CPE14*, 527–30, Document 336.

259 *According to Heisenberg* Holton, Gerald. Werner Heisenberg and Albert Einstein. *Physics Today*, 2000, 53: 38–42.

259 *In these months* Heisenberg, W. Oral History Interviews: Werner Heisenberg, Session VI. *American Institute of Physics* [Online]. February 1963. https://www.aip.org/history-programs/niels-bohr-library/oral-histories/4661-6.

259 *In July 1925* Heisenberg, W. Über quantentheoretische Umdeutung kinematischer und mechanischer Beziehungen. *Zeitschrift für Physik*, 1925, 33: 879–93.

260 *They reproduced Heisenberg's results* Born, M, Jordan, P. Zur Quantenmechanik. *Zeitschrift für Physik*, 1925, 34: 858–88.

260 *They also applied the new theory* Born, M, Heisenberg, W, Jordan, P. Zur Quantenmechanik II. *Zeitschrift für Physik*, 1926, 35: 557–615.

260 *He went out of his way* Einstein, Albert, et al. *CPE15*, 70–3, Document 23; 201–2, Document 114; 208–11, Document 119.

260 *Bohr also attended* Einstein, Albert, Planck, Max. *CPE15*, 203–4, Document 116; 212–15, Documents 121, 122.

260 *Heisenberg is not an author* McCubbin, N. Beauty in Physics: The Legacy of Paul Dirac. *Contemporary Physics*, 2004, 45: 319–33.

261 *'Everybody (every physicist)'* Stachel, John. *Einstein from B to Z* (Berlin, Springer Science and Business Media, 2001), 526–37.

261 *Einstein certainly thought* Heisenberg, Werner. *Physics and Beyond* (New York, Harper Torchbook, 1972), 62–9.

262 *He wrote to Ehrenfest* Einstein, Albert. *CPE15*, 427–31, Document 233.

262 *Despite his worries* —. *CPE15*, 433–4, Document 256.

262 *No meeting of minds* Heisenberg, Werner. *Physics and Beyond* (New York, Harper Torchbook, 1972), 70–6.

262 *When interpreted in this way* —. Oral History Interviews: Werner Heisenberg, Session VII. *American Institute of Physics* [Online]. February 1963. https://www.aip.org/history-programs/niels-bohr-library/oral-histories/4661-7.

262 *Schrödinger himself published* Schrödinger, Erwin. Über das Verhältnis der Heisenberg–Born–Jordanschen Quantenmechanik zu der meinem. *Annalen der Physik*, 1926, 384: 734–56.

262 *In fact, Schrödinger's paper* Casado, C. M. A Brief History of the Mathematical Equivalence between the Two Quantum Mechanics. *Latin American Journal of Physics Education*, 2008, 2: 152–5.

263 *When Bohr returned* Heisenberg, Werner. Über den anschaulichen Inhalt der quantentheoretischen Kinematik und Mechanik. *Zeitschrift für Physik*, 1927, 43: 172–98.

264 *After a period* Camilleri, K. Indeterminacy and the Limits of Classical Concepts: The Transformation of Heisenberg's Thought. *Perspectives on Science*, 2007, 15: 178–201.

264 *Nevertheless, Bohr agreed* Heisenberg, Werner. Oral Histories: Werner Heisenberg, Session VIII. *American Institute of Physics* [Online]. February 1963. https://www.aip.org/history-programs/niels-bohr-library/oral-histories/4661-8.

264 *Rising after a rather obscure* Cassidy, David. C. *Uncertainty* (New York, W.H. Freeman & Co., 1992), 248–9.

264 *Indeed, it is likely* Beller, Mara. *Quantum Dialogue: The Making of a Revolution* (Chicago, University of Chicago Press, 1999), 138–44.

264 *Heisenberg had already heard* Heisenberg, Werner. *CPE15*, 823–4, Document 524; 810–14, Document 516.

264 *This episode reveals* Pauling, Linus. Heisenberg 1927. Fisicafundamental [Online]. 1927. http://www.fisicafundamental.net/relicario/doc/Heisenberg1927.pdf.

264 *Much Later Einstein* Einstein, Albert. AEA [Online] 29 November 1952. Cat. no. 61–845. http://alberteinstein.info/

266 *Ehrenfest told his Leiden students* Mehra, Jagdish, Rechenberg, Helmut. *The Historical Development of Quantum Theory*, Vol. 6, Part 1 (New York, Springer, 2000), 251–4.

266 *The following morning* Heisenberg, Werner. Quantum Theory and its Interpretation. In *Niels Bohr: His Life and Work as Seen by His Friends and Colleagues*, S. Rozental, ed. (Amsterdam, North Holland, 1967), 107.

266 *Certainly, this was Heisenberg's* —. *Werner Heisenberg: Liebe Eltern*, A.-M. Hirsch-Heisenberg, ed. (Munich, Langen Mueller, 2003), 126.

266 *The following account* Bacciagaluppi, G., Valentini, A. *Quantum Theory at the Crossroads: Reconsidering the 1927 Solvay Conference* (Cambridge, Cambridge University Press, 2009), 175–83.

267 *'But still, it cannot be'* Heisenberg, Werner. *Encounters with Einstein* (Princeton, Princeton University Press: Princeton Paperbacks, 1989), 116–17.

267 *It is notable that* Zangger, Heinrich. *Seelenverwandte*, Robert Schulmann, ed. (Zurich, Verlag Neue Zürcher Zeitung, 2012), 454–5, Document 290.

267 *Now all of equal academic status* Cassidy, David C. *Uncertainty* (New York, W.H. Freeman & Co., 1992), 246.

267 *The first time that Einstein* Einstein, Albert. *CPE15*, 654, Document 426.

268 *Einstein published his idea* Einstein, Albert. *CPE15*, 377–9, Document 223.

268 *Rupp set to work* Einstein, Albert, Rupp, Emil. *CPE15*, 387–627, Documents 231, 240, 252, 259, 270, 275, 277, 279, 284, 288, 290, 291, 296, 298, 299, 306, 313, 315, 323, 350, 354, 361, 384, 391, 395, 396, 404, 410.

268 *Jeroen van Dongen, on whose researches* van Dongen, Jeroen. Emil Rupp, Albert Einstein and the Canal Ray Experiments on Wave–Particle Duality: Scientific Fraud and Theoretical Bias. *Historical Studies in the Physical and Biological Sciences*, 2007, 37(Suppl): 73–120.

269 *It is particularly strong* Einstein, Albert. *CPE14*, 340–6, Document 220.

269 *Einstein's article is a long* —. *CPE14*, 498–503, Document 321.

269 *With a confidence entirely justified* —. *CPE14*, 514–25, Document 332.

269 *Einstein stood in awe* —. *CPE15*, 783–91, Document 503, 504; 792–7, Document 506.

269 *If only he had taken* —. *CPE15*, 165–71, Document 92.

266 *Here Einstein gave up* —. *CPE15*, 435–7, Document 258.

269 *In 1926, Einstein again became* —. *CPE15*, 498, Document 302; 569–71, Document 358.

269 *Klein wrote to Einstein* —. *CPE15*, 575–7, Document 363.

270 *However, he had written to Ehrenfest* —. *CPE15*, 716–21, Document 459; 752–9, Document 480; 578–9, Document 365.

270 *Hermann Weyl wrote to Einstein* —. *CPE15*, 682–98, Document 443; 741–3, Document 473.

270 *In December 1927* Anon. *The Jewish Chronicle*. 5 January 1925, 25; 4 March 1927, 24; 9 December 1927; 16 December 1927, 26; see also *Judische Rundschau*, 9 December 1927, 698.

270 *In January 1924* Einstein, Albert. *CPE14*, 313–16, Document 194.

270 *Thus, their relationship* Einstein, Albert, Judah Magnes. *CPE12*, 164–6, Documents 122, 124.

271 *Einstein's outrage and astonishment* Einstein, Albert. *CPE15*, 249–50, Document 142.

272 *In March 1926* Einstein, Albert, Judah Magnes. *CPE15*, 312–13, Document 172; 328–30, Document 186; 333–4, Document 189; 366, Document 214.

272 *Einstein once again wrote* Kohn, Leo. *CPE15*, 818–20, Document 521; see also *CPE16*, 121, Document 59.

272 *Although generally exasperated* Einstein, Albert. *CPE16*, 375–6, Document 231.
272 *In UK universities* Einstein, Albert. *CPE15*, 324–30, Document 201.
273 *This complex conflict* Kotzin, Daniel P. *Judah L. Magnes: An American Jewish Nonconformist* (Syracuse, Syracuse University Press, 2010), 169–211. See also Rosenkranz, Z. *Einstein before Israel* (Princeton, Princeton University Press, 2011), 181–96.

Chapter 19

274 *Sadly, political developments* Wharton, James. *CPE16*, 911–12, Appendix A.
275 *'[Einstein] adored music'* Barjansky, Catherine. *Portraits with Backgrounds* (New York, Macmillan, 1947), 174–6.
275 *Einstein proposed to come* Haldane, Lord. *AEA* [Online] 29 June 1927. Cat. no. 32–654. See also Cat. no. 45–452. http://alberteinstein.info/.
275 *Einstein returned to Berlin* Einstein, Albert. *CPE16*, 71–2, Documents 23, 24; 75–85, Documents 26–28, 31, 32, 34.
275 *Alfred sent congratulations* Einstein, Alfred. *CPE16*, 194, Document 100; 668, Document 464.
275 *'When I reached my ninth'* Fink, Michael, Bess, Hieronymus. The Autobiography and Early Diary of Alfred Einstein (1880–1952). *The Musical Quarterly*, 1980, 66: 366; see also Anselm Gerhard. Die Vorgeblicher Verwandte. In *Musizieren, Leben – und Maulhalten*, Ivana Rentsch & Anselm Gerhard, eds., (Basel, Schwabe Ag, 2006), 80.
276 *Together they wrote* Winternitz, Josef, Albert Einstein. *CPE16*, 201–2, Documents 105, 107.
276 *Although the details are unclear* Zannger, Heinrich. *CPE16*, 220, Document 120; 222, Document 123; 257–9, Documents 149–51. See also *AEA* [Online] 30 December 1927, Cat. no. 89–541. http://alberteinstein.info/.
276 *On the following day* Einstein, Albert. *CPE16*, 923–32, Appendix G.
276 *'I was in Holland'* —. *CPE16*, 258, Document 150.
276 *Einstein had remarked on this* — *CPE16*, 249–50, Document 144.
276 *They did, Einstein playing* Seelig, Carl. *BIOG*, 18–19.
277 *Byland's son and daughter* Einstein, Albert. *CPE16*, 456–7, Document 298.
277 *He was apparently carrying* Plesch, Janos. *The Story of a Doctor*, [trans.] Edward Fitzgerald (London, Victor Gollancz, 1947), 216.
277 *Although he concluded* Einstein, Albert. *CPE16*, 271–6, Document 156.
277 *A sketch of Einstein playing* Grandjean, Martin. *Les cours universitaires de Davos 1928–1931. Au centre de l'Europe intellectuelle. Lausanne, Switzerland.* Unpublished master's thesis (University of Lausanne, 2011). https://core.ac.uk/download/pdf/77147672.pdf. See also Davoser Blätter, 30 March 1928.
277 *The only source* Grüning, Michael. *Ein Haus für Albert Einstein* (Berlin, Verlag der Nation, 1990), 150.
277 *Whatever happenned* Einstein, Elsa. AEA [Online] 20 April 1928. Cat. no. 72–355. http://alberteinstein.info/
278 *However, he felt well enough* Seelig, Carl. *Albert Einstein: Leben und Werk eines Genies unserer Zeit* (Zurich, Europa Verlag, 1960), 320–1.
278 *Despite this, and protests* Heinrich Zangger, Albert Einstein. *CPE16*, 285–8, Documents 161, 163, 164; 290–1, Documents 168, 169.
278 *Only by 24 May* Heinrich Zangger, Elsa Einstein, Albert Einstein. *CPE16*, 302, Document 176; *AEA* [Online] 20 April 1928. Cat. no. 72–355. http://alberteinstein.info/; 324, Document 200.

278 *Einstein also stopped drinking coffee* Herneck, Friedrich. *Einstein privat* (Berlin, 1978), 38–9, 108.

278 *During the 1920s, Marianoff* Marianoff, Dimitri. *AEA* [Online] 1931. Cat. no. 50–331. http://alberteinstein.info/.

278 *The process had dragged on* Zangger, Heinrich. *Seelenverwandte*, Robert Schulmann, ed. (Zurich, Verlag Neue Zürcher Zeitung, 2012), 474–5, Document 306.

279 *Einstein had said he wanted* Marianoff, Dimitri, Palma Wayne. *Einstein: An Intimate Study of a Great Man* (New York, Doubleday, Doran and Co., Inc., 1944), 4.

279 *It is also possible that* Rösch, Juliane. 'Di berliner hobn nisht gegloybt zeyere oygn.': Das jiddischsprachige Theater GOSET und der Regisseur Alexander Granovsky. *PARDES- Zeitschrift der Vereinigung für Jüdische Studien E.V*, 2008, 14: 55–7.

279 *However, the first (indirect) mention* Einstein, Albert. *CPE16*, 554–5, Document 374. See also See also Einstein, Eduard. *AEA* [Online] 28 October 1928. Cat. no. 144–501. http://alberteinstein.info/.

279 *If any further evidence* Goldschmidt, Heinrich. *AEA* [Online] 29 November 1930. Cat. no. 67–50. http://alberteinstein.info/.

279 *The accuracy of his characterisations* Marianoff, Dimitri, Palma Wayne. *Einstein: An Intimate Study of a Great Man* (New York, Doubleday, Doran and Co., Inc., 1944), 5–10.

279 *Einstein and Lunatscharski* Hoffmann, Dieter. *Einsteins Berlin: Auf den Spuren eines Genies* (Weinheim, Wiley-VCH Verlag GmbH & Co., 2006), 154–5.

280 *Plesch's autobiography demonstrates* Plesch, Janos. *The Story of a Doctor*, [trans.] Edward Fitzgerald (London, Victor Gollancz, 1947), 255–8.

280 *Their suspicion was that Marianoff* Grundmann, Siegfried. *The Einstein Dossiers* (Berlin, Springer-Verlag, 2004), 347–8.

280 *The* volte face *that Einstein executed* Fölsing, Albrecht. *Albert Einstein: Eine Biographie* (Frankfurt am Main, Suhrkamp Verlag, 1993), 727.

280 *He decided not to take the job* Einstein, Albert. *CPE16*, 310–11, Document 183.

280 *Einstein agreed to join* Rabkin, Anna. *Antonina Vallentin: A European of Foreign Affairs.* Master's thesis, California State University (Hayward, 2003) [Cited: 12 June 2024]. https://scholarworks.calstate.edu/concern/theses/bv73c1387. See also Grüning, M. *Ein Haus für Albert Einstein* (Berlin, Verlag der Nation, 1990), 40; Szilard, L. *AEA* [Online] 2 April 1929. Cat. no. 45–649. http://alberteinstein.info/.

280 *The editors of the illustrated weekly* Einstein, Albert. *CPE16*, 286, Document 162; 465, Document 308.

280 *The families parted in harmony* —. *CPE16*, 344, Document 211.

281 *They went sailing together* Rogger, Franziska. *ES*, 132; see also Einstein, Elsa. *CPE16*, 421–2, Document 265.

281 *It seems that Tete joined Einstein* Einstein, Albert, Einstein, Elsa. *CPE16*, 380–1, Document 236; 421–2, Document 265; 457–8, Document 299; see also *AEA* [Online] 1–10 September 1928. Cat. no. 123–439. http://alberteinstein.info/.

281 *Chaim Weizmann and Leo Kohn arrived* Einstein, Elsa. *AEA* [Online] 20 July 1928. Cat. no. 65–563. http://alberteinstein.info/.

281 *For the next few months* Einstein, Albert. *CPE16*, 226–30, Document 128; 246–7, Document 140; 269–70, Document 155; 324–34, Documents 201, 202; 342–3, Document 209; 370, Document 227. See also Rosenkranz, Z. *Einstein before Israel* (Princeton, Princeton University Press, 2011), 196–206.

281 *She is presumably referring* Einstein, Albert, Einstein, Elsa. *CPE16*, 424, Document 268; 467–8, Document 311; Albert Einstein, *AEA* [Online] July–September 1928. Cat. no. 48–12, Cat. no. 65–560. http://alberteinstein.info/.

282 *'Contemporary estimates are'* Russell, Bertrand. *The Collected Works of Bertrand Russell*, Peter Köllner, J. G. Slater, eds., Vol. X (London, Routledge, 1996), 38–42.

282 *The distance by which* Einstein, Albert. *CPE16*, 570–5, Document 387.

282 *Another was Chaim Herman Müntz* —. *CPE16*, 113–15, Documents 53, 54; 387–8, Document 244.

282 *Sadly, Schwarz died early* Schwarz, Curt, Einstein, Albert, Schwarz, Otto. *CPE16*, 371–4, Document 229; see also *AEA* [Online] August 1927–January 1929. Cat. nos. 48–442 through 48–449, 48–451, 48–453 through 48–456, 48–458, 47–508, 48–460. http://alberteinstein.info/.

283 *Lanczos's compatriot Szilard* Gellai, Barbara. *The Intrinsic Nature of Things* (Providence, American Mathematical Society, 2010), 19–28. See also Seelig, C. *Albert Einstein: Leben und Werk eines Genies unserer Zeit* (Zurich, Bertelsmann Lesering, 1960), 312.

283 *Lanczos nevertheless arrived early* Einstein, Albert, Lanczos, Cornel. *CPE16*, 429–34, Documents 273–7; 437, Document 280; 527, Document 353.

283 *Although they met occasionally* Einstein, Albert. *CPE16*, 710–11, Document 505.

283 *He may instead have remembered* Lanczos, Cornelius. *The Einstein Decade: 1905–1915* (London, Elek Science, 1974), x, 24–5.

283 *However, Einstein seems to have brought* Einstein, Albert, Müller, Ferdinand. *AEA* [Online] 1928–1931. Cat. nos. 47–679 through 47–692. http://alberteinstein.info/.

284 *This appeared on the front page* Miller, Paul D. Einstein on Verge of Great Disovery: Resents Intrusion. *The New York Times*. 4 November 1928, 1,15. Also available in *CPE16*, 933–5, Appendix H.

284 *'Music does not* directly affect *research work'* Einstein, Albert. *CPE16*, 450, Document 293.

285 *Plesch recounted that* Plesch, Janos. *The Story of a Doctor*, [trans.] Edward Fitzgerald (London, Victor Gollancz, 1947), 214.

285 *He first stayed in Gatow* Einstein, Albert, Ehrenfest, Paul. *CPE16*, 456, Document 309; see also *AEA* [Online] 31 January 1929. Cat. no. 10–188. http://alberteinstein.info/.

285 *A few weeks later* Einstein, Albert. *CPE16*, 531–3, Documents 358, 359; 579, Document 392.

285 *'The success of this attempt'* —. *CPE16*, 298–304, Document 313.

285 *However, he admitted* — *CPE16*, 537–44, Document 365.

285 *He also drafted a fuller correction* —. *CPE16*, 545–51, Documents 366–8.

286 *There was nothing to be excited about* Anon. Five Pages of Einstein Manuscript Give World New Conception of Inter-relation between Mechanics and Electro-dynamics. *Jewish Daily Bulletin*. 14 January 1929, VI: 3.

286 *Reichenbach grasped it* Einstein, Albert, Reichenbach, Hans. *CPE16*, 563–7, Documents 381, 383, 384; 577–8, Documents 390, 391.

286 *Arthur Eddington reported that crowds* Einstein, Albert, Eddington, Arthur. *CPE16*, 570–5, Document 387; 598, Document 403.

286 *This could be associated* Einstein, Albert, Weitzenböck, Ronald, Cartan, Elie. *CPE16*, 391–403, Documents 246, 248, 251, 254; 580, Document 394; 600, Document 407; 606, Document 413; 731–2, Document 520; 740–1, Document 525; 745, Document 531; 751, Document 536.

286 *These articles were published* Einstein, Albert, Cartan, Eli. Auf die Riemann-Metrik und den Fern-Parallelismus gegründete einheitliche Feldtheorie. *Mathematische Annalen*, 1930, 102: 685–97, 698–706. Cartan's paper entitled 'Notice historique sur la notion de parallélisme absolu' follows Einstein's directly.

287 *He therefore indicated* Fölsing, Albrecht. *Albert Einstein: Eine Biographie* (Frankfurt am Main, Suhrkamp Verlag, 1993), 695–6.

287 *Einstein told Besso that* Einstein, Albert. *CPE16*, 531–2, Document 358.

287 *'Dear Mr Einstein'* Pauli, Wolfgang. The Pauli Letter Collection. *CERN Scientific Information Service* [Online]. 1929: einstein_0079; jordan_0165; ehrenfest_0077-6; weyl_0498. https://scientific-info.cern/archives/Pauli_archive/guide/letters.

288 *Weyl, in laying the foundations* Weyl, Hermann. Elektron und Gravitation. *Zeitschrift für Physik*, 1929, 56: 330–52.

288 *This viewpoint* Dongen, Jeroen van. *Einstein's Unification* (Cambridge, Cambridge University Press, 2010), 64–74.

288 *'It is indeed a bold statement'* Pauli, Wolfgang. *Collected Scientific Papers*, Vol. 2, V. F. Weisskopf and R. Kronig, eds. (New York, Interscience, 1964), 1399. Appeared in *Naturwissenschaften*, 1932, 20: 186.

288 *'You were right after all, you rascal!'* Einstein, Albert. *AEA* [Online] 22 January 1932. Cat. no. 19–169. http://alberteinstein.info/.

289 *Although they initially produced* Sauer, Tilman. Field equations in teleparallel space–time: Einstein's Fernparallelismus approach toward unified field theory. *Historia Mathematica*, 2006, 33: 399–439.

289 *By March, he was once again* Einstein, Albert. *CPE16*, 576–7, Document 389.

289 *Unsurprisingly, it was Bach* Anon. *Manchester Guardian*. 27 February 1929, 10.

289 *Images of himself* Grüning, Michael. *Ein Haus für Albert Einstein* (Berlin, Verlag der Nation, 1990), 241.

289 *The family departed* Anon. *CPE16*, 947–8, Appendix L. Published in *The New York Times*, 15 March 1929, p. 3.

289 *Her love-life was legendary* Fry, Helen. *Music and Men* (Cheltenham, History Press, 2008)

290 *'Think of me as one'* Freud, Sigmund, Einstein, Albert. *CPE16*, 627, Document 432; 669, Document 465; 674–5, Document 471.

290 *'I have on the whole'* Einstein, Albert, Weizmann, Chaim. *CPE16*, 686, Document 482; 592, Document 397; 636, Document 441.

290 *Otto Neustätter, friend of Einstein* Neustätter, Otto. *CPE16*, 625–6, Document 430.

290 *He mentioned that music* Byland, Hans. *AEA* [Online] 20 March 1929. Cat. no. 30–386. http://alberteinstein.info/.

290 *Besso sent a brief but eloquent letter* Besso, Michele, Einstein, Albert. *CPE16*, 667, Document 462; 669, Document 466.

290 *'The speeches delivered'* Anon. *The Jewish Chronicle*. 3 May 1929, 19.

290 *He received with great joy* —. *CPE16*, 670–1, Document 467; see also *AEA* [Online]. Cat. nos. 93–457, 98–031, 98–039. http://alberteinstein.info/.

291 *It was in use* Einstein, Albert. *CPE16*, 638, Document 446; 664, Document 460; Elsa Einstein to Maja Winteler-Einstein, 19 August 1929, in Michael Gruning, *Ein Haus fur Albert Einstein* (Berlin, Verlag der Nation, 1990), 304–5.

291 *The city authorities then proposed* Plesch, Janos. *The Story of a Doctor*, [trans.] Edward Fitzgerald (London, Victor Gollancz, 1947), 224.

291 *At this point, a young architect* Grüning, Michael. *Ein Haus für Albert Einstein* (Berlin, Verlag der Nation, 1990), 32–127; see also https://einstein-website.de/initiativkreis/.

292 *Natalia Saz* Saz, Natalia. *Novellen meines Lebens* (Berlin, Henschel Verlag, 1973). Section reprinted in Michael Grüning, *Ein Haus für Albert Einstein* (Berlin, Verlag der Nation, 1990), 494–507 and photograph on 508. See also Friedrich Herneck, *Einstein privat* (Berlin, Buchverlag der Morgen, 1978), 128, 137.

292 *'. . . false in its construction'* Anon. *Jewish Daily Bulletin*. 9 April 1929, 3; 14 April 1929, 2.

293 *Quite what O'Connor made* Lemaître, Georges. Un Univers homogène de masse constante et de rayon croissant rendant compte de la vitesse radiale des nébuleuses extragalactiques. *Annales de la Société Scientifique de Bruxelles*, 1927, A47: 49–59.

293 *Einstein replied that he believed* Anon. *CPE16*, 715–17, Document 508.

293 *Walter, watching him* Walter, Bruno. *Theme and Variations* (New York, Alfred Knopf, 1946), 233.

293 *Lemaître had first met Einstein* Lemaître, Georges. Recontres avec Einstein. *Revue de Questions Scientifiques: Actualité, histoire et philosophie des sciences*, 1958, 129: 129–32. https://inters.org/lemaitre-einstein. See also H. Krage. *Masters of the Universe* (Oxford, Oxford University Press, 2014), 74–75.

294 *'When we left the concert hall'* Magidoff, Robert. *Yehudi Menuhin* (Westport, Greenwood Press, 1975), 136–7. See also Einstein, Elsa. *AEA* [Online] 12 March 1934. Cat. no. 95–814. http://alberteinstein.info/.

294 *'I feel rather sorry for the boy'* Plesch, Janos. *The Story of a Doctor*, [trans.] Edward Fitzgerald (London, Victor Gollancz, 1947), 339.

294 *Kreisler was famous* Horkan, Vincent. Fritz & Harriet Kreisler. *Crisis Magazine*, 1 May 1995 [Online]. https://www.crisismagazine.com/1995/fritz-harriet-kreisler.

294 *She obviously made a great fuss* Einstein, Albert. *AEA* [Online] April–December 1929. Cat. nos. 143–224, 84–101. http://alberteinstein.info/.

294 *However, he would take advantage* de Dijn, Rosine. *Albert Einstein und Elisabeth von Belgien* (Regensburg, Verlag Friedrich Pustet, 2015), 64–65.

Chapter 20

295 *An insight into his life* Herneck, Friedrich. *Einstein Privat* (Berlin, Buchverlag der Morgen, 1978), 114.

296 *'His day usually began'* Marianoff, Dimitri, Wayne, Palma. *Einstein: An Intimate Study of a Great Man* (New York, Doubleday, Doran and Co., Inc., 1944), 20–22.

296 *'He considers Shakespeare's mighty work'* Reiser, Anton. *Albert Einstein: A Biographical Portrait* (London, Thornton Butterworth Limited, 1931), 196–9.

297 *'Shortly after my first return'* Eisner, Bruno. Gedenken und Gedanken aus des Leben eines Musikers. *The Leo Baeck Institute* [Online], 132–41. [Cited: 5 August 2021]. https://www.lbi.org/.

298 *'He was not a very good'* Plesch, Janos. *The Story of a Doctor*, [trans.] Edward Fitzgerald (London, Victor Gollancz, 1947), 214–15.

298 *Other opinions of his playing* Herneck, Friedrich. *Einstein Privat* (Berlin, Buchverlag der Morgen, 1978), 129.

298 *Einstein told the author that* Einstein, Albert. *AEA* [Online]. 20 November 1929. Cat. no. 34–328. http://alberteinstein.info/.

298 *'Everything is determined'* Viereck, George Sylvester. What Life Means to Einstein. *The Saturday Evening Post*, 26 October 1929), 17, 110, 113, 114, 117.

299 *Unfortunately, the combination* Anon. Einstein's Message Spoken to Edison. *The New York Times*, 23 October 1929, 3.

299 *Einstein stayed with Mileva* Einstein, Albert. *AEA* [Online]. August 1929. Cat. no. 75–765. http://alberteinstein.info/.

299 *By 19 August* Anon. *Jewish Daily Bulletin*, 13 August 1929, Vol. VI, 1–2; Also 16 August 1929, 1, 3.

299 *Einstein replied to Maja's report* Rogger, Franziska. *ES*, 123.

299 *He had signed up* Popović, Milan. *AS*, 159.

299 *By mid-December* Trubuhović-Gjurić, Desanka. *ISAE*, 171–2.

300 *His final words* Zangger, Heinrich. *Seelenverwandte*, Robert Schulmann, ed. (Zurich, Verlag Neue Zürcher Zeitung, 2012), 484–5, Document 313.

300 *Zangger thought that* Einstein, Albert, Zangger, Heinrich. *Seelenverwandte*, Robert Schulmann, ed. (Zurich, Verlag Neue Zürcher Zeitung, 2012), 485–90, Documents 315–17. See also Einstein, Albert. *AEA* [Online]. 22 December 1929. Cat. no. 75–763. http://alberteinstein.info/.

300 *... an honorary degree from ETH* Anon. E. T. S-Jubiläum. *Schaffauser Nachrichten*. 8 November 1930.

300 *Einstein's extended family* Vallentin, Antonina. *Le Drame d'Albert Einstein* (Paris, Les Petits-fils de Plon et Nourrit, 1954), 164–5. Note that the French edition is used in preference to the English translation, which Vallentin considered unreliable.

301 *With Einstein by then* Wolff, Barbara. 'Derartige kolossale Opfer ...' Der Nobelpreis für Physik für das Jahr 1921 – was geschah mit dem Preisgeld? *Max-Planck-Institute für Wissenschaftsgeschichte*, 2019. Preprint 493.

301 *Yourgrau's fascinating life* van de Merve, Alwyn. *Old and New Questions in Physics, Cosmology, Philosophy, and Theoretical Biology – Essays in Honor of Wolfgang Yourgrau* (New York, Plenum, 1983), 1–37.

302 *'It is true that the honoured'* Yourgrau, Wolfgang. Einstein und der Akademische Dünkel. [ed.] Peter Bergmann, Peter Aichelburg. *Albert Einstein: Sein Einfluß auf Physik, Philosophie und Politik* (Berlin, Springer Verlag, 2013), 223–31.

302 *A newspaper reported* Anon. *AEA* [Online]. 29 January 1930. Cat. nos. 75–54, 65–596. http://alberteinstein.info/.

302 *'I don't think this can'* Marianoff, Dimitri, Wayne, Palma. *Einstein: An Intimate Study of a Great Man* (New York, Doubleday, Doran and Co., Inc., 1944), 97–105.

302 *This is interesting* Herneck, Friedrich. *Einstein Privat* (Berlin, Buchverlag der Morgen, 1978), 42–3. The photograph can be found on the Getty Images website, https://www.gettyimages.co.uk/

303 *They were moving out* Einstein, Albert. *AEA* [Online]. 11 April 1930. Cat. no. 34–353. http://alberteinstein.info/.

303 *'Planck loves you like a son'* Grüning, Michael. *Ein Haus für Albert Einstein* (Berlin: Verlag der Nation, 1990), 177.

303 *A cat was also a great favourite* —. *Ein Haus für Albert Einstein* (Berlin, Verlag der Nation, 1990), 215, 305; see also Herneck, Friedrich. *Einstein Privat* (Berlin, Buchverlag der Morgen, 1978), 123–6.

303 *By a happy chance* Anon. *The London Gazette - Supplement*. 30 May 1930, 33611. https://www.thegazette.co.uk/London/issue/33611/supplement/3473.

304 *Einstein would have enjoyed* Babbage, Dennis. *Magdalen College Magazine*. 1983. See also Anon. *AEA* [Online]. 5 June 1930. Cat. no. 30–247. http://alberteinstein.info/.

304 *'Bertrand Russell and I each met'* Littlewood, John. *Littlewood's Miscellany*. Bela Bollobas, ed. (Cambridge, Cambridge University Press, 1953), 130. See also Anon. *AEA* [Online] 5 June 1930. Cat. no. 30–248. http://alberteinstein.info/.

304 *The men wined and dined* Salaman, Esther. Memories of Einstein. *Endeavour*, 1979, 52: 23.

304 *He lectured in German* Einstein, Albert. *Professor Einstein's Address at the University of Nottingham. Science*, 1930, 71: 608–10 [trans.] Professor Henry Brose. See https://www.youtube.com/watch?v=161UNSza_qk for a film of Einstein speaking at this event.

305 *'Einstein was a remarkably serene'* Anscombe, Charlotte. Remembering when . . . Albert Einstein visited the University – and was late! *University of Nottingham.* [Online]. 5 June 2015. https://blogs.nottingham.ac.uk/newsroom/2015/06/05/remembering-whenalbert-einstein-visited-the-university-and-was-late/. See also Brose, Henry. *AEA* [Online]. 1927. Cat. no. 45–675. http://alberteinstein.info.

305 *After a government reception* Einstein, Albert. *CPAE15*, 585, Document 369.

306 *In response to Einstein's questioning* Singer, Wendy. 'Endless Dawns' of Imagination. *The Kenyon Review*, 2001, 23: 7–33.

306 *Maillot commented that* Harry, Graf Kessler. *Tagebücher, 1918–1937.* Wolfgang Pfeiffer-Belli, ed. (Frankfurt-am-Main, Insel-Verlag, 1961), 15 July 1930.

306 *Marianoff also accompanied* Herneck, Friedrich. *Einstein Privat* (Berlin, Buchverlag Der Morgen, 1978), 101–2. See also *The New York Times*, 30 December 1930, 2, and Marianoff, Dimitri, Wayne, Palma. *Einstein: An Intimate Study of a Great Man* (New York, Doubleday, Doran & Company, 1944), 90–4.

307 *Despite having several violins* Einstein, Albert. *AEA* [Online]. July 1930. Cat. no. 75–692. http://alberteinstein.info/.

307 *When Maja left to return* Rogger, Franziska. *ES*, 132–3.

307 *Naturally his speech was broadcast* Küpper, Hans-Josef. Einstein's speech on occasion of the opening of the Radio Show in 1930. *Albert Einstein in the World Wide Web.* [Online] [Cited: 20 August 2021.] A recording of parts of Einstein's speech can be heard from a link on this web page. https://einstein-website.de/tondokument/.

307 *Bentwich also noted that* Bentwich, Norman. *My Seventy-Seven Years.* London: Routledge and Keegan Paul, 1962. pp. 95, 99.

307 *He stayed until Einstein left* Grüning, Michael. *Ein Haus für Einstein* (Berlin, Verlag der Nation, 1930), 334. Elsa Einstein to Hedwig Born, 18th September 1930. See also Einstein, Eduard. *AEA* [Online] October 1930. Cat no. 86–527. http://alberteinstein.info/.

307 *Indeed, Einstein made very little Le Magnétism - Rapports et Discussions de Sixième Conseil de Physique*. Various. Brussels: Institut International de Physique Solvay, 1932.

307 *It was the latter who recalled* Rosenfeld, Leon. Oral History Interviews, Leon Rosenfeld, Session III. *American Institute of Physics* [Online]. February 1963. https://www.aip.org/history-programs/niels-bohr-library/oral-histories/4847-3. See also Mehra, Jagdish. *The Solvay Conferences on Physics* (Dordrecht, D. Reidel, 2012), xvii–xviii.

308 *Bohr calculated that the uncertainty* Bohr, Niels. Discussion with Einstein on Epistemological Problems in Atomic Physics. Paul Arthur Schlipp, ed. *Albert Einstein: Philosopher-Scientist*, Vol. VII (Evanston, Open Court, Library of Living Philosophers, 1947) 224–30.

308 *In July 1931* Howard, Don. Revisiting the Einstein–Bohr Dialogue. *Iyyun: The Jerusalem Philosophical Quarterly*, 2007, 56: 74–5.

308 *Bohr mixed uncertainties* de la Torre, A. C., Daleo, A., Garcia-Mata, I. The Photon-Box Bohr–Einstein Debate Demythologized. *European Journal of Physics*, 2000, 21: 253–60.

308 *Indeed, general relativity is unnecessary* Unruh, W. G., Opat, G. I. On the Bohr–Einstein 'Weighing of Energy' debate. *American Journal of Physics*, 1979, 47: 743.

308 *It is now clear that* Nikolić, Hrvoje. EPR before EPR: A 1930 Einstein–Bohr thought experiment revisited. *European Journal of Physics*, 2012, 33: 1089–97. See also Dieks, Dennis, Lam, Sander. Complementarity in the Einstein–Bohr Photon Box. *American Journal of Physics* 2008, 76: 838.

308 *Others attending included Bohr* de Dijn, Rosine. *Albert Einstein und Elisabeth von Belgien* (Regensburg, Verlag Friedrich Pustet, 2015), 66–9.

308 *Weizmann was anxious to discuss* Weizmann, Chaim. CHAIM, Vol. XIV (New Brunswick, Transaction Books, 1978), 90–2, Letter 97; 122–4, Letter 122; 389–90, Letter 366.

309 *'There is not a superfluous word'* McRory, Desmond J. Shaw, Einstein and Physics. *Shaw: The Journal of George Bernard Shaw Studies*, 1986, 6: 33–67. Available on http://www.jstor.com/stable/40681170.

309 *After their lunch* Patch, Blanche. *Thirty Years with G.B.S.* (London, Victor Gollancz, 1951), 105.

310 *The evening was a fundraising success* Anon. *The Jewish Chronicle*. 31 October 1930, VI, 17, 23.

310 *'Luncheon was served'* Barjansky, Catherine. *Portraits with Backgrounds* (New York, Macmillan, 1947), 135.

310 *At some point in the visit* de Dijn, Rosine. *Albert Einstein und Elisabeth von Belgien* (Regensburg, Verlag Friedrich Pustet, 2015), 67–9. See also *AEA* [Online]. Cat. no. 74–886. http://alberteinstein.info/.

310 *He wrote asking Michele* Einstein, Albert. AEA [Online] 1 November 1930. Cat. no. 7–120. http://alberteinstein.info/.

310 *Although he did not explicitly* Einstein, Albert. Religion and Science. *The New York Times*. 9 November 1930. Section 5, 1.

311 *On the other hand* Fölsing, Albrecht. *Albert Einstein: Eine Biographie* (Frankfurt am Main, Suhrkamp Verlag, 1993), 715.

311 *Arnold Schoenberg, who was* Schoenberg, Arnold, Einstein, Albert. AEA [Online]. 11 November 1930. Cat. nos. 34–382, 34–383. http://alberteinstein.info/.

311 *Nothing came of the initiative* Rathjen, Friedhelm, Weigel, Andreas. A Portrait of the Artist as an Adolf Loos Campaigner. *James Joyce Quarterly*, 2004, 42/43: 315–19. Available on https://www.jstor.org/stable/25570969.

311 *On 29 November, Dima Marianoff* Marianoff, Dimitri, Wayne, Palma. *Einstein: An Intimate Study of a Great Man* (New York, Doubleday, Doran and Co., Inc., 1944), 11–12. See also *The New York Times*, 30 November 1930, 2; and *The Jewish Chronicle*, 5 December 1930, 8.

312 *Nevertheless, despite repeated urgings* Epstein, Paul. *CPE14*, 598–9, Document 390.

312 *On the evening of 30 November* Goenner, Hubert. *Einstein in Berlin 1914–1933* (Munich, C. H. Beck, 2005), 138–40.

313 *He had also translated* van Loon, Gerard Willem. *The Story of Hendrik Willem van Loon* (Philadelphia, Lippincott, 1972), 238–9.

313 The New York Times *reported that* Einstein, Albert. *AEA* [Online] December–March 1930–1931. Cat. no. 29–134. http://alberteinstein.info/. The description of the New York visit is based on this document, which is Einstein's Travel Diary, and reports from *The New York Times* on each day of the visit. See also Eisinger, Josef. *Einstein on the Road* (Buffalo, Prometheus, 2011), 102–6.

313 *Einstein was the only savant* Anon. Einstein Saw His Statue in Church Here; Comments That He Must Be Careful Now. *The New York Times*. 28 December 1930, 1.

314 *The party did not return* Dukas, Helen. *AEA* [Online] 13 December 1930. Cat. no. 75–322. http://alberteinstein.info/.

315 *Toscanini's interpretation* Downes, Olin. Einstein and Toscanini Meet Backstage and Afford Demonstrations of Famous Theory. *The New York Times*, 21 December 1930. Section X, 8; for the concert programme, see https://www.omnia.ie/index.php?navigation_function=2&navigation_item=4bc201eccce2872c1fbf9c681c5a98da&repid=2.

315 *It certainly proved too strong* Einstein, Albert. *Einstein on Peace*, Otto Nathan, Heinz Norden, eds. (New York, Schocken Books, 1968), 116–18.

315 *Einstein found his discussion* Michelmore, Peter. *Einstein, Profile of the Man* (New York, Dodd, Mead & Company, 1962), 152; see also *AEA* [Online] 14 December 1930. Cat. no. 75–322. http://alberteinstein.info/.

315 *The exchange was catalysed* Sauer, Tilman, Trautman, Andrzej. Myron Mathisson: What Little We Know of His Life. *Acta Physica Polonica B Proceedings Supplement*, 2008, 1: 7–26.

316 *His colleagues, the musicians* Dukas, Helen. *AEA* [Online] 24 December 1930. Cat. no. 75–322. http://alberteinstein.info/.

Chapter 21

318 *Lili Schober had been born* Rodrigues, Ruth Elizabeth. *Selected Students of Leopold Auer – A Study in Violin Performance-Practice* (Birmingham, University of Birmingham, 2009), 62. Available on https://core.ac.uk/download/pdf/76634.pdf.

318 *She and Elsa became close* Petschnikoff, Lili. *The World at our Feet* (New York, Vantage Press, 1968).

319 *Sinclair and Einstein enjoyed* Sinclair, Upton. *The Autobiography of Upton Sinclair* (New York, Harcourt, Brace & World, Inc., 1962), 254–9. See also Einstein, Albert. *Einstein on Peace*, Otto Nathan, Heinz Norden, eds. (New York, Schocken Books, 1968), 120–1.

319 *He ingratiated himself* Dukas, Helen. *AEA* [Online]. 12 January 1931. Cat. no. 75–322. http://alberteinstein.info/.

319 *As Hilb remarked to Elsa* Hilb, Emil, Einstein, Elsa, Marianoff, Dimitri. *AEA* [Online]. January–August 1931. Cat. nos. 50–323 through 50–329. http://alberteinstein.info/.

320 *The latter, who was staying* Kellen, Konrad. *Mein Boss, Der Zauberer* (Hamburg, Rowohlt Verlag, 2011), 109.

320 *'I thank you for all'* Petschnikoff, Lili. *The World at Our Feet*. New York, Vantage Press, 1968), 203–5.

320 *After a striking performance* Dukas, Helen. *AEA* [Online]. Cat. no. 74–840. http://alberteinstein.info/.

321 *In his autobiography* Chaplin, Charles. *My Autobiography* (Harmondsworth, Penguin Books, 1966), 317, 125.

322 *Chaplin himself immediately dashed* —. *My Autobiography* (Harmondsworth, Penguin Books, 1966), 326–7.

322 *Einstein's speech emphasised* Anon. Einstein Lauded by LA Jewish Community. *B'nai B'rith Messenger*, 1931, XXXIII: 1,5.

323 *The view of Moszkowski's relatives* —. *The Detroit Jewish Chronicle*, 19 June 1931, XXXII: 1, 7. See also *AEA* [Online]. 18 February 1931, Cat. nos. 75–332; 69–568. http://alberteinstein.info/.

323 *Sadly, Michelson was by then* —. Dr Einstein comes to Santa Anna. *Santa Anna Register*, 20 February 1931, 3.

324 *However, it does indicate* Einstein, Albert, Tolman, Richard, Podolsky, Boris. Knowledge of Past and Future in Quantum Mechanics. *Physical Review Journal*, 1931, 37: 780–1. See also Pais, A, *LORD*, 448.

326 *When he asked for one* Marianoff, Dimitri, Wayne, Palma. *Einstein: An Intimate Study of a Great Man* (New York, Doubleday, Doran and Co., Inc., 1944), 84–9.

326 *He restarted his erratic diary* Einstein, Albert. *AEA* [Online] December–March 1930–1931. Cat. no. 29–134; the description of both the Berlin period and the Oxford visit is based on this document, which is Einstein's Travel Diary. See also Eisinger, J. *Einstein on the Road* (New York, Prometheus, 2011), 102–6. http://alberteinstein.info/.

326 *After dinner on the following day* Mendlesohn, Erich. The Erich Mendelsohn Archive - Correspondence of Erich and Luise Mendelsohn 1910–1953. *Kunstbibliothek Staatliche Museen zu Berlin* [Online] [Cited: 22 October 2021]. Evidence that Mendelsohn was a pianist can be found in letter 732 of 19 August 1918, 2; see also letters 597, 1163. http://ema.smb.museum/.

326 *After obtaining estimates* Einstein, Albert. Zum kosmologischen Problem der allgemeinen Relativitätstheorie. *Sitzungsberichte der Preussischen Akademie der Wissenschaften, Physikalisch-mathematische Klasse*, 1931, XII: 235–7. See also O'Raifeartaigh, C, McCann, B. Einstein's Cosmic Model of 1931 Revisited: An Analysis and Translation of a Forgotten Model of the Universe. *European Physics Journal H*, 2014, 39: 63–85.

327 *When it arrived in Florence* Rogger, Franziska. *ES*, 88–9.

327 *That evening they arrived* Fox, Robert. Einstein in Oxford. *Notes and Records of the Royal Society*, 2018, 72: 293–318. See also Papers of Lindemann, F. A., Viscount Cherwell. Nuffield College Library, D54/1, 22 June 1927.

327 *In a letter to her* Einstein, Albert. *AEA* [Online]. May 1931. Cat. nos. 143–242, 84–104. http://alberteinstein.info/.

327 *After Maja fled Italy* Palla, Francesco. Einstein's Piano. *Il Colle Di Galileo*, 2016, 5: 9–27. See also https://www.staude.it/biografia/einsteins-piano/?lang=en.

327 *Their relationship began* Einstein, Albert. AEA [Online] May 1931. Cat. nos. 143–242, 143–243, 143–292. http://alberteinstein.info/

328 *Much to the lecturer's amusement* Fox, Robert. Einstein in Oxford. *Notes and Records of the Royal Society*, 2018, 72: 293–318.

330 *'To Professor Albert Einstein'* Deneke, Margaret. *Dedication. [Sheet music]*. Oxford, UK: Fantasia String Quartet, May 1931. AEA Hebrew University of Jerusalem: Einstein Sheet Music Collection.

330 *'After dinner we were established'* —. *Memoirs of Margaret Deneke*. Bodleian Library, University of Oxford. Oxford: unpublished typescript. Deneke Deposit, MS. Eng. misc. d. 1065.

331 *The lecturer began* Newlandsmith, Ernest. 'The Ancient Music of the Coptic Church.' A Lecture Delivered at the University Church, Oxford, on 21 May 1931. *Library of Congress*. [Online] 21 May 1931 [Cited: 12 November 2021]. https://www.loc.gov/resource/ihas.200155827.0/?sp=1.

331 *Rapidly returning to familiar ground* Anon. Professor Einstein. *The Jewish Chronicle*, 29 May 1931, p. 16.

331 *Einstein accepted with alacrity* Clark, Ronald. *Einstein: The Life and Times* (New York, Avon Books, 1972), 536.

332 *It may be this* Einstein, Albert, Einstein, Elsa. *AEA* [Online] 1931–1932. Cat. nos. 47–422, 79–719. http://alberteinstein.info/.

332 *With typical concern for his friends* Jammer, Max. *The Philosophy of Quantum Mechanics: The Interpretations of Quantum Mechanics in Historical Perspective* (New York, Wiley, 1974), 171–2.

332 *However, the thought experiment* Howard, Donald. 'Nicht Sein Kann was Nicht Sein Darf,' or the Prehistory of EPR, 1909–1935: Einstein's Early Worries about the Quantum Mechanics of Composite Systems. In *Sixty-Two Years of Uncertainty: Historical, Philosophical, and Physical Inquiries into the Foundations of Quantum Mechanics. Proceedings of the 1989 Conference*, 'Ettore Majorana' Centre for Scientific Culture, International School of History of Science, Erice, Italy, 5–14 August. Arthur Miller, ed. (New York, Plenum, 1990), 61–111.

332 *He was no doubt encouraged* Hentschel, Klaus. Erwin Finlay Freundlich and Testing Einstein's Theory of Relativity. *Archive for History of Exact Sciences*, 1994, 47: 143–201.
333 *In two papers* Einstein, Albert, Meyer, Walther. Einheitliche Theorie von Gravitation und Elektrizität. *Sitzungsberichte der Königlich Preussischen Akademie der Wissenschaften*, 1931, 541–57. See also *Sitzungsberichte der Königlich Preussischen Akademie der Wissenschaften*, 1932, 130–7.
333 *'This theory does not yet'* Einstein, Albert. Gravitational and Electromagnetic Fields. *Science*, 1931, 74: 438–9.
333 *Sadly, none of these statements* Goenner, Hubert F. M. *On the History of Unified Field Theories*, Living Reviews in Relativity, Vol. 7. (Berlin, Springer, 2004).
333 *The Viennese authorities* Anon. *American Jewish World*, 23 October 1931, 1, 5.
334 *The opening to subvert* Vallentin, Antonina. *Le Drame d'Albert Einstein* (Paris, Les Petits-fils de Plon et Nourrit, 1954), 164–5.
334 *Given the background* Anon. Einstein Defines Aim of Physicists. *The New York Times*, 5 October 1931, 11.
334 *MacDonald, flanked on one side* MacDonald, James Ramsay. *Hansard*. 31 July 1931, 255: cc2636. https://api.parliament.uk/historic-hansard/commons/1931/jul/31/berlin-conversations#S5CV0255P0_19310731_HOC_99.
334 *The news of the Japanese invasion* Einstein, Albert. *Einstein on Peace*, Otto Nathan, Heinz Norden, eds. (New York, Schocken Books, 1968), 145–50.
335 *Einstein relented and signed* Clark, Ronald. *Einstein: The Life and Times* (New York, Avon Books, 1972), 536–9.
335 *Sadly, Einstein's suggestion* Einstein, Albert. The Road to Peace – by Einstein. *The New York Times Magazine*, 22 November 1931, 1–2.
336 *He also recommended that Einstein* Einstein, Eduard. *AEA* [Online] August 1931. Cat. no. 144–523. http://alberteinstein.info/.
336 *He had intended to visit* Rogger, Franziska. *ES*, 124.
336 *The Burghölzli doctors* Huonker, Thomas. *Diagnose: 'moralisch defekt'* (Zurich, Orell Füssli Verlag AG, 2003), 223.
336 *It is likely that an episode* Trbuhović-Gjurić, Desanka. *ISAE*, 169. The dates and even the order of events in this part of the book can be seen from other sources to be very confused.
336 *On 14 November, Einstein left* Einstein, Albert. *AEA* [Online] 1931–32. Cat. no. 29–136. http://alberteinstein.info/.
336 *His father had however* Einstein, Albert. AEA [Online] May/June 1930. Cat. nos. 75–778, 75–781, 75–993, 143–292. http://alberteinstein.info/
337 *He had also presented this* Einstein, Albert. Über die Unbestimmtheitsrelationen. *Zeitschrift für Angewandte Chemie*, 1931, 45: 23. Editor's report on the Colloquium.
337 Casimir, Hendrik. *Haphazard Reality* (New York, Harper & Row, 1983), 315–18. The author is grateful to Dirk van Delft, Director of the Boerhaave Museum, Leiden, the Netherlands, for information on these colloquia.

Chapter 22

338 *Indeed, Einstein kept much more* Einstein, Albert. *AEA* [Online]. 30 December 1931. Cat. no. 29–136. http://alberteinstein.info/.When not otherwise specified, information on the stays in

Pasadena is taken from Einstein's Travel Diaries and from reports in the *Los Angeles Times*. See also Eisinger, J. *Einstein on the Road* (New York, Prometheus Books, 2011).

339 *The subsequent discussions* Einstein, Albert, de Sitter, Willem. On the Relation between the Expansion and the Mean Density of the Universe. *Proceedings of the National Academy of Sciences of the United States of America*, 1932, 18: 213–14.

339 *Georges Lemaître, who was shortly* O'Raifeartaigh, Cormac, O'Keeffe, Michael, Mitton, Simon. Historical and Philosophical Reflections on the Einstein–de Sitter Model. *European Physical Journal H*, 2021, 46: 4.

339 *This endorsement, together with the foreword* Grattan, Hartley. Why, Dr. Einstein! *New Republic*, 9 March 1932. https://newrepublic.com/article/119292/controversy-einsteins-endorsement-psychic-upton-sinclair-defends.

340 *Although he spoke no German* Burke, Tony. History: El Mirador Hotel Did Well during the 1930s. *Desert Sun* [Online]. 31 January 1932. https://eu.desertsun.com/story/news/2015/08/15/history-tony-burke-nd-letter-palm-springs/31642399/.

340 *Pieces by Bach, Mozart* Petschnikoff, Lili. *The World at our Feet* (New York, Vantage Press, 1968), 203–4.

341 *This could be interpreted* Sauer, Tilman. Einstein's Unified Field Theory Program. In *The Cambridge Companion to Einstein*, Christoph Lehner, Michel Janssen, eds. (New York: Cambridge University Press, 2014), 301–5.

342 *'Einstein played his violin'* Marianoff, Dimitri, Wayne, Palma. *Einstein: An Intimate Study of a Great Man* (New York, Doubleday, Doran and Co., Inc., 1944), 144–5.

342 *Einstein enjoyed the performance* Several, Michael. Pearl Hassler. *Flame: Newsletter for the Pasadena Jewish Temple*. 2012, 7(August): 7. See also *Flame*, 2012, 7(June): 11.

343 *The pianist was Sidney Cutler* Anon. Prof. Einstein Guest at Pasadena Tea. *B'nai B'rith Messenger*, 19 February 1932, XXXIV: 6.

343 *Despite fleeing back* Petschnikoff, Lili. *The World at our Feet* (New York: Vantage Press, 1968), 205–7.

344 *He was once again moved* Anon. Dr. Einstein Makes Plea for Harmony between Jew and Arab in Palestine. *B'nai B'rith Messenger*, 4 March 1932, XXXIV: 1, 8.

344 *He suggested that Tete* Einstein, Albert. *AEA* [Online] 29 February 1932. Cat. no. 75–671. http://alberteinstein.info/.

344 *He argued endlessly with Maja* Rogger, Franziska. *ES*, 124. See also Einstein, Eduard. *AEA* [Online] 25 April 1932. Cat. no. 144–510. http://alberteinstein.info/.

345 *In June she wrote again* Einstein, Mileva. *AEA* [Online]. May and June 1932. Cat. nos. 144–416, 144–417. http://alberteinstein.info/.

345 *It was only after 2 am* Zangger, Heinrich. *Seelenverwandte*, Robert Schulmann, ed. (Zurich: Verlag Neue Zürcher Zeitung, 2012), 501–7, Documents 327–31.

345 *This was something that Besso* Besso, Michele. *Albert Einstein Michele Besso: Correspondance*, Pierre Speziali, ed. (Paris, Hermann, 1972), 285–6 (Document 112, 18 September 1932).

345 *The letter to Mileva* Einstein, Albert. *AEA* [Online]. 6 October 1932. Cat. no. 75–719. http://alberteinstein.info/.

345 *When Tete did not give* Einstein, Albert, Einstein, Eduard. *AEA* [Online] August–September 1932. Cat. nos. 75–686, 75–669, 144–515 through 144–517. http://alberteinstein.info/.

345 *'You will have to admit'* Ettema, Robert, Mutel, Cornelia F. *Hans Albert Einstein* (Reston, ASCE Press, 2014), 50. See also *AEA* [Online]. October–November 1932. Cat. no. 75–684. http://alberteinstein.info/.

345 *She concluded by begging* Einstein, Mileva. *AEA* [Online]. 18 November 1932. Cat. no. 144–418. http://alberteinstein.info/.

346 *In fact, he did not* Einstein, Albert. *AEA* [Online]. 1932. Cat. nos. 75–668, 7–370. http://alberteinstein.info/.

346 *Einstein had organised this* Zangger, Heinrich. *Seelenverwandte*, Robert Schulmann, ed. (Zurich: Verlag Neue Zürcher Zeitung, 2012), 506–12, Documents 331–333 and notes therein; see also *AEA* [Online]. Cat. nos. 74–992, 75–799. http://alberteinstein.info/.

346 *Disembarking at the final destination* Einstein, Albert. *AEA* [Online]. 3–5 April 1932. Cat. nos. 10–227, 10–231, 10–233. http://alberteinstein.info/.

346 *He ended his eulogy* Franke, Kurt. *Jüdische Miniaturen: Moritz Katzenstein* (Berlin, Hentrich & Hentrich, 2005), 40–9.

346 *However, from what was presumably* Grüning, Michael. *Ein Haus für Albert Einstein* (Berlin, Verlag der Nation, 1990), 266–7, 384–5. See also https://einstein-website.de/kaiserlich-deutsche-akademie-der-naturforscher-zu-halle-leopoldina/

348 *It is unlikely that Einstein* Reichinstein, David, Einstein, Albert. *AEA* [Online]. 1932–1935. Cat. nos. 20–185, 20–195, 20–171, 20–173, 20–198. http://alberteinstein.info/. For letters from June–July 1932, see https://library.ethz.ch/en/locations-and-media/platforms/einstein-online/die-berliner-zeit-1914-1933.html.

348 *Margaret lent him* Deneke, Margaret. *Ernest Walker* (Oxford, Oxford University Press, 1951), 91–2. See also *Memoirs of Margaret Deneke*. Bodleian Library, University of Oxford. Oxford: unpublished. typescript. Deneke Deposit, MS. Eng. misc. d. 1065.

349 *Indeed,* The Cambridge Review *speculated* Anon. Lecture by Professor Einstein. *Cambridge Review*, 13 May 1932, LIII: 382.

349 *It is probably with this trip* —. Einstein Talks to Savants. *The New York Times*, 7 May 1932, 13. See also Grazer, Walter. Foreword. In Eisinger, Josef. *Einstein on the Road* (Amherst, Prometheus Books, 2011), 1.

349 *He even at times* Fox, Robert. Einstein in Oxford. *Notes and Records of the Royal Society*, 2018, 72: 305.

349 *Einstein, however, was still reluctant* Eisinger, Josef. *Einstein on the Road* (Amherst, Prometheus Books, 2011), 152.

350 *Although some characterised both* Einstein, Albert. *Einstein on Peace*, Heinz Norden, Otto Nathan, eds. (New York, Schoken, 1968), 167–72.

350 *'The hotel clerk told me'* Bercovici, Konrad. The Comedy of Peace. *Pictorial Review*, March 1933.

350 *Flexner's recollection of his visit* Einstein, Albert. AEA [Online]. 8 June 1932. Cat. no. 38–3. http://alberteinstein.info/.

352 *The following morning, she telephoned* Vallentin, Antonina. *Le Drame d'Albert Einstein* (Paris: Les Petits-fils de Plon et Nourrit, 1954), 170–6.

352 *By August, it was already* Einstein, Albert, Flexner, Abraham. *AEA* [Online]. June–July 1932. Cat. nos. 38–1 through 38–8. http://alberteinstein.info/. See also *The New York Times*, 24 August 1932, 19.

352 *Ominously, they added that* Anon. Nazis Engage in Fight at Berlin University. *The New York Times*, 1 July 1932, 4.

352 *He told Tete that he would forbid* Einstein, Albert. *AEA* [Online]. 20 August 1932. 75–688. http://alberteinstein.info/.

352 *She closed by wishing him* de Dijn, Rosine. *Albert Einstein und Elisabeth von Belgien* (Regensburg, Verlag Friedrich Pustet, 2015), 72, 75.

353 *'When we clean house'* Robinson, Andrew. *Einstein on the Run* (New Haven, Yale University Press, 2019), 192.

354 *Here he had meetings* Einstein, Albert, Einstein, Elsa, Ehrenfest, Paul. *AEA* [Online]. July–August 1932. Cat. nos. 10–235, 10–236, 17–330. http://alberteinstein.info/; D. van Delft, private communication; see also van Delft, D. Paul Ehrenfest's Final Years. *Physics Today*, 2014, 67: 41–7.

354 *He was surprised to be asked* Einstein, Albert. *AEA* [Online]. 13 August 1932. Cat. no. 123–59. http://alberteinstein.info/.

354 *He did however send a message* —. *Einstein on Peace*, Heinz Norden, Otto Nathan, eds. (New York, Schoken, 1968), 175–83.

355 *'The struggle for complete'* —. *AEA* [Online] 23 August 1932. Cat. no. 48–30. http://alberteinstein.info/.

355 *It can be assumed that;* Various. Gästebuch vom Hause Einstein: dem grossen Nachbar in Verehrung gewidmet. *Centre for Jewish History, Leo Baeck Institute* [Online]. September 1932 [Cited: 11 April 2022.] https://www.lbi.org/. See also *The New York Times*, 23 September 1932, 22; and *Jüdische Rundschau*, 27 September 1932, 374.

355 *It is worth quoting it in full* Grüning, Michael. *Ein Haus für Albert Einstein* (Berlin, Verlag der Nation, 1990), 283–8.

356 *'It is a particular blessing'* Einstein, Albert. *AEA* [Online]. August 1932. Cat. no. 28–218. http://alberteinstein.info/.

356 *If, however, the improvisation* Grüning, Michael. *Ein Haus für Albert Einstein* (Berlin, Verlag der Nation, 1990), 251–2.

356 *Einstein gave another talk* Anon. *Jüdische Rundschau*. 21 October 1932, 407; see also *The New York Times*, 17 October 1932, 1, 11; and 18 October 1932, 6.

357 *Her views on Palestine and Zionism* —. Lady Erleighs Berliner Besuch. *Jüdische Rundschau*. 15 November 1932: 441.

357 *The play, written in 1907* Graf Kessler, Harry. *Tagebücher, 1918–1937*, Wolfgang Pfeiffer-Belli, ed. (Frankfurt-am-Main, Insel Verlag, 1961). See entry for 15 November 1932.

357 *'Einstein is seen to be a great man'* Various. Gästebuch vom Hause Einstein: dem grossen Nachbar in Verehrung gewidmet. Centre for Jewish History, Leo Baeck Institute [Online]. September 1932 [Cited: 11 April 2022.] https://www.lbl.org/

357 *However, this initiative seems* Einstein, Albert. *Einstein on Peace*, Heinz Norden, Otto Nathan, eds. (New York, Schoken, 1968), 183–5. See also Goenner, Herbert. *Einstein in Berlin* (Munich, Verlag C. H. Beck, 2005), 335.

357 *The young Bucky* Bucky, Thomas. Einstein: An Intimate Memoir. *Harper's Magazine*, 1964, September, 43–4.

357 *'Never before has any attempt'* Anon. Einstein Ridicules Women's Fight on Him Here. *The New York Times*, 4 December 1932: 1.

358 *They returned to Caputh* Jerome, Fred. *The Einstein File* (New York: St Martin's Press, 2002), 4–13.

358 *It was this aspect that nettled* Anon. Dr Einstein Gets a Visa. *B'nai B'rith Messenger*. 1932, XXXVI: 4.

358 *He obliged and provided Frau Kielow* Kielow, Margaret, Einstein, Albert. *AEA* [Online]. 1946, 1932. Cat. nos. 57–8, 50–698. http://alberteinstein.info/.

358 *He, together with Ilse* Einstein, Albert. *AEA* [Online]. 10 December 1932. Cat. no. 29–138. http://alberteinstein.info/.

359 *Philipp Frank wrote that* Frank, Philipp. *LZ*, 363–64.

359 *Elsa didn't believe him* Vallentin, Antonina. *Le Drame d'Albert Einstein* (Paris: Les Petits-fils de Plon et Nourrit, 1954), 179.

Chapter 23

360 *He also remarked that* Anon. Einstein Embarks; Jests About Quiz. *The New York Times*, 11 December 1932, 3.
360 *Modern history on his reading* Einstein, Albert. *AEA* [Online]. 27 December 1932. Cat. no. 75–675. http://alberteinstein.info/.
360 *'To Charlie Chaplin'* Anon. Einstein Guest of Southland. *The Los Angeles Times*, 10 January 1933, LII, 1–2. See also *Jewish Daily Bulletin*, 11 January 1933, 1.
361 *The two had not resolved* Lambert, Dominique. Einstein and Lemaître: Two friends, two cosmologies . . . *Interdisciplinary Encyclopedia of Religion and Science* [Online] [Cited: August 1, 2022]. https://inters.org/einstein-lemaitre.
361 *This partial record* Einstein, Elsa. *AEA* [Online]. 9 January 1933. Cat. no. 89–274. http://alberteinstein.info/.
361 *Millikan's communications* Clark, Ronald. *Einstein: The Life and Times* (New York, Avon Books, 1972), 552. See also *Illustrated Daily News*, 21 January 1933, 3; *Los Angeles Times*, 22 January 1933, 1, 2, and 24 January 1933, 4.
361 *While recognising the geographical* Einstein, Albert. *Out of My Later Years* (New York, Philosophical Library, 1950), 215.
362 *Since he was wearing* Anon. World Thinkers Survey Cause of Depression. *The Los Angeles Times*, 24 January 1933, LII, 1–2.
362 *It was probably this idea* Anon. *Jewish Daily Bulletin*, 22 January 1933, X, 1, 4. See also Einstein, A. *Einstein and Peace*, O. Nathan, H. Norden, eds. (New York, Schoken Books, 1968), 184.
362 *It was addressed by US* Anon. Mack Biltmore Lunch is Singular Affair. *B'nai B'rith Messenger*, 3 February 1933, XXXVI, 3.
362 *In fact, within two months* Clark, Christopher. *The Iron Kingdom* (London, Penguin, 2007), 648–50.
362 *One of them, child star* Barnes, Eleanor. Kreisler Hailed Undisputed King of Violin World. *Illustrated Daily News (Los Angeles)*, 1 February 1933, 13. See also *Los Angeles Times*, 1 February 1933, 7.
363 *The Hertzs drove* Anon. Einstein Hears New Idea. *Los Angeles Times*, 7 February 1933, 7.
364 *'[Einstein] warned me'* Chaplin, Charles. *My Autobiography* (Harmondsworth, Penguin Books, 1966), 318–19.
364 *Eddington considered Paul* Kragh, Helga. Eddington's Dream: A Failed Theory of Everything. In *Information and Interaction*, I. Rickles, D. Durham, eds. (Cham: Springer International, 2017), 45–58.
364 *It was during this stay* Einstein, Albert. *Albert Einstein: The Human Side*, Banesh Hoffmann, Helen Dukas, eds. (Princeton: Princeton University Press, 1979), 48–9.
365 *Also on 27 February* Anon. Prof. Piccard will make Stratosphere Trip in Southwest. *Pasadena Post*, 1 March 1933, 1. See also Anon. 'Mystery Woman' is Einstein. *Los Angeles Times*, 27 February 1933, 8.
365 *It and other right-of-centre* Sapinsley, Barbara. *From Kaiser to Hitler* (New York, Grosset and Dunlap, 1968), 141–8.

365 *He finished the letter* Einstein, Albert. *AEA* [Online] 27 February 1933. Cat. no. 50–834. http://alberteinstein.info/.

366 *Einstein and Elsa seem* Anon. Einstein Delays Trip to Germany as Harm Feared. *Santa Rose Republican*, 3 March 1933, 2. See also *Hollywood Citizens News*, 3 March 1933, 1; and *Chicago Daily Tribune*, 13 March 1933, 1.

366 *Although more than 100* —. Einstein, Ready to Go, Hits at Hitler Regime. *Los Angeles Times*. 11 March 1933, 1, 2. See also Knopoff, Leon. Beno Gutenberg. In *Biographical Memoirs of the National Academy of Sciences*, Vol. 76 (Washington, DC, National Academies Press, 1999), 29.

366 *That evening the Einsteins departed* Kinsley, Philip. Einstein Tells 3 Problems of World Today. *Chicago Daily Tribune*, 15 March 1933, 1. See also Rosenkranz, Ze'ev. Albert Einstein and the Hebrew University. *Proceedings of the World Congress of Jewish Studies, Division B: The History of the Jewish People*, Volume III: *Modern Times* (Jerusalem, World Union of Jewish Studies, 1993), 231.

366 *He was nevertheless hopeful* Anon. Einstein Honored at a Dinner Here. *The New York Times*, 16 March 1933, 1, 10.

367 *'I am sorry to say'* —. Dr. Einstein urges Hitler Protests. *The New York Times*, 17 March 1933, 15.

367 *When this was questioned* —. Dr. Einstein Acts as Boy's Godfather. *The New York Times*, 18 March 1933, 15. See also *The New York Times*, 19 March 1933, 7.

368 *'The raid on the home'* Einstein, Albert. *Einstein on Peace*, Heinz Norden, Otto Nathan, eds. (New York, Schoken, 1968), 212–13. See also Anon. Nazis Hunt Arms in Einstein Home. *The New York Times*, 21 March 1933, 10.

368 *He continued that the political* —. *AEA* [Online]. 21 March 1933. Cat. no. 97–234. http://alberteinstein.info/.

368 *'Brutal acts of violence'* Anon. Prof. Einstein Hopes For Curbs On Nazis. *The New York Times*, 28 March 1933, 11.

368 *There is some controversy* Herneck, Friedrich. *Einstein privat* (Berlin, Buchverlag Der Morgen, 1978), 148–9. See also Einstein, Elsa. AEA [Online]. 11 April 1933. Cat. no. 83–160. http://alberteinstein.info/.

369 *He then abruptly changed* Einstein, Elsa, Einstein, Albert, Einstein, Eduard. *AEA* [Online] 1932–1933. Cat. nos. 99–219, 144–521. http://alberteinstein.info/.

369 *In the letter, he acknowledged* Einstein, Albert. *AEA* [Online]. 28 March 1933. Cat. nos. 29–179.25, 36–55. http://alberteinstein.info/.

369 *Music was no doubt* de Dijn, Rosine. *Albert Einstein und Elisabeth von Belgien* (Regensburg, Verlag Friedrich Pustet, 2015), 93.

369 *This must be the violin* Einstein, Albert. AEA [Online]. 28 July 1933. Cat no. 50–904. http://alberteinstein.info/.

370 *At this point there were cries* Birchall, Frederick. Whips Crowd into Frenzy. *The New York Times*, 1 April 1933, 10.

370 *'Feuchtwanger's idiotic articles'* —. Hitlerites Order Boycott Against Jews in Business, Professions and Schools. *The New York Times*, 29 March 1933, 1, 8.

370 *He was joined at the start* Marianoff, Dimitri, Wayne, Palma. *Einstein: An Intimate Study of a Great Man* (New York, Doubleday, Doran and Co., Inc., 1944), 141–4.

370 *By 1 June the three women* Herneck, Friedrich. *Einstein Privat* (Berlin, Buchverlag der Morgen, 1978), 148–56.

370 *It is impossible to reconcile* —. *Einstein Privat* (Berlin, Buchverlag Der Morgen, 1978), 150.

371 *'. . . I am tired and at the end'* Einstein, Elsa. *AEA* [Online]. 11 April 1933. Cat. no. 83–160. http://alberteinstein.info/. See also *The New York Times*, 6 April 1933, 9.

371 *On 15 August* Anon. Versuch einer Klärung. *Jüdische Rundschau*, 31 March 1933, XXXVIII, 1. See also *Jüdische Rundschau*, 4 April 1933, 131–2; *Jüdisch-liberale Zeitung*, 15 August 1933, 10.

372 *His reason was that* Einstein, Albert. *Mein Weltbild*, Carl Seelig, ed. (Frankfurt-am-Main, Ullstein Materialien, 1980), 81–6.

372 *It may be that von Laue* Gerstengarbe, Sybille. Die Leopoldina und ihre jüdischen Mitglieder. *Acta Historica Leopoldina*, 2014, 64: 428.

372 *When Planck insisted that* Planck, Max. Mein Besuch bei Adolf Hitler. *Physikalische Blätter*, 1947, 3: 143. See also O'Flaherty, J. Max Planck and Adolf Hitler. *AAUP Bulletin*, 1956, 42: 437–44.

372 *On 8 April 1933* Anon. Albert Einstein et le Collège de France: Quelques Points d'Histoire. *Archives du Collège de France* [Online] [Cited: 12 August 2022]. https://www.college-de-france.fr/media/lettre-du-college-de-france/UPL54220_J13EINSTEIN.pdf.

373 *'I now have more Professorships'* Einstein, Albert. *AEA* [Online]. April 1933. Cat. nos. 21–223, 121–707. http://alberteinstein.info/.

373 *This seems to have mollified* Einstein, Albert, Flexner, Abraham. *AEA* [Online]. April–July 1933. Cat. nos. 38–19, 38–22 through 38–29, 38–36. http://alberteinstein.info/.

374 *It was probably her* Elisabeth, Queen of the Belgians. *AEA* [Online]. 10 April 1933. 32–358. http://alberteinstein.info/.

374 *On the latter date* de Dijn, Rosine. *Albert Einstein und Elisabeth von Belgien* (Regensburg, Verlag Friedrich Pustet, 2015), 96.

374 *'Einstein's playing of Mozart'* Bucky, Peter. *The Private Albert Einstein* (Kansas City, Andrews and McMeel, 1992), 150.

374 *They had been recommended* Einstein, Albert, Lindemann, Frederick. *AEA* [Online]. May 1933. Cat. no. 16–372–16–374. http://alberteinstein.info/. See also Fort, Adrian. *The Prof* (London, Jonathan Cape, 2003), 119–27.

374 *Born spent two years* Einstein, Albert. *AEA* [Online] 30 May 1933. Cat. no. 8–192. http://alberteinstein.info/. See also Born, G. V. R. The Wide-Ranging Family History of Max Born. *Notes and Records of the Royal Society*, 2002, 56: 219–62.

374 *Einstein compared it* Ehrenfest, Paul, Einstein, Albert. *AEA* [Online]. March–May 1933. Cat. nos. 10–245, 10–246, 10–251, 12–378. http://alberteinstein.info/.

375 *'The significance of the University'* Anon. Fifteenth Anniversary and Banquet of the Jewish Telegraph Agency. *Jewish Daily Bulletin*, 16 March 1933, 29.

375 *However, this fell* Weizmann, Chaim. *CHAIM*, Vol. XV, 170–1, Letter 164; 316–21, Letter 283; 401, Letter 360.

375 *He was clearly deeply hurt* —. *CHAIM*, Vol. XV, 404–8, Letter 366; 415–18, Letter 377.

375 *In letters later that month* Einstein, Albert. *AEA* [Online]. April–May 1933. Cat. nos. 37–65, 33–426. http://alberteinstein.info/. See also *CHAIM*, 423, 432.

375 *That this reluctance had suddenly* —. *AEA* [Online] 9 May 1933. 16–378. http://alberteinstein.info/.

376 *However, he continued to oscillate* Einstein, Mileva. *AEA* [Online]. May 1933. Cat. nos. 144–419, 144–420. http://alberteinstein.info/.

377 *Certainly by the following month* Einstein, Albert, Zangger, Heinrich. *Seelenverwandte*, Robert Schulmann, ed. (Zurich, Verlag Neue Zürcher Zeitung, 2012), 519–21, 525–7, Documents 340, 341, 344. See also *AEA* [Online]. 5 November 1932. Cat. no. 75–685; Marić, Mileva. 20 June 1933. Cat. no. 144–421. http://alberteinstein.info/.

377 *He was never to see* —. *AEA* [Online]. May 1933. 10–253,75–663,143–252. http://alberteinstein.info/.

Chapter 24

378 *This year he had been moved* Deneke, Margaret. *Memoirs of Margaret Deneke.* Bodleian Library, University of Oxford. Oxford: unpublished. typescript. Deneke Deposit, MS. Eng. misc. d. 1065.

378 *'I realise that I have'* Einstein, Albert. *AEA* [Online] 30 May 1933. Cat. no. 75–663. http://alberteinstein.info/.

379 *He thought it a good performance* Deneke, Margaret. *Memoirs of Margaret Deneke.* Bodleian Library, University of Oxford. Oxford: unpublished. typescript. Deneke Deposit, MS. Eng. misc. d. 1065.

379 *He mentioned the discovery* Anon. Probing the Atom. *The Times*, 3 June 1933, 9.

379 *'As he delivered his speech'* Clark, Ronald. *Einstein: The Life and Times* (New York, Avon Books, 1972), 582.

380 *'Our experience up to date'* Einstein, Albert. On the Method of Theoretical Physics. *Philosophy of Science*, 1934, 1: 163–9. The original German source for the lecture can be found in *AEA* [Online]. June 1932. Cat. no. 1-114. http://alberteinstein.info/.

380 *In May 1931* Anon. News and Views. *Nature*, 1931, 127: 826–7. No written version of the lectures exists. Only an English summary, prepared for distribution beforehand, and reports in the press survive. Although *The New York Times* and *The Times* covered the first two lectures, only *Nature* summarised all three.

380 *'I cannot believe that'* Pais, Abraham. *LORD*, 347.

381 *'Dear Elsa, My time here'* Einstein, Albert. *AEA* [Online] 11 June 1933. 143–257. http://alberteinstein.info/.

381 The Times *report quoted him* —. *AEA* [Online] 13 April 1933. Cat. no. 5–2. http://alberteinstein.info/.See also Anon. Dr Einstein at Oxford. *The Times*, 14 June 1933, 12.

382 *'My biggest regret is that'* —. *AEA* [Online] 14 June 1933. Cat. no. 10–255. http://alberteinstein.info/.

382 *Einstein, however, refused* Weizmann, Chaim. *CHAIM*, Vol. XV, 439–42, Letter 409.

382 *Einstein liked him very much* Anon. The English Association. *The Times*, 19 June 1933, 17. See also Einstein, Albert. *AEA* [Online]. 17 June 1933. Cat. no. 143–258. http://alberteinstein.info/.

383 *As Weizmann had just left* Weizmann, Chaim. *CHAIM*, Vol. XVI, 23–5, Letter 23; 6, Letter 6.

383 *He stated that if* Magnes, Judah. *Dissenter in Zion*, Arthur Goren, ed. (Cambridge, MA: Harvard University Press, 1982), 295–6.

383 *Hartog cabled a report* Weizmann, Chaim. *CHAIM*, Vol. XVI, 11, Letter 12.

383 *'It gives me great pleasure'* Anon. Dr Einstein to Head Physics Faculty at Hebrew University, says Weizmann. *Jewish Daily Bulletin*, 6 July 1933, 3.

383 *He also asked Haber* Einstein, Albert. *AEA* [Online] 8 August 1933. Cat. no. 12–387. http://alberteinstein.info/.

383 *He had the impression* Weizmann, Chaim. *CHAIM*, Vol. XVI, 85–7, Letter 81.

384 *Einstein arrived by train* Anon. *The Manchester Guardian*, 21 June 1933, 10.

384 *'Once we have recognised'* Einstein, Albert. *The Origins of the General Theory of Relativity* (Glasgow, Jackson, Wylie & Co., 1933), 1–11. See also *The Manchester Guardian*, 21 June 1933, 10. *The Glasgow Herald*, 21 June 1933, 11. The German original can be found in *AEA* [Online]. June 1933. Cat. no. 5–3. http://alberteinstein.info/.

384 *He intimated to Langevin* —. *AEA* [Online]. 26 June 1933. Cat. no. 15–400. http://alberteinstein.info/.

385 *In no circumstances would he* —. *AEA* [Online] 14 July 1933. Cat. no. 32–361. http://alberteinstein.info/. See also de Dijn, R. *Albert Einstein und Elisabeth von Belgien* (Regensburg, Verlag Friedrich Pustel, 2016), 96–8.

385 *'To prevent the great evil'* —. *Einstein on Peace*, Heinz Norden, Otto Nathan, eds. (New York: Schoken, 1968), 228–36.

385 *Einstein expressed himself* —. *AEA* [Online]. 16 July 1933. Cat. no. 50–632. http://alberteinstein.info/.

386 *In his apology* Robinson, Andrew. *Einstein on the Run* (New Haven, Yale University Press, 2019), 197–202, 230–1. See also *AEA* [Online] June–July 1933. Cat. nos. 37–412, 121–753, 123–402, 32–364. http://alberteinstein.info/.

386 *He was also impressed* Einstein, Albert. *AEA* [Online]. 22 July 1933. Cat. no. 143–250. http://alberteinstein.info/.

386 *Against 'Address', after a pause* Clark, Ronald. *Einstein: The Life and Times* (New York: Avon Books, 1972), 598–9. See also Robinson, Andrew. *Einstein on the Run* (New Haven, Yale University Press, 2019), 233–4.

386 *Locker-Lampson was a friend* McLaren, Stuart. *Saving Einstein* (Lowestoft, Poppyland Publishing, 2021), 84–5.

387 *'I am not personally a Jew'* Locker-Lampson, Oliver. Nationality Of Jews. *Hansard*, 26 July 1933, Vol. 280. cc2604–6. https://hansard.parliament.uk/Commons/1933-07-26/debates/98f2b556-282d-459e-8c15-d16d71402f23/NationalityOfJews.

387 *By the end of August* Einstein, Albert, Flexner, Abraham. *AEA* [Online]. 27 July–August 1933. Cat. nos. 38–31–38–45, 39–560. http://alberteinstein.info/. See also *L'Intrangiseant*, 13 September 1933, 3.

387 *However, Einstein apparently enjoyed* Anon. La Simplicité et les Goutes. *Le Carillon*, 5 August 1933, 1, 2. See also Ensor à Einstein. *Le Littoral*, 26 August 1933, 3.

388 *It would be pleasant to think* de Dijn, Rosine. *Albert Einstein und Elisabeth von Belgien* (Regensburg, Verlag Friedrich Pustet, 2015), 93. See also *L'Echo de Ostende*, 9 August 1933, 2.

388 *Einstein congratulated him* Einstein, Albert. *AEA* [Online]. August 1933. Cat. no. 75–557. http://alberteinstein.info/.

388 *His health gradually broke down* Einstein, Albert, Einstein, Elsa. *AEA* [Online]. 1933–1934. Cat. nos. 122–916 through 122–930. http://alberteinstein.info/.

388 *Just before departing* Hill, Alfred. AEA [Online]. 4 October 1933. Cat. no. 50–914. http://alberteinstein.info/.

389 *One was the stars* Vallentin, Antonina, Einstein, Albert. *AEA* [Online]. August 1933. Cat. nos. 83–165, 29–416. http://alberteinstein.info/.

389 *'Out of all the terrible things'* Einstein, Albert. *AEA* [Online]. 22 August 1933. Cat. no. 50–807. http://alberteinstein.info/.

389 *Furthermore, it reprinted his statement* Anon. *The Brown Book of the Hitler Terror* (London, Victor Gollancz, 1933). See also Robinson, Andrew. *Einstein on the Run* (New Haven, Yale University Press, 2019), 242–5.

389 *He did not go outside* Einstein, Albert, Einstein, Elsa. *AEA* [Online]. September 1933. Cat. nos. 37–688, 83–166, 50–909. http://alberteinstein.info/.

389 *On 10 September* Einstein, Albert. *AEA* [Online]. 1935, 1933. Cat. nos. 70–789, 143–259. http://alberteinstein.info/.

390 *They continued to correspond* —. *AEA* [Online]. June–October 1933 [Cited: 27 May 2024]. Cat. nos. 82–902, 82–903. See also 82–904 through 82–906, 91–316. http://alberteinstein.info/.

390 *His first move* Robinson, Andrew. *Einstein on the Run* (New Haven, Yale University Press, 2019), 249–56. See also McLaren, Stuart. *Saving Einstein* (Lowestoft, Poppyland Publishing, 2021), 104–8; Einstein, Albert. *AEA* [Online]. September 1933. Cat. nos. 121–834, 143–260, 143–262. http://alberteinstein.info/.

390 *He advised Einstein to remove* Einstein, Albert. *AEA* [Online]. September 1933. 143–259,143–262. http://alberteinstein.info/.

391 *He drew a pistol* van Delft, Dirk. Paul Ehrenfest's Final Years. *Physics Today*, 2014, 67: 41. See also Ehrenfest, P. *AEA* [Online]. August–September 1933. Cat. no. 10–264. http://alberteinstein.info/.

391 *He knew about it by 1 October* Einstein, Albert. *Seelenverwandte*, Robert Schulmann, ed. (Zurich, Verlag Neue Zürcher Zeitung, 2012), 538–40, Document 351.

391 *'My essential problem, he said'* Marso. Une Interview d'Einstein. *L'Intransigeant*, 11 November 1933, 2. See also Marianoff, Dimitri, Wayne, Palma. *Einstein: An Intimate Study of a Great Man* (New York, Doubleday, Doran and Co., 1944), 160–3.

391 *A copy was donated* Epstein, Jacob. *Let There Be Sculpture* (New York, G.P. Putnam's Sons, 1940), 67–9. See also *The Times*, 12 January 1934, 10; 15 January 1934, 12; and 23 January 1934, 8.

392 *Einstein does not seem* McLaren, Stuart. *Saving Einstein* (Lowestoft, Poppyland Publishing, 2021), 152–5.

392 *He only heard about it* Einstein, Albert. *AEA* [Online] December 1933. Cat. no. 38–72. http://alberteinstein.info/.

392 *'In my opinion any Power'* Anon. Professor Einstein's Political Views. *The Times*, 16 September 1933, 12.

392 *He outlined the seriousness* Robinson, Andrew. *Einstein on the Run* (New Haven, Yale University Press, 2019), 258–70.

393 *'Only through perils and upheavals'* Einstein, Albert. Prof. Einstein's Address at Albert Hall, London. *The New York Times*, 4 October 1933, 17. For the video recording, see https://youtu.be/ZBage5Ff57E.

393 *'I know very many people'* —. *AEA* [Online] 1933. Cat. no. 120–516. http://alberteinstein.info/. Note that the published version of the pamphlet is very different from that given in Einstein, Albert. *Einstein and Peace*, O. Nathan, H. Norden, eds. (New York, Schocken Books, 1968), 241–3.

393 *Yahuda had indeed written* Yahuda, Abraham. Papers of F.A. Lindemann, Viscount Cherwell. Nuffield College Library, 26 August 1933. D57/24.

393 *Elsa, Dukas, and Mayer* —. *AEA* [Online]. September 1933. Cat. nos. 121–839, 143–262. http://alberteinstein.info/. See also *The Times*, 9 October 1933, 16.

394 *They were preoccupied with trying* Einstein, Albert, Einstein, Elsa. *AEA* [Online]. October 1933. 143-302, 99-536, 37-691. http://alberteinstein.info/.

394 *He suggested that* Einstein, Albert. *AEA* [Online]. 14 October 1933. 29–327. http://alberteinstein.info/.

394 *On 16 October* Flexner, Abraham. Chronological files: 1933 cards. *Institute for Advanced Study* [Online]. 16 October 1933 [Cited: 5 December 2022], 248. https://albert.ias.edu/handle/20.500.12111/2507.

394 *His life in Princeton had begun* Anon. Einstein Arrives; Pleads for Quiet. *The New York Times*, 18 October 1933, 23. See also Einstein, Albert, Einstein, Elsa. *AEA* [Online], October 17–18 1933. Cat. nos. 37–691, 83–168, 84–191; 23 April 1934. Cat. no. 51–41. http://alberteinstein.info/.

Chapter 25

395 *His fellow executor* Sayen, Jamie. *Einstein in America* (New York, Crown Publishers, Inc., 1985), 62, 75. See also Einstein, Elsa. *AEA* [Online]. 17 October 1933. Cat. no. 37–691. http://alberteinstein.info/.

395 *Indeed, she was concerned* Einstein, Elsa. *AEA* [Online]. October 1933. Cat. nos. 37–691, 83–169. http://alberteinstein.info/. See also *Paterson Morning Call*, 31 October 1933, 1.

395 *There was an unofficial* Sanua, Marianna. Stages in the Development of Jewish Life at Princeton University. *American Jewish History*, 1987, 76: 391–415.

396 *Eventually he gave way* Sayen, Jamie. *Einstein in America* (New York, Crown Publishers, Inc., 1985), 63–4, 80. See also *The Princeton Herald*, 20 October 1933, 1; *The Daily Princetonian*, 19 October 1933, 1. The Einstein photograph and date can be found at https://upload.wikimedia.org/wikipedia/commons/archive/1/16/20160820062811%21Einstein_1933.jpg.

396 *James Bamberger and Caroline* Kennedy, George H. *Paterson Evening News*, 4 November 1933, 23.

396 *'Princeton is a wonderful corner'* Einstein, Albert. *AEA* [Online]. 20 November 1933. Cat. no. 32–369. http://alberteinstein.info/.

396 *He hoped that they would* —. *AEA* [Online]. 1 December 1933. Cat. no. 75–665. http://alberteinstein.info/.

396 *Of course, they agreed* Sayen, Jamie. *Einstein in America* (New York, Crown Publishers, Inc., 1985), 64,74.

397 *Hilb knew Seidel* Anon. Einstein has Musicale. *The New York Times*, 10 November 1933, 24.

398 *'Herr Seidel was kind enough'* Einstein, Elsa. *AEA* [Online]. 20 November 1933. Cat. no. 50–342. http://alberteinstein.info/.

398 *He tried to circumvent* Einstein, Albert. *Papers of Lord Cherwell*. Library & Archives, Nuffield College Oxford. Oxford: unpublished, 1933, 1934. typescript. D57/26, D57/27, D63/1.

398 *It is the German spirit* Anon. Germanischer and jüdischer Geist in der Wissenschaft. *Jüdische Rundschau*, 28 November 1933, XXXVIII, 867.

398 *Although not having yet* Macrae, Norman. *John von Neumann* (New York, Pantheon Books, 1992), 166–8.

399 *'I am beginning to weary'* Flexner, Abraham. Chronological files: 1933 cards. *Institute for Advanced Study* [Online]. 13 November 1933 [Cited: 5 December 2022], 268. https://albert.ias.edu/handle/20.500.12111/2507.

399 *Einstein probably shook his head* Einstein, Albert, Einstein, Elsa, Flexner, Abraham. *AEA* [Online]. 14–15 November 1933. Cat. nos. 38–55 through 38–61. http://alberteinstein.info/.

400 *Surely part of the reason* Sayen, Jamie. *Einstein in America* (New York, Crown Publishers, Inc., 1985), 65–6. See also Chronological files: 1933 cards, 276–7. *Institute of Advanced Study* [Online]. https://albert.ias.edu/handle/20.500.12111/2507.

400 *They were invited to spend* Rosenthau Sr, Henry, Einstein, Albert, Roosevelt, Eleanor. *AEA* [Online]. 1933–1934. Cat. nos. 33–69, 33–75, 33–78, 33–79, 33–81. http://alberteinstein.info/.

400 *Einstein concluded the draft* Einstein, Albert. *AEA* [Online]. November–December 1933. Cat. nos. 38–62, 38–72, 38–73, 38–75, 38–76. http://alberteinstein.info/.

401 *Despite desultory sniping* Flexner, Abraham. Chronological files: 1933 cards. *Institute for Advanced Study* [Online]. 11 December 1933 [Cited: 5 December 2022], 286. https://albert.ias.edu/handle/20.500.12111/2507.

401 *Recovering his strength* Hartog, Mabel. *Philip Hartog* (London, Constable, 1949), 127.

401 *'Magnes himself proposed'* Weizmann, Chaim. *CHAIM*, Vol. XV, 319–20, Letter 292. See also *AEA* [Online]. 17 May 1934. Cat. no. 39–571. http://alberteinstein.info/.

401 *Magnes apparently accused* Weizmann, Chaim. *CHAIM*, Vol. XV, 322, Note 3 to Letter 295; 322–5, Letter 296.

402 *'Ironically, Einstein, for whom'* Rosenkranz, Ze'ev. Albert Einstein's Involvement in the Affairs of the Hebrew University, 1919–1935. *Proceedings of the World Congress of Jewish Studies: Division B: The History of the Jewish People*, Vol. III: *Modern Times* (Jerusalem: World Union of Jewish Studies, 1993), 232. Available at https://www.jstor.org/stable/23536848.

402 *Friends of long standing* Weizmann, Chaim. *CHAIM*, Vol. XVI, 28–9, Letter 25; 36–7, Letter 32; 44–5, Letter 38.

402 *'During a speech'* Pais, Abraham. *LORD*, 454.

402 *There is no physics involved* Einstein, Albert, Mayer, Walter. Darstellung der Semi-Vektoren als gewöhnliche Vektoren von besonderem Differentiations Charakter. *Annals of Mathematics*, 1934, 35: 104–10.

403 *'Nobel's soul might have been'* Anon. Einstein Eulogizes Nobel as Idealist. *The New York Times*, 19 December 1933, p.18.

403 *Einstein was reported to* —. Einstein Attends Choir Recital at Princeton. *Paterson Evening News*, 14 December 1933, p. 3.

404 *The plan was eventually* Wakin, Daniel. The Maestro and the Money. *The New York Times*, 16 January 2009, CY9.

404 *The remainder of his recital* Anon. *The New York Times*, 5 November 1933, 82. See also *The New York Times*, 27 October 1933, 16; 28 October 1933, 20; 10 December 1933, 87.

404 *'What a splendid man'* Einstein, Elsa. *AEA* [Online]. 21 December 1933. Cat. no. 89–277. http://alberteinstein.info/.

405 *'It wasn't until 1 October'* Auner, Joseph. *A Schoenberg Reader* (New Haven, Yale University Press, 2011), 244. See also http://schoenblog.com/wp-content/uploads/2021/09/Screen-Shot-2021-09-18-at-11.28.55-PM.jpg.

405 *They finally met* McDonald, Malcolm. *Schoenberg*, 2nd edn. (Oxford, Oxford University Press, 2008), 74–5.

405 *Godowsky was enchanted* Strachstein, Harriet. Godowski's Pet Enthusiasm is Purely, Simply – Einstein. *Jewish Daily Bulletin*, 30 March 1934, 6.

405 *'My husband played wonderfully'* Einstein, Elsa. *AEA* [Online]. 9 January 1934. Cat. no. 50–355. http://alberteinstein.info/.

405 *On the third ring* Burton, Humphrey. *Menuhin: A Life* (London, Faber and Faber, 2000), 150–2.

407 *Menuhin also played* Downes, Olin. Menuhin Soloist at Philharmonic. *The New York Times*, 19 January 1934, 25. See also *The New York Times*, 19 March 1934, 12. Einstein, Albert, Einstein, Elsa, Menuhin, Yehudi *AEA* [Online]. March 1934. Cat. nos. 95–812, 95–815, 34–366, 34–367. http://alberteinstein.info/.

Chapter 26

408 *My husband is completely comfortable* Einstein, Elsa. *AEA* [Online]. 3 November 1933. Cat. no. 50–338. http://alberteinstein.info/.

408 *With typical tactlessness* —. *AEA* [Online]. 30 May 1933. 50–341. http://alberteinstein.info/.

409 *Hilb defended Knabe* Einstein, Elsa, Hilb, Emil. *AEA* [Online]. November 1933. Cat. nos. 50–340, 38–63. http://alberteinstein.info/.

409 *This fiery exchange however* Flexner, Abraham. *AEA* [Online]. 4 December 1933. Cat. no. 38–64. http://alberteinstein.info/.

409 *Hilb wrote to her* Hilb, Emil. *AEA* [Online]. December 1933. Cat. nos. 50–343, 50–345. http://alberteinstein.info/.

409 *Einstein had been exerting* Anon. Lehmann Is Hailed at Maccabee Fete; Dr Kayser Is Invited Here. *The New York Times*, 24 December 1933, 15, 44.

410 *'Professor Albert Einstein'* Hilb, Emil. *AEA* [Online]. 12 December 1933. Cat. no. 50–346. http://alberteinstein.info/.

410 *'Tonight we should have gone'* Einstein, Elsa, Hilb, Emil. *AEA* [Online]. December 1933, January 1934. Cat. nos. 50–349 through 50–357. http://alberteinstein.info/.

411 *This reference relates* Gwynne, Helen. Yesterday in New York. *The Hollywood Reporter*, 18 January 1934, 3.

411 *The rehearsal for the concert* Einstein, Elsa. *AEA* [Online]. 12 January 1934. 50–357. http://alberteinstein.info/. See also the *New York Times*, January 15th, p. 6; January 16th, p. 23.

411 The New York Times *reported* Anon. Einstein in Debut as Violinist here. *The New York Times*, 18 January 1934, 23.

412 *'I think that it can be said'* Salpeter, Harry. The Human Touch. *Jewish Daily Bulletin*, 21 January 1934, 6.

413 *'Now to the Movies'* Einstein, Elsa. *AEA* [Online]. 19 January 1934. 50–364. http://alberteinstein.info/.

414 *Irrespective of arguments* Einstein, Elsa, Hilb, Emil. *AEA* [Online]. January 1934. Cat. nos. 50–363, 50–367 through 50–369. http://alberteinstein.info/.

414 *Einstein sent the following poem* Einstein, Elsa, Einstein, Albert. *AEA* [Online]. January 1934. Cat. nos. 50–370, 32–371. http://alberteinstein.info/.

414 *'Seldom in my life'* Einstein, Albert. *AEA* [Online]. 20 February 1934. Cat. no. 32–383. http://alberteinstein.info/.

414 *Einstein promised to write* Anon. *Jewish Daily Bulletin*, 16 February 1934, 3.

414 *On 19 February* Anon. 12000 at Jewish Pageant. *The New York Times*, 20 February 1934, 25.

415 *They had lunch at Library Place* —. *Monmouth Democrat*, 1 March 1934, 6.

415 *He concluded the lecture* Spies, Claudio. Vortrag / 12 T K / Princeton. *Perspectives of New Music*, 1974, 13: 58–136.

415 *'. . . if Mozart could hear this'* McDonald, Malcolm. *Schoenberg*, 2nd edn. (Oxford, Oxford University Press, 2008), 75. See also Hilb, Emil, Einstein, Elsa. *AEA* [Online]. 9 and 11 March 1934. Cat. nos. 50–387, 50–389. http://alberteinstein.info/.

415 *The* Jewish Daily Bulletin *took* Anon. *Jewish Daily Bulletin*, 5 March 1934, 1, 8.

415 *While there is no record* Stern, Beatrice. *A History of the Institute of Advanced Study, 1930–1950* (Princeton: unpublished typescript, 1964), 182. Available from https://albert.ias.edu/handle/20.500.12111/110.

415 *Elsa Hilger's Pietro Guarnerius* Anon. A Tribute to Elsa Hilger. *The Cello Museum* [Online]. [Cited: 05 December 2022]. https://cellomuseum.org/a-tribute-to-elsa-hilger/.

416 *As drily remarked* Gribbin, John. *Erwin Schrödinger and the Quantum Revolution* (Hoboken, John Wiley & Sons, 2012), 175.

416 *'He [Schrödinger] disliked college'* Born, Max. *My Life* (New York, Charles Scribners Sons, 1978), 270.

416 *Thus, Flexner's reaction* Flexner, Abraham. Biographical files: S cards. *Institute for Advanced Study* [Online]. 1934–1935 [Cited: 30 December 2022], 40–51. https://albert.ias.edu/handle/20.500.12111/2826.

416 *Einstein does not seem to* —. Institute of Advanced Study Regular Meeting Minutes. *Institute for Advanced Study*. [Online]. 29 January 1934 [Cited: 30 December 2022]. https://albert.ias.edu/

handle/20.500.12111/2447. See also Prof. Dirac Appointed. *The New York Times*, 16 March 1934, 19.

416 *Despite Dirac's office* Farmelo, Graham. *The Strangest Man* (London, Faber & Faber, 2009), 254–9.

417 *Watters put his chauffeur-driven limousine* Watters, Leon. Comments on the 'Letters of Professor and Mrs Albert Einstein to Dr Leon L. Watters'. The Leon L. Watters Collection, California Institute of Technology Archives.

417 *She seems to have spent* Anon. Dinner to the Einsteins. *The New York Times*, 6 March 1934, 27. See also the *Jewish Daily Bulletin*, 15 March 1934, 2.

417 *He gave a speech* Anon. *Jewish Daily Bulletin*. 20 March 1934, 1.

417 *The concert ended* Anon. Music Notes. *The New York Times*, 26 February 1934, 20. See also *The New York Times*, 28 February 1934, 23.

417 *Einstein chased after* —. Einsteins Hailed at Jersey Fetes. *The New York Times*, 26 March 1934, 8.

418 *'If at all possible'* Born, Max. *The Born–Einstein Letters*, [trans.] Irene Born (London, Macmillan, 1971), 121–2. See also Einstein, Elsa. *AEA* [Online]. 24 February 1934. Cat. no. 37–693; 22 April 1934. Cat. no. 37–699. http://alberteinstein.info/.

419 *'His long devotion'* Anon. Einstein Cancels his Trip Abroad. *The New York Times*, 2 April 1934, 9. See also the *Jewish Daily Bulletin*, 27 March 1934, 7; Carnegie Hall - https://www.carnegiehall.org/About/History/Performance-History-Search?q=&dex=prod_PHS&event=18145&start=-1128279600.

419 *Another bizarre occurrence* Downes, Olin. AEA [Online]. 1 April 1934. Cat. no. 34–355. http://alberteinstein.info/.

421 *'One might almost say'* Goldschmidt, Alfons. In Einstein, Music and Mathematics Blend. *The New York Times*, 1 April 1934, 18.

421 *'It seems to me that'* Anon. Musical Activities. *The New York Times*, 15 April 1934, 156.

421 *He considered that this* —. Einstein Greeted in Jersey Capitol. *The New York Times*, 11 April 1934, 23.

421 *In his view* Einstein, Albert. *Einstein on Peace*, Heinz Norden, Otto Nathan, eds. (New York, Schoken, 1968), 252–3.

421 *However, this message did not* Einstein, Albert, Einstein, Elsa. *AEA* [Online]. April 1934. Cat. nos. 37–694, 37–699. http://alberteinstein.info/.

422 *This turned out to be fortunate* Infeld, Leopold. *Quest* (New York, Chelsea, 1980), 10, 206–17. See also Einstein, Albert. *AEA* [Online]. 5 May 1934. Cat. no. 39–568; 10 June 1934. Cat. no. 39–575. http://alberteinstein.info/.

422 *She sailed on the luxurious* Anon. Shipping and Mails. *The New York Times*. 20 May 1934, p. 76. See also Elsa Einstein, *AEA* [Online]. May 1934. 38-478. http://alberteinstein.info/.

422 *Einstein replied that* Watters, Leon. Comments on the 'Letters of Professor and Mrs Albert Einstein to Dr Leon L. Watters', 16. The Watters Papers, The Archives, California Institute of Technology.

422 *The cause of death* Einstein, Albert. *Seelenverwandte*, Robert Schulmann, ed. (Zurich, Verlag Neue Zürcher Zeitung, 2012), 541–4, Document 353 and Note 1; Document 354. See also Einstein, Elsa. *AEA* [Online]. 10 January 1935. Cat. no. 37–701. http://alberteinstein.info/.

423 *After a heart-rending farewell* Anon. SS Pennland Passenger List. *GG Archives* [Online]. 24 August 1934 [Cited: 11 January 2023]. https://www.ggarchives.com/OceanTravel/Passengers/RedStarLine/Pennland-PassengerList-1934-08-24.html. See also Einstein, Albert, Einstein, Elsa. *AEA* [Online]. July–August 1934. Cat. nos. 143–277, 83–174, 75–243. http://alberteinstein info/.

423 *Elsa recounted to Hilb* Einstein, Elsa, Dukas, Helen. *AEA* [Online]. April, June 1934. Cat. nos. 50–395, 50–398. http://alberteinstein.info/.

423 *Perhaps if he had owned* Einstein, Albert. *AEA* [Online]. 22 July 1934. Cat. no. 76–817. http://alberteinstein.info/.

423 *Einstein inscribed a photograph* Einstein, Elsa. *AEA* [Online]. 13 September 1934. Cat. no. 89–281. http://alberteinstein.info/. See also https://tarisio.com/cozio-archive/cozio-carteggio/historic-women-performers-leonora-jackson-mckim/.

424 *As his vacation wore on* Watters, Leon. Comments on the 'Letters of Professor and Mrs. Albert Einstein to Dr. Leon L. Watters'. 21–24. The Watters Papers, The Archives, California Institute of Technology. See also *Los Angeles Times*, 10 June 1943, 21.

424 *Einstein however refused* Brian, Denis. *Einstein: A Life* (New York, J. Wiley, 1996), 260–3. See also *Los Angeles Times*, 10 June 1934, 21.

424 *However, he was clearly suffering* Rogger, Franziska. *ES*, 125.

424 *In August, the faithful Zangger* Einstein, Albert, Einstein, Eduard, Zangger, Heinrich. *AEA* [Online]. June–August 1934. 143–266 through 143–274, 143–278, 143–279, 144–533, 75–989, 75–966, 87–114. http://alberteinstein.info/.

425 *'One of the most remarkable'* Anon. Scientists Mourne Mme. Curie's Death. *The New York Times*, 5 July 1934, 17. See also *The Los Angeles Times*, 5 July 1934, 3.

425 *The three Einsteins returned* Einstein, Elsa. *AEA* [Online]. 10 January 1935. 37–701. http://alberteinstein.info/. See also *The New York Times*, 9 October 1934, 21.

Chapter 27

426 *Supported by Einstein* Pais, Abraham. *LORD*, 494. See also Dirac, Paul, Fock, Vladmir, Podolsky, Boris. On Quantum Electrodynamics. *Physikalische Zeitschrift der Sowjetunion*, 1932, 2: 468–79; and Rosen, Nathan. The Normal State of the Hydrogen Molecule. *Physical Review*, 1931, 38: 2099–2114.

427 *By considering the collisions* Einstein, Albert. An Elementary Proof of the Theorem Concerning the Equivalence of Mass and Energy. *Bulletin of the American Mathematical Society*, 1935, 41: 223–30.

427 *Einstein attended a dinner* Topper, David, Vincent, Dwight. Einstein's 1934 Two-blackboard Derivation of Energy–Mass Equivalence. *American Journal of Physics*, 2007, 75: 978–83.

427 *The exact genesis* Peres, A. Einstein, Podolsky, Rosen, and Shannon. *ArXiv Quantum Physics* [Online]. 2 October 2003. https://arxiv.org/abs/quant-ph/0310010.

428 *'One could object to this'* Einstein, Albert, Podolsky, Boris, Rosen, Nathan. Can Quantum-Mechanical Description of Physical Reality Be Considered Complete? *Physical Review*, 1935, 47: 777–80.

429 *He and Einstein never collaborated* Anon. Einstein Attacks Quantum Theory. *The New York Times*, 4 May 1935, 11. See also *The New York Times*, 7 May 1935, 21.

429 *Here he demonstrated that* Einstein, Albert. *AEA* [Online]. 19 June 1935. Cat. no. 124–816. http://alberteinstein.info/.

429 *In his reply on 19 August* Einstein, Albert, Schrödinger, Erwin. *AEA* [Online]. July–August 1935. Cat. nos. 22–48 through 22–51. http://alberteinstein.info/.

429 *This triad of concepts* Schrödinger, Erwin. Die Gegenwärtige Situation in der Quantenmechanik. *Die Naturwissenschaften*, 1935, 23: 807–12; 824–8; 844–9.

429 *A discussion of the genesis* Fine, Arthur. The Einstein–Podolsky–Rosen Argument in Quantum Theory. Stanford Encyclopedia of Philosophy [Online]. https://plato.stanford.edu/entries/qt-epr/.

430 *He submitted a paper* Bohr, Niels. Can Quantum-Mechanical Description of Physical Reality Be Considered Complete? *Physical Review*, 1935, 48: 696–702. See also Bohr, Niels. *Nature*, 1935, 136: 65.

430 *Schrödinger, in a letter* Fine, Arthur. The Einstein–Podolsky–Rosen Argument in Quantum Theory. *Stanford Encyclopedia of Philosophy* [Online]. https://plato.stanford.edu/entries/qt-epr/.

430 *'I am now having fun'* —. *The Shaky Game* (Chicago, Chicago University Press, 1996), 64–85.

430 *'Dirac came here after'* Bohr, Niels. Oral History Interviews, Niels Bohr, Session V. *American Institute of Physics* [Online]. 17 November 1962. https://www.aip.org/history-programs/niels-bohrlibrary/oral-histories/4517-5. See also Mehra, Jagdish. *The Solvay Conferences on Physics* (Dordrecht, D. Reidel, 2012), xvii–xviii.

431 *It is therefore unsurprising* Pais, Abraham. *LORD*, 456. See also Bohm, David *Quantum Theory* (New York, Dover, 1989), 583–624; Bohm, David, Aharanov, Yakir. *Physical Review*, 1957, 108: 1070–6; Bell, John *Physics*, 1964, 1: 195–290.

431 *'Spooky action at a distance'* Einstein, Albert. Autobiographical Notes; Reply to Criticisms. In *Albert Einstein: Philosopher-Scientist*, Vol. VII, Paul Arthur Schlipp, ed. (Evanston, Open Court, Library of Living Philosophers, 1947), 87, 682. See also Einstein, Albert. *AEA* [Online]. 3 December 1947. Cat. no. 8–210. http://alberteinstein.info/. For a contrary opinion, see Bell, J. S. Einstein Podolsky Rosen Experiments. CERN, 1976. https://cds.cern.ch/record/610098/files/cer-002369328.pdf.

431 *Einstein however was as busy* Hilb, Emil, Einstein, Elsa. *AEA* [Online]. January 1935. Cat. nos. 50–404, 50–405, 50–412. http://alberteinstein.info/.

432 *On the evening of 12 March* Anon. List or Manifest of Alien Passengers. *New York Passenger Lists, 1909, 1927–1957*. 6 March 1935. See also *The New York Times*, 12 March 1935, 18.

432 *No doubt he clarified that* —. About People. *The Detroit Jewish Chronicle*, 19 April 1935, 5. See also *Jewish Daily Bulletin*, 24 May 1935, 4; Leo Baeck Institute, https://www.lbi.org/griffinger/record/213935.

432 *His muse deserting him* —. Einstein is Silent as He Gets Medal. *The New York Times*, 16 May 1935, 1, 21. See also *The New York Times*, 14 May 1935, 19.

432 *Marianoff's description* Marianoff, Dimitri, Wayne, Palma. *Einstein: An Intimate Study of a Great Man* (New York: Doubleday, Doran and Co., Inc., 1944), 144–5. See also New York Passenger Lists, 1909, 1925–1957.

433 *Her attitude reinforced Einstein's* Einstein, Elsa. *AEA* [Online]. 1934-1935. 83–175, 37–701, 83–176. http://alberteinstein.info/.

433 *The possibility thereby inherent* Einstein, Albert, Rosen, Nathan. The Particle Problem in the General Theory of Relativity. *Physical Review*, 1935, 48: 73–7.

434 *'Finally, in 1935'* Lemaître, Georges. Recontres avec Einstein. *Revue de Questions Scientifiques: Actualité, histoire et philosophie des sciences*, 1958, 129: 129–32. See also Silberstein, Ludwik. *Physical Review*, 1936, 49: 268–70; Einstein, Albert, Rosen, Nathan. *Physical Review*, 1936, 49: 404–5.

434 *The Einstein party stayed* Anon. History: When Albert Einstein Called Bermuda Home. *ForeverBermuda.com* [Online]. 13 November 2019. See also *The New York Times*, 26 May 1935, 3; 28 May 1935, 27. http://www.foreverbermuda.com/history-when-albert-einstein-einstein-called-bermuda-home/.

435 *'Bermuda was unsurpassably beautiful'* Einstein, Elsa. *AEA* [Online]. 8 June 1935. Cat. no. 38–490. http://alberteinstein.info/.

435 *'... [would come in] to the music room'* Brian, Denis. *Einstein: A Life* (New York, John Wiley, 1996), 280. See also *The New York Times*, 4 August 1935, 1.

435 *Meier sailed from Cherbourg* Einstein, Albert. *AEA* [Online]. 5 June 1935. Cat. no. 51–415. http://alberteinstein.info/. See also New York Passenger Lists, 1909, 1925–1957.

435 *In 1936 he married Eva Urgiss* Anon. Rudolf Kayer, German Scholar. *The New York Times*, 7 February 1964, 32. See also Einstein, Elsa. *AEA* [Online]. 24 April 1936. Cat. no. 37–703. http://alberteinstein.info/ and New York Passenger Lists, 1909, 1925–1957.

435 *Schwarz replied enthusiastically* Einstein, Elsa. AEA [Online]. 1935–1936. Cat. nos. 123–673, 98–248. http://alberteinstein.info/.

436 *However, Einstein did have* Anon. Francesko. *The Detroit Jewish Chronicle*, 20 September 1935, 5. See also *The New York Times*, 29 June 1935, 16; 4 October 1935, 25; 9 October 1935, 27; and the New York Passenger Lists, 1909, 1925–1957.

436 *The fact that this was written* Einstein, Elsa, Marianoff, Dimitri. *AEA* [Online]. 22 November 1936. Cat. nos. 83–181, 144–825. http://alberteinstein.info/.

436 *Reinhard's colleagues* Grosch, Nils. Albert Einstein und Der Weg der Verheissung. In *Musizieren, Leben – und Maulhalten!*, Ivana Rentsch, Anselm Gerhard, eds. (Zürich, Schwaber Verlag Basel, 2006), 66.

437 *Another visitor in September* Einstein, Elsa, Einstein, Albert, et al. *AEA* [Online]. June–July 1935. Cat. nos. 38–493, 38–494, 35–431, 35–455 through 35–461, 51–590, 49–658. http://alberteinstein.info/.

437 *Such sentiments would have* Anon. Pirandello backs war with Ethiopia. *The New York Times*, 21 July 1935, 21. See also Woolf, Samuel. *AEA* [Online]. 18 July 1935. Cat. no. 52–325. http://alberteinstein.info/.

437 *For lighter reading* Einstein, Albert, Chase, Stuart. *AEA* [Online]. July–September 1935. Cat. nos. 50–689, 50–688, 49–677, 36–537, 49–393, 49–394. http://alberteinstein.info/.

437 *She hoped that Einstein* Busch, Frieda. *AEA* [Online]. 14 July 1935. Cat. no. 34–339. http://alberteinstein.info/.

438 *Although Austria, as Einstein feared* Einstein, Albert. *AEA* [Online]. 5 July 1935. Cat. no. 70–789. http://alberteinstein.info/.

438 *These were required* —. *AEA* [Online]. 1935–1936. Cat. nos. 29–418, 29–420, 97–237. http://alberteinstein.info/.

438 *Elsa and Einstein were to remain* Einstein, Elsa. *AEA* [Online]. 1934–1935. Cat. nos. 38–485, 37–703. http://alberteinstein.info/.

439 *She returned to Princeton* Einstein, Elsa, Einstein, Albert. *AEA* [Online]. June–September 1935. Cat. nos. 83–180, 50–188, 52–205. http://alberteinstein.info/. See also Watters, Leon. Comments on the 'Letters of Professor and Mrs Albert Einstein to Dr Leon L. Watters'. The Watters Papers, The Archives, California Institute of Technology.

439 *As Elsa was supervising* Einstein, Elsa. *AEA* [Online]. 22 November 1935. 83–181. http://alberteinstein.info/.

439 *In November, similar remarks* Anon. Einstein and Smith Ask Aid for Exiles. *The New York Times*, 23 October 1935, 22. See also *The New York Times*, 25 November 1935, 9.

439 *Von Laue therefore stayed* Flexner, Abraham. Abraham Flexner and Richard Courant correspondence. *Institute for Advanced Study*. 1934–1936. Flexner to Courant, 23 September 1935. See also Einstein, Elsa. *AEA* [Online]. 21 October 1935. Cat. no. 16–108. http://alberteinstein.info.

439 *Pauli bought a new car* Enz, Charles. *No Time to be Brief* (Oxford, Oxford University Press, 2002), 294–5. See also Pauli, Wolfgang, Rose, M. E. *Physical Review*, 1936, 49: 462–5.

439 *Such a predilection* Pauli, Wolfgang. heisenberg_0216. *Pauli Letter Collection* [Online]. 9 June 1936. http://cds.cern.ch/record/84998/files/heisenberg_0216.pdf.

440 *Von Ossietzky died in 1938* Einstein, Albert. *Einstein on Peace*, Heinz Norden, Otto Nathan, eds. (New York, Schoken, 1968), 264–7.

440 *She was moved by Einstein's* Einstein, Elsa. *AEA* [Online]. January 1936. Cat. nos. 49–747, 83–183. http://alberteinstein.info/. See also *The New York Times*, 18 December 1935, 28; and 19 December 1935, 30.

440 *It may be that he waited* Einstein, Albert. Declaration of Intention for Albert Einstein. *Library of Congress* [Online]. 15 January 1936 [Cited: 29 May 2023]. https://www.loc.gov/item/2021667583/. See also *The New York Times*, 16 January 1936, 23; and 1 October 1940, 5.

441 *He offered to come* Huberman, Bronislaw. *AEA* [Online]. 1 February 1936. Cat. no. 34–358. http://alberteinstein.info/. See also https://www.porta-polonica.de/en/atlas-of-remembrance-places/bronis%C5%82aw-huberman.

441 *It is estimated that* Einstein, Albert. *AEA* [Online]. February/March 1936. 80–792. http://alberteinstein.info/. See also *Jewish Telegraph Agency*, 28 December 1936, 1–2.

442 *Other speakers included* Anon. London Candle Ray Opens Museum Here. *The New York Times*, 12 February 1936, 1, 19.

442 *'Mozart', he mused,* Einstein, Albert. *AEA* [Online]. 20 March 1936. Cat. no. 32–387. http://alberteinstein.info/.

443 *Einstein subsequently wrote to Freud* —. *AEA* [Online]. April 1936. Cat. nos. 75–941, 32–565. http://alberteinstein.info/.

443 *Electroshock therapy* Huonker, Thomas. *Diagnose: 'moralisch defekt'* (Zurich, Orell Füssli Verlag AG, 2003). pp. 221–24. See also Einstein, Albert. *AEA* [Online]. 11 January 1935, 75-964. http://alberteinstein.info/.

443 *Intrigued but not competent* Seigfried, Tom. The Amateur Who Helped Einstein See the Light. *Science News* [Online]. 1 October 2015 [Cited: 18 June 2023]. https://www.sciencenews.org/blog/context/amateur-who-helped-einstein-see-light.

443 *He concluded that* Einstein, Albert. Lens-like Action of a Star by the Deviation of Light in the Gravitational Field. *Science*, 1936, 84, pp. 506–7.

443 *Gravitational lensing observations* Renn, Jürgen, Sauer, Tilman. Eclipses of the Stars. In *Revisiting the Foundations of Relativistic Physics – Festschrift in Honour of John Stachel*, Jürgen Renn, Lindy Divarci, Petra Schröter, Abhay Ashtekar, Robert S. Cohen, Don Howard, et al., eds. (Dordrecht, Springer, 2003), 69–92. https://www.mpiwg-berlin.mpg.de/Preprints/P160.PDF.

444 *Convinced that he had made* Einstein, Albert. *AEA* [Online]. July–August 1936. Cat. nos. 18–55, 16–112. http://alberteinstein.info/.

444 *Einstein never submitted* —. *AEA* [Online]. 7 July 1936. Cat. no. 19–86. http://alberteinstein.info/.

444 *'If you ask me'* Infeld, Leopold. *Quest* (New York, Chelsea Publishing Company, 1980), 254–69.

445 *He was even more disgruntled* Einstein, Albert, Rosen, Nathan. On Gravitational Waves. *Journal of the Franklin Institute*, 1937, 223: 43–54.

445 *At least, however* Kennefick, Daniel. *Travelling at the Speed of Thought* (Princeton: Princeton University Press, 2007), 81–99.

445 *The advice must have been good* Einstein, Elsa. *AEA* [Online]. July–September 1936. Cat. nos. 89–285, 89–286. http://alberteinstein.info/.

445 *He demonstrated by giving* Anon. *Kingston Daily Freeman*, 3 July 1936, 10. See also *The New York Times*, 7 August 1936, 9.
446 *'I think you will be'* Einstein, Elsa. *AEA* [Online]. 10 September 1936. Cat. no. 52–215. http://alberteinstein.info/.
446 *'But one thing is really nice'* Einstein, Albert. *AEA* [Online]. 26 September 1936. Cat. no. 11–515. http://alberteinstein.info/.
446 *Presumably Schwarz and Einstein* Schwarz, Boris. Musical and Personal Reminiscences of Albert Einstein. In *Albert Einstein, Historical and Cultural Perspectives: The Centennial Symposium in Jerusalem*, Yehuda Elkana, Gerald Holton, eds. (Princeton: Princeton University Press, 1982), 414.
446 *On 20 December 1936* Einstein, Albert. *AEA* [Online]. December 1936. Cat. nos. 29–168, 65–603, 97–238. http://alberteinstein.info/. See also Infeld, Leopold. *Quest* (New York, Chelsea, 1980), 244.

Chapter 28

447 *'About 14 days ago'* Einstein, Albert. *AEA* [Online]. 5 January 1937. Cat. nos. 75–926, 97–238. http://alberteinstein.info/.
448 *'Life in a small circle'* —. *AEA* [Online]. 11 August 1937. Cat. no. 32–390. http://alberteinstein.info/.
448 *Bloch's Violin Concerto* Anon. Honour for Jewish Composer. *The Jewish Chronicle*, 19 November 1937, 48. See also *The New York Times*, 15 May 1938, 153. Cohen, Alex. *AEA* [Online]. 16 May 1936. Cat. no. 34–333. http://alberteinstein.info/.
449 *The first night on 7 January* Einstein, Elsa. *AEA* [Online]. 1936. 83–183,83–184,37–703. http://alberteinstein.info/. See also *The New York Times*, 8 January 1937, 14.
449 *By then Marianoff had disappeared* Einstein, Albert. *AEA* [Online]. August 1936. Cat. nos. 144–905, 144–916. http://alberteinstein.info/. See also *The New York Times*, 7 December 1937, 27.
449 *He was a regular visitor* Schwarz, Boris. Musical and Personal Reminiscences of Albert Einstein. In *Albert Einstein, Historical and Cultural Perspectives: The Centennial Symposium in Jerusalem*, Yehuda Elkana Gerald Holton, eds. (Princeton, Princeton University Press, 1982), 414–16. See also *The New York Times*, 8 April 1940, 19.
450 *Eisner found work initially* Eisner, Bruno. Gedenken und Gedanken aus des Leben eines Musikers. *The Leo Baeck Institute*, 136–7. See also Einstein, Albert. *AEA* [Online]. 26 September 1936. Cat. no. 99–453. http://alberteinstein.info/.
450 *Even more remarkably* Hoffman, Banesh, Dukas, Helen. *Einstein: Creator and Rebel* (London, Rupert Hart-Davis, 1973), 228–31.
451 *It was published* Infeld, Leopold. *Quest* (New York, Chelsea Publishing Company, 1980), 301–21.
451 *'Will the further development'* Einstein, Albert, Infeld, Leopold. *The Evolution of Physics* (New York, Simon & Schuster, 1938).
452 *He was offered a temporary appointment* Crowther, J. A. From Galileo to Einstein. *Nature*, 1938, 141: 891–2. See also Kaempffert, W. *The New York Times*, 10 April 1938, 86.
452 *'Nowadays I am greatly valued'* Einstein, Albert, Flexner, Abraham. *AEA* [Online]. 1937–1938. Cat. nos. 38–88, 38–90, 21–230. http://alberteinstein.info/.
453 *That October, the IAS* Sayen, Jamie. *Einstein in America* (New York, Crown Publishers, Inc., 1985), 92–6. See also *AEA* [Online]. Cat. nos. 53–763, 53–766, 38–94, 38–95. http://alberteinstein.info/.

453 *Whenever she performed nearby* Jerome, Fred, Taylor, Rodger. *Einstein on Race and Racism* (New Brunswick, Rutgers University Press, 2006), 42, 43, 57. See also Sanborn III, D. H. A Magnificent Voice. *Princetonmagazine.com* [Online]. https://www.princetonmagazine.com/a-magnificent-voice/; Anon. Albert Einstein Used His Fame to Speak Out against Racism. *iloveancestry.com* [Online]. https://tinyurl.com/36p7ue99; Anderson, M. *My Lord, What a Morning* (New York, Viking Press, 1956), 266.

454 *'I often think about our old days'* Einstein, Albert. *AEA* [Online]. 9 June 1937. Cat. no. 7–142. http://alberteinstein.info/.

455 *However, before everything could be finalised* Lonner, Thomas D. My Blood Strangers. *Palais des Beaux Arts* [Online]. 2018. https://www.palaisdesbeauxarts.at/collection/thomas-lonner/my-blood-strangers. See also *AEA* [Online]. Cat. nos. 28–390, 53–747 through 53–758; 87–19. http://alberteinstein.info/.

456 *'I was a friend of G. H. Hardy'* Snow, Charles Percival. *Variety of Men* (London, Macmillan, 1967), 84–8.

456 *'I live now as a lonely old man'* Einstein, Albert. *AEA* [Online]. 25 May 1937. Cat. no. 75–933. http://alberteinstein.info/.

457 *Hans Moos was Einstein's* Ettema, Robert, Mutel, Cornelia F. *Hans Albert Einstein* (Reston, ASCE Press, 2014), 94–8. See also *The New York Times*, 13 October 1937, 3; *AEA* [Online]. 1937–1938. Cat. nos. 75–926, 75–928, 75–939, 75–950, 88–341. http://alberteinstein.info/.

457 *Zangger also gave Einstein news* Einstein, Albert. *Seelenverwandte*, Robert Schulmann, ed. (Zurich, Verlag Neue Zürcher Zeitung, 2012), 548–64, 577, Documents 358–69; Notes to Document 379.

457 *To general laughter* Mann, Erica, Mann, Klaus. *Escape to Life* (Boston, Houghton-Mifflin, 1939), 258–64. See also Mann, Klaus. Tagebuch entry for 8 February 1938. https://www.monacensia-digital.de/mann/content/titleinfo/30161.

458 *'While urging continued effort'* Anon. Einstein is Guest of Jewish Stars. *The New York Times*, 7 April 1938, 19. See also 3,000 Hear Einstein at Seder Service. *The New York Times*, April 18 1938, 15.

458 *However, there is no indication* Neumann, Betty. Subgroup V: Other Family Members, undated, 1883–1982. *Muehsam Family Collection, Centre for Jewish History* [Online]. 1938–1944 [Cited: 17 October 2023]. Series 2J. See also Neumann, B. *AEA* [Online]. 1938–1939. Cat. nos. 54–103, 30–630, 54–107, 54–108. http://alberteinstein.info/. See also The Statue of Liberty – Ellis Island; Passenger Search. https://archives.cjh.org/repositories/5/archival_objects/842229.

458 *He criticised the isolationist* Davies, Lawrence. Einstein Criticizes 'Barbarity' Abroad. *The New York Times*, 7 June 1938, 16.

459 *This may have been related* Einstein, Bernhard. Recollections of My Grandfather, Albert Einstein. In *The Fascinating Life and Theory of Albert Einstein*, Walther C. Mih, ed. (New York, Kroshka Books, 2000), ix–x. See also *AEA* [Online]. June 1938. Cat. no. 29–425. http://alberteinstein.info/.

459 *In September he wrote once more* Einstein, Albert. *AEA* [Online]. June–September 1938. Cat. nos. 29–425, 97–242, 29–427. http://alberteinstein.info/.

459 *She wrote that she could* Winteler, Marie. *AEA* [Online]. 1938–1939. Cat. nos. 54–634, 54–635. http://alberteinstein.info/.

460 *Obviously discouraged* Schoenberg, Arnold, Einstein, Albert. *AEA* [Online]. 1938. Cat. nos. 52–461, 93–911. http://alberteinstein.info/. See also Tonietti, Tito M. Albert Einstein and Arnold Schoenberg Correspondence, *NTM Zeitschrift für Geschichte der Wissenschaften, Technik und Medizin*, 1997, 5: 1–22.

460 The New York Times *headlined this* Anon. Einstein Hopeful for Better World. *The New York Times*, 16 September 1938, 22. See also Einstein, Albert. *AEA* [Online]. 27 October 1938. Cat. no. 75–950. http://alberteinstein.info/.

460 *Unfortunately, the initiative* Barati, George. George Barati: A Life in Music. *escholarship* [Online] [Cited: 13 September 2025]. 37–40. https://escholarship.org/uc/item/5zs7n9hv.

460 *These evenings became regular* Loeb Jr, James, Einstein, Albert. *AEA* [Online]. 7 March 1937. Cat. nos. 34–344, 34–345. http://alberteinstein.info/.

461 *The level of her accomplishment* Anon. The Silent Drama. *The Sunday Oregonian*, 5 September 1920, 4:2.

461 *'I remember one evening'* Sayen, Jamie. *Einstein in America* (New York, Crown Publishers, Inc., 1985), 137.

461 *Despite this and his deep* Einstein, Albert. *AEA* [Online]. 19 November 1938. Cat. no. 34–391. http://alberteinstein.info/. See also Grosch, Nils. Albert Einstein und Der Weg der Verheissung. In *Musizieren, Leben – und Maulhalten!* (Basel, Schwaber Verlag, 2006), 75–7.

461 *In his letter, Einstein suggested* —. *AEA* [Online]. 14 December 1938. Cat. no. 29–428. http://alberteinstein.info/.

461 *Einstein immediately wrote* Einstein, Albert, Einstein, Hans Albert. *AEA* [Online]. January–February 1939. Cat. nos. 143–304.1, 144–86, 75–904. http://alberteinstein.info/.

462 *Within four months* Rogger, Franziska. *ES*, 140–2. See also *The New York Times*, 21 April 1939, 28.

462 *After the concert* Potter, Tully. *Adolf Busch*, Vol. I (London, Toccata Press, 2010), 694–5. See also *The New York Times*, 29 January 1939, 144.

462 *Although he loved Godowsky* Einstein, Albert. *AEA* [Online]. 1939. Cat. no. 34–322. http://alberteinstein.info/.

462 *'If you really like to fiddle'* —. *AEA* [Online]. 18 April 1939. Cat. no. 31–557. http://alberteinstein.info/.

463 *He concluded by hailing* von Laue, Max. *AEA* [Online]. 10 January 1939. Cat. no. 30–579. http://alberteinstein.info/.

463 *Einstein was not at home* Anon. Einstein at Sixty. *The New York Times*, 14 March 1939, 14, 16. See also *The New York Times*, 25 November 1939, 23.

464 *Bohr and Fermi discussed* Pais, Abraham. *Niels Bohr's Times* (Oxford, Oxford University Press, 1991), 453–6.

464 *After a few minutes* Anon. Nuclear Fission Announcement. *George Washington University: Libraries & Academic Innovation* [Online] [Cited: 21 August 2023]. https://library.gwu.edu/nuclear-fission-announcement. See also *The New York Times*, 31 January 1939, 18; Badash, Lawrence, Hodes, Elizabeth, Tiddens, Adolph. Nuclear Fission: Reaction to the Discovery in 1939. *Proceedings of the American Philosophical Society*, 1986, 130: 196–231.

464 *By June, Bohr and Wheeler* Bohr, Niels. Resonance in Uranium and Thorium Disintegrations and the Phenomenon of Nuclear Fission. *Physical Review*, 1939, 55: 418–19. See also Bohr, Niels, and Wheeler, John A. The Mechanism of Nuclear Fission. *Physical Review*, 1939, 56: 426–50; von Halban, Hans, Joliot, Frédéric, Kowarski, Lew. Number of Neutrons Liberated in the Nuclear Fission of Uranium. *Nature*, 1939, 143: 680; Anderson, Herbert L,. Fermi, Enrico, Szilard, Leo. Neutron Production and Absorption in Uranium. *Physical Review*, 1939, 56: 284–6.

465 *The atomic age was approaching* Oppenheimer, J. R. Celebration of the Sixtieth Birthday of Albert Einstein. *Science*, 1939, 89: 335–6.

465 *The most likely explanation* Garbedian, H. G. *Albert Einstein: Maker of Universes* (New York, Funk & Wagnalls, 1939). See also T*he New York Times*, 19 March 1939, 93.

465 *Why Einstein should choose* Anon. Einstein Will Mark 60th Year Tomorrow. *The New York Times*, 13 March 1939, 14.

465 *Einstein's careful phrasing* Laurence, W. L. Einstein Sees Key to Universe Near. *The New York Times*, 14 March 1939, 1, 15.

466 *'I urge my listeners'* Anon. Notables Attend Benefit Concert. *The New York Times*, 22 March 1939, 27. See also *The New York Times*, 22 March 1939, 10.

466 *The expert was not to know* Illy, Jozsef. *The Practical Einstein* (Baltimore, Johns Hopkins University Press, 2012), 131–3. See also *The New York Times*, 23 March 1939, 22.

466 *He liked it because he thought* Rubinstein, Arthur. *Mein Gluckliches Leben* (Frankfurt am Main, S. Fischer Verlag, 1980), 587–9. See also https://tinyurl.com/47tx37cf.

466 *Although several of the speakers* Kuznick, P. J. Losing the World of Tomorrow: The Battle Over the Presentation of Science at the 1939 New York World's Fair. *American Quarterly*, 1994, 46: 341–73. See also *Jewish Telegraph Agency News*, 29 May 1939, 1–3; *The New York Times*, 1 May 1939, 6; 29 May 1939, 7; 10 July 1939, 9.

467 *In addition to the distractions* Anon. Dr. Einstein Visits the Fair. *The New York Times*, 12 June 1939, 10. See also The Fair Today. *The New York Times*, 21 June 1939, 20; 2 July 1939, 13.

Chapter 29

468 *Here Fermi took up* Szilard, Leo. *Reminiscences*,K. R. Windsor, G. W. Szilard, eds. (Cambridge, MA, Harvard University Press, 1969), 105–9. See also Pais, Abraham. *Niels Bohr's Times* (Oxford, Oxford University Press, 1991), 453–4.

468 *Nevertheless, he was able to convince* Wigner, Eugene Paul. *The Collected Works of Eugene Paul Wigner*, Vol. 5 (Berlin, Springer, 1992), 23–8. See also Strauss, Lewis. *Men and Decisions* (New York, Doubleday, 1962), 171–5.

469 *It was in fact the possibility* Badash, L., Hodes, E., Tiddens, A. Nuclear Fission: Reaction to the Discovery in 1939. *Proceedings of the American Philosophical Society*, 1986, 130: 196–231. See also Ahern, J. J. *Historical Studies in the Physical and Biological Sciences*, 2003, 33: 217–36.

469 *Teller agreed that the President* Szilard, Leo. *AEA* [Online]. 19 July 1939. Cat. no. 39–461. http://alberteinstein.info/.

470 *He gave both to Sachs* Rhodes, Richard. *The Making of the Atomic Bomb* (New York, Simon & Schuster, 1986), 296–308.

470 *When they arrived in Peconic* Ettema, Robert, Mutel, Cornelia F. *Hans Albert Einstein* (Reston, ASCE Press, 2014), 114–15.

470 *He ran a local orchestra* Bucky, Peter. *The Private Albert Einstein* (Kansas City, Andrews and McMeel, 1992), 66–8.

471 *'You know, they are very talented'* Britten, Benjamin. *Letters from a Life: The Selected Letters and Diaries of Benjamin Britten 1913–1976*, Donald Mitchell, ed. (London, Faber & Faber, 1991), 723–7. Particularly Note 4. See also Palmer, Tony. Benjamin Britten – a Time There Was. https://www.youtube.com/watch?v=ECQqWrn1-i4 (28:30 – 32.00); and White, P. Albert Einstein – the Violinist. *The Physics Teacher*, 2005, 43: 287–8.

471 *Rothman's memories are* Rattiner, Dan. *In the Hamptons Too* (New York, State University of New York Press, 2014), 182–3.

472 *'Alex, what you are after'* Rhodes, Richard. *The Making of the Atomic Bomb* (New York, Simon & Schuster, 1986), 312–15.

472 *If he was satisfied* Schweber, Silvan. Einstein and Nuclear Weapons. In *Einstein for the 21st Century,*Gerald Holton, Silvan S. Schweber, Peter L. Galison, eds. (Princeton, Princeton University Press, 2008), 76–7.

473 *A briefing for this committee* Sachs, Alexander. *AEA* [Online]. December 1936. Cat. nos. 39–488.2 through 39–488.20. http://alberteinstein.info/.

473 *Debye's position* van Ginkel, Gijs. Prof. Peter J. W. Debye (1884–1966) in 1935–1945. Radboud Universiteit [Online]. 2006 [Cited: November 20, 2023]. https://www.theochem.ru.nl/~pwormer/Historical%20sources%20Debye%201935-1945.pdf.

474 *In ten minutes, Bush was leaving* Rhodes, Richard. *The Making of the Atomic Bomb* (New York, Simon & Schuster, 1986), 336–8.

474 *Lawrence was using cyclotrons* Conant, Jennet. *Tuxedo Park* (New York, Simon & Schuster, 2002), 157–69, 247.

474 *The army, presumably* Anon. Federal Bureau of Investigation: Albert Einstein, Part 1 of 9 (BUFILE NUMBER: 61-7099. FBI [Online]. [Cited: 20 November 2023]. 25–6. https://vault.fbi.gov/Albert%20Einstein/Albert%20Einstein%20Part%201%20of%2014/view.

475 *On the following day* Rhodes, Richard. *The Making of the Atomic Bomb* (New York, Simon & Schuster, 1986), 395–8.

475 *Although pressed by several physicists* Clark, Ronald. *Einstein: The Life and Times* (New York, Avon Books, 1972), 684–6.

475 *After 1944, however* Brunauer, Stephen. Einstein in the U.S. Navy. *Journal of the Washington Academy of Sciences*, 1979, 69: 108–13.

475 *Probably Gamow did take* Gamow, George. *My World Line* (New York, Viking Press, 1970), 148–51.

475 *Indeed, Bush either forgot* Einstein, Albert. *Einstein on Peace*, Heinz Norden, Otto Nathan, eds. (New York, Schoken, 1968), 664, Note 9.

476 *'At Princeton, my grandfather Albert'* Einstein, Bernhard. Recollections of My Grandfather, Albert Einstein. In *The Fascinating Life and Theory of Albert Einstein*, Walther C. Mih, ed. (New York, Kroshka Books, 2000), x.

476 *On 7 May* Potter, Tully. *Adolf Busch*, Vol. I (London, Toccata Press, 2010), 719.

476 *It would be nice to think* Anon. 17 Scientists Score Isolationist Stand. *The New York Times*, 21 May 1940, 12. See also *The New York Times*, 21 May 1940, 33; 26 May 1940, 147.

477 *However, its continued reliance* Einstein, Albert. *Considerations Concerning the Fundaments of Theoretical Physics. Science*, 1940, 91: 487–92. See also *The New York Times*, 13 May 1940, 16.

477 *'. . . the most beautiful cultural contribution'* Einstein, Albert. *AEA* [Online]. 22 September 1940 [Cited: 23 May 2024]. Cat. no. 28–527. http://alberteinstein.info/.

478 *'[Einstein] went through playing'* Sayen, Jamie *Einstein in America* (New York, Crown Publishers, Inc., 1985), 137. See also *Time*, 3 February 1941. https://time.com/archive/6764587/music-einstein-fiddles/.

478 *However, when Foldes pressed him* Foldes, Andor. *Errinerungen* (Berlin, Limes, 1993), 86–9. See also Foldes, Lili. Two on a Continent (New York, E.P. Dutton, 1947), 193–6.

479 *In January 1941* Safranek, Milos. *Bohuslav Martinu: The Man and his Music* (New York, Alfred A. Knopf, 1944), 2–80.

479 *It is likely that one* Rybka, F. James. *Bohuslav Martinů: The Compulsion to Compose* (Lanham, Scarecrow Press, 2011), 99–123. See also Safranek, M. *Bohuslav Martinu: The Man And His Music* (New York, Alfred A. Knopf, 1944), 97–101.

480 *Given Einstein's lack of interest* Rentsch, Ivana. Der Klang der 'inneren Harmonie'. *Opus Musicum*, 2005, 37: 4–9. See also Rentsch, Ivana. A Gift for Einstein. *Bohuslav Martinu Newsletter*, 2005, 5(3): 12–14.

480 *In concluding she sent* Winteler, Marie. *AEA* [Online]. 31 December 1939. Cat. no. 54–636. http://alberteinstein.info/.

481 *She spent her final years* —. *AEA* [Online]. 1940–1944. Cat. nos. 56–354 through 56–357. http://alberteinstein.info/.

481 *He also told Rothman* Einstein, Albert. *AEA* [Online]. 19 June 1944. Cat. no. 56–62. http://alberteinstein.info/.

481 *These are detailed* —. Letters Received from Albert Einstein, 17 May 1943–15 October 1944. *National Archives* [Online]. 1943–1944 [Cited: November 12, 2023]. ID 305247–305255. https://catalog.archives.gov/id/305246.

481 *On some matters* Dry, Sarah. *The Newton Papers* (New York, Oxford University Press, 2014), 161–73. See also Einstein, Albert. *AEA* [Online]. 1940–1944. Cat. nos. 39–607, 39–611, 39–614, 69–57, 69–58, 72–906, 93–300, 123–25, 123–590, 121–947 through 121–968. http://alberteinstein.info/.

482 *None of her letters* Brian, Dennis. *The Unexpected Einstein* (New York, John Wiley, 2005), 219–22. See also Pogrebin, R. *The New York Times*, 1 June 1998, 10; Einstein, Albert. *AEA* [Online]. 1943–1946. Cat. nos. 85–154 through 85–167. http://alberteinstein.info/.

482 *'I visited my grandfather'* Einstein, Bernhard. Recollections of My Grandfather, Albert Einstein. In *The Fascinating Life and Theory of Albert Einstein*, Walther C. Mih, ed. (New York, Kroshka Books, 2000), xi. See also Einstein, Frieda. *AEA* [Online]. 26 July 1942. Cat. nos. 75–813. http://alberteinstein.info/.

482 *Indeed, already in May* Einstein, Albert, Einstein, Mileva. AEA [Online]. 1940–1944 [Cited: 20 November 2023]. Cat. nos. 144–96,144–205,75–987,75–807. http://alberteinstein.info/.

483 *He eventually became a Swiss* Enz, Charles P. *No Time to be Brief* (Oxford, Oxford University Press, 2002), 338–61. See also https://etheritage.ethz.ch/2021/12/10/die-eth-als-eine-art-unterabteilung-der-fremdenpolizei-der-streit-zwischen-einem-nobelpreistraeger-und-der-eth/.

483 *This, although interesting* Einstein, Albert, Pauli, Wolfgang. On the Non-Existence of Regular Stationary Solutions of Relativistic Field Equations. *Annals of Mathematics*, 1943, 44: 131–7. See also C.P. Enz, Charles P. *No Time to be Brief* (Oxford, Oxford University Press, 2002), 388–9.

483 *Einstein was outraged* Brian, Dennis. *The Unexpected Einstein* (New York, John Wiley, 2005), 156. See also Einstein, Albert. *AEA* [Online]. 19 March 1940. Cat. no. 33–163. http://alberteinstein.info/.

484 *A condensed version of Einstein* Einstein, Albert, Kahler, Erich. *AEA* [Online]. 1944 [Cited: 17 November 2023]. Cat. nos. 38–266, 122–242. http://alberteinstein.info/.

484 *In his final letter* Anon. Legge antisemite e deportazione. *Liceo Parini* [Online] 2015. [Cited: 10 December 2023] 91–6. https://liceoparini.edu.it/wp-content/uploads/sites/82/2017/01/Il-dolore-di-avervi-dovuto-lasciare-ebook.pdf.

484 *After Christmas, Bohr and his son* Schweber, Silvan. Einstein and Nuclear Weapons. In *Einstein for the 21st Century*, Gerald Holton, Silvan S. Schweber, Peter L. Galison, eds. (Princeton, Princeton University Press, 2008), 79.

485 *Despite Frederick Lindemann* Pais, Abraham. *Niels Bohr's Times* (Oxford, Oxford University Press, 1991), 485–500.

485 *Such a frightening misconception* Jones, R.V. Churchill and Science. In *Churchill*, W.R. Louis, R. Blake, eds. (Oxford, Oxford University Press, 1993), 438.

485 *'It seems to me that'* Pais, Abraham. *Niels Bohr's Times* (Oxford, Oxford University Press, 1991), 501–2.
486 *Einstein wrote to Stern* Einstein, Albert. *AEA* [Online]. December 1944. Cat. nos. 89–901; 22–240. http://alberteinstein.info/. See also Clark, Ronald. *Einstein: The Life and Times* (New York, Avon Books, 1972), 577.
486 *The German stocks of uranium* Speer, Albert. *Inside the Third Reich*, [trans.] R. Winston, C. Winston (New York, Macmillan, 1970), 269–74.
486 *Quite what the President* Einstein, Albert, Szilard, Leo. *AEA* [Online]. 25 March 1945 [Cited: 17 November 2023]. Cat nos. 93–334, 93–358. http://alberteinstein.info/.
486 *Szilard, who had arrived* Szilard, Leo. *Reminiscences*, Perspectives in American History, Vol. II, K. R. Windsor, G. W. Szilard, eds. (Cambridge, MA, Harvard University Press, 1968), 123–8.
487 *Byrnes was just as caustic* Byrnes, James. *All in One Lifetime* (New York, Harper, 1958), 284.
487 *'[It] warned of Doomsday'* Strauss, Lewis. *Men and Decisions* (New York, Doubleday, 1962), 184–98.
488 *'Einstein is often blamed'* Oppenheimer, J. Robert. On Albert Einstein. *The New York Review of Books*, 17 March 1966. https://www.nybooks.com/articles/1966/03/17/on-albert-einstein/.
488 *Similar remarks were made* Bush, Vannevar. *Pieces of the Action* (London, Cassel, 1972), 58. See also Lerner, A. B. *Einstein and Newton*. (Minneapolis, Lerner Publications, 1973), 212–15.
488 *It sets the tone* Einstein, Albert. *Einstein on Peace*, Heinz Norden, Otto Nathan, eds. (New York, Schoken, 1968), 308.

Chapter 30

489 *Einstein agreed with this analysis* Smyth, Henry. *Atomic Energy for Military Purposes* (Princeton, Princeton University Press, 1945). See also Kaempffert, W. *The New York Times*, 16 August, 1945, 8; 19 August 1945, 67.
489 *In August 1946* Sayen, Jamie. *Einstein in America* (New York, Crown Publishers, Inc., 1985), 181–6.
490 *A poll in August 1945* Erskine, Hazel Gaudet. The Polls: Atomic Weapons and Nuclear Energy. *The Public Opinion Quarterly*, 1963, 27: 155–90. See also Sayen, J. *Einstein in America* (New York, Crown, 1985), 183; Einstein, Albert. *AEA* [Online]. 27 August 1945. Cat. no. 57–443. http://alberteinstein.info/.
490 *Both of these changes* Historian, Foreign Service Institute. The Acheson-Lilienthal & Baruch Plans, 1946. *Office of the Historian, U.S. Department of State* [Online] [Cited: April 21, 2024]. https://history.state.gov/milestones/1945-1952/baruch-plans.
491 *Einstein famously replied that* Einstein, Albert. *Einstein on Peace*, Heinz Norden, Otto Nathan, eds. (New York, Schoken, 1968), 369–99. See also Brian, Dennis. *The Unexpected Einstein* (New York, John Wiley, 2005), 50; *Alfred Werner Collection*, https://archive.org/stream/alfredwernercoll06wern/alfredwernercoll06wern_djvu.txt; Einstein, Albert. *AEA* [Online]. Cat. no. 40–470.
491 *'Dear Born, Einstein's death'* Pauli, Wolfgang. born_0027. *Pauli Letter Collection* [Online]. 24 April 1955. http://cds.cern.ch/record/83440/files/born_0027.pdf?version=1. See also Enz, Charles P. *No Time to Be Brief* (Oxford, Oxford University Press, 2002), 391.
492 *'Professor Einstein talked to me'* Brian, Denis. *Einstein: A Life* (New York, John Wiley, 1996), 353, 406. See also Commins, Dorothy. *Journal of Historical Studies*, 1967–68, Winter: 160.

492 *The remaining 80%* Jerome, Fred. *Einstein, Race, and the Myth of the Cultural Icon. Isis*, 2004, 95: 627–39. See also *The New York Times*, 4 May 1946, 7.

492 *The Ku Klux Klan* Jerome, Fred, Taylor, Rodger. *Einstein on Race and Racism* (New Brunswick, Rutgers University Press, 2006), 92–7.

493 *'Despite being something'* Einstein, Albert. *AEA* [Online]. September 1946 [Cited: 15 May 2024]. Cat. nos. 56–866, 37–468. http://alberteinstein.info/. See also Tam, F. M. Einstein in Western Maryland. *The Physics Teacher*, 2005, 43: 206–8.

493 *'We now hear so many'* Einstein, Albert, Einstein, Eduard. *AEA* [Online]. 1946–1947 [Cited: 19 May 2024]. Cat. nos. 75–848, 144–547. http://alberteinstein.info/.

493 *Solovine remarked on this* Einstein, Albert. *Letters to Solovine* (New York, Citadel Press, 1993), 94–7, 110.

494 *'. . . even in these times'* —. *From My Later Years* (New York, Philosophical Library, 1950), 229–30.

495 *The name of the distinguished* Goldstein, Israel. *Brandeis University* (New York, Bloch Publishing Company, 1951). For name question, see 79–81. See also Einstein, Albert. Cat. no. 40–373; Goldstein, Israel. Cat. nos. 40–374, 40–375, 40–377, 40-385-7; Wise, Stephen. Cat. no. 35–264. All *AEA* [Online]. 1946. http://alberteinstein.info/.

495 *However, on 2 September* Einstein, Albert, Young, Boris. *AEA* [Online]. 1946. Cat. nos. 40–403, 40–407; 40–409. http://alberteinstein.info/.

495 *He further remarked that* Schweber, Silvan. *Einstein and Oppenheimer* (Cambridge, Harvard University Press, 2008), 116–17.

495 *In this, Wise proved* Wise, Stephen, Goldstein, Israel. *AEA* [Online]. 1946–47. Cat. nos. 122–341, 40–416, 122–342, 122–345. http://alberteinstein.info/.

496 *At least he seems to have forgiven* Einstein, Albert, Sachar, Abram, Laski, Harold, Goldstein, Israel. *AEA* [Online]. 1947–1954. Cat. nos. 40–432, 40–450, 61–212, 61–214, 38–590, 37–318. http://alberteinstein.info/. See detailed account in Schweber, S. S. *Einstein and Oppenheimer* (Cambridge, MA, Harvard University Press, 2008), 101–35; see also Sachar, A. L. *A Host at Last* (Boston, Little, Brown, 1976), 12–30.

496 *Schrödinger's latest effort* Einstein, Albert. *AEA* [Online]. February 1947. [Cited: 19 May 2024.] 22–133,22–146. http://alberteinstein.info/.

497 *However, they continued to discuss* Anon. Unifying the Cosmos. *The New York Times*, 16 February 1947, 8E. See also New Einstein Theory Gives a Master Key to the Universe. *The New York Times*, 27 December 1949, 1; Halpern, P. *Einstein's Dice and Schrödinger's Cat* (New York, Basic Books, 2016), Chapter 7.

497 *He thought that it ought* Einstein, Albert. *AEA* [Online]. 4 November 1947 [Cited: 19 May 2024]. Object no. 58–488. http://alberteinstein.info/.

497 *Albert Einstein was however* Truman, Harry S. Commencement Address at Princeton University. *Truman Library* [Online]. 18 June 1947 [Cited: 21 May 2024]. https://www.trumanlibrary.gov/library/public-papers/117/commencement-address-princeton-university. See also Einstein, E. *Mozart Society of America Newsletter*, 2003, VII, No. 2.

498 *Einstein replied in an article* Einstein, Albert. *Einstein on Peace*, Heinz Norden, Otto Nathan, eds. (New York, Schoken, 1968), 400–57.

498 *Vallentin's book appeared* Vallentin, Antonina. *Le Drame d'Albert Einstein* (Paris: Les Petits-fils de Plon et Nourrit, 1954), 248–52.

499 *'Perhaps it would be possible'* Dünki, Robert, Maissen, Anna Pia. ' … damit das traurige Dasein unseres Sohnes etwas besser gesichert wird.' In *Stadtarchiv Zürich - Jahresbericht 2007/2008*, Robert

Dünki, Anna Pia Maissen, eds. (Zurich, Stadtarchiv Zürich, 2009), 341–59. See also Trbuhović-Gjurić, Desanka, *ISAE*, 196–7; Einstein, Albert. *AEA* [Online]. 1947–1948 [Cited: 16 September 2025]. Cat. nos. 71–49, 71–51. http://alberteinstein.info/.

499 *The letter quoted above* Einstein, Mileva. *AEA* [Online]. 21 May 1940 [Cited: 2 June 2024]. Cat. no. 144–96. http://alberteinstein.info/.

500 *It was intended for Einstein's grandson* Einstein, Albert. *AEA* [Online]. 21 January 1948 [Cited: 20 May 2024]. Cat. no. 75–959. http://alberteinstein.info/.

500 *Her words were transcribed* Trubuhović-Gjurić, Desanka. *ISAE*, 199–200. See also Dünki, Robert, Maissen, Anna Pia. *Stadtarchiv Zürich - Jahresbericht 2007/2008*, Robert Dünki, Anna Pia Maissen, eds. (Zurich, Stadtarchiv Zürich, 2009), 341–59; Einstein, Mileva, Einstein, Eduard, Nathan, Otto. *AEA* [Online]. 1948. Cat. nos. 99–855, 144–536, 58–694. http://alberteinstein.info/.

500 *He also relayed Mileva's plea* Einstein, Eduard. *AEA* [Online]. May–June 1948 [Cited: 17 November 2023]. Cat. no. 144–546. http://alberteinstein.info/.

500 *In the end, Hans Albert* Einstein, Hans Albert, Einstein, Albert. *AEA* [Online]. June–August 1948 [Cited: 17 November 2023]. Cat. nos. 144–107, 75–957, 75–958, 144–111, 75–830, 75.553.2, 75–553.1. http://alberteinstein.info/.

500 *On 4 August 1948* Trbuhović-Gjurić, Desanka. *ISAE*, 200–2.

501 *'Tetel is the most'* Einstein, Albert. *AEA* [Online]. 4 August 1948. Cat. nos. 83–485, 75–836. http://alberteinstein.info/.

501 *In fact, Hans Albert was Tete's heir* —. *AEA* [Online]. 9 September 1948 [Cited: 22 May 2024]. Cat. no. 38–575. http://alberteinstein.info/. See also Wolff, B. The Nobel Prize in Physics 1921 – What happened to the prize money?' *einstein-website-de* [Online]. 2016. https://einstein-website.de/what-happened-to-the-nobel-prize-money/.

501 *He closed by remarking that* Einstein, Eduard. *AEA* [Online]. September–November 1949. Cat. nos. 144–548, 144–549. http://alberteinstein.info/.

501 *Einstein did not answer* Schweitzer, Albert, Einstein, Albert. *AEA* [Online]. February–September 1951 [Cited: 30 June 2024]. Cat. nos. 33–218, 33–220. http://alberteinstein.info/.

502 *The situation de-escalated* Sayen, Jamie. *Einstein in America* (New York, Crown Publishers, Inc., 1985), 208–10. See also Einstein, Albert. *AEA* [Online]. 11 February 1948. Cat. no. 5–151. http://alberteinstein.info/.

502 *On 18 May* Einstein, Albert. *AEA* [Online]. 1948–1949 [Cited: 24 May 2024]. Cat nos. 33–448, 86–975, 120–979. http://alberteinstein.info/.

502 *He ensured that his real text* —. *Einstein on Peace.* Heinz Norden, Otto Nathan, eds. (New York: Schoken, 1968), 458–99.

503 *The cap to this* annus horribilis —. *Einstein on Peace*, Heinz Norden, Otto Nathan, eds. (New York, Schoken, 1968), 504–7, 558. See also Einstein, Albert. *AEA* [Online]. June 12 1951. Cat. no. 40–623. http://alberteinstein.info.

503 *Indeed, when warned by his doctors* —. *AEA* [Online]. 1947–1949. Cat. nos. 22–136, 38–578, 21–260, 75–828. http://alberteinstein.info/. See also Plesch, Janos, Plesch, Peter H. Some Reminiscences of Albert Einstein. *Notes and Records of the Royal Society of London*, 1995, 49(2): 307.

503 *In fact, the committee* Szilard, Leo. AEA [Online]. February 24 1950 [Cited: 12 June 2024]. Cat. no. 93–347. http://alberteinstein.info/.

504 *He also went to a local drugstore* LaHurd, Jeff. Sarasota's Rich and Famous. *Sarasota Herald Tribune*, 7 October 2010. https://eu.heraldtribune.com/story/news/2010/10/07/sarasotas-rich-and-famous/28971003007/. See also Einstein, Albert. *AEA* [Online]. 18–19 February 1949. Cat. nos. 58–899, 38–579. http://alberteinstein.info/.

504 *Sadly, this exchange of verses* Toller, Else, Einstein, Albert. *AEA* [Online]. March 1949 [Cited: 23 May 2024]. Cat. nos. 30–1043; 31–636. http://alberteinstein.info/.

504 *Another letter in May* Stern, Antonia, Einstein, Albert. *AEA* [Online]. February–May 1949 [Cited: 31 May 2024]. Cat. nos. 58–843, 58–848, 58–849, 58–850, 58–853, 58–854, 123–661. http://alberteinstein.info/. See also Scherrer, L. Der mysteriöse Tod eines Helden. *Neue Zürcher Zeitung*, 10 April 2018. https://www.nzz.ch/schweiz/der-mysterioese-tod-eines-helden-ld.1375839.

505 *'To [the pleasure of intimate conversations]'* Lawrence, W. L. World Scientists Hail Einstein at 70. *The New York Times*, 13 March 1949, 34.

505 *The last is a somewhat acerbic* Anon. Einstein Honored. *Princeton Alumni Weekly*, 25 March 1949. https://paw.princeton.edu/article/einstein-honored. See also https://albert.ias.edu/entities/archivalmaterial/c5828f4f-00a7-4fdc-a1a4-d7bb2bbe712d.

506 *He ignored the peace overtures* Sachar, Abram, Einstein, Albert. *AEA* [Online]. May 1949 [Cited: 27 May 2024]. Cat. nos. 58–737, 58–738. http://alberteinstein.info/.

506 *Around this time* Goldstein, Rebecca. *Incompleteness* (New York, W.W. Norton & Co., 2005), 33.

506 *Von Neumann explained* Feinberg, A. Robot Brain Plots Orbits of Planets. *The New York Times*, 12 September 1949, 23–4. See also Plumb, R. K. *The New York Times*, 18 November 1949, 30; 19 November 1949, 15.

506 *The conditions in which* United Press. Atom Blast in Russia Disclosed. *The New York Times*, 24 September 1949, 1.

506 *The topic of their conversation* Einstein, Albert. *AEA* [Online]. 13 June 1947 [Cited: 30 May 2024]. Cat. no. 32–725. http://alberteinstein.info/. See also *The New York Times*, 6 November 1949, 34.

506 *He was to give* Einstein, Albert. AEA [Online]. December 1 1951 [Cited: 1 July 2024]. Cat. no. 20–247. http://alberteinstein.info/.

507 *He undoubtedly saw the future* Spiegel, Irving. 1950 Fund Ratified for Jewish Appeal. *The New York Times*, 28 November 1949, 22.

Chapter 31

508 *The new decade began* Brian, Denis. *Einstein*: A *Life* (New York, John Wiley, 1996), 382–92.

508 *He warned that an arms race* Einstein, Albert. *Einstein on Peace*, Heinz Norden, Otto Nathan, eds. (New York, Schoken, 1968), 520–2.

508 *He agreed to write a statement* —. *AEA* [Online]. 2 February 1950 [Cited: 30 May 2024]. Cat. nos. 61–608, 61–609. http://alberteinstein.info/.

509 *'I have retained my good sense of humour'* Moos, Paul, Einstein, Albert. *AEA* [Online]. March 1950 [Cited: 9 June 2024]. Cat. nos. 60–586, 36–543. http://alberteinstein.info/.

509 *They loved each other very much* Brian, Denis. *The Unexpected Einstein* (New York, John Wiley, 2005), 51–2. Interview with Alice Kahler.

509 *Her sight began to fail* Rogger, Franziska. *ES*, 156.

509 *In the first one* Einstein, Albert. *AEA* [Online]. 22 June 1946 [Cited: 8 June 2024]. Cat. nos. 29–433, 29–434. http://alberteinstein.info/.

510 *He described Russell* —. *AEA* [Online]. 1947–1951 [Cited: 1 July 2024]. Cat. nos. 80–863, 75–808, 80–865, 33–256,7–401. http://alberteinstein.info/.

510 *By March of 1951* —. *AEA* [Online]. 1950–1951 [Cited: 15 May 2024]. Cat. nos. 36–543, 21–271, 21–280. See also entries for 108 and 112 Mercer Street, US Census, 1940 and 1950. http://alberteinstein.info/.

510 *A few months after Maja's death* —. *AEA* [Online]. June–December 1951 [Cited: 15 May 2024]. Cat. nos. 38–303, 29–435, 75–794, 7–401. http://alberteinstein.info/.

511 *Fantova was also often a witness* Fantova, Johanna. Introduction to 'Gespräche mit Einstein'. *The Princeton University Library Chronicle*, 2003, 65: 51–64. See also Einstein, Albert. *AEA* [Online]. 25 June 1952. Cat. no. 87–350. http://alberteinstein.info; and *Hanna Fantova Collection of Albert Einstein*, Folder 1 (Princeton, Princeton University Library).

511 *Finally there was a flurry* Schrödinger, Erwin, Einstein, Albert, von Laue, Max, Besso, Michele. *AEA* [Online]. March–December 1950 [Cited: 7 June 2024]. Cat. nos. 22–158 through 22–162; 22–174; 16–141, 16–142, 7–196 through 7–199, 7–396 through 7–398. http://alberteinstein.info/.

512 *Einstein concluded his article* Einstein, Albert. On the Generalized Theory of Gravitation. *Scientific American*, 1950, 182: 13–17.

512 *Sadly, he was proven* Highfield, Roger, Carter, Paul. *The Private Lives of Albert Einstein* (London, Faber and Faber, 1993), 255. See also Einstein, Hans Albert, Einstein, Albert. *AEA* [Online]. (1949–1950). Cat. nos. 144–108, 75–552.2–3, 75–552.6; 75–796, 75–810. http://alberteinstein.info/.

512 *He explained that he was afraid* Einstein, Albert. *AEA* [Online]. 1950–1954 [Cited: 10 June 2024]. Cat. nos. 83–486, 80–141, 39–59. http://alberteinstein.info/.

512 *The aim was to raise* —. *AEA* [Online]. May–June 1950 [Cited: 10 June 2024]. Cat. nos. 120–587, 122–374. http://alberteinstein.info/. See also *The New York Times*, 27 May 1950, 21.

513 *Einstein wrote to the family* —. *AEA* [Online]. 1950–1952 [Cited: 8 June 2024]. Cat. nos. 37–715, 37–716, 124–369. http://alberteinstein.info/. See also *The New York Times*, 14 April 1950, 26.

513 *These were to no avail* —. *AEA* [Online]. July–December 1950 [Cited: 13 June 2025]. Cat. nos. 32–736, 32–738, 32–741, 32–743, 32–744. http://alberteinstein.info/.

513 *'My knowledge of modern music'* Neilson, Francis. *My Life in Two Worlds*, Vol. 2 (Appleton, Dutton, 1953), 142. See also Einstein, Albert. *AEA* [Online]. Cat. no. 74–180 (1939); Cat. nos. 71–341, 34–332 (1950); Zeisler, Ernest. *AEA* [Online]. Cat. no. 120–792 (1936). http://alberteinstein.info/.

513 *To both correspondents* Einstein, Albert. *AEA* [Online]. 1950–1951 [Cited: 13 June 2024]. Cat. nos. 60–434, 32–400. http://alberteinstein.info/.

513 *Fantova remembered him* Fantova, Johanna. Introduction to 'Gespräche mit Einstein'. *The Princeton University Library Chronicle*, 2003, 65: 58. See also Fantova, Johanna. *Hanna Fantova Collection of Albert Einstein*, Folder 6 (Princeton, Princeton University Library), entry for 24 March 1954.

514 *He arrived with his violin* Samuels, Gertrude. Einstein, at 75, Is still a Rebel. *The New York Times*, 14 March 1954, 306–7.

514 *'If you don't agree with me'* Einstein, Albert. *AEA* [Online]. 1 January 1951 [Cited: 1 July 2024]. Cat. no. 21–274. http://alberteinstein.info/.

514 *In a similarly light-hearted vein* —. *AEA* [Online]. 16 January 1951 [Cited: 30 June 2024]. Cat. nos. 34–427, 61–419, 61–569. http://alberteinstein.info/.

514 *He concluded another letter* Sorokin, Pitirim. War or New Foreign Policy. *The New York Times*, 17 January 1951, 22. See also Einstein, Albert. *AEA* [Online]. January–March 1951, Cat. nos. 61–354, 65–960, 65–737, 60–819. http://alberteinstein.info/.

515 *Einstein commented that* Schweitzer, Albert, Einstein, Albert. *AEA* [Online]. 1951–1954 [Cited: 30 June 2024]. Cat. nos. 33–221, 33–225. http://alberteinstein.info/.

515 *'The works of the greatest masters'* Einstein, Albert. *AEA* [Online]. 21 March 1951 [Cited: 30 June 2024]. Cat. no. 34–400. http://alberteinstein.info/.

515 *On the day following* —. *AEA* [Online]. 23 March 1951 [Cited: 30 June 2024]. Cat. no. 61–812. http://alberteinstein.info/. See also Yeshiva University's Albert Einstein College of Medicine. *Yeshiva University Library* [Online]. 2005. https://library.yu.edu/c.php?g=1073982&p=7871665.

515 *However, Einstein would never have* White, Peregrine. Albert Einstein: The Violinist. *The Physics Teacher*, 2005, 43: 286–8.

515 *The programme began appropriately* Anon. Program. Princeton Symphony Orchestra, First Concert. *Historical Society of Princeton* [Online]. 1995–2016 [Cited: 1 July 2024]. https://princeton.pastperfectonline.com/Archive/62583685-3E62-4970-9F04-371027207030. Also the late Dr Paul Kent, private communication. Dr Kent spent the academic year 1950–51 in Princeton and took part in the chamber music sessions in Mercer Street.

516 *Eventually this would result* Peat, F. D. *Infinite Potential: the Life and Times of David Bohm* (Reading, Addison-Wesley, 1997), 56–119. See also Einstein, Albert. *AEA* [Online]. 17 April and 5 August 1951. Cat. nos. 120–545, 8–6. http://alberteinstein.info/.

516 *'Late in his life'* Portnow, Allison. *Einstein, Modernism, and Musical Life in America, 1921–1945.* Unpublished PhD Thesis (University of North Carolina at Chapel Hill, 2011), 253.

517 *Waking at 11.45 p.m.* McDonald, Malcolm. *Schoenberg* (Oxford, Oxford University Press, 2008), 86–7.

517 *Yes, he concluded* Einstein, Albert. *AEA* [Online]. 3 January 1952 [Cited: 3 July 2024]. Cat. no. 32–403. http://alberteinstein.info/.

517 *Einstein thought that the idea* —. *AEA* [Online]. 2 January 1952 [Cited: 3 July 2024]. Cat. no. 26–74. http://alberteinstein.info/.

517 *Fortunately for posterity* Einstein, Albert, Besso, Vero. *AEA* [Online]. 1952 [Cited: 3 July 2024]. Cat. nos. 61–141, 29–436 through 29–439, 7–408. http://alberteinstein.info/.

517 *This is such a good* Stuckenschmidt, Hans H. *Schoenberg: His Life, World and Work* (New York, Schirmer, 1978), 521.

518 *'I myself, on the other hand'* Pelizza Marangoni, Ernesta, Einstein, Albert. *AEA* [Online]. 1952 [Cited: 15 July 2024]. Cat. nos. 60–403, 60–409. http://alberteinstein.info/.

518 *Since people are not machines* Einstein, Albert. *AEA* [Online]. 8 January 1952 [Cited: 3 July 2024]. Cat. no. 59–62. http://alberteinstein.info/.

518 *Einstein was charmed by the book* —. *AEA* [Online]. 1952 [Cited: 13 July 2024]. Cat. nos. 39–10 through 39–23, 39–29, 39–37, 39–42, 124–412. http://alberteinstein.info/.

518 *It was only in October 1954* Pauling, Linus, Einstein, Albert. AEA [Online]. May 1952 [Cited: 14 July 2024]. Cat. nos. 60–864 through 60–870. http://alberteinstein.info/.

519 *Despite Einstein's support* Einstein, Albert. *AEA* [Online]. 15 May 1952 [Cited: 14 July 2024]. Cat. nos. 60–504, 60–505. http://alberteinstein.info/.

519 *Einstein replied in similar vein* Ehrat, Jakob, Einstein, Albert. *AEA* [Online]. April–May 1954 [Cited: 14 July 2024]. Cat. nos. 59–553, 59–554. http://alberteinstein.info/.

519 *'I am deeply moved'* Einstein, Albert, Eban, Abba. *AEA* [Online]. November 1952 [Cited: 15 July 2024]. Cat. nos. 33–454, 41–87, 41–89. http://alberteinstein.info/. See also *The New York Times*, 15 November 1952, 9; *The New York Times*, 19 November 1952, 1; *JTA Daily News Bulletin*, 20 November 1952, 1–2; Isaacson, W. E*instein: His Life and Universe* (New York, Simon and Schuster, 2007), 521–2.

520 *He went on to opine* —. *AEA* [Online]. October–November 1952 [Cited: 15 July 2024]. Cat. nos. 80–314, 42–528. http://alberteinstein.info/. See also *Hanna Fantova Collection of Albert Einstein*, Folder 6 (Princeton, Princeton University Library, Transcription of 17 June 1954).

520 *A few months later* Seelig, Carl. *BIOG*, 243. See also *The New York Times*, 29 April 1953, 53.

520 *Undeterred by this threatening atmosphere* Jerome, Fred. Taylor, Rodger. *Einstein on Race and Racism* (New Brunswick, Rutgers University Press, 2006), 122–5.

520 *In fact is was Bucky* Anon. Einstein Concedes Even He Can Err. *The New York Times*, 29 November 1952, 1, 13.
521 *The Rosenbergs were executed* Einstein, Albert. *AEA* [Online]. 1952–1953 [Cited: 16 July 2024]. Cat. nos. 41–547, 41–551, 41–615. http://alberteinstein.info/.
521 *A glance at the surnames* Einstein, Albert, Frauenglass, William. AEA [Online]. April–June 1953 [Cited: 22 July 2024]. Cat. nos. 34–635, 34–639; 34–641. See also *The New York Times*, 5 May 1953, 18; 12 June 1953, 9; 13 June 1953, 8; 19 June 1953, 9. http://alberteinstein.info/.
521 *'In your issue of June 13th'* Russell, Bertrand. Obeying Law in Testifying. *The New York Times*, 26 June 1953, 18.
522 *In his last letter in March* Jerome, Fred. *The Einstein File* (New York, St Martin's Griffin, 2003), 238–55. See also Einstein, Albert, Frauenglass, W. *AEA* [Online]. (1954–1955). Cat. nos. 59–767 through 59–769, 59–772, 85–104; 59–760, 59–771. http://alberteinstein.info/.
522 *He remarked on the strange phenomenon* Einstein, Albert. *AEA* [Online]. 12 January 1953 [Cited: 16 July 2024]. Cat. no. 32–405. http://alberteinstein.info/.
522 *'In receiving it I compared'* —. *AEA* [Online]. March 1953 [Cited: 16 July 2024]. Cat. nos. 73–282, 124–719. http://alberteinstein.info/.
522 *'The evaluation of Pablo Casals'* —. *AEA* [Online]. 30 March 1953 [Cited: 16 July 2024]. Cat. no. 34–347. http://alberteinstein.info/.
523 *Similarly, in a letter* Einstein, Albert, Ott, Emil. *AEA* [Online]. 1952–1953 [Cited: 20 July 2024]. Cat. nos. 21–294, 60–802, 60–803, 60–606. http://alberteinstein.info/.
523 *Lebach outlived Einstein* Einstein, Albert. *AEA* [Online]. 12 May 1953 [Cited: 20 July 2024]. Cat. no. 60–221. http://alberteinstein.info/.
523 *In May 1953* —. *AEA* [Online]. 18 May 1953 [Cited: 20 July 2024]. Cat. no. 65–858. http://alberteinstein.info/. See also Manfred Clynes. *Wikipedia*. https://en.wikipedia.org/wiki/Manfred_Clynes.
523 *Instead, he responded that* —. *AEA* [Online]. 1953–1954 [Cited: 23 July 2024]. Cat. nos. 122–234, 59–416, 59–417. http://alberteinstein.info/.
524 *According to the groom* Fantova, Johanna. *Hanna Fantova Collection of Albert Einstein*, Folder 6 (Princeton, Princeton University Library, Transcriptions October–December 1953). See also Brian, Denis. *Einstein: A Life* (New York, Wiley, 1996), 413–14; Einstein, Albert. *AEA* [Online]. May 22 1954. Cat. no. 23–207. http://alberteinstein.info/.
524 *Borden wrote that* Rhodes, Richard. *Dark Sun* (London, Simon & Schuster, 1996), 532–4.

Chapter 32

525 *Equally angry, Strauss rejoined* Bird, Kai, Sherwin, Martin J. *American Prometheus* (New York, Knopf, 2005), 437–90.
526 *He considered Oppenheimer's reaction* Einstein, Albert. *AEA* [Online]. 12 August 1954 [Cited: 29 July 2024]. Cat. no. 124–422. http://alberteinstein.info/.
526 *However, the anticipated press scrum* Pais, Abraham. *LORD*, 10–11. See also *The New York Times*, 13 April 1954, 1, 15, 16–18, 20.
526 *Many years later* Bird, Kai, Sherwin, Martin J. *American Prometheus* (New York: Knopf, 2005), 537–50.
527 *Realising the situation* Allison, Samuel K., Condon, Edward, Urey, Harold C., Daniels, Farrington, Loomis, F. W., Shils, Edward, et al. Scientists Affirm Faith in Oppenheimer. *Bulletin of the*

Atomic Scientists, 10(5): 188–90. See also Einstein, Albert. *AEA* [Online]. 12 August 1954. Cat. no. 124–422; Sayen, Jamie. *Einstein in America* (New York, Crown, 1985), 289; *The New York Times*, 14 April 1954, 19; and 2 October 1954, 1, 6.

527 *He concluded that* Einstein, Albert. *AEA* [Online]. 3 January 1954 [Cited: 31 July 2024]. Cat. no. 32–408. http://alberteinstein.info/.

527 *'I feel somewhat unhappy'* —. *AEA* [Online]. 5 January 1954 [Cited: 4 August 2024]. Cat. no. 59–67. http://alberteinstein.info/.

527 *His support was no doubt* —. *AEA* [Online]. 12 January 1954 [Cited: 3 August 2024]. Cat. no. 30–89. http://alberteinstein.info/.

528 *This visit was the last time* Pauli, Wolfgang. born_0025. *Pauli Letter Collection* [Online]. March 1954 [Cited: 8 August 2024]. https://cds.cern.ch/record/83428/files/born_0025.pdf. See also born_0024.

528 *In his letter of thanks* Einstein, Albert. *AEA* [Online]. 20 February 1954 [Cited: 4 August 2024]. Cat. no. 34–354. http://alberteinstein.info/.

528 *The Schwarz family visited* —. *AEA* [Online]. March 1954 [Cited: 4 August 2024]. Cat. nos. 32–411, 30–1162, 30–1166, 61–204, 84–202 (p. 1), 70–785, 97–441, 81–653, 61–680, 8–242, 79–684. http://alberteinstein.info/. See also *Hanna Fantova Collection of Albert Einstein*, Folder 6, entry for 16 April 1954. Isaacson, Walter. *Einstein: His Life and Universe* (New York, Simon & Schuster, 2007), 536.

528 *Pauli died of pancreatic* Enz, Charles P. *No Time to be Brief* (Oxford, Oxford University Press, 2002), 533–4.

528 *This replaced the previous* Potter, Tully. *Adolf Busch*, Vol. II (London, Toccata Press, 2010), 1112.

529 *In return it gave him an infection* Fantova, Johanna. *Hanna Fantova Collection of Albert Einstein*, Folder 6, entry for 20 February 1955 (Princeton, Princeton University Library)

529 *Still, he mused, all things pass* Einstein, Albert. *AEA* [Online]. 30 March 1954 [Cited: 4 August 2024]. Cat. no. 120–847. http://alberteinstein.info/.

529 *'If only everything was as clear'* —. *AEA* [Online]. 14 May 1954 [Cited: 5 August 2024]. Cat. no. 7–236. http://alberteinstein.info/.

529 *Unusually, he reminded Hans Albert* —. *AEA* [Online]. 11 May 1954 [Cited: 5 August 2024]. Cat. no. 75–918. http://alberteinstein.info/.

529 *However, they must have been successful* Van, Jon. An Old Friend Rembers: Einstein, a Genius with Heart. *Chicago Tribune*, 11 March 1979, 40. See also Einstein, Albert. *AEA* [Online]. Cat. nos. 70–597, 70–598, 59–558, 59–559, 60–250, 59–560 through 59–562, 87–332 (1939-1954). [Online] http://alberteinstein.info.

530 *He thought that he was condemned* Fantova, Johanna. *Hanna Fantova Collection of Albert Einstein*, 29 July 1954. Folder 1, Letters 27, 28, and Folder 6, entry for 19 June 1954 (Princeton, Princeton University Library). See also Einstein, Albert. *AEA* [Online]. 28 December 1954, 38–310. http://alberteinstein.info/.

530 *Dr Wise, President* Spiegel, Irving. Einstein Extols Israel as Refuge. *The New York Times*, 20 September 1954, 16. See also Einstein, Albert. *AEA* [Online]. 10 August 1954, Cat. no. 120–608. http://alberteinstein.info/.

531 *Einstein wrote an appreciative obituary* Einstein, Albert. *AEA* [Online]. December–February 1954–1955 [Cited: 8 August 2024]. Cat. nos. 120–850, 75–917, 2–14, 5–162. http://alberteinstein.info/. See also *Hanna Fantova Collection of Albert Einstein*, Folder 6, entries from October–November 1954. (Princeton, Princeton University Library).

531 *He remarked that as he grew older* —. *AEA* [Online]. 2 January 1955 [Cited: 8 August 2024]. Cat. no. 32–414. http://alberteinstein.info/.

531 *'In January 1955'* Anderson, Marian. *My Lord, What a Morning* (New York, Viking, 1956), 267. See also Jerome, Fred, Taylor, Rodger. *Einstein on Race and Racism* (New Brunswick, Rutgers University Press, 2006), 43–4.

531 *Nevertheless, on 31 January* von Laue, Max, Einstein, Albert. *AEA* [Online]. January–February 1955 [Cited: 13 August 2024]. Cat. nos. 16–207, 16–207.1, 5–98. http://alberteinstein.info/.

531 *Einstein thought that Casals* Einstein, Albert. *AEA* [Online]. 19 February 1955 [Cited: 14 August 2024]. Cat. no. 124–425. http://alberteinstein.info/.

532 *His final words to her* Queen Elisabeth of the Belgians, Einstein, Albert. *AEA* [Online]. February–March 1955 [Cited: 16 August 2024]. Cat. nos. 32–415, 32–416. http://alberteinstein.info/.

532 *Here he was at pains to emphasise* Einstein, Albert, von Laue, Max. *AEA* [Online]. March 1955 [Cited: 16 August 2024]. Cat. nos. 120–931, 16–201, 16–203. http://alberteinstein.info/.

532 *'Forty years have passed'* Einstein, Albert. *AEA* [Online]. March 1955 [Cited: 14 August 2024]. Cat. no. 65–457. http://alberteinstein.info/.

532 *'Anyway, I have found'* —. *AEA* [Online]. 1953–1955 [Cited: 16 August 2024]. Cat. nos. 21–301, 122–135. http://alberteinstein.info/.

533 *'His end was as harmonious'* Besso, Vero, Besso-Rusconi, Bice. Einstein, Albert. *AEA* [Online]. March 1955 [Cited: 16 August 2024]. Cat. nos. 7–246, 7–247, 7–245. http://alberteinstein.info/.

533 *Among modern operas* Fantova, Johanna. *Hanna Fantova Collection of Albert Einstein*, Folder 6, Transcriptions March 1955 (Princeton, Princeton University Library).

533 *Einstein got as far as* Einstein, Albert. *AEA* [Online]. April 1955 [Cited: 16 August 2024]. Cat. nos. 120–576, 122–95. http://alberteinstein.info/. See also Einstein, Albert. *Einstein on Peace*, Heinz Norden, Otto Nathan, eds. (New York, Schoken, 1968), 638–40.

533 *In fact, Enstein had explained* Einstein, Albert. AEA [Online]. 29 August 1953 [Cited: 14 August 2024]. Cat. no. 7–416. http://alberteinstein.info/.

534 *He closed his letter* —. *AEA* [Online]. 6 April 1955 [Cited: 16 August 2024]. Cat. no. 32–750. http://alberteinstein.info/.

534 *'. . . a nuclear physicist'* Nickerson, Sylvia. Taking a Stand: Exploring the Role of the Scientists prior to the First Pugwash Conference on Science and World Affairs, 1957. *Scientia Canadensis*, 2013, 36: 63–87. See also Russell, Bertrand. *AEA* [Online]. 11 February 1955. Cat. no. 33–199. http://alberteinstein.info/.

534 *He suggested Whitehead and Urey* Russell, Bertrand, Einstein, Albert. *AEA* [Online]. February 1955 [Cited: 13 August 2024]. Cat. nos. 33–199, 33–201. http://alberteinstein.info/.

535 *Einstein gave Russell carte blanche* Bohr, Niels. Letter to Bertrand Russell, 23 March 1955. *The Bertrand Russell Archives*, Vol. File 1, Box 1–36 (Hamilton, Canada, McMaster University). See also Einstein, Albert. *AEA* [Online]. March 1955. Cat. nos. 33–204, 33–206. http://alberteinstein.info/.

535 *This letter and his signature* Einstein, Albert. *Einstein on Peace*, Heinz Norden, Otto Nathan, eds. (New York, Schoken, 1968), 631–5.

536 *On the evening of Sunday 17 April* Dukas, Helen. *AEA* [Online]. 12–18 April 1955 [Cited: 17 August 2024]. Cat. nos. 75–324, 39–71. http://alberteinstein.info/. See also Plesch, Janos, Plesch, Peter H. Some Reminiscences of Albert Einstein. *Notes and Records of the Royal Society of London*, 1995, 49(2): 303–28; Brian, Denis. *The Unexpected Einstein* (New York, Wiley, 2005), 81.

536 *'Twice I was allowed to see him'* Born, Max. *Physics in My Generation* (Oxford, Pergamon Press, 1956), 164–5.

537 *The programme contained the instruction* Anon. Program. Princeton Symphony Orchestra, In Memory of Albert Einstein. *Historical Society of Princeton* [Online]. 17 December 1955 [Cited:

18 August 2024]. https://princeton.pastperfectonline.com/archive/153C4EB0-E3F0-43EB-B935-143367573347.

537 *'In 1952, the Juilliard String Quartet'* Mann, Robert. *A Passionate Journey* (New York, East End Press, 2018), 136–8.

Chapter 33

538 *'Biography, the most interesting perhaps'* Lockhart, John G. *Memoires of the Life of Sir Walter Scott, Bart*, 2nd edn. (Edinburgh, R. Cadell, 1839), 123.

538 *'One has to know something about history'* Einstein, Albert. *AEA* [Online]. April 1954 [Cited: 21 August 2024]. Cat. no. 26–45. http://alberteinstein.info/.

538 *The above quotations* Cohen, I. Bernard. An Interview with Einstein. *Scientific American*, 1955, 193(2): 68–73.

538 *'By the way, it also proves'* Einstein, Albert. *AEA* [Online]. 2 November 1952 [Cited: 21 August 2024]. Cat. no. 39–42. http://alberteinstein.info/.

540 *'His violin playing was not as good'* Brian, Denis. Interview with Vicky Weisskopf. *The Unexpected Einstein* (New York, John Wiley, 2005), 146.

541 *He loved women* —. Interview with Alice Kahler. *The Unexpected Einstein* (New York, John Wiley, 2005), 50.

541 *It can perhaps also be mentioned* Moore, Ealter J. *Schrödinger: Life and Thought* (Cambridge, Cambridge University Press, 1992), 363.

542 *'This sorrow is eating up Albert'* Vallentin, Antonina. *Le Drame d'Albert Einstein* (Paris, Les Petits-fils de Plon et Nourrit, 1954), 166–7.

542 *'One can't analyse him too much'* Einstein, Elsa. *AEA* [Online]. 1 January 1929 [Cited: 27 August 2024]. Cat. no. 73–338. http://alberteinstein.info/.

542 *Michelmore talked to* Michelmore, Peter. *Einstein, Profile of the Man.* (New York, Dodd, Mead & Company, 1962). 124.

543 *'Being with this man'* Dukas, Helen. Entry from Dukas' diary for 11 September 1930. *AEA* [Online]. 11 September 1930 [Cited: 29 August 2024]. Cat. no. 38–592.7. http://alberteinstein.info/.

543 *'[Einstein's] sense of humour'* Viscount Cherwell (attr). Rhodes Lecture Biography. *The Cherwell Papers* (Oxford, Nuffield College Library, 1931), D55/5.

544 *'Through all the stream of events'* Infeld, Leopold. *Quest* (New York, Chelsea Publishing Company, 1980), 185.

544 *'I do not socialize'* Plesch, Janos, Plesch, Peter. Some Reminiscences of Albert Einstein. *Notes and Records of the Royal Society of London*, 1995, 49(2): 307.

544 *His stepson-in-law Rudi Kayser* Reiser, Anton. *Albert Einstein: A Biographical Portrait* (London, Thornton Butterworth Limited, 1931), 165.

544 *'Einstein is always impersonal'* Marianoff, Dimitri, Wayne, Palma. *Einstein: An Intimate Study of a Great Man* (New York, Doubleday, Doran & Co., Inc., 1944), 95.

544 *A recipient of Einstein's most passionate* Newmann, Betty. *AEA* [Online]. 1956 [Cited: 2 September 2024]. Cat. no. 90–781, 9. http://alberteinstein.info/.

545 *Born had felt close enough to Einstein* Born, Max. *The Born–Einstein Letters*, [trans.] Irene Born (London, Macmillan, 1971), 130.

545 *'My passionate sense of social justice'* Einstein, Albert. *Ideas and Opinions* (New York, Bonanza Books, 1954), 9.

546 *He is buried nearby* Ettema, Robert, Mutel, Cornelia F. *Hans Albert Einstein* (Reston, ASCE Press, 2014), 286–7.

547 *'I am not a bit interested'* Bogolub, Alexander, Einstein, Albert. *AEA* [Online]. June 1950 [Cited: 29 August 2024]. Cat. nos. 59–288, 59–289. http://alberteinstein.info/.

547 *'And I believe that although'* Einstein, Albert. *AEA* [Online]. 14 March 1954 [Cited: 2 September 2024]. Cat. no. 61–380, 2. http://alberteinstein.info/.

548 *'Women are able to please'* Neumann, Betty. *AEA* [Online]. 1956 [Cited: 2 September 2024]. Cat. no. 90–781, 7. http://alberteinstein.info/.

548 *'Einstein was not only a scientist'* Brian, Dennis. *The Unexpected Einstein* (New York, John Wiley, 2005), 156. Bertrand Russell, BBC Broadcast, 1966.

549 *'My dear Mr Salit'* Salit, Norman, Einstein, Albert. *AEA* [Online]. February 1950 [Cited: 4 September 2024]. Cat. nos. 61–225, 61–226. http://alberteinstein.info/.

550 *'... is an important document'* Lemaître, Georges. Recontres avec Einstein. *Revue de Questions Scientifiques: Actualité, histoire et philosophie des sciences*, 1958, 129: 132. https://inters.org/lemaitre-einsten.

550 *'I believe that Nature'* Einstein, Albert. *AEA* [Online]. November 1954 [Cited: 4 September 2024]. Cat. no. 61–380. http://alberteinstein.info/.

551 *'I believe that ideas'* Born, G.V.R. The Wide-Ranging Family History of Max Born. *Notes and Records of the Royal Society of London*, 2002, 56(2): 261.

551 *'The appeal of General Relativity'* Whittaker, Edmund. Albert Einstein. 1879–1955. *Biographical Memoirs of Fellows of the Royal Society*, 1955, I: 59–60.

551 *'In this sense'* Einstein, Albert. *AEA* [Online]. 10 August 1954 [Cited: 4 September 2024]. Cat. no. 7–421. http://alberteinstein.info/.

552 *The discovery of gravitational waves* Abbott, B. P., Abbott, R., Abbott, T. D., Abernathy, M. R., Acernese, F., Ackley, K., et al. Observation of Gravitational Waves from a Binary Black Hole Merger. *Physical Review Letters*, 2016, 116: 061102.

553 *After the cremation* Brian, Denis. *Einstein: A Life* (New York, John Wiley, 1996), 427.

Index

For the benefit of digital users, indexed terms that span two pages (e.g., 52–53) may, on occasion, appear on only one of those pages.

Note that entries in italics refer to names or places primarily associated with music and the arts. Note also that the contents of footnotes have not been indexed, nor have names of some individuals peripheral to the narrative unless themselves noteworthy. Names in the main entry are given in full, with frequently used nicknames or contractions, while subentries are shortened to the names normally used in the main text. Place names in Germany are given without qualification, as are large, well known cities; others are placed in the appropriate country, or, for USA, state.

C

D

F

G

H

M

N

O

Q

R

T

U

V

W